NUTRIENT REQUIREMENTS OF FISH AND SHRIMP

Committee on Nutrient Requirements of Fish and Shrimp

Board on Agriculture and Natural Resources

Division on Earth and Life Studies

NATIONAL RESEARCH COUNCIL
OF THE NATIONAL ACADEMIES

THE NATIONAL ACADEMIES PRESS
Washington, D.C.
www.nap.edu

THE NATIONAL ACADEMIES PRESS 500 Fifth Street, N.W. Washington, DC 20001

NOTICE: The project that is the subject of this report was approved by the Governing Board of the National Research Council, whose members are drawn from the councils of the National Academy of Sciences, the National Academy of Engineering, and the Institute of Medicine. The members of the committee responsible for the report were chosen for their special competences and with regard for appropriate balance.

This study was supported by a grant from the Agricultural Research Service of the United States Department of Agriculture under Contract No. 59-0790-5-186, the National Oceanic and Atmospheric Administration under Award No. NA08OAR4170833; the United Soybean Board under Project No. 8490; and internal National Research Council funds derived from sales of publications in the Animal Nutrition Series. Any opinions, findings, conclusions, or recommendations expressed in this publication are those of the author(s) and do not necessarily reflect the views of the organizations or agencies that provided support for the project.

Library of Congress Cataloging-in-Publication Data

Nutrient requirements of fish and shrimp / Committee on the Nutrient Requirements of Fish and Shrimp, Board on Agriculture and Natural Resources, Division on Earth and Life Studies, National Research Council of the National Academies.
p. cm.
Includes bibliographical references and index.
ISBN-13: 978-0-309-16338-5 (cloth)
ISBN-10: 0-309-16338-2 (cloth)
ISBN-13: 978-0-309-16339-2 (pdf)
ISBN-10: 0-309-16339-0 (pdf)
1. Fishes—Nutrition—Requirements. 2. Shrimps—Nutrition—Requirements. 3. Fishes—Feeding and feeds. 4. Shrimps—Feeding and feeds. I. National Research Council (U.S.). Committee on the Nutrient Requirements of Fish and Shrimp.
SH156.N865 2011
595.3'88—dc22
2011008752

Additional copies of this report are available from the National Academies Press, 500 Fifth Street, N.W., Lockbox 285, Washington, DC 20055; (800) 624-6242 or (202) 334-3313 (in the Washington metropolitan area); Internet, *http://www.nap.edu.*

Copyright 2011 by the National Academy of Sciences. All rights reserved.

Printed in the United States of America

THE NATIONAL ACADEMIES

Advisers to the Nation on Science, Engineering, and Medicine

The **National Academy of Sciences** is a private, nonprofit, self-perpetuating society of distinguished scholars engaged in scientific and engineering research, dedicated to the furtherance of science and technology and to their use for the general welfare. Upon the authority of the charter granted to it by the Congress in 1863, the Academy has a mandate that requires it to advise the federal government on scientific and technical matters. Dr. Ralph J. Cicerone is president of the National Academy of Sciences.

The **National Academy of Engineering** was established in 1964, under the charter of the National Academy of Sciences, as a parallel organization of outstanding engineers. It is autonomous in its administration and in the selection of its members, sharing with the National Academy of Sciences the responsibility for advising the federal government. The National Academy of Engineering also sponsors engineering programs aimed at meeting national needs, encourages education and research, and recognizes the superior achievements of engineers. Dr. Charles M. Vest is president of the National Academy of Engineering.

The **Institute of Medicine** was established in 1970 by the National Academy of Sciences to secure the services of eminent members of appropriate professions in the examination of policy matters pertaining to the health of the public. The Institute acts under the responsibility given to the National Academy of Sciences by its congressional charter to be an adviser to the federal government and, upon its own initiative, to identify issues of medical care, research, and education. Dr. Harvey V. Fineberg is president of the Institute of Medicine.

The **National Research Council** was organized by the National Academy of Sciences in 1916 to associate the broad community of science and technology with the Academy's purposes of furthering knowledge and advising the federal government. Functioning in accordance with general policies determined by the Academy, the Council has become the principal operating agency of both the National Academy of Sciences and the National Academy of Engineering in providing services to the government, the public, and the scientific and engineering communities. The Council is administered jointly by both Academies and the Institute of Medicine. Dr. Ralph J. Cicerone and Dr. Charles M. Vest are chair and vice chair, respectively, of the National Research Council.

www.national-academies.org

COMMITTEE ON NUTRIENT REQUIREMENTS OF FISH AND SHRIMP

RONALD W. HARDY, *Chair,* University of Idaho, Hagerman
DELBERT M. GATLIN, III, *Vice-Chair,* Texas A&M University, College Station
DOMINIQUE P. BUREAU, University of Guelph, Ontario
LOUIS R. D'ABRAMO, Mississippi State University, Mississippi State
D. ALLEN DAVIS, Auburn University, Auburn, Alabama
JOHN E. HALVER, University of Washington, Seattle
ÅSHILD KROGDAHL, Norwegian School of Veterinary Science, Oslo, Norway
FRANÇOISE MÉDALE, French National Institute for Agricultural Research (INRA), Pee-Sur-Nivelle, France
SHI-YEN SHIAU, National Taiwan Ocean University, Keelung
DOUGLAS R. TOCHER, University of Stirling, Scotland

Staff

AUSTIN J. LEWIS, Study Director
RUTHIE S. ARIETI, Research Associate
ERIN P. MULCAHY, Senior Program Assistant (through August 2010)

External Support

PAULA T. WHITACRE (Full Circle Communications), Editor

BOARD ON AGRICULTURE AND NATURAL RESOURCES

NORMAN R. SCOTT, *Chair,* Cornell University, Ithaca, New York
PEGGY F. BARLETT, Emory University, Atlanta, Georgia
HAROLD L. BERGMAN, University of Wyoming, Laramie
RICHARD A. DIXON, Samuel Roberts Noble Foundation, Ardmore, Oklahoma
DANIEL M. DOOLEY, University of California, Oakland
JOAN H. EISEMANN, North Carolina State University, Raleigh
GARY F. HARTNELL, Monsanto Company, St. Louis, Missouri
GENE HUGOSON, Minnesota Department of Agriculture, St. Paul
KIRK C. KLASING, University of California, Davis
VICTOR L. LECHTENBERG, Purdue University, West Lafayette, Indiana
PHILIP E. NELSON, Purdue University, West Lafayette, Indiana
ROBERT PAARLBERG, Wellesley College, Watertown, Massachusetts
KEITH PITTS, Curragh Oaks Consulting, Fair Oaks, California
CHARLES W. RICE, Kansas State University, Manhattan
HAL SALWASSER, Oregon State University, Corvallis
PEDRO A. SANCHEZ, The Earth Institute, Columbia University, Palisades, New York
ROGER A. SEDJO, Resources for the Future, Washington, DC
KATHLEEN SEGERSON, University of Connecticut, Storrs
MERCEDES VÁZQUEZ-AÑÓN, Novus International, Inc., St. Charles, Missouri

Staff

ROBIN A. SCHOEN, Director
KAREN L. IMHOF, Administrative Assistant
AUSTIN J. LEWIS, Senior Program Officer
EVONNE P. Y. TANG, Senior Program Officer
PEGGY TSAI, Program Officer
CAMILLA YANDOC ABLES, Associate Program Officer
KARA N. LANEY, Associate Program Officer
RUTH S. ARIETI, Research Associate
JANET M. MULLIGAN, Research Associate
KAMWETI MUTU, Research Associate
ERIN P. MULCAHY, Senior Program Assistant (through August 2010)
KATHLEEN REIMER, Program Assistant (from August 2010)

Acknowledgments

This report has been reviewed in draft form by persons chosen for their diverse perspectives and technical expertise in accordance with procedures approved by the National Research Council's Report Review Committee. The purpose of this independent review is to provide candid and critical comments that will assist the institution in making its published report as sound as possible and to ensure that the report meets institutional standards of objectivity, evidence, and responsiveness to the study charge. The review comments and draft manuscript remain confidential to protect the integrity of the deliberative process. We wish to thank the following for their review of this report:

Geoff L. Allan, Port Stephens Fisheries Centre, New South Wales, Australia
Ian P. Forster, Fisheries and Oceans Canada, British Columbia, Canada
Menghe H. Li, Mississippi State University, Mississippi State
Ingrid Lupatsch, Centre for Sustainable Aquaculture Research, Swansea University, UK
Wing Keong Ng, Universiti Sains, Malaysia
Marty Alan Riche, Harry K. Dupree Stuttgart National Aquaculture Research Center, Agricultural Research Service, U.S. Department of Agriculture, FL
Michael B. Rust, National Marine Fisheries Service, National Oceanic and Atmospheric Administration, WA
Wendy M. Sealey, Bozeman Fish Technology Center, U.S. Fish and Wildlife Service, MT
Albert G. J. Tacon, Aquaculture Consultant, Vista, CA
Carl D. Webster, Aquaculture Research Center, Kentucky State University, Frankfort
Robert P. Wilson, *Emeritus*, Mississippi State University, Mississippi State
Thomas R. Zeigler, Zeigler Bros, Inc., PA

Although the reviewers listed above have provided many constructive comments and suggestions, they were not asked to endorse the conclusions or recommendations, nor did they see the final draft of the report before its release. The review of this report was overseen by Robert R. Stickney, Texas A&M University. Appointed by the National Research Council, he was responsible for making certain that an independent examination of this report was carried out in accordance with institutional procedures and that all review comments were carefully considered. Responsibility for the final content of this report rests entirely with the author committee and the institution.

The Committee on Nutrient Requirements of Fish and Shrimp thanks the U.S. Department of Agriculture, the National Oceanic and Atmospheric Administration, and the United Soybean Board for contributing funds to support the committee's work. The committee has been fortunate to have Dr. Austin Lewis, Senior Program Officer, Ruthie Arieti, Research Associate, and Erin Mulcahy, Senior Program Assistant, assigned to the committee. Dr. Lewis has provided excellent guidance, advice, and encouragement throughout the development of the report, and the committee is grateful for his sustained support and friendship. Ms. Arieti has been extremely effective at keeping the process of writing, revising, and editing sections moving along smoothly as well as keeping committee members informed and on track. The committee thanks Robin Schoen, Director of the Board on Agriculture and Natural Resources, for her efforts to get the revision underway and for her support and encouragement during its preparation. The committee members wish to thank their colleagues Arlene Ali, Brett Glencross, Katheline Hua, Kyeong-Jun Lee, Yu-Hung Lin, Biswamitra Patro, and Guillaume Salze, whose assistance was essential to complete the publication and Victoria Blondin for producing original drawings of fish anatomy. Finally, the committee wishes to thank the National Research Council for giving the members the opportunity to produce a new revised version of the report that they hope will guide the aquaculture feed industry, scientists, students, and others who share their passion for fish and shrimp nutrition for many years to come.

Contents

Tables and Figures

TABLES

FIGURES

Summary

The National Research Council (NRC) has published several previous reports on the nutrition and feeding of fish, the most recent of which is the *Nutrient Requirements of Fish* (1993). Since 1993 a large amount of information on fish and shrimp nutrition has been published. Consequently, many of the requirements and recommendations set forth in the 1993 report are no longer relevant or appropriate. Since publication of the previous report, aquaculture production has expanded more than 10-fold and has become of much greater national and international significance, as a food supplier and source of income. In fact, aquaculture now supplies half of the seafood and fisheries products consumed worldwide. Given the state of the world's fisheries, future demand for seafood and fisheries products can only be met by expanded aquacultural production. Such production will likely become more intensive and increasingly depend on nutritious and efficient aquaculture feeds containing ingredients from sustainable sources. This challenge can only be met by applying the latest nutritional and feed production information.

The need for an updated nutrient requirement publication for aquatic species has been apparent for several years. With the support of industry leaders and federal agencies, funding was obtained for a new report. In developing the NRC study, it was decided that one publication should address both coldwater and warmwater fish and shrimp. It was recognized that the audience for the new report would be varied, ranging from research scientists and those involved in regulation to people working with commercial aspects of fish and shrimp feeding, and therefore the report should attempt to address topics of importance for each audience. Also, because aquaculture has grown rapidly throughout the world, international representation was considered crucial.

The task given to the committee is presented in Appendix A. In brief, the committee was asked to prepare a report that evaluates the scientific literature on the nutrient requirements of fish and shrimp in all stages of life. The report was to focus primarily on the species that are most important commercially, but other emerging species could be included. Other elements of the task included: a review of strategies to increase nutrient retention and thus reduce fecal and metabolic excretions that contribute to environmental pollution; a discussion of the benefits and detriments of including marine products in fish feeds; and consideration of the relationship between diet formulation and the nutrient content of fish, especially omega-3 fatty acid levels.

The project was sponsored by the Agricultural Research Service of the United States Department of Agriculture, the National Oceanic and Atmospheric Administration, the United Soybean Board, and the NRC. To ensure international representation, the committee was composed of scientists from the United States, Canada, France, Norway, Taiwan, and the United Kingdom.

A primary realization of the committee is that continued growth of aquaculture production depends on precision feed formulation using information on nutritional requirements, as well as nutrient levels and availability in feed ingredients, to produce efficient feeds that maximize fish growth and health while minimizing environmental effects. Thus, the report contains nutritional recommendations condensed from the scientific literature as well as substantial other information to provide a context for understanding how to use the information in preparing feeds and applying appropriate feeding regimes to support efficient aquacultural production. The committee recognized the global shift in aquaculture feeds toward higher use levels of ingredients derived from grains, oilseeds, and other alternative sources to replace ingredients produced from marine resources and the nutritional challenges this will create.

This publication is expanded considerably from the 1993 report and contains several new chapters and sections. It begins with an Introduction that documents the expansion of aquaculture during the past two decades and the rapid growth in the number of research reports on the nutrient requirements of fish and shrimp. The topic of finding alternatives to fish meal and fish oil derived from marine resources in aquatic feeds is also introduced, along with a brief overview of the various topics covered in the report.

Chapter 2 discusses basic concepts and methodology used in experimental studies to determine the nutrient requirements of aquatic animals. Conducting nutritional studies using aquatic animals presents challenges to nutritionists compared to conducting studies with livestock and poultry. Several challenges are associated with the species' aquatic existence, including differences in husbandry and feeding and the fact that aquatic animals are poikilothermic (their body temperatures vary with the temperature of their surroundings). Another challenge is the high degree of variability among individuals, strains, and stocks compared to livestock and poultry. These complications make it necessary to use appropriate experimental designs with adequate replication such that treatment effects can be detected. The need for proper interpretation of results is also discussed, along with the importance of choosing a valid response criterion.

Chapter 3 examines the digestive physiology of fish and shrimp. This material, which was not included in previous NRC publications on fish, begins with a description of the anatomical features of the digestive organs of fish and the composition and role of digestive secretions. The processes of digestion and nutrient absorption are then discussed. The second part of the chapter covers shrimp.

Dietary energy utilization and metabolic integration are the topics of Chapter 4. It begins with the classical energy-partitioning scheme, which describes various losses of energy between the feed and the energy ultimately retained or recovered. Factors that affect the partitioning of energy in practical feeding systems for fish and shrimp are then addressed. The discussion of energy utilization has been expanded considerably from the 1993 report.

Chapters 5, 6, 7, 8, and 9 cover proteins and amino acids, lipids, carbohydrates and fiber, minerals, and vitamins, respectively. Constituting more that 40% of the total report, these chapters contain an extensive discussion of each of the nutrient classes and a review of experiments to determine nutrient requirements. Each chapter contains a comprehensive list of references. The material in these chapters forms the basis for the nutrient requirements listed later in the publication.

Chapter 5, on proteins and amino acids, begins with an overview of the biochemistry and roles of proteins and amino acids. The essential amino acids are then discussed in more detail and the principles involved in quantifying amino acid requirements are reviewed. This includes a discussion of the ideal protein concept (optimal amino acid proportions) and factorial approaches to determining amino acid requirements.

Knowledge about lipid functions and requirements has expanded considerably, and this is reflected in a much longer section than in the 1993 report. After general comments about the structures and functions of lipids, Chapter 6 discusses dietary lipid levels and requirements for specific fatty acids, phospholipids, and cholesterol by both fish and shrimp. The chapter ends with a discussion of other issues relevant to aquaculture.

Although fish and shrimp do not have specific requirements for dietary carbohydrates or fiber, these are major components of most commercial fish feeds. Chapter 7 reviews types of carbohydrates, the role of starch and non-starch polysaccharides, the metabolic fate of glucose, and the nutritional role of digestible carbohydrates in fish and shrimp. Also covered is the digestibility of various carbohydrate classes by fish.

Mineral nutrition of aquatic species is more complicated than that of terrestrial species because the former can absorb some minerals from the aquatic medium in which they live. Chapter 8 discusses six macrominerals and seven trace minerals. Other minerals such as cobalt and molybdenum are mentioned briefly. The chapter concludes with comments about the sources and forms of minerals and their interactions with other dietary components.

An update of the 1993 report's review of vitamin requirements is provided in Chapter 9. The chapter is divided into fat-soluble and water-soluble vitamins and also includes a brief discussion of other vitamin-like compounds. Sources and stability of vitamins used in feeds are also covered.

Chapter 10, titled "Feed Additives," is an update of the chapter titled "Other Dietary Components" in the 1993 publication. It includes a discussion of substances such as antimicrobial agents, enzymes, and other compounds that are commonly found in, or added to, feeds for fish and shrimp. Chapter 11, titled "Antinutritional Factors and Adventitious Toxins in Feeds," presents tables of antinutrients and other undesirable substances found in fish feeds, as well as an extensive publication list.

Chapter 12 addresses nutrient digestibility and availability. In addition to a discussion of topics such as methodology and potential errors, the digestibility of proteins, carbohydrates and fiber, lipids, and minerals is reviewed. This chapter also contains five tables with digestibility values for various fish species and shrimp that can be used in formulating feeds for various aquatic species.

Chapter 13 reviews some of the more applied aspects of fish nutrition and feeding. Subjects addressed are feeding early life stages, production diets, and feed management. Coverage includes important topics such as feeding in intensive production systems and pollution loading and waste management.

Feeding larval fish is a topic covered briefly in the 1993 report, but now has a chapter of its own. Chapter 14 reviews the practical feeding aspects as well as the limited data on nutritional requirements of larval fish and shrimp.

Chapter 15 addresses ingredients used in fish nutrition, formulation, and processing of feeds for aquatic animals. These topics were covered in the previous report, but this chapter provides updated information as well as more details on various considerations associated with feed production for aquatic animals.

Chapter 16 contains new material and covers two topics that have attracted considerable public and consumer attention during the past decade. The first issue is the limitations to supply and use of marine resources, particularly fish meal and fish oil. Many commercial aquaculture feeds contain appreciable amounts of fish meal and fish oil, and there is concern that the increasing demand on these resources will outstrip supply unless levels in feeds are reduced. As a consequence, a considerable amount of research is being devoted to identifying, developing, and evaluating alternative ingredients, and this research is discussed within the chapter. The second issue is the nutritional value of fish and aquatic products in human nutrition. Fish are unique and rich sources of long-chain, polyunsaturated fatty acids that are important components of the human diet. These topics are also covered in the chapter.

In Chapter 17, the committee identifies critical research needs for defining nutrient requirements. It is hoped that this chapter will aid researchers, administrators, and others as future research agendas are developed.

Chapter 18 contains tables of nutrient requirements for fish and shrimp. Requirements are expressed on a dry-matter basis. These are minimum requirements that assume 100% bioavailability and do not contain "margins of safety" or other adjustments for specific practical feeding situations. For the most part, requirement values in the literature were obtained with young, rapidly growing fish or shrimp. The committee critically evaluated published studies to arrive at the estimates presented. As such, values in these tables are the best estimates of the committee rather than an average of literature values.

Chapter 19 consists of tables of feed ingredients for feedstuffs commonly fed to fish and shrimp, including average composition values. Readers should be aware that values among different products available in the marketplace may differ from the average values presented in these tables.

Aquaculture production is sure to increase, both in quantity and in the range of organisms being produced, and increasing aquaculture production should be conducted in a manner that lowers the environmental effects of various production systems and that utilizes sustainably produced feed ingredients. These goals are both connected to nutrient requirements; without solid information on nutrient requirements of the range of farmed aquatic species, feeds cannot be formulated using alternative feed ingredients. The committee designed this report to be a comprehensive summary of extant knowledge on nutrient requirements of fish and shrimp and also to be forward-looking by including information to explain the nutritional science that underpins nutrient requirements. This approach allows the reader to understand better both the strengths and weaknesses of current information, and thus use it appropriately. The reader will also understand the importance of nutrient requirements to the production of efficient, economical, and sustainable feeds for use in aquaculture. The committee also hopes that the information assembled in this report inspires scientists to strive to develop better estimates of nutrient requirements of farmed fish and shrimp using both conventional and new approaches.

1

Introduction

More than 17 years have elapsed since the last National Research Council report on *Nutrient Requirements of Fish* was published. During the intervening years, global aquaculture production has grown at a rate of nearly 10% per year and now provides approximately 50% by weight of fish and shellfish consumed. The consensus among agencies and experts is that seafood supply from capture fisheries cannot increase to meet expected demand arising from growing populations, increasing incomes in developing countries and changing food preferences in developed countries (FAO, 2008). Therefore, aquaculture production is predicted to continue to grow.

Approximately half of global aquaculture production is from species that rely on feed inputs. The number of cultured species also has increased to well over 100, with the largest increase in farmed species being in the marine sector. Aquaculture production has become a major global industry and an important source of income and food in many countries.

Similar to livestock and poultry production, nutrition plays a key role in the aquaculture industry by influencing fish growth, health, product quality, and waste generation. Feed costs often account for over 50% of variable costs of an aquaculture enterprise and thus commonly influence economic returns. Development of nutritious, efficiently delivered, and cost-effective diets depends on knowing a species' nutritional requirements and meeting those requirements with balanced diet formulations and appropriate feeding practices. Therefore, this publication covers important aspects of nutrient provision, including digestive physiology (Chapter 2) and nutrient utilization of aquatic species (Chapter 12), as well as comprehensive information about nutritional energetics (Chapter 4) and all the major nutrient groups (Chapters 5–9). Reviews of feed additives (Chapter 10) and antinutritional factors (Chapter 11) also are provided.

In contrast to terrestrial livestock, the diversity is much greater among aquatic species and the environments in which they are produced in aquaculture. In fact, numerous representatives of the phyla Chordata, Arthropoda, and Mollusca are cultured in freshwater, brackish, and marine environments. The organisms within each of these phyla also exhibit considerable diversity within their different life stages, which influences or dictates nutritional and environmental requirements. These aquatic species also possess tremendous genetic diversity to allow broad heritability of traits to ensure sufficient recruitment in a stochastic natural environment. The type of culture system in which they are produced also varies considerably relative to production intensity. In many culture systems, natural productivity within the system contributes nutrients to the fish or shrimp. Unlike terrestrial animals, these various factors result in considerable complexity in estimating the quantitative nutrient requirements of aquatic species. Therefore, the ideal of universally applicable nutrient requirement values for all aquacultured species cannot be realized.

The unique considerations relative to fish and shrimp nutrition are addressed in this report, including updated information about the major feedstuffs used in aquatic feed formulations (Chapter 19), and feed processing and manufacture (Chapter 15). Special attention is devoted to fish meal and fish oil due the increased demand for these feedstuffs in aquatic feeds such that global supplies have become both limiting and costly. Chapter 16 of this report addresses development of alternatives to fish meal and fish oil in aquatic feeds while maintaining the desirable characteristics conferred by these ingredients to the diet, cultured organism and ultimately the consumer (e.g., fatty acid composition). Minimizing any potential negative environmental effects from aquaculture through improvements in diet formulations and feeding practices is also addressed in several chapters of this publication (Chapters 8, 10, 12, and 13). Providing nutritious diets and conducting efficient feeding practices in an aquatic environment are complicated by a number of factors that include quantifying feed intake of organisms and accounting for the magnitude of nutrient loss (leaching) from feeds into the water before consumption. These specific feed management issues also are addressed in this report (Chapters 13 and 14).

In response to the expansion of global aquaculture during the past two decades, research efforts have increased to support production practices of established and emerging species throughout the world. Consequently, information about the nutritional requirements of cultured fish and shrimp has correspondingly increased rapidly. The quantity of products generated by aquaculture has increased substantially over this time. Additionally, the number of aquatic species evaluated for food production, stock enhancement, and as research models has expanded. Aquatic species are used as model systems for understanding specific nutrient needs, metabolism, environmental challenges and animal health. The major consumers of prepared feeds are herbivorous and omnivorous species, with 42% of the total volume consumed by carps; however, marine salmonids and shrimp are the major consumers of feeds containing marine-derived ingredients. These quantitative and qualitative feed requirements were important considerations for the committee, and part of the charge to the committee was the challenge to present a current review of the nutritional requirements of prominent fish and crustacean species in this report.

The nutrient requirement values presented in this publication (Chapter 18) represent the committees' recommendations based on the background material presented in the proceeding chapters. They are minimum dietary concentrations required to support normal growth or other physiological responses. These values assume 100% bioavailability and therefore do not include a margin of safety. Nutrient requirements of those aquatic species that do not rely heavily on manufactured feeds were not included in this report. This publication is designed to provide the most comprehensive compilation of data from fish and crustacean species that are well established in commercial aquaculture. Another overriding objective of the committee was to provide information on methodology for quantifying nutrient requirements (Chapter 2) to guide future research efforts devoted to the determination of nutrient requirements of similar or related species whose potential for aquaculture is emerging. In addition, modeling tools for predicting requirements at various life stages are also considered (Chapter 2). Accordingly, this report should serve as a useful guide for the feed industry as well as researchers in aquatic animal nutrition.

Based on the previously stated criteria, major fish groups receiving particular attention in this publication are the warmwater omnivorous species, such as carps, tilapia and catfish; coldwater carnivorous species, such as the salmonids; and warmwater marine species, such as sea bass and sea bream. The major crustacean group includes various marine penaeid shrimp species.

REFERENCES

FAO (Food and Agriculture Organization of the United Nations). 2008. FISHSTAT Plus: Universal software for fishery statistical time series, Version 2.32: Aquaculture production 1950–2006; Commodities production and trade 1976–2006, Fishery Information, Data and Statistics Unit, Fisheries and Aquaculture Department. Rome, Italy: FAO. Available online at: *http://www.fao.org/fi/oldsite/STATIST/FISOFT/FISHPlus.asp*. Accessed on August 23, 2010.

2

Basic Concepts and Methodology

Aquatic animal nutrition is a relatively new science. Although it has many similarities to terrestrial animal nutrition, fish nutrition researchers are confronted with a wide variety of challenges not encountered in terrestrial animal nutrition research. These challenges are primarily due to the aquatic media that require special considerations related to the delivery of nutrients, monitoring of intake, collection and quantification of waste products, and the unique physiology of various species of fish and shrimp. The number of species currently under investigation and the widely different ingredients and corresponding nutrient composition of available diets further confound the goal of collecting accurate data and interpreting the results for practical application.

To the extent possible, aquatic animal nutrition should continue to refine research protocol, adopt concepts and methodological approaches from terrestrial species, build on previous research results, and critically examine the results of studies relative to current knowledge and scientific quality. Obtaining reliable information about the nutrient requirements of fish and shrimp and other crustaceans is based on an array of recommendations presented in this chapter. These recommendations are intended to provide guidance that underlies a common understanding of experimental design and data analysis. Proper experimental protocols and corresponding statistical analyses in the goal of estimating nutrient requirements will positively advance knowledge of fish and shrimp nutrition.

DETERMINATION OF NUTRIENT REQUIREMENTS

It is important that the objectives of any experiment are carefully defined before planning an experimental design. Experimental design considers many factors that include species, culture system, diet design, sample and data collection, chemical analyses, and data analyses. It is important that the design is developed in conjunction with the testing of a clearly defined, concise, and testable hypothesis.

Progress in nutritional research with fish and shrimp requires hypothesis testing using conventional experimental procedures. Using established experimental protocols is essential for experimental results to be compared with those already published. However, not all procedures are commonly applicable to all species or developmental stages of a particular species. Nonetheless, fundamental knowledge of guidelines for experimental designs and procedures for feeding trial execution as well as analysis of data is essential. Therefore, the objective of this chapter is to provide recommendations to control important variables or sources of variation that make it difficult to detect responses to the nutrient or ingredient being studied. These recommendations are designed to establish a level of standardization while recognizing that better approaches to measure and evaluate responses of fish and shrimp to nutritional manipulation will be established and adopted in the future.

EXPERIMENTAL DESIGN AND CONDITIONS

A common goal of any research investigation is to draw a conclusion about treatment effects. Therefore, the aim of experimental design is to minimize variation attributed to experimental error so that treatment effects will be detected if they exist. Although the basic principles of experimental design in fish nutrition studies are similar to those that guide investigations with terrestrial animals, including swine, beef cattle, and poultry, nutritional experiments with aquatic animals have unique considerations. Water, the culture medium and environment in which aquatic organisms live, presents some particular challenges because it may limit observation of the cultured organism and may adversely affect the integrity of prepared feed and the nutrients within if not consumed promptly. The water itself also may provide nutrients such as dissolved minerals and possibly planktonic organisms that also contain nutrients. Thus, it is important to pay attention to the culture system and water supply used in nutrition studies with aquatic animals so that favorable water quality conditions are maintained throughout the duration of the experiment. Maintaining good water quality becomes particularly challenging and critical as the biomass of organisms and feeding rates increase during the course of a feeding trial. Social interactions inherent in many aquatic species, such as density-dependent growth or hierarchical dominance, need to be avoided by keeping the rearing density (numbers and associated biomass) at appropriate levels.

The genetic diversity of fish and shrimp used in experimental studies generally leads to a wider variation in traits and responses when compared to that observed for poultry and rats, the main species on which many aspects of experimental design in nutrition studies are based. The principal causes of variation in response among cohorts are genetic and environmental. Lack of controlled breeding (domestication) of aquatic species over a comparatively long time is a major contrast between many fish and shrimp species and livestock or laboratory animals.

Experimental Culture Systems

The design of experimental culture systems is an important factor in achieving confidence in the evaluation of responses in nutritional experiments. Ultimately, the design of the system allows the organism under investigation to reach its growth potential through effective utilization of experimental diets.

A variety of culture systems, e.g., flow-through, recirculating, semirecirculating, and static, have been used to conduct nutritional experiments. Each system has shortcomings, but the goal is to use a rearing system that allows normal growth of the species under investigation given a set of physical conditions. Use of systems in which there is no continuous flow requires careful management of water quality to prevent accumulation of metabolites that may reduce growth or affect other response variables. Therefore, in systems without flow (static), routine replacement of water is necessary, and this practice can disrupt to the extent that response variables are adversely affected. Recirculating systems need appropriate biological and physical filters to prevent accumulation of metabolites and particulate material (an indirect increase in bacterial load) that can adversely affect growth or other response variables. Flow-through systems need to be managed whereby water quality conditions do not vacillate relative to spatial and temporal origin.

An ideally designed culture system isolates replicates from one another: that is, water in contact with an organism or group of organisms from one replicate does not come in contact with another replicate. This design avoids the possible negative effect of a metabolite in a culture unit(s) being transferred and possibly concentrated as water sequentially passes through other culture units. Although such metabolites may be ephemeral in nature, even small accumulations can negatively influence the growth of organisms exposed to the water. Therefore, systems using a serial delivery of water to replicates of different treatments are to be avoided in nutritional experiments.

When one organism or a group of organisms in an experimental unit such as a tank is the replicate, then the volume of the experimental unit to be used becomes important. It is essential that each experimental culture unit (replicate) has a sufficiently large volume whereby the increase in biomass occurring during the feeding trial does not reach a level that becomes a variable affecting responses to different levels of nutrients. This effect has been demonstrated by D'Abramo et al. (2000) for juvenile prawns *Macrobrachium rosenbergii* individually held in culture containers. Therefore, the stocking density of experimental organisms is planned with an anticipation of what the final density (biomass) or organisms will be. Sufficient tank volume, combined with an appropriate water flow, is needed to produce a tank turnover rate that will eliminate any possible tank effects contributing to variation in treatment-based response variables.

Individual units that compose an experimental system, whether containers, tanks, or aquaria, generally represent the units of observation from which measured responses are obtained for statistical analysis. To conduct an analysis of variance (ANOVA) or other parametric statistical analyses, the assumption is that each unit of observation is independent, including water supply. The number of organisms that are assigned to each culture unit will principally depend on the species being investigated. The number of organisms placed into each experimental unit must be sufficiently low to limit hierarchical feeding behavior and limit the potential effects of unequal sex ratios. If there is more than one organism per experimental unit, the response of the group rather than that of each individual within the group is the appropriate unit for the statistical analysis. In some cases, if the experimental organisms within an experimental unit can be individually identified, their individual responses can be measured. If only two treatments are compared, then a simple "t" test is the preferred statistical test. If the weight of each organism for each dietary treatment is recorded at different times over the duration of an experiment, including initial and final weights, then a growth rate exponential (GRE) can be determined. The GRE is based on the calculated slope of a straight line regression plot of the logarithm of the individual weight of each organism in each treatment determined at different times. The slopes (growth rates) of replicate organisms within each treatment can then be statistically analyzed by ANOVA to determine whether there are any significant differences among treatments. However, repeated weighings of certain species can cause stress leading to mortality that is not treatment dependent.

Experimental Organisms

Source, strain, age, previous nutritional history, and condition of the fish or shrimp can all influence responses in nutritional studies. The source and genetic background of the fish or shrimp used in the experiments should be known and reported. Otherwise, an important variable for any future comparison of responses is not controlled. Genetic background influences growth. An adequate growth rate is necessary to detect responses to nutrient variables. Knowing the age of the experimental organisms avoids undesirable variation resulting from age-dependent differences in nutrient requirements and the corresponding response variables that are measured.

Depending on the goal of the experiment, experimental organisms derived from a single mating (full sibs) or several matings can be used in the interest of possibly reducing the level of variation in a response variable. If the organisms have been derived from several independent spawns, proportional representation among treatment groups may be helpful to minimize variation associated with multiple parentages.

Experimental Diets

To determine nutrient requirements, a diet that ideally permits an independent change (increase, decrease, removal) in the level or concentration of a particular nutrient is required. Research studies often use basal diets that consist of refined ingredients with defined chemical compositions. Such diets are termed semipurified or chemically defined and provide maximum control over the nutrient composition of the experimental diets, particularly the nutrient under investigation. However, for some species, feed intake and growth may be significantly less than optimal when semipurified diets are fed. Therefore, as part of the experimental design, a reference diet of known composition and a demonstrated ability to support normal weight gain of the targeted species is recommended to estimate the growth potential that can ultimately be realized under the conditions of the experiment. Control diets can be composed of practical ingredients (e.g., corn and fish meal) or semipurified ingredients (e.g., starch and casein). In addition, the high and consistent quality of many practical ingredients has led to their use in experiments involving assessment of nutrient requirements. Use of practical ingredients generally requires greater care in maintaining equal levels of all nutrients besides the one under investigation. By careful control over the quality and quantity of practical ingredients, a suitable range in the level of the nutrient under investigation can be achieved. For some species, a live food diet may be used as a reference diet, providing important baseline information about the adequacy of the experimental animals and culture system in the performance of the experiment. The reference diet could be included in a statistical analysis to compare performance to other experimental diets. However, data derived from the reference treatment are not to be included in an experiment in which treatments consisting of chemically defined diets are being used to quantify a nutrient requirement.

With the use of either semipurified or practical diets, adequate preparation is crucial to ensure that desired physical characteristics are controlled and maintained. Each ingredient for all diets should be sourced from one homogeneous lot. After mixing, all diets should be ground to a standard mesh size and then remixed to ensure homogeneity before use in preparing a diet. After the diet is prepared in a form that can be readily consumed by the experimental organism, care also needs to be exercised to avoid the possible influence of heat in reducing the biological availability of the nutrients, particularly those under investigation. Storage under conditions (usually frozen) that maintain the freshness (nutrient integrity) for the duration of the experiment is highly recommended. Small quantities of diet can be stored at 4°C shortly before feeding.

Prior to the beginning of a feeding trial, the diet ought to be analyzed to confirm that the targeted nutrient levels were achieved. The results of the feeding trial should be reported according to analyzed levels rather than the intended (formulated) levels. In the absence of knowledge of whether the intended levels of the nutrient under investigation were achieved, possible confusing and misleading results may occur.

Accurate determination of the dietary requirement for a nutrient requires a sufficient number of treatments (diets) formulated to contain graded amounts of the nutrient under investigation and that all other nutrients in the experimental diets are provided at levels equal to or in excess of their requirements. A basal diet, not supplemented with the targeted nutrient, provides a baseline level for evaluation of the response variable(s). The other experimental diets would satisfactorily span a graded range of concentrations of the nutrient, from significantly deficient to in excess of the anticipated requirement.

The design of experiments to determine nutrient requirements commonly begins with a wide range of equally spaced levels for initial evaluation of the response. The results are then used to select an appropriately smaller interval that will satisfactorily serve to establish a precise estimate of the requirement. Generally, the total number of treatments within a selected interval is at least five.

Feed Management and Duration of Experiment

Experimental organisms are normally fed at rates approaching apparent satiation to ensure maximum growth so potential dietary differences can be detected, if they exist. Maximum growth can be achieved by hand feeding to satiation in an appropriate number of meals or by providing a slight excess of the amount of feed that the organisms will consume in a specified period of time. Any effort to calculate feed utilization will focus on minimizing the amount of uneaten feed or recovering uneaten feed and subtracting that amount recovered from the amount delivered. The number of feedings that yield maximum growth or feed efficiency can be species- or life-stage specific and should be an important consideration in the design of an experiment. Young, rapidly growing organisms generally benefit from being fed several times per day, and feeding frequency will be adjusted accordingly.

Most species of fish used in research studies exhibit reduced feed intake for a short period when their diet is changed. Normally, experiments do not begin until fish or shrimp in the study have consumed the same diet for a period of time to avoid influencing response variables. A period of time is necessary to establish a "baseline" and may involve the feeding of a conditioning or transition diet. In some cases, the ingredient and nutrient composition is similar to the experimental diets to be investigated. This approach serves to establish a nutritional baseline among experimental organisms so that each treatment group contains organisms that presumably are nutritionally equivalent before any assessment can occur.

Juvenile organisms grow faster than preadults or adults and therefore commonly manifest responses to different dietary levels of a nutrient more quickly. When weight gain is used as a response variable, care needs to be exercised to ensure that the magnitude of the response achieves a level that is sufficiently great for comparison among treatments before an experiment is terminated. For rapidly growing organisms, observed changes in weight gain of up to 1,000% of initial weight (10 ×) are recommended, but for larger fish and shrimp an increase of 200 to 300% (2–3 ×) may be acceptable. A standard often used is a 300% increase in body weight (which represents the twice doubling of the initial weight). The aforementioned increases are guidelines, and ultimately acceptable weight gain responses may be lower as determined by the objective of the experiment as well as the life stage of the organisms under consideration. The time required (duration of the experiment) to observe these recommended weight increases is dependent on the rate of response to the nutrient under investigation and the temperature of the water. A readily observable difference in monitored responses among treatments may permit early termination of the feeding trial. However, if experiments are terminated prematurely, potentially significant responses to the range of nutrient levels under investigation may not be observed.

It has been suggested that the duration of feeding normally encompass at least 8 to 12 weeks, but it is not appropriate to arbitrarily assign a specific time period (Cowey, 1992). The important point is that if a dose-response experiment is conducted, then a sufficient increase in weight will be required to observe recognizable and perhaps significant dose-response relationships. Weight gain is the most common response variable recorded, but it may not always be the most appropriate (Baker, 1986).

REPLICATES IN THE EXPERIMENTAL DESIGN

Requirement estimates are typically based on experiments with regression design. Knowledge of the variation in response of the organism under investigation is the basis for determining the number of replicates needed per dietary treatment. Response of experimental fish and shrimp is a product of the treatment applied combined with experimental/biological variability. Variation due to dietary treatments must be detectable beyond the experimental/biological variation in responses exhibited by the cultured organism in replicate units. A combination of an inherently wide variation of responses coupled with an insufficient number of replicates can result in failing to identify a significant treatment response when one actually exists. The number of replicates used in requirement studies will vary with the experimental design, species, and strain of fish or shrimp, and aims of the study. A statistical power analysis can be used to determine the number of replicates that will result in statistical judgments that are accurate and reliable relative to the response variable measured (Bausell and Li, 2002). It is important that the power that is established be sufficiently high to detect reasonable departures from the null hypothesis. Recognition of the importance of power analysis and calculation of sample size is growing and becoming integral to the proper design of experiments.

MEASURED RESPONSES

Nutrients in a well-balanced diet are intended to support all physiological requirements and growth of the culture species. Dietary ingredients may contain chemical compounds that positively, negatively, or neutrally influence growth. Researchers need to design experiments with the understanding of what physiological responses will confidently and accurately provide an appropriate measure of the animal's integrated response to the desired nutrient, additive or metabolic modifier.

Weight Gain

Growth, the deposition of new tissue, is the most commonly used response to evaluate modifications to the dietary content of a nutrient. The growth potential of an aquatic animal species is influenced by life stage, genetic strain, environmental conditions, and nutrient intake. Growth can be measured as a change in weight or length of the organism, but weight gain is most commonly measured in aquatic animal nutrition studies. If all of the experimental organisms begin at the same weight, then weight gain responses to different dietary treatments can be expressed in various manners. Common examples of calculated measures of weight gain responses include:

Mean weight = total biomass of animals/number of animals

Weight gain = final weight (W_t) – initial weight (W_0)

Percent weight gain = (final weight – initial weight) × 100 / initial weight

Instantaneous growth rate (IGR) = $W_t = W_0 (1 + \alpha / 100) t$

where: $\alpha = (\ln W_t - \ln W_0) \times 100 / t$ and t = time (days).

Growth rates should be reported based on a model that fits the response of the specific stage of the organism (Dumas et al., 2010). Animals do not grow geometrically (i.e., exponentially, across life stages). For example, larval growth is commonly exponential, so IGR is an appropriate measure. However, the exponential growth function is not suitable for predicting or describing the growth trajectory of other life stages of fish and other animals, although it has been incorrectly reported for such (Dumas et al., 2010). In addition, transformation is highly dependent on the initial weight and yields unavoidably systematic deviations.

Although IGR remains in use as an expression of growth rate, it has severe limitations because it generally does not properly represent the growth trajectory of fish. The IGR has no comparative application between and within studies or even across the duration of an experiment. This mathematical model is inappropriate and should not be used.

An accurate and useful coefficient to express or predict fish growth relative to water temperature is based on the exponent 1/3 power of body weight (Iwama and Tautz, 1981).

$$\text{Thermal-unit growth coefficient (TGC)} = (\text{FBW}^{1/3} - \text{IBW}^{1/3}) / \Sigma (\text{T} \times \text{D}) \times 100$$

$$\text{Predicted final body weight} = (\text{IBW}^{1/3} + \Sigma (\text{TGC} / 100 \times \text{T} \times \text{D}))^3$$

where: IBW is initial body weight (g/fish), FBW is final body weight (g/fish), D is the number of days, and T is water temperature (°C).

There is convincing evidence that this model equation truly represents the actual growth curves of rainbow trout, lake trout, brown trout, Chinook salmon, and Atlantic salmon over a wide range of temperatures. The TGC model has since been widely used in the aquaculture literature (e.g., Einen et al., 1995; Kaushik, 1998; Willoughby, 1999; Stead and Laird, 2000; Hardy and Barrows, 2002). Lupatsch et al. (2003) used a growth model that was similar to TGC but differed in that different weight exponents were specifically applied to describe the growth of each of the species (seabream, sea bass, and grouper) under investigation.

The TGC values and growth rates are dependent on species, stock (genetics), nutrition, environment, husbandry, and others factors; therefore, the TGC for a given aquaculture condition is calculated using past growth records or records obtained from similar stocks and husbandry conditions. This simple model has been adapted recently to the different growth stanzas of rainbow trout across life stages (Dumas et al., 2007) and is considered useful for estimation of the growth of fish over an extended period of time. Despite its convenient application, the thermal unit approach can result in systematic errors arising from situations where the temperature moves too far away from the optimum for growth (Krogh, 1914; Hayes, 1949; Ricker, 1979; Jobling, 2003).

Feed Utilization

New tissue deposition is the net result of the retention of nutrients by the animal as influenced by diet composition, feed consumption, and physiological state, and is consequently not constant. The most common terms used to describe the conversion of feed to weight gain is feed conversion ratio (FCR, total feed fed divided by body weight gain) or feed efficiency (FE, body weight gain divided by total feed fed × 100). These descriptive responses are quantitative measures of efficiency but semidescriptive because the nutrient composition of both gain and feed may vary. Therefore, these terms provide limited, descriptive information about nutrient utilization and are primarily used to assess economic practicality. Accurate measures of conversion efficiencies are only achieved through accurate estimates of uneaten feed combined with high survival.

Other calculated responses have been adopted from investigations with other species, but are not always properly applied. For example, protein efficiency ratio (PER) is derived from published literature using the rat to evaluate the quality of a protein and is defined as weight gain per unit of protein fed, when the dietary protein level and daily ration are fixed and below requirement. This assay is applied for comparison of treatment responses within an individual feeding trial, but has limited comparative value among different trials because conditions are rarely the same. More informative measures of an experimental organism's responses to dietary treatments generally entail the determination of the composition of their gained weight.

Tissue Composition

Lean tissue synthesis as protein rather than lipid deposition is typically the preferred form of weight gain in aquatic organisms produced for food. Thus, in most nutrition experiments, the proximate composition (protein, lipid, ash, and moisture content) of the organism's whole-body tissue is determined before and after the feeding trial. Total energy may also be determined. Retention of dietary protein, lipid, or energy is computed as:

$$\% \text{ retention} = (\text{final weight} \times \text{final nutrient (or energy) content}) - (\text{initial weight} \times \text{initial nutrient (or energy) content}) \times 100 / \text{nutrient (or energy) intake}$$

Animals seek to eat a sufficient amount of a nutritionally balanced feed in order to follow a genetically determined growth path (Oldham et al., 1997), where maximal protein accretion and associated carcass lean growth rate determine nutrient requirements for growth and composition of growth (Schinkel and de Lange, 1996). Thus, as protein deposition is the main determinant of amino acid requirements, information on the requirement level is ideally based in relationship to protein or lysine retention, as proposed by Rodehutscord et al. (1997), Susenbeth et al. (1999), and Hauler and Carter (2001). Estimation of nutrient requirements is significantly impacted by what response parameter(s) is selected. For example, the requirement for maximizing protein gain is higher than that required to maximize weight gain (Bureau and Encarnação, 2006).

Weight gain and protein deposition often serve as poor indicators in requirement studies of some micronutrients, such as minerals. For example, Wilson and El Naggar (1992) reported that weight gain of channel catfish did not respond quantitatively to supplementation of dietary potassium. The reported requirement of this species was therefore based on the whole-body potassium retention. The use of whole-body mineral retention has also been demonstrated in requirements of potassium (Shiau and Hsieh, 2001), sodium (Shiau and Lu, 2004), and manganese (Lin et al., 2008a) for hybrid tilapia, as well as selenium (Lin and Shiau, 2005) and copper (Lin et al., 2008b) for grouper. For these and other nutrients such as vitamins, subclinical measures such as enzyme activities or tissue levels of the nutrient or its metabolites are likely to be more sensitive measures than weight gain. Nonetheless, growth data obtained in long-term studies remain an important way to determine a dietary requirement for some trace elements. Thus, when dealing with micronutrients, tissue studies are an important accompaniment to growth and feed intake data.

Beside growth responses, measurements of enzyme activity in tissues also may be useful in nutrient requirement studies with some minerals and vitamins because a functional measure of the nutritional status of an organism is provided with respect to those specific nutrients (Cowey, 1976). However, attempts to determine requirements based on activity of a key enzyme of which the vitamin is a component or cofactor have not succeeded (Yen et al., 1976; Anderson et al., 1978). This approach was questioned by Baker (1986), who contended that maximal enzymatic activity is not necessarily desirable in a growing animal. Nevertheless, enzyme activity data are often useful as indicators of incipient deficiency. Recent investigations have shown that dietary nutrients can influence the product of gene expression, and this knowledge establishes a basis for the determination of nutrient requirements (Cahu et al., 2003; Villeneuve et al., 2006). Vitamin deficiency and excess and other dietary components (phospholipids) have exerted effects on gene expression for skeletal development.

ESTIMATING QUANTITATIVE NUTRIENT REQUIREMENTS

A variety of statistical procedures have been used to quantify nutrient requirements of aquatic organisms including mean separation tests after ANOVA, nonlinear regression, and various linear kinetic models. Mean separation is technically not a correct statistical analysis for quantitative data such as that generated by feeding graded levels of a nutrient. In addition, this approach has limited value as it does not accurately estimate the minimum level of nutrient required to maximize the growth response. In contrast to mean separation techniques, various types of regression analysis or fitting response curves is the preferred way of evaluating the responses to graded nutrient levels if the dietary levels are sufficiently varied. However, fitting response curves has been handled quite differently and several nonlinear models as well as linear (broken-line model) approaches have been utilized. Shearer (2000) emphasized that any reported nutrient requirements should be scrutinized by examination of experimental design and the corresponding statistical procedures. Using regression analysis, he demonstrated that previously published estimates of nutrient requirements of fish, derived from ANOVA and broken-line analysis, were underestimates, primarily because of inappropriate choice of statistical analyses.

Broken-Line Model

The broken-line model has been the most widely used method of evaluating dose-response data in nutrient requirement studies with aquatic species. This technique involves using two straight lines to model the dose-response relationship (Robbins et al., 1979).

This linear model assumes that the growth response of an animal to increasing dietary intake of a limiting indispensable nutrient will increase linearly until the requirement is met, after which no increase in response, represented by a horizontal line (slope = 0) (Figure 2-1a), or a negative response will be observed. The break point corresponds to the nutrient requirement or minimum nutrient level that will produce the maximum response.

The general equation of the broken-line model is:

$$Y = f + 0.5\, g\, (x - h - | x - h |)$$

where Y is the measured response (body weight gain g/organism), x is the independent variable (nutrient concentration in g/kg diet), f is the ordinate, h is the abscissa of the breakpoint, and g is the slope of the line for $x < h$.

This model is fitted by obtaining the least squares estimates of f and g for several values of x using the ordinary least squares technique. The maximum likelihood estimate of h is accepted as the value of h that maximizes the model sum of squares (Robbins et al., 1979).

For several amino acids, intake and weight gain are apparently linearly related. Therefore, broken-line regression became the model of choice for requirement studies, mainly because of the convenient and straightforward method of objectively defining the point of intersection of two linear functions as "the requirement." Thus, this approach simplified the interpretation of response curves. However, it is generally recognized that this model tends to provide lower

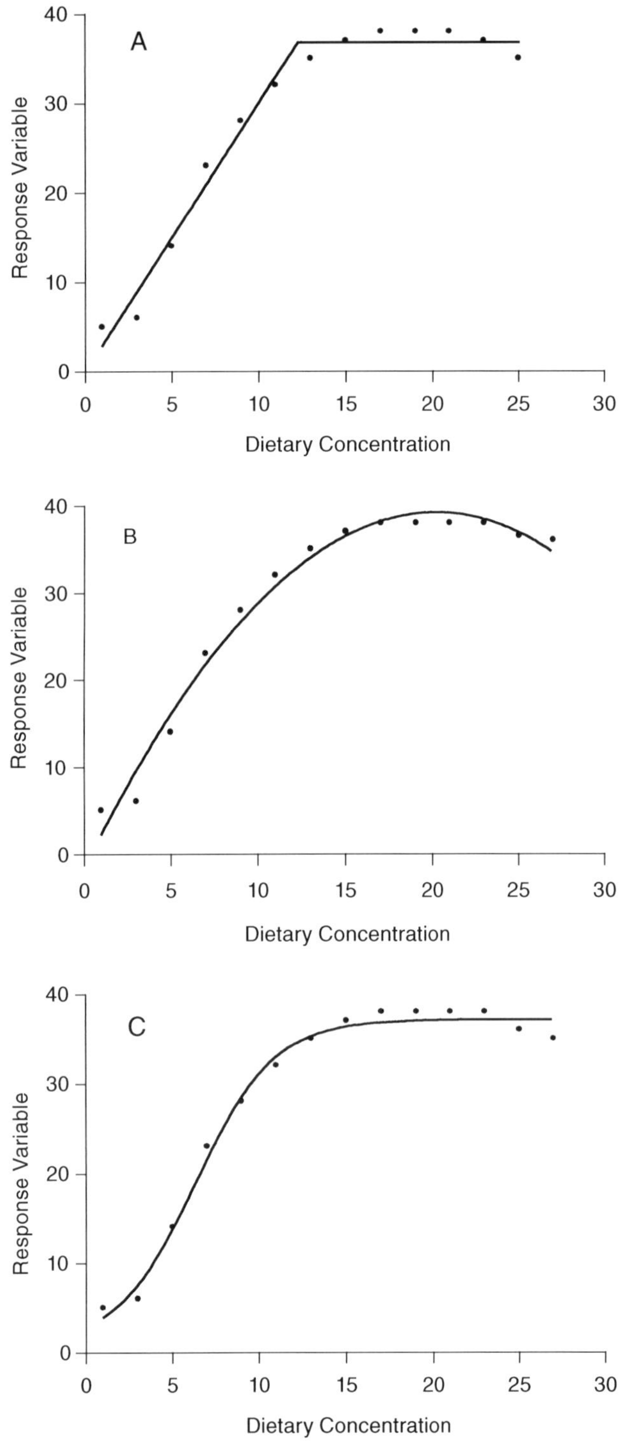

FIGURE 2-1 Models for the interpretation of dose-response experiments. A: broken-line regression. B: exponential model. C: sigmoidal model.

requirement estimates than other models described below. When regression analyses are used, all data points derived from each replicate of each treatment need to be included rather than the common methodological shortcoming of using the overall mean data point for each treatment.

Nonlinear Models

Nonlinear models began to be used in an effort to address situations where the data were not properly fitting the broken-line model. In these cases, weight gain of the animal reached a maximum and decreased significantly. Nonlinear models are therefore based on the biological principle that is often referred to as the "law of diminishing returns." Among all the nonlinear approaches, the exponential (Rodehutscord et al., 1995, 1997) and sigmoidal (Mercer et al., 1989; Gahl et al., 1991) models have been commonly used in requirement studies with aquatic species (Figure 2-1b and c). The limitation of these models is the arbitrary assignment of 95% as being the maximum response.

For the exponential model (Figure 2-1b), the response to increasing nutrient intake is not linear and is defined as a decreasing increment in gain as equal increments of the dietary nutrient are added near the level at which maximum gain is realized (Gahl et al., 1991). The exponential model derived from Rodehutscord et al. (1995) is:

$$Y = Y_{max} \left(1 - e^{-b(x-c)}\right)$$

where Y_{max} is the plateau of the curve (upper asymptote), b is the parameter describing the steepness of the curve, and c is x at y = 0.

At highly deficient levels, the first levels of supplementation of the limiting nutrient will result in a significant increase in performance (linear response). The effects will diminish progressively with further increased levels of supplementation, evidenced by the slope of the curve progressively decreasing, until no further increase in response occurs. The requirement estimated from nonlinear curves is generally calculated as that nutrient level required to achieve 95% of maximum response, such as weight gain, protein gain, and survival.

The sigmoidal response model (Figure 2-1c) consists of four parameters to describe the sigmoidal shape of the response curve (Mercer et al., 1989). They defined the different sections of this nutrient response curve as threshold, deficient, adequate, and optimal. An inhibition constant was then included for the development of the Saturation Kinetics Model (SKM).

The sigmoidal model as derived from Gahl et al. (1991) is:

$$Y = \frac{Y_{max} + (d\,(1 + m) - Y_{max})\, e^{-kx}}{1 + m\, e^{-kx}}$$

where Y_{max} is the plateau of the curve (upper asymptote), d is the intercept of the y axis, k is the scaling parameter for x, and m is the shaping parameter that locates the inflection point.

A nonlinear approach model has been suggested to be the most appropriate to describe biological responses (Mercer et al., 1989; Gahl et al., 1991; NRC, 1993; Cowey, 1994; Rodehutscord and Pack, 1999). However, if different models are used, it is critical that the experimental design, specifically the range of nutrient concentrations under evaluation, is considered. If there is a very wide range in dietary concentration, from extremely deficient to clear inhibition, then the SKM with inhibition appears to be the most appropriate approach (Rodehutscord and Pack, 1999). However, most of the nutrient requirement studies that have been published rarely include extremely high concentrations of the investigated nutrient in the experimental design.

The sigmoidal approach generally appears superior to the exponential approach in those requirement studies where the dietary concentration of the nutrient varies from severely deficient to high above the optimum level because the linear or near linear part of the ascending section of the response curve is better described (Rodehutscord and Pack 1999). However, if the nutrient is evaluated within a restricted range, then use of the simple, easy to interpret exponential approach is well justified (Rodehutscord and Pack 1999). This approach is particularly suitable for those investigations where the dietary ingredients preclude formulation of diets that contain severely deficient levels of the nutrient under investigation.

Essential amino acid (EAA) requirements of rainbow trout were reevaluated using the exponential model (Rodehutscord et al., 1995), the four parameter logistic function (Gahl et al., 1991), the SKM (Mercer et al., 1989), and the broken-line model (adaptation of Robbins et al., 1979). Estimates of the requirements were considerably different and dependent on the model used (Rodehutscord and Pack, 1999). For the EAA analyzed, the lowest requirement values were derived from use of the broken-line model, while estimated requirements using the exponential model were the highest. Encarnação et al. (2004) observed that use of the broken-line model resulted in an estimated dietary requirement for lysine (1.8% of diet) that was lower than that derived from both the exponential and four-parameter logistic equation models (2.3% of diet). The estimated requirement value of 2.3% of diet agreed well with that reported by Rodehutscord et al. (1997) who used a similar dietary and data analysis model. The comparative results of the aforementioned investigations suggest that the requirement value for rainbow trout (1.8% of diet), as proposed by NRC (1993) and derived from the advocated use of the broken-line model, falls significantly below what is required to maximize weight gain. The intercept of the two linear functions is clearly below the lysine level maximizing weight gain and therefore fails to accurately represent the response of the animal to increasing levels of this amino acid.

Factorial Model

Shearer (1995) introduced a unique look at the determination of nutrient requirements of elements for fish through a factorial model. The model required the determination of factors such as bioavailability of the nutrient within the feed, availability of the nutrient from the water, the requirement for new tissue synthesis, and endogenous loss. This proposed approach was based on the argument that nutrient requirements derived via this model would have broader applicability than requirements generated from a variety of other empirical approaches.

Bioenergetic Modeling

Another increasingly popular method is the use of bioenergetic modeling to predict requirements for nutrients, primarily protein and energy. This approach is described in Chapter 13.

CONCLUSIONS

Although many methodological principles currently used in nutritional studies of fish and shrimp have been derived from nutritional investigations of terrestrial animals, new and different approaches are also warranted because of the variety of species under investigation and because of an aquatic culture medium. Experimental protocol, such as whether restrictive or satiation feeding will be used, is dictated by the objective of the research. Nonetheless, recommendations provided in this chapter cut across all species relative to the important features to be maintained in experimental design, collection of data, and analysis of results. Continued improvements in methodology, sometimes species-specific, will increase confidence in the results that are obtained. Future work needs to focus more on obtaining a better understanding of interactions among nutrients themselves and antinutritional factors that affect their availability. Caution needs to be exercised in the application of knowledge of nutrient requirements because differences may exist based on the culture system that is used, the specific stage of development of the organism, the response variable that is measured, or whether the species is freshwater, marine, or estuarine.

Nutrient requirement information provides the foundation for the development of cost-effective, practical formulations. Knowledge of the nutrient composition and availability of practical feedstuffs will be necessary to ensure that requirements are met. The challenge does not stop there. For those species grown in ponds, primary and secondary productivity may contribute to the satisfaction of nutrient requirements. High-density (intensive) culture will require nutritionally balanced feeds, whereas in semi-intensive and extensive culture with reduced densities, formulations may take advantage of nutrient supplements from the consumption of natural food. Therefore, based upon the level of contribution of natural food, requirement levels can be conservatively reduced. Combining the knowledge of nutrient requirements with an understanding of the relative contributions of nutrients from natural productivity will lead to the development of effective formulated feeds for pond culture.

REFERENCES

Anderson, P. A., D. H. Baker and S. P. Mistry. 1978. Bioassay determination of the biotin content of corn, barley, sorghum and wheat. J. Anim. Sci. 47:654–659.

Baker, D. H. 1986. Problems and pitfalls in animal experiments designed to establish dietary requirements of essential nutrients. J. Nutr. 116:223–2249.

Bausell, R. B., and Y. Li. 2002. Power Analysis for Experimental Research. A Practical Guide for the Biological, Medical and Social Sciences. Cambridge, UK: Cambridge University Press.

Bureau, D. P., and P. M. Encarnação. 2006. Adequately determining the amino acid requirements of fish: The case example of lysine. Pp. 29–51 in Avances en Nutricion Acuicola VIII, L. E. Cruz-Suarez, D. Ricque-Marie, M. Tapia-Salazar, M. G. Nieto-Lopez, D. A. Villareal Cavazos, A. C. Puello Cruz, and A. Garcia-Ortega, eds. VIII Simposium Internacional de Nutricion Acuicola, 15–17 November 2006. Mazatlan, Sinaloa, Mexico. Monterrey, Nuevo Leon, Mexico: Universidad Autonoma de Nuevo Leon.

Cahu, C., J. Z. Infante, and T. Takeuchi. 2003. Nutritional components affecting skeletal development in fish larvae. Aquaculture 227(1–4):245–258.

Cowey, C. B. 1976. Use of synthetic diets and biochemical criteria in the assessment of nutrient requirement of fish. J. Fish. Res. Board Can. 33:1040–1045.

Cowey, C. B. 1992. Nutrition: Estimating requirements of rainbow trout. Aquaculture 100:177–189.

Cowey, C. B. 1994. Amino acid requirements of fish: A critical appraisal of present values. Aquaculture 124:1–11.

D'Abramo, L. R., W. H. Daniels, P. Gerard, W. H. Jun, and C. G. Summerlin. 2000. Influence of water volume, surface area, and water replacement rate on weight gain of juvenile freshwater prawns Macrobrachium rosenbergii. Aquaculture 182:161–171.

Dumas, A., J. France, and D. P. Bureau. 2007. Evidence of three growth stanzas in rainbow trout (*Oncorhynchus mykiss*) across life stages and adaptation of the thermal-unit growth coefficient. Aquaculture 267:139–146.

Dumas, A., J. France, and D. P. Bureau. 2010. Modelling growth and body composition in fish nutrition: Where have we been and where are we going? Aquac. Res. 41:161–181.

Einen, O., I. Holmefjord, T. Åsgård, and C. Talbot. 1995. Auditing nutrient discharges from fish farms: Theoretical and practical considerations. Aquac. Res. 26:701–713.

Encarnação, P., C. de Lange, M. Rodehutscord, D. Hoehler, W. Bureau, and D. P. Bureau. 2004. Diet digestible energy content affects lysine utilization, but not dietary lysine requirements of rainbow trout (*Oncorhynchus mykiss*) for maximum growth. Aquaculture 235:569–586.

Gahl, M. J., M. D. Finke, T. D. Crenshaw, and N. J. Benevenga. 1991. Use of four-parameter logistic equation to evaluate the response of growing rats to ten levels of each indispensable amino acid. J. Nutr. 121:1720–1729.

Hardy, R. W., and F. T. Barrows. 2002. Diet formulation and manufacture. Pp. 505–600 in Fish Nutrition, 3rd Edition, J. E. Halver and R. W. Hardy, eds. San Diego, CA: Academic Press.

Hauler, R., and C. Carter. 2001. Lysine deposition responds linearly to marginal lysine intake in Atlantic salmon (*Salmo salar* L) parr. Aquac. Res. 32(suppl. 1):147–156.

Hayes, F. R. 1949. The growth, general chemistry, and temperature relations of salmonid eggs. Q. Rev. Biol. 24:281–308.

Iwama, G. K., and A. F. Tautz. 1981. A simple growth model for salmonids in hatcheries. Can. J. Fish. Aquat. Sci. 38:649–656.

Jobling, M. 2003. The thermal growth coefficient (TGC) model of fish growth: A cautionary note. Aquac. Res. 34:581–584.

Kaushik, S. J. 1998. Nutritional bioenergetics and estimation of waste production in non-salmonids. Aquat. Living Resour. 11:211–217.

Krogh, A. 1914. On the influence of the temperature on the rate of embryonic development. Z. Allg. Physiol. 16:163–177.

Lin, Y.-H., and S. Y. Shiau. 2005. Dietary selenium requirements of juvenile grouper, *Epinephelus malabaricus*. Aquaculture 250:356–363.

Lin, Y.-H., S.-M. Lin, and S.-Y. Shiau. 2008a. Dietary manganese requirements of juvenile tilapia, *Oreochromis niloticus* × *O. aureus*. Aquaculture 284:207–210.

Lin, Y.-H., S.-M. Lin, and S.-Y. Shiau. 2008b. Dietary copper requirements of juvenile grouper, *Epinephelus malabaricus*. Aquaculture 274:161–165.

Lupatsch, I., G. Wm. Kissil, and D. Sklan. 2003. Comparison of energy and protein efficiency among three fish species: gilthead seabream (*Sparus aurata*), European seabass (*Dicentrarchus labrax*) and white grouper (Epinephelus aeneus): Energy expenditure for protein and lipid deposition. Aquaculture 225:175–189.

Mercer, L. P., H. E. May, and S. J. Dodds. 1989. The determination of nutritional requirements in rats: Mathematical modeling of sigmoidal, inhibited nutrient-response curves. J. Nutr. 119:1465–1471.

NRC (National Research Council). 1993. Nutrient Requirements of Fish. Washington, DC: National Academy Press.

Oldham, J. D., G. C. Emmans, and I. Kyriazakis.1997. Limits and limitations to nitrogen use in farm animals. Proc. Nutr. Soc. 56:525–534.

Ricker, W. E. 1979. Growth rates and models. Fish Physiol. 8:677–743.

Robbins, K. R., H. W. Norton, and D. H. Baker. 1979. A flexible growth function for empirical use. J. Exp. Bot. 10:290–300.

Rodehutscord, M., and M. Pack. 1999. Estimates of essential amino acid requirements from dose-response studies with rainbow trout and broiler chicken: Effect of mathematical model. Arch. Anim. Nutr. 52:223–244.

Rodehutscord, M., S. Jacobs, M. Pack, and E. Pfeffer. 1995. Response of rainbow trout (*Oncorhynchus mykiss*) growing from 50 to 150 g to supplements of DL-methionine in a semipurified diet containing low or high levels of cysteine. J. Nutr. 125:964–969.

Rodehutscord, M., A. Becker, M. Pack, and E. Pfeffer. 1997. Response of rainbow trout (*Oncorhynchus mykiss*) to supplements of individual essential amino acids in a semipurified diet, including an estimate of the maintenance requirement for essential amino acids. J. Nutr. 127:1166–1175.

Schinkel, A. P., and C. F. M. de Lange. 1996. Characterization of growth parameters needed as inputs for pig growth models. J. Anim. Sci. 74:2021–2036.

Shearer, K. D. 1995. The use of factorial modeling to determine the dietary requirements for essential elements in fishes. Aquaculture 133:57–72.

Shearer, K. D. 2000. Experimental design, statistical analysis and modeling of dietary nutrient requirements studies for fish: A critical review. Aquacult. Nutr. 6:91–102.

Shiau, S.-Y., and J. F. Hsieh. 2001. Dietary potassium requirement of juvenile grass shrimp *Penaeus monodon*. Fish. Res. 67:592–595.

Shiau, S.-Y., and L.-S. Lu. 2004. Dietary sodium requirement determined for juvenile hybrid tilapia (*Oreochromis niloticus* × *O. aureus*) reared in fresh water and seawater. Brit. J. Nutr. 91:585–590.

Stead, S. M., and L. Laird. 2000. Handbook of Salmon Farming. New York: Springer.

Susenbeth, A., T. Dickel, A. Diekenhorst, and D. Hohler. 1999. The effect of energy intake, genotype, and body weight on protein retention in pigs when dietary lysine is the first-limiting factor. J. Anim. Sci. 77:2985–2989.

Villeneuve, L. A. N., E. Gisbert, J. Moriceau, C. L.Cahu, and J. L. Zambonino Infante. 2006. Intake of high levels of vitamin A and polyunsaturated fatty acids during different developmental periods modifies the expression of morphogenesis genes in European sea bass (*Dicentrarchus labrax*). Br. J. Nutr. 95(4):677–687.

Willoughby, S. 1999. *Manual of Salmonid Farming*. Oxford, UK: Fishing News Books, Blackwell Science.

Wilson, R. P., and G. El Naggar. 1992. Potassium requirement of fingerling channel catfish, *Ictalurus punctatus*. Aquaculture 108:169–175.

Yen, J. T., A. H. Jensen, and D. H. Baker. 1976. Assessment of the concentrations of biologically available vitamin B-6 in corn and soybean meal. J. Anim. Sci. 42:866–870.

3

Digestive Physiology of Fish and Shrimp

Digestion is the process of solubilizing and degrading nutrients into smaller components and elements that can be transported across the intestinal wall to support physiological processes. This chapter reviews comparative aspects of digestive function in fish and shrimp relevant to aquaculture. An understanding of the digestive processes and their limitations is necessary for the formulation of diets that can fulfill nutrient requirements. First, some considerations about feeding habits of species are presented. Then, knowledge about the structure of digestive organs, secretions of different parts of the digestive tract, enzyme activities, hydrolytic processes, and nutrient transport are addressed. The chapter focuses on the adult stages of fish and shrimp and their ability to digest the macronutrients proteins, lipids, and carbohydrates. Earlier stages of development are partly covered in other chapters. Sources of information for the present chapter include recent original papers as well as books, book chapters, and review papers such as Ceccaldi (1997), Carrillo-Farnes et al. (2007), Cyrino et al. (2008), Kuz'mina (2008), and Holmgren and Olsson (2009).

FISH

Knowledge about what an organism eats aids in understanding the diversity of the anatomical and physiological characteristics of the organism. Some fish species feed on dead items (scavengers), others on living material, some feed solely on microorganisms, others on larger plants and animals, and some are opportunistic eating whatever they can find. Food for fish in the wild comprises detritus, phytoplankton, zooplankton, micro- and macroalgae, aquatic plants, meiofauna, insects, crustaceans, mollusks, shellfish, fish, seeds and fruits, and even birds and mammals (Platell and Potter, 2001; Lundstedt et al., 2004; De Almeida et al., 2006). One way to classify fish is according to the primary ingredients of their natural diet; herbivores (milkfish and some carps), omnivores (channel catfish and some tilapia), and carnivores (salmonids, basses, seabreams, flounders, and groupers). Species that have a similar dietary selection may show great variation in intestinal anatomy, and within the same species there are differences among developmental stages.

Structural and Functional Aspects of Digestive Organs

Variation in anatomy and histomorphology of the digestive tract among fish species is greater than for any other phylum (Buddington and Kuz'mina, 2000a,b). The tract can be subdivided into the foregut with mouth, pharynx, esophagus, and stomach; the midgut with pyloric ceca; and the distal or hindgut terminating in the rectum. Figure 3-1 illustrates general characteristics of the digestive tract of fish grouped in four categories according to the anatomy of the tract. Figures 3-2 and 3-3 illustrate the organization of internal organs in a generalized fish with stomach and pyloric ceca and in Atlantic cod, respectively. The anatomy, particularly of the foregut, has presumably developed through evolution and been influenced by the nature of the food of the species to allow efficient intake and digestion. Bottom feeders have downward orientation of the mouth, whereas species eating food in the water column have the mouth oriented at the tip of the body (Jobling, 1995). There seems to be a relationship between mouth size of the fish and size of the food. However, this is not always the case (Platell and Potter, 2001). The second largest fish in the world, the basking shark (*Cetorhinus maximus*), is a filter feeder feeding on planktonic prey.

Most fish species start out at hatching with a straight simple digestive tract without a stomach. Through the larval and juvenile stages, the gastrointestinal (GI) tract develops into more complicated structures. Some fish continue to have a short and relatively simple tract, whereas others have long, more complex tracts. Some fish species, mostly herbivores, have no stomachs even as adults. Most fish without a true stomach belong to the microphagous, detrivorous,

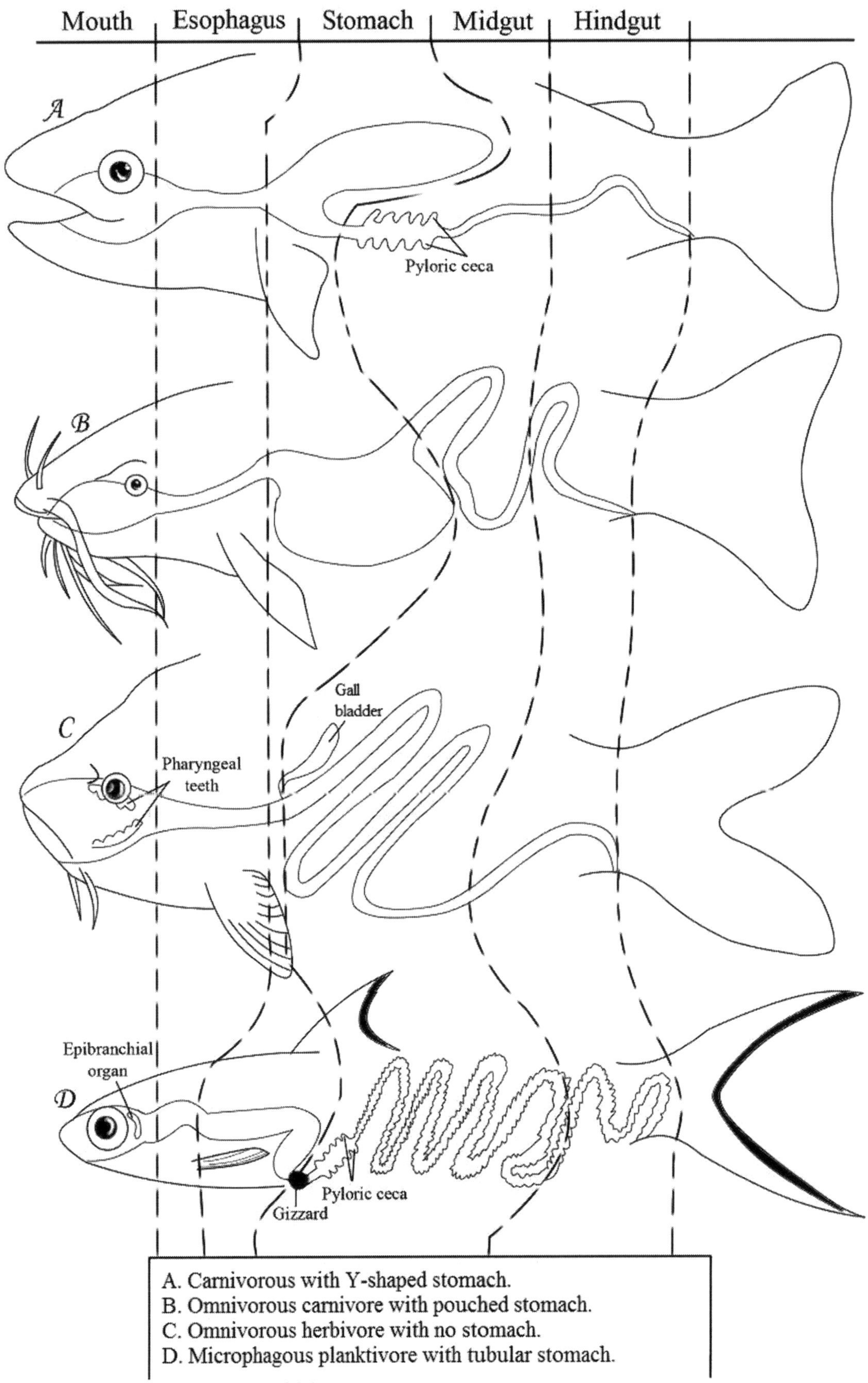

FIGURE 3-1 Comparative digestive anatomy of fish.
Illustration courtesy of Victoria Blondin, University of Guelph, Ontario.

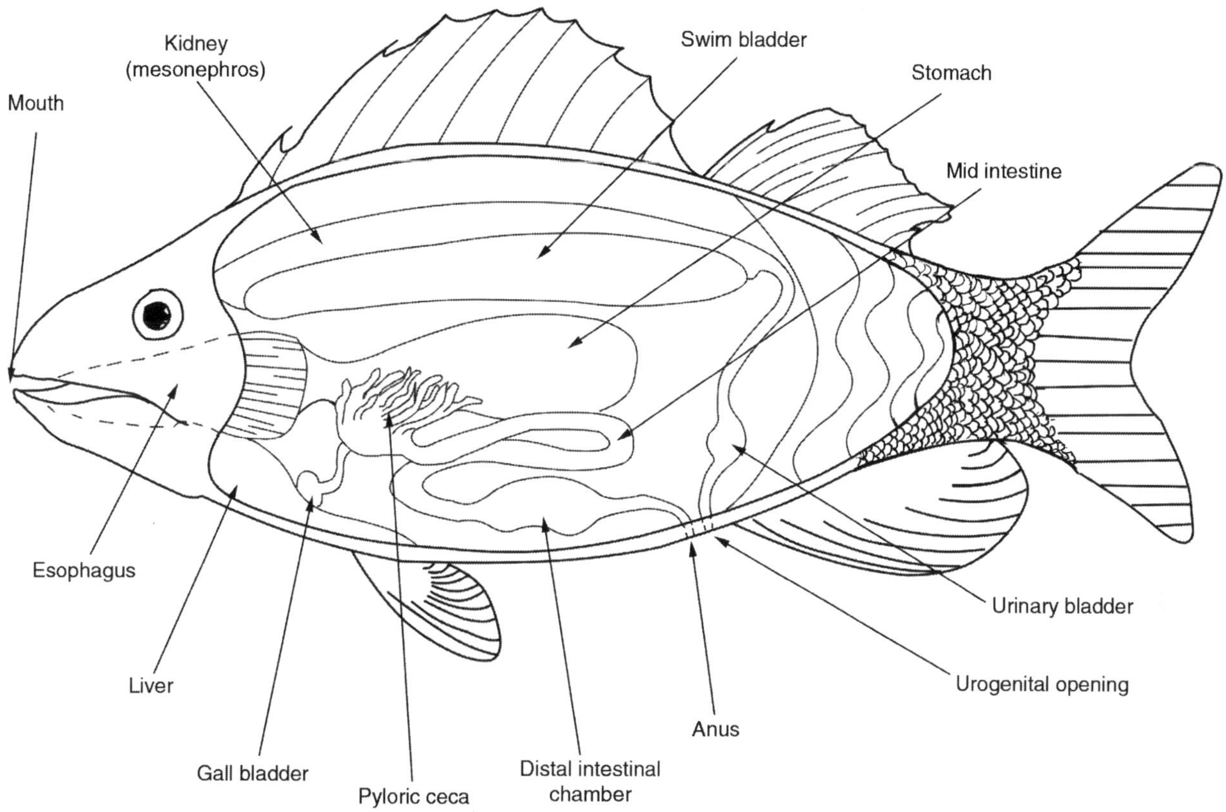

FIGURE 3-2 Organization of internal organs in a generalized fish.
Illustration courtesy of Victoria Blondin, University of Guelph, Ontario.

and herbivorous species. There are, however, examples of carnivorous fish without a stomach such as the cyprinid Colorado squawfish (*Ptychocheilus oregoensis*) (Jobling, 1995).

Fish vary greatly in the manner they catch food. Some, such as the great white shark (*Carcharodon carcharias*), use their teeth to hold and tear their prey. Salmonids suck in their prey with water into the foregut, whereas pacu, such as the cultivated *Piaractus mesopotamicus*, chew and grind plants with teeth that resemble human teeth. Many species, such as the silver carp (*Hypophthalmichthys molitrix*), blue tilapia (*Oreochromic aurea*), and Nile tilapia (*Oreochromis niloticus*), are filter feeders, collecting plankton by filtering large volumes of water and collecting food with gill rakers (Sims, 2008).

Fish stomachs show variation in anatomy, ranging from straight, to U- and T-shaped (Suyehiro, 1941). Some fish without a stomach have a gizzard-like structure in the foregut that aids in grinding of the food. Likewise, the intestines vary from short and straight to long and complex. The long intestines can have different three-dimensional organizations such as spirals or balls with various twists and turns. In some species of fish with short intestines, such as cartilaginous species, the surface is increased by luminal spiral valve formations. Some species are equipped with pyloric appendages that can number from one to several hundred. The distal intestinal structures usually differ from the more proximal compartments and can be very complex (Suyehiro, 1941). The functional morphology and biochemistry of the distal structures indicate that, not only water and minerals, but also protein, lipids, and carbohydrates are hydrolyzed and absorbed in this region—in contrast to the situation in mammals. Absorption of macromolecules also takes place in this region, which seems to be of great importance in antigen presentation and immune function (McLean and Ash, 1986, 1987a,b; Sire and Vernier, 1992; Amthauer et al., 2000a,b). In some species (e.g., the Atlantic cod, *Gadus morhua*) the distal intestinal regions have a section with a holding capacity allowing fermentation of dietary fiber components. However, in most fish species passage rate through the intestine is rapid, limiting the quantitative importance of fermentation for the nutrient supply of the fish (Kuz'mina, 2008).

The anatomy and histomorphology of the accessory organs of the digestive tract, the exocrine pancreas, and the liver also show high variability among fish species. Greatest variation is seen for pancreatic tissue. The pancreas is a distinct organ in some species, such as the sturgeon (*Acipenser*

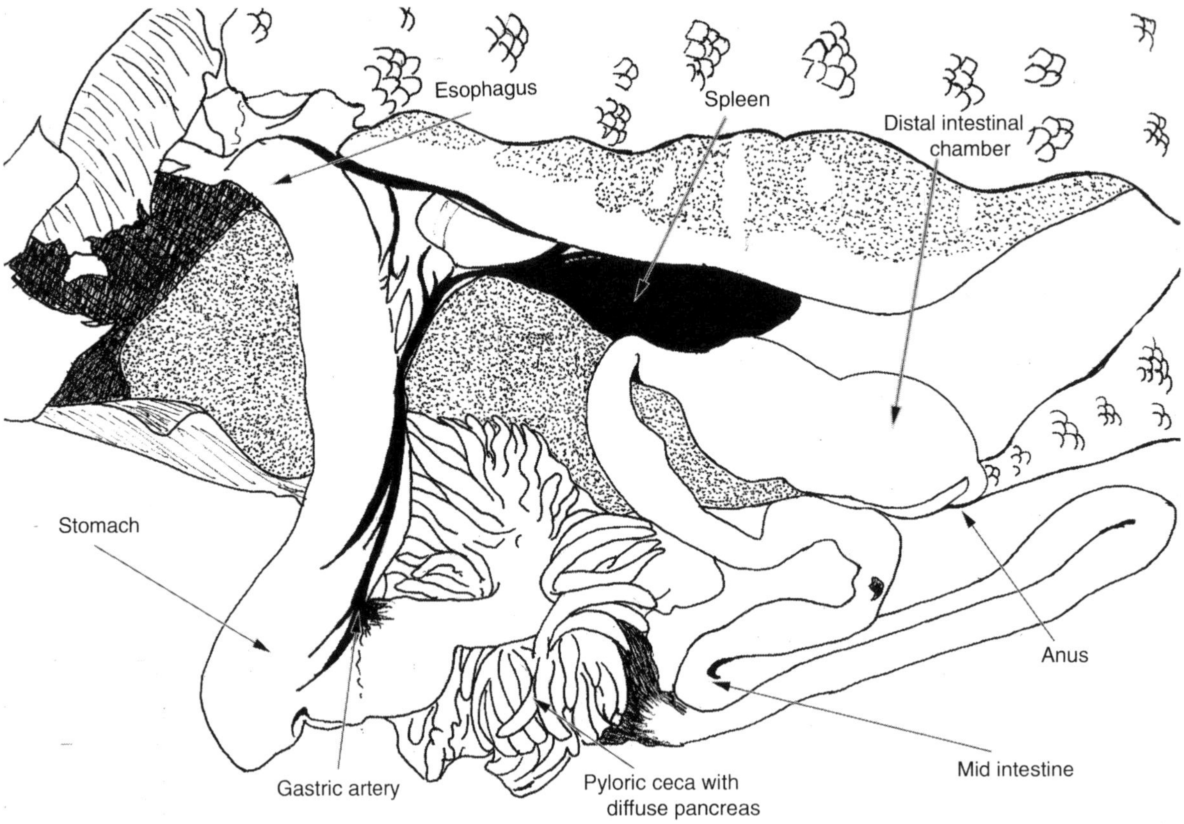

FIGURE 3-3 Drawing of stomach and pyloric ceca in Atlantic cod (*Gadus morhua*). Illustration courtesy of Victoria Blondin, University of Guelph, Ontario.

spp.), but in most species it is diffusely dispersed in mesenteric tissue along the intestines and blood vessels, as in salmonids and/or in the liver (hepatopancreas), as in breams. The structure and function of the diffuse fish pancreas is difficult to study and not well known for any species. It seems clear, however, that the acinar cells of the pancreas, found in clusters, produce and store digestive enzymes that, upon signaling from the intestine, are secreted into tubules that converge into the pancreatic duct (Kurokawa and Suzuki, 1996; Morrison et al., 2004). Water and bicarbonate are added from cells along the tubules. In some fish the tubules converge into main pancreatic ducts leading to the intestine, bile duct, or both, whereas in others the large number of fine tubules opens directly into the pyloric ceca or intestine (Einarsson and Davies, 1997).

Livers in most fish species are organized as a single organ, some with two or more lobes. In contrast to other vertebrates, fish livers are not organized into well-defined lobules of acinar units (Rust, 2002). They have a complex network of blood vessels, tubules, sinusoids, and ducts. Bile ducts drain bile into the gall bladder, and, in fish with a hepatopancreas, pancreatic ducts drain pancreatic juice to the intestinal tract. Hepatocytes comprise the main volume of the liver, storing glycogen and lipid. Great variations exist among fish species regarding liver lipid levels. Some fish, such as the cod, store lipid almost solely in the liver and accumulate large amounts depending on feed composition and intake (Rosenlund et al., 2004). In the liver of Atlantic cod (*Gadus morhua*) lipid may exceed 70% (Karlsen et al., 2006), whereas Atlantic salmon generally show levels below 10% (Pratoomyot et al., 2010). Also the European sea bass (*Dichentrarcus labrax*) liver has the capacity for high lipid accumulation, whereas the sea bream accumulates lipid at levels comparable to the salmon (Peres et al., 1999). In contrast to mammals, but similar to birds, fish seem to transport lipids from the intestine to the body via the portal vein and the liver. Lacteals, which are ducts that transport lipids and larger molecules and complexes, have not been observed in fish. Fish fed high-carbohydrate diets often accumulate large glycogen depots in the liver. The function of such livers may be compromised (Hilton and Dixon, 1982). The mechanism behind this high glycogen accumulation is not well understood (Enes et al., 2009).

Secretions

Mucus

All along the digestive tract, from the mouth to the anus, mucus is secreted from specialized cells. Water, ions, and mucins (i.e., highly glycated proteins with large water binding capacity) are the main components of the mucus, which also contains bicarbonate and may contain antibodies. The density of mucus-producing cells in the mucosa varies among intestinal compartments and is typically lower in the mouth region and high in the intestinal regions (Kuperman and Kuzmina, 1994; Sklan et al., 2004; Abate et al., 2006; Diaz et al., 2008a; Dezfuli et al., 2009; Manjakasy et al., 2009). The components of mucus protect the surface of the tract from mechanical damage by rough dietary components and from chemical damage by endogenous acid, alkali, and digestive enzymes. The mucus is also important for the protection of the body against microbes and chemicals that may challenge the health and well-being of the animal (Shephard, 1994). The number of mucus-producing cells and mucus flow may be affected by feeding habits and feed composition. An increased number has been observed with increasing dietary inclusion of plant ingredients (Olsen et al., 2007).

Gastric Juice

The principal digestive components secreted in the stomach are pepsinogen and hydrochloric acid (HCl), both secreted from cells embedded in the stomach wall with the gastric juice. In most fish, pepsinogen and acid are secreted from the same cell type, the oxynticopeptic cells. In some species, however, such as in some elasmobranchs, cells of the gastric mucosa that seem specialized to secrete either pepsinogen or HCl are found among oxynticopeptic cells (Holmgren and Olsson, 2009). In winter flounder (*Pleuronectes americanus*), mucosal mucus-producing cells expressing a proton pump have been observed, indicating that these cells also secrete HCl (Gawlicka et al., 2001). Feed intake stimulates the secretion of both pepsinogen and acid. However, at least partly independent regulation seems to occur as some stimuli with strong effects on pepsinogen secretion weakly affect acid secretion and vice versa (Holmgren and Olsson, 2009). Depending on the species, feeding rate, diet composition, and time after a meal, stomach pH varies between 1 and 6 (Pérez-Jiménez et al., 2009). The pH in stomach chyme is the result of acid secreted from the stomach wall and buffering capacity of the feed components and elements in the drinking water. Proteins, carbohydrates, and minerals are among the buffering components (Thompson and Weber, 1979; Lawlor et al., 2005). A study of effects on chyme pH after ingestion of a single meal in rainbow trout showed a pH just below 3 in chyme from the stomach before the meal, rising to a pH between 5 and 6 immediately after feed intake (Bucking and Wood, 2006). The pH was stable until about 6 hours after the meal and then gradually decreased to pH 4 at 24 hours. Similar changes in pH have been observed with gilthead sea bream (*Sparus aurata*) (Deguara et al., 2003). Stomach pH in fish seems to be quite regulated and most likely, as in other species, the regulation involves stimulants such as gastrin, acetylcholine, ghrelin, and orexin and inhibitors such as somatostatin, nitric oxide, and dopamine (Schubert, 2009). Whether histamine, a potent stimulator of acid secretion in humans and other nonruminant land animals, is also present in fish is not clear. Histamine has not been observed in fish. The integration of mechanisms that regulate pH is not well understood, either in fish or in other animal species (Schubert, 2009).

The proenzyme pepsinogen is activated to pepsin in the stomach catalyzed by the HCl. Fish seem to secrete more than one form of pepsinogen, and the different forms show different activation rates, pH optima, specific activities, and specificities regarding which peptide bonds they hydrolyze most efficiently (Wu et al., 2009). Most fish pepsins show more than one pH optimum between pH 1 and 5, and some show appreciable activity even at higher pH. Pepsins are endopeptidases (i.e., an enzyme that hydrolyzes peptide bonds at some distance from the terminal amino acids) with specificity for peptide bonds adjacent to aromatic amino acids (i.e., phenylalanine and tyrosine). Secretion of pepsinogen from the gastric mucosa seems to be regulated according to dietary protein level in Atlantic cod (Krogdahl et al., 2009) and accumulates in the mucosa during fasting (Einarsson and Davies, 1996).

There are yet no clear indications that gastric juice of fish contains lipase and amylase. Reports of high specific activity of lipase and amylase in homogenates of stomach tissue from the Brazilian catfish "pintado" (*Pseudoplanty stoma corruscans*) and tambaqui (*Colossoma macropomum*) are suggestive that the stomachs of these fish are a source of the enzymes (Lundstedt et al., 2004; De Almeida et al., 2006). However, the enzyme activities were observed in extracts of tissue samples from fasted fish and not tested in contents of the stomach. A possible explanation for the high activity is the presence of pancreatic tissue in or on the stomach wall.

Whether fish have an endogenous mechanism for the digestion of the exoskeleton of crustaceans is unclear. Chitinolytic activity has been measured in the stomach of many fish species (Divakaran et al., 1999; Gutowska et al., 2004; Ikeda et al., 2009; Fange et al., 2010). The enzyme seems to be associated with fish that consume chitinous prey but do not have mechanical structures breaking down the crustacean exoskeleton. Genomic studies of a species in the pufferfish family, *Takifugu rubripes* (Altschul et al., 1997), show the presence of gene sequences with high similarity to sequences coding for chitinase. However, whether these genes code for proteins to be secreted into the digestive tract is unclear. Chitinase activity in the digestive tract may originate from the prey and/or microbiota. Several species of microbes common to fish intestine have the potential of producing chitinases (Sugita and Ito, 2006).

Bile

The important digestive components of bile are bile acids, phospholipids, and bicarbonate. Bile also contains cholesterol, fatty acids, bile pigments from catabolism of heme, and other inorganic salts. Bile acids, produced from cholesterol or taken up from the blood by the hepatocytes, and phospholipids are excreted continuously from the cells into the bile canaliculi and transported to the gallbladder. From the gallbladder the bile is emptied into the proximal midgut upon stimuli from the intestine triggered by the entering chyme. Cholecystokinin seems to be an important mediator of the contraction of the gallbladder in fish (Aldman et al., 1992; Aldman and Holmgren, 1995; Einarsson et al., 1997). However, knowledge on regulation of bile output in fish is very limited. Between meals, bile accumulates and is concentrated in the gallbladder, which becomes distended and darkly colored. In the intestine the bile acids stabilize lipid droplets and form micelles for the dispersion of lipid components produced in the chyme by lipolytic activities. The bile acids are also essential cofactors for the action of the main lipases acting in the intestine and possibly for the stability of other digestive enzymes and mucosal integrity (Ogata et al., 2003). Taurocholic, taurolithocholic, and taurochenodeoxy cholic acids seem to be the major forms of the bile acids in fish, but glycocholic acid has also been reported (Haslewood, 1967; Une et al., 1991; Bogevik et al., 2009; Velez et al., 2009). Accordingly, taurine, either supplied with the feed or produced from cysteine, is essential for efficient nutrient digestion. Bile can also be classified as an excretory secretion because it serves as the major excretory route for many components that have no known physiological function in the biliary or intestinal tract, including cholesterol; bilirubin; conjugates or hormonal steroids; lipophilic xenobiotics in conjugated or unconjugated form; polyvalent cations, such as iron and copper; and cobalamins. In humans, bile also carries immunoglobulins. Whether this is the case for fish is not known (Hofmann, 1994). An investigation with Atlantic salmon (*Salmo salar*) fed diets with wax esters replacing fish oil showed that wax esters increased bile volume and concentrations of bile acid and phospholipids (Bogevik et al., 2009). However, there seemed to be a limit to the compensatory mechanism. Limitations in compensatory responses were also observed in rainbow trout (*Oncorhynchus mykiss*) in a study of effects of dietary soybean meal inclusion on bile salt concentration in intestinal segments. The study showed a rapid decrease in the bile salt concentration up to 40 days of feeding (Romarheim et al., 2008).

Pancreatic Juice

Pancreatic secretions carry water and bicarbonate, adding to the solubilizing and buffering capacity of the intestine. The digestive enzymes are considered the most important components in the pancreatic juice. The diffuse nature of the exocrine pancreas in most fish makes investigation particularly difficult, and present knowledge needs to be strengthened to understand its function, capacity, and limitation in different feeding situations. The fish pancreas seems to produce most of the same digestive enzymes as the pancreas in mammals and birds, some in proenzyme forms. They comprise the proenzymes of the endopeptidases trypsin, chymotrypsin, and elastase I and II; the exopeptidases carboxypeptidase A and B; the active forms of lipase; phospholipase; α-amylase; and DNase and RNase (Kurtovic et al., 2009). Very little information is available on fish colipase. It seems that digestive lipases in most fish species are independent of a colipase, although colipase is present in some species (Kurtovic et al., 2009). The exocrine pancreatic cells store the digestive enzymes in granules and excrete them with the pancreatic juice upon signals from the intestine (Einarsson et al., 1996; Einarsson et al., 1997).

Digestive enzymes secreted from the pancreas are produced in several isoforms that show variation in molecular weights, molecular structure, pH optima, efficiency, and stability, both within and between species (Krogdahl et al., 2005; Asgeirsson and Cekan, 2006; Ogiwara and Takahashi, 2007). Table 3-1 summarizes some of the general biochemical aspects. Molecular characteristics of digestive enzymes from fish are under investigation, and new nucleotide and amino acid sequences are being published with increasing frequency (Froystad et al., 2006; Psochiou et al., 2007; Manchado et al., 2008; Kurtovic et al., 2009). Pancreatic enzymes mainly act freely mixed in the chyme. However, these enzymes also seem to be associated with the brush border membrane of the enterocytes, exerting their action in close vicinity to the nutrient transporters of these cells (see review by Kuz'mina, 2008).

Species differences exist regarding enzyme output and activity, particular for α-amylase (Krogdahl et al., 2005). In general, herbivorous species seem to produce higher levels of amylase than do omnivores. The lowest activities are observed in carnivores such as eel (*Anguilla anguilla*), which have been shown to have amylase activities less than 1/100th of the activity observed in carp (*Cyprinus carpio*) (Hidalgo et al., 1999). In a comparison of capacity for starch hydrolysis in intestinal contents among Atlantic salmon, Atlantic cod, and rainbow trout, Atlantic salmon ranked the lowest (Froystad et al., 2006). The cod had an intermediate level. The low activity of the salmon amylase was suggested to be due to a defect in a substrate-anchoring structure of the molecule. Species differences have also been described for the proteases in terms of their molecular structure, pH optima, and thermal stability (Glass et al., 1989).

In general, fish seem to be able to adjust their secretion of the pancreatic digestive enzymes according to dietary level and quality of the corresponding nutrient (Buddington et al., 1997). Increasing lipase activity with increasing lipid level has been shown for both rainbow trout and yellowtail (*Seriola quinqueradiata*) (Morais et al., 2004; Ducasse-

TABLE 3-1 Digestive Enzymes of the Digestive Tract[a,b]

Source	Enzyme	Substrate	Specificity or Products
Stomach	Pepsins (pepsinogens)	Proteins and polypeptides	Peptide bonds adjacent to aromatic amino acids
Exocrine pancreas	Trypsins (trypsinogens)	Proteins and polypeptides	Peptide bonds adjacent to arginine or lysine
	Chymotrypsins (chymotrypsinogens)	Proteins and polypeptides	Peptide bonds adjacent to aromatic amino acids
	Elastase I (proelastase I)	Elastin, some other proteins	Peptide bonds adjacent to aliphatic and neutral amino acids
	Elastase II (proelastase II)	Elastin, some other proteins	Peptide bonds adjacent to aliphatic and neutral amino acids
	Carboxypeptidase A (procarboxypeptidase A)	Proteins and polypeptides	Carboxy terminal amino acids that have aromatic or branched aliphatic side chains
	Carboxypeptidase B (procarboxypeptidase B)	Proteins and polypeptides	Carboxy terminal amino acids that have basic side chains
	Colipase (procolipase)	Fat droplets	Binds to bile salt-triglyceride-water interface, making anchor for lipase
	Pancreatic lipase	Triglycerides	Monoglycerides and fatty acids
	Cholesteryl ester hydrolase	Cholesteryl esters	Cholesterol and fatty acids
	Pancreatic α-amylase	Starch	1,4,α-linkages, producing α-limit dextrins, maltotriose, and maltose
	Ribonuclease	RNA	Nucleotides
	Deoxyribonuclease	DNA	Nucleotides
	Phospholipase A (prophospholipase A)	Phospholipids	Fatty acids, lysophospholipids
	Enteropeptidase	Trypsinogen	Trypsin
	Aminopeptidases	Polypeptides	N-terminal amino acid from peptide
Intestinal mucosa	Dipeptidases	Dipeptides	Two amino acids
	Glucoamylase	Maltose, maltotriose	Glucose
	Sucrase	Sucrose	Fructose and glucose
	Nuclease and related enzymes	Nucleic acids	Pentoses and purine and pyrimidine bases
Cytoplasm of mucosal cells	Various peptidases	Di-, tri-, and tetrapeptides	Amino acids

[a]Adapted from Ganong (2009).
[b]The corresponding proenzymes are shown in parentheses.

Cabanot et al., 2007; Murashita et al., 2007). Replacing fish oil with wax esters from *Calanus finmarchicus* increased specific activity of lipolytic enzymes in the intestinal contents (Bogevik et al., 2009). However, at the highest inclusion level of wax ester (25% of the diet), reduced lipid digestibility was observed. Similarly, dietary protein and amino acids stimulate pancreatic secretion of proteolytic enzymes (Cahu et al., 2004). Protease secretion seems to respond to substrate level in the diet up to a limit. The mechanism behind the response may be related to the level of free proteases in the chyme. Protease inhibitors mixed at increasing levels into diets of rainbow trout have been found to cause a curvilinear increase in total trypsin protein concentration in the intestinal content although trypsin activity decreased (Berg-Lea et al., 1989). However, at an inclusion level of about 5 g/kg diet, the capacity for trypsin synthesis seemed to be exceeded. Strain differences in responses to dietary protein level have been described for the winter flounder (*Pseudopleuronectes americanus*) (Gawlicka et al., 2001).

The regulatory mechanisms behind exocrine pancreatic secretion are not well known. Cholecystokinin is involved in the endocrine regulation of pancreatic secretion (Koven et al., 2002), but other peptide hormones as well as neurological signals play roles in the regulation (Volkoff, 2006; Koji et al., 2008; Holmgren and Olsson, 2009).

Bicarbonate Secretion and pH of the Intestine

In fish with a functional stomach the acid chyme entering the proximal intestine seems to be quickly neutralized supposedly by HCO_3^- in bile and pancreatic juice. Secretion from epithelial cells may also add to the pH adjustment of the chyme (Cooper et al., 2010). Only limited information has been published on variation in intestinal pH and the effects of diet composition. Reduction in luminal pH in European flounder intestine has been found to stimulate HCO_3^- secretion (Wilson and Grosell, 2003; Cooper et al., 2010). Buffering capacity in the intestine seems well adjusted in light of the constant liberation of amino and fatty acids. The pH is observed to be above 7 all along the intestinal tract, for example in rainbow trout, with an increasing trend toward the distal sections (Bucking and Wood, 2006). In the distal most compartments, in which the microbial activity is higher than in the more proximal sections, pH would be expected to be lower, such as in Atlantic cod (Seppola et al., 2005). Secretion of HCO_3^- from epithelial cells appears to play an important role also in preventing excessive uptake of Ca^+ ingested by marine fish via drinking water and prey fish. Bicarbonate precipitates Ca^+ as $CaCO_3$, which is unavailable for absorption. This process seems to be an important element in intestinal water absorption (Whittamore et al., 2010).

Membrane Bound Digestive Enzymes

The brush border of the absorptive cells is equipped with membrane bound peptidases that complete the hydrolysis of peptides before transport into the cells. The peptidases act on bonds at the amino terminal end of the peptides. They are numerous, with different specificities. Peptidases from different fish species show different characteristics in terms of pH optima, thermostability, and distribution along the intestinal tract (Kuz'mina, 2008). Dietary protein level affects brush border aminopeptidase activity in herbivorous, omnivorous, and carnivorous fish (Buddington et al., 1997), with moderate differences between the groups of fish (Cahu and Infante, 1995).

Brush border disaccharidases hydrolyze low molecular carbohydrates with 2–4 units producing free forms of their respective monosaccharides. The highest hydrolytic capacity of intestinal homogenates is found for maltose. Glucose is produced from maltose at rates several times higher than from sucrose and trehalose (Buddington and Hilton, 1987; Krogdahl et al., 1999; Kuz'mina, 2008). A homogenate of fish intestinal mucosa also shows the ability to hydrolyze lactose. The enzymes responsible for this activity seem to be cytosolic because the activity remains in the homogenate when the brush border membranes are extracted (Krogdahl et al., personal communication). Herbivorous and omnivorous fish species have several-fold higher disaccharidase activities in the intestinal brush border compared to carnivorous species (Kuz'mina, 2008). Present knowledge indicates that disaccharidases from fish living in cold waters have higher specific activities than the same enzymes from fish in warmer waters (Maffia et al., 1993). In most species the highest activities are observed in the proximal intestine with decreasing activities toward the anus. Whether disaccharidase activity of the brush border is affected by dietary carbohydrate level seems to depend on the feeding situation of the fish. A comparative study of effects of starch level in diets for rainbow trout and Atlantic salmon showed that both species increased their disaccharidase capacity with increasing starch level (Krogdahl et al., 2004). However, in other studies on salmonids, varying dietary starch level did not alter disaccharidase activity (Buddington and Hilton, 1987; Krogdahl et al., 1999; Kuz'mina, 2008). The conflicting results may be related to differences between the studies regarding dietary starch level, starch processing, technical qualities of the feed, feed intake, and/or environmental factors such as temperature and salinity. It should be kept in mind that fish can adjust intestinal brush border enzyme capacity either by increasing enzyme concentration in the tissue or by increasing brush border area or by both methods. Both possibilities should be taken into account in studies of effects on brush border enzyme capacity. No apparatus for hydrolysis of lipids has been identified in the intestinal brush border.

Intestinal Transit Time

Intestinal passage rate and transit time vary with diet composition, meal size, and feed structure in many animal species (Guilloteau, 1979; Hill, 2007). Increased flow of digestible carbohydrates, proteins, and lipids into the distal regions of the small intestine inhibits intestinal motility. Lipids elicit the strongest signals (Hasler, 2006). These observations are in accordance with the results of investigations of gastric emptying rate in fish (dos Santos and Jobling, 1988). Most investigations on intestinal passage rate in fish have focused on effects of fibers and bulking agents (Storebakken, 1985; Storebakken and Austreng, 1997; Dias et al., 1998). Soluble indigestible carbohydrates such as alginates and guar gum as well as bulking agent such as silica and zeolite in general seem to cause reduced transit rates in fish. Intestinal passage rate may be suggested to be regulated to optimize nutrient utilization and to prevent overload of nutrients in the distal intestinal compartments. However, soluble fibers and bulking agents often reduce nutrient digestibilities, particularly of dietary lipids. Insoluble fibers, such as cellulose, on the other hand, may speed up passage rate (Dohnalek, 2004). Intestinal passage rate is expected to vary among fish species. However, comparative studies are not available. The studies conducted with rainbow trout and sea bass indicate similar transit times for these species with presence in feces of markers from a meal between 5 and 35 hours after the meal (Storebakken, 1985; Dias et al., 1998).

Digestion

Stomach Digestion

A condition for efficient digestion and absorption of a nutrient is solubility in water. The concerted action of hydrochloric acid and pepsin in the stomach denatures and degrades most proteins and increases their solubility. The process also increases the solubility of other nutrients such as carbohydrates and minerals bound or trapped in the feed matrix. The low pH increases the solubility of many minerals and transforms them to their chloride forms, which often are more water-soluble than their native form in the feed. Lipids are also released. The hydrophobicity of lipids gives them a tendency to aggregate into droplets. Under normal circumstances, emulsifiers from the feed and stomach, such as phospholipids and certain proteins, will limit the size of the lipid droplets. However, if the rate of lipid release is too fast or the supply of emulsifiers is limited, accumulations of lipid will form. The result may be fat belching as seen in some farming situations (Baeverfjord et al., 2006). The condition seems to be multifactorial and is influenced by rate of pellet disintegration, rapid changes in salinity, and temperature.

The importance of enzymes present in food organisms for the digestive process of fish has been an issue discussed by several scientists (see review by Kuz'mina, 2008). Unarguably, live prey animals are eaten with their own intestinal digestive apparatus providing a range of gastric, pancreatic, and membrane bound enzymes. Moreover, each cell of the prey has lysosomes that contain enzymes for the degradation of proteins, lipids, carbohydrates, nucleic acids, and other cell components at acid pH. They may be activated when stressed, for example when exposed to the host gastric juice that contains acid and enzymes. The term "induced autolysis" has been suggested for the process (Kuz'mina, 2008). Researchers have argued (Kuz'mina, 2008) that the fast degradation of whole prey animals observed in fish involves activation of digestive apparatus of the cells of the prey by H^+ ions from the gastric juice. The ions have been estimated to diffuse 1,000 times more rapidly into the prey than the digestive enzymes. In this view, gastric digestion proceeds from three starting points: the digestive tract of both the host and the prey and from within the tissue of the prey. The quantitative importance of the prey enzymes for nutrient digestion is not established, but is suggested to vary depending on the nutrient in question, the fish and its stage of development, the prey and its physiological status, environmental temperature, and oxygen level (Kuz'mina, 2008). For fish fed dry pellets, enzymes from the feed are certainly of no importance unless specifically supplemented.

Intestinal Digestion

Once the mixing and churning action of the stomach muscles and structures has processed the feed to the appropriate particle size and moisture level sufficient for further transport and processing, the partially digested feed, now called chyme, is passed on to the midgut, the intestine's pyloric or hepatopancreatic region. The product of stomach processes is a mixture of dissolved nutrients, mainly proteins and large peptides; mono-, di-, oligo-, and polysaccharides; water-soluble vitamins; emulsified lipids, including lipid-soluble vitamins; dissolved minerals and vitamins; and fine particles of any undissolved and insoluble feed material. In fish without a stomach, particle size of feed is reduced in some species by various structures such as the gizzard, and the enzymatic breakdown of nutrients starts in the midgut.

Protein and peptide hydrolysis take place in the chyme by the concerted action of the endo- and exopeptidases (for characteristics of the enzymes, see Table 3-1). Trypsin hydrolyzes internal peptide bonds adjacent to lysine and arginine, leaving them as carboxyterminal peptide ends, which are substrates for carboxypeptidase B (i.e., basic amino acids). Chymotrypsin preferentially hydrolyzes bonds to branched-chain amino acids, giving rise to carboxyterminal ends suitable for the action of carboxypeptidase A. The elastases preferentially hydrolyze peptides adjacent to aliphatic and neutral amino acids and are particularly efficient in initiating elastin hydrolysis. After the action of the pancreatic enzymes, the peptide chains are short, usually with less than five amino acids. After further hydrolysis by the aminopeptidases, a large proportion of the amino acids are absorbed as free amino acids. However, it is likely that substantial amounts are taken up as small peptides for further hydrolysis within the cell.

Lipid digestion requires emulsification of the lipids released from the feed in the stomach and intestine in the initial steps of digestion. The bile salt-dependent, carboxyl ester lipase is the dominating lipase in most fish species and the only lipase in many species. This carboxyl ester lipase seems to have broad substrate specificity, preferentially hydrolyzing bonds involving long, highly unsaturated fatty acids in the 1 and 3 positions of triacylglycerols. This lipase also has the ability to hydrolyze wax esters (Tocher and Sargent, 1984; Gjellesvik et al., 1989; Tocher, 2003; Kurtovic et al., 2009). Fish hydrolyze phospholipids quite efficiently, but a specific phospholipase has not been described (Tocher, 2003). Whether the final hydrolysis products, the results of concerted action of more than one lipolytic enzyme, are free fatty acids and monoglycerides or glycerol is not known. However, the enzymes responsible for resynthesis of triacylglycerols in the intestinal mucosa of this species seem to prefer monoglycerides before glycerol (Oxley et al., 2007), an indication that monoglycerols are important endproducts. The fatty acid products of lipolysis in the chyme are incorporated into primary micelles formed by bile acids and phospholipids. As the micelles enlarge, they are transformed into secondary micelles that have the capacity to include the more lipophilic compounds such as long-chain saturated fatty acids, cholesterol esters, and fat-soluble vitamins. The further process of absorption of lipids is not well known, but is believed to proceed as in mammals. As the secondary micelles reach the so-called unstirred water layer covering the intestinal brush border, they disintegrate because of the lower pH of this layer. The fatty acids cross the brush border membrane by diffusion or facilitated transport aided by proteins.

Low molecular weight carbohydrates, such as glucose, maltose, and sucrose, seem to be digested efficiently in all fish (Singh and Nose, 1967; Hilton et al., 1982; Hilton and Atkinson, 1982; Storebakken et al., 1998). They are all highly water-soluble and their hydrolysis is dependent only on glucosidases located in the brush border. The products of the hydrolysis are mainly glucose and fructose. Digestion of starch and chitin takes place by the action of α-amylase and chitinase, respectively. However, starch in most feedstuffs is contained in granules that are mostly insoluble and therefore not hydrolyzed by the amylase in the fish intestine unless well heat-treated in the presence of moisture (Krogdahl et al., 2005). The exception from this general pattern is starch in oats, which can be digested without heating, even in Atlantic salmon (Arnesen et al., 1990; Arnesen and Krogdahl, 1995; Krogdahl et al., 2005). Chitin seems to be quite poorly hydrolyzed even in fish species having crustaceans in their natural diet (Krogdahl et al., 2005). The reason may be low solubility of this polysaccharide, very low (or no) chitinase production, or low uptake efficiency of N-acetyl glucosamine, the product of chitinase activity (Gutowska et al., 2004).

Microbial Digestion

The digestive tract of all animals, including fish, is inhabited by microorganisms of many kinds, aerobic, facultative aerobic, as well as anaerobic. The numbers of bacteria in fish are in general lower than in homoeothermic animals, but great differences exist among fish species. Some bacteria, the allochthonous, are transient and present in the chyme; others, the autochthonous bacteria, are inhabitants of the mucosal surface and reproduce in situ. Until the past decade, studies of intestinal microbiota were largely hampered by methodological limitations because only live bacteria that were able to grow on the available media could be studied. The development of molecular tools and collection of data in international databases have changed the situation, and the number of studies is increasing. It is evident that the fish digestive tract harbors microbes with the ability to secrete enzymes that are able to hydrolyze and metabolize proteins, starch, cellulose, other nonstarch polysaccharides, chitin, and lignin (Kuz'mina, 2008; Ray et al., 2009). Higher concentrations of bacteria are found in the distal intestinal compartments than in the proximal. The variation is related to variation of feed sources. Also the microbiota of the environment has a great impact on intestinal microbiota. Herbivorous species normally have higher bacterial numbers than do carnivores. But also within these groups, variation is seen due to differences in intestinal anatomy. Fish such as cod, which have a chamber-like compartment that is closed by sphincters, have higher bacterial numbers than Atlantic salmon (Seppola et al., 2005). The products of bacterial fermentation of dietary nutrients are amino acids, glucose, acetate, propionate, and butyrate, all compounds that apparently can be absorbed efficiently by the distal intestinal tract of fish. However, the quantitative contribution from microbial fermentation to total nutrient supply is most likely small even in herbivorous species.

Nutrient Absorption

Products of the action of digestive enzymes can enter the organism across the brush border by diffusion or facilitated transport down a concentration gradient or by active and energy-dependent transport against a concentration gradient. Passage via paracellular pathways is also possible, but considered to be of minor importance in fish (Ferraris et al., 1990; Oxley et al., 2007). Facilitated and active transport takes place via specialized transporters unique for the nutrient or a group of nutrients with similar chemical characteristics. Both are saturable mechanisms. Fish have the apparatus for nutrient absorption all along the intestinal tract including the distal most areas (Ferraris and Ahearn, 1984; Collie, 1985; Buddington and Diamond, 1987; Bakke-McKellep et al., 2000). Distribution of the transporters along the intestinal tract differs, however, among species, although most fish show decreasing absorption rates toward the distal segments (Buddington et al., 1987; Bakke et al., 2010). Thus, the basic mechanisms of nutrient absorption seem to be similar to those found in mammals. However, for most, but perhaps not all transporters, the rate of nutrient absorption is lower in fish (Reshkin and Ahearn, 1987; Buddington et al., 1997).

The active transporters are generally dependent on ions such as Na^+, Cl^-, K^+, or H^+, and the energy for transport is needed to maintain necessary ion gradients across the cell membrane. The nutrient transporters show a high degree of conservation through evolution. However, variation among fish species has been observed in terms of traits such as substrate affinity (K_m) and maximum velocity (V_{max}). An apparent tendency for higher substrate affinity of amino acid transporters in herbivorous fish compared to carnivorous fish and an opposite trend regarding V_{max} of the glucose transporters has been suggested (Ferraris and Ahearn, 1984; Buddington and Diamond, 1987). Higher influx of nutrients per unit of tissue in freshwater than saltwater fish has also been indicated (Ferraris and Ahearn, 1984; Collie, 1985; Buddington and Diamond, 1987; Collie and Ferraris, 1995; Lionetto et al., 1996). As expected, transporter capacity tends to increase with increasing water temperature (Houpe et al., 1996).

Based mainly on studies with the European eel (*Anguilla anguilla*), it seems that fish have at least four distinct Na^+-dependent transporters for amino acids, one transporter for each of acidic, neutral, N-methylated amino acids, and proline (Storelli et al., 1989). Sodium-independent transporters seem to be present for the absorption of neutral and basic amino acids (i.e., for glycine, alanine, and lysine) as in mammals. For histidine, a highly specific transporter has been suggested because the transport seems to be independent of the presence of other amino acids (Glover and Wood, 2008). However, differences among fish species exist regarding substrate specificity of various amino acids (Collie and Ferraris, 1995).

Fish are also equipped with peptide transporters as demonstrated in herbivorous and carnivorous fish species such as the tilapia (*Oreochromis mossambicus*), European eel, rockfish (*Sebastes caurinus*), sea bass (*Dichentrarcus labrax*), rainbow trout, and Atlantic salmon (Thamotharan et al., 1996a,b; Maffia et al., 1997; Bakke-McKellep et al., 2000; Nordrum et al., 2000; Verri et al., 2000; Terova et al., 2009; Ostaszewska et al., 2010). The molecular structures of both the PepT1 and PepT2 transporters have been characterized for zebrafish (*Danio rerio*) and cod (Buddington et al., 1997; Verri et al., 2003; Romano et al., 2006). Diet composition seems to affect the expression of PepT1 as demonstrated for rainbow trout (Ostaszewska et al., 2010). A comparative study of amino acid and peptides transport has been carried out with rainbow trout and Atlantic salmon, showing species differences in transport activity along the intestinal tract. In both species, transport decreased along the intestine. In the distal intestine, transport seemed higher in the trout than in the salmon for lysine and methionine, equal or lower for phenylalanine and proline. Soybean feeding decreased transporter-mediated uptake and increased permeability. In both species, nutrient transport was also influenced by water salinity. The results indicate that transporter-mediated uptake is of greater importance in saltwater than in freshwater (Bakke-McKellep et al., 2000; Nordrum et al., 2000).

Some dietary and endogenous proteins escape proteolytic digestion in the proximal sections of the intestine. Such proteins may be absorbed as macromolecules. Uptake of human gamma globulin, horseradish peroxidase, ferritin, prion proteins, and oral vaccines has been demonstrated in various fish species (Lavelle and Harris, 1997; Hernandez-Blazquez and da Silva, 1998; Amthauer et al., 2000a; Concha et al., 2002; Quentel et al., 2007; Uran et al., 2008; Valle et al., 2008). The distal intestine seems to be the most important site of absorption of larger peptides and proteins, and uptake of intact proteins is considered essential for development of the defense apparatus against exogenous proteins and pathogens. The nutritional importance of macromolecular uptake, however, is considered minor. Indication of an enteropancreatic circulation of proteins exists based on macromolecular uptake studies. However, despite efforts to gain information on such recirculation of proteins, data that can support the concept are scarce (Rothman et al., 2002).

Information about lipid transport across the intestinal mucosa in fish is limited, but present knowledge indicates that the processes are quite similar to those in other vertebrates. The proximal intestine seems to absorb most dietary lipids. Medium-chain and longer highly unsaturated fatty acids are absorbed in more proximal regions compared to the longer and more saturated, which are absorbed in more distal regions (Røsjø et al., 2000). It is believed that fatty acids as well as the fatty alcohols pass the brush border membrane by diffusion. However, demonstration of the presence of fatty acid binding proteins (FABP), also in the fish intestine, indicates that facilitated transport may take place (Andre et al., 2000; Concha et al., 2002; Iqbal and Hussain, 2009). Uptake of fatty alcohols from wax esters, abundantly present in some marine organisms such as copepods, is slower than uptake of fatty acids (Bogevik et al., 2008). Both fatty acids and alcohols are reesterified in the enterocytes. The triacylglycerols are produced from fatty acids and monoglycerols or glycerol-3-phosphate (Caballero et al., 2006). The monoglycerols seem to dominate as substrate for the production of triacylglycerols, whereas phospholipid synthesis utilizes glycerol-3-phosphate. The efficiency of production and partitioning between the two seems to depend on the source of lipid in the diet. Triacylglycerols produced by the mucosal cells are incorporated into lipoproteins that accumulate in lipid droplets in the cells and exit the cells via exocytosis (Hernandez-Blazquez and da Silva, 1998; Kjaer et al., 2009). The lipoproteins produced by the enterocytes should perhaps be named portomicrons rather than chylomicrons because they are not conveyed by chylus into collective lymph ducts, or so-called lacteals (Tocher and Sargent, 1984; Bogevik et al., 2008). However, the movement of lipids between the gut and the general circulation is not well known.

A dietary supply of phospholipids is essential for efficient lipid digestion, absorption, intracellular metabolism, and further transport in the body (Tocher et al., 2008). A deficiency of phospholipids has been observed to cause lipid accumulation within the intestinal absorptive cells and histological alterations in carp and salmonids (Fontagne et al., 1998; Olsen et al., 2003)

Glucose uptake has been studied in several fish species and seems qualitatively similar to that in other vertebrates. D-glucose and galactose are taken up by the same brush border transporter, SGLT1, which is electrogenic and dependent on Na^+ and energy (Krogdahl et al., 2005; Geurden et al., 2007). Fructose is also absorbed in fish. However, a putative facilitative transporter for fructose, such as GLUT5 in other vertebrates, has not been identified in fish. Other transporters may supply additional transport capacity for monosaccharides but have not yet been described in fish.

Carbohydrate absorption also takes place mainly in the proximal intestinal compartments of the fish intestine, as shown for Atlantic salmon (Krogdahl et al., 1999; Bakke-McKellep et al., 2000). This is in agreement with the observation that brush border enzymes for hydrolysis of disaccharides have the highest activities in the proximal regions. Differences exist among fish species, but most fish absorb mono- and disaccharides with high efficiency (Singh and Nose, 1967; Hilton et al., 1982; Hilton and Atkinson, 1982; Storebakken et al., 1998). No information has been found in the scientific literature on uptake mechanisms for N-acetylglucosamines, the hydrolysis products of chitinolytic activity.

Knowledge on mechanisms behind increases and decreases in nutrient transport capacities in the brush border and basolateral membranes and their regulation is very limited. Transporter concentration in the brush border membrane can change quickly, for example by the introduction of transporters stored intracellularly. The signals may be mediated by endocrine and/or neurological signals (Holmgren and Olsson, 2009).

SHRIMP

Shrimp can be filter feeders, scavengers, and predators and are classified as herbivores, carnivores, and omnivores. Investigations of stomach contents of shrimp have shown that they eat other species of crustaceans, annelids, mollusks, echinoderms, nematodes, fish tissue, insects, seeds, algae, macrophytes, vegetable matter, and detritus (Focken et al., 1998; Figueiredo and Anderson, 2009). Some species have developed more carnivorous feeding habits than others. In extensive and semi-intensive pond-cultured shrimp, the naturally available food organisms can dominate over the exogenously supplied feed (Nunes et al., 1997; Nunes and Parsons, 2000), whereas in more intensive systems the contribution from natural feed is reduced or eliminated.

Structural and Functional Aspects of Digestive Organs

The anatomy of the digestive tract of shrimp is often divided into three major parts: foregut, midgut, and hindgut. A further division can be made of the foregut: esophagus, cardiac stomach, and pyloric stomach, a chamber where feed particles are ground and filtered (Mantel, 1983; Ceccaldi, 1997). A drawing of the structure of the tract is shown in Figure 3-4. Lamellae of various sizes, brushes and needles, and dents make the stomach structure rather complicated. Most shrimp possess a calcified structure in the stomach, known as the gastric mill. The hepatopancreas (midgut gland), the major digestive organ of shrimp, is a large multilobate structure, a diverticulum of the midgut. A blind tubule covered by a single epithelial layer with digestive characteristics is the basic unit. The tubules vary in length and fuse into larger collective ducts and end in one or two major ducts opening into the midgut. The hindgut of the intestine is straight and widens into the rectum before termination at the anus. A layer of a chitin-protein complex, which is part of the shrimp exoskeleton, covers the external surface of the foregut and hindgut. The midgut is not lined by this complex and is the only section of the intestine with characteristics of intestinal absorptive surface. Shedding and replacement of the chitin protein layer takes place at each molt. In some shrimp species, chewing structures located in the stomach are also replaced at each molt (Ceccaldi, 1997). Shrimp larvae start out with simple intestines that develop into more complicated structures as they progress through distinct stages.

Shrimp catch and preprocess feed with their mouth pieces with specialized prehensile appendages. The feed is passed through the relatively short esophagus to the stomach. Feed disintegration takes place mainly in the stomach by the action of the various lamellae, appendages, and calcified parts, including the gastric mill (Ceccaldi, 1997). The feed is turned into very fine particles that are passed on to the midgut. Larger particles are conveyed by fluid streaming retrograde to the more proximal parts of the stomach for further degradation. Another sorting of feed particles takes place in the midgut by the glandular filter (ampulla). Indigestible particles are passed on to the distal compartments of the tract, whereas nutritionally valuable material enters the hepatopancreas.

The hepatopancreas combines the functions of the pancreas, intestine, and liver and is responsible for processes such as synthesis and secretion of digestive enzymes, absorption of digested material, and metabolism of lipids, carbohydrates, and minerals. It is the center for the production of materials required for the temporally distinct events of molt and vitellogenesis. The glandular tissue also serves as a detoxification organ for heavy metals and toxic organic compounds (Ceccaldi, 1997). Several shrimp species have midgut ceca of different lengths and numbers located close to the stomach, at the opening of the hepatopancreatic canal, or at the entrance to the hindgut. Cells of the ceca have microvilli indicating absorptive functions. The epithelium of the hindgut of shrimp is involved in osmoregulation (i.e., transport and metabolism of water and ions) and in condensing the material for excretion in feces (Ceccaldi, 1997).

Shrimp encase their feces in a peritrophic membrane, an acellular layer that separates ingested materials from the gut

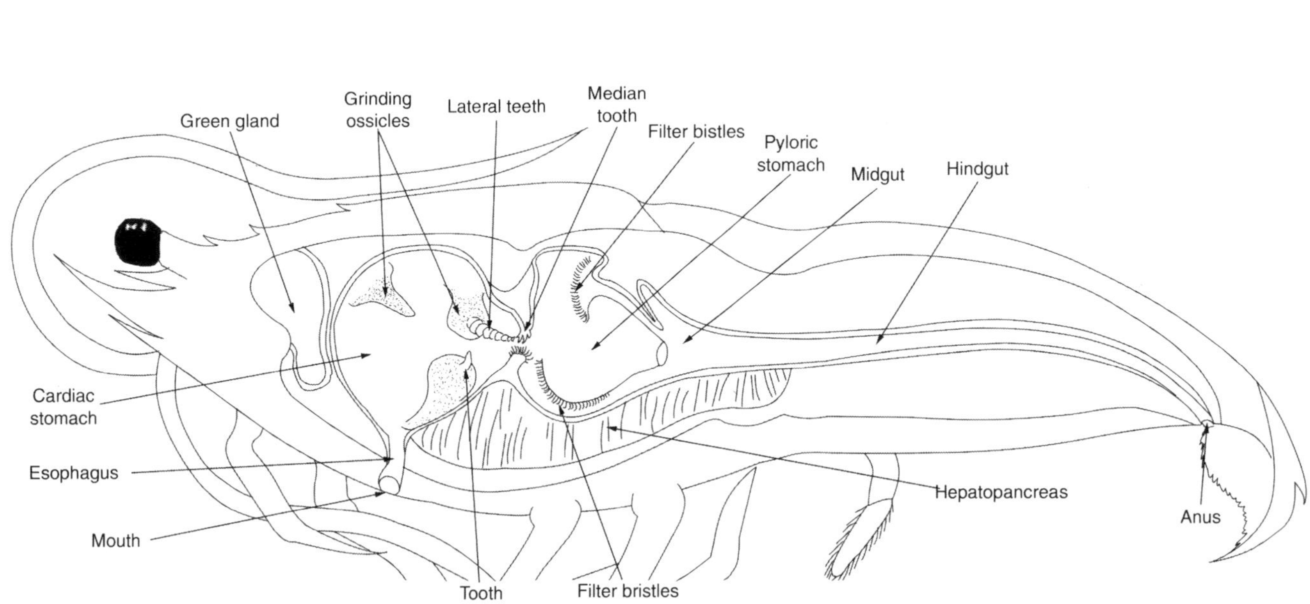

FIGURE 3-4 Anatomy of the digestive tract of shrimp.
Illustration courtesy of Victoria Blondin, University of Guelph, Ontario.

epithelium. The membrane is secreted from the midgut and contains chitin and protein (Martin et al., 2006). This has implications for the measurement of apparent digestibility. On one hand, nutrients in the membrane are lost, and on the other hand there is reduced loss of nutrients in the feces from leaching.

Digestive Enzymes

The shrimp stomach, covered with a chitin level, does not secrete acid and enzymes, but its contents often show digestive enzyme activities, some of which may originate from the hepatopancreas and some from food animals. At least in some crustaceans, digestive enzymes produced in the hepatopancreas pass from the midgut into the stomach (Vogt et al., 1989). Hence, the digestive processes can be initiated before the feed enters the midgut.

The hepatopancreas is the main secretory organ of shrimp. Digestive enzymes are synthesized in the F-cells (fibrillar cells) and accumulate in the B-cells (blister-like cells) (Vogt et al., 1989; Sousa et al., 2005). In some but not all species, the B-cells contain granules, presumably containing enzymes in active or proenzyme form (Babu and Manjulatha, 1995; Sainz et al., 2004; Ong and Johnston, 2006). The hepatopancreas secretes a wide range of digestive enzymes; proteases, including specific collagenases; lipolytic enzymes; chitinase; cellulase; laminarinase; α/β-glucosidase and/or α-amylase to be able to make use of the cellulose from plant cell walls; laminarin from brown algae; and other nonstarch polysaccharides (Xue et al., 1999; Johnston and Freeman, 2005; Carrillo-Farnes et al., 2007; Figueiredo and Anderson, 2009). Species that have high protein content in their diet show high proteinase activities. Species that feed on crustaceans synthesize chitinase. Herbivorous species tend to produce high amounts of the various carbohydrases to be able to disrupt cell walls and make use of the cellulose from plant cell walls, laminarin from brown algae, and other nonstarch polysaccharides (Xue et al., 1999; Johnston and Freeman, 2005). Omnivorous opportunistic feeders have high activities of several of the enzymes mentioned above and are able to utilize a wide range of food sources.

The main endopeptidases for most crustaceans are trypsin and chymotrypsin. Some species, however, synthesize cathepsin L as the main proteolytic enzyme (Teschke and Saborowski, 2005; Carrillo-Farnes et al., 2007; Chisty et al., 2009). Very few studies have been carried out on purified enzymes, but gene sequences are available for many (Carrillo-Farnes et al., 2007). Characteristics such as inhibition patterns, pH optima, and heat tolerance of enzymes are mainly the results of studies on crude hepatopancreatic extracts (Carrillo-Farnes et al., 2007). Shrimp trypsinogen seems to lack the enterokinase-recognizing sequence of trypsin from vertebrates. It is however activated by extracts of hepatopancreas and so is chymotrypsinogen (Babu and Manjulatha, 1995; Viader-Salvado et al., 2007). Information on exopeptidases, such as carboxypeptidase and aminopeptidases, is difficult to find in shrimp (Carrillo-Farnes et al., 2007).

Enzymes capable of hydrolyzing triglycerides and phospholipids have been observed in several shrimp species (Carrillo-Farnes et al., 2007). There is, however, some debate about whether the hydrolysis of triglycerides is catalyzed by a triglyceride lipase, a phospholipase, or both. *Litopenaeus vannamei* seem to have both. From this species, two fractions with lipolytic activity have been isolated; one with preference for triglyceride substrate, and the other for phospholipid (Carrillo-Farnes et al., 2007). A comparison of substrate specificity of lipases from *Litopenaeus schmitti* indicated a strong preference for n-3 and n-6 fatty acids. Studies of lipases from *L. vannamei*, *Farfantepenaeus californiensis*, and *Farfantepenaeus notialis* showed similar preferences. Several shrimp lipases have shown two pH optima in the range 5–11 (Carrillo-Farnes et al., 2007). The lipases and esterases are found associated with the microvilli of R-cells (resorptive cells), as well as in vacuoles of B-cells, supranuclear vacuoles of F-cells, lumen of the hepatopancreas tubule, and in intertubular connective tissue (Lopez-Lopez et al., 2003). The R-cells can take up fatty acids from the lumen and store them intracellularly.

Shrimp are able to hydrolyze a great variety of oligo- and polysaccharides and seem to surpass greatly even herbivorous fish. Hepatopancreatic and/or tissues from other sections of the digestive tract of various shrimp species have shown a wide range of enzyme activities characterized as α- and β-galactosidase, α-fucosidase, laminarinase, α-mannosidase, β-glucuronidase, β-glucosaminidase, xylanase and α-xylosidase, raffinase, β-fructofuranosidase, and cellulase (reviewed by Carrillo-Farnes et al., 2007). Whether these enzymes are endogenous to the shrimp or to the food ingested by the shrimp or both is not clear. In accordance with these observations, many shrimp species seem to utilize starch and other polysaccharides very efficiently. Three α-amylases have been cloned from *L. vannamei* and show great sequence similarities with mammalian α-amylase (Van Wormhoudt and Sellos, 2003). Activities described as α- and β-galactosidases, chitobiase, α-fucosidase, laminarinase, α-mannosidases, β-glucuronidase, β-glucosaminidase, xylanase, and α-xylosidase have been observed in one or more species (Van Wormhoudt and Sellos, 2003). It may be suggested that the great ability of shrimp to hydrolyze polysaccharides is related to the fact that they all start as an herbivore or omnivore with phytoplankton as a major nutrient source (Le Vay et al., 2001; Diaz et al., 2008b).

Two of the cell types of the hepatopancreas, the R and F, are equipped with microvilli, indicating absorptive functions. Also, epithelial cells of the intestinal ceca, present in several shrimp species, have well-organized microvilli (Ceccaldi, 1997). Whether they are equipped with digestive enzymes such as aminopeptidases and disaccharidases is not clear (Ceccaldi, 1997). Homogenates of hepatopancreas show α-glucosidase activity, but the enzymes may be intracellular.

Larger changes in enzyme content of the hepatopancreas are seen during molting periods (e.g., for trypsin and chitinase) (Hernandez and Murueta, 2009). Chitinase digests the old exoskeleton so it can be resorbed and replaced by newly synthesized chitin. The production of digestive enzymes also seems to vary throughout the year and even within the species, depending on the available nutrient sources. The Caridean shrimp (*Crangon crangon*) has high trypsin activity during the summer and low activity during winter periods (Pöhlmann, 2007; Sahlmann, 2008). The latter work also indicated that shrimp may recirculate digestive enzymes. Even enzymes from ingested prey can survive the hydrolytic conditions in the intestine and be recycled via the hepatopancreas (Sahlmann, 2008).

Enzymes seem to be emptied from the hepatopancreas upon feeding (Ong and Johnston, 2006). Passage of enzymes from the midgut to the stomach has been found to induce additional synthesis and secretion of enzymes (Vogt et al., 1989). Adjustments to diet composition for proteases, lipolytic enzymes, and amylase have been shown for many species. The responses vary among species. For some species, responses are seen in proteolytic and amylase activity but not in lipolytic activity, but for other species, amylase and/or protease activity seem unresponsive (Moss et al., 2001; Gamboa-Delgado et al., 2003; Lopez-Lopez et al., 2005). A high dietary starch level was found to increase the specific activity of α-amylase and an α-glucosidase in *L. vannamei* (Le Moullac et al., 1997; Gaxiola et al., 2005). The same species has shown variation in trypsin and chymotrypsin activity with variation in protein level (Le Moullac et al., 1997; Lemos et al., 2000; Muhlia Almazan et al., 2003, 2008). The magnitude of the stimulation seems to differ among species and to depend on the protein source of the diet.

The regulatory mechanisms behind adaptation to dietary composition are not well understood. Intestinal hormones are likely to be involved in this regulation (Santos et al., 1997). Gastrin-cholecystokinin-like peptides isolated from the stomach of the marine crustacean *Nephrops norvegicus* were found to stimulate isolated midgut gland cells (Favrel et al., 1991). Moreover, GI hormones from vertebrates, CCK-8 (desulfated form), gastrin, bombesin, secretin, and substance P were all stimulating the release of proteases and amylase from the hepatopancreas (Resch-Sedlmeier and Sedlmeier, 1999). Also, hormones from the eyestalk, such as the hyperglycaemic hormone, have been suggested to be involved in regulation of digestive functions (Carrillo-Farnes et al., 2007).

Digestion

Qualitatively, digestive processes seem quite similar in shrimp and fish. Even though shrimp do not have a secretory stomach, nutrient hydrolysis seems to be initiated in the foregut in many species by the action of enzymes delivered from the hepatopancreas or from food animals. All macronutrients may be partially hydrolyzed when they reach the midgut as the juice from the hepatopancreas contains proteases, lipases, and amylase. The breakdown of macronutrients continues in the hepatopancreatic chamber and the endproducts are supposedly small peptides and amino acids, fatty acids, and monoglycerol or possibly free glycerol. Digestion of lipids in crustaceans is similar to that of fish, and lipid digestibility is typically > 90%. The midgut gland of *L. vannamei* shows lipase activity from the very early stages of development, indicating a capacity for lipid digestion as well as the importance of lipid in development (Rivera-Pérez et al., 2010). A major difference between shrimp and fish is the fact that crustaceans do not produce bile and do not utilize bile salts in their lipid digestion and metabolism (Cherif et al., 2007). Demand for other emulsifiers may therefore be higher in shrimp than in vertebrates.

The intestinal microbiota of shrimp may play a role in some shrimp species feeding on high-carbohydrate diets. However, transit time is high and prevents extensive microbial fermentation. The microbes may supply vitamins and possibly add some digestive enzymes, but neither the proximal compartment nor the distal compartment of the intestine has a surface facilitating colonization (Ceccaldi, 1997).

Absorption

The main absorption of nutrients in crustaceans takes place in the hepatopancreas. This tubular system has a single-cell layer of epithelial cells that facilitates rapid transcellular nutrient transport to the haemolymph. However, the absorptive functions of the different cell types of the hepatopancreatic tissue have not been fully investigated. The activity of several brush-border membrane transporters in the hepatopancreas is reported to be pH-dependent (Verri et al., 2001). This has been demonstrated for the Na^+/D-glucose cotransporter, the Na^+/Cl^-/L-alanine cotransporter, the $Na^+/2Cl^-$/L-leucine cotransporter, and the Na^+/Cl^-/L-glutamate cotransporter. The low pH in the hepatopancreatic lumen facilitates nutrient influx into the epithelial cells.

CONCLUSIONS

Fish and shrimp differ greatly in the anatomical characteristics of the digestive tract, which seems to be more complicated in shrimp than in fish. However, variation in structure is greater for fish. The digestive processes, on the other hand, are less variable and generally follow the same principles as found in higher animals. Present knowledge on digestive physiology of fish far exceeds that of shrimp, but many details still require further investigation, even in fish. Better understanding is needed of the fate of the feed in the digestive tract and limitations of the digestive processes to be able to formulate and process diets optimally so that they can fulfill the nutrient requirements and secure health and wellbeing of the cultivated organisms. The processes depend on the stages of development of the animals and vary with environmental conditions. Regulatory aspects are in particular weakly described, even in fish. Interactions between diet compounds and defense mechanisms in the gut also need greater attention.

REFERENCES

Abate, F., G. Germana, F. De Carlos, G. Motalbano, R. Laura, M. Levanti, and A. Germana. 2006. The oral cavity of the adult zebrafish (*Danio rerio*). Anat. Histol. Embryol. 35:304.

Aldman, G., and S. Holmgren. 1995. Intraduodenal fat and amino-acids activate gallbladder motility in the rainbow-trout, *Oncorhynchus-mykiss*. Gen. Comp. Endocrinol. 100:27–32.

Aldman, G., D. Grove, and S. Holmgren. 1992. Duodenal acidification and intraarterial injection of Cck8 increase gallbladder motility in the rainbow-trout, *Oncorhynchus-mykiss*. Gen. Comp. Endocrinol. 86:20–25.

Altschul, S. F., T. L. Madden, A. A. Schaffer, J. H. Zhang, Z. Zhang, W. Miller, and D. J. Lipman. 1997. Gapped BLAST and PSI-BLAST: A new generation of protein database search programs. Nucleic Acids Res. 25:3389–3402.

Amthauer, R., C. Santander, J. Villanueva, and M. Concha. 2000a. Specific binding of horseradish peroxidase to carp (*Cyprinus carpio*) intestinal apical membrane. Mol. Biol. Cell 11:557.

Amthauer, R., H. Tobar, H. Molina, M. Concha, and J. Villanueva. 2000b. Horseradish peroxidase binding to intestinal brush-border membranes of *Cyprinus carpio*. Identification of a putative receptor. J. Cell Biochem. 80:274–284.

Andre, M., S. Ando, C. Ballagny, M. Durliat, G. Poupard, C. Briancon, and P. J. Babin. 2000. Intestinal fatty acid binding protein gene expression reveals the cephalocaudal patterning during zebrafish gut morphogenesis. Int. J. Dev. Biol. 44:249–252.

Arnesen, P., and Å. Krogdahl. 1995. Nutrient digestibilities, weight gain and plasma and liver levels of carbohydrate in rainbow trout, *Oncorhynchus mykiss* (Walbaum), fed diets with oats. Aquacult. Nutr. 1:201–211.

Arnesen, P., L. E. Brattås, J. J. Olli, and Å. Krogdahl. 1990. Soybean carbohydrates appear to restrict the utilization of nutrients by Atlantic salmon (*Salmo salar* L.). P. 273 in Proc. Third Int. Symp. on Feeding and Nutr. in Fish, Toba, Japan, 1989, M. Takeda and T. Watanabe, eds. Tokyo, Japan: Laboratory of Fish Nutrition, Department of Aquatic Biosciences.

Asgeirsson, B., and P. Cekan. 2006. Microscopic rate-constants for substrate binding and acylation in cold-adaptation of trypsin I from Atlantic cod. FEBS Lett. 580:4639–4644.

Babu, D., and C. Manjulatha. 1995. Zymogen secreting cells in the hepatopancreatic duct of *Pagurus bernhardus* (Linnaeus, 1758) and *Clibanarius longitarsus* (de Haan, 1849) (Anomura). Crustaceana 68:616–628.

Baeverfjord, G., S. Refstie, P. Krogedal, and T. Asgard. 2006. Low feed pellet water stability and fluctuating water salinity cause separation and accumulation of dietary oil in the stomach of rainbow trout (*Oncorhynchus mykiss*). Aquaculture 261:1335–1345.

Bakke, A. M., D. Tashjian, C. Wang, S. Bai, and S. Hong. 2010. Competition between selenomethionine and methionine absorption in the intestinal tract of green sturgeon (*Acipenser medirostris*). Aquat. Toxicol. 96:62–69.

Bakke-McKellep, A. M., S. Nordrum, Å. Krogdahl, and R. K. Buddington. 2000. Absorption of glucose, amino acids, and dipeptides by the intestines of Atlantic salmon (*Salmo salar* L.). Fish Physiol. Biochem. 22:33–44.

Berg-Lea, T., L.-E. Brattås, and A. Krogdahl. 1989. Soybean proteinase inhibitors affect nutrient digestion in rainbow trout. P. 99 in Recent Advances of Research in Antinutritional Factors in Legume Seeds, J. Huisman, T. F. B. van der Pool, and I. Liener, eds. Wageningen, the Netherlands: Pudoc.

Bogevik, A. S., R. E. Olsen, and D. R. Tocher. 2008. Utilization and metabolism of palmityl and oleoyl fatty acids and alcohols in caecal enterocytes of Atlantic salmon (*Salmo salar* L.). Aquacult. Nutr. 14:270–280.

Bogevik, A. S., D. R. Tocher, E. Langmyhr, R. Waagbo, and R. E. Olsen. 2009. Atlantic salmon (*Salmo salar*) postsmolts adapt lipid digestion according to elevated dietary wax esters from Calanus finmarchicus. Aquacult. Nutr. 15:94–103.

Bucking, C., and C. M. Wood. 2006. Water dynamics in the digestive tract of the freshwater rainbow trout during the processing of a single meal. J. Exp. Biol. 209:1883–1893.

Buddington, R. K., and J. M. Diamond. 1987. Pyloric ceca of fish—a new absorptive organ. Am. J. Physiol. Gastr. L. 252:G65–G76.

Buddington, R. K., and J. W. Hilton. 1987. Intestinal adaptations of rainbow-trout to changes in dietary carbohydrate. Am. J. Physiol. Gastr. L. 253:G489–G496.

Buddington, K. K., and V. Kuz'mina. 2000a. Digestive system. Part 4 Microscopic functional anatomy. P. 379 in The Laboratory Fish, G. K. Ostrander, ed. San Diego, CA: Academic Press.

Buddington, K. K., and V. Kuz'mina. 2000b. Digestive system. Part 3 Gross functional anatomy. P. 173 in The Laboratory Fish, G. K. Ostrander, ed. San Diego, CA: Academic Press.

Buddington, R. K., J. W. Chen, and J. Diamond. 1987. Genetic and phenotypic adaptation of intestinal nutrient transport to diet in fish. J. Physiol.-London 393:261–281.

Buddington, R. K., Å. Krogdahl, and A. M. Bakke-McKellep. 1997. The intestine of carnivorous fish: Structure and functions and the relations with diet. Acta Physiol. Scand. 161:67–80.

Caballero, M. J., G. Gallardo, L. Robaina, D. Montero, A. Fernandez, and M. Izquierdo. 2006. Vegetable lipid sources affect in vitro biosynthesis of triacylglycerols and phospholipids in the intestine of sea bream (*Sparus aurata*). Brit. J. Nutr. 95:448–454.

Cahu, C. L., and J. L. Z. Infante. 1995. Maturation of the pancreatic and intestinal digestive functions in sea bass (*Dicentrarchus labrax*): Effect of weaning with different protein sources. Fish Physiol. Biochem. 14:431–437.

Cahu, C., I. Ronnestad, V. Grangier, and J. L. Z. Infante. 2004. Expression and activities of pancreatic enzymes in developing sea bass larvae (*Dicentrarchus labrax*) in relation to intact and hydrolyzed dietary protein; involvement of cholecystokinin. Aquaculture 238:295–308.

Carrillo-Farnes, O., A. Forrellat-Barrios, S. Guerrero-Galvan, and F. Vega-Villasante. 2007. A review of digestive enzyme activity in penaeid shrimps. Crustaceana 80:257–275.

Ceccaldi, H. 1997. Anatomy and physiology of the digestive system. P. 261 in Crustaceaen Nutrition, L. R. D'Abramo, D. Conklin, and D. M. Akiyama, eds. Baton Rouge, LA: The World Aquaculture Society.

Cherif, S., A. Fendri, N. Miled, H. Mejodoub, and Y. Gargouri. 2007. Crab digestive lipase acting at high temperature: Purification and biochemical characterization. Biochimie 89:1018.

Chisty, M. A. H., R. Hashim, and A. S. C. Chong. 2009. Identification and partial characterization of selected proteolytic enzymes in the digestive system of giant freshwater Prawn Macrobrachium rosenbergii (De Man) Postlarvae. Aquac. Res. 40:519–525.

Collie, N. L. 1985. Intestinal nutrient transport in coho salmon (*Oncorhynchus kisutch*) and the effects of development, starvation, and seawater adaptation. J. Comp. Physiol. B 156:163–174.

Collie, N. L., and R. P. Ferraris. 1995. Nutrient fluxes and regulation in fish intestine. P. 221 in Metabolic Biochemistry, P. W. Hochachka and T. P. Mommsen, eds. Amsterdam, the Netherlands: Elsevier Science.

Concha, M. I., C. N. Santander, J. Villanueva, and R. Amthauer. 2002. Specific binding of the endocytosis tracer horseradish peroxidase to intestinal fatty acid-binding protein (I-FABP) in apical membranes of carp enterocytes. J. Exp. Zool. 293:541–550.

Cooper, C. A., J. M. Whittamore, and R. W. Wilson. 2010. Ca_2^+-driven intestinal HCO_3^- secretion and $CaCO_3$ precipitation in the European flounder in vivo: Influences on acid-base regulation and blood gas transport. Am. J. Physiol. Reg. I 298:R870–R876.

Cyrino, J. E. P., D. P. Bureau, and B. G. Kapoor. 2008. Feeding and Digestive Functions of Fishes. Enfield, NH: Science Publishers.

De Almeida, L. C., L. M. Lundstedt, and G. Moraes. 2006. Digestive enzyme responses of tambaqui (Colossoma macropomum) fed on different levels of protein and lipid. Aquacult. Nutr. 12:443–450.

Deguara, S., K. Jauncey, and C. Agius. 2003. Enzyme activities and pH variations in the digestive tract of gilthead sea bream. J. Fish Biol. 62:1033–1043.

Dezfuli, B. S., C. Szekely, G. Giovinazzo, K. Hills, and L. Giari. 2009. Inflammatory response to parasitic helminths in the digestive tract of Anguilla anguilla (L.). Aquaculture 296:1–6.

Dias, J., C. Huelvan, M. T. Dinis, and R. Metailler. 1998. Influence of dietary bulk agents (silica, cellulose and a natural zeolite) on protein digestibility, growth, feed intake and feed transit time in European seabass (*Dicentrarchus labrax*) juveniles. Aquat. Living Resour. 11:219–226.

Diaz, A. O., A. M. Garcia, D. E. Figueroa, and A. L. Goldemberg. 2008a. The mucosa of the digestive tract in Micropogonias furnieri: A light and electron microscope approach. Anat. Histol. Embryol. 37:251–256.

Diaz, A. C., A. V. F. Gimenez, S. M. Velurtas, and J. L. Fenucci. 2008b. Ontogenetic changes in the digestive system of Pleoticus muelleri (Decapoda, Penaeoidea). Invertebr. Reprod. Dev. 52:1–12.

Divakaran, S., B. G. Kim, and A. C. Ostrowski. 1999. Digestive enzymes present in Pacific threadfin Polydactylus sexfilis (Bloch & Schneider 1801) and bluefin trevally Caranx melampygus (Cuvier 1833). Aquac. Res. 30:781–787.

Dohnalek, M. H. 2004. The role of fibre in clinical nutrition. P. 271 in Dietary Fibre, J. W. van der Kamp, N.-G. Asp, J. Miller Jones, and G. Schaafsma, eds. Wageningen, the Netherlands: Academic Publishers.

dos Santos, J., and M. Jobling. 1988. Gastric-emptying in cod, gadusmorhua l—effects of food particle-size and dietary energy content. J. Fish Biol. 33:511–516.

Ducasse-Cabanot, S., J. Zambonino-Infante, N. Richard, F. Medale, G. Corraze, M. Mambrini, J. Robin, C. Cahu, S. Kaushik, and S. Panserat. 2007. Reduced lipid intake leads to changes in digestive enzymes in the intestine but has minor effect on key enzymes of hepatic intermediary metabolism in rainbow trout (*Oncorhynchus mykiss*). Animal 1:1272–1282.

Einarsson, S., and P. S. Davies. 1996. On the localisation and ultrastructure of pepsinogen, trypsinogen and chymotrypsinogen secreting cells in the atlantic salmon, *Salmo salar* L. Comp. Biochem. Phys. B 114B:295–301.

Einarsson, S., and P. S. Davies. 1997. A multiductal system conveys digestive enzymes from the pancreas into the intestine in the Atlantic salmon. J. Fish Biol. 50:1120–1123.

Einarsson, S., P. S. Davies, and C. Talbot. 1996. The effect of feeding on the secretion of pepsin, trypsin and chymotrypsin in the Atlantic salmon, *Salmo salar* L. Fish Physiol. Biochem. 15:439–446.

Einarsson, S., P. S. Davies, and C. Talbot. 1997. Effect of exogenous cholecystokinin on the discharge of the gallbladder and the secretion of trypsin and chymotrypsin from the pancreas of the Atlantic salmon, *Salmo salar* L. Comp. Biochem. Phys. C. 117:63–67.

Enes, P., S. Panserat, S. Kaushik, and A. Oliva-Teles. 2009. Nutritional regulation of hepatic glucose metabolism in fish. Fish Physiol. Biochem. 35:519–539.

Fange, R., G. Lundblad, J. Lind, and K. Slettengren. 2010. Chitinolytic enzymes in the digestive-system of marine fishes. Mar. Biol. 53:317–321.

Favrel, P., G. Kegel, D. Sedlmeier, R. Keller, and A. VanWormhoudt. 1991. Structure and biological-activity of crustacean gastrointestinal peptides identified with antibodies to gastrin cholecystokinin. Biochimie 73:1233–1239.

Ferraris, R. P., and G. A. Ahearn. 1984. Sugar and amino-acid-transport in fish intestine. Comp. Biochem. Phys. A 77:397–413.

Ferraris, R. P., S. Yasharpour, K. C. K. Lloid, R. Mirzayan, and J. M. Diamond. 1990. Luminal glucose concentration in the gut under normal conditions. Am. J. Physiol. Gastr. L. 259:G820–G837.

Figueiredo, M., and A. J. Anderson. 2009. Digestive enzyme spectra in crustacean decapods (Paleomonidae, Portunidae and Penaeidae) feeding in the natural habitat. Aquac. Res. 40:282–291.

Focken, U., A. Groth, R. M. Coloso, and K. Becker. 1998. Contribution of natural food and supplemental feed to the gut content of *Penaeus monodon* Fabricius in a semi-intensive pond system in the Philippines. Aquaculture 164:105–116.

Fontagne, S., I. Geurden, A. M. Escaffre, and P. Bergot. 1998. Histological changes induced by dietary phospholipids in intestine and liver of common carp (*Cyprinus carpio* L.) larvae. Aquaculture 161:213–223.

Froystad, M. K., E. Lilleeng, A. Sundby, and A. Krogdahl. 2006. Cloning and characterization of alpha-amylase from Atlantic salmon (*Salmo salar* L.). Comp. Biochem. Phys. A 145:479–492.

Gamboa-Delgado, J., C. Molina-Poveda, and C. Cahu. 2003. Digestive enzyme activity and food ingesta in juvenile shrimp *Litopenaeus vannamei* (Boone, 1931) as a function of body weight. Aquac. Res. 34:1403–1411.

Ganong, W. 2009. Digestion and absorption. P. 398 in Review of Medical Physiology, W. Ganong, ed. London, UK: Appelton & Lange.

Gawlicka, A., C. T. Leggiadro, J. W. Gallant, and S. E. Douglas. 2001. Cellular expression of the pepsinogen and gastric proton pump genes in the stomach of winter flounder as determined by *in situ* hybridization. J. Fish Biol. 58:529–536.

Gaxiola, G., G. Cuzon, T. Garcia, G. Taboada, R. Brito, M. Chimal, A. Paredes, L. Soto, C. Rosas, and A. van Wormhoudt. 2005. Factorial effects of salinity, dietary carbohydrate and moult cycle on digestive carbohydrases and hexokinases in *Litopenaeus vannamei* (Boone, 1931). Comp. Biochem. Phys. A 140:29–39.

Geurden, I., M. Aramendi, J. Zambonino-Infante, and S. Panserat. 2007. Early feeding of carnivorous rainbow trout (*Oncorhynchus mykiss*) with a hyperglucidic diet during a short period: Effect on dietary glucose utilization in juveniles. Am. J. Physiol. Reg. I. 292:R2275–R2283.

Gjellesvik, D. R., A. J. Raae, and B. T. Walther. 1989. Partial-purification and characterization of a triglyceride lipase from cod (*Gadus morhua*). Aquaculture 79:177–184.

Glass, H. J., N. L. MacDonald, R. M. Moran, and J. R. Stark. 1989. Digestion of protein in different marine species. Comp. Biochem. Phys. B 94:607–611.

Glover, C. N., and C. M. Wood. 2008. Histidine absorption across apical surfaces of freshwater rainbow trout intestine: Mechanistic characterization and the influence of copper. J. Membrane Biol. 221:87–95.

Guilloteau, P., R. Toullec, and P. Patureau-Miraud. 1979. Influence de la vitesse d'évacuation gastrique des protéines et des lipides sur l'utilisation digestive chez le veau préruminant. Ann. Biol. Anim. Bioch. 19:955–958.

Gutowska, M. A., J. C. Drazen, and B. H. Robison. 2004. Digestive chitinolytic activity in marine fishes of Monterey Bay, California. Comp. Biochem. Phys. A 139:351–358.

Hasler, W. 2006. Small intestinal motility. P. 935 in Physiology of the intestinal tract, L. R. Johnsen, ed. New York: Academic Press.

Haslewood, G. A. 1967. Bile salt evolution. J. Lipid Res. 8:535–551.

Hernandez, J., and J. Murueta. 2009. Activity of trypsin from *Litopenaeus vannamei*. Aquaculture 290:190–195.

Hernandez-Blazquez, F. J., and J. R. M. C. da Silva. 1998. Absorption of macromolecular proteins by the rectal epithelium of the Antarctic fish Notothenia neglecta. Can. J. Zool. 76:1247–1253.

Hidalgo, M. C., E. Urea, and A. Sanz. 1999. Comparative study of digestive enzymes in fish with different nutritional habits. Proteolytic and amylase activities. Aquaculture 170:267–283.

Hill, J. 2007. Impacts of nutritional technology on feeds offered to horses: A review of effects of processing on voluntary intake, digesta characteristics and feed utilisation. Anim. Feed Sci. Tech. 138:92–117.

Hilton, J. W., and J. L. Atkinson. 1982. Response of rainbow trout (*Salmo gairdneri*) to increased levels of available carbohydrate in practical trout diets. Brit. J. Nutr. 47:597–607.

Hilton, J. W., and D. G. Dixon. 1982. Effect of increased liver-glycogen and liver weight on liver-function in rainbow-trout, *Salmo gairdneri* Richardson: Recovery from anesthesia and plasma S-35-Sulphobromophthalein clearance. J. Fish Dis. 5:185–195.

Hilton, J. W., J. L. Atkinson, and S. J. Slinger. 1982. Maximum tolerable level, digestion, and metabolism of D-glucose (cerelose) in rainbow trout (*Salmo gairdneri*) reared on a practical trout diet. Can. J. Fish Aquat. Sci. 39:1229–1234.

Hofmann, A. F. 1994. Biliary secretion and excretion. P. 1555 in Physiology of the Gastrointesinal Tract, L. R. Johnson, ed. New York: Raven Press.

Holmgren, S., and C. Olsson. 2009. The neuronal and endocrine regulation of gut function. P. 467 in Fish Neuroendocrinology, N. Bernier, G. Van Der Krak, A. Farrell, and C. Brauner, eds. New York: Elsevier.

Houpe, K. L., C. Malo, P. B. Oldham, and R. K. Buddington. 1996. Thermal modulation of channel catfish intestinal dimensions, BBM fluidity, and glucose transport. Am. J. Physiol. Reg. I. 270:R1037–R1043.

Ikeda, M., K. Miyauchi, A. Mochizuki, and M. Matsumiya. 2009. Purification and characterization of chitinase from the stomach of silver croaker Pennahia argentatus. Protein Expres. Purif. 65:214–222.

Iqbal, J., and M. M. Hussain. 2009. Intestinal lipid absorption. Am. J. Physiol. Endoc. M. 296:E1183–E1194.

Jobling, M. 1995. Environmental Biology of Fishes. London, UK: Chapman & Hall.

Johnston, D., and J. Freeman. 2005. Dietary preference and digestive enzyme activities as indicators of trophic resource utilization by six species of crab. Biol. Bull. 208:46.

Karlsen, Ø., G. I. Hemre, K. Tveit, and G. Rosenlund G. 2006. Effect of varying levels of macro-nutrients and continuous light on growth, energy deposits and maturation in farmed Atlantic cod (*Gadus morhua* L.). Aquaculture 255:242–254.

Kjaer, M. A., A. Vegusdal, G. M. Berge, T. F. Galloway, M. Hillestad, A. Krogdahl, H. Holm, and B. Ruyter. 2009. Characterisation of lipid transport in Atlantic cod (*Gadus morhua*) when fasted and fed high or low fat diets. Aquaculture 288:325–336.

Koji, M., H. Fukada, I. Ronnestad, T. Kurokawa, and T. Masumoto. 2008. Nutrient control of release of pancreatic enzymes in yellowtail (Seriola quinqueradiata): Involvement of CCK and PY in the regulatory loop. Comp. Biochem. Phys. A 150:438–443.

Koven, W., C. R. Rojas-Garcia, R. N. Finn, A. Tandler, and I. Ronnestad. 2002. Stimulatory effect of ingested protein and/or free amino acids on the secretion of the gastro-endocrine hormone cholecystokinin and on tryptic activity, in early-feeding herring larvae, Clupea harengus. Mar. Biol. 140:1241–1247.

Krogdahl, Å., S. Nordrum, M. Sørensen, L. Brudeseth, and C. Røsjø. 1999. Effects of diet composition on apparent nutrient absorption along the intestinal tract and of subsequent fasting on mucosal disaccharidase activities and plasma nutrient concentration in Atlantic salmon, *Salmo salar* L. Aquacult. Nutr. 5:121–133.

Krogdahl, Å., A. Sundby, and J. J. Olli. 2004. Atlantic salmon (*Salmo salar*) and rainbow trout (*Oncorhynchus mykiss*) digest and metabolize nutrients differently. Effects of water salinity and dietary starch level. Aquaculture 229:335–360.

Krogdahl, A., G. I. Hemre, and T. P. Mommsen. 2005. Carbohydrates in fish nutrition: Digestion and absorption in postlarval stages. Aquacult. Nutr. 11:103–122.

Krogdahl, Å., H. Holm, B. Ruyter, T. F. Galloway, G. M. Berge, and M. Hillestad. 2009. Cod (*Gadus morhua*) adjust digestive processes to variation in diet composition. P. 333 in Aquaculture Europe 2009. Brussels, Belgium: Aquaculture Europe.

Kuperman, B. I., and V. V. Kuzmina. 1994. The ultrastructure of the intestinal epithelium in fishes with different types of feeding. J. Fish Biol. 44:181–193.

Kurokawa, T., and T. Suzuki. 1996. Formation of the diffuse pancreas and the development of digestive enzyme synthesis in larvae of the Japanese flounder *Paralichthys olivaceus*. Aquaculture 141:267–276.

Kurtovic, I., S. N. Marshall, X. Zhao, and B. K. Simpson. 2009. Lipases from mammals and fishes. Rev. Fish. Sci. 17:18–40.

Kuz'mina, V. 2008. Classical and modern concepts in fish digestion. P. 85 in Feeding and Digestive Functions of Fishes, J. E. P. Cyrino, D. P. Bureau, and B. G. Kapoor, eds. Enfield, NH: Science Publishers.

Lavelle, E. C., and J. E. Harris. 1997. The processing of an orally administered protein antigen in the digestive tract of rainbow trout, *Oncorhynchus mykiss*. Comp. Biochem. Phys. A 117:263–275.

Lawlor, P. G., P. B. Lynch, P. J. Caffrey, J. J. O'Reilly, and M. K. O'Connell. 2005. Measurements of the acid-binding capacity of ingredients used in pig diets. Irish Vet. J. 58:447–452.

Le Vay, L., D. A. Jones, A. C. Puello-Cruz, R. S. Sangha, and C. Ngamphongsai. 2001. Digestion in relation to feeding strategies exhibited by crustacean larvae. Comp. Biochem. Phys. A 128:623–630.

Lemos, D., J. M. Ezquerra, and F. L. Garcia-Carreno. 2000. Protein digestion in penaeid shrimp: Digestive proteinases, proteinase inhibitors and feed digestibility. Aquaculture 186:89–105.

Le Moullac, G., B. Klein, D. Sellos, and A. VanWormhoudt. 1997. Adaptation of trypsin, chymotrypsin and alpha-amylase to casein level and protein source in Penaeus vannamei (Crustacea Decapoda). J. Exp. Mar. Biol. Ecol. 208:107–125.

Lionetto, M. G., M. Maffia, F. Vignes, C. Storelli, and T. Schettino. 1996. Differences in intestinal electrophysiological parameters and nutrient transport rates between eels (*Anguilla anguilla*) at yellow and silver stages. J. Exp. Zool. 275:399–405.

Lopez-Lopez, S., H. Nolasco, and F. Vega-Villasante. 2003. Characterization of digestive gland esterase–lipase activity of juvenile redclaw crayfish *Cherax quadricarinatus*. Comp. Biochem. Phys. B 135:347.

Lopez-Lopez, S., H. Nolasco, H. Villarreal-Colmenares, and Civera-Cerecedo R. 2005. Digestive enzyme response to supplemental ingredients in practical diets for juvenile freshwater crayfish *Cherax quadricarinatus*. Aquacult. Nutr. 11:79–85.

Lundstedt, L. M., J. Melo, and G. Moraes. 2004. Digestive enzymes and metabolic profile of Pseudoplatystoma corruscans (Teleostei: Siluriformes) in response to diet composition. Comp. Biochem. Phys. B 137:331–339.

Maffia, M., R. Acierno, G. Deceglie, S. Vilella, and C. Storelli. 1993. Adaptation of intestinal cell membrane enzymes to low temperatures in the Antarctic teleost *Pagothenia bernacchii*. J. Comp. Physiol. B 163:265–270.

Maffia, M., T. Verri, A. Danieli, M. Thamotharan, M. Pastore, G. A. Ahearn, and C. Storelli. 1997. H+-glycyl-L-proline cotransport in brush-border membrane vesicles of eel (Anguilla anguilla) intestine. Am. J. Physio. Reg. I. 272:R217–R225.

Manchado, M., C. Infante, E. Asensio, A. Crespo, E. Zuasti, and J. P. Cañavate. 2008. Molecular characterization and gene expression of six trypsinogens in the flatfish Senegalese sole (*Solea senegalensis* Kaup) during larval development and in tissues. Comp. Biochem. Phys. B 149:334–344.

Manjakasy, J. M., R. D. Day, A. Kemp, and I. R. Tibbetts. 2009. Functional morphology of digestion in the stomachless, piscivorous needlefishes Tylosurus gavialoides and Strongylura leiura ferox (Teleostei: Beloniformes). J. Morphol. 270:1155–1165.

Mantel, L. H. 1983. Internal anatomy and physiological regulation. Pp. 1–53 in The Biology of Crustacea, Vol. 5, D. E. Bliss, ed. New York: Academic Press.

Martin G. G., R. Simcox, A. Nguyen, and A. Chilingaryan. 2006. Peritrophic membrane of the penaeid shrimp Sicyonia ingentis: Structure, formation, and permeability. Biol. Bull. 211:275–285.

McLean, E., and R. Ash. 1986. The time-course of appearance and net accumulation of horseradish peroxidase (HRP) presented orally to juvenile carp *Cyprinus carpio* (L.). Comp. Biochem. Phys. A 84:687–690.

McLean, E., and R. Ash. 1987a. Intact protein (antigen) absorption in fishes: Mechanism and physiological significance. J. Fish Biol. 31:219–223.

McLean, E., and R. Ash. 1987b. The time-course of appearance and net accumulation of horseradish peroxidase (HRP) presented orally to rainbow trout *Salmo gairdneri* (Richardson). Comp. Biochem. Phys. A 88:507–510.

Morais, S., C. Cahu, J. L. Zambonino-Infante, J. Robin, I. Ronnestad, M. T. Dinis, and L. E. C. Conceicao. 2004. Dietary TAG source and level affect performance and lipase expression in larval sea bass (*Dicentrarchus labrax*). Lipids 39:449–458.

Morrison, C. M., B. Pohajdak, J. Tam, and J. R. Wright. 2004. Development of the islets, exocrine pancreas, and related ducts in the Nile tilapia, *Oreochromis niloticus* (Pisces:Cichlidae). J. Morphol. 261:377–389.

Moss, S., S. Divakaran, and B. Kim. 2001. Stimulating effects of pond water on digestive enzyme activity in the Pacific white shrimp, *Litopenaeus vannamei* (Boone). Aquac. Res. 32:125–131.

Muhlia-Almazan, A., F. L. Garcia-Carreno, J. A. Sanchez-Paz, G. Yepiz-Plascencia, and A. B. Peregrino-Uriarte. 2003. Effects of dietary protein on the activity and mRNA level of trypsin in the midgut gland of the white shrimp Penaeus vannamei. Comp. Biochem. Phys. B 135:373–383.

Muhlia-Almazan, A., A. Sanchez-Paz, and F. L. Garcia-Carreno. 2008. Invertebrate trypsins: A review. J. Comp. Physiol. B 178:655–672.

Murashita, K., H. Fukada, H. Hosokawa, and T. Masumoto. 2007. Changes in cholecystokinin and peptide Y gene expression with feeding in yellowtail (Seriola quinqueradiata): Relation to pancreatic exocrine regulation. Comp. Biochem. Phys. B 146:318–325.

Nordrum, S., A. M. Bakke-McKellep, A. Krogdahl, and R. K. Buddington. 2000. Effects of soybean meal and salinity on intestinal transport of nutrients in Atlantic salmon (*Salmo salar* L.) and rainbow trout (*Oncorhynchus mykiss*). Comp. Biochem. Phys. B 125:317–335.

Nunes, A. J. P., and G. J. Parsons. 2000. Effects of the southern brown shrimp, Penaeus subtilis, predation and artificial feeding on the population dynamics of benthic polychaetes in tropical pond enclosures. Aquaculture 183(1–2):125–147.

Nunes, A. J. P., T. C. V. Gesteira, and S. Goddard. 1997. Food consumption and assimilation by the Southern brown shrimp *Penaeus subtilis* under semi-intensive culture in NE Brazil. Aquaculture 149:121–136.

Ogata, Y., M. Nishi, H. Nakayama, T. Kuwahara, Y. Ohnishi, and S. Tashiro. 2003. Role of bile in intestinal barrier function and its inhibitory effect on bacterial translocation in obstructive jaundice in rats. J. Surg. Res. 115:18–23.

Ogiwara, K., and T. Takahashi. 2007. Specificity of the medaka enteropeptidase serine protease and its usefulness as a biotechnological tool for fusion-protein cleavage. P. Natl. Acad. Sci. USA 104:7021–7026.

Olsen, R. E., B. T. Dragnes, R. Myklebust, and E. Ringo. 2003. Effect of soybean oil and soybean lecithin on intestinal lipid composition and lipid droplet accumulation of rainbow trout, *Oncorhynchus mykiss* Walbaum. Fish Physiol. Biochem. 29:181–192.

Olsen, R. E., A. C. Hansen, G. Rosenlund, G. I. Hernre, T. M. Mayhew, D. L. Knudsen, O. T. Eroldogan, R. Myklebust, and O. Karlsen. 2007. Total replacement of fish meal with plant proteins in diets for Atlantic cod (*Gadus morhua* L.) II—Health aspects. Aquaculture 272:612–624.

Ong, B. L., and D. Johnston. 2006. Influence of feeding on hepatopancreas structure and digestive enzyme activities in Penaeus monodon. J. Shellfish Res. 25:113–121.

Ostaszewska, T., M. Kamaszewski, P. Grochowski, K. Dabrowski, T. Verri, E. Aksakal, I. Szatkowska, Z. Nowak, and S. Dobosz. 2010. The effect of peptide absorption on PepT1 gene expression and digestive system hormones in rainbow trout (*Oncorhynchus mykiss*). Comp. Biochem. Phys. A 155:107–114.

Oxley, A., F. Jutfelt, K. Sundell, and R. E. Olsen. 2007. Sn-2-monoacylglycerol, not glycerol, is preferentially utilised for triacylglycerol and phosphatidylcholine biosynthesis in Atlantic salmon (*Salmo salar* L.) intestine. Comp. Biochem. Phys. B 146:115–123.

Peres, H., P. Goncalves, and A. Oliva-Teles. 1999. Glucose tolerance in gilthead seabream (*Sparus aurata*) and European seabass (*Dicentrarchus labrax*). Aquaculture 179:415–423.

Pérez-Jiménez, A., G. Cardenete, A. E. Morales, A. García-Alcázar, E. Abellán, and M. C. Hidalgo. 2009. Digestive enzymatic profile of Dentex dentex and response to different dietary formulations. Comp. Biochem. Phys. A 154:157–164.

Platell, M. E., and I. C. Potter. 2001. Partitioning of food resources amongst 18 abundant benthic carnivorous fish species in marine waters on the lower west coast of Australia. J. Exp. Mar. Biol. Ecol. 261:31–54.

Pöhlmann, K. 2007. Regional and seasonal differences in triacylglyceride and digestive enzymes in North sea brown shrimp. Masters Thesis, University of Bremen, Germany.

Pratoomyot, J., E. A. Bendiksen, J. G. Bell, and D. R. Tocher. 2010. Effects of increasing replacement of dietary fishmeal with plant protein sources on growth performance and body lipid composition of Atlantic salmon (*Salmo salar* L.). Aquaculture 305:124–132.

Psochiou, E., E. Sarropoulou, Z. Mamuris, and K. A. Mouton. 2007. Sequence analysis and tissue expression pattern of *Sparus aurata* chymotrypsinogens and trypsinogen. Comp. Biochem. Phys. B147:367–377.

Quentel, C., M. Bremont, and H. Pouliquen. 2007. Farm fish vaccination. Prod. Anim. 20:233–238.

Ray, A. K., T. Roy, S. Mondale, and E. Ringø. 2009. Identification of gut-associated amylase, cellulase and protease-producing bacteria in three species of Indian major carps. Aquac. Res. 41:1462–1469.

Resch-Sedlmeier, G., and D. Sedlmeier. 1999. Release of digestive enzymes from the crustacean hepatopancreas: Effect of vertebrate gastrointestinal hormones. Comp. Biochem. Phys. B 123:187–192.

Reshkin, S. J., and G. A. Ahearn. 1987. Basolateral glucose transport by intestine of teleost, *Oreochromis mossambicus*. Am. J. Physiol. Reg. I. 252:R579–R586.

Rivera-Pérez, C., M. d. l. Á. N. del Toro, and F. L. García-Carreño. 2010. Digestive lipase activity through development and after fasting and re-feeding in the whiteleg shrimp Penaeus vannamei. Aquaculture 300:163–168.

Romano, A., G. Kottra, A. Barca, N. Tiso, M. Maffia, F. Argenton, H. Daniel, C. Storelli, and T. Verri. 2006. High-affinity peptide transporter PEPT2 (SLC15A2) of the zebrafish Danio rerio: Functional properties, genomic organization, and expression analysis. Physiol. Genomics 24:207–217.

Romarheim, O. H., A. Skrede, M. Penn, L. T. Mydland, A. Krogdahl, and T. Storebakken. 2008. Lipid digestibility, bile drainage and development of morphological intestinal changes in rainbow trout (*Oncorhynchus mykiss*) fed diets containing defatted soybean meal. Aquaculture 274:329–338.

Rosenlund, G., O. Karlsen, K. Tveit, A. Mangor-Jensen, and G. I. Hemre. 2004. Effect of feed composition and feeding frequency on growth, feed utilization and nutrient retention in juvenile Atlantic cod, *Gadus morhua* L. Aquacult. Nutr. 10:371–378.

Røsjø, C., S. Nordrum, J. J. Olli, Å. Krogdahl, B. Ruyter, and H. Holm. 2000. Lipid digestibility and metabolism in Atlantic salmon (*Salmo salar*) fed medium-chain triglycerides. Aquaculture 190:65–76.

Rothman, S., C. Liebow, and L. Isenman. 2002. Conservation of digestive enzymes. Physiol. Rev. 82:1–18.

Rust, M. 2002. Nutritional physiology. P. 367 in Fish Nutrition, J. E. Halver and R. M. Hardy, eds. Amsterdam, the Netherlands: Academic Press.

Sahlmann, C. 2008. Seasonal and regional aspects of the nutritional ecology and physiology of the brown shrimp, *Crangon crangon* (Decapoda: Caridea). Masters Thesis, University of Bremen, Germany.

Sainz, J. C., F. L. Garcia-Carreno, and P. Hernandez-Cortes. 2004. Penaeus vannamei isotrypsins: Purification and characterization. Comp. Biochem. Phys. B 138:155–162.

Santos, E. A., L. E. M. Nery, R. Keller, and A. A. Goncalves. 1997. Evidence for the involvement of the crustacean hyperglycemic hormone in the regulation of lipid metabolism. Physiol. Zool. 70:415–420.

Schubert, M. L. 2009. Gastric exocrine and endocrine secretion. Curr. Opin. Gastroen. 25:529–536.

Seppola, M., R. E. Olsen, E. Sandaker, P. Kanapathippillai, W. Holzapfel, and E. Ringø. 2005. Random amplification of polymorphic DNA (RAPD) typing of Carnobacteria isolated from hindgut chamber and large intestine of Atlantic cod (*Gadus morhua l.*). Syst. Appl. Microbiol. 29:131–137.

Shephard, K. L. 1994. Functions for fish mucus. Rev. Fish Biol. Fisher. 4:401–429.

Sims, D. 2008. Sieving a living: A review on the biology, ecology and conservation status of the plankton-feeding basking shark Cetorhinus maximus. P. 171 in Advances in Marine Biology, D. W. Sim, ed. London, UK: Elsevier.

Singh, R. P., and T. Nose. 1967. Digestibility of carbohydrates in young rainbow trout. Bull. Freshwater Fish. Res. Lab. (Japan) 17:21–25.

Sire, M. F., and J. M. Vernier. 1992. Intestinal absorption of protein in teleost fish. Comp. Biochem. Phys. A 103:771–781.

Sklan, D., T. Prag, and I. Lupatsch. 2004. Structure and function of the small intestine of the tilapia *Oreochromis niloticus* × *Oreochromis aureus* (Teleostei, Cichlidae). Aquac. Res. 35:350–357.

Sousa, L., E. Cuartas, and A. Petrielle. 2005. Fine structural analysis of the epithelial cells in the hepatopancreas of *Palaemonetes argentinus* (Crustacea, Decapoda, Caridea) in intermoult. Biocell 29:25–31.

Storebakken, T. 1985. Binders in fish feeds. I. Effect of alginate and guar gum on growth, digestibility, feed intake and passage through the gastrointestinal tract of rainbow trout. Aquaculture 47:11–26.

Storebakken, T., and E. Austreng. 1987. Binders in fish feeds. II. Effect of different alginates on the digestibility of macronutrients in rainbow trout. Aquaculture 60:121–131.

Storebakken, T., K. D. Shearer, S. Refstie, S. Lagocki, and J. McCool. 1998. Interactions between salinity, dietary carbohydrate source and carbohydrate concentration on the digestibility of macronutrients and energy in rainbow trout (*Oncorhynchus mykiss*). Aquaculture 163:347–359.

Storelli, C., S. Vilella, M. P. Romano, M. Maffia, and G. Cassano. 1989. Brush-border amino acid transport mechanisms in carnivorous eel intestine. Am. J. Physiol. Reg. I. 257:R506–R510.

Sugita, H., and Y. Ito. 2006. Identification of intestinal bacteria from Japanese flounder (*Paralichthys olivaceus*) and their ability to digest chitin. Lett. Appl. Microbiol. 43:336–342.

Suyehiro, Y. 1941. A study of the digestive system and feeding habits of fish. Jpn. J. Zool. 10:1–303.

Terova, G., S. Cora, T. Verri, S. Rimoldi, G. Bernardini, and M. Saroglia. 2009. Impact of feed availability on PepT1 mRNA expression levels in sea bass (*Dicentrarchus labrax*). Aquaculture 294:288–299.

Teschke, M., and R. Saborowski. 2005. Cysteine proteinases substitute for serine proteinases in the midgut glands of Crangon crangon and Crangon allmani (Decapoda:Caridea). J. Exp. Mar. Biol. Ecol. 316:213–229.

Thamotharan, M., J. Gomme, G. V. Zonno, M. Maffia, C. Storelli, and G. A. Ahearn. 1996a. Electrogenic, proton-coupled, intestinal dipeptide transport in herbivorous and carnivorous teleosts. Am. J. Physiol. Reg. I. 270:R939–R947.

Thamotharan, M., V. Zonno, C. Storelli, and G. A. Ahearn. 1996b. Basolateral dipeptide transport by the intestine of the teleost *Oreochromis mossambicus*. Am. J. Physiol. Reg. I. 270:R948–R954.

Thompson, S. A., and C. W. Weber. 1979. Influence of Ph on the binding of copper, zinc and iron in 6 fiber sources. J. Food Sci. 44:752–754.

Tocher, D. R. 2003. Metabolism and functions of lipids and fatty acids in teleost fish. Rev. Fish. Sci. 11:107–184.

Tocher, D. R., and J. R. Sargent. 1984. Studies on triacylglycerol, wax ester and sterol ester hydrolases in intestinal caeca of rainbow trout (*Salmo gairdneri*) fed diets rich in triacylglycerols and wax esters. Comp. Biochem. Phys. B 77:561–571.

Tocher, D. R., E. A. Bendiksen, P. J. Campbell, and J. G. Bell. 2008. The role of phospholipids in nutrition and metabolism of teleost fish. Aquaculture 280:21–34.

Une, M., T. Goto, K. Kihira, T. Kuramoto, K. Hagiwara, T. Nakajima, and T. Hoshita. 1991. Isolation and identification of bile—Salts conjugated with cysteinolic acid from bile of the red seabream, Pagrosomus major. J. Lipid Res. 32:1619–1623.

Uran, P. A., R. Aydin, J. W. Schrama, J. A. J. Verreth, and J. H. W. M. Rombout. 2008. Soybean meal-induced uptake block in Atlantic salmon *Salmo salar* distal enterocytes. J. Fish Biol. 73:2571–2579.

Valle, A. Z. D., M. Iriti, F. Faoro, C. Berti, and S. Ciappellano. 2008. In vivo prion protein intestinal uptake in fish. APMIS 116:173–180.

Van Wormhoudt, A., and D. Sellos. 2003. Highly variable polymorphism of the alpha-amylase gene family in *Litopenaeus vannamei* (Crustacea decapoda). J. Mol. Evol. 57:659–671.

Velez, Z., P. C. Hubbard, K. Welham, J. D. Hardege, E. N. Barata, and A. V. M. Canario. 2009. Identification, release and olfactory detection of bile salts in the intestinal fluid of the Senegalese sole (*Solea senegalensis*). J. Comp Physiol. A. 195:691–698.

Verri, T., M. Maffia, A. Danieli, M. Herget, U. Wenzel, H. Daniel, and C. Storelli. 2000. Characterisation of the H+/peptide cotransporter of eel intestinal brush-border membranes. J. Exp. Biol. 203:2991–3001.

Verri, T., A. Mandal, L. Zilli, D. Bossa, P. K. Mandal, L. Ingrosso, V. Zonno, S. Vilella, G. A. Ahearn, and C. Storelli. 2001. D-glucose transport in decapod crustacean hepatopancreas. Comp. Biochem. Phys. A 130:585–606.

Verri, T., G. Kottra, A. Romano, N. Tiso, M. Peric, M. Maffia, M. Boll, F. Argenton, H. Daniel, and C. Storelli. 2003. Molecular and functional characterisation of the zebrafish (Danio rerio) PEPT1-type peptide transporter. FEBS Lett. 549:115–122.

Viader-Salvado, J., M. Castillo-Galvan, L. Galan-Wong, and M. Guerrero-Olazaran. 2007. Biochemical characterization of recombinant shrimp trypsinogen. J. Biotechnol. 131:S233.

Vogt, G., W. Söcker, V. Storch, and R. Zwillinger. 1989. Biosynthesis of *Astacus* protease, a digestive enzyme from crayfish. Histochemistry 91:373–383.

Volkoff, H. 2006. The role of neuropeptide Y, orexins, cocaine and amphetamine-related transcript, cholecystokinin, amylin and leptin in the regulation of feeding in fish. Comp. Biochem. Phys. A 144:325–331.

Whittamore, J. M., C. A. Cooper, and R. W. Wilson. 2010. HCO_3^- secretion and $CaCO_3$ precipitation play major roles in intestinal water absorption in marine teleost fish in vivo. Am. J. Physiol. Reg. I. 298:R877–R886.

Wilson, R. W., and M. Grosell. 2003. Intestinal bicarbonate secretion in marine teleost fish-source of bicarbonate, pH sensitivity, and consequences for whole animal acid-base and calcium homeostasis. BBA-Biomembranes 1618:163–174.

Wu, T., L. C. Sun, C. H. Du, Q. F. Cai, Q. B. Zhang, W. J. Su, and M. J. Cao. 2009. Identification of pepsinogens and pepsins from the stomach of European eel (Anguilla anguilla). Food Chem. 115:137–142.

Xue, X. M., A. J. Anderson, N. A. Richardson, A. J. Anderson, G. P. Xue, and P. B. Mather. 1999. Characterisation of cellulase activity in the digestive system of the redclaw crayfish (*Cherax quadricarinatus*). Aquaculture 180:272–386.

4

Dietary Energy Utilization and Metabolic Integration

Nutrients are required by animals to sustain life processes and allow activity, growth, and reproduction. Nutrients serve as precursors for the biosynthesis of structural or storage molecules, enzymes, metabolic intermediates, and a plethora of other molecules. A proportion of the nutrients consumed is catabolized to harness chemical (free) energy, which is required for use in anabolic and other life-sustaining processes (Blaxter, 1989; Mayes, 2000). Animals do not simply metabolize energy *per se*. Instead they metabolize specific nutrients, each with specific roles and metabolic fates (Van Milgen, 2002). Several nutrients or metabolic intermediates derived from nutrients are used simultaneously in the same process, and interactions between nutrients are numerous. A large number of endogenous (genetics, sex, physiological state, nutritional history, etc.) and exogenous (temperature, stressors) factors also affect the fate of nutrients in animals (Blaxter, 1989). To quantitatively examine the utilization of all dietary components in a detailed and integrative fashion is highly desirable. However, it also is an extremely complex undertaking. Numerous frameworks have been developed to describe and predict the utilization of dietary nutrients by animals in a practical fashion (Dumas et al., 2008). Bioenergetics or biochemical thermodynamics, the study of the energy changes accompanying biochemical reactions in biological systems (Patton, 1965; Mayes, 2000), has been the foundation of several of the more popular frameworks. Life processes (e.g., anabolic reactions, muscular contraction, active transport) obtain energy by chemical linkage with some energy being transferred to synthetic reaction and some energy lost as heat. According to the first law of thermodynamics, the partition of energy-yielding components between catabolism as fuels and anabolism as storage in tissues can be tracked by the study of the balance between dietary energy intake and expenditure.

Ege and Krogh (1914) were possibly the first to apply the principles of bioenergetics to fish. Since then, hundreds of reports on studies of energy utilization and expenditure for a range of species of fish have been produced. Numerous reviews have also been written on nutritional energetics (bioenergetics applied in a nutritional context), including those of Phillips (1972), Brett and Groves (1979), Cho et al. (1982), Elliott (1982), Cho and Kaushik (1985, 1990), Tytler and Calow (1985), Smith (1989), Jobling (1994), Kaushik and Médale (1994), Cho and Bureau (1995), Cui and Xie (1999), Médale and Guillaume (1999), and Bureau et al. (2002).

Nutritional energetic frameworks have progressively evolved over the past five decades to include some considerations for the types of macronutrients consumed and/or body tissue components deposited (e.g., body protein and body lipids) or more or less explicit representations of the digestion of feed components, metabolism of absorbed nutrients, and partition of nutrients among tissues and functions within the animal (Kielanowski 1965; Baldwin and Bywater, 1984; Emmans and Fisher, 1986; Emmans, 1994; Noblet et al., 1994; DeLange, 1997; Lupatsch et al., 1998, 2003; Birkett and de Lange, 2001; Van Milgen, 2002). This chapter's objective is to present key principles of nutritional energetics and their underlying metabolic and physiological mechanisms, and review estimates of energy expenditure in different fish and shrimp species. This chapter also aims to identify the gaps in knowledge, highlights some of the limitations of common nutritional energetics frameworks, and fosters a reflection about the need for aquaculture nutritionists to examine growth and nutrient utilizations in a more explicit, mechanistic and integrative fashion in the future.

STANDARD ENERGY PARTITIONING SCHEME—NRC 1981 NOMENCLATURE

Different systems of nomenclatures that describe the partitioning of energy in animals have been used. This is especially apparent in fish biology where the nomenclatures and modes of expression of energy transaction used are extremely diverse. In 1981, a subcommittee of the Committee on Animal Nutrition of the National Research Council was appointed to develop a systematic terminology for descrip-

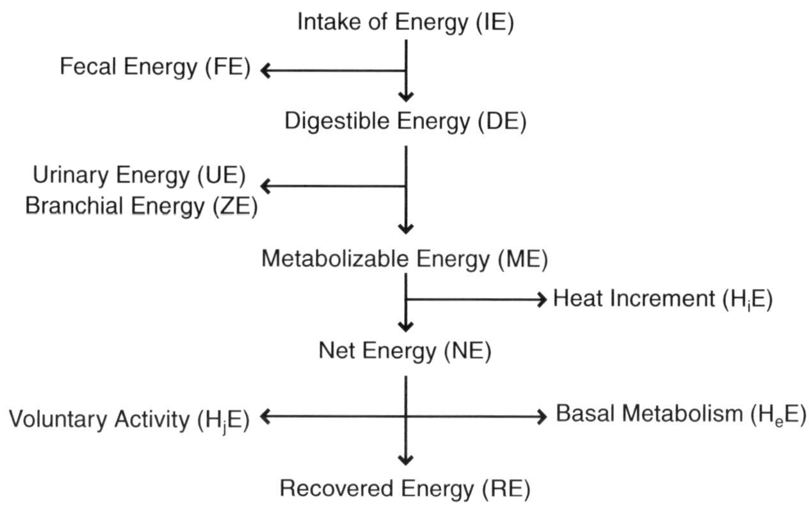

FIGURE 4-1 Schematic representation of the energy flow through an animal (NRC, 1981).

tion of energy utilization by domestic animals, including fish (NRC, 1981). This system is presented schematically in Figure 4-1 and has been widely adopted by animal nutritionists around the world, sometimes with some modifications. This pragmatic nomenclature (Table 4-1) has also been adopted by a number of fish nutrition researchers. This nomenclature was adopted in this document with minor modifications and additions.

Classically, animal nutritionists have expressed all measurements of energy transactions in terms of calories. The calorie used in nutrition is termed the 15°C calorie (cal), which is the energy required to raise the temperature of 1 g water from 14.5 to 15.5°C. One thousand calories is a kilocalorie (kcal). The kcal is a common unit of expressing dietary energy in animal nutrition. The joule (J) was adopted in the Système International des Unités (SIU, International System of Units) as the unit for expression of electrical, mechanical, and chemical energy. One J is defined as 1 kg m^2/sec^2 or 10^7 erg. One 15°C cal is equivalent to 4.184 J. The joule, like most other SIU units, has gained in popularity as the unit for expressing dietary energy in scientific literature. However, it is the National Research Council (NRC) policy to use the kcal as the unit of reference in nutritional energetics. Values in this document are presented in both units where feasible and practical.

GROSS ENERGY AND INTAKE OF ENERGY

Gross energy (GE) is the commonly used term for enthalpy (ΔH) of combustion in nutrition. However, as opposed to enthalpy, GE is generally represented by a + sign (Mayes, 2000). GE content of a substance is usually measured by its combustion in a heavily walled metal container (bomb) under an atmosphere of compressed oxygen. The method of determination is referred to as bomb calorimetry. Under these conditions, the carbon and hydrogen are fully oxidized to carbon dioxide and water, as they are in vivo. However, the nitrogen is converted to oxides, which is not the case in vivo. The oxides of nitrogen interact with water to produce strong acids, an endergonic reaction. These acids can be estimated by titration, allowing a correction to be applied for the difference between combustion in an atmosphere of oxygen and catabolism in vivo (Blaxter, 1989).

The GE content of an ingredient or a compounded diet

TABLE 4-1 Terminology of Types of Dietary Energy and Energy Budget Components

Dietary Energy Types	Abbreviation
Gross energy	GE
Digestible energy	DE
Metabolizable energy	ME
Net energy	NE
Energy Budget Components/Terms	
Intake of energy	IE
Fecal energy losses	FE
Digestible energy intake	DEI
Urinary and branchial (nonfecal) energy losses	UE + ZE
Metabolizable energy intake	MEI
Surface losses	SE
Heat losses (heat production)	HE
Basal metabolism	H_eE
Fasting heat losses	HE_f
Maintenance energy	HE_m
Voluntary activity energy losses	H_jE
Heat increment of feeding	H_iE
Heat of digestion and absorption processes	H_dE
Heat of formation and excretion of metabolic wastes	H_wE
Heat of transformation and retention of substrates	H_rE
Recovered energy	RE

depends upon its chemical composition. The mean values of GE of carbohydrates, proteins, and lipids are on average 4.11, 5.64, and 9.44 kcal/g (17.2, 23.6, and 39.5 kJ/g), respectively (Blaxter, 1989). Intake of energy (IE) is the notation adopted by NRC (1981) for the intake of GE of an animal (Figure 4-1). Intake of energy is simply the product of feed consumption (g) and GE (kcal/g).

FECAL ENERGY LOSSES—DIGESTIBLE ENERGY

Before the feed components can serve as fuels for animals, they must be digested and absorbed. Some feed components resist digestion and pass through the digestive tract to be voided as fecal material. Egestion (excretion as feces) of components containing GE is referred to as fecal energy losses (FE). The difference between the GE and FE of a unit quantity of this diet is termed the digestible energy (DE). Digestible energy intake (DEI) was adopted by NRC (1981) to represent intake of digestible energy, the product of feed intake (g/fish) and DE (kcal/g) of the feed or IE minus FE (Figure 4-1, Table 4-1).

Variation in the digestibility of dietary components is generally a major factor affecting the variation in their usefulness as energy sources to the animal. The FE often represents about 15–30% of IE for fish and shrimp fed practical diets and is a significant loss of energy. The DE values are better estimates of levels of "available" energy to the animal than are GE values of feeds and ingredients (Cho and Kaushik, 1990). Consequently, formulation on a DE (and digestible nutrients) basis is more practical and logical than formulating on GE or crude nutrients (e.g., crude protein) basis. Formulation based on a DE basis has gained popularity in fish and crustacean nutrition over the past 30 years. Methods for determining digestibility and the factors that affect the digestibility of nutrients and energy are reviewed later in this document (Chapter 12).

Digestible energy content is thought to be one of the major factors controlling feed intake in fish (Lee and Putnam, 1972; Jobling and Wandsvik, 1983; Kentouri et al., 1995; Paspatis and Boujard, 1996; Lupatsch et al., 2001a). This assumption is derived from evidence in the literature showing that when offered diets with various DE levels, fish appear to adjust their feed intake to maintain a particular (daily) energy intake (Jobling and Wandsvik, 1983; Boujard and Médale, 1994; Kaushik and Médale, 1994; Yamamoto et al., 2000; Lupatsch et al., 2001b; Yamamoto et al., 2002, 2005). The capacity of fish to adjust to diets of different DE density is believed to be determined by the physical capacity of the digestive tract (Lupatsch et al., 2001b). Nevertheless, the expected feed intake adjustments of the fish to dietary DE was not observed in several other studies (Alanärä, 1994; Alanärä and Kiessling, 1996; Helland and Grisdale-Helland, 1998; Koskela et al., 1998; Peres and Oliva-Tele, 1999; Encarnação et al., 2004; Geurden et al., 2006). Feeding trials with lipid-rich diets did not indicate a negative feedback from extra dietary DE on feed intake in rainbow trout (Geurden et al., 2006). There is increasing evidence that feed intake of animals is regulated in part by the lean growth potential of animals (Encarnação et al., 2004; Geurden et al., 2006; Dumas et al., 2007). Animals will seek to eat a sufficient amount of a nutritional adequate diet to allow them to achieve their target or preferred performance unless limited by constraints or overridden by an externally managed intervention (Oldham et al., 1997). The differences in feed intake of fish fed diets with different DE levels observed in several studies is likely a reflection of the animals trying to adjust their feed intakes to consume sufficient amounts of different digestible nutrients to enable them to meet their growth and/or body composition targets, not simply a response to DE level of the diet per se. The DE density of the diet in itself has limited effect on feed intake regulation of fish. However, it represents a practical and valuable mode of expression of the digestible nutrient density of the diet.

NONFECAL LOSSES—METABOLIZABLE ENERGY

Catabolism of certain nutrients results in the production of metabolic wastes (e.g., ammonia) that must be excreted by the animal. Fish and shrimp excrete metabolic wastes through their gills and in urine. Excess of some nutrients, such as water-soluble vitamins, glucose, and amino acids, and some metabolites are also excreted in the urine as the result of glomerular filtration, which is present in most fish species (Dantzler, 1989). The excretion of ammonia and other types of combustible materials, such as urea, creatinine, glucose, amino acids, trimethylamine (TMA), and trimethylamine oxide (TMAO), through the gills and in urine results in energy losses that must be accounted for in an energy budget. Excretion of combustible products through the gills is termed branchial energy loss (ZE) and through the urine, urinary energy loss (UE). Subtracting these nonfecal losses from DE results in an estimate of the metabolizable energy (ME) value of the diet:

$$ME = IE - (FE + UE + ZE)$$

Direct determination of the ME values of diets for fish and shrimp is very difficult. Smith (1971) developed a metabolic chamber and experimental procedure that collect combustible products excreted from the gills and urine and quantify UE + ZE and allow estimation of ME of feedstuffs. However, this method requires restraint of the fish in a sealed vessel with a diaphragm separating the front from the rear portion of the body. This is a source of considerable stress in most fish species. Fish do not feed freely under such conditions and need to be force fed. These force-fed fish frequently vomit. Due to the stress, the animal generally exhibits much lower, and often negative, nitrogen balance than a free-swimming animal feeding normally. As a result, the estimates of UE + ZE obtained with this method are much greater than would be the case for unrestrained fish feeding normally, and the estimates of ME of diets and feed ingredients is much lower than expected (Cho and Kaushik, 1990).

Branchial (ZN) and urinary (UN) nitrogenous wastes represent the bulk of the nonfecal energy losses of fish and crustaceans. Ammonia consists of approximately 85% of the nitrogenous wastes excreted by fish, whereas urea usually consists of less than 15% (Kaushik and Cowey, 1991). Monitoring production of N wastes in water of the rearing environment is an approach that has been commonly used (Brett and Zala, 1975; Kaushik, 1980a,b; Kaushik et al., 1982; Dosdat et al., 1996; Chakraborty and Chakraborty, 1998), but it requires an elaborate sampling protocol and considerable care.

Given the limitations of direct measurement of UE + ZE and UN + ZN, the use of an indirect method to estimate UE + ZE has been recommended as a means of obtaining realistic estimates (Cho and Kaushik, 1985). Cho and Kaushik (1990) proposed that the branchial and urinary excretion of 1 g of nitrogen by fish under normal conditions could be equated to an energy loss of 5.95 kcal (24.9 kJ), based on an energy of combustion value of ammonia (82.3% N by weight) of 4.90 kcal/g (20.5 kJ/g) (Bradfield and Llewellyn, 1982). Using this approach, the sum of branchial and urinary N excretion (ZN + UN) can be estimated by the difference between digestible nitrogen intake (DNI) and recovered nitrogen (RN) as follows:

$$ZN + UN = DNI - RN,$$
$$ZE + UE = (ZN + UN)\ 5.95\ \text{kcal/g N},$$
$$ME = DE - (ZE + UE)$$

Estimates of nonfecal losses are variable, but their contribution to the energy budget of fish is commonly no more than 3–6% of ME (Kaushik, 1998; Bureau et al., 2002). The main factors affecting nonfecal energy losses are those that influence the retention of protein/amino acids by the body and hence govern the loss of nitrogenous endproducts through the gills or in the urine.

Excretion of other combustible compounds may occasionally contribute significantly to UE + ZE of animals. Estimates of UE + ZE based on nitrogenous waste compounds excretion may occasionally underestimate actual nonfecal energy losses (Bureau et al., 1998). For example, excretion of glucose in the urine (Yokote, 1970; Furuichi, 1988; Kakuta and Namba, 1989; Bureau et al., 1998; Deng et al., 2000), as well as through the gills (Hemre and Kahrs, 1997), has been detected in fish made hyperglycemic by feeding a diet containing high levels of digestible carbohydrate or injected with glucose. The energy lost as urinary glucose is, nonetheless, relatively small and has since been estimated to be less than 5% of the ME intake of the animal (Bureau, 1997).

SURFACE ENERGY LOSSES

Shedding of combustible components through losses of mucus, scales, and epithelial cells represents loss of energy that is termed surface energy loss (SE). These losses are difficult to quantify in fish and are probably small. However, molting is an essential part of the growth processes of crustaceans. The exuvia produced by crustaceans results in the loss of combustible material and can also be classified as SE. Limited information is available on the SE losses of crustaceans. Based on a brief review of available information from unpublished information from past trials, Bureau et al. (2000) estimated SE associated with molting in growing penaeid shrimp to be small, approximately equivalent to 3% of ME intake.

HEAT LOSSES

Combustion of organic molecules results in the release of heat. For example, the combustion of one mole of glucose in a bomb calorimeter results in the liberation of 670 kcal (2,803 kJ) as heat (Blaxter, 1989). When oxidation of glucose occurs in the tissues, some of the energy is not lost immediately as heat but is captured in high-energy phosphate bonds through coupling reactions. Under aerobic conditions, glucose is completely oxidized to CO_2 and water, and the equivalent of 36 high-energy phosphate bonds are generated per molecule. The total energy captured in ATP per mole of glucose oxidized is 334 kcal (1,398 kJ), or the equivalent to approximately 50% of the enthalpy of combustion (or GE) (Blaxter, 1989). The remainder is dissipated as heat. In turn, when ATP generated by the catabolism of glucose is hydrolyzed during coupling with endergonic reaction, only a fraction of the free energy may be retained in the synthesized compounds and the rest is liberated as heat. Therefore, ultimately free energy liberated by exergonic reactions that is not captured in the products of anabolism (e.g., protein, lipids, carbohydrates, and nucleic acids) is liberated as heat by biological organisms.

The first law of thermodynamics states that heat produced by a chemical reaction is always the same, regardless of whether the process occurred directly or proceeded through a number of intermediate steps (Blaxter, 1989). Therefore, the amount of heat liberated depends on the chemical nature (energy content) of the compounds catabolized and of the overall reaction rather than the chemical reactions pathways by which this catabolism occurred.

According to the NRC (1981) nomenclature, HE is the total heat losses of an animal. It is also commonly designated as "metabolic rate" (Kleiber, 1975), which actually represents a much broader term. The HE is an indication of the intensity of ongoing metabolic reactions in the animal. A relatively large number of reviews have discussed at length the merits of various methodological approaches for measurement of HE in fish (Cho and Kaushik, 1985; Tytler and Calow, 1985; Cho and Kaushik,1990; Cho and Bureau, 1995; Bureau et al., 2002).

Three components of animal metabolism lead to the release of energy as heat. Heat liberated by animals as a consequence of the need to sustain the structure and function of the body tissues is termed basal metabolism (H_eE) according to NRC (1981) nomenclature or minimal metabolism according to the nomenclature of Blaxter (1989). Physical activity also increases metabolic rate because of work done, and it is termed heat of voluntary activity (H_jE). The ingestion of feed increases the metabolic rate as a consequence of the extra work needed to ingest, digest, and metabolically utilize the components of the diet. This increase is termed the "heat increment of feeding" (H_iE). Standard dynamic action (SDA) is another term commonly used in the literature for this type of heat loss.

BASAL/MINIMAL METABOLISM

Animals require a continuous supply of free energy for those functions of the body immediately necessary for maintaining life regardless of whether or not food or feed is consumed. Basal/minimal metabolism (H_eE) represents use of energy for such things as the circulation of the blood, pulmonary ventilation, repair and replacement of cells, homeostasis, transport of ions (especially of sodium and potassium), and muscle tone. In fish and shrimp, H_eE is clearly related to temperature because environmental temperature has a determinant effect on the internal temperature, rate of biochemical reactions, and metabolic rate of the animal.

Meaningful assessment of H_eE requires the conditions by which standardized measurements are made. This objective is achieved by attempting to measure a minimum rate of heat production free of any controlling factors known to increase it. Such factors include exercise (voluntary movement), the consumption of feed and its subsequent metabolism, and the physical environment. The object of standardization is to ensure comparability of estimates rather than to establish some absolute minimum value of metabolism that is compatible with life. A number of terms have thus arisen to describe these presumably standardized measurements of "minimal metabolism." With domesticated animals, and hence fish under aquaculture conditions, what is usually measured is the fasting heat production (HE_f) (Blaxter, 1989). HE_f is also known as standard metabolism in the fish biology literature (Elliott, 1982). HE_f of different species of fish measured under various conditions has been reported in a very large number of studies. Unfortunately, significant variability in the estimates of HE_f or H_eE of fish reported in the literature exists and is probably mostly due to very significant differences in the methodological approaches and experimental conditions (Cho and Bureau, 1995).

It is difficult to ensure that the fish are in a state of muscular repose (complete rest) because they need to maintain their orientation in the water, which requires some muscular activity. It has been suggested that H_eE could be estimated using fasted fish swimming at different rates by extrapolation to zero activity (Smith, 1989). However, fish of many species spend considerable periods resting on the bottom of tanks or maintaining their position in quiet water with minimal activity. Consequently, HE_f of free-swimming animals has been regarded as a close approximation of minimal or basal metabolism (Cho and Kaushik, 1990). Oxygen consumption of free-swimming fish fasted for 3 to 7 days to eliminate the effect of the feed consumed, and its subsequent metabolism is the most common approach for measuring HE_f (Kaushik and Médale, 1994; Cho and Bureau, 1995; Glencross and Felsing, 2006). Measuring carcass energy losses during fasting is another common method of estimating HE_f and, consequently, H_eE (Cho and Kaushik, 1985; Lupatsch et al., 1998, 2003; Glencross et al., 2010). Oxygen consumption of fasting fish and carcass energy losses during fasting have been shown to result in similar HE_f estimates for rainbow trout (Bureau, 1997) and Asian sea bass (Glencross and Felsing, 2006; Glencross, 2008).

Available data on the HE_f of fish species show that, for a given weight, the rates are 1/5 to 1/20 of terrestrial vertebrates. Data from Kaushik and Gomes (1988), Cho (1991), and Bureau (1997) suggest HE_f of approximately 7.17–9.56 kcal/kg $BW^{0.80}$ (30–40 kJ/kg $BW^{0.80}$) per day for rainbow trout between 15 and 18°C. Based on carcass energy loss during starvation, Kaushik et al. (1995) calculated that Nile tilapia (*Oreochromis niloticus*) lost 16.73 kcal/kg $BW^{0.80}$ (70 kJ/kg $BW^{0.80}$) at water temperatures of 28ºC. Using the same approach, Lupatsch et al. (2003) estimated the HE_f of gilthead sea bream (*Sparus aurata*) to be 10.04 kcal/kg $BW^{0.82}$ (42 kJ/kg $sBW^{0.82}$), that of European sea bass (*Dicentrarchus labrax*) to be about 8.37 kcal/kg $BW^{0.80}$ (35 kJ/kg $BW^{0.80}$), and that of white grouper (*Epinephelus aeneus*) to be 5.98 kcal/kg $BW^{0.79}$ (25 kJ/kg $BW^{0.79}$) at a water temperature of 23ºC, while Glencross (2008) estimated the HE_f of Asian sea bass (*Lates calcarifer*) to be 10.28 kcal/kg $BW^{0.80}$ (43 kJ/kg $BW^{0.80}$) at 30ºC. By comparison, HE_f has been reported to vary between 40.6 to 141.0 kcal/kg $BW^{0.75}$ (170 to 590 kJ/kg $BW^{0.75}$) per day in homeothermic domestic animals (Blaxter, 1989). The low HE_f of fish compared to homeotherms can be attributed to the lack of expenditure for thermoregulation, lower sodium pump activity, their buoyancy, and the mode of nitrogen excretion (ammoniotelism).

EFFECT OF BODY WEIGHT ON BASAL METABOLISM

In poikilotherms as well as in homeotherms, H_eE in absolute terms (kcal/animal per day) increases as the mass of the animal increases. The relationship of body weight to metabolic rate in animals can be described by the general equation $Y = aW^b$, where Y is the metabolic rate, W is the body weight, and "a" is a coefficient dependent on species and temperature.

The logarithm of H_eE increases linearly with the logarithm of the body mass (Blaxter, 1989). However, the slope of this relation is lower than 1, which means that in all species, animals of smaller size spend more energy per unit of mass than animals of larger size. The value of the exponent for fish has been described as ranging from 0.50 to 1.00. Hepher (1988), who reviewed experimental data from the literature, concluded that the exponent 0.8 describes, with reasonable accuracy, the change in metabolic rate with body mass of several fish species. Detailed observations by Brett and Groves (1979), Hogendoorn (1983) for African catfish (*Clarias gariepinus*), Cui and Liu (1990) for six different teleost species (*Cyprinus carpio*, *Oreochromis mossambicus*, *Pseudobagrus fulvidraco*, *Carassium auratus*, *Macropodus chinensis*, and *Pseudorasbora parva*), Cho (1991) for rainbow trout, Sanchez et al. (1993) for turbot (*Scophthalmus maximus*), Lupatsch et al. (1998) for gilthead sea bream, Liu et al. (2000) for mandarin fish (*Siniperca chuatsi*) and Chinese snakehead (*Channa argus*), Lupatsch et al. (2003) for European sea bass and white grouper, and Glencross (2008) for barramundi (also known as Asian seabass) suggest that across species the exponent is greater than 0.7 and less than 1.0. Thus it appears reasonable to assume metabolic body weight (MBW) can in practice be calculated using $kg^{0.8}$ for most fish species. However, recent evidence suggests that for penaeid shrimp, the appropriate scaling coefficient may to be closer to 1.0 (Lupatsch et al., 2008).

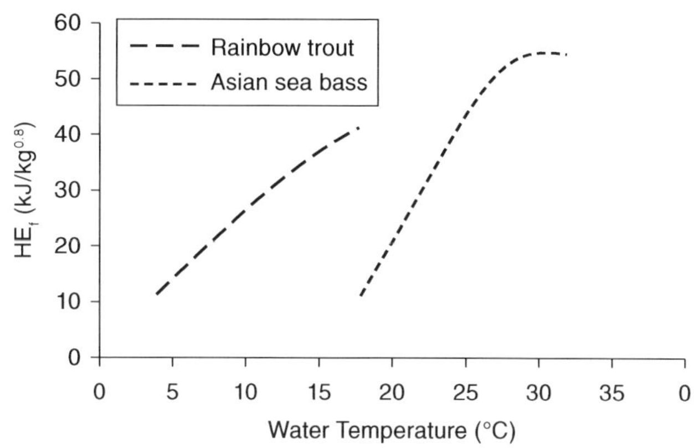

FIGURE 4-2 Fasting heat losses of rainbow trout, *Oncorhynchus mykiss*, and Asian sea bass, *Lates calcarifer* (expressed as HE_f, kJ per $kg^{0.8}$ per day and as a function of water temperature).

EFFECT OF TEMPERATURE ON BASAL METABOLISM

Water temperature is the major factor determining the metabolic rate and energy expenditure of poikilothermic animals, such as teleosts and crustaceans.

Based on mathematical analysis of oxygen consumption data of fasting rainbow trout reared at water temperatures ranging from 5 to 16°C, Cho and Kaushik (1990) concluded that H_eE as a function of water temperature could be described as:

$$H_eE = ((-1.04 + 3.26(T) - 0.05(T)^2) / (BW^{0.824})) / d$$

where: H_eE is basal metabolism (kilojoules), T is water temperature (°C), and BW is body weight (kg).

Glencross (2008) developed the following equation to estimate the H_eE of barramundi (Asian sea bass):

$$H_eE = (0.4462426 - 0.0848448(T) + 0.0048282(T)^2 - 0.0000750(T)^3) \times BW^{0.80}$$

where: BW is body weight (g).

Within species and certain temperature ranges, increasing water temperature results in a curvilinear (almost linear) increase of H_eE (Figure 4-2). Studies with Asian sea bass (barramundi) have reported a significant increase in H_eE above thermal optimum for this species (Bermudes et al., 2010). Increases in temperature above thermal optima result in metabolic disorders that can affect H_eE. As temperature continues to be elevated, fish reduce feeding activity and metabolic perturbations lead to death (upper lethal temperature). The upper lethal temperature varies, and the effect of temperature on H_eE may vary with species and strains within a species (Jonsson and Jonsson, 2009). Conversely, the metabolic rate of fish is reduced when water temperature and consequently body temperature of the fish is reduced. This reduction continues until the lower lethal limit is reached and the fish dies. This lower limit differs with species and for some species, such as some Antarctic fish species, can be slightly below 0°C (Clarke and Johnson, 1999).

Studies with temperate and tropical species show no clear relationship between preferred environmental temperature and H_eE across species (Médale and Guillaume, 1999). However, Clarke and Johnson (1999) observed a curvilinear relationship between metabolic rate and temperature based on analysis of data from 69 teleost fish species. These different conclusions may be related to the fact that the analysis of Clarke and Johnson (1999) was based on a survey of published data from 69 species with only one temperature per species, defined as the "experimental temperature most representative of that experienced in the wild." Using this approach, a statistically significant curvilinear relationship is seen but is mostly the results of low metabolic rate for polar species (water temperature < 5°C) and higher metabolic rate for certain fish species between 35 and 40°C (Bureau et al., 2002). At their optimal growth temperature, H_eE of salmonids (Cho and Kaushik, 1990), mandarin fish (Liu et al., 2000), Chinese snakehead (Liu et al., 2000), gilthead sea bream (Lupatsch et al., 1998), European sea bass (Kaushik, 1998; Lupatsch et al., 2003), grouper (Lupatsch et al., 2003) and barramundi (Glencross, 2008) appear to be fairly similar, at about 5.98–11.95 kcal/kg $BW^{0.80}$ (25–50 kJ/kg $BW^{0.80}$) per day.

BASAL METABOLISM OF SHRIMP

The fasting oxygen consumption of the 30 g crayfish *Cherax tenuimanus* (Smith) at 22°C was estimated to be about 0.04 mg O_2 per minute (Villarreal, 1990), corresponding to H_eE of about 26 kJ/kg BW per day. Tchung (1995) observed that the HE_f of blue shrimp (*Penaeus stylirostris*) weighing between 20–28 g at 28°C was about 14.82 kcal/kg $BW^{0.66}$ (62 kJ/kg $BW^{0.66}$) per day during the intermolt period. Data from Gauquelin (1996) suggest that fasting oxygen consumption of *Penaeus stylirostris* weighing between 20–30 g was 3.3 g O_2/kg BW per day, which corresponds to HE_f of 10.76 kcal/kg BW (45 kJ/kg BW) per day. Lupatsch et al. (2008) estimated the HE_f (based on carcass energy losses during fasting) of Pacific white shrimp, *Litopenaeus vannamei* (1–35 g) kept at 28°C to be about 32 cal/g $BW^{0.95}$ (134 J/g $BW^{0.95}$), which is equivalent to 11.7 kcal/kg $BW^{0.8}$ (49 kJ/kg $BW^{0.8}$) per day. Oxygen consumption data from Maldonaldo et al. (2009) suggest an HE_f of about 2.4–8.4 kcal/kg $BW^{0.8}$ (10–35 kJ/kg $BW^{0.8}$) per day of Pacific white shrimp weighing between 0.2 and 6.0 g live weight reared at 28–32°C. Available data suggest that at their respective optimal temperatures, H_eE of different shrimp and other crustacean species is similar to that of fish species.

The effect of temperature on fasting oxygen consumption has been studied by Ocampo (1998) with *Penaeus californiensis* in the intermolt stage. Fasting oxygen consumption increased from 0.19 to 0.35 to 0.43 mg/g per hour when temperature increased from 19 to 23 to 27°C, respectively. Taken together, these results suggest that the effect of temperature on H_eE in shrimp is similar to that seen in fish. Results of several studies on oxygen consumption of fasting crustaceans are scattered throughout the literature. There is a need to review and analyze the available experimental evidence using the approaches that have been applied to higher vertebrates.

MAINTENANCE ENERGY REQUIREMENT

Although frequently confused, maintenance energy requirement (HE_m) and basal metabolism (H_eE) are two closely related but distinct concepts. Figure 4-3 contrasts the concepts of maintenance and basal metabolism. HE_m is generally defined as the amount of ME required for an animal to maintain zero energy balance (zero energy gain, RE = 0). The most commonly used method for estimating HE_m consists in feeding fish at different levels and using regression of the results of RE as a function of ME intake and extrapolating to zero carcass energy gain (i.e., RE = 0) (Figure 4-3). In theory, HE_m should be equal to H_eE plus the H_iE associated with feeding a maintenance ration. Consequently, HE_m values would be expected to be 20–60% greater than H_eE. Estimates of HE_m obtained across studies with the same species often have relatively large variances. Several factors, such as methodological approach, scaling factor used to calculate metabolic weight, regression model, and composition of the diet used, may have significant impacts on the estimate of HE_m. Evidence from a large number of

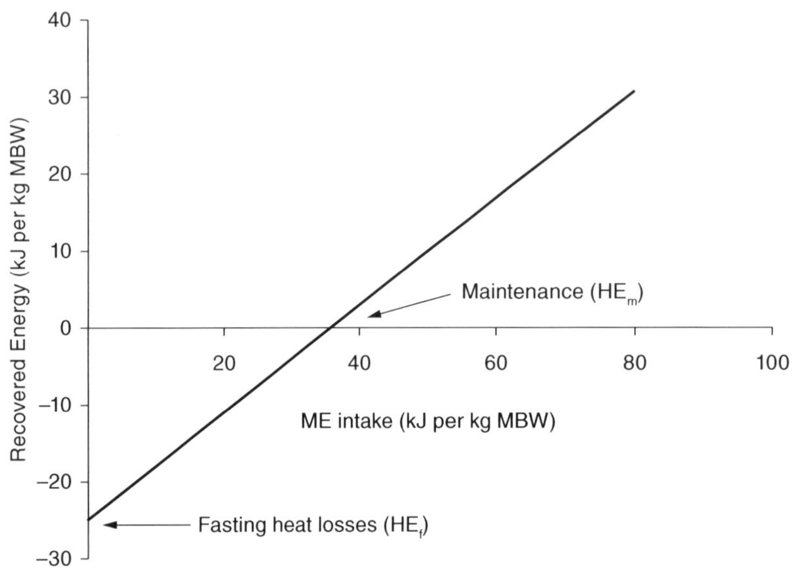

FIGURE 4-3 Illustration of the concept of maintenance and fasting heat losses (recovered energy [RE] as a function of metabolizable energy [ME] intake of fish and illustration of the concept of maintenance [HE_m] and fasting heat losses [HE_f] [an estimate of basal metabolism (H_eE)]).

published studies suggests that HE_m of fish reared at their optimum growth temperature is approximately 9.6–19.1 kcal/kg $BW^{0.8}$ (40–80 kJ ME / ($BW^{0.8}$)) per day. A summary of the results of a number of these studies is presented in Table 4-2. Due to the difficulty associated with measuring ME and because UE + ZE is generally small, it is increasingly common (although theoretically incorrect) to estimate HE_m on a DE basis for fish and shrimp.

It is worth noting that at zero carcass energy gain (RE = 0), fish fed a nutritionally adequate diet still deposit body protein (positive "protein-energy" gain) and mobilize body lipids (negative "nonprotein" energy gain) and still gain live weight (Figure 4-4). This phenomenon is found in all young animals fed a maintenance ration that is adequate in protein (Blaxter, 1989). Many have argued that the concept of maintenance is an irrational concept for growing animals and accordingly should be phased out. Others have argued that the concept of maintenance, although far from perfect, is still very useful in practice (Baldwin and Bywater, 1984).

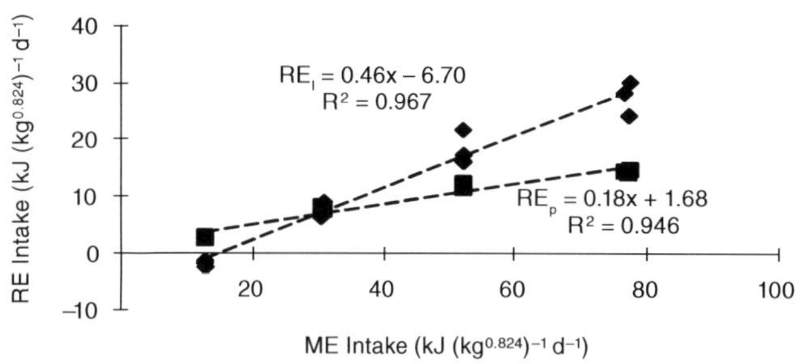

FIGURE 4-4 Recovered energy and metabolizable energy in rainbow trout, *Oncorhynchus mykiss*. (Recovered energy [RE] as protein [RE_p, squares] and lipid [RE_l, diamonds] as a function of metabolizable energy [ME] intake of rainbow trout reared at 8.5°C [data from Bureau et al., 2006]).

TABLE 4-2 Estimate of Maintenance[a] Energy Requirement of Different Fish and Shrimp Species Obtained Through Feeding Trials

Species	Weight (g/fish)	Temperature (°C)	HE_m (kcal ME/$kg^{0.80}$ per day)	HE_m (kJ ME/$kg^{0.80}$ per day)	Reference
Atlantic salmon (*Salmo salar*)	5	15	4.06	17	Bureau et al. (1999)
Asian sea bass/Barramundi (*Lates calcarifer*)	15	30	8.37[b]	35[b]	Glencross (2008)
	410	30	11.00[b]	46[b]	Glencross (2008)
Channel catfish (*Ictalurus punctatus*)	8–10	27	5.98	25	Gatlin et al. (1986)
Chinese sucker (*Myxocyprinus asiaticus*)	12	27	9.08	38	Yuan et al. (2009)
European sea bass (*Dicentrarchus labrax*)	15–140	24	10.76[b]	45[b]	Lupatsch et al. (2001a, 2003)
Gilthead sea bream (*Sparus aurata*)	30–160	24	11.47[b]	48[b]	Lupatsch et al. (1998, 2003)
Pacific white shrimp (*Litopenaeus vannamei*)	1.5–7.5	28	26.29[b]	110[b]	Lupatsch et al. (2008)
Rainbow trout (*Oncorhynchus mykiss*)	150	18	13.86–17.93	58–75	Kaushik and Gomes (1988)
	158	9	4.54	19	Bureau et al. (2006)
	300	15	7.89	33	Storebakken et al. (1991)
	96	14	10.04[b]	42[b]	Glencross (2008)
	55	16	10.52[b]	44[b]	Glencross (2009)
Red drum (*Sciaenops ocellatus*)	5.5	25	11.47	48	McGoogan and Gatlin (1998)
Nile tilapia (*Oreochromis niloticus*)	330	28	14.34	60	Lupatsch et al. (2010b)
Tra catfish (*Pangasianodon hypothalamus*)	40	32	9.56[b]	40[b]	Glencross et al. (2010)
Yellowtail (*Seriola quinqueradiata*)	750	15	18.40	77	Watanabe et al. (2000)

[a]HE_m.
[b]Expressed on a digestible energy (DE) basis.

HEAT LOSSES FOR VOLUNTARY ACTIVITY

Fish have an efficient mode of transportation. Their bodies are supported by water, and they do not need to expend energy against gravity like terrestrial animals. A streamlined body moving through the water is one of the most efficient forms of transportation. A large number of studies have focused on the metabolic cost of swimming for fish (Brett and Groves, 1979). Krohn and Boisclair (1994) suggested that the metabolic costs of turning and accelerating may be substantially more than the cost of swimming at constant speed in a straight line. Estimation of the energetic cost of activity may be very significant for wild fish due to their need to expend considerable amounts of energy to acquire food and escape predators. H_jE associated from activity is difficult to estimate separately from H_eE because there is always a certain amount of voluntary activity in any group of fish (Cho et al., 1982). It has been assumed that when constructing an energy budget of free-swimming fish under normal aquaculture conditions, the cost of activity is rather negligible and is already included in the estimate of H_eE (Bureau et al., 2002). This assumption may be an oversimplification of reality. Cooke et al. (2000) used electromyogram telemetry to observe a 60% increase in voluntary swimming activity and a 26% increase in oxygen consumption in rainbow trout held at high stocking density compared to those held at low stocking density. These authors hypothesized that differences in feed efficiency observed in fish held at different stocking densities may be related to increase in energy losses due to activity. Conversely, a recent study with European sea bass (Lupatsch et al., 2010a) found that oxygen consumption as well as HE_m of fish were higher at low stocking density, but no difference was found in feed efficiency or growth rate, apart from a slightly reduced body lipid content of fish kept at the low stocking density. It may be concluded that H_jE may be a significant contribution to HE of fish under certain conditions but that there are likely significant differences between species, life stages, rearing environments, and environmental conditions. More work needs to be carried out to quantify H_jE of the numerous fish species cultured under a great variety of rearing environments. The broad range of technologies available today (e.g., mesocosms, underwater camera, sonar, radio transmitters, global positioning system [GPS], internal temperature loggers, electromyogram telemetry, image analysis software) combined with traditional techniques (e.g., respirometry, comparative carcass analysis) could enable accurate quantification of H_jE of fish reared under practical conditions.

HEAT INCREMENT OF FEEDING

Ingestion of feed by an animal that has been fasting results in an increase in the animal's HE. This expenditure of energy due to feeding is referred to as heat increment of feeding (H_iE). This component of the energy budget is also referred to as extra heat, specific dynamic action (SDA), calorigenic effect, and dietary thermogenesis in the literature. The factors that contribute to H_iE have traditionally been separated into three categories: (1) digestion and absorption processes (H_dE), (2) formation and excretion of metabolic wastes (H_wE), and (3) transformation and interconversion of the substrates and their retention in tissues (H_rE). For crustaceans, H_iE should also include cost of molting (H_xE).

ESTIMATES OF HEAT INCREMENT OF FEEDING

The length of time for which consumption of diet influences HE depends on many factors; chief among these factors are the quantity and quality of the diet, the water temperature, and growth (nutrient deposition) of the animal. The rise in oxygen uptake corresponds more or less to the rate of transit of feedstuffs through the digestive tract (Kaushik and Dabrowski, 1983). The H_iE principally depends on the balance of dietary nutrients and the plane of nutrition (Brody, 1945). Therefore, attempts to measure the H_iE of individual feed ingredients that are a nutritionally unbalanced diet (Smith et al., 1978; Tandler and Beamish, 1979) or measurements of effect of fish size (Beamish, 1974) and fish density (Medland and Beamish, 1985) performed under forced activity conditions have very doubtful meaning. Similarly, the estimation of H_iE of an animal without reference to its growth and nutrient deposition (energy or protein and lipid deposition) (e.g., Ross et al., 1992) is also inadequate.

Many studies have shown highly significant linear (or largely linear) relationships between ME intake and RE (Figure 4-5). The slope is often identified as the "efficiency of metabolizable utilization for production," K_g or K_{pf}, and has been reported to vary between 0.26–0.70 for various fish species fed practical diets (Meyer-Burgorff et al., 1989; Cui and Liu, 1990; Azevedo et al., 1998; Lupatsch et al., 1998; Médale et al., 1998; Ohta and Watanabe, 1998; Rodehutscord and Pfeffer, 1999; Peres and Oliva-Teles, 2000; Lupatsch et al., 2003; Bureau et al., 2006; Glencross, 2008; Yuan et al., 2009). Consequently, in most of the species studied so far, H_iE appears to be equivalent to 30–75% of ME intake in fish fed nutritionally adequate diets. Although significant interspecific differences exist, a large proportion of the variability in the estimates of H_iE among studies can likely be attributed to differences in diet composition and composition of weight gain (protein vs. lipid deposition), as well as a variety of methodological issues (such as experimental protocol, range of data, stress conditions, assessment of feed intake and nutrient digestibility, and statistical model used). Estimate of H_iE may only be applicable to a certain set of conditions (same species, life stage, and diet composition). However, for a given diet and species, H_iE expressed as a proportion of ME intake, DE intake, or RE does not appear to significantly vary with water temperature (Azevedo et al., 1998; Lupatsch et al., 1998; Rodehutscord and Pfeffer, 1999; Lupatsch et al., 2003; Lupatsch and Kissil, 2005), at

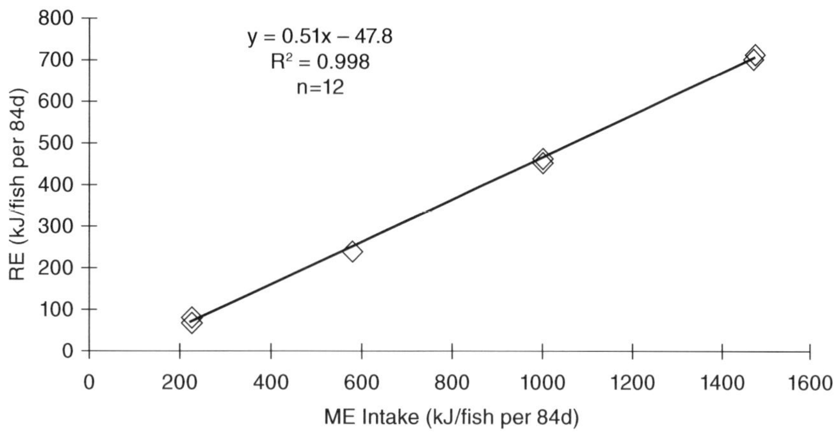

FIGURE 4-5 Recovered energy and metabolizable energy in Atlantic salmon, *Salmo salar* (recovered energy [RE] as a function of metabolizable energy [ME] intake in Atlantic salmon fed at different feeding levels. The slope indicates the "efficiency of ME utilization for production [K_{pf}]" and 1-slope is an estimate of the heat increment of feeding [H_iE] of the animal. [Source of data: Bureau et al., 1999, n = 12]).

least within a certain range of temperatures (which depends on species, strains, and rearing environment). Outside this thermal range metabolic perturbations occur, and these negatively affect efficiency of ME and DE utilization (Bermudes et al., 2010). The effect of feeding levels is less consistent across studies. Some studies have shown no effect of feeding level on efficiency of ME utilization (K_{pf}) (Azevedo et al., 1998; Lupatsch et al., 2001a,b; Bureau et al., 2006; see also Figure 4-5). However, results from other studies (Glencross, 2008; Lupatsch et al., 2008; Helland et al., 2010) suggest that at high feeding levels, efficiency of energy utilization tends to decrease, and H_iE, as a proportion of ME or RE, appears to increase significantly. This decrease in efficiency of energy utilization or increase in H_iE may be attributable to the curvilinear response in protein deposition that is observed at high feeding levels as animals approach their so-called maximal protein deposition rate or PD_{max} (Bureau et al., 2006; Dumas et al., 2007). The relationship between ME and RE and H_iE are the results of complex metabolic processes, and properly contrasting results of studies requires a more comprehensive analysis of nutrient utilization and metabolism, as opposed to a simple relation of ME intake and energy retention.

There are limited data on H_iE in shrimp. Interpretation of data from Warukamkul et al. (2000) suggest K_{pf} of the black tiger shrimp (*Penaeus monodon*) can be estimated at about 0.5; therefore, for every 0.12 kcal (0.5 kJ) of RE or H_eE, 0.12 kcal (0.5 kJ) is expended as H_iE. Lupatsch et al. (2008) observed that the efficiency of DE utilization by Pacific white shrimp was only about 30%, indicating that for every 0.07 kcal (0.3 kJ) of RE or H_eE, 0.17 kcal (0.7 kJ) is expended as H_iE. Estimation of H_iE in crustaceans is complicated by the difficulty in properly estimating ME intake of the animal and by contribution of molting processes to energy losses. Estimates of cost of molting (H_xE) are very scarce. Read and Caulton (1980) estimated that as much as 25% of RE accumulated in intermolt may be expended due to molting. This high estimate is very difficult to corroborate, and more work is needed to estimate heat losses in shrimp throughout their growth cycle.

DIGESTION AND ABSORPTION PROCESSES

Digestion and absorption processes (H_dE) refer to the heat losses related to biochemical and "mechanical" aspects of feeding and digestion in fish. Early studies using either "sham feeding" or feeding nondigestible materials such as kaolin or cellulose indicated that "mechanical SDA" approached 10–30% of total H_iE (Tandler and Beamish, 1979). However, other studies found that neither sham feeding nor kaolin feeding significantly increased the metabolic rate of the fish (Jobling and Davies, 1980). Emmans (1994) estimated the heat losses associated with egesting indigestible material to be about 0.91 kcal (3.8 kJ) per g of fecal organic matter (FOM) in terrestrial animals. If this value is applicable to fish, H_dE would probably represent less than 10% of H_iE of fish fed high-quality practical diets.

Heat losses associated to the enzymatic hydrolyses of lipids, polysaccharides, and proteins in the lumen of the gut have been estimated, in theory, to be about 0.1–0.2% of the GE of the substrate hydrolyzed (Blaxter, 1989). The absorption of certain products of digestion, such as amino acids, peptides, and glucose by the intestinal mucosa often occurs through an energy-dependent transport system known as active transport. Carrier proteins simultaneously transport the target molecule and a cotransported ion. The maintenance of a sodium gradient across the membrane is achieved by an ATP-dependent sodium transporter working in the opposite direction. This transporter hydrolyses one ATP molecule per every three sodium ions extruded. Theoretical cost of transport of glucose through active transport is one-third of an

ATP, which is equivalent to less than 1% of the GE of glucose or about 1% of the potential amount of ATP generated by the aerobic metabolism of glucose (i.e., 36 ATP). Absorption of lipid digestion products differ significantly: triacylglycerides (TAG) are hydrolyzed to free fatty acids (FFA) and monoacylglycerol (MAG) in the lumen. The FFA and MAG are absorbed passively, and TAG are resynthesized in the mucosa and exported as chylomicron to the circulation. Synthesis of TAG and chylomicrons requires a certain amount of energy, but again, this amount represents only a small proportion of the GE content of these molecules (Blaxter, 1989).

Heat losses arising from anaerobic fermentation in the gut is another factor contributing to H_dE. However, fermentation in most fish species is very limited (Leenhouwers et al., 2008), perhaps with the exception of certain marine herbivorous fish species (Clement, 1996). Very few quantitative studies on fate of volatile fatty acids and energy lost in the fermentation process have been conducted for fish. However, available estimates of fermentation suggest that heat losses associated with fermentation are small (Leenhouwers et al., 2008).

Overall, the heat losses associated with diet ingestion and digestion (H_dE) are probably small compared to that associated with metabolic work (H_rE + H_wE) (Brody, 1945). The physiological basis of this increased heat production is the postabsorptive processes related to ingested diet. These processes are primarily the metabolic work required for the synthesis of proteins and lipids in the tissues from the newly absorbed, metabolized amino acids, fatty acids, and glucose.

FORMATION AND EXCRETION OF METABOLIC WASTE

Deamination and catabolism of amino acids lead to ammonia production. As ammonia is toxic and cannot be rapidly eliminated by mammals and birds; these animals synthesize urea and uric acid, which are less toxic. The energy cost of synthesis for these products is 3.11 and 2.39 kcal/g N (13 and 10 kJ/g N), respectively, for urea and uric acid (Martin and Blaxter, 1965). The concentration of urea and uric acid for further excretion by the kidneys in terrestrial animals requires additional expenditure of energy. In contrast, ammonia is the primary waste product of protein catabolism in fish (Kaushik and Cowey, 1991). Urea is mainly the product of degradation of purines and arginine catabolism. As ammonia is efficiently excreted by the gills, fish generally do not require energy to detoxify or concentrate this waste. As a result, heat of formation and excretion of metabolic waste (H_wE) should represent only a very small fraction of H_iE of fish.

TRANSFORMATION OF SUBSTRATES AND RETENTION IN TISSUES

The heat losses associated with transformation of the substrates and their retention in tissues (H_rE) should represent a very large proportion of H_iE in animals. Much evidence suggests that the efficiency of utilization of ME varies with the chemical nature of the energy-yielding nutrients absorbed (Blaxter, 1989). When a fasting animal is refed, nutrients absorbed by the animal replace body constituents as the source of energy. The efficiency of utilization of ME is in proportion to the ATP yield of the nutrients absorbed (Blaxter, 1989; Van Milgen, 2002).

Growing animals accrete new tissues where part of the energy supplied is stored mainly as protein, lipid, and glycogen. Theoretical efficiency of transformation to or retention of substrates in tissue has been calculated for higher vertebrates (Blaxter, 1989; Flatt, 1992; van Milgen, 2002), and these theoretical costs are also valid for fish given the great similarity of the intermediate metabolism of fish and higher vertebrates. According to the calculations of Blaxter (1989) and van Milgen (2002), converting glucose into glycogen costs 5% of the energy of glucose as H_iE whereas converting glucose into lipids entails an increase of H_iE equal to about 30% of its GE. Conversion of dietary lipids into body lipids is, in theory, about 96%; therefore 4% of GE of lipids is dissipated as H_iE. The maximum theoretical efficiency of the conversion of dietary amino acids into body proteins is 85% efficient, entailing an H_iE of 15% of the GE value of proteins (Blaxter, 1989; van Milgen, 2002). Conversion of amino acids into body lipids is, in theory, only 66% efficient so approximately 34% energy would be lost as H_iE.

Protein and lipid deposition is the result of both synthesis and degradation rates of either protein or lipid, respectively, i.e., their turnover rates. Energy is lost as heat in the biochemical reactions that lead to protein synthesis and degradation, lipogenesis, and lipolysis, and in regulating and integrating the various cellular metabolic activities involved in protein and lipid deposition (van Milgen, 2002). Calculation of the theoretical costs of protein and lipid deposition is extremely complicated and fraught with uncertainties. Alternatively, these costs can be estimated in an empirical manner based on statistical analysis of energy expenditure and protein and lipid depositions.

PRACTICAL NET ENERGY SYSTEMS

Many studies have attempted to relate ME to RE (or H_iE) and then tried to delineate the various determinants of H_iE. The most popular approach is a factorial one and was first proposed by Kielanowski (1965). Factorial approaches have been at the foundation of popular energy requirement systems for pigs (NRC, 1998), chickens (NRC, 1994), beef cattle (NRC, 2000), and dairy cattle (NRC, 2001; Kebreab et al., 2003).

In the classic factorial approach, the partial energy costs for protein and lipid deposition are determined through a multiple regression approach using ME intake as the independent variable and protein and lipid energy deposition rates as the dependent variables to determine (Reeds, 1991).

The energy cost for lipid and protein deposition is simply defined as ME required to promote a defined increment in body protein or lipid. The partial efficiency of ME utilization for whole body growth (K_{pf}), protein deposition (K_p), and lipid deposition (K_f) is the ratio of net energy retained to the corresponding ME intake components:

$$ME = HE_m + RE_p / K_p + RE_f / K_f$$

Using this type of approach, Emmans (1994) concluded that the net energy cost for protein retention in terrestrial livestock species was 2.54 kcal per kcal of protein retained (that is, 1.54 kcal of heat expended for each 1 kcal of protein deposited) equivalent to a K_p of 40%. The calculated energy cost for lipid retention was 1.4 kcal and 1.1 kcal per kcal lipid deposited (i.e., heat losses of 0.4 or 0.1 kcal per each 1 kcal lipid deposited) when deposited from nonlipid or lipid, respectively. These are equivalent to a K_f = 90% when deposited from lipid and K_f = 70% when deposited from nonfat substrates.

An increasing number of studies have used the factorial approach to estimate HE_m, K_p, and K_f of fish. Results from these studies are summarized in Table 4-3. The estimates of K_p ranged between 0.49 and 0.81 and those of K_f between 0.66–0.91. These values appear to be similar to that observed with mammals and birds.

The factorial approach of Kielanowski (1965) has been criticized because there is, in general, a strong correlation between protein and lipid depositions and that it is much easier to control ME intake than it is to control protein and lipid depositions (Emmans, 1995). If multicolinearity is present to a harmful degree in physiological data, multiple linear regressions often yield nonsensical results (Slinker and Glantz, 1985; Birkett and de Lange, 2001; Azevedo et al., 2005). To overcome some of the limitations of the factorial approach, a multivariate approach has been proposed by van Milgen and Noblet (1999). In this multivariate approach, protein deposition (PD) and lipid deposition (LD) (dependent variables) are considered a function of ME intake. Azevedo et al. (2005) investigated the utilization of ME for growth vs. maintenance in rainbow trout and Atlantic salmon using both the factorial and multivariate approaches. The estimates

TABLE 4-3 Estimates of Maintenance[a], Cost of Protein[b] and Lipid Deposition[c] Determined Using the Factorial Approach[d]

Species	Temp. (°C)	HE_m (per day)	HE_m (per day)	K_p	K_f	Reference
Common carp (*Cyprinus carpio*)	18	10.04 kcal/kg$^{0.75}$	42 kJ/kg$^{0.75}$	0.56	0.72	Schwartz and Kirchgessner (1995)
European sea bass (*Dicentrarchus labrax*)	23	9.56 kcal/kg$^{0.79}$	40 kJ/kg$^{0.79}$	0.53	0.90	Lupatsch et al. (2003)
Gilthead sea bream (*Sparus aurata*)	23	10.28 kcal/kg$^{0.83}$	43 kJ/kg$^{0.83}$	0.53	0.76	Lupatsch et al. (2003)
Grouper (*Epinephelus* spp.)	23	7.41 kcal/kg$^{0.83}$	31 kJ/kg$^{0.83}$	0.56	0.91	Lupatch et al. (2003)
Rainbow trout (*Oncorhynchus mykiss*)	15	0.33 kcal/g$^{0.39}$	1.37 kJ/g$^{0.39}$	0.54	0.90[e]	Rodehutscord and Pfeffer (1999)
Rainbow trout (*O. mykiss*)	8.5	8.37 kcal/kg$^{0.80}$	35 kJ/kg$^{0.80}$	0.53	0.90[e]	Azevedo et al. (2005)
Rainbow trout[f] (*O. mykiss*)	8.5	4.78 kcal/kg$^{0.80}$	20 kJ/kg$^{0.80}$	0.43	0.81	Azevedo et al. (2005)
Rainbow trout (*O. mykiss*)	8.5	4.54 kcal/kg$^{0.80}$	19 kJ/kg$^{0.80}$	0.63	0.72	Bureau et al. (2006)
Atlantic salmon (*Salmo salar*)	8.5	8.37 kcal/kg$^{0.80}$	35 kJ/kg$^{0.80}$	0.81	0.90[e]	Azevedo et al. (2005)
Atlantic salmon[f] (*S. salar*)	8.5	4.78 kcal/kg$^{0.80}$	20 kJ/kg$^{0.80}$	0.52	0.81	Azevedo et al. (2005)
Mulloway (*Argyrosomus japonicus*)	20–26	10.76 kcal/kg$^{0.80}$	45 kJ/kg$^{0.80}$	0.49	0.75	Pirozzi et al. (2010)

[a] HE_m.
[b] K_p.
[c] K_f.
[d] Using the method of Kielanowski (1965).
[e] Fixed.
[f] Multivariate analysis.

obtained with the factorial and multivariate approaches were different (Table 4-3) but highly dependent on the model and corresponding assumptions (Azevedo et al., 2005).

Significant differences in the energy cost of protein deposition exist between theoretical assumptions (86% efficient) and that calculated using the factorial and multivariate approaches (43–81% efficient). Both the factorial and multivariate approaches are based on statistical relationships between inputs and outputs, without a true representation of the underlying biological principles. Results obtained with the factorial and multivariate approaches may be statistical artifacts. Protein and lipid depositions are not merely the deposition of energy but rather the results of a highly complicated array of metabolic processes. Rates and efficiencies (hence turnover) of deposition are governed by dietary factors (nutrient balance and utilization) and biological factors (e.g., genetics, physiological state of the animal, and types of tissue made).

In growing animals, the rates of protein synthesis greatly exceed those of protein deposition (Reeds et al., 1981). The efficiency of retention of protein deposition in fish ranges from 30 to 70% (Langar and Guillaume, 1994). Therefore, changes in protein turnover are a possible explanation for the variable energy cost of protein deposition (Reeds et al., 1985; Milligan and Summers, 1986). On the other hand, when problems associated with structural and kinetic heterogeneity of amino acid pools are involved (Watt et al., 1991), in addition to the difficulty of measuring the synthesis of rapidly turning over proteins (Wheatley et al., 1988), one is led to conclude that current estimates of protein synthesis in vivo may underestimate actual rates. Consequently, the theoretical energy cost for protein deposition will also be underestimated. The nutrient composition of the diet may also contribute to K_p. Catabolism of amino acids due to amino acid excess or imbalances in the diet or low nonprotein energy content will result in waste of energy (H_iE associated with catabolism of amino acid higher than that of lipid) and a decrease in K_p. Diets with unbalanced amino acid composition will result in lower retained energy as protein and consequently result in K_p values lower than that achievable if the diet were perfectly balanced. The type of amino acid catabolized also plays a role in loss of energy (van Milgen et al., 2002).

Nonetheless, heat losses associated with lipid deposition estimated using the factorial and multivariate approaches appear to be close to the theoretical chemical cost. Since K_f differs depending on the origin of the lipid deposited, the composition of the diet will also affect the efficiency of lipid deposition. Dietary intake of preformed lipids will lead to a high efficiency of utilization for lipid deposition, whereas de novo synthesis of lipid from dietary carbohydrates will lead to a slightly lower efficiency. The pathways of lipogenesis in fish are qualitatively similar to those in other vertebrates. A review of the literature reveals significant differences amongst species and studies on the same species in terms of de novo lipogenesis. Some studies have indicated that de novo fatty acid synthesis is generally limited in fish fed high-lipid diets (Brauge, 1994; Dias et al., 1998). Consequently, in fish fed high-lipid diets, where almost all the lipid deposited is of dietary origin (Brauge, 1994), the cost of lipid deposition is apparently very low (Table 4-3).

The early studies of Cho et al. (1976) showed that an increase in dietary fat levels led to a decrease in H_iE. LeGrow and Beamish (1986) confirmed that the increase in oxygen uptake as dietary protein levels increased was consistent, regardless of the dietary lipid level, thus highlighting the importance of dietary digestible protein/digestible energy (DP/DE) ratios. In addition, the effectiveness of the energy derived from the catabolism of different amino acids by the fish is unclear (Encarnação et al., 2006). It has been suggested that digestible carbohydrates are possibly also significant contributors to H_iE (Beamish et al., 1986; Hilton et al., 1987). Beamish et al. (1986) observed that fish fed a diet with high glucose content consumed more oxygen than fish fed a diet with the same protein level but rich in lipid, suggesting that an increase in digestible carbohydrate intake results in an increase in H_iE. The fish fed the diet high in glucose had a lower nitrogen retention efficiency (N gain:N intake) than the fish fed the diet high in lipid (Beamish et al., 1986). These data suggest that the effect on H_iE observed was in fact more related to the variation in dietary lipid and the sparing of protein than to the digestible carbohydrate itself. However, Helland and Grisdale-Helland (1998) observed that, at low intake levels, an increase in digestible starch at the expense of digestible protein resulted in an increase in oxygen consumption of Atlantic salmon but no change in the N gain:N intake. Bureau et al. (1998) observed a very poor retention of ME of digestible starch fed to rainbow trout, suggesting that at high levels of intakes, the utilization of digestible carbohydrates can be associated with very significant H_iE. However, these authors observed no significant increase in oxygen consumption in fish fed diets with very high digestible starch levels. They speculated that anaerobic catabolism of glucose may be a source of energy loss for fish fed high carbohydrate diets. This hypothesis should be investigated.

The discussion above illustrates the value of using net energy (NE) systems (factorial and multivariate approaches) as simple and practical means of estimating the cost of protein and lipid depositions in animals. However, more accurate and detailed delineation of the cost of growth requires detailed analysis of digestion, metabolism, and retention of individual nutrients, and thus, the use of framework based on nutrient utilization as opposed to energy utilization.

RECOVERED ENERGY

Protein and lipids are the main energy-yielding components of the bodies of animals. Glycogen, another energy-yielding body component, generally represents only a small proportion of the body of the animal (< 1%). Based on detailed allometric analysis, Shearer (1994) and Dumas et al.

(2007) concluded that the protein concentration in the whole body of growing salmonid was relatively constant, but that their lipid concentration was highly variable and affected by both endogenous (fish size, growth rate) and exogenous (dietary, environmental) factors. Studies with numerous other species also support this conclusion (Lupatsch et al., 2003; Glencross, 2008; Glencross et al., 2010).

Studies have shown that the composition of biomass gain includes more lipids and less water in a large fish than in a small fish (Shul'man, 1974; Dumas et al., 2007). Consequently, more energy is contained in one unit of biomass gain for a large fish (e.g., 2.39 kcal/g BW or 10 kJ/g BW) than for a small fish (e.g., 0.96–1.20 kcal/g BW or 4–5 kJ/g BW). The protein of the whole body appears to vary little with growth of a given species of fish, whereas whole-body GE content varies considerably over time. There are about 4 g of water associated with each gram of protein tissue deposited (Dumas et al., 2007). On a wet-weight basis, tissue contains about 16% protein. Since protein has a GE of 5.64 kcal/g (23.6 kJ/g), protein contributes about 0.88 kcal GE/g (3.7 kJ GE/g) of wet tissue. Lipids are stored in tissues generally substituting water (Dumas et al., 2007), although experimental evidence suggests that lipid deposits can significantly contribute to live weight gain of fish. Differences exist in the lipid deposition dynamic amongst species, strains, life stages, and animals with different nutritional histories. Whole-body GE content increase seen in fish of increasing weight is mainly due to increasing lipid content.

The difference between the enthalpy of combustion (i.e., GE) of the body at the beginning and at the end of a period of time is referred to as RE. The most direct way of estimating RE is to determine GE (by bomb calorimetry) of representative whole-body samples from a group of experimental animals at the beginning and at the end of a growth assay. This method of determining RE is termed "comparative carcass analysis" or "slaughter technique." Alternatively, RE can be estimated by difference between IE and FE, UE + ZE, and HE (Blaxter, 1989) and is also known as the "energy balance technique." The RE can either be positive or negative and represents the enthalpy of combustion of organic compounds stored or lost by the body of the animal.

Protein and lipid deposition are two distinct biological processes regulated by different mechanisms. The relative importance of protein and lipid deposition depends upon a great number of nutritional factors. Severe feed restriction results in significant alteration of the protein to lipid deposition ratio in fish. As mentioned earlier, fish fed a maintenance ration (ration resulting in RE = 0) can still deposit body protein (positive protein-energy gain) while mobilizing body lipids (Figure 4-4). There is clearly evidence that live weight gain is largely driven by protein deposition (Dumas et al., 2007). Studies on the effect of feeding levels on fish have shown that protein and lipid deposition increase linearly with feed allocation, but that they have different slopes and intercepts (Figure 4-4). At severe feed restriction, protein deposition greatly exceeds lipid deposition. However, energy deposition as lipids often exceeds that as protein at moderate to high feeding levels (Lupatsch et al., 2001b; Dumas et al., 2007). A number of studies have also shown that protein deposition tends to plateau off at high feeding levels (Lupatsch et al., 2003; Glencross, 2008; Glencross et al., 2010), whereas lipid deposition does not appear to level off (increases linearly). A decreasing protein to lipid deposition ratio can be observed at a feeding level approaching maximum protein deposition (Dumas et al., 2007). Differences in the GE content of the carcass and RE of fish are largely determined by variations in lipid depositions.

High DE intake and feeding diets with improper balance of DP to DE and protein sources with poorer essential amino acid profile generally result in the deposition of a larger proportion of RE as lipid. Seasonal changes in body composition, in relation to specific physiological stages or endocrine status are also known to occur. There are also considerable interspecific differences in lipid deposition and tissue distribution. Nutrient deposition and temporal changes in body composition of fish and effects of all the factors mentioned above need to be investigated more systematically than has been the case in the past (Dumas et al., 2010).

REPRODUCTION AND GONADS—OVUM ENERGY

Reproduction is a demanding period of life for many organisms. Resources need to be diverted from somatic growth into processes necessary for successful breeding (Thorpe, 1992; Hendry and Berg, 1999). The chemical composition and energy content of gonads and gametes have been fairly well characterized for a number of fish species (Kaushik and Médale, 1994). The energy content of the eggs, termed ovum energy (OE) in the nomenclature proposed by NRC (1981), of rainbow trout has been estimated to be about 6.45 kcal/g (27 kJ/g) dry matter, irrespective of the size of the eggs. The average energy content in eggs, measured in about 50 teleosts, was 5.62 kcal/g (23.5 kJ/g) dry matter whatever the size of eggs. Lupatsch et al. (2010b) determine that eggs of tilapia had a GE of 2.51 kcal/g (10.5 kJ/g) wet weight. The total amount of energy stored in the eggs would represent 8–15% of gross energy of the whole body, very much correlated to the gonado-somatic index (Kaushik and Médale, 1994). For the majority of the species, the male gonads with maturity represent only one small proportion of the body mass. On the other hand, the ovaries with maturity can represent up to 30% of the body mass of certain species. In some multiple spawners, such as the gilthead seabream, total egg production over a single season can even reach 100% of body mass (Tandler et al., 1995). Lupatsch et al. (2010b) determined that adult female tilapia produced about 1 g of egg per kg BW per day.

There are few studies on the actual cost of gonad formation (e.g., H_iE associated with nutrient deposition in gonads). Efficiency of DE utilization (K_{pf}) for both gonad formation and somatic growth in adult female tilapia was recently estimated to be 0.63, which is very similar to estimates of K_{pf} for somatic growth in the same species (Lupatsch et al., 2010b).

The total cost of reproduction exceeds that of only production of the gametes. The development of secondary sexual characters, the production of mucus, nuptial behaviors and activities, migration, and other aspects are all processes that involve expenditure of energy. Under aquaculture conditions, expenses associated with reproduction may certainly not be as dramatic as those incurred in wild fish. However, reproduction is a critical part of the production cycle for farmed species and involves fairly dramatic changes in the energy partitioning by the animal. Reproduction involves the synthesis and temporary storage of new tissues that are formed almost regardless of the level of dietary energy intake, the necessary energy being withdrawn from other body tissues if the dietary supply is insufficient (Jonsson et al., 1991, 1997). Consequently the redistribution of tissue energy that takes place in the breeding season can complicate measurements of energy balance and estimation of the cost of growth (Lupatsch et al., 2010b).

CALCULATION OF ENERGY REQUIREMENT FOR GROWTH

The essential thrust of studies on bioenergetics of animals is to provide a convenient and accurate system to predict feed requirements or efficiency of feed utilization based on body weight, growth rate, sex, activity, physiological state, environment, and the composition of feed provided to the animal (Baldwin and Bywater, 1984). A large number of bioenergetics models of different formats have been developed to predict growth, feed ration, FCR, and waste outputs of various fish species under a variety of conditions (Cho, 1992; Cho and Bureau, 1998; Kaushik, 1998; Lupatsch and Kissil, 1998; Cui and Xie, 1999; Bureau et al., 2002, 2003; Zhou et al., 2005; Glencross, 2008; Lupatsch et al., 2008; Glencross et al., 2010; Pirozzi et al., 2010).

Fish growth has usually been predicted using two different approaches in bioenergetics models. One approach assumes that IE drives weight gain. This assumption is encountered mostly in fisheries and ecology studies because the availability of food in natural ecosystems often limits fish growth (Elliott, 1976; Kitchell et al., 1977; From and Rasmussen, 1989). An alternative approach considers genetic or potential/desired growth rate rather than nutrition as the factor limiting animal growth (Hubbell, 1971; Calow, 1973; Oldham et al., 1997). Here, IE is a function of the requirements of the individual to achieve a given growth potential or growth target. This approach was suggested by Winberg (1956) and is mostly used in aquaculture where fish are generally fed to satiation with nutritionally complete diets (Cho, 1990, 1992; Lupatsch, et al., 2001b; Zhou et al., 2005; Glencross, 2008).

In several bioenergetics models, FE, UE + ZE, H_iE, and H_eE, as well as GE content of the carcass, are considered to be fixed fractions of IE, regardless of the composition of the feed and performance of the fish (e.g., Hanson et al., 1997). Basic understanding of nutrition should indicate that these are unreasonable assumptions. It is also common to observe energy requirement expressed as absolute amount of DE required per kg body weight per day for "maximal production" or energy expenditure and deposition expressed as a proportion of "maximal feed consumption" (C_{max}) in numerous fish bioenergetics studies (e.g., Gatlin et al., 1986; McGoogan and Gatlin, 1998; Ohta and Watanabe, 1998; Cui and Xie, 1999; Elliott and Hurley, 1999; Watanabe et al., 2000). Maximal production or C_{max} of an animal are factors that are highly dependent on genetics, diet composition, environmental conditions (e.g., temperature), husbandry practices, health status, and other variables. Maximum production and C_{max} are, therefore, highly variable parameters. Consequently, the energy requirement for maximum production calculated in some studies (i.e., energy requirement expressed in absolute term such as kcal per fish per day) can only be valid for the extremely specific conditions (e.g., diet composition, strain, temperature, and culture conditions) encountered in the study. Fish growing at different rates will deposit nutrients at different rates and, consequently, have different energy and feed requirements. Energy requirement should therefore be calculated for explicitly expressed levels of performance (e.g., expected or achievable level of performance), feed composition, and life stage (Cho, 1991, 1992; Cho and Bureau, 1998; Kaushik, 1998; Guillaume et al., 1999). This is at the basis of a large number of factorial models that have been developed in recent years for different aquaculture species (Cho and Bureau, 1998; Lupatsch et al., 1998, 2003, 2008; Bureau et al., 2003; Glencross, 2008). These factorial models divide energy requirement into its different components or fractions.

Cho (1991) proposed factorial models to determine energy requirement of fish based on expected level of performance, diet composition, and expected body composition. The approach of Cho (1991) was slightly modified by Cho (1992), Cho et al. (1994), Cho and Bureau (1998), and Bureau et al. (2002, 2003) and used to estimate feed requirement and waste outputs of different salmonid fish species reared under commercial-like conditions. Lupatsch et al. (1998) proposed a similar approach and subsequently used this approach to estimate energy, protein, and feed requirements of a variety of fish species (Lupatsch et al., 1998, 2001a, 2003, 2008).

Using the approach of Cho (1991), estimation of DE requirements and, consequently, feed requirements (or allocation) of fish can be determined as follows:

1. Characterization of diet (including DE content)
2. Calculation of expected live weight gain and RE
3. Allocation of H_eE based on fish size and water temperature
4. Allocation of H_iE for maintenance and energy deposition
5. Allocation of UE + ZE

6. Calculation of minimum DE requirement
7. Calculation of feed requirement.

Using the approach similar to that of Cho (1991), energy, oxygen, and feed requirements and expected feed efficiency of fish of different sizes reared under different conditions or rearing periods can be calculated (Cho and Bureau, 1998). Table 4-4 presents energy and oxygen requirements and theoretical feed efficiency of rainbow trout reared at 12°C and fed a practical diet (44% digestible protein and 4.54 Mcal [19 MJ] DE) at different weights or growing from 1 g to 1,000 g. The DE requirements to produce 1 kg biomass of rainbow trout were estimated to vary from about 2.51 Mcal (10.5 MJ) for 1 g fish to 6.41 (26.8 MJ) for 1 kg fish.

Table 4-5 similarly presents estimates of energy and oxygen requirements and theoretical feed efficiency of European sea bass but calculated using the approach of Lupatsch et al. (1998). Glencross (2008) used a very similar approach for Asian sea bass (Table 4-6). It is of utmost importance to understand that these estimates are only valid for the

TABLE 4-4 Energy and Oxygen Requirements[a] and Expected Feed Efficiency of Rainbow Trout[b] (*Oncorhynchus mykiss*)

Live Weight	RE[c]	H_eE[d]	H_iE[e]	UE + ZE[f]	DE[g]	Oxygen[h]	
(g/fish)	Mcal/kg (MJ/kg) Weight Gain					(g/kg weight gain)	Feed Efficiency[i]
1	1.34 (5.6)	0.24 (1.0)	0.81 (3.4)	0.12 (0.5)	2.51 (10.5)	320	1.8
5	1.34 (5.6)	0.33 (1.4)	0.81 (3.4)	0.12 (0.5)	2.63 (11.0)	350	1.7
10	1.36 (5.7)	0.41 (1.7)	0.81 (3.4)	0.12 (0.5)	2.68 (11.2)	370	1.7
50	1.43 (6.0)	0.53 (2.2)	0.86 (3.6)	0.14 (0.6)	2.99 (12.5)	430	1.5
100	1.55 (6.5)	0.60 (2.5)	0.93 (3.9)	0.14 (0.6)	3.21 (13.4)	470	1.4
500	1.79 (7.5)	0.76 (3.2)	1.39 (5.8)	0.22 (0.9)	4.54 (19.0)	670	1.0
1,000	3.27 (13.7)	0.86 (3.6)	1.96 (8.2)	0.31 (1.3)	4.41 (26.8)	880	0.7
1–1,000	2.51 (10.5)	0.79 (3.3)	1.51 (6.3)	0.24 (1.0)	5.04 (21.1)	710	0.9

[a]Mcal (MJ) or g/kg weight gain.
[b]At various sizes or growing from 1 to 1,000 g based on assumption that the fish are reared at 12°C, growing with a thermal-unit growth coefficient (TGC) = 0.220, and fed diet with 420 g DP, 220 g lipids, and 4.54 Mcal (19 MJ) digestible energy (DE) per kg.
[c]RE (kJ/fish) = (–0.0039 (live body weight at $T_{(i+1)}$, g/fish)^2 + 5.5812 (live body weight at $T_{(i+1)}$, g/fish)) – (–0.0039 (live body weight at $T_{(i)}$, g/fish)^2 + 5.5812 (live body weight at $T_{(i)}$)).
[d]$H_eE = ((-1.04 + 3.26(T) - 0.05(T)^2) / (0.020^{0.824})) / d$.
[e]$H_iE = 0.6 \times RE$.
[f]$UE + ZE = 0.05 \times (RE + H_eE + H_iE)$.
[g]DE requirement = $RE + H_eE + H_iE + (UE + ZE)$.
[h]Oxygen requirement = $H_eE + H_iE$ / oxycalorific coefficient (13.6 kJ/g O_2 consumed).
[i]Expected feed efficiency (gain/feed).

TABLE 4-5 Energy and Oxygen Requirements[a] and Expected Feed Efficiency of European Sea Bass (*Dicentrarchus labrax*)[b]

Live Weight	RE[c]	H_eE[d]	H_iE + (UE + ZE)[e]	DE[f]	Oxygen[g]	
(g/fish)	Mcal/kg (MJ/kg) Weight Gain				(g/kg weight gain)	Feed Efficiency[h]
1	1.15 (4.8)	0.62 (2.6)	0.57 (2.4)	2.34 (9.8)	360	2.0
5	1.43 (6.0)	0.84 (3.5)	0.72 (3.0)	2.99 (12.5)	480	1.6
10	1.58 (6.6)	1.00 (4.1)	0.81 (3.4)	3.37 (14.1)	550	1.4
50	1.98 (8.3)	1.41 (5.9)	1.08 (4.5)	4.47 (18.7)	770	1.1
100	2.15 (9.0)	1.65 (6.9)	1.22 (5.1)	5.07 (21.2)	880	0.9
250	2.41 (10.1)	2.01 (8.4)	1.41 (5.9)	5.83 (24.4)	1,050	0.8
400	2.56 (10.7)	2.22 (9.3)	1.53 (6.4)	6.26 (26.2)	1,160	0.8
1–400	2.34 (9.8)	1.89 (7.9)	1.34 (5.6)	5.59 (23.4)	1,039	0.8

[a]Mcal (MJ) or g/kg weight gain.
[b]At various sizes or growing from 1 to 400 g based on assumption that the fish are reared at 22°C, growing according to $y = 0.64 \times BW(kg)^{0.587} \times exp^{0.07T}$ (Lupatsch et al., 2001a), and fed a diet with 500 g digestible protein, 180 g lipids, and 4.78 Mcal (20 MJ) DE per kg.
[c]RE (kJ/fish) = (5.17 (live body weight (g) at $T_{(I+1)})^{0.107}$) – (5.17 kJ (live body weight (g) at $T_{(i)})^{0.107}$.
[d]$H_eE = 35 \times$ (live body weight (kg) at $T_{(i)})^{0.8} / d$.
[e]$H_iE + (UE + ZE) = 0.32 \times (RE + H_eE)$.
[f]DE requirement = $RE + H_eE + H_iE$.
[g]Oxygen requirement = $H_eE + H_iE$ / oxycalorific coefficient (13.6 kJ/g O_2 consumed).
[h]Expected feed efficiency (gain/feed).

TABLE 4-6 Energy and Oxygen Requirements[a] and Expected Feed Efficiency of Asian Sea Bass (*Lates calcarifer*)[b]

Live Weight (g/fish)	Growth Rate[c] (g/fish per day)	RE[d]	H_cE[e]	H_iE + (UE + ZE)[f]	DE[g]	Oxygen[h] (g/kg weight gain)	Feed Efficiency[i]
		Mcal/kg (MJ/kg) weight gain					
10	1.1	1.08 (4.5)	0.29 (1.2)	0.74 (3.1)	2.13 (8.9)	319	2.6
50	2.2	1.36 (5.7)	0.55 (2.3)	0.93 (3.9)	2.84 (11.9)	454	1.9
100	3.0	1.51 (6.3)	0.69 (2.9)	1.03 (4.3)	3.25 (13.6)	533	1.7
250	4.4	1.72 (7.2)	0.98 (4.1)	1.17 (4.9)	3.90 (16.3)	666	1.4
500	5.9	1.91 (8.0)	1.29 (5.4)	1.29 (5.4)	4.49 (18.8)	794	1.2
1,000	8.0	2.10 (8.8)	1.67 (7.0)	1.43 (6.0)	5.21 (21.8)	953	1.0
2,000	10.7	2.32 (9.7)	2.15 (9.0)	1.58 (6.6)	6.07 (25.4)	1,152	0.9
3,000	12.7	2.46 (10.3)	2.51 (10.5)	1.67 (7.0)	6.64 (27.8)	1,290	0.8

[a]Mcal (MJ) or g/kg weight gain.
[b]At various sizes or growing from 10 to 3,000 g based on assumption that the fish are reared at 30°C, growing according to a growth rate (g/fish per day) = $(0.54 - 0.1199T + 0.0074T^2 - 0.0001T^3) \times BW(kg)^{0.424} \times \exp^{0.07T}$ (Glencross, 2008) and fed a diet with 5.38 Mcal (22.5 MJ) DE per kg.
[c]Growth rate g/fish per day = $(0.54 - 0.1199T + 0.0074T^2 - 0.0001T^3)\ BW^{0.424}$.
[d]RE (kJ/fish) = 3.273 (live body weight gain (g/fish)) × live weight (g/fish)$^{0.143}$.
[e]H_cE (kJ/fish per day) = $(0.44624 - 0.08484T + 0.00483T^2 + 0.0008T^3) \times BW^{0.8035}$.
[f]H_iE + (UE + ZE) = 0.684 × (RE).
[g]DE requirement = RE + H_cE + H_iE.
[h]Oxygen requirement = H_cE + H_iE / oxycalorific coefficient (13.6 kJ/g O_2 consumed).
[i]Expected feed efficiency (gain/feed).

given set of conditions (such as species, water temperature, growth rates, and diet composition) and should not be applied blindly.

The DE requirement (kcal per fish per day) is largely dependent on growth rate of the animal. In general, as temperature increases, the growth rate, DE intake, and RE increase, but efficiency of ME and DE utilization (RE/ME or RE/DE) does not change (Azevedo et al., 1998). Total energy requirement should ideally be expressed as DE because FE and, consequently, IE are highly dependent on the composition of diet fed. FE losses by animals are largely a factor of the composition of the diet and not always greatly affected by biological factors.

LIMITATIONS OF NUTRITIONAL ENERGETICS APPROACHES

Nutritional energetics models are simple and practical, but their limitations have been increasingly recognized (Birkett and de Lange, 2001; Bureau et al., 2002; Bajer et al., 2004; Dijkstra et al., 2007; Dumas et al., 2008). Energy systems simplify the partitioning of dietary components into body deposition of nutrients on the basis of their heats of combustion, disregard specific metabolic roles of different nutrients and their interaction, and ignore significant differences in contribution of protein and lipid toward live weight gain. More than a century ago, Rubner (1902) described bioenergetics as the "heat doctrine" (Dumas et al., 2008). The lumping of nutrients together solely on the basis of their energy values is largely irrational. It does not allow a complete evaluation of the effect of chemical composition of the feed and role or efficiency of use of specific nutrients. "Energy requirement" and "dietary energy utilization" are parameters and tools that help deal with the complexity of animal metabolism and nutritional requirements, not immutable concepts.

One of the practical limitations of bioenergetics models is that they assume that energy is allocated in a hierarchical fashion and that growth is the surplus of energy after all other components of the energy budget have been covered or satisfied (Elliot and Hurley, 1999). Models based on bioenergetic principles assume that growth and feed efficiency will be nil when animals are fed a maintenance ration (RE = 0). That assumption has been proven inaccurate in fish, as well as in other animals, where positive weight gain was still observed even though animals were fed on a maintenance ration of a nutritionally adequate diet (Le Dividich et al., 1980; Bureau et al., 2006). Live weight gain is mainly (but not solely) driven by protein deposition due to the close association of water with protein deposition (Shearer, 1994; Dumas et al., 2007). Lipid reserves can be mobilized to support protein deposition (Black, 1974; Campbell, 1988; Bureau et al., 2006; Figure 4-4). Azevedo et al. (2004a,b) observed dramatic changes in the feed efficiency in rainbow trout of increasing weight. Efficiency of protein retention decreased significantly in these animals but efficiency of ME utilization did not change.

Bioenergetics models rely on estimates of the cost of growth calculated empirically based on statistical interpretations of experimental data. How much of the energy cost is truly due to "biological inefficiencies" or simply due to the fact that animals are fed "imperfect" diets is not known. Furthermore, these empirical estimates cannot be legitimately extrapolated to conditions beyond which data are collected (France and Thornley, 1984; Baldwin, 1995). Although nutritional energetics frameworks are including increasingly explicit representations of metabolic use of dietary nutrients, there is increasing evidence that current bioenergetics models are not sufficiently rational and flexible to be applied to the wide range of conditions encountered in fish culture (Bureau and Hua, 2008; Dumas et al., 2008). There is a necessity to move to an approach based on a clearer understanding of the role of specific absorbed nutrients and their metabolism in determining productive responses of the animal (Reynolds, 1999; Dumas et al., 2008). A more scientifically correct feed evaluation and requirement system should be based on characterization of nutrient fractions relevant to their actual digestion, metabolism, and deposition in the animal under varying practical conditions (Boisen and Verstegen, 1998).

Mechanistic models, based on more or less explicit representation of biochemical reactions and metabolic use of amino acids, fatty acids, and glucose, have been developed for various fish species, such as African catfish (Machiels and Henken, 1986; Conceição et al., 1998), Nile tilapia (van Dam and De Vries, 1995), rainbow trout (van Dam and De Vries, 1995; Hua and Bureau, 2010), turbot (Conceição et al., 1998), and Atlantic salmon (*Salmo salar*) (Bar et al., 2007). Construction of mechanistic models requires adequate knowledge of the system, and relies on sufficient and accurate data to quantify the perceived system (Baldwin, 1995). The process of parameterization can be a major bottleneck in the development and application of complex mechanistic models (Kyriazakis 1999; McNamara, 2004; Dumas et al., 2010). Consequently, all biochemical models have been developed with some degree of simplification of metabolic pathways, have included numerous assumptions, and have been generally driven by more or less transparent partitioning rules. However, metabolic models are often more complex than what is required to represent growth at the whole animal level (van der Honing, 1998; Birkett and de Lange, 2001). These highly detailed models can work well within the narrow range of conditions for which they are parametrized and calibrated. However, they generally fail to accurately describe nutrient utilization by fish as influenced by a wide range of conditions (including differences in feed composition, environmental conditions, husbandry practices, life stages, and genetic background of animals) encountered in fish culture (Dumas et al., 2010; Hua and Bureau, 2010). Despite these limitations, there is a need for the fish and shrimp nutrition community to increasingly embrace approaches based on explicit and integrative utilization of nutrients in order to enable the development of more rational systems aimed at better describing and predicting growth and nutrient utilization and requirements of fish and shrimp.

REFERENCES

Alanärä, A. 1994. The effect of dietary temperature, dietary energy content and the reward level on the demand feeding activity of rainbow trout (*Oncorhynchus mykiss*). Aquaculture 126:349–359.

Alanärä, A., and A. Kiessling. 1996. Changes in demand feeding behaviour in Arctic charr, *Salvelinus alpinus* L., caused by differences in dietary energy content and reward level. Aquac. Res. 27:479–486.

Azevedo, P. A., C. Y. Cho, S. Leeson, and D. P. Bureau. 1998. Effects of feeding level and water temperature on growth, nutrient and energy utilization and waste outputs of rainbow trout (*Oncorhynchus mykiss*). Aquat. Living Resour. 11:227–238.

Azevedo, P. A., S. Leeson, C. Y. Cho, and D. P. Bureau. 2004a. Growth, nitrogen and energy utilization of juveniles from four salmonid species: Diet, species and size effects. Aquaculture 234:393–414.

Azevedo, P. A., S. Leeson, C. Y. Cho, and D. P. Bureau. 2004b. Growth and feed utilization of large size rainbow trout (*Oncorhynchus mykiss*) and Atlantic salmon (*Salmo salar*) reared in freshwater: Diet and species effects, and responses over time. Aquacult. Nutr. 10:401–411.

Azevedo, P. A., J. van Milgen, S. Leeson, and D. P. Bureau. 2005. Comparing efficiency of metabolizable energy utilization by rainbow trout (*Oncorhynchus mykiss*) and Atlantic salmon (*Salmo salar*) using factorial and multivariate approaches. J. Anim. Sci. 83:842–851.

Bajer, P. G., G. W. Whitledge, and R. S. Hayward. 2004. Widespread consumption-dependent systematic error in fish bioenergetics models and its implications. Can. J. Fish. Aquat. Sci. 61:2158–2167.

Baldwin, R. L. 1995. Modeling Ruminant Digestion and Metabolism. New York: Chapman and Hall.

Baldwin, R. L., and A. C. Bywater. 1984. Nutritional energetics of animals. Annu. Rev. Nutr. 4:101–114.

Bar, N. S., T. Sigholt, K. D. Shearer, and Å. Krogdahl. 2007. A dynamic model of nutrient pathways, growth, and body composition in fish. Can. J. Fish. Aquat. Sci. 64:1669–1682.

Beamish, F. W. H. 1974. Apparent specific dynamic action of largemouth bass (*Micropterus salmonides*). J. Fish. Res. Board Can. 31:1763–1769.

Beamish, F. W. H., J. W. Hilton, E. Niimi, and S. J. Slinger. 1986. Dietary carbohydrate and growth, body composition and heat increment in rainbow trout (*Salmo gairdneri*). Fish Physiol. Biochem. 1:85–91.

Bermudes, M., B. D. Glencross, K. Austen, and W. Hawkins. 2010. Effect of high water temperatures on nutrient and energy retention in barramundi (*Lates calcarifer*). Aquaculture 306:160–166.

Birkett, S., and K. de Lange. 2001. Limitations of conventional models and a conceptual framework for a nutrient flow representation of energy utilization by animals. Brit. J. Nutr. 86:647–659.

Black, J. L. 1974. Manipulation of body composition through nutrition. P. Aus. S. Ani. 10:211–218.

Blaxter, K. L. 1989. Energy Metabolism in Animals and Man. Cambridge, UK: Cambridge University Press.

Boisen, S., and M. W. A. Verstegen. 1998. Evaluation of feedstuffs and pig diets. Energy or nutrient-based evaluation systems? I. Limitations of present energy evaluation systems. Acta Agr. Scand. A-An. 48:86–94.

Boujard, T., and F. M. Médale. 1994. Regulation of voluntary feed intake in juvenile rainbow trout fed by hand or by self-feeders with diets containing two different protein/energy ratios. Aquat. Living Resour. 7:211–215.

Bradfield, A. E., and M. J. Llewellyn. 1982. Animal Energetics. Glasgow, UK: Blackie and Son, Ltd.

Brauge, C. 1994. Influences des Sources Non Protéiques d'Énergie et des Facteurs Environnementaux sur les Métabolismes Énergétique et Lipique de la Truite Arc-en-Ciel, *Oncorhynchus mykiss*. Ph.D. thesis, L'Université de Bordeaux, Bordeaux, France.

Brett, J. R., and T. D. D. Groves. 1979. Physiological energetics. Pp. 279–352 in Fish Physiology, W. S. Hoar, D. J. Randall, and J. R. Brett, eds. New York: Academic Press, Inc.

Brett, J. R., and C. A. Zala. 1975. Daily pattern of nitrogen excretion and oxygen consumption of sockeye salmon (*Oneorhynchus nerka*) under controlled condition. J. Fish. Res. Board Can. 32:2479–2486.

Brody, S. 1945. Bioenergetics and Growth with Special Reference to the Efficiency Complex in Domestic Animals. New York: Hafner Publishing Company Inc.

Bureau, D. P. 1997. The partitioning of energy from digestible carbohydrate by rainbow trout (*Oncorhynchus mykiss*). Ph.D. thesis, University of Guelph, Guelph, Ontario.

Bureau D. P., and K. Hua. 2008. Models of nutrient utilization by fish and potential applications for fish culture operations. Pp. 442–461 in Mathematical Modelling in Animal Nutrition, J. France and E. Kebreab, eds. Wallingford, UK: CAB International Press.

Bureau, D. P., J. B. Kirkland, and C. Y. Cho. 1998. The partitioning of energy from digestible carbohydrate by rainbow trout (*Oncorhynchus mykiss*). Pp. 163–166 in Energy Metabolism of Farm Animals, K. J. McCracken, E. F. Unsworth, and A. R. G. Wylie, eds. Wallingford, UK: CAB International Press.

Bureau, D. P., P. A. Azevedo, and C. Y. Cho. 1999. Growth and nutrient deposition as a function of feeding level in salmonids. P. 41 in International Workshop on Fish Nutrition and Growth, Wuhan, Hubei, P.R. China, 10–14 October 1999.

Bureau, D. P., P. A. Azevedo, M. Tapia-Salazar, and G. Cuzon. 2000. Pattern and cost of growth and nutrient deposition in fish and shrimp: Potential implications and applications. Pp. 111–140 in Avances en Nutricion Acuicola V. Memorias del V Simposium Internacional de Nutricion Acuicola, L. E. Cruz-Suarez, D. Ricque-Marie, M. Tapia-Salazar, M. A. Olvera-Novoa, and R. Civera-Cerecedo, eds. 19–22 Noviembre, 2000. Merida, Yucatan, Mexico.

Bureau, D. P., S. J. Kaushik, and C. Y. Cho. 2002. Bioenergetics. Pp. 1–56 in Fish Nutrition, J. E. Halver and R. W. Hardy, eds. San Diego, CA: Academic Press.

Bureau, D. P., S. J. Gunther, and C. Y. Cho. 2003. Chemical composition and preliminary theoretical estimates of waste outputs of rainbow trout reared on commercial cage culture operations in Ontario. N. Am. J. Aquacult. 65:33–38.

Bureau, D. P., K. Hua, and C. Y. Cho. 2006. Effect of feeding level on growth and nutrient deposition in rainbow trout (*Oncorhynchus mykiss* Walbaum) growing from 150 to 600 g. Aquac. Res. 37:1090–1098.

Calow, P. 1973. The relationship between fecundity, phenology, and longevity: A systems approach. Am. Nat. 107:559–574.

Campbell, R. G. 1988. Nutritional constraints to lean tissue accretion in farm animals. Nutr. Res. Rev. 1:233–253.

Chakraborty, S. C., and S. Chakraborty. 1998. Effect of dietary protein level on excretion of ammonia in Indian major carp, *Labeo rohita*, fingerlings. Aquacult. Nutr. 4:47–51.

Cho, C. Y. 1990. Fish nutrition, feeds, and feeding: With special emphasis on salmonid aquaculture. Food Rev. Int. 6:333–357.

Cho, C. Y. 1991. Prediction model for energy requirement of salmonid fishes. Pp. 446–449 in Energy Metabolism of Farm Animals, C. Wenk and M. Boessinger, eds. Zürich, Switzerland: Institut für Nutztierwissenschaften, Gruppe Ernährung.

Cho, C. Y. 1992. Feeding systems for rainbow trout and other salmonids with reference to current estimates of energy and protein requirements. Aquaculture 100:107–123.

Cho, C. Y., and D. P. Bureau. 1995. Determination of the energy requirements of fish with particular reference to salmonids. J. Appl. Ichthyol. 11:141–163.

Cho, C. Y., and D. P. Bureau. 1998. Development of bioenergetics models and the Fish-PrFEQ software to estimate production, feeding ration and waste output in aquaculture. Aquat. Living Resour. 11:199–210.

Cho, C. Y., and S. J. Kaushik. 1985. Effects of protein intake on metabolizable and net energy values of fish diets. Pp. 95–117 in Nutrition and Feeding in Fish, C. B. Cowey, A. M. Mackie, and J. G. Bell, eds. London, UK: Academic Press.

Cho, C. Y., and S. J. Kaushik. 1990. Nutritional energetics in fish: Energy and protein utilization in rainbow trout (*Salmo gairdneri*). World Rev. Nutr. Diet. 61:132–172.

Cho, C. Y., H. S. Bayley, and S. J. Slinger. 1976. Energy metabolism in growing rainbow trout: Partition of dietary energy in high protein and high fat diets. Pp. 299–302 in Energy Metabolism of Farm Animals, M. Vermorel, ed. Clermont-Ferrand, France: G. de Bussac.

Cho, C. Y., S. J. Slinger, and H. S. Bayley. 1982. Bioenergetics of salmonid fishes: Energy intake, expenditure and productivity. Comp. Biochem. Phys. B 73B:25–41.

Cho, C. Y., J. D. Hynes, K. R. Wood, and H. K. Yoshida. 1994. Development of high nutrient-dense, low pollution diets and prediction of aquaculture wastes using biological approaches. Aquaculture 124:293–305.

Clarke, A., and N. M. Johnson. 1999. Scaling of metabolic rate with body mass and temperature in teleost fish. J. Anim. Ecol. 68:893–905.

Clement, K. D. 1996. Fermentation and gastrointestinal microorganisms in fishes. Pp. 156–198 in Gastrointestinal Microbiology, R. I. Mackie and B. A. White, eds. New York: Chapman and Hall.

Conceição, L. E. C., J. A. J. Verreth, M. W. A. Verstegen, and E. A. Huisman. 1998. A preliminary model for dynamic simulation of growth in fish larvae: Application to the african catfish (*Clarias gariepinus*) and turbot (*Scophthalmus maximus*). Aquaculture 163(3–4):215–235.

Cooke, S. J., K. P. Chandroo, T. A. Beddow, R. D. Moccia, and R. S. McKinley. 2000. Swimming activity and energetic expenditure of captive rainbow trout *Oncorhynchus mykiss* (Walbaum) estimated by electromyogram telemetry. Aquac. Res. 31:495–505.

Cui, Y., and J. Liu. 1990. Comparison of energy budget among six teleosts. II. Metabolic rates. Comp. Biochem. Phys. A 97A:169–174.

Cui, Y., and S. Xie. 1999. Modelling growth in fish. Pp. 413–434 in Feeding Systems and Feed Evaluation Models, M. K. Theodorou and J. France, eds. Wallingford, UK: CAB International Press.

Dantzler, W. H. 1989. Comparative Physiology of the Vertebrate Kidney. Zoophysiology, Volume 22. Berlin, Germany: Springer-Verlag.

De Lange, C. F. M. 1997. Modelling energy intake and utilization in grower-finisher pigs. Pp. 96–108 in Proceedings of Canadian Feed Industry Association Eastern Nutrition Conference, May 6–7, 1997, Guelph, Ontario, Canada.

Deng, D. F., S. Refstie, H. I. Hemre, C. E. Crocker, H. Y. Chen, J. J. Cech, Jr., and S. S. O. Hung. 2000. A new technique of feeding, repeated sampling of blood and continuous collection of urine in white sturgeon. Fish Physiol. Biochem. 22:191–197.

Dias, J., M. J. Alvarez, A. Diez, J. Arzel, G. Corraze, J. M. Bautista, and S. J. Kaushik. 1998. Regulation of hepatic lipogenesis by dietary protein/energy in juvenile European seabass (*Dicentrarchus labrax*). Aquaculture 161:169–186.

Dijkstra, J., E. Kebreab, J. A. N. Mills, W. F. Pellikaan, S. López, A. Bannink, and J. France. 2007. Predicting the profile of nutrients available for absorption: From nutrient requirement to animal response and environmental impact. Animal 1:99–111.

Dosdat, A., F. Servais, R. Métailler, C. Huelvan, and E. Desbruyères. 1996. Comparison of nitrogenous losses in five teleost fish species. Aquaculture 141:107–127.

Dumas, A., C. F. M. de Lange, J. France, and D. P. Bureau. 2007. Quantitative description of body composition and rates of nutrient deposition in rainbow trout (*Oncorhynchus mykiss*). Aquaculture 273:165–181.

Dumas, A., J. Dijkstra, and J. France. 2008. Mathematical modelling in animal nutrition: A centenary review. J. Agr. Sci. 146:123–142.

Dumas, A., J. France, and D. P. Bureau. 2010. Modelling growth and body composition in fish nutrition: Where have we been and where are we going? Aquac. Res. 41:161–181.

Ege, R., and A. Krogh. 1914. On the relation between the temperature and the respiratory exchange in fishes. Internationale Revue der Gesamten Hydrobiologie und Hydrographie 1:48–55.

Elliott, J. M. 1976. The energetics of feeding, metabolism and growth of brown trout (*Salmo trutta* L.) in relation to body weight, water temperature and ration size. J. Anim. Ecol. 45:923–948.

Elliott, J. M. 1982. The effects of temperature and ration size on the growth and energetics of salmonids in captivity. Comp. Biochem. Phys. B 73B:81–92.

Elliott, J. M., and M. A. Hurley. 1999. A new energetics model for brown trout, *Salmo trutta*. Freshwater Biol. 42:235–246.

Emmans, G. C. 1994. Effective energy: A concept of energy utilization applied across species. Brit. J. Nutr. 71:801–821.

Emmans, G. C. 1995. Models of the relationship between animal performance and dietary energy. Pp. 47–58 in Modelling Growth in the Pig, P. J. Moughan, M. W. A. Verstegen, and M. I. Visser-Reyneveld, eds. Wageningen, the Netherlands: Wageningen Pers.

Emmans, G. C., and C. Fisher. 1986. Problems in nutritional theory. Pp. 9–39 in Nutrient Requirements of Poultry and Nutritional Research, C. Fisher and K. N. Boorman, eds., London, UK: Butterworths.

Encarnação P., C. F. M. de Lange, M. Rodehutscord, D. Hoehler, W. Bureau, and D. P. Bureau. 2004. Diet digestible energy content affects lysine utilisation, but not dietary lysine requirements of rainbow trout (*Oncorhynchus mykiss*) for maximum growth. Aquaculture 235:569–586.

Encarnação, P., C. F. M. de Lange, and D. P. Bureau. 2006. Diet energy source affect lysine utilization for protein deposition in rainbow trout (*Oncorhynchus mykiss*). Aquaculture 261:1371–1381.

Flatt, J. P. 1992. Energy cost of ATP synthesis. Pp. 319–343 in Energy Metabolism: Tissue Determinants and Cellular Corollaries, J. M. Kinney, and H. N. Tucker, eds. New York: Raven Press Ltd.

France, J., and J. H. M. Thornley. 1984. Mathematical Models in Agriculture. London, UK: Butterworths.

From, J., and G. Rasmussen. 1989. Fish growth. Pp. 331–339 in Mathematical Submodels in Water Quality Systems, S. E. Jørgensen and M. J. Gromiec, eds. Amsterdam, the Netherlands: Elsevier Science Publishers.

Furuichi, M. 1988. Carbohydrates. Pp. 44–55 in Fish Nutrition and Mariculture, T. Watanabe, ed. Tokyo, Japan: Kanagawa International Fisheries Training Centre, Japan International Cooperation Agency.

Gatlin, D. M., W. E. Poe, and R. P. Wilson. 1986. Protein and energy requirements of fingerling channel catfish for maintenance and maximum growth. J. Nutr. 116:2121–2131.

Gauquelin, F. 1996. Etude de l'excrétion azotée et de l'influence de régimes alimentaires chez *P. stylirostris*. Issy-Les-Moulineaux, France: Rapport Interne Ifremer.

Geurden, I., E. Gondouin, M. Rimbach, W. Koppe, S. Kaushik, and T. Boujard. 2006. The evaluation of energy intake adjustments and preferences in juvenile rainbow trout fed increasing amounts of lipid. Physiol. Behav. 88:325–332.

Glencross, B. D. 2008. A factorial growth and feed utilization model for barramundi, *Lates calcarifer*, based on Australian production conditions. Aquacult. Nutr. 14:360–373.

Glencross, B. D., and M. Felsing. 2006. Influence of fish size and water temperature on the metabolic demand for oxygen by barramundi, *Lates calcarifer* (bloch), in freshwater. Aquac. Res. 37(11):1055–1062.

Glencross, B. D., T. T. T. Hien, N. T. Phuong, and T. L. Cam Tu. 2010.A factorial approach to defining the energy and protein requirements of Tra catfish, *Pangasianodon hypothalamus*. Aquacult. Nutr. doi 10.111/j.1365-2095.2010.00774.x.

Guillaume, J., S. J. Kaushik, P. Bergot, and R. Métailler. 1999. Nutrition et Alimentation des Poissons et des Crustacés, Collections Du Labo au Terrain. Paris, France: Éditions INRA.

Hanson, P., T. Johnson, J. Kitchell, and D. E. Schindler. 1997. Fish bioenergetics 3.0. University of Wisconsin Sea Grant Institute, Report WISCU-T-97-001, Madison, Wisconsin.

Helland, S. J., and B. Grisdale-Helland. 1998. The influence of dietary carbohydrate and protein levels on energy and nitrogen utilization of Atlantic salmon in seawater. Pp. 391–394 in Energy Metabolism of Farm Animals, K. J. McCracken, E. F. Unsworth, and A. R. G. Wylie, eds. Wallingford, UK: CAB International Press.

Helland, S. J., B. Hatlen, and B. Grisdale-Helland. 2010. Energy, protein and amino acid requirements for maintenance and efficiency of utilization for growth of Atlantic salmon post-smolts determined using increasing ration levels. Aquaculture 305:150–158.

Hemre, G. I., and F. Kahrs. 1997. 14C-glucose injection in Atlantic cod, *Gadus morhua*, metabolic responses and excretion via the gill membrane. Aquacult. Nutr. 3:3–8.

Hendry, A. P., and O. K. Berg. 1999. Secondary sexual characters, energy use, senescence, and the cost of reproduction in sockeye salmon. Can. J. Zoolog. 77:1663–1675.

Hepher, B. 1988. Nutrition of Pond Fishes. Cambridge, UK: Cambridge University Press.

Hilton, J. W., J. L. Atkinson, and S. J. Slinger. 1987. Evaluation of the net energy value of glucose (cerelose) and maize starch in diets for rainbow trout (*Salmo gairdneri*). Brit. J. Nutr. 58:453–461.

Hogendoorn, H. 1983. Growth and production of the African catfish, *Clarias lazera*. III. Bioenergetic relations of body weight and feeding level. Aquaculture 35:1–17.

Hua, K., and D. P. Bureau. 2010. Construction of a growth and nutrient utilization model for salmonid fish and its application in meta-analysis of fish meal replacement by plant proteins. Technical Report prepared for the United Soybean Board.

Hubbell, S. P. 1971. Of sowbugs and systems: The ecological bioenergetics of a terrestrial isopod. Pp. 269–324 in Systems Analysis and Simulation in Ecology, Volume 1, B. C. Patten, ed. New York: Academic Press.

Jobling, M. 1994. Fish Bioenergetics, Fish and Fisheries Series 13. London, UK: Chapman and Hall.

Jobling, M., and P. S. Davies. 1980. Effects of feeding on metabolic rate, and the specific dynamic action in plaice, *Pleuronectes platessa* L. J. Fish Biol. 16:629–638.

Jobling, M., and A. Wandsvik. 1983. An investigation of factors controlling food intake in Arctic charr, *Salvelinus alpinus* L. J. Fish Biol. 23:397–404.

Jonsson, B., and N. Jonsson. 2009. A review of the likely effects of climate change on anadromous Atlantic salmon *Salmo salar* and brown trout *Salmo trutta*, with particular reference to water temperature and flow. J. Fish Biol. 75:2381–2447.

Jonsson, N., B. Jonsson, and L. P. Hansen. 1991. Energetic cost of spawning in male and female Atlantic salmon (*Salmo salar* L.). J. Fish Biol. 39:739–744.

Jonsson, N., B. Jonsson, and L. P. Hansen. 1997. Changes in proximate composition and estimates of energetic costs during upstream migration and spawning in Atlantic salmon *Salmo salar*. J. Anim. Ecol. 66(3):425–436.

Kakuta, I., and K. Namba. 1989. Urine properties in carp, *Cyprinus carpio* L., with Sekoke disease. J. Fish Biol. 12:241–248.

Kaushik, S. J. 1980a. Influence of a rise in temperature on the nitrogen excretion of rainbow trout (*Salmo gairdneri* R.). Pp. 77–89 in Aquaculture in Heated Effluents and Recirculated Systems, Volume 1, K. Tiews, ed., Berlin, Germany: Heenemann GmbH.

Kaushik, S. J. 1980b. Influence of nutritional status on the daily patterns of nitrogen excretion in the carp (*Cyprinus carpio* L.) and the rainbow trout (*Salmo gairdneri* R.). Reprod. Nutr. Dev. 20:1751–1765.

Kaushik, S. J. 1998. Nutritional bioenergetics and estimation of waste production in non-salmonids. Aquat. Living Resour. 11:211–217.

Kaushik, S. J., and C. B. Cowey. 1991. Dietary factors affecting nitrogen excretion by fish. Pp. 3–19 in Nutritional Strategies and Aquaculture Waste, Proceedings of the 1st International Symposium on Nutritional Strategies in Management of Aquaculture Waste, C. B. Cowey and C. Y. Cho, eds. Guelph, Ontario, Canada.

Kaushik S. J., and K. Dabrowski. 1983. Postprandial metabolic changes in larval and juvenile carp (*Cyprinus carpio*). Reprod. Nutr. Dev. 23:223–234.

Kaushik, S. J., and E. F. Gomes. 1988. Effect of frequency of feeding on nitrogen and energy balance in rainbow trout under maintenance conditions. Aquaculture 73:207–216.

Kaushik, S. J., and F. Médale. 1994. Energy requirements, utilization and dietary supply to salmonids. Aquaculture 124:81–97.

Kaushik, S. J., K. Dabrowski, and P. Luquet. 1982. Patterns of nitrogen excretion and oxygen consumption during ontogenesis of common carp (*Cyprinus carpio*). Can. J. Fish. Aquat. Sci. 39:1095–1105.

Kaushik, S. J., T. Doudet, F. Medale, P. Aguirre, and D. Blanc. 1995. Protein and energy needs for maintenance and growth of Nile tilapia (*Oreochromis niloticus*) using different criteria. J. Appl. Ichthyol. 11:290–296.

Kebreab, E., J. France, R. E. Agnew, T. Yan, M. S. Dhanoa, J. Dijkstra, D. E. Beever, and C. K. Reynolds. 2003. Alternatives to linear analysis of energy balance data from lactating dairy cows. J. Dairy Sci. 86:2904–2913.

Kentouri, M., P. Divanach, I. Geurden, and M. Anthouard. 1995. Evidence of adaptative behavior in gilthead sea bream (*Sparus aurata* L.) in relation to diet composition in a self-feeding condition. Ictyophysiol. Acta 18:125–143.

Kielanowski, J. 1965. Estimates of the energy cost of protein deposition in growing animals. Pp. 13–20 in Proceedings of the 3rd Symposium on Energy Metabolism, K. L. Blaxter, ed. London, UK: Academic Press.

Kitchell, J. F., D. J. Stewart, and D. Weininger. 1977. Applications of bioenergetics model to yellow perch (*Perca flavescens*) and walleye (*Stizostedion vitreum vitreum*). J. Fish. Res. Board Can. 34:1922–1935.

Kleiber, M. 1975. The Fire of Life. New York: Robert E. Kleiber Publishing Co.

Koskela, J., M. Jobling, and R. Savolainen. 1998. Influence of dietary fat level on feed intake, growth and fat deposition in the whitefish *Coregonus lavaretus*. Aquacult. Int. 6:95–102.

Krohn, M. M., and D. Boisclair. 1994. Use of a stereo-video system to estimate the energy expenditure of free swimming fish. Can. J. Fish. Aquat. Sci. 51:1119–1127.

Kyriazakis, I. 1999. A Quantitative Biology of the Pig. Oxon, UK: CABI Publishing.

Langar, H., and J. Guillaume. 1994. Effect of feeding pattern and dietary protein source on protein synthesis in European sea bass (*Dicentrarchus labrax*). Comp. Biochem. Phys. A 108A:461–466.

Le Dividich, J., M. Vermorel, J. Noblet, J. C. Bouvier, and A. Aumaitre. 1980. Effects of environmental temperature on heat production, energy retention, protein and fat gain in early weaned piglets. Brit. J. Nutr. 44:313–323.

Lee, D. J., and G. B. Putnam. 1972. The response of rainbow trout to varying protein/energy ratios in a test diet. J. Nutr. 103:916–922.

Leenhouwers, J. I., W. F. Pellikaan, H. F. A. Huizing, R. O. M. Coolen, J. A. J. Verreth, and J. W. Schrama. 2008. Fermentability of carbohydrates in an *in vitro* batch culture method using inocula from Nile tilapia (*Oreochromis niloticus*) and European sea bass (*Dicentrarchus labrax*). Aquacult. Nutr. 14:523–532.

LeGrow, S. M., and F. W. H. Beamish. 1986. Influence of dietary protein and lipid on apparent heat increment of rainbow trout, *Salmo gairdneri*. Can. J. Fish. Aquat. Sci. 43:19–25.

Liu, J., Y. Cui, and J. Liu. 2000. Resting metabolism and specific dynamic action in the mandarin fish *Siniperca chuatsi* and the Chinese snakehead *Channa argus*. Comp. Biochem. Physiol. 127A, 131–138.

Lupatsch, I., G. W. and Kissil. 1998. Predicting aquaculture waste from gilthead seabream (*Sparus aurata*) culture using a nutritional approach. Aquat. Living Resour. 11:265–268.

Lupatsch, I., and G. W. Kissil. 2005. Feed formulations based on energy and protein demands in white grouper (*Epinephelus aeneus*). Aquaculture 248:83–95.

Lupatsch, I., G. W. Kissil, D. Sklan, and E. Pfeffer. 1998. Energy and protein requirements for maintenance and growth in gilthead seabream (*Sparus aurata* L.). Aquacult. Nutr. 4:165–173.

Lupatsch, I., G. W. Kissil, and D. Sklan. 2001a. Optimization of feeding regimes for European sea bass *Dicentrarchus labrax*: A factorial approach. Aquaculture 202:289–302.

Lupatsch, I., G. W. Kissil, D. Sklan, and E. Pfeffer. 2001b. Effects of varying dietary protein and energy supply on growth, body composition and protein utilization in gilthead seabream (*Sparus aurata* L.). Aquacult. Nutr. 7:71–80.

Lupatsch, I., G. W. Kissil, and D. Sklan. 2003. Comparison of energy and protein efficiency among three fish species, gilthead sea bream (*Sparus aurata*), European sea bass (*Dicentrarchus labrax*) and white grouper (*Epinephelus aeneus*): Energy expenditure for protein and lipid deposition. Aquaculture 225:175–189.

Lupatsch, I., L. Cuthbertson, S. Davies, and R. J. Shields. 2008. Studies on energy and protein requirements to improve feed management of the Pacific white shrimp, Litopenaeus vannamei. Pp. 281–295 in Avances en Nutricion Acuicola IX, L. E. Cruz Suarez, D. Ricque Marie, M. Tapia Salazar, M. G. Nieto Lopez, D. A. Villarreal Cavazos, J. Pablo Lazo, and M. Teresa Viana, eds. IX Symposio Internacional de Nutricion Acuicola, Universidad Autonoma de Nuevo Leon, Monterrey, Nuevo Leon, Mexico, November 24–27, 2008.

Lupatsch, I., G. A. Santos, J. W. Schrama, and J. A. J. Verreth. 2010a. Effect of stocking density and feeding level on energy expenditure and stress responsiveness in European sea bass (*Dicentrarchus labrax*). Aquaculture 298:245–250.

Lupatsch, I., R. Deshev, and I. Magen. 2010b. Energy and protein demands for optimal egg production including maintenance requirements of female tilapia *Oreochromis niloticus*. Aquac. Res. 41:763–769.

Machiels, M. A. M., and A. M. Henken. 1986. A dynamic simulation model for growth of the African catfish, *Clarias gariepinus* (Burchell 1822) I. Effect of feeding level on growth and energy metabolism. Aquaculture 56:29–52.

Maldonado, C., G. Cuzon, E. Guzman, R. Brito, L. Soto, L. Arena, and G. Gaxiola. 2009. Effect of an herbivorous diet on energy balance of *Litopenaeus vannamei* at selected ontogenetic stages. Aquaculture 296:123–128.

Martin, A. K., and K. L. Blaxter. 1965. The energy cost of urea synthesis in sheep. Pp. 83–91 in Energy Metabolism, K. L. Blaxter, ed. New York: Academic Press.

Mayes, P. A. 2000. Bioenergetics and the metabolism of carbohydrates and lipids. Pp. 123–298 in Harper's Biochemistry, R. K. Murray, D. K. Graner, P. A. Mayes, and V. W. Rodwell, eds. Stamford, CT: Appleton and Lange.

McGoogan, B. B., and D. M. Gatlin. 1998. Metabolic requirements of red drum, *Sciaenops ocellatus*, for protein and energy based on weight gain and body composition. J. Nutr. 128:123–129.

McNamara, J. P. 2004. Research, improvement and application of mechanistic, biochemical, dynamic models of metabolism in lactating dairy cattle. Anim. Feed Sci. Tech. 112:155–176.

Médale, F., and J. Guillaume. 1999. Nutrition énergétique. Pp. 87–111 in Nutrition et alimentation des poissons et des crustacés, Collections Du Labo au Terrain, J. Guillaume, S. J. Kaushik, P. Bergot, and R. Métailler, eds. Paris, France: Éditions INRA.

Médale, F., T. Boujard, F. Vallée, D. Blanc, M. Mambrini, A. Roem, and S. J. Kaushik. 1998. Voluntary feed intake, nitrogen and phosphorus losses in rainbow trout (*Oncorhynchus mykiss*) fed increasing dietary levels of soy protein concentrate. Aquat. Living Resour. 11:239–246.

Medland, T. E., and F. W. H. Beamish. 1985. The influence of diet and fish density on apparent heat increment in rainbow trout, *Salmo gairdneri*. Aquaculture 4T:1–10.

Meyer-Burgdorff, K.-H., M. F. Osman, and K. D. Günther. 1989. Energy metabolism in *Oreochromis niloticus*. Aquaculture 79:283–291.

Milligan, L. P., and M. Summers. 1986. The biological basis of maintenance and its relevance to assessing responses to nutrients. P. Nutr. Soc. 45:185–193.

NRC (National Research Council). 1981. Nutritional Energetics of Domestic Animals and Glossary of Energy Terms. Washington, DC: National Academy Press.

NRC. 1994. Nutrient Requirements of Poultry. Washington, DC: National Academy Press.

NRC. 1998. Nutrient Requirements of Swine. Washington, DC: National Academy Press.

NRC. 2000. Nutrient Requirements of Beef Cattle. Washington, DC: National Academy Press.

NRC. 2001. Nutrient Requirements of Dairy Cattle. Washington, DC: National Academy Press.

Noblet, J., H. Fortune, X. S. Shi, and S. Dubois. 1994. Prediction of net energy value of feeds for growing pigs. J. Anim. Sci. 72:344–354.

Ocampo. V. L. 1998. Metabolizable energy and growth efficiency of the shrimp under the effect of environmental variations. Pp. 44–60 in Simposium Internacional de Nutricion Acuicola, La Paz, Baja California Sur, Mexico, November 15–18, 1998.

Ohta, M., and T. Watanabe. 1998. Effect of feed preparation methods on dietary energy budgets in carp and rainbow trout. Fish. Sci. 64:99–114.

Oldham, J. D., G. C. Emmans, and I. Kyriazakis. 1997. Limits and limitations to nitrogen use in farm animals. P. Nutr. Soc. 56:525–534.

Paspatis, M., and T. Boujard. 1996. A comparative study of automatic feeding and self-feeding in juvenile Atlantic salmon (*Salmo salar*) fed diets of different energy levels. Aquaculture 145:245–257.

Patton, A. R. 1965. Biochemical Energetics and Kinetics. Philadelphia, PA: W.B. Saunders Co.

Peres, H., and A. Oliva-Teles. 1999. Effect of dietary lipid level on growth performance and feed utilization by European sea bass juveniles (*Dicentrarchus labrax*). Aquaculture 179:325–334.

Peres, H., and A. Oliva-Teles. 2000. Maintenance requirements and efficiency of protein and energy utilization in European sea bass (*Dicentrarchus labrax*) (abstract). Second Workshop on Energy Metabolism of Fish; Efficiency of Energy Utilization: Methodological Considerations and Effects of Dietary, Biological and Environmental Factors, Miyazaki, Japan, May 22, 2000.

Phillips, A. J. 1972. Calorie and energy requirement. Pp. 2–28 in Fish Nutrition, J. E. Halver, ed. New York: Academic Press.

Pirozzi, I., M. A. Booth, and G. L. Allan. 2010. Protein and energy utilization and the requirements in juvenile mulloway (*Argyrosomus japonicus*). Fish Physiol. Biochem. 36:109–121.

Read, G. H. L., and M. S. Caulton. 1980. Changes in mass and chemical composition during the moult cycle and ovarian development in immature and mature *Penaeus indicus* Milne Edwards. Comp. Biochem. Phys. A 66A:431–437.

Reeds, P. J. 1991. The energy cost of protein deposition. Pp. 473–479 in Energy Metabolism of Farm Animals, C. Wenk and M. Boessinger, eds. Proceedings of the 12th Symposium, EAAP Publication 58:473–479.

Reeds, P. J., M. F. Fuller, A. Cadenhead, G. E. Lobley, and J. D. McDonald. 1981. Effects of changes in the intakes of protein and non-protein energy on whole-body protein turnover in growing pigs, nitrogen retention and body protein synthesis. Brit. J. Nutr. 45:539–546.

Reeds, P. J., M. F. Fuller, and B. A. Nicholson. 1985. Metabolic basis of energy expenditure with particular reference to protein. Pp. 46–67 in Substrate and Energy Metabolism in Man, J. S. Garrow and D.Halliday, eds. London, UK: John Libbey and Son.

Reynolds, C. K. 1999. Measurement of energy metabolism. Pp. 87–107 in Feeding Systems and Feed Evaluation Models, M. K. Theodorou and J. France, eds. Wallingford, UK: CABI Publishing.

Rodehutscord, M., and E. Pfeffer. 1999. Maintenance requirement for digestible energy and efficiency of utilisation of digestible energy for retention in rainbow trout, *Oncorhynchus mykiss*. Aquaculture 179:95–107.

Ross, L. G., R. W. McKinney, S. K. Cardwell, J. G. Fullarton, S. E. J. Roberts, and B. Ross. 1992. The effects of dietary protein content, lipid and ratio level on oxygen consumption and specific dynamic action in *Oreochromis niloticus* L. Comp. Biochem. Phys. A 103A:573–578.

Rubner, M. 1902. Die Gesetze des Energieverbrauchs bei der Erndhrung. Leipzig, Germany: F. Deuticke.

Sanchez, F. J., R. M. Cal, and C. Guisande. 1993. Oxygen consumption of turbot *Scophthalmus maximus* juveniles between 15°C and 22°C, under experimental conditions. P. 459 in Proceedings of the World Aquaculture Society Meeting, Torremolinos, Spain, May 26–28, 1993.

Schwartz, F. J., and M. Kirchgessner. 1995. Effects of different diets and levels of feeding on retention and efficiency of utilization of energy and protein by carp. J. Appl. Ichthyol. 11:363–366.

Shearer, K. D. 1994. Factors affecting the proximate composition of cultured fishes with emphasis on salmonids. Aquaculture 119:63–88.

Shul'man, G. E. 1974. Life Cycles of Fish, Physiology and Biochemistry. Jerusalem, Israel: Halsted Press.

Slinker, B. K., and S. A. Glantz. 1985. Multiple regression for physiological data analysis: The problem of multicollinearity. Am. J. Physiol. Reg. I 249(1):1–12.

Smith, R. R. 1971. A method for measuring digestibility and metabolizable energy of feeds. Prog. Fish Cult. 33:132–134.

Smith, R. R. 1989. Nutritional energetics. Pp. 2–29 in Fish Nutrition, J. E. Halver, ed. San Diego, CA: Academic Press.

Smith, R. R., G. L. Rumsey, and M. L. Scott. 1978. Heat increment associated with dietary protein, fat and carbohydrate and complete diets in salmonids: Comparative energetic efficiency. J. Nutr. 108:1025–1032.

Storebakken, T., S. S. O. Hung, C. C. Calvert, and E. M. Plisetskaya. 1991. Nutrient partitioning in rainbow trout at different feeding rates. Aquaculture 96:191–203.

Tandler, A., and F. W. H. Beamish. 1979. Mechanical and biochemical components of apparent specific dynamic action in largemouth bass, *Micropterm salmoides Laceepede*. J. Fish Biol. 14:343–350

Tandler, A., M. Harel, W. M. Koven, and S. Kolkovski. 1995. Broodstock and larvae nutrition in gilthead seabream (*Sparus aurata*). New findings on its mode involvement in improving growth, survival and swimbladder inflation. Isr. J. Aquacult. Bamidgeh 47:95–111.

Tchung, B. 1995. Effet de la Nourriture sur le Métabolisme de la Crevette *Penaeus stylirostris*. Mémoire de Diplôme d'Études Approfondies, Université Française du Pacifique, Polynésie Française.

Thorpe, J. E. 1992. Reproductive strategies in Atlantic salmon, *Salmo salar* L. Aquacult. Fish. Manage. 25(1):77–87.

Tytler, P., and P. Calow. 1985. Fish Energetics: New Perspectives. London, UK: Croom Helm.

van Dam, A. A., and F. W. T. P. De Vries. 1995. Parameterization and calibration of a model to simulate effects of feeding level and feed composition on growth of *Oreochromis niloticus* (L.) and *Oncorhynchus mykiss* (Walbaum). Aquac. Res. 26:415–425.

van der Honing, Y. 1998. Perspectives of future feed information based on energy and nutrient availability. Arch. Tierernahr. 51:135–153.

Van Milgen, J. 2002. Modeling biochemical aspects of energy metabolism in mammals. J. Nutr. 132(10):3195–3202.

Van Milgen, J., and J. Noblet. 1999. Energy partitioning in growing pigs: The use of a multivariate model as an alternative for the factorial analysis. J. Anim. Sci. 77:2154–2162.

Villarreal, H. 1990. Effect of temperature on oxygen consumption and heart rate of the Australian crayfish *Cherax tenuimanus* (Smith). Comp. Biochem. Phys. 95A:189–193.

Warukamkul, P., V. Viyakarn, C. Nitithamyong, and S. Piyatiratitivorakul. 2000. Effects of salinity and protein levels on energy budget of juvenile black tiger prawn, *Penaeus monodon* (abstract O56). IX International Symposium on Nutrition and Feeding of Fish, May 21–25, 2000, Miyazaki, Japan.

Watanabe, K., H. Aoki, Y. Yamagata, V. Kiron, S. Satoh, and T. Watanabe. 2000. Energy and protein requirements of yellowtail during winter season. Fish. Sci. 66:521–527.

Watt, P. W., Y. Lindsay, C. M. Scrimgeour, P. A. F. Chien, J. N. A. Gibson, D. J. Taylor, and M. J. Rennie. 1991. Isolation of aminoacyl-tRNA and its labeling with stable-isotope tracers: Use in studies of human tissue protein synthesis. P. Natl. Acad. Sci. USA 88:5892–5896.

Wheatley, D. N., A. Glenesk, and M. S. Inglis. 1988. Protein synthesis and mitosis. 1. Influence of anti-tubulins on protein synthetic activity in synchronous M-phase and asynchronous populations of HeLa S-3 cells. Cytobios 55(220):41–50.

Winberg, G. G. 1956. Rate of Metabolism and Food Requirements of Fishes. Fisheries Resources Board of Canada Translation Series No. 194. Nanaimo, Canada: Fisheries Resources Board of Canada.

Yamamoto, T., T. Shima, T. Unuma, M. Shiraishi, T. Akiyama, and M. Tabata. 2000. Voluntary intake of diets with varying digestible energy contents and energy sources, by juvenile rainbow trout *Oncorhynchus mykiss*, using self feeders. Fish. Sci. 66:528–534.

Yamamoto, T., T. Shima, H. Furuita, and N. Suzuki. 2002. Influence of dietary fat level and whole-body adiposity on voluntary energy intake by juvenile rainbow trout *Oncorhynchus mykiss* (Walbaum) under self-feeding conditions. Aquac. Res. 33:715–723.

Yamamoto, T., T. Sugita, and H. Furuita. 2005. Essential amino acid supplementation to fish meal-based diets with low protein to energy ratios improves the protein utilization in juvenile rainbow trout *Oncorhynchus mykiss*. Aquaculture 246(1–4):379–391.

Yokote, M. 1970. Sekoke disease, spontaneous diabetes in carp found in fish farms—II. Some metabolic aspects. B. Jpn. Soc. Sci. Fish. 36:1214–1218.

Yuan, Y.-C., H. J. Yang, S. Y. Gong, Z. Luo, H. W. Yuan, and X. K. Chen. 2009. Effects of feeding levels on growth performance, feed utilization, body composition and apparent digestibility coefficients of nutrients for juvenile Chinese sucker, *Myxocyprinus asiaticus*. Aquac. Res. 1–13; doi:10.1111/j.1365-2109.2009.02387.x.

Zhou, Z., S. Xie, W. Lei, X. Zhu, and Y. Yang. 2005. A bioenergetic model to estimate feed requirement of gibel carp, *Carassius auratus gibelio*. Aquaculture 248:287–297.

5

Proteins and Amino Acids

Proteins and amino acids are critical molecules because of the role they play in the structure and metabolism of all living organisms. Fish and shrimp cannot synthesize all amino acids and must acquire several in their diet, through the consumption of protein or mixtures of amino acids.

François Magendie (1783–1855), in his textbook *Précis Élémentaire de Physiologie* (1817), was the first to report the importance of the type of nitrogenous compounds in the diet of animals. A century later, in 1914, L. B. Mendel and T. B. Osborne studied the protein requirements of rats and demonstrated nutritional requirements for individual amino acids (Carpenter, 2003). Studies in the early 1940s by W. Rose of the University of Illinois found that 10 amino acids were essential for rats. The removal of any one of these essential amino acids from the diet of growing rats led to profound nutritive failure, accompanied by a rapid decline in weight, loss of appetite, and eventually death (Carpenter, 2003). The first definitive studies on protein and amino acid nutrition of fish were conducted by Halver and collaborators in the late 1950s and early 1960s with Chinook salmon (*Oncorhynchus tshawytscha*). Since this seminal work, hundreds of studies involving a very large number of fish and shrimp species have been carried out, and the body of literature is continuously expanding.

This chapter provides an overview of the nutritional biochemistry of proteins and amino acids and discusses the roles and basis of essentiality of amino acids and some of the factors affecting efficiency of utilization of these nutrients. The results of published studies on quantitative amino acid requirements of a large number of commercially important fish and shrimp species are compiled, and methodological approaches, strategies, and challenges associated with defining and meeting essential amino acid requirements of fish and shrimp are also discussed.

PROTEINS AND AMINO ACIDS: BIOCHEMISTRY, ROLES, AND OVERVIEW OF METABOLISM

Proteins and their building blocks, amino acids, are organic compounds that are essential components of all living organisms. Amino acids can link together by a covalent peptide bond between the α-carboxyl end of one amino acid and the α-amino end of the other (Brody, 1999). Any number of amino acids can be joined by successive peptide linkages, forming a peptide chain. An oligomer consisting of two amino acids is called a dipeptide. Peptides of 2 to around 20 amino acid residues are termed polypeptides. Protein typically contains about 300 amino acids. The polypeptide chains that constitute proteins are linear and contain no branching (Brody, 1999). Amino acids can be linked in varying sequences to form a vast variety of proteins. Proteins are defined by their unique sequence of amino acid residues, which is encoded in the genetic material of the organism. The amino acid sequence is the primary structure of protein. Peptide chains are cross-linked by disulfide bridges, hydrogen bonds, and van der Waals forces that result in the formation of the secondary, tertiary, and quaternary structures of proteins (Buxbaum, 2007).

Proteins have numerous structural and metabolic functions. Proteins, such as actin and tubulin, confer stiffness and rigidity to otherwise fluid biological components (Buxbaum, 2007). Collagen and elastin are critical components of connective tissue, such as cartilage. Other proteins, such as myosin, also have a mechanical function and are capable of generating mechanical forces, such as those exerted by contracting muscles (Buxbaum, 2007). Many proteins are enzymes that catalyze biochemical reactions or transporters that allow the entry and exit of molecules through cells. Some proteins are important in cell signaling, immune responses, cell adhesion, and functioning of the cell cycle (Buxbaum, 2007).

Protein is thus an essential component for every type of cell in the body, including muscles, bones, organs, tendons, and ligaments. Body tissues are continuously being formed and broken down. In growing animals, protein synthesis exceeds degradation, and the balance between these two processes results in protein deposition or accretion (Millward, 1989). Protein deposition appears to be the main determinant of live weight (biomass) gain in fish (Dumas et al., 2007). The close association between live weight gain and protein mass (Figure 5-1) is due to the close association of water with protein (Figure 5-2). Protein hydration is important for their structure and activity, and proteins help maintain the aqueous intracellular milieu in a "gel" state (Chaplin, 2006). Conversely, lipid deposition does not always appear to contribute substantially to live weight gain because a large proportion of triglycerides is stored in tissues by substituting for water (Dumas et al., 2007). However, other experimental evidence suggests that lipid deposits can significantly contribute to live weight gain of fish, and differences exist among species, strains, life stages, and animals with different nutritional history.

Protein deposition in organisms is dictated by specific templates determined by genetic and epigenetic "codes" of the animal and specific targets determined by endogenous (genetic, life stage) and exogenous (environment, diet) factors. Thousands of different proteins are produced by biological organisms, and each one of these different proteins has a specific structure, function and/or a unique amino acid sequence (Buxbaum, 2007; Finn and Fyhn, 2010)

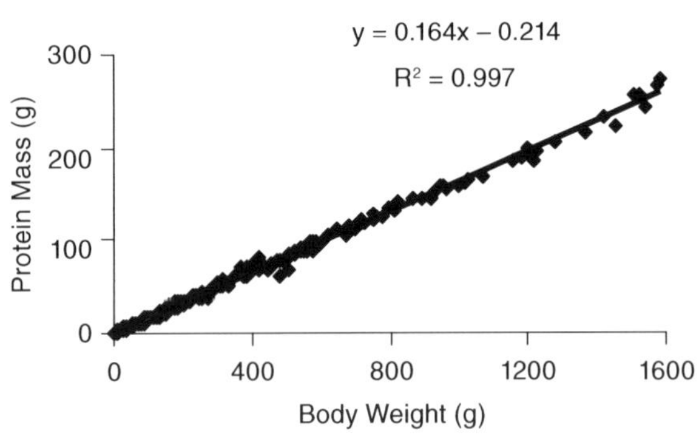

FIGURE 5-1 Relationship between protein mass and live weight of rainbow trout (*Oncorhynchus mykiss*) (Dumas et al., 2007).

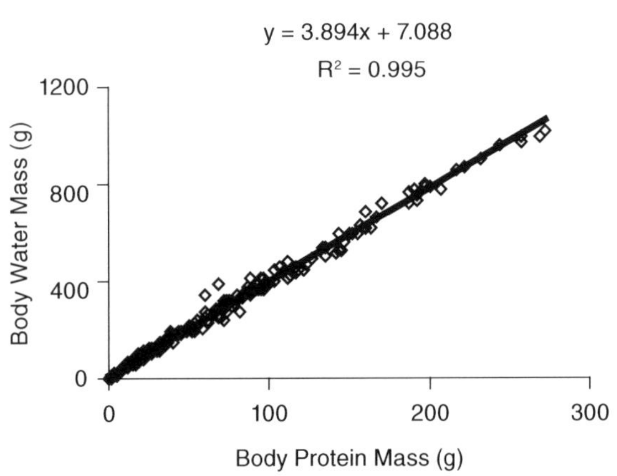

FIGURE 5-2 Relationship between water mass and protein mass of rainbow trout (*Oncorhynchus mykiss*) (Dumas et al., 2007).

(Table 5-1). Different tissues contain different proteins or the same proteins in different proportions. For example, in muscle cells, actin represents about 20% of total protein content, whereas in nonmuscle cells it represents only 5–10%. The amino acid composition of different body proteins also differ significantly. Collagen, a main component of connective tissues, contains only about 3% lysine, whereas myosin and tropomyosin, major components of muscle tissues, contain more than 14% lysine (Pellett and Young, 1984). Interestingly, there are only limited differences in the whole-body amino acid compositions among fish species (Wilson and Cowey, 1985; Kaushik, 1998; Kaushik and Seiliez, 2010). Some differences appear to exist between the whole-body amino acid profiles of shrimp and fish, but overall these differences appear to be minor (Table 5-2). Studies have also shown that the whole-body amino acid composition is minimally affected by body size of fish, at least in juvenile (immature) fish (Kaushik, 1998; Portz and Cyrino, 2003). The relative proportions of the different body components, and proteins these compartments contain, are apparently relatively constant in fish over a wide range of live weight (Dumas et al., 2007).

Amino acids are molecules containing both amine and carboxyl functional groups, with the general formula $H_2NCHRCOOH$, where R is a side chain. The most naturally abundant and metabolically important amino acids are the L-α-amino acids, in which the amino and carboxylate groups are attached to the same carbon atom, called the α-carbon from organic chemistry nomenclature (Brody, 1999). The various α-amino acids differ in which R group, often referred to as the "side-chain," is attached to the α-carbon. The nature and size of the R group can vary from a single hydrogen atom in glycine through a methyl group in alanine to a large heterocyclic group in tryptophan.

The properties of amino acids result from variations in the structures of different R groups, which influence the size, shape, electrical charge, and other characteristics. Amino acids can exist as D- or L-isomers or mixtures of the products. Amino acids are generally found in nature in the L configuration, which are, with a few exceptions, the most biologically active forms. D-amino acids occur in small quantities in certain molecules synthesized by invertebrates and bacteria. Heat-processed ingredients can also contain

TABLE 5-1 Amino Acid Composition of Different Body Proteins of Animals

Protein	Lysine (%)	Leucine (%)	Methionine (%)	Threonine (%)
Collagen	3.0	2.8	0.8	1.8
Myosin	14.6	10.8	3.3	5.4
Actin	7.0	8.2	4.2	7.7
Tropomyosin	18.4	12.1	2.5	3.1
Elastin	0.4	7.2	Traces	0.9

SOURCE: Adapted from Pellett and Young (1984).

TABLE 5-2 Amino Acid Composition (g/16 g N) of Various Fish and Shrimp Species

Amino Acid	Rainbow Trout (*Oncorhynchus mykiss*)	Atlantic Salmon (*Salmo salar*)	Channel Catfish (*Ictalurus punctatus*)	Largemouth Bass (*Micropterus salmoides*)	European Sea Bass (*Dicentrarchus labrax*)	Gilthead Sea Bream (*Sparus auratus*)	Turbot (*Scophthalmus maximus*)	Penaeid Shrimp
Alanine	6.6	6.5	6.3	6.0	6.8	6.8	7.3	5.6
Arginine	6.4	6.6	6.7	8.5	7.5	8.8	7.7	7.4
Asparate	9.9	9.9	9.7	11.8	9.5	9.4	10.3	8.8
Cysteine	0.8	1.0	0.9	0.8	1.0	1.0	1.1	0.8
Glutamate	14.2	14.3	14.4	13.3	15.5	15.1	16.5	16.2
Glycine	7.8	7.4	8.1	7.8	8.1	7.9	9.7	9.0
Histidine	3.0	3.0	2.2	2.1	2.6	2.7	2.5	2.5
Isoleucine	4.3	4.4	4.3	4.0	4.3	4.3	4.3	3.6
Leucine	7.6	7.7	7.4	8.0	7.1	7.3	7.5	6.5
Lysine	8.5	9.3	8.5	8.1	7.9	8.1	8.1	7.8
Methionine	2.9	1.8	2.9	2.6	2.7	3.0	3.4	2.3
Phenylalanine	4.4	4.4	4.1	4.0	4.3	4.7	4.5	3.6
Proline	4.9	4.6	6.0	6.0	5.3	5.3	5.5	8.0
Serine	4.7	4.6	4.9	4.2	4.5	4.5	5.2	3.6
Threonine	4.8	5.0	4.4	4.4	4.4	4.6	4.6	3.8
Tryptophan	1.0	0.9	0.8	0.9	N/A	N/A	N/A	N/A
Tyrosine	3.4	3.5	3.3	2.8	3.9	4.0	4.1	7.5
Valine	5.1	5.1	5.2	4.6	4.7	4.8	4.7	5.1

NOTE: N/A: not available, not reported.
SOURCES: Wilson and Cowey (1985), Kaushik (1998), Portz and Cyrino (2003), Alam (2004), Sara (2007).

D-amino acids because of heat-induced racemization, and chemical synthesis of amino acids result in the production of racemic mixtures of L and D-isomers.

Amino acids are generally represented by three-letter or single-letter abbreviations (Table 5-3) (IUPAC-IUB-JCBN, 1984; Buxbaum, 2007). The three-letter abbreviations are commonly used in animal nutrition, whereas single-letter abbreviations are commonly used in molecular biology and bioinformatics. The convergence of nutritional sciences with molecular genetics and the increasing use of molecular biology techniques and bioinformatics in nutrition studies will likely lead to greater use of single-letter abbreviations in the not too distant future in the animal nutrition literature.

TABLE 5-3 Essential and Nonessential Amino Acids

Essential	Abbreviations	
Arginine	Arg	R
Histidine	His	H
Isoleucine	Ile	I
Leucine	Leu	L
Lysine	Lys	K
Methionine	Met	M
Phenylalanine	Phe	F
Threonine	Thr	T
Tryptophan	Trp	W
Valine	Val	V
Nonessential		
Alanine	Ala	A
Asparagine	Asn	N
Aspartate	Asp	D
Cysteine[a]	Cys	C
Glycine	Gly	G
Glutamate	Glu	E
Glutamine	Gln	Q
Proline	Pro	P
Serine	Ser	S
Tyrosine[a]	Tyr	Y

[a]Conditionally essential.

Twenty primary amino acids are used by cells in protein biosynthesis (Table 5-3). Aside from the primary amino acids found in proteins, there are a vast number of other amino acids, formed by posttranslational modification. These modifications are often essential for the function or regulation of a protein; for example, the carboxylation of glutamate allows for better binding of calcium cations, and the hydroxylation of proline is critical for maintaining connective tissues.

Beside their role as the building blocks of protein, amino acids also have a variety of roles in metabolism. Amino acids are important in many other biological molecules, such as forming parts of coenzymes, precursors for the biosynthesis of structural molecules (e.g., heme, chitin, and purine bases), metabolic intermediates (e.g., acetate and pyruvate), and neurotransmitters, hormones, biogenic amines, or numerous other molecules (e.g., serotonin, gamma-aminobutyric acid, melamine, nitric oxide, and histamine) important in the response of the organism to different stimuli. However, the conversion of amino acids to these specific compounds is considered to be quantitatively minor compared to that used for protein synthesis or catabolized by organisms (Cowey and Walton, 1989; Moughan, 1999).

Most microorganisms and plants can biosynthesize all 20 primary amino acids, while animals must obtain some of the amino acids from their diet. The amino acids that an organism cannot synthesize on its own (or is incapable of synthesizing sufficient amounts) are referred to as "essential amino acids" (EAAs) (Table 5-3). In contrast, the nonessential amino acids (NEAAs) can be synthesized from precursors, for example by addition of an amino group to a tricarboxylic acid (TCA)-cycle intermediate, such as α-ketoglutarate or oxaloacetate (Cowey and Walton, 1989; Zubay, 1993).

The essentiality of various amino acids for fish and shrimp has been determined either by feeding trials involving the successive deletion of each amino acid in the diet or by isotopic-labeling studies (Wilson, 1989). In isotopic-labeling studies, a radio-labeled substrate (e.g., ^{14}C glucose) is injected and the radioisotope label is incorporated into those amino acids that the animal is able to synthesize. All amino acids that have not incorporated ^{14}C are considered to be essential (Coloso and Cruz, 1980; Kanazawa and Teshima, 1981).

Distinction between EAAs and NEAAs also can be made in trials where the individual amino acids are deleted from the diet and growth performance analyzed (Wilson, 1989; Cowey, 1994). The deletion of an EAA would significantly reduce animal growth performance, while a NEAA would not affect growth, suggesting that it could be synthesized by the animal. Nose et al. (1974) used this approach to test the essentiality of 18 amino acids for common carp. They identified that 10 of these amino acids resulted in significant reduction in growth performance after 4 weeks.

It is clear from the available evidence published to date that all fish and shrimp require the same 10 EAAs (Table 5-2) required by most other animals (Ketola, 1982; Wilson, 1989; NRC, 1993). These include arginine, histidine, isoleucine, leucine, lysine, methionine, phenylalanine, threonine, tryptophan, and valine. Tyrosine is synthesized from phenylalanine, and cysteine is synthesized from methionine. Therefore, these two amino acids are considered semiessential or conditionally essential. These amino acids are frequently included in estimates of requirements (e.g., total sulfur amino acid requirement, or TSAA). There is growing evidence that some amino acids or related compounds, such as taurine, may be essential or conditionally essential for some, but not all, fish species or for certain life stages in certain species, notably the larval stage of marine fish species.

Along with these 10 EAAs, fish have a nonspecific requirement for a source of amino groups (also known as nonspecific nitrogen) for the synthesis of NEAAs. Most NEAAs, except tyrosine, can be synthesized by simple pathways leading from one of four common metabolic intermediates: pyruvate, oxaloacetate, α-ketoglutarate, or 3-phosphoglycerate (Zubay, 1993). Alanine, asparagine, aspartate, glutamate, and glutamine are synthesized by simple one- or two-step amination reactions from their organic precursors, pyruvate, oxaloacetate, or α-ketoglutarate. Cysteine and glycine are derived from serine, which is in turn synthesized from 3-phosphoglycerate.

An Overview of the Dynamic of Amino Acid Utilization

Ingested proteins are broken down during digestion through hydrolysis to free amino acids, dipeptides, and tripeptides by digestive enzymes secreted into the gastrointestinal tract. These products are absorbed by the mucosal cells where intracellular digestion of small peptides occurs; thus only amino acids appear to be released into the portal vein as products of protein digestion (Murai et al., 1987). Some evidence has shown that small amounts of certain whole proteins may be absorbed through the wall of the gastrointestinal tract (McLean et al., 1999). The amounts involved are not quantitatively significant but this process may be a physiologically important mechanism, possibly for modulation of the immune system through antigen sampling (McLean et al., 1999).

Amino acids supplied by the digestion of dietary protein (exogenous source) or breakdown of body protein (endogenous source) enter the free amino acid pool, also called the metabolic pool (Cowey and Walton, 1989; Kaushik and Seiliez, 2010). From this pool, amino acids can be used for body protein synthesis and as precursors for other substances. Utilization of the amino acids is a function of the metabolic demands of the organism, and the efficiency with which other nutrients absorbed by the animal can be utilized to meet those demands. The metabolism of amino acids is complex and highly integrated with continuous flux within and between cells (Wilson and Cowey, 1985; Cowey and Walton, 1989; Kaushik and Seiliez, 2010). Factors, such as nutrient transport across cell membranes, rate of blood flow, organ uptake, and rates of enzyme activity associated with different biochemical pathways, all interact to control metabolism of amino acids, and they are influenced by complex and sensitive hormonal and neural systems (Cowey and Walton, 1989).

Many factors cause amino acid oxidation. However, a common feature is the imbalance between amino acid supply and amino acid utilization for protein synthesis (Weijs, 1993). The principal endproducts of amino acid catabolism are ammonia, carbon dioxide, and bicarbonate. Ammonia is highly toxic and in order to prevent toxicity, higher vertebrates convert ammonia to urea for excretion in the urine. Fish and crustaceans have an extremely efficient transfer mechanism for ammonia across the gills; thus, they have no need to expend energy converting ammonia into urea (Cowey and Walton, 1989). The great majority of fish and shrimp excrete more than 80% of their nitrogenous wastes in the form of ammonia (Kaushik and Cowey, 1991; Mambrini and Guillaume, 1999). Gills are the main excretory organs, accounting for more than 75–90% of total nitrogen excretion (Cowey and Walton, 1989; Kaushik and Cowey, 1991).

The breakdown of amino acids generally occurs in two steps. The first is generally deamination and comprises the removal of the amino group, which is either converted to ammonia or transferred to become the amino group of a glutamic acid molecule (Cowey and Walton, 1989; Zubay, 1993). The second stage is the conversion of the carbon skeletons (the α-keto acids produced by deamination) to citric acid cycle intermediates (Cowey and Walton, 1989; Zubay, 1993). The carbon backbone of amino acids contain usuable (free) energy that can be harnessed by the TCA cycle or converted into fatty acids and/or glycogen by the animal for future use (Cowey and Walton, 1989). Different amino acids have different carbon skeletons, and their conversions to citric acid intermediates follow correspondingly diverse pathways, which can be grouped according to the intermediate at which they enter the citric acid cycle (Cowey and Walton, 1989). In this respect, amino acids fall into three categories: glucogenic, ketogenic, or glucogenic and ketogenic. Glucogenic amino acids are those that give rise to a net production of pyruvate or TCA-cycle intermediates, such as α-ketoglutarate or oxaloacetate, all of which are precursors of glucose via gluconeogenesis (Guillaume et al., 1999). All amino acids except lysine and leucine are at least partly glucogenic (Cowey and Walton, 1989). Lysine and leucine are the only amino acids that are solely ketogenic, giving rise only to acetyl-CoA or acetoacetyl-CoA, neither of which can bring about net glucose production, although they can be used for ketone body synthesis. Their pathways are similar in the final steps and resemble the steps in the β-oxidation of fatty acids (Cowey and Walton, 1989). Isoleucine, phenylalanine, threonine, tryptophan, and tyrosine give rise to both glucose and fatty acid precursors and are thus characterized as being both glucogenic and ketogenic (Cowey and Walton, 1989).

Evidence suggests that invertebrates and lower vertebrates, including teleost fish, have a better ability to utilize D-isomers of amino acids than do mammals (Cowey and Walton, 1989; Deng et al., 2010). Specific D-amino acid transaminase (D-AAO) may catalyse deamination of D-isomers to α-keto acids, which, in turn, can be reaminated to the natural L-form (Deng et al., 2010). Alternatively, racemases and epimerases can convert D-isomers to racemic mixtures. Significant activity of these enzymatic processes appear to be induced in invertebrates and fish tissues upon feeding D-isomers of amino acids (Deng et al., 2010).

Utilization of amino acids is affected by numerous factors, such as diet composition, chemical form of the amino acids supplied, and a number of biological factors (including species and life stage). In light of the complexity of this issue, the main biological processes that determine efficiency of EAA utilization for body protein deposition (PD) are frequently described (or estimated) in a factorial (or categorized) fashion. Factorial models have been extensively used in poultry and swine nutrition for about four decades and take various forms (D'Mello, 2003; Moughan, 2003).

In the most commonly used framework, EAA utilization is explicitly represented in terms of deposition as body protein, maintenance requirement, inevitable catabolism, preferential catabolism for energy use, and catabolism associated with intakes exceeding requirement (Moughan, 2003). This type of factorial framework has been used to predict EAA requirement of fish (e.g., Hauler and Carter, 2001b; Bodin et al., 2008) and, although it has a number of limitations when applied to fish (reviewed by Bureau and Encarnação, 2006), it provides a relatively straightforward framework of describing some of the determinants of amino acid catabolism and/or retention.

Amino Acid Deposition as Body Protein

A review of the literature indicates that for a large majority of fish and shrimp species fed high-quality diets, deposition of amino acids into body protein represents between 25 and 55% of total amino acids consumed. The deposition of protein is consequently a major determinant of amino acid utilization and requirements of fish and shrimp (Cowey and Walton, 1989).

Because there is a very strong association between live weight gain and PD (Shearer, 1994; Dumas et al., 2007), there also is a very close association between live weight gain and amino acid requirements in absolute terms (g/fish per day). The EAA needs for protein accretion correspond to the amino acid content of tissue protein gain (Kaushik and Seiliez, 2010).

Maintenance Amino Acid Requirements

Maintenance amino acid requirement is defined as the amount of dietary amino acid required to maintain the protein pool of the animal in equilibrium. Amino acid needs for maintenance include a certain amount for loss of endogenous gut proteins, mucins, and other secretions. Amino acids are also required as precursors for various metabolites, neurotransmitters, hormones, cofactors, and the like. Maintenance requirements of certain amino acids may account for a greater proportion of total requirement (maintenance + growth) because they can be involved in a wide variety of other metabolic reactions beside protein synthesis or be subject to significant endogenous losses (Rodehutscord et al., 1997; Nichols and Bertolo, 2008).

Estimate of maintenance requirement for amino acids are generally obtained by linear regression. Diets containing graded levels of protein and essential amino acids are fed, and protein (N) gain is monitored (Rodehutscord et al., 1997; Richard et al., 2010). The intake of amino acid resulting in no net protein (N) gain in the animal is assumed to be the maintenance requirement of the animal.

Rodehutscord et al. (1997) estimated the maintenance EAA requirement of rainbow trout (live weight = 50 g/fish) to be as follows (in mg per $kg^{0.75}$ per day): lysine, 4; tryptophan, 2; histidine, 2; valine, 5; leucine, 16, and isoleucine, 2. Bodin et al. (2009) obtained a significantly higher estimate of maintenance lysine requirement (24 mg per $kg^{0.75}$ per day) for rainbow trout. Rollin et al. (2006) and Abboudi et al. (2007) estimated the maintenance Thr requirement of Atlantic salmon fry (live weight = 1–2 g/fish) to be between 5 to 7 mg per $kg^{0.75}$ per day. Richard et al. (2010) estimated maintenance lysine and methionine requirement of *Penaeus monodon* (2 g live weight) to be about 40 and 20 mg per $kg^{0.75}$ per day.

Studies have suggested that the minimum maintenance requirement EAA pattern is quite different from the patterns of EAA in tissue proteins (Rodehutscord et al., 1997). Significant differences between species may also exist. Based on data from Fournier et al. (2003), Kaushik and Seiliez (2010) concluded that there are significant quantitative and qualitative differences in endogenous amino acid losses between rainbow trout and turbot.

Estimates of maintenance requirements should be viewed with a healthy degree of skepticism. The confidence intervals for estimates of maintenance derived from regression analyses are generally not provided but are likely very broad. Amino acid composition of protein gain can also be variable in fish fed diet deficient in EAA (Encarnação et al., 2004), which suggests that N gain may not be an adequate indicator of the maintenance requirement of individual amino acids. In addition, composition of the diet can affect catabolism of some EAA even when the diet is highly deficient in this amino acid (Encarnação et al., 2004, 2006). Finally, there may be significant differences in the nutrient partitioning of animals fed diets highly deficient in EAA compared to fish fed a nutritionally adequate diet. The relevance of "maintenance requirement" to fast-growing animals has often been questioned in the animal nutrition literature (Baldwin and Bywater, 1984; Bureau et al., 2002).

Overall, the maintenance amino acid requirement of domesticated fish and shrimp represents a small proportion (generally between 5 to 20%) of their total amino acid requirement (Rodehutscord et al., 1997; Abboudi et al., 2007, 2009; Richard et al., 2010). The relative contributions of maintenance to the total amino acid needs is likely greater in slow-growing than in fast-growing animals.

Inevitable Amino Acid Catabolism

Inevitable catabolism is defined as the degradation of amino acids through active catabolic pathways, which still occurs when energy supply is not limiting for protein synthesis (Moughan, 1995). This degradation appears to be an "inevitable" consequence of the presence of catabolic systems in cells that cannot be completely inactivated. As a result, a fraction of any absorbed amino acid is catabolized, even when the intake of this amino acid is below requirements for maximum protein deposition (Moughan, 1995; de Lange et al., 2001). This inevitable amino acid catabolism affects the marginal efficiency of utilization of amino acid for PD when amino acid intake is below requirements to achieve maximum protein deposition (PD_{max}). Expressed as a fraction of available amino acid intake, the rate of inevitable amino acid catabolism is assumed to be constant when amino acid intake is between about 70 and 100% of that required to achieve PD_{max} (de Lange et al., 2001; Moehn et al., 2004). Gahl et al. (1995), however, indicated a decrease in efficiency of utilization as limiting EAA intakes increased above maintenance (diminishing returns). Studies in which efficiencies of EAA utilization have been compared suggest that the efficiencies are different for each EAA (Fuller, 1994; Adeola, 1995; Gahl et al., 1996).

Based on a metaanalysis of data from different studies, Hauler and Carter (2001b) observed that lysine utilization for lysine gain (lysine retention/lysine intake) was constant despite differences in dietary lysine concentrations and lysine intakes by the fish. Based on this observation, Hauler and Carter (2001b) suggested that lysine utilization remains constant at marginal lysine intakes over different dietary formulations and life stages. Rodehutscord et al. (2000), Hauler and Carter (2001a), Encarnação et al. (2004, 2006), and Bodin et al. (2009) reported a maximal marginal efficiency of lysine retention of 71 and 78% for rainbow trout and Atlantic salmon, respectively. These observations suggest that inevitable catabolism represents 20 to 30% of digestible lysine consumed above maintenance in these fish species. Efficiency of lysine utilization appears to remain constant at marginal lysine intake over different dietary formulations and life stages in Atlantic salmon (Hauler and Carter, 2001a). There is a general lack of specific data for other amino acids and other fish species. However, based on available evidence, inevitable catabolism can in practice be estimated at about 20 to 40% of digestible amino acid consumed by animals above maintenance requirement.

The results of Encarnação et al. (2004, 2006) and Encarnação (2005), however, provided evidence that, at marginal levels of intake the efficiency of lysine utilization for PD is not constant, but is affected by dietary lysine concentration and intake of digestible energy (DE) supplied as lipids, but not DE supplied by other energy-yielding nutrients, such as amino acids (e.g., mixture of NEAA, leucine). Figure 5-3 illustrates the effect of DE content of the diet on efficiency of lysine utilization in rainbow trout (Encarnação et al., 2004). Studies also showed that the source of dietary amino acids (protein bound vs. free amino acids) (Tantikitti and March, 1995; Zarate and Lovell, 1997; Williams et al., 2001; El Haroun and Bureau, 2006) and the feed ingredient matrix (Nang Thu et al., 2007) can also affect efficiency of EAA utilization for PD. There is still considerable debate about the rate of inevitable catabolism and the factors affecting it, both in terrestrial animals (Moughan, 2003) and fish (Bureau and Encarnação, 2006). Compared to other nonruminant animals, fish seem to have a more "elastic" inevitable EAA catabolism, and the border between "inevitable" and "preferential" catabolism may be blurred in these animals (Bureau and Encarnação, 2006).

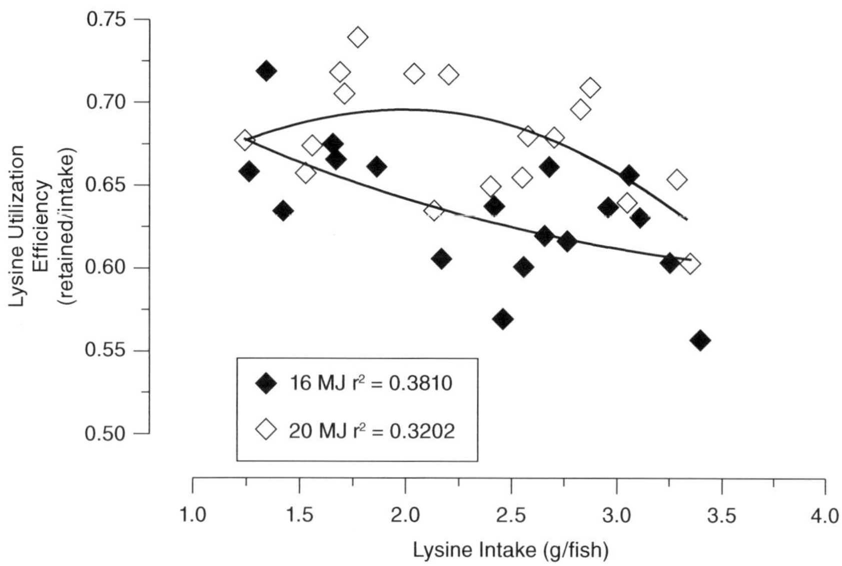

FIGURE 5-3 Effect of lysine and DE content of the diet (3.82 vs. 4.78 Mcal DE or 16 or 20 MJ DE/kg) on efficiency of lysine utilization of rainbow trout (*Oncorhynchus mykiss*) (Encarnação et al., 2004).

Preferential Catabolism

Preferential catabolism refers to catabolism of amino acids to provide energy, when dietary energy intake is limiting PD, which implies that the animal is portioning EAA away from protein synthesis toward catabolism to meet a specific metabolic need (Moughan, 2003). Distinction between preferential and inevitable catabolism can then be derived from the slope of the relationship between metabolizable energy (ME) intake and PD (Mohn and de Lange, 1998).

In fish, where amino acids appear to provide a significant proportion of the total energy (ATP) requirement (Ronnestad et al., 1999, 2003; Finn et al., 2002; Kaushik and Seiliez, 2010), inevitable and preferential catabolism may be difficult to separate. It is not clear to what extent the significant catabolism of amino acids despite adequate ME and net energy (NE) intakes (Encarnação et al., 2006) is related to inevitable losses (maintenance requirement, inevitable catabolism) of amino acids or catabolism of amino acids that are in excess of requirements, or preferential catabolism of amino acid as energy sources. The determinants of amino acid catabolism in fish deserve to be studied more systematically than it has been the case in the past.

Catabolism of Excess Amino Acids

Intake of amino acids in excess of the amounts required for protein deposition, maintenance requirements, inevitable catabolism, and preferential catabolism will result in additional catabolism of these amino acids. Feeding a diet in which the amino acid profile is deficient in one or multiple amino acids compared to dietary requirements will limit protein deposition, limit the retention of the other amino acids, and force their deamination and catabolism.

ESSENTIAL AMINO ACIDS—BIOCHEMISTRY, ROLES, AND DEFICIENCY SIGNS

Lysine

Lysine (abbreviated as Lys or K) is an α-amino acid with the chemical formula $HO_2CCH(NH_2)(CH_2)_4NH_2$. Lysine contains two amino groups with the ε-amino group and is an amino acid that often participates in hydrogen bonding and as a general base in catalysis. Common posttranslational modifications of lysine include methylation of the ε-amino group, giving methyl-, dimethyl-, and trimethyllysine as well as acetylation (Buxbaum, 2007). Collagen contains hydroxylysine, which is derived from lysine by lysyl hydroxylase (Sassi, 2001). Allysine and hydroxyallysine are produced by the actions of the enzyme lysyl oxidase on lysine and hydrolysine in the extracellular matrix. These molecules are essential in the crosslink formation that stabilizes the structure of collagen (Sassi, 2001).

Lysine is abundant in body protein of fish and other animals. Lysine is found in high concentration in ingredients such as fish meal and blood meal and in low concentrations in some plant protein ingredients, notably cereal grain byproducts, such as corn gluten meal and wheat gluten. Due to the reactive nature of its ε-amino group, lysine is sensitive to heat damage and to nonenzymatic glycolyzation reactions resulting in production of Maillard reaction products (Moughan and Rutherfurd, 1996). These result in irreversible chemical damage to lysine and reduce the amount of available lysine in feedstuffs and feeds (Carpenter, 1960; Moughan and Rutherfurd, 1996). Consequently, lysine is commonly the first limiting amino acid in feeds, particularly those formulated with high levels of plant protein ingredients or with protein ingredients processed under harsh conditions.

Beside reduced growth and feed efficiency, lysine deficiency has been shown to cause some health issues such as dorsal and caudal fin erosion in rainbow trout and common carp (Ketola, 1983; Guillaume et al., 1999).

Sulfur Amino Acids and Related Molecules

Methionine (abbreviated as Met or M) and cysteine (abbreviated Cys or C) are two sulfur-containing amino acids. Methionine is a nonpolar amino acid. Like other hydrophobic amino acids, it can play a role in binding/recognition of hydrophobic ligands such as lipids (Brosnan and Brosnan, 2006b). With a thiol side chain, cysteine is classified as a hydrophilic amino acid. The high reactivity of this thiol means cysteine is an important structural and functional component of many proteins and enzymes, forming disulphide bridges (dimers) in some proteins, which are important in folding the protein chain—RSH + RSH → RS-RS (Brosnan and Brosnan, 2006b; Buxbaum, 2007). Cysteine is readily oxidized to cystine (Cys-Cys) in the environment and is promptly reduced to the two cysteine molecules by the organism.

In addition to its role as a precursor in protein synthesis, L-methionine participates in a wide range of other metabolic reactions including the production of S-adenosylmethionine (SAM) (Baker, 2006), L-cysteine, glutathione, taurine, sulphate, phosphatidylcholine, and other phospholipids (Brosnan and Brosnan, 2006b). SAM itself is involved in the synthesis of creatine, epinephrine, melatonin, and the polyamines spermine and spermidine, among several other substances (Brosnan and Brosnan, 2006b). L-methionine is also a glycogenic amino acid and may participate in the formation of D-glucose and glycogen.

Methionine is converted to SAM by methionine adenosyltransferase. SAM serves as a methyl-donor in many methyltransferase reactions and is converted to S-adenosylhomocysteine (SAH). SAM, a remarkably versatile molecule, is said to be second, only to ATP, in the number of enzymes that require it (Brosnan and Brosnan, 2006b). Adenosylhomocysteinase converts SAH to homocysteine, which can be used to regenerate methionine or to form cysteine. Methionine can be regenerated from homocysteine via methionine synthase (Brosnan and Brosnan, 2006b). It can also be remethylated using glycine betaine (NNN-trimethyl glycine) to methionine via the enzyme betaine-homocysteine methyltransferase (BHMT). To a certain extent, betaine and choline can both help spare methionine in terms of the methyl donor function (Baker, 2006). Homocysteine can also be converted to cysteine. Cystathionine-β-synthase combines homocysteine and serine to produce cystathionine. Instead of degrading cystathionine via cystathionine-β-lyase, as in the biosynthetic pathway, cystathionine is broken down to cysteine and α-ketobutyrate via cystathionine-γ-lyase (Brosnan and Brosnan, 2006b). The enzyme α-ketoacid dehydrogenase converts α-ketobutyrate to propionyl-CoA, which is metabolized to succinyl-CoA (Brosnan and Brosnan, 2006b).

Walton et al. (1982) demonstrated a reduction in the methionine requirement in rainbow trout as dietary cysteine increased from 0 to 2% of the diet. These results have subsequently been supported by observations by Rumsey et al. (1983) and Cowey et al. (1992) and with numerous other fish species (NRC, 1993). When cystine is included in the diet, the need for methionine is reduced because cystine replaces methionine in the synthesis of cysteine and its derivatives. Cysteine is incapable of meeting the entire methionine requirement because the irreversibility of the cystathionine synthase reaction prevents cysteine from being converted back to homocysteine and serine (Brosnan and Brosnan, 2006b).

The conversion of methionine into cysteine complicates the estimation of a precise methionine requirement as the level of dietary cysteine will vary the amount of dietary methionine required by the fish. Some researchers have therefore expressed the requirement as a single total sulfur amino acid (TSAA) requirement, or as a "Met + Cys" requirement. Estimates of the TSAA requirement of rainbow trout are between 0.8 and 1.1% of the diet (Rumsey et al., 1983; Cowey et al., 1992). Dietary cysteine is estimated to have replacement values for methionine of 40 to 60% for various species of fish (Wilson, 2002). Estimates include 60% replacement value for channel catfish (Harding et al., 1977), 44% for blue tilapia (Liou, 1989), 42% for rainbow trout (Kim et al., 1992a), and 40% for red drum and hybrid striped bass (Moon and Gatlin, 1991; Griffin et al., 1992). Available evidence suggests that cysteine can provide almost half of the TSAA needs for protein accretion in several fish species that have been evaluated (NRC, 1993; Goff and Gatlin, 2004). Some studies have suggested that no more than 3 g/kg of dietary cysteine (0.3%) is an effective source of sulfur amino acids in fish, meaning that dietary cysteine in excess of this amount has no methionine sparing effect (Kim et al., 1992a; Pack et al., 1995). It is recommended that diets be formulated with separate methionine and cysteine requirements to ensure that the animal needs are effectively met (Rodehutscord et al., 1995a). In practice, many feedstuffs contain an excess of cysteine to methionine, which allow nutritionists to focus mainly on meeting the methionine requirement.

As in the case for other essential nutrients, feeding diets with low methionine content results in poor growth and feed efficiency. In addition, salmonids, including rainbow trout, Atlantic salmon (*Salmo salar*), and lake trout (*Salvelinus namaycush*), also suffer from cataracts when given a diet deficient in methionine (Poston et al., 1977). The lens begins to become opaque after 2 to 3 months, depending on the extent of the deficiency. Riboflavin, thiamine, vitamin A, zinc, tryptophan, and, more recently, histidine deficiencies have also been implicated in the development of ocular pathologies in fish (Poston et al., 1977; Simmons et al., 1999; Waagbo, 2010).

The mechanism linking methionine deficiency to the development of cataract is not completely understood. However, it is likely related to the role of methionine in the formation of glutathione. Oxidative damage is known to occur to protein-bound thiol groups in the lens of older mammals. Glutathione, which requires methionine or N-acetylcysteine for synthesis (Ferrer et al., 1990), probably has a role in protecting the lens from such damage (Cowey et al., 1992). Low dietary methionine supply might prevent the turnover of oxidized methionine molecules in the lens, leading to the formation of disulfide bonds and the development of ocular opacity (Simmons et al., 1999).

Methionine is the most toxic of the primary amino acids. Dietary levels in excess of two or three times the requirement level has been shown to affect growth in mammals (Edmonds and Baker, 1987; Baker, 2006). Growth of rainbow trout also seems to be negatively affected by very high methionine intake (Poppi et al., 2010). Methionine toxicity is thought to be due to hepatic accumulation of SAM (Regina et al., 1993). Alleviation of the effects of methionine toxicity can be achieved in rats with supplemental glycine and serine. Glycine facilitates the catabolism of excess SAM by the glycine-N-methyltransferase enzyme (Baker, 2006). Serine is combined with homecysteine in the transulfuration pathway to form cystathionine and then cysteine. This pathway can also eventually produce taurine, which can be easily excreted by the animal. Yokoyama and Nakazoa (1992) found elevated hepatic concentrations of cystathionine and reduced hepatic levels of serine when diets containing excess methionine were fed to rainbow trout. They postulated that cystathionine synthesis via the transulfuration pathway proceeds more rapidly than cysteine biosynthesis from cystathionine in this species.

Methionine Isomers

Large quantities of methionine are produced industrially by chemical synthesis for use in animal feeds. The chemical synthesis process produces a racemic mixture of D- and L-isomers of methionine (DL-Met). Most animals utilize D- and L-isomers of methionine with similar efficiencies (Baker, 2006). Utilization of D-methionine requires its deamination by D-amino acid oxidase and subsequently reamination to L-methionine. Rainbow trout can use D-methionine to replace L-methionine on an equimolar basis (Kim et al., 1992a). This metabolic capacity is probably also characteristic of other fish as well as crustacean species (Guillaume et al., 1999).

Methionine Hydroxy Analog

An hydroxy analog of methionine, 2-hydroxy-4-(methylthio)butanoic acid (HMB or OH-Met), also known as DL-methionine-hydroxyanalog (MHA), is commercially available and has found widespread use in animal feeds (Baker, 2006). The HMB differs from methionine by having a hydroxyl group on the alpha carbon rather than an amino group (Dibner, 2003). Like synthetic DL-met, HMB has one asymmetrical carbon atom and therefore occurs as a racemic mixture of L-isomer and D-isomer (Baker, 2006). Because HMB bears a hydroxyl group instead of an amino group, it is an organic acid (Dibner, 2003). Uptake of DL-met and HMB across the brush border membrane takes place by two different transport systems (Brachet and Puigserver, 1987; Richards et al., 2005). Methionine is transported by the system B amino acid transporter, whereas MHA-FA is transported via an H^+-dependent transporter, Monocarboxylate Transporter 1, which is also involved in the transport of lactic acid (Martın-Venegas et al., 2007). Upon absorption by the animal, HMB is rapidly converted to L-met in the liver using two different enzymes for D- and L-isomers, a dehydrogenase for the D-isomer and an oxidase for the L-isomer (Baker, 2006).

The biological efficiency of various forms of HMB, including liquid free acid form and crystalline calcium salt of HMB has been the focus of numerous trials with different terrestrial livestock species. There has been considerable debate over the biological efficacity of HMB in comparison to DL-met (Jansman et al., 2003). Various metaanalyses of data from a large number of trials with poultry have suggested that the relative biological efficiency of HMB in comparison with DL-met is about 75 to 80% on an equimolar basis (Jansman et al., 2003; Baker, 2006; Sauer et al., 2008). The results of studies carried out with different fish species (Robinson et al., 1978; Keembiyehetty and Gatlin, 1995, 1997a; Cheng et al., 2003a; Goff and Gatlin, 2004; Kelly et al., 2006) also support the notion that HMB can be used as a source of methionine by fish but that its relative biological efficiency is lower than that of DL-met on an equimolar basis. On the basis of available experimental evidence, the committee considers it reasonable to assume that the biological efficacy of HMB for fish is about 75 to 80% that of DL-met on an equimolar basis. The difference in biological efficiency between HMB and DL-met may be related to the differences in the absorption dynamic of these two compounds, a phenomenon that has been well documented in poultry (Drew et al., 2003) but has yet to be examined in fish and crustaceans.

Taurine

Taurine, or 2-aminoethanesulfonic acid, is an organic acid. It is also a major constituent of bile and can be found in the lower intestine and in small amounts in the tissues of animals. Taurine is a derivative of cysteine, and it is synthesized by the transsulfuration pathway (Goto et al., 2002). Taurine is often referred to as an amino acid in the literature although strictly speaking, it is not an amino acid because it lacks a carboxyl group.

Functional roles of taurine include conjugation of bile acids, osmoregulation, putative neurotransmitter function, cell membrane stabilization, and antioxidant effects (Lombardini et al., 1979; Gaull and Wright, 1986; Huxtable, 1992). Small polypeptides have been identified that contain taurine, but to date no aminoacyl tRNA synthetase has been identified as specifically recognizing taurine and capable of incorporating it into a protein.

Taurine appears to be a major contributor to osmotic pressure balance in some animals. It can be found at levels as high as the total pool of free amino acids in marine fish and invertebrates. Studies suggest significant interspecific differences in both the pathway and capacity of taurine biosynthesis in fish (Goto et al., 2002; Kim et al., 2008). Taurine biosynthesis appears to be very low in some species (Tagaki et al., 2005). A relatively large number of studies suggest that taurine may be conditionally essential for some juvenile marine fish and shrimp species, such as Japanese flounder (*Paralichtys olivaceus*) (Park et al., 2002; Kim et al., 2003, 2005a,b), European sea bass (Martinez et al., 2004), red sea bream (*Pagrus major*) (Goto et al., 2001), yellowtail (*Seriola quinqueradiata*) (Matsunari et al., 2005; Tagaki et al., 2005; Kim et al., 2008), cobia (*Rachycentron canadum*) (Lunger et al., 2007), and black tiger shrimp (*Penaeus monodon*) (Shiau and Chou, 1994), but not for common carp (*Cyprinus carpio*) and Atlantic salmon (Espe et al., 2008). One study reported a positive effect of taurine supplementation on growth of rainbow trout (Gaylord et al., 2006). Evidence suggests that taurine may be conditionally essential in freshwater fish during early life stages (e.g., larval stage) (Zhang et al., 2006). Taurine supplementation effectively reduced the severity of green liver disease in yellowtail fed a diet devoid of fish meal and containing less than 0.1% taurine (Tagaki et al., 2005). This suggests that taurine deficiency might be at the origin of this nutritional disease. Differences in the rate of synthesis of taurine from cysteine and metabolic and physiological demands for taurine may explain differences observed between species, life stages, and studies. Relatively few studies have examined the effect of graded levels of taurine on growth and health of different fish species reared in different environmental conditions and fed diets of different compositions. This makes it difficult to determine the taurine requirement of different species.

Selenocysteine and Selenomethionine

Selenocysteine (abbreviated as Sec or U) is an amino acid that is present in several enzymes, notably glutathione peroxidases, tetraiodothyronine 5' deiodinases, thioredoxin reductases, formate dehydrogenases, glycine reductases, and some hydrogenases. It is unique among amino acids because it is the only one synthesized directly on a tRNA (Ganichkin et al., 2008; Castellano et al., 2009). Selenocysteine is structurally similar to cysteine except the sulfur in cysteine is replaced with selenium, making the side chain-CH2-SeH. However, selenocysteine is not synthesized from Cys, but rather from serine (sidechain-CH2-OH), covalently linked to tRNA, by the replacement of oxygen with Se. Selenium is converted to selenophosphate, a high-energy molecule, at the cost of one ATP molecule, which then reacts with serine to form selenocysteinyl-tRNA (Ganichkin et al., 2008). Overall, Sec and Cys residues do not seem to be functionally exchangeable in proteins, suggesting that selenoproteins have important specific roles in the metabolism of vertebrates (Castellano et al., 2009). Proteins containing Sec (selenoproteins) apparently account for the essentiality of Se to vertebrates and have an important role in several pathologies associated with Se deficiency (Behne and Kyriakopoulos, 2001; Castellano et al., 2009). In mammals, there are 19 selenoproteins with known functions, and all of them are enzymes (Behne and Kyriakopoulos, 2001). The glutathione peroxidase (GPx) family of selenoproteins catalyzes the reduction of lipid and hydrogen peroxides to lipid alcohols and water, respectively, with glutathione as a reductant (Miranda et al., 2009). The GPx family is an essential component of the antioxidative system protecting membrane lipids and macromolecules from oxidative damage (Miranda et al., 2009).

Selenomethionine (Se-met and Sem) is another L-amino acid containing selenium. L-Se-met is synthesized by plants, marine algae, and yeast along with Met in quantities depending on the amount of Se available (Schrauzer, 2000). Selenomethionine was detected in organs in their intact form, suggesting that selenoamino acids are absorbed and delivered to organs in their intact forms, at least in mammals (Suzuki et al., 2006). There is no available evidence suggesting that Se-met is an essential nutrient. However, it seems to be a highly bioavailable source of Se. In vivo, Sem is randomly incorporated instead of methionine in protein The replacement of Met by Se-met does not significantly alter protein structure, but may influence the activity of enzymes if Se-met replaces Met in the vicinity of the active site (Scrauzer, 2000). Selenomethionine in protein is readily oxidized to Se-met oxide, which is easily reduced back to Se-met by glutathione (GSH). For Se-met to stimulate synthesis of selenoproteins such as GPx, the Se must be released by enzymatic degradation and converted to selenophosphate, which is the substrate for cotranslational Sec synthesis (Allmang and Krol, 2006; Miranda et al., 2009).

Branched-Chain Amino Acids: Leucine, Isoleucine, and Valine

Branched-chain amino acids (BCAAs) refer to the three amino acids (leucine, isoleucine, and valine) having aliphatic side chains that are nonlinear. Leucine (abbreviated as Leu or L) is an α-amino acid with an isobutyl R group. Isoleucine (abbreviated as Ile or I) has a sec-butyl side chain that is a large aliphatic hydrophobic chiral side chain. Four stereoisomers of isoleucine are possible although, in nature, only one enantiomeric form, (2S,3S)-2-amino-3-methylpentanoic acid, exists. Valine (abbreviated as Val or V) has an isopropyl side chain.

The key property of the three BCAAs is their hydrophobicity. Therefore, in proteins, these amino acids are largely excluded from aqueous environments, but they interact well with other hydrophobic molecules (Brosnan and Brosnan, 2006a). They are largely found in the hydrophobic interior core of globular proteins where their interactions with other similar amino acids play a key role in determining the three-dimensional shapes of these proteins and, hence, their functions (Brosnan and Brosnan, 2006a). These three EAAs play important structural roles and are primarily deposited in body protein, notably in skeletal muscles (Cowey and Walton, 1989; Brosnan and Brosnan, 2006a). Valine is also involved in the synthesis of the myelin covering of the nerves, and valine deficiency can cause degenerative neurological conditions in mammals. Because of their critical roles in the protein structure, most proteins have a relatively high proportion of BCAAs, and these represent a significant proportion of amino acids consumed by animals.

The increase in circulating BCAAs that occurs after a protein-containing meal is "sensed" by a number of different tissues and has important effects in these tissues (Yang et al., 2008). Thus, the BCAAs serve as important signals to other tissues; among the tissues that respond to BCAA concentrations are brain and skeletal muscle (Brosnan and Brosnan, 2006a). Leucine is increasingly recognized as an anabolic nutrient signal, communicating the presence of an ingested protein-containing meal to peripheral tissues, and stimulating insulin secretion by the β-cells of the pancreas and protein synthesis in muscle and adipose tissue through the target of rapamycin signalling pathway (Yang et al., 2008).

The metabolism of BCAAs differs from that of the other amino acids in three important respects. First, rather than being restricted to the liver as for most EAAs, the catabolic enzymes for BCAAs are distributed widely in body tissues, including the kidney, muscle, and even the central nervous system (Cowey and Walton, 1989; Brosnan and Brosnan, 2006a). Second, all three BCAAs share the same common transporter for intestinal absorption. Finally, the first steps in the oxidation of each of these three amino acids are catalyzed by two common enzymes, and so the organism metabolizes these three amino acids using the same enzymatic system (Cowey and Walton, 1989; Brosnan and Brosnan, 2006a). The first step in BCAA catabolism is transamination catalyzed by BCAT (branched-chain aminotransferase) isozymes. In this reaction, the amino group is transferred from a BCAA to α-ketoglutarate to form glutamate and the respective branched-chain α-keto acid (BCKA). The keto acid products are irreversibly oxidized by the second enzyme in the catabolic pathway, the mitochondrial BCKA dehydrogenase enzyme complex (Brosnan and Brosnan, 2006a).

Antagonisms Among Branched-Chain Amino Acids

Interactions between the BCAAs, leucine, isoleucine, and valine are known to produce antagonistic effects in chicks, pigs, rats, and humans (D'Mello, 1994). Reduction of plasma isoleucine and valine concentration after consumption of an excessive amount of leucine has been reported in rats, chicks, pigs, and humans (Block and Harper, 1991; Langer et al., 2000). Leucine-induced changes in plasma levels of isoleucine and valine have mainly been attributed to competitive inhibition during intestinal absorption and increased oxidation through BCKA dehydrogenase activation (Block and Harper, 1991; D'Mello, 1994; Langer et al., 2000).

In fish, antagonism involving BCAAs have not been fully assessed, and the results obtained have shown some inconsistencies. No effect of excess leucine on the other BCAAs was observed by Robinson et al. (1984), Choo et al. (1991), and Rodehutscord et al. (1997). However, Chance et al. (1964) observed that the isoleucine requirement of Chinook salmon (*Oncorhynchus tshawytscha*) increased slightly with increasing concentrations of dietary leucine. Hughes et al. (1983) observed changes in concentrations of BCAAs in lake trout (*Salvelinus namaychus*) given diets containing increasing amounts of valine. Plasma isoleucine and leucine were both elevated in valine-deficient fish, and their concentrations decreased as dietary valine was increased. An antagonist effect of excess leucine on plasma and muscle levels of other BCAAs has also been reported by Hughes et al. (1984) in lake trout and Yamamoto et al. (2004) in rainbow trout. Yamamoto et al. (2004) reported a negative effect of feeding diets formulated to low Ile:Leu and Val:Leu ratios. Choo et al. (1991) and Encarnação (2005) did not observe any effect of excess dietary leucine on plasma valine and isoleucine concentrations. Rainbow trout showed a high tolerance for dietary leucine; no growth depression occurred with concentrations as high as 9.2% of diet (Choo et al., 1991). Even with excessive dietary leucine concentrations (13.4% of diet), which were overtly toxic, the concentrations of free valine and isoleucine in plasma, liver, and muscle were not depressed (Choo et al., 1991).

Arginine

Arginine (abbreviated as Arg or R) is an α-amino acid with a side chain consisting of a 3-carbon aliphatic straight chain, the distal end of which is capped by a complex guanidinium group. Arginine is in zwitterionic form at neutral pH. The guanidinium group is positively charged in neutral, acidic, and even most basic environments and thus imparts basic chemical properties to arginine. Because of the conjugation between the double bond and the nitrogen lone pairs, the positive charge is delocalized, enabling the formation of multiple H-bonds. The distributing basics of the moderate structure found in geometry, charge distribution, and ability to form multiple H-bonds make arginine ideal for binding negatively charged groups. In mammals, arginine is classified as a semiessential or conditionally essential amino acid, depending on the developmental stage and health status of the individual (Baker, 2007). Arginine is an intermediate of the urea cycle and can be synthesized from citrulline (Wan et al., 2006). However, in fish and shrimp, arginine has been shown to be an EAA due to the very poor activity of the urea cycle. Huggins et al. (1969) demonstrated that all five enzymes of the urea cycle exist in teleosts. The urea cycle enzymes appear to be expressed during embryogenesis in a number of teleosts, but their activity is highly down-regulated during later life stages (Chadwick and Wright, 1999). There is evidence that teleosts can utilize ornithine and citrulline to synthesize arginine (Chiu et al., 1986). Some teleosts, such as *Heteropneustes fossilis* and *Oreochromis alcalicus grahami*, also upregulate the activity of the urea cycle during certain environmental conditions (Randall et al., 1989). The ability of these animals to produce arginine from the urea cycle to meet their arginine requirement has not been investigated.

Arginine is a precursor for creatine and nitric oxide synthesis and serves as a potent stimulant of insulin and growth hormone so that it may play an important role in anabolic processes (Wan et al., 2006).

Antagonism of arginine by excess dietary lysine is a phenomenon that has been characterized in a number of animal species, including chicks, rats, guinea pigs, and dogs (Baker, 2007). Excessive levels of dietary lysine have been shown to cause growth depression, which can be alleviated with additional dietary arginine (Austic and Scott, 1975). Lysine and arginine are transported on the same dibasic amino acid carrier, and competitive inhibition between these two amino acids can affect their absorption, transport, and metabolism (Kaushik and Fauconneau, 1984). In birds, excess lysine enhances arginine catabolism and increases the arginine requirement by inducing renal arginase (Jones et al., 1967; D'Mello and Lewis, 1970). In mammals, competitive inhibition of arginase by L-Lys has been observed and this can result in reduced arginine catabolism and urea production (Statter et al., 1978; Fico et al., 1982).

Studies with different fish species have yielded no convincing evidence of antagonism between lysine and arginine. Robinson et al. (1981), Tibaldi et al. (1994), and Alam et al. (2002a) have found no negative effect of feeding excess lysine on growth and/or plasma arginine levels in channel catfish, European sea bass, and Japanese flounder. Conversely, Kaushik and Fauconneau (1994) observed a significant decrease in plasma arginine in rainbow trout in response to increasing the dietary lysine level from 1.8 to 3.0% of the diet. However, the results of Kaushik and Fauconneau (1994) can probably be explained by the fact that the control diet (1.8% lysine) was marginally deficient in lysine (Encarnação et al., 2004). Increasing dietary lysine levels may have improved protein deposition and increased demand for other EAA, including arginine, thereby reducing plasma levels and oxidation of arginine.

Threonine

Threonine (abbreviated as Thr or T), together with serine and tyrosine, is one of three primary amino acids bearing an alcohol group. The threonine residue is susceptible to numerous posttranslational modifications. The hydroxy side chain can undergo O-linked glycosylation. In addition, threonine residues undergo phosphorylation through the action of a threonine kinase. In its phosphorylated form, it can be referred to as phosphothreonine.

Threonine is abundant in mucins. Evidence with mammals suggests that gut mucin synthesis may comprise a significant proportion of the whole-body threonine requirement (Nichols and Bertolo, 2008). Fish produce significant amounts of mucus, notably during stressful conditions, transfer to seawater, or exposure to heavy metals, ammonia, and pollutants (Eddy and Fraser, 1982). Mucus production may represent a non-negligible portion of the threonine requirement under certain conditions.

Threonine is metabolized to pyruvate via threonine dehydrogenase. An intermediate in this pathway can undergo thiolysis with CoA to produce acetyl-CoA and glycine. Beside a depression of growth and feed efficiency, threonine deficiency does not appear to cause specific deficiency signs (Ahmed, 2007).

Tryptophan

Tryptophan (abbreviated as Trp or W) is an α-amino acid containing an indole functional group. Only the L-stereoisomer of tryptophan is used in structural or enzyme proteins, but the D-stereoisomer is occasionally found in naturally produced peptides. Tryptophan functions as a biochemical precursor for several compounds, including 5-HT (serotonin), a neurotransmitter, which is synthesized via tryptophan hydroxylase. Serotonin, in turn, can be converted to melatonin (a neurohormone), via N-acetyltransferase and 5-hydroxyindole-O-methyltransferase activities.

Beside reduced growth and feed efficiency, tryptophan deficiency leads to scoliosis (lateral curvature of the vertebral column) and to a derangement of mineral metabolism in certain salmonids, including rainbow trout (Walton et al., 1984a), sockeye salmon (*Oncorhynchus nerka*) (Halver and Shanks, 1960), and chum salmon (*Oncorhynchus keta*) (Akiyama et al., 1986). Significantly greater concentrations of calcium (Ca) (a fourfold increase over control), sodium (Na), and potassium (K) were found in the kidneys of tryptophan-deficient trout (Walton et al., 1984a). Concentrations of

Ca, magnesium (Mg), Na, and K in the livers of tryptophan-deficient trout were also significantly greater than in normal trout. The underlying mechanisms for this accumulation of minerals in liver and kidney have not been resolved.

Scoliosis induced by Trp deficiency may be reversed by restoring tryptophan to an adequate concentration or by inclusion of serotonin in the diet (Akiyama et al., 1986). Thus, the experimental evidence suggests that this condition may be related to a decline in circulating level of serotonin (Akiyama et al., 1986).

The synthesis of serotonin in the brain is dependent on availability of tryptophan. This occurs because the rate-limiting enzyme in the biosynthetic pathway of 5-HT, tryptophan hydroxylase, is unsaturated with Trp under normal physiological conditions (Hoglund et al., 2005). Feeding high dietary levels of Trp resulted in elevated body and brain Trp content, increased brain 5-HTergic activity, affected the behavioral response to stress, and reduced aggressiveness in different fish species (Winberg et al., 2001; Lepage et al., 2002; Hseu et al., 2003; Hoglund et al., 2005, 2007).

Niacin can also be synthesized from tryptophan via kynurenine and quinolinic acids as key biosynthetic intermediates, although there is evidence that fish are unable to convert any significant amount of tryptophan into niacin (Ng et al., 1997).

Histidine

Like arginine and lysine, histidine (abbreviated as His or H) is classified as a basic amino acid. However, histidine has a positively charged imidazole functional group that can act as both an acid and a base, i.e., it can both donate and accept protons under some conditions. This amino acid side chain has important roles as a coordinating ligand in metalloproteins, and also as a catalytic site in certain enzymes, such as aiding the catalytic functions of chymotrypsin (digestive enzyme) and those enzymes involved with metabolism of proteins and carbohydrates. The residue can also serve a role in stabilizing the folded structures of proteins. Histidine is also found abundantly in hemoglobin, is the direct precursor of histamine, and is an important source of carbon atoms in the synthesis of purines.

The enzyme histidine ammonia-lyase converts histidine into ammonia and urocanic acid (Cowey and Walton, 1989) in the liver. In the liver, urocanate is further processed into ammonia, glutamate, and a 1-C fragment that it used in the folate coenzyme system. Decarboxylation into histamine plays important roles in immune function and as a paracrine agent acting on the stomach.

Following removal of blood meal from feeds in the mid-1990s, a high incidence of cataracts was observed in Atlantic salmon smolts in Northern Europe (Breck et al., 2003; Waagbo et al., 2010). Studies confirmed the mitigating effect of blood meal on development of cataract in these fish (Breck et al., 2003). Blood meal is considerably richer in histidine (4–6% histidine) than fish meal (1–2% histidine) and other common fish-feed ingredients. A series of studies published by Bjerkas and Sveier (2004), Breck et al. (2003; 2005a,b), and Tröße et al., (2009) present evidence suggesting that feeding diets containing 0.9 to 1.0% total histidine support maximal growth but could result in high incidence and severity of cataract in fast growing Atlantic salmon smolts. These studies report a significant interaction among genotype, salinity variation/saltwater transition, and water temperature and dietary histidine level on the development of cataract in salmon. The exact mechanism by which histidine mitigates cataract development is not completely understood (Waagbo et al., 2010). Histidine and related compounds (imidazoles) may play important biochemical roles, such as osmoregulation, muscle pH buffering, and detoxification of reactive carbonyl species (Waagbo et al., 2010). Breck et al. (2005b) found that increasing dietary histidine inclusion at levels from 0.9 to 1.4% of diet in salmon undergoing smoltification had a significantly positive effect on eye lens protein turnover and n-acetyl histidine (NAH) content of the lens. NAH may play a role in protecting the lens against variation in osmotic pressure and oxidative stress associated with seawater transfer (Breck et al., 2005b). Protein turnover is likely an essential mechanism for repairing damage to eye lens proteins and maintaining clarity of the eye lens (Breck et al., 2005b). The "metabolic" requirement for histidine may increase during the smoltification process and the subsequent period of fast growth in seawater. Histidine requirement for optimal ocular health appears to be significantly higher than that for maximal growth in fast-growing Atlantic salmon undergoing transition to seawater (Waagbo et al., 2010). A higher requirement for optimal ocular health compared to that for maximizing growth was also reported for methionine in Arctic charr (*Salvelinus alpinus*) (Simmons et al., 1999).

Phenylalanine and Tyrosine

Phenylalanine (abbreviated as Phe or F) is a nonpolar α-amino acid because of the hydrophobic nature of the benzyl side chain. Tyrosine (abbreviated as Tyr or Y) or 4-hydroxyphenylalanine is synthesized in the body from phenylalanine and is considered a semiessential, or conditionally essential, amino acid. Fish readily convert phenylalanine to tyrosine so that phenylalanine alone can meet requirements for aromatic amino acids (Wilson, 1989; Guillaume et al., 1999). However, the presence of tyrosine in the diet will reduce some of the requirement for phenylalanine. Phenylalanine sparing by tyrosine is believed to be between 40–60% in the species studied to date (e.g., NRC, 1993; Guillaume et al., 1999).

Aside from being a proteinogenic amino acid, tyrosine has a special role by virtue of the phenol functionality. It occurs in proteins that are part of signal transduction processes and functions as a receiver of phosphate groups that are transferred to the hydroxyl group by protein kinases (so-called receptor tyrosine kinases), with phosphorylation changing the activity of the target protein. L-tyrosine also can be converted into the catecholamines, norepinephrine, and epinephrine, dopamine (a neurotransmitter), and thyroxin (Cowey and Walton, 1989).

The catabolism of L-tyrosine involves a series of reactions that yield fumarate and acetoacetate (3-ketobutyroate). Acetoacetate is a ketone body that can be converted into acetyl-CoA, which in turn can be oxidized by the TCA cycle or be used for fatty acid synthesis (Cowey and Walton, 1989).

QUANTITATIVE PROTEIN AND ESSENTIAL AMINO ACID REQUIREMENTS

Protein Requirements

The protein requirement of animals corresponds to the increasingly well-understood requirements for specific EAA and to a nonspecific need for amino groups (nonspecific nitrogen) for the synthesis of NEAA, as well as the contribution of amino acids to meeting the energy and other metabolic needs of the animal.

Protein generally refers to crude protein (CP); that is, N × 6.25, a definition based on the assumption that proteins contain 16% N. This assumption is simplistic and rarely appropriate because the percentage of N in pure proteins is known to vary significantly, depending on their amino acid composition (Mariotti et al., 2008). Using a fixed conversion factor (e.g., 6.25) can lead to a 10–20% error in estimation of the true protein content of certain protein-rich ingredients (Mariotti et al., 2008). Nitrogenous compounds in feeds do not only comprise protein and amino acids; they also include numerous compounds such as nucleic acids, amines, urea, ammonia, nitrates, nitrites, phospholipids, and nitrogenous glycosides (Mariotti et al., 2008). The contribution of these different compounds to the total N content of ingredients is highly variable. Consequently, crude protein (as well as digestible protein) refers to a mixture of very different substances differing in terms of their biochemical nature and nutritive value (Mariotti et al., 2008). The sum of individual amino acids in ingredients (as analyzed by reference laboratories using rigorous methodological approaches and modern equipment) only seems to account for 80 to 90% of the "protein" content of feed ingredients (Helland et al., 2010). This suggests that N-containing compounds other than amino acids may account for as much as 10–20% of the crude protein content of feed ingredients.

All dietary proteins are not identical in their nutritive value, which is a function of their digestibility and amino acid profile. The amino acid composition of proteins in different ingredients differs markedly (NRC, 1993). Digestibility of protein and availability of individual amino acids also vary considerably among ingredients. The capacity of different dietary protein sources to meet the EAA and other metabolic needs of animals is expected to differ considerably.

Amino acids also play an important role in meeting the energy (metabolic) requirements of fish and shrimp and serve as metabolic fuels for most fish and crustacean species (Kaushik and Seiliez, 2010). The persistent use of the dietary crude protein concept by most nutritionists and feed manufacturers is understandable because it is a simple and practical parameter but it is also to some extent perplexing.

The protein requirements have been examined for a very large number of fish and shrimp species at different life

TABLE 5-4 Recommended Dietary Protein Levels (%) for Various Fish Species of Commercial Importance (As-Fed Basis)

	Weight Range				
Species	< 20 g	20–200 g	200–600 g	600–1,500 g	> 1,500 g
Atlantic salmon (*Salmo salar*)	48	44	40	38	34
Channel catfish (*Ictalurus punctatus*)	44	36	32	32	28
Common carp (*Cyprinus carpio*)	45	38	32	28	28
Nile tilapia (*Oreochromis niloticus*)	40	34	30	28	26
Pacific salmon (*Oncorhynchus* spp.)	55	45	40	38	38
Rainbow trout (*Oncorhynchus mykiss*)	48	40	38	38	36
European sea bass[a] (*Dicentrarchus labrax*)	55	50	45	45	—
Gilthead sea bream[b] (*Sparus auratus*)	50	45	40	40	—

[a]This recommendation applies to other sea bass species.
[b]This recommendation applies to other sea bream species.

TABLE 5-5 Recommended Dietary Protein Levels (%) of Different Shrimp Species

Species	Weight Range		
	0.1–5 g	5–20 g	> 30 g
Tiger shrimp (*Penaeus monodon*)	45	40	40
Pacific white shrimp (*Litopenaeus vannamei*)	40	35–40	35
Kuruma prawn (*Marsupenaeus japonicus*)	50	45	40

stages. Tables 5-4 and 5-5 provide practical estimates of dietary protein requirements of fish and shrimp as a percentage of the diet. Understanding the nutritional constraints and limitations used in arriving at the protein requirement is important for their proper application. This information provides a simple basis for formulation of practical feeds for different species, notably when limited information is available on the EAA requirements and optimal levels (or the effectiveness) of lipid and carbohydrate as energy sources are not well defined for a given species. Estimates of protein requirements should be considered highly approximate, especially in the current context in which an increasingly wide variety of feedstuffs is used in the formulation of feeds for different fish and shrimp species.

As a dietary concentration, the dietary protein requirements of fish and shrimp seem to be appreciably higher than those of terrestrial warm-blooded animals. However, although the protein requirement in terms of dietary concentration (percentage of the diet) is high, the absolute requirement (grams of protein intake per kilogram of body weight gain) is highly comparable, although significant differences seem to exist among species (Figure 5-4). This is because fish have a lower maintenance energy requirement than warm-blooded animals, which results in a similar amount of body weight gain per unit of protein ingested as in warm-blooded animals but better feed efficiency (gain:feed). Some fish species, such as the Atlantic salmon, are typically more efficient converters of protein than domesticated warm-blooded animals and "omnivorous" fish species, such as tilapia and common carp (Figure 5-4). Direct or indirect comparisons of species indicate that carnivorous species often have higher protein retention efficiencies (N gain/N intake) than do omnivorous fish (e.g., Pei et al., 2004). These observations cast doubt on the oft-voiced "opinion" that "carnivorous" fish species rely more heavily on amino acids to meet their energy/metabolic requirements than do omnivorous fish species.

Digestible Protein to Digestible Energy Ratio

Amino acids play an important role in meeting the energy (metabolic) requirements of fish and shrimp and are generally efficient metabolic fuels for most fish and crustacean species. The use of protein as a dietary source of energy by animals is considered undesirable because of the relatively high cost of protein compared to the cost of other energy-yielding nutrients (starch and lipids) and because of the release of ammonia associated with the catabolism of amino

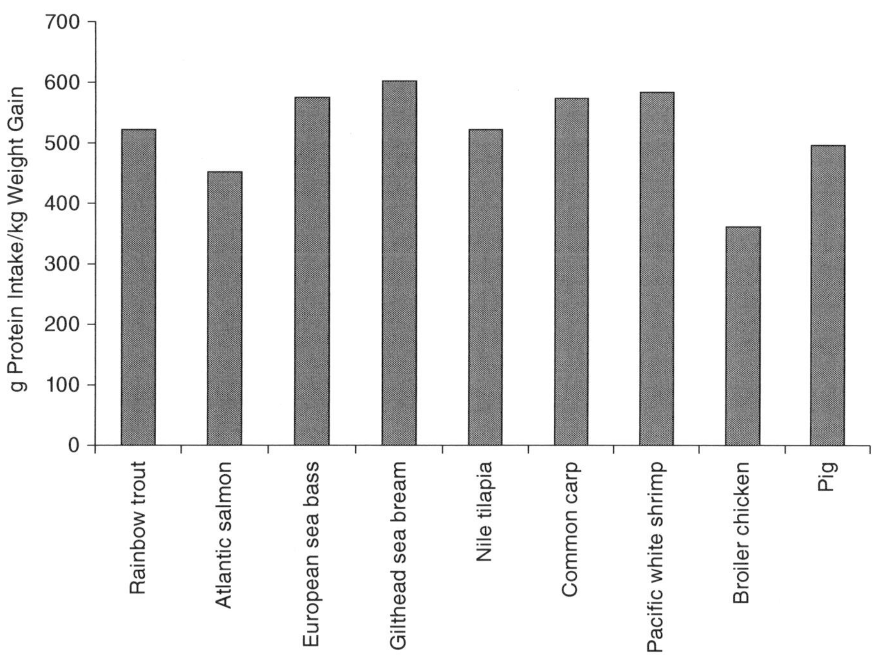

FIGURE 5-4 Protein intake per kilogram of live weight gain in different fish and shrimp species, chicken, and swine.

acids. In general, some economy can be made if other dietary energy-yielding nutrients are present in adequate amounts to reduce dietary amino acid catabolism, an effect commonly referred to as "protein-sparing." There is a strong argument that digestible protein (DP) to DE ratio is a more rational way of expressing protein requirement than the dietary "crude protein" requirement (percentage CP of diet).

A decrease in digestible protein to digestible energy ratio (DP:DE) achieved by reducing the dietary DP levels with or without concomitant increase in the dietary non-protein DE supply has proven to be extremely efficient in improving protein (nitrogen) utilization and decreasing nitrogenous loses in numerous farmed fish species (Lee and Putnam, 1973; Cho and Kaushik, 1985; Kaushik and Oliva-Teles, 1985; Cho and Woodward, 1989; Hillestad and Johnsen, 1994; Vergara et al., 1996; Einen and Roem, 1997; Grisdale-Helland and Helland, 1997, Helland and Grisdale-Helland, 1998; Hillestad et al., 1998; Steffens et al., 1999; Azevedo et al., 2004a,b; Satoh et al., 2004). Variation in the DE content of the diet also involves other energy-yielding nutrients that may not have the same ability to be utilized by fish and spare dietary amino acids (Encarnação et al., 2006). In general, increasing the lipid content of the diet can help reduce dietary protein (amino acid) catabolism in a number of species. A number of studies also have shown that extra DE provided as lipid can have a limited impact on efficiency of protein utilization under certain conditions (Azevedo et al., 2004a,b). Many fish species have a limited ability to utilize carbohydrate (e.g., starch) efficiently even when gelatinized. In excess of certain levels, the metabolic utilization of absorbed glucose is limited in most fish, and the amount of net energy that can be derived from digestible carbohydrate is limited (Bureau, 1997), although there are differences between species (Kaushik and Seiliez, 2010).

Overall, differences seem to exist among species and life stages in terms of optimal DP/DE. For most species studied so far, optimal DP/DE seems to range from about 84 to 105 g DP/Mcal DE (20 to 25 g DP/MJ DE) (Winfree and Stickney, 1981; Cho and Woodward, 1989; Azevedo et al., 2004a,b; Wang et al., 2006). Optimal DP/DE for species that show high capacity to utilize high dietary levels of lipids (e.g., Atlantic salmon) can be below 84 g DP/Mcal DE (20 g DP/MJ DE) (Einen and Roem, 1997). The optimal DP/DE of carnivorous fish species that are relatively poorly tolerant of dietary lipids (or "lean" species), such as Asian sea bass, haddock, and cobia, seems to be closer to 117 to 134 g DP/Mcal DE (28 to 32 g DP/MJ DE) (Catacutan and Coloso, 1995; Kim et al., 2004; Tibbett et al., 2005; Webb et al., 2010). In some species, feeding a low DP/DE diet (e.g., high-lipid diets) can improve protein utilization but result in undesirable levels of lipid deposition (Tibbett et al., 2005), which may have long-term implications in terms of final product quality or health of the animal.

Evidence suggests that the protein requirements, as a proportion of the mass of the diet, decrease as fish increase in size. For example, 25% protein was adequate in the diet of channel catfish of 114 to 500 g, but 35% protein produced faster gains than did 25% protein in fish weighing between 14 to 100 g (Page and Andrews, 1973). Similar results have been obtained with salmonids, common carp, and tilapia (Wilson and Halver, 1986). The optimal DP:DE ratio for young Atlantic salmon has been estimated to be between 84–100 g DP/Mcal DE (20–24 g DP/MJ DE), whereas evidence suggest that for large salmon (> 2.5 kg) it may decrease to 67–71 g DP/Mcal DE (16–17 g DP/MJ DE) (Einen and Roem, 1997). Results from a study conducted on gilthead seabream (*Sparus aurata* L.) predicts that optimum DP:DE ratio could decrease from 119 down to 82 g DP/Mcal DE (28.5 down to 19.5 g DP/MJ DE) as the animal grows from 10 to 250 g (Lupatsch et al., 2001b).

Despite requiring (and/or tolerating) lower dietary protein concentrations, larger fish frequently show lower protein retention efficiency (N gain/N intake) than do smaller fish. Studies in rainbow trout indicated that protein retention efficiency decreased very significantly and the ratio of lipid to protein deposited (LD:PD) increased dramatically with an increase in live body weight of these animals (Azevedo et al., 2004a,b; Peña-Ortega and Bureau, 2004; Azevedo et al., 2005; Dumas et al., 2007). A higher absolute amount of protein (grams of protein per kilogram of biomass gain) can therefore be required by some species as the fish grow larger despite a significant decrease in their protein concentration or optimal DP:DE ratio for their diet.

Effect of Environmental Factors

Because the metabolism of fish and shrimp is affected by temperature and other environmental factors (Bureau et al., 2002), it is often assumed that environmental conditions may have a significant effect on nutritional requirement of fish and crustaceans. However, so far no convincing evidence exists to show that protein requirement or optimal DP:DE ratio is affected by water temperature or other environmental factors (e.g., salinity) (NRC, 1981; Lupatsch and Kissil, 2005), at least within a "normal" range of conditions (range specific to each species and/or strain). A small decrease in protein and energy digestibility is observed with decreasing water temperature but the efficiency of DP and DE utilization of rainbow trout and grouper fed various rations at different water temperatures does not seem to be affected (Azevedo et al., 1998; Rodehutscord and Pfeffer, 1999; Lupatsch et al., 2001; Lupatsch and Kissil, 2005). In general, all feeding and growth functions increase in parallel as water temperature rises. A significant increase in protein requirement may occur in Asian sea bass (Barramundi) reared at temperatures that are significantly above the thermal optima, possibly as a means for compensating for lower feed intake of the animal, a reduction in the efficiency of utilization of protein, and an increase in the rate of protein losses (Bermudes et al., 2010).

Factorial models have been proposed as straightforward

means to estimate protein requirements of fish and shrimp to account for the potential impacts of diet composition, life stage, growth rate, and environmental factors (Lupatsch et al., 1998, 2001a, 2003, 2008, 2010; Glencross, 2008, Glencross et al., 2010; Richard et al., 2010). Using factorial approaches, the requirement is generally estimated as an absolute amount of protein (or digestible protein) required per kilogram of body weight per day for explicitly expressed levels of performance (e.g., expected or achievable level of performance) and life stages. Information on feed intake and diet composition (e.g., digestible energy levels) allow the backcalculation of "optimal" dietary protein concentration. In a number of studies, an absolute protein requirement (grams of DP required per kilogram of body weight per day) has been estimated for "maximal production" (Gatlin et al., 1986; Watanabe et al., 2000; Richard et al., 2010). Maximal production of an animal is highly dependent on genetics, diet composition, environmental conditions (e.g., temperature), husbandry practices, health status, and other variables and the absolute protein requirements for maximum production calculated consequently can only be valid for the specific conditions encountered in the study. Consequently, estimates of protein (or amino acid) requirements expressed in absolute terms (grams of DP required per kilogram of body weight per day) need to be presented within a production or growth modeling context as opposed to stand alone estimates (Lupatsch et al., 1998; Bureau et al., 2002; Glencross et al., 2010).

QUANTIFYING ESSENTIAL AMINO ACID REQUIREMENTS

Methodological Approaches

The EAA requirements of fish have been a topic of investigation for more than 50 years. Initial studies with Chinook salmon were conducted by Halver and coworkers in the late 1950s, when they evaluated various amino acid test diets. Most successful diets were based on the profile of whole hen egg, and this diet was used to determine qualitative amino acid requirements of Chinook salmon (Halver, 1957). Subsequently, quantitative requirements for the 10 EAA were investigated by the same group, providing a model for much of the work on quantitative EAA requirements of other fish species (Wilson, 1989). Most quantitative estimates of EAA requirements have been established by dose-response studies, although a number of other methods have also been applied. These methodological approaches are based on analyses such as plasma or tissue amino acid concentration and rates of amino acid oxidation. However, these methods have always been subsidiary to growth studies (Kim et al., 1992c; Cowey, 1994).

Growth Response Assays

Most estimates of EAA requirements have been determined based on conventional growth response assays. In such studies, a basal diet, deficient in a single EAA but meeting all other known nutrient requirements of the animal, is supplemented with graded levels of the EAA studied. Various types of chemically defined, purified, and natural ingredients have been used to provide graded increments of the amino acid under test. Most studies have used test diets in which the nitrogen component consisted of either amino acids or a mixture of amino acids, casein, and gelatin formulated to provide an indispensable amino acid composition identical with some reference protein (such as whole hen's egg protein or fish body protein) minus the amino acid under test (Cowey, 1994). These diets are generally fed to young, fast-growing fish, although the method can also be applied to other life stages depending on the objective of the investigation.

Quantification of EAA requirements is generally based on analysis of dose-response curves with weight gain used as a response criterion. The lowest level of EAA maximizing live weight gain is then identified as the minimum dietary requirement. Protein and EAA depositions are also increasingly used as response parameters along with weight gain (e.g., Pfeffer et al., 1992; Rodehutscord et al., 1995a, 1997; Hauler and Carter, 2001b; Encarnação et al., 2004). Evidence suggests that protein and EAA depositions may be a more robust and rational criteria. Estimates of requirements based on protein deposition have been shown to be higher than those obtained based on weight gain in some studies (Encarnação et al., 2004).

Alternative dietary and experimental designs have been used to estimate protein and EAA requirements in some studies (Gatlin et al., 1986; Gurure, 1997; Hung et al., 2004; Abboudi et al., 2006, 2007; Liebert and Benkendorff, 2007; Helland et al., 2010). These approaches include the requirement at ration level (RRL) method, in which a diet that may or may not be nutritionally complete is fed at graded levels, thus achieving graded intake of protein and/or amino acids to estimate requirements for maintenance and maximum growth or protein gain (a type of factorial method). The diet dilution technique, in which serial dilution of a "summit diet" also known to be first limiting in a specific EAA are fed to animals is another approach to achieve intake of incremental amino acid levels. The main advantage of these approaches is that the amino acid balance of the diet does not change in the different diets with graded levels of EAA fed to the animal (Gous and Morris, 1985; D'Mello, 2003). However, the supply of several nutrients varies at once and/or there is a substitution of energy-yielding nutrients that arguably are more important confounding factors than small imbalances in the profile of dietary amino acids (D'Mello, 2003). The estimate of requirements obtained with the RRL method is generally expressed as absolute values (e.g., g/fish BW per day) and may be applicable only to the specific conditions (performance level, growth rate and protein deposition) observed in the study.

Whole-Body Amino Acid Profile

A review of the literature indicates that for a large majority of fish species fed high-quality diets, deposition of amino acids into body protein represents between 25 to 55% of total amino acids consumed. The deposition of protein, consequently, is one of the determinants of amino acid requirements by fish, and composition of body protein deposited can be used to provide an indication of the diet EAA profile required by the animal. Early estimates of amino acid requirements of fish were based on the amino acid profiles of fish, egg, and whole-body proteins (Wilson, 1989; NRC, 1993). This approach is still popular to develop initial estimates of amino acid requirements of fish and shrimp species for which limited information is available on their nutritional requirements (Kaushik, 1998; Kaushik and Seiliez, 2010). However, because a minor proportion (typically less than 50%) of the digested amino acids is generally retained, the profile of retained EAA may not fully correspond to the profile of EAA required by the animal. Certain amino acids are known to be preferentially retained in tissues, whereas others appear to play more active metabolic roles and may be less efficiently retained.

In most cases, a reasonably good agreement between body EAA profile and EAA requirement (% protein) has been observed. Wilson and Poe (1987) observed very good agreement between EAA requirement pattern and whole-body EAA profile of channel catfish and between EAA requirement pattern and catfish egg EAA profile. Nose (1979) also found good agreement between body EAA and EAA requirements of common carp.

Amino acid profile of whole body or muscle protein (Ogino, 1980; Mambrini and Kaushik, 1995) or whole body A:E ratios (EAA content/total EAA × 1,000) (Arai, 1981; Moon and Gatlin, 1991; Wilson, 1994; Brown, 1995) have also been employed to generate simple quantitative estimates of EAA requirements of fish. These are at the basis of many of the estimates of the "ideal protein" pattern of amino acids published for many studies (as discussed later in this chapter). However, it is generally recognized that EAA requirement estimates based on body protein overestimate the requirement for amino acid preferentially deposited in body protein, such as leucine and lysine, and underestimate the requirements of amino acids playing important metabolic roles, such as methionine, threonine, histidine, and arginine.

Factorial Models of Amino Acid Requirements

The mathematical representation of the main biological processes that determine EAA utilization for body protein deposition has been an increasingly used methodological approach to estimate EAA requirements of fish and shrimp (Hauler and Carter, 2001b; Teshima et al., 2002; Bodin et al., 2008, 2009; Richard et al., 2010). These models are often extensions of protein and energy requirement models discussed earlier (Lupatsch et al., 1998; Bureau et al., 2002; Glencross et al., 2010). Factorial EAA requirement models are generally based on integration of information derived from different methodological approaches (e.g., whole-body EAA profile, estimates of maintenance requirements and inevitable catabolism of amino acids obtained by experimentation [growth assays] or metaanalyses of published studies, growth, and bioenergetic modeling).

Factorial models are practical because they enable the estimation of EAA requirements as a function of diet composition, life stage, and growth rate. The main limitations of current factorial models are that they estimate EAA requirement by generating independent estimates for each EAA, they assume that there are no interactions among EAA and other nutrients (e.g., fatty acids and glucose), and assume constant efficiency of utilization of amino acids regardless of physiological state. They also assume that "feed intake" or "energy requirement" or "feed conversion ratio" are "determinant"; that is, they are "independent" from diet composition, nutrient intake, interaction between nutrients, changes in target protein, and lipid deposition or physiological state. More efforts need to be invested in developing a more rational, yet practical, framework of EAA and nutrient requirements of fish and shrimp (Bureau and Encarnação, 2006; Hua and Bureau, 2010).

Blood and Muscle Amino Acid Levels

At subrequirement intake levels, the serum or tissue content of the tested EAA should remain low until the requirement for the EAA is met and then increase to high levels when excessive amounts of the amino acid are fed (Wilson, 1989). This technique has proven to be useful in corroborating the EAA requirements, but only in a few cases. In channel catfish, serum lysine data (Wilson et al., 1977) were useful in confirming the requirement values estimated by weight gain data. However, this technique has not always been reliable for assessing EAA requirements (Kaushik and Luquet, 1979; Hughes et al., 1983; Walton et al., 1986; Were, 1989; Kim et al., 1992b). Its validity seems to be linked to (1) the nature of the EAA tested, (2) interactions between the different amino acids, and (3) time elapsed between meal and blood sampling (Mambrini and Kaushik, 1995)

Amino Acid Oxidation Studies

Direct and indirect oxidation studies are based on tissue-free amino acid concentrations and give a measure of the partitioning of the EAA between protein synthesis and oxidation (Kim et al., 1983). The direct oxidation method is based on the principles that, at limiting levels, rates of oxidation of the EAA under study should be low because concentration in the free amino acid pool should be small, the major portion being utilized for protein synthesis, and little would be oxidized (Kim et al., 1983). Thus, the oxidation rate of the tested amino acid should remain low until the requirement level is reached and then it would increase sharply. The intake level that produces a marked increase in amino acid oxidation should then be a direct indicator of the requirement value for that specific amino acid (Kim et al., 1983).

Indirect oxidation studies measure the oxidation of an EAA other than the one under study. In this instance, incorporation of this other amino acid into tissue protein is limited by the intake level of the amino acid under study; consequently, high rates of oxidation of this other amino acid will occur. As dietary concentration of the amino acid under study increases, tissue protein synthesis will increase progressively, and the amounts of other amino acids being oxidized will decrease as proportionally larger amounts are used for protein synthesis (Kim et al., 1983; Cowey, 1994).

These techniques have been evaluated in rainbow trout with limited success. Compared to growth studies, oxidation studies gave similar estimates for lysine (Walton et al., 1984b) and tryptophan (Were, 1989) requirements, but apparently resulted in unreliable estimates of arginine requirements (Fauconneau et al., 1992; Lall et al., 1994). These discrepancies could be explained by the variable rates of oxidation of the different EAA, differences in the techniques of administration of the labeled amino acids, and the inherent variability associated with the technical complexity of the method (Mambrini and Kaushik, 1995).

SUMMARY OF PUBLISHED ESTIMATES OF ESSENTIAL AMINO ACID REQUIREMENTS OF FISH AND SHRIMP

Estimation of EAA requirements of fish and shrimp has been the objective of a large number of published studies (> 200 papers). These studies have been very diverse in terms of scope and methodological approaches used. Quantitative estimates have been generated for all 10 essential amino acids in a number of species (or highly related group of species), including channel catfish, common carp, Indian major carp (rohu and mrigal), Nile tilapia, Pacific salmon (Chinook, chum, coho), and rainbow trout. Increasing information is available on marine fish species (cobia, croaker, drum, sea bass, sea bream, turbot, flounder, and others). However, the significant numbers (10) of EAA and the very large number of fish and shrimp species cultivated around the world result in significant dilution of research efforts. In addition, the experimental design of many studies is deficient, and few studies can be considered sufficiently robust to generate reliable estimates of EAA requirements. An improvement of the quality of efforts invested in the estimation of EAA requirements of different fish and shrimp species would be a valuable contribution to the aquaculture nutrition community.

Tables 5-6 to 5-15 provide a summary of the experimental conditions and conclusion of studies on EAA requirements of fish. Table 5-16 summarizes information from studies on penaeid shrimp. Estimates of EAA requirements vary significantly among studies. Part of this variability can be attributed to differences among species but very significant variability exists within species. This variability may be attributable to experimental design and conditions (including weight of fish, composition and physical characteristics of the experimental diets, number of treatment and replicates, feeding method, and growth rate achieved) but also to the wide range of mathematical and statistical approaches used to estimate requirements, which may have a significant effect on estimate of EAA requirements (Rodehutscord and Pack, 1999; Encarnação et al., 2004; Wang et al., 2010).

ESSENTIAL AMINO ACID REQUIREMENTS IN THE CONTEXT OF FEED FORMULATION

Nutrient requirements find their usefulness in their translation into nutritional recommendations for feed formulations. Formulating cost-effective feeds meeting essential amino acid requirements of fish and shrimp can represent a challenge. Aquaculture feeds are different from other livestock feeds due to the wide variability of their composition in terms of digestible protein, lipid, carbohydrate, and DE contents (Encarnação et al., 2004; Bureau and Encarnação, 2006). The impact of diet composition on EAA utilization and requirements of fish and shrimp has been the focus of a limited number of studies and thus remains poorly understood and controversial (Encarnação et al., 2004, 2006; Bureau and Encarnação, 2006). Significant differences in opinion exist as to how EAA requirement data should be expressed and EAA levels deemed adequate in feed formulations should be calculated (Bureau and Encarnação, 2006). This situation limits the ability to review and interpret information on EAA requirements of fish and make recommendations that are widely applicable to practical conditions, for example, the wide variability in protein, and digestible energy (or energy-yielding nutrient) levels to which commercial feeds for a given species are formulated (Encarnação et al., 2004; Bureau and Encarnação, 2006).

Basic Approaches and Considerations

When formulating feeds, an appropriate safety margin is allocated on top of established EAA requirement values to compensate for putative processing and storage losses, variation in composition, digestibility, and bioavailability of nutrients in feed ingredients, as well as account for variations in requirements caused by environmental and biotic factors (NRC, 1993). Part of this safety margin is to account for lower digestibility of nutrients in feed ingredients compared to the high-quality ingredients used in laboratory diets (NRC, 1993). Most estimates of EAA requirements have been determined with diets in which EAA supplied were near 100%

TABLE 5-6 Arginine

Species	IBW (g/fish)	Diet CP (%)	Diet DE (kcal/g)	Diet DE (kJ/g)	Arg Levels (% diet DM)	Estimated Arg Requirement	TGC (%)	Response Variable	Model	Reference
Atlantic salmon (*Salmo salar*)	383	42	3.51	14.7	1.13–2.86	2.0–2.2% of diet 4.8% of CP	0.10–0.21	Weight gain	Logistic model	Berge et al. (1997)
	110	40	4.06	17.0	1.1–3.2	1.6% of diet 4.1% of CP	0.18–0.22	Weight gain	Broken-line regression	Lall et al. (1994)
Asian sea bass (*Lates calcarifer*)	2.6	48	4.21	17.6	0–3.00	1.8% of diet 3.8% of CP	0.04–0.05	Weight gain	Broken-line regression	Murillo-Gurrea et al. (2001)
Black sea bream (*Sparus macrocephalus*)	10.5	38	4.18	17.5	1.85–3.46	2.8–3.1% of diet 7.7–8.1% of CP	0.06	Weight gain	Broken-line and quadratic regressions	Zhou et al. (2010a)
Channel catfish (*Ictalurus punctatus*)	195–205	24	2.29	9.6	0.20–1.60	1.0% of diet 4.3% of CP	0.02–0.12	Weight gain	Broken-line regression	Robinson et al. (1981)
Chinook salmon (*Oncorhynchus tshawytscha*)	3.5	40	3.78	15.8	1.60–3.60	2.4% of diet 6.0% of CP	0.07–0.09	Weight gain	ANOVA, least squares regression	Klein and Halver (1970)
Coho salmon (*Oncorhynchus kisutch*)	0.9	45	3.73–3.97	15.6–16.6	2.28–4.32	2.2–2.5% of diet 4.9–5.5% of CP	0.07–0.09	Weight gain	Broken-line regression	Luzzana et al. (1998)
	2.9–2.9	40	3.78	15.8	1.20–3.60	2.3% of diet 5.8% of CP	0.06–0.10	Weight gain	ANOVA, least squares regression	Klein and Halver (1970)
Common carp (*Cyprinus carpio*)	0.6	48	3.51	14.7	0–2.89	1.7% of diet 4.3% of CP	–0.03–0.04	Weight gain	Broken-line regression	Nose (1979)
European sea bass (*Dicentrarchus labrax*)	2.1	46	4.30	18	1.00–2.87	1.8% of diet 3.9% of CP	0.05–0.06	Weight gain	Broken-line regression	Tibaldi et al. (1994)
Hybrid catfish (*Clarias gariepinus* × *Clarias macrocephalus*)	0.6	40	3.59–3.61	15.0–15.1	1.00–2.25	1.8–2.0% of diet 4.5–5.0% of CP	0.12–0.22	Weight gain, feed conversion	Quadratic regression	Singh and Khan (2007)
Hybrid striped bass (*Morone saxatilis* × *Morone chrysops*)	3.1	33	3.47	14.5	1.0–2.4	1.5% of diet 4.4% of CP	0.16–0.17	Feed efficiency	Broken-line regression	Griffin et al. (1992)
	7.1	33	3.47	14.5	0.6–2.0	1.5% of diet 4.4% of CP	0.16–0.21	Weight gain, feed efficiency	ANOVA, broken-line regression	Griffin et al. (1992)
Mrigal carp (*Cirrhinus mrigala*)	0.6	40	3.51	14.7	1.00–2.25	1.8% of diet 4.6% of CP	0.02–0.04	Weight gain	ANOVA, quadratic regression	Ahmed and Khan (2004a)
Japanese flounder (*Paralichthys olivaceus*)	1.9	50	3.75–3.87	15.7–16.2	1.25–3.25	2.0–2.5% of diet 4.1–4.9% of CP	0.05–0.09	Weight gain, feed conversion	Broken-line regression	Alam et al. (2002a,b)
Nile tilapia (*Oreochromis niloticus*)	0.02	28	2.01	8.4	0.60–1.80	1.2% of diet 4.2% of CP	0.05	Weight gain	Broken-line regression	Santiago and Lovell (1988)

TABLE 5-6 Continued

Species	IBW (g/fish)	Diet CP (%)	Diet DE (kcal/g)	Diet DE (kJ/g)	Arg Levels (% diet DM)	Estimated Arg Requirement	TGC (%)	Response Variable	Model	Reference
Rainbow trout (*Oncorhynchus mykiss*)	2.5	30	4.28	17.9	1.01–5.65	1.4% of diet 4.2% of CP	0.12–0.15	Weight gain	Saturation kinetic model	Cho et al. (1992)
	2.6–4.7	40	3.35	14.0	1.30–2.40	1.4–1.7% of diet 3.5–4.2% of CP	0.14–015	Weight gain	Polynomial regression	Chiu et al. (1988)
	7.0	45	4.02	16.8	0.80–2.80	1.6–1.8% of diet 3.6–4.0% of CP	0.08–0.14	Weight gain	Broken-line regression	Walton et al. (1986)
	9.3	43	4.76–4.85	19.9–20.3	1.60–4.00	1.6% of diet 4.0% of CP	0.13–0.16	Weight gain	ANOVA Fisher's test	Fournier et al. (2003)
	12.4	50	3.59	15.0	0.47–2.50	1.4% of diet 4.0% of CP	0.11–0.18	Weight gain	Broken-line regression	Kim et al. (1992b)
Tilapia (*Sarotherodon mossambicus*)	1.7	40	3.01	12.6	1.59–3.18	1.6% of diet 4.0% of CP	0.04–0.05	Weight gain	ANOVA Duncan's multiple range test	Jackson and Capper (1982)
Turbot (*Psetta maxima*)	7.4	55	4.80–4.92	20.1–20.6	1.60–4.00	1.6–3.0% of diet 3.0–5.4% of CP	0.07–0.14	Weight gain	ANOVA Fisher's test	Fournier et al. (2003)
Yellow perch (*Perca flavescens*)	11.0	33	4.06	17.0	0.44–1.84	1.6% of diet 4.9% of CP	0.04–0.29	Weight gain	Quadratic regression	Twibell and Brown (1997)

digestible to the fish. EAA requirements should be assumed to be digestible amino acid requirements (as opposed to "total" amino acid requirements) (NRC, 1993). The digestibility of amino acid in practical feedstuffs may be markedly less than that in the laboratory diets (Cho and Bureau, 1997). Estimates of digestibility of protein and amino acids of common feed ingredient are presented in detail in Chapter 12.

Three basic approaches exist for meeting digestible EAA requirements of fish when feed formulations are deficient in amino acids (e.g., feed is formulated with proteins of low biological value): (1) increasing the total protein level of the diet (Figure 5-5a), (2) supplementation with crystalline amino acids (Figure 5-5b), or (3) use of a combination of different protein sources with different amino acid profiles (Figure 5-5c) (e.g., Guillaume et al., 1999). These three approaches are considered acceptable, although some disagreements exist among nutritionists (some of these issues are reviewed below).

Effectiveness of Crystalline Amino Acids

Crystalline amino acids (CAA) have been used commercially to meet EAA needs of animals for more than 40 years. DL-Methionine, produced by chemical synthesis, and L-Lysine, produced by fermentation, began to find their way into animal feeds in the early 1960s, L-Threonine and L-Tryptophan in the 1980s, and L-Isoleucine and L-Valine in the 1990s. Progress in biotechnology allowed reduction in the cost of large-scale production of amino acids, which has been one of the key factors in expanded use of supplemental amino acids in animal feeds. Because of the high cost of fish meal and concurrent to the increased utilization of more economical protein sources with "imperfect" EAA profiles, CAA are being increasingly used to meet EAA requirements of the fish. CAA are, as such, becoming key components of cost-effective fish-feed formulations. Nonetheless, some segments of the aquaculture industry have been relatively slow to adopt widespread use of CAA to meet EAA require-

TABLE 5-7 Histidine[a]

Species	IBW (g/fish)	Diet CP (%)	Diet DE (kcal/g)	Diet DE (kJ/g)	His Levels (% diet)	Estimated His Requirement	TGC (%)	Response Variable	Model	Reference
Channel catfish (*Ictalurus punctatus*)	200	24	2.87	12.0	0.10–0.80	0.4% of diet 1.5% of CP	0.03–0.18	Weight gain	Broken-line regression	Wilson et al. (1980)
Chinook salmon (*Oncorhynchus tshawytscha*)	2.9	40	3.73	15.6	0.70–1.30	0.7% of diet 1.8% of CP	0.09–0.10	Weight gain	Least squares regression	Klein and Halver (1970)
Chum salmon (*Oncorhynchus keta*)	1.5	40	3.85	16.1	0–2.11	0.7% of diet 1.6% of CP	0.05–0.10	Weight gain	Duncan's multiple range test	Akiyama et al. (1985)
Coho salmon (*Oncorhynchus kisutch*)	3.2	40	3.73	15.6	0.50–1.30	0.7% of diet 1.8% of CP	0.08–0.09	Weight gain	Least squares regression	Klein and Halver (1970)
Common carp (*Cyprinus carpio*)	1.7	48	3.51	14.7	0–1.00	0.8% of diet 2.1% of CP	0.01–0.04	Weight gain	Broken-line regression	Nose (1979)
Mrigal carp (*Cirrhinus mrigala*)	0.6	40	ND	ND	0.25–1.50	0.9% of diet 2.1% of CP	—	Weight gain	Quadratic regression	Ahmed and Khan (2005a)
Nile tilapia (*Oreochromis niloticus*)	0.08	28	2.01	8.4	0.20–0.80	1.0% of diet 1.7% of CP	0.04–0.05	Weight gain	Broken-line regression	Santiago and Lovell (1988)
Rainbow trout (*Oncorhynchus mykiss*)	40.0	34	4.80	20.1	0.26–1.35	0.5–0.6% of diet 1.0–1.2% of CP	—	Weight gain, protein deposition	Exponential function	Rodehutscord et al. (1997)

[a]ND = not determined.

TABLE 5-8 Isoleucine

Species	IBW (g/fish)	Diet CP (%)	Diet DE (kcal/g)	Diet DE (kJ/g)	Ile Levels (% diet)	Estimated Ile Requirement	TGC (%)	Response Variable	Model	Reference
Channel catfish (*Ictalurus punctatus*)	200	28	2.87	12.0	0.3–1.30	0.6% of diet 2.6% of CP	0.04–0.18	Weight gain	Broken-line regression	Wilson et al. (1980)
Chinook salmon (*Oncorhynchus tshawytscha*)	1.0	41	3.54	14.8	0.50–2.00	1.0% of diet 2.6% of CP	0.03–0.08	Weight gain	Not reported	Chance et al. (1964)
Common carp (*Cyprinus carpio*)	3.5	48	3.51	14.7	0–4.29	1.0% of diet 2.5% of CP	–0.01–0.04	Weight gain	Broken-line regression	Nose (1979)
Mrigal carp (*Cirrhinus mrigala*)	0.6	40	3.51	14.7	0.50–1.75	1.3% of diet 3.2% of CP	0.02–0.03	Weight gain	Polynomial regression	Ahmed and Khan (2006)
Rohu carp (*Labeo rohita*)	0.4	40	3.51	14.7	0.75–2.00	1.5–1.6% of diet 3.8–4.0% of CP	0.02–0.04	Weight gain	Polynomial regression	Khan and Abidi (2007a)
Lake trout (*Salvelinus namaycush*)	3.2	35	3.49	14.6	0.54–1.26	0.7–0.9% of diet 1.5–2.1% of CP	0.12	Weight gain	Duncan's multiple range test	Hughes et al. (1983)
Nile tilapia (*Oreochromis niloticus*)	0.09	28	2.01	8.4	0.40–1.10	1.8% of diet 3.1% of CP	0.04–0.05	Weight gain	Broken-line regression	Santiago and Lovell (1988)
Rainbow trout (*Oncorhynchus mykiss*)	47.0	33	4.80	20.1	0.50–1.53	0.7–1.4% of diet 1.5–2.8% of CP	0.10–0.14	Weight gain, protein deposition	Exponential function	Rodehutscord et al. (1997)

TABLE 5-9 Leucine[a]

Species	IBW (g/fish)	Diet CP (%)	Diet DE (kcal/g)	Diet DE (kJ/g)	Leu Levels (% diet)	Estimated Leu Requirement	TGC (%)	Response Variable	Model	Reference
Channel catfish (*Ictalurus punctatus*)	200	24	2.87	12.0	0.60–2.00	0.8% of diet 3.5% of CP	0.11–0.16	Weight gain	Broken-line regression	Wilson et al. (1980)
Chinook salmon (*Oncorhynchus tshawytscha*)	2.5	41	3.54	14.8	1.0–3.1	1.6% of diet 3.9% of CP	0.02–0.06	Weight gain	N/A	Chance et al. (1964)
Common carp (*Cyprinus carpio*)	0.5	48	3.51	14.7	0–2.50	1.3% of diet (with 1.0% isoleucine) 3.3% of CP	0.01–0.05	Weight gain	Broken-line regression	Nose (1979)
Mrigal carp (*Cirrhinus mrigala*)	0.6	40	3.51	14.7	0.75–2.00	1.5% of diet 3.9% of CP	0.02–0.03	Weight gain	Polynomial regression	Ahmed and Khan (2006)
Rohu carp (*Labeo rohita*)	0.4	40	3.51	14.7	0.75–2.00	1.5–1.6% of diet 3.8–3.9% of CP	0.01–0.03	Weight gain	Polynomial regression	Abidi and Khan (2007)
Lake trout (*Salvelinus namaycush*)	3.2	35	3.49	14.6	0.96–2.24	1.3–1.7% of diet 2.7–3.7% of CP	0.11–0.13	Weight gain	Duncan's multiple range test	Hughes et al. (1983)
Nile tilapia (*Oreochromis niloticus*)	0.06	28	2.01	8.4	0.60–1.20	1.9% of diet 3.4% of CP	0.05	Weight gain	Broken-line regression	Santiago and Lovell (1988)
Rainbow trout (*Oncorhynchus mykiss*)	6.0	43	ND	ND	1.10–13.4	3.4 of diet 9.2% of CP	0.12–0.16	Weight gain	Tukey's HSD test difference test	Choo et al. (1991)
	49.0	34	4.80	20.1	1.00–4.20	1.1–1.4% of diet 2.3–2.9% of CP	0.09–0.15	Weight gain, protein deposition	Exponential function	Rodehutscord et al. (1997)

[a]ND = not determined.

ments of fish and shrimp, notably because of concerns with the efficiency with which dietary CAA are used by fish and shrimp. Several studies have shown that CAA were utilized as efficiently as those of intact protein origin in meeting EAA requirements of fish (Murai et al., 1987; Kim et al., 1991; Espe and Lied, 1994; Rodehutscord et al., 1995b; Rollin, 1999; Williams et al., 2001; Rollin et al., 2003; Espe et al., 2006). Conversely, other studies have indicated that CAA appear to be utilized with a lower efficiency than EAA supplied by intact protein (Yamada et al., 1981b; Murai et al., 1987; Espe and Njaa, 1991; Schuhmacher et al., 1997; Zarate and Lovell, 1997; de la Higuera et al., 1998; Refstie et al., 2001; Sveier et al., 2001; Liu et al., 2002; Dabrowski et al., 2003; Peres and Oliva-Teles, 2005; El Haroun and Bureau, 2006; Hauler et al., 2007; X. Q. Zhou et al., 2007; Dabrowski et al., 2010).

Several studies in different fish and shrimp species have shown quite convincingly that CAA may be absorbed slightly more rapidly and/or earlier in the gastrointestinal tract than protein-bound amino acids (Deshimaru, 1976; Yamada et al., 1981a; Kaushik and Dabrowski, 1983; Murai et al., 1987; Cowey and Walton, 1988; Tantikitti and March, 1995; Zarate and Lovell, 1997; Zarate et al., 1999). This faster and/or earlier absorption may result in temporary higher tissue or plasma concentrations of amino acids provided as CAA, because of a slight metabolic dyssynchrony with amino acids derived from protein digestion and a greater proportion of the CAA being catabolized (Batterham, 1984; Cowey and Walton, 1988; Tantikitti and March, 1995; Schumacher et al., 1997; Zarate et al., 1999; Fox et al., 2006). This hypothesis is supported by evidence of better metabolic utilization of CAA in animals fed more frequently (Batterham and Morrison, 1981; Batterham, 1984; Tacon and Cowey, 1985; Tantikitti and March, 1995; Zarate et al., 1999). Reducing the solubility and absorption rate of CAA using coating, encapsulation, or polymerization techniques reportedly improves the efficiency of utilization of CAA in fish and shrimp (Murai et al., 1981; Teshima et al., 1990, 1992; Chen et al., 1992; Cho et al., 1992; Fox et al., 1995; Dabrowski et al., 2003; Alam et al., 2004; Segovia-Quintero and Reigh, 2004; X. Q. Zhou et al., 2007; Dabrowski et al., 2010). Experimental evidence also suggests that the dietary ingredient matrix (e.g., wheat gluten- vs. corn gluten-based diets) and life stage of the animal may influence the efficiency of CAA utilization in some species (Dabrowski et al., 2003; Nang Thu et al., 2007). In both cyprinids (Nose et al., 1974; Murai et al.,

TABLE 5-10 Lysine[a]

Species	IBW (g/fish)	Diet CP (%)	Diet DE (kcal/g)	Diet DE (kJ/g)	Lys Levels (% diet)	Estimated Lys Requirement	TGC (%)	Response Variable	Model	Reference
Atlantic salmon (*Salmo salar*)	4.7	50	4.78	20.0	1.24–2.94	2.0% of diet 4.0% of CP	0.03–0.06	Weight gain	Broken-line regression	Anderson et al. (1993)
	642	43.8	3.51	14.7	0.4–5.6	2.2% of diet 5.0% of CP	0.15–0.26	Weight gain	Exponential equation	Espe et al. (2007)
Asian sea bass (*Lates calcarifer*)	13.1	45	4.09	17.1	0–1.50	2.1% of diet 4.5% of CP	0.05–0.07	Weight gain	Broken-line regression	Murillo-Gurrea et al. (2001)
Black sea bream (*Sparus macrocephalus*)	9.1	38	3.75–3.78	15.7–15.8	2.08–4.05	3.3% of diet 8.6% of CP	0.07–0.08	Weight gain	Polynomial regression	Zhou et al. (2010b)
Channel catfish (*Ictalurus punctatus*)	200	24	2.87	12.0	0.75–1.65	1.2% of diet 5.1% of CP	0.09–0.18	Weight gain	Duncan's multiple range test	Wilson et al. (1977)
	200	30	2.87	12.0	0.75–2.00	1.5% of CP 5.0% of CP	0.05–0.15	Weight gain	Duncan's multiple range test	Robinson et al. (1980a)
Chum salmon (*Oncorhynchus keta*)	1.5	40	3.85	16.1	0–4.47	1.9% of diet 4.8% of CP	0.04–0.10	Weight gain	Duncan's multiple range test	Akiyama et al. (1985)
Cobia (*Rachycentron canadum*)	1.3	44	4.09	17.1	1.15–3.25	2.3% of diet 5.3% of CP	0.09–0.11	Weight gain	Broken-line regression	Q.-C. Zhou et al. (2007)
Common carp (*Cyprinus carpio*)	1.5	48	3.51	14.7	0–2.40	2.2% of diet 5.7% of CP	0.00–0.03	Weight gain	Broken-line regression	Nose (1979)
European sea bass (*Dicentrarchus labrax* L.)	0.9	50	4.18	17.5	1.20–2.45	2.2% of diet 4.4% of CP	0.04–0.05	Weight gain	Broken-line regression	Tibaldi and Lanari (1991)
Freshwater catfish (*Mystus nemurus*)	150	35	ND	ND	0.70–3.10	1.2% of diet 3.5% of CP	0.04–0.06	Weight gain	Nonlinear regression	Tantikitti and Chimsung (2001)
Grass carp (*Ctenopharyngodon idella*)	3.2	38	3.30–3.42	13.8–14.3	0.69–3.08	2.1% of diet 5.4% of CP	0.02–0.04	Weight gain	Broken-line regression	Wang et al. (2005)
Grouper (*Epinephelus coloides*)	15.8	48	4.02	16.8	1.92–3.95	2.8% of diet 5.6% of CP	0.07–0.14	Weight gain	Broken-line regression	Luo et al. (2006)
Hybrid striped bass (*Morone chrysops* × *Morone saxatilis*)	8.0	35	3.25	13.6	1.14–2.37	1.4% of diet 4.0% of CP	0.09–0.11	Weight gain, plasma free lysine	Broken-line regression	Keembiyehetty et al. (1992)
	5–12	35	3.59	15.0	1.2–2.6	1.4% of diet 4.0% of CP	0.07–0.08	Weight gain	Broken-line regression	Griffin et al. (1992)
Mrigal carp (*Cirrhinus mrigala*)	0.6	40	ND	ND	1.50–2.75	2.3% of diet 5.8% of CP	0.02–0.04	Weight gain	Quadratic regression	Ahmed and Khan (2004b)
Japanese flounder (*Paralichthys olivaceusa*)	3.0	46	3.68–3.70	15.4–15.5	0–2.50	1.5–2.1% of diet 3.3–4.6% of CP	0.04–0.09	Weight gain, feed efficiency, N retention	Broken-line regression	Forster and Ogata (1998)
Japanese sea bass (*Lateolabrax japonicus*)	5.5	43	—	—	1.28–4.25	2.5–2.6% of diet 5.8–6.1% of CP	0.10–0.12	Weight gain, FER, PER	Broken-line regression	Mai et al. (2006)

TABLE 5-10 Continued

Species	IBW (g/fish)	Diet CP (%)	Diet DE (kcal/g)	Diet DE (kJ/g)	Lys Levels (% diet)	Estimated Lys Requirement	TGC (%)	Response Variable	Model	Reference
Jundia (*Rhamdia quelen*)	1.4	33	—	—	3.00–6.50	1.5–1.7% of diet 4.5–5.1% of CP	0.02–0.03	Weight gain	Broken-line and polynomial regression	Montes-Girao and Fracalossi (2006)
Nile tilapia (*Oreochromis niloticus*)	0.04	28	2.01	8.4	1.10–1.90	1.4% of diet 5.1% of CP	0.04–0.05	Weight gain	Broken-line regression	Santiago and Lovell (1988)
	118	25	3.08	12.9	1.13–1.57	1.3–1.4% of diet 5.4–5.7% of CP	0.07–0.09	Weight gain	Polynomial regression	Furuya et al. (2004a)
	5.7	30	3.23	13.5	1.04–1.74	1.44% of diet 5.23% of CP	0.12–0.15	Weight gain	Broken-line regression	Furuya et al. (2006)
Rainbow trout (*Oncorhynchus mykiss*)	5.0	45	4.04	16.9	1.00–2.60	1.9% of diet 5.3% of CP	0.11–0.17	Weight gain	Duncan's multiple range test	Walton et al. (1984b)
	5.2	50	4.80–4.95	20.1–20.7	1.30–3.40	2.1–2.7% of diet 6.6–8.4% of CP	0.16–0.21	Weight gain, protein gain, lysine gain	Exponential, polynomial and broken-line regression	Wang et al. (2010)
	7.0	45	4.04	16.9	1.00–2.60	1.9% of diet 4.3% of CP	0.17–0.28	Weight gain	Broken-line regression	Walton et al. (1986)
	13.7	35	3.59	15.0	0.72–1.60	1.3% of diet 3.7% of CP	0.12–0.20	Weight gain	ANOVA	Kim et al. (1992b)
	51.0	34	4.80	20.1	0.45–5.80	1.4–2.5% of diet 3.0–5.3% of CP	0.09–0.15	Weight gain, protein gain	Exponential function	Rodehutscord et al. (1997)
	14.9	45	3.51–4.02	14.7–16.8	1.50–2.25	1.8–2.3% of diet 4.1–5.1% of CP	0.21–0.26	Weight gain	ANOVA	Cheng et al. (2003b)
	24.0	40	4.21–5.07	17.6–21.2	1.20–2.50	1.8–2.3% of diet 4.5–5.8% of CP	0.16–0.22	Weight gain	Broken-line regression, four-parameter logistic equation, exponential equation	Encarnação et al. (2004)
Red sea bream (*Pagrus major*)	1.70	46	3.92–4.21	16.4–17.6	0–2.50	1.7–2.1% of diet 3.6–4.4% of CP	0.06–0.10	Weight gain, feed efficiency, N gain	Broken-line regression	Forster and Ogata (1998)
Tilapia (*Sarotherodon mossambicus*)	1.70	40	3.01	12.6	1.42–2.44	1.6% of diet 4.1% of CP	0.03–0.05	Weight gain	Duncan's multiple range test	Jackson and Capper (1982)
Turbot (*Psetta maxima*)	18.1	50	4.42	18.5	1.19–3.11	2.5% of diet 5.0% of CP	0.04–0.06	Weight gain	Broken-line regression	Peres and Oliva-Teles (2008)

[a]ND = not determined.

TABLE 5-11 Methionine[a]

Species	IBW (g/fish)	Diet CP (%)	Diet DE (kcal/g)	Diet DE (kJ/g)	Met Levels (% diet)	Estimated Met Requirement	TGC (%)	Response Variable	Model	Reference
Arctic charr (*Salvelinus alpinus*)	20.5	40	4.52–4.68	18.9–19.6	0.90–2.40	0.7% of diet (with 0.2% Cys) 1.8% of CP	0.04–0.21	Weight gain	Quadratic regression	Simmons et al. (1999)
Atlantic salmon (*Salmo salar*)	493	42.9	4.9	ND	0.7–1.28	0.7% of diet (with 0.6% Cys) 1.73% of CP	0.32–0.40	Weight gain	ANOVA	Espe et al. (2008)
Asian sea bass (*Lates calcarifer*)	2.59	46	3.82	16.0	0.62–1.26	1.0% of diet (with 0.3% Cys) 2.2% of CP	0.07–0.08	Weight gain	Duncan's multiple range test, break point analysis	Coloso et al. (1999)
Channel catfish (*Ictalurus punctatus*)	200	24	2.53	10.6	0.25–0.81	0.6% of diet 2.3% of CP	0.10–0.17	Weight gain	Linear regression	Harding et al. (1977)
Cobia (*Rachycentron canadum*)	11.6	44	4.11	17.2	0.61–1.68	1.2% of diet (with 0.7% Cys) 2.6% of CP	0.09–0.13	Weight gain	Quadratic regression	Q. C. Zhou et al. (2006)
Common carp (*Cyprinus carpio*)	2.4	48	3.51	14.7	0–1.20	0.8% of diet (with 2.0% Cys) 2.0% of CP	0.00–0.02	Weight gain	Broken-line regression	Nose (1979)
European sea bass (*Dicentrarchus labrax*)	13.4	44	4.06	17.0	0.49–1.62	0.8–0.9% of diet (with 0.4% Cys) 1.8–1.9% of CP	0.03–0.05	Weight gain N gain	Broken-line regression	Tulli et al. (2010)
Hybrid striped bass (*Morone chrysops* × *Morone saxatilis*)	52.7	35	3.25	13.6	0.60–1.60	0.8–0.9% of diet (with 1.3% Cys) 1.8–2.0% of CP	0.03–0.09	Weight gain	Broken-line regression	Keembiyehetty and Gatlin (1993)
	94.4	35	3.25	13.6	0.60–1.00	0.9% of diet (with 1.3% Cys) 1.9% of CP	0.06–0.11	Weight gain	Broken-line regression	Keembiyehetty and Gatlin (1993)
Japanese flounder (*Paralichthys olivaceus*)	2.8	50	3.78–3.82	15.8–16.0	0.53–2.03	1.4–1.5% of diet (with 0.06% Cys) 2.9–3.0% of CP	0.02–0.11	Weight gain, feed efficiency	Broken-line regression	Alam et al. (2000)
Mrigal carp (*Cirrhinus mrigala*)	3.9	40	2.86	12.0	0.50–2.00	1.0–1.2% of diet (with 1.0% Cys) 2.0–3.0% of CP	—	Weight gain	Polynomial regression	Ahmed et al. (2003)
Nile tilapia (*Oreochromis niloticus*)	0.06	28	2.01	8.4	0.15–1.35	0.8% of diet (with 0.2% Cys) 2.7% of CP (with 0.5% Cys)	0.04–0.05	Weight gain	Broken-line regression	Santiago and Lovell (1988)
	1.3	30.6	3.06	12.8	0.89–1.29	1.1% of diet (total sulfur amino acids) 2.1% of CP	0.07–0.08	Weight gain	Polynomial regression	Furuya et al. (2001)
	1.28	28	3.35	14.0	0.31–1.01	0.85% of the diet (total sulfur amino acids) 2.8% of CP	0.02–0.04	Weight gain	Broken-line regression	Nguyen and Davis (2009)

TABLE 5-11 Continued

Species	IBW (g/fish)	Diet CP (%)	Diet DE (kcal/g)	Diet DE (kJ/g)	Met Levels (% diet)	Estimated Met Requirement	TGC (%)	Response Variable	Model	Reference
Rainbow trout (*Oncorhynchus mykiss*)	2.5	40	4.35	18.2	0–2.20	0.8% of diet 1.9% of CP	0.11–0.13	Weight gain	ANOVA	Cowey et al. (1992)
	11.0–15.0	45	3.59	15.0	0.23–0.80	0.5% of diet (with 0.5% Cys) 1.5% of CP	0.11–0.18	Weight gain	Broken-line regression	Kim et al. (1992a)
	51.0	34	4.80	20.1	0–1.20	0.4–0.9% of diet 0.7–1.9% of CP	0.09–0.15	Weight gain, protein deposition	Regression analysis	Rodehutscord et al. (1995a)
Red drum (*Sciaenops ocellatus*)	0.9	35	3.25	13.6	0.35–1.85	0.9% of diet (with 0.12% Cys) 2.7% of CP (with 0.3% Cys)	0.02–0.09	Weight gain	Broken-line regression	Moon and Gatlin (1991)
Mossambique tilapia (*Oreochromis mossambicus*)	1.7	40	3.01	12.6	0.53–0.86	0.5% of diet (with 0.7% Cys) 1.3% of CP	0.04–0.05	Weight gain	Duncan's multiple range test	Jackson and Capper (1982)
Yellow croaker (*Pseudosciaena crocea*)	1.2	43	3.82	16	0.66–1.89	1.3–1.4% of diet (with 0.29% Cys) 3.2–3.3% of CP (with 0.78% Cys)	0.06–0.08	Specific growth rate, feed conversion efficiency	Polynomial regression	Mai et al. (2006)
Yellow perch (*Perca flavescens*)	4.7	34	3.51	14.7	0.37–1.77	1.0% of diet (with 0.03% Cys) 3.1% of CP	0.01–0.03	Weight gain	Broken-line regression	Tibwell et al. (2000)
	8.2	34	3.51	14.7	0.5–1.2 (TSAA)	0.85% of diet (TSAA) 2.5% of CP	0.01–0.05	Weight gain	Broken-line regression	Tibwell et al. (2000)

[a]ND = not determined.

1981) and shrimp (Lim, 1993), adjustment of dietary pH has been shown to improve the utilization of diets with high levels of CAA. It is also clear that water stability of the diet and feeding behavior of the animal can affect efficiency of CAA utilization. In crustaceans, the leaching of dietary CAA may explain, at least in part, the reported poor ability of these animals to utilize CAA (Fox et al., 2006). The efficiency of utilization of CAA appears to be a dynamic function on how dietary and physiological factors interact (El Haroun and Bureau, 2006). This complex issue deserves to be investigated in a systematic and comprehensive fashion.

On the basis of available evidence, the committee suggests that it may be advisable to recommend a slightly higher "safety margin" when meeting a significant proportion of the requirement for an amino acid using CAA compared to using a highly digestible protein-bound amino acid sources. However, estimates of EAA requirements of fish and shrimp are generally derived from growth response assays in which a deficient basal diet is supplemented with graded levels of the deficient EAA supplied in the free form (as a CAA). Therefore, estimates of EAA requirements should already have the appropriate safety margin to account for the potientially reduced efficiency of utilization of CAA by fish and shrimp.

Diet Composition and Essential Amino Acid Requirements

Different modes of expressing EAA requirement are used interchangeably in the scientific literature. These modes of expression are based on different, often diametrically opposed assumptions with regards to how diet composition affects EAA requirements. In practice, the use of different modes of expression will result in dramatically different recommendations, depending on which "school of thought" one adheres to (Table 5-17). Individual EAA levels deemed adequate in the diet may be very different depending on the mode of expression adopted, the composition of the diet formulated, and the amino acid composition of the ingredients used in the formulation (Table 5-18).

The EAA requirements of fish and other animals have been expressed as a percentage of diet by many nutrition-

TABLE 5-12 Phenylalanine

Species	IBW (g/fish)	Diet CP (%)	Diet DE (kcal/g)	Diet DE (kJ/g)	Phe Levels (% diet)	Estimated Phe Requirement	TGC (%)	Response Variable	Model	Reference
Channel catfish (*Ictalurus punctatus*)	195–205	24	2.29	9.6	0.20–0.80	0.5% of diet (with 0.6% Tyr) 2.1% of CP	0.05–0.12	Weight gain, feed efficiency	Broken-line regression	Robinson et al. (1980b)
Chinook salmon (*Oncorhynchus tshawytscha*)	2.5	41	3.54	14.8	0.96–3.50	1.7% of diet (with 0.4% Tyr) 4.4% of CP	0.04–0.05	Weight gain	Not reported	Chance et al. (1964)
Common carp (*Cyprinus carpio*)	1.5	48	3.51	14.7	0–2.75	1.3% of diet (with 2.9% Tyr) 3.3% of CP (with 2.9% Tyr) 2.5% of diet (with 0% Tyr) 6.5% of CP (with 0% Tyr)	–0.01–0.04	Weight gain	Broken-line regression	Nose (1979)
Rohu carp (*Labeo rohita*)	0.2	40	3.51	14.7	0.40–1.65	1.2% of diet (with 1.0% Tyr) 2.9–3.1% of CP (with 2.5% Tyr)	0.02–0.03	Weight gain, feed efficiency	Polynomial and quadratic regression	Khan and Abidi (2007b)
Mrigal carp (*Cirrhinus mrigala*)	0.6	40	3.35	14	0.50–1.75	1.3% of diet (with 0.1% Tyr) 3.3% of CP	0.02–0.03	Weight gain	Quadratic regression	Ahmed (2009)
Nile tilapia (*Oreochromis niloticus*)	0.01	28	2.01	8.4	0.60–1.80	1.1% of diet (with 0.5% Tyr) 3.8% of CP	0.04–0.05	Weight gain	Broken-line regression	Santiago and Lovell (1988)
Rainbow trout (*Oncorhynchus mykiss*)	12.7	35	3.82	16.0	0.26–1.75	0.7% of diet 2.0% of CP	0.06–0.13	Weight gain	ANOVA	Kim (1993)

ists and in the reference literature (e.g., NRC, 1993). What is assumed with this approach is that the concentration of EAA the diet needed to meet metabolic requirements is not influenced by the digestible nutrient composition of the diet. This assumption is difficult to rationalize because optimal concentrations of essential nutrients are expected to be different between lower digestible nutrient density feeds and higher nutrient density feeds.

In order to deal with this issue, Pfeffer et al. (1992) and Rodehutscord et al. (1997) hypothesized that EAA requirements of fish should be expressed per unit of DE (g/Mcal DE). These researchers suggested that dietary DE determines feed intake of animals and that high DE diets should be formulated to higher EAA levels to compensate for a lower feed intake by the fish. This assumption is derived from evidence in the literature showing that when offered diets with various DE levels, fish appear to adjust their feed intake to maintain a particular (daily) energy intake (Jobling and Wandsvik, 1983; Boujard and Médale, 1994; Kaushik and Médale, 1994; Yamamoto et al., 2000, 2002, 2005). However, Encarnação et al. (2004) obtained very similar estimates of requirement for lysine (2.28 vs. 2.33% diet) with diets containing different dietary DE levels (3.82 vs. 4.78 Mcal DE or 16 vs. 20 MJ DE). The observations of Encarnação et al. (2004) suggest that expressing EAA requirements in relation to DE content of the diet is not entirely appropriate. A very similar conclusion was drawn based on a critical analysis of studies on EAA requirements of broiler chickens (Lemme, 2007). In the study of Encarnação et al. (2004), fish fed high-DE diets achieved higher weight gain and similar protein deposition at lower levels of feed intake as compared to fish fed low-DE diets. The lower feed intake observed with increasing diet DE density was offset by a higher efficiency of utilizing lysine for body protein deposition, especially at marginally deficient levels of lysine intake (Figure 5-3). Subsequent studies, however, suggest that this "lysine-sparing" effect is dependent on the dietary DE source used (Encarnação et al., 2006). Fatty acids appear to be more effective in improving lysine and protein deposition when compared to other energy-yielding nutrients, such as amino acids (Encarnação et al., 2006). This further supports the conclusion that expressing EAA requirements in relation to diet DE content is probably not adequate.

Cowey and Cho (1993) and Mambrini and Guillaume (1999) contended that EAA requirements should be expressed as a proportion of dietary protein (g/16 g N) on the assumptions that EAA requirements are dependent on the protein content of the diet, and that a certain balance between amino acids needs to be respected (Cowey and Cho, 1993). Cowey and Cho (1993) assumed that efficiency of utilization

TABLE 5-13 Threonine[a]

Species	IBW (g/fish)	Diet CP (%)	Diet DE (kcal/g)	Diet DE (kJ/g)	Thr Levels (% diet)	Estimated Thr Requirement	TGC (%)	Response Variable	Model	Reference
Atlantic salmon (*Salmo salar*)	1.8	40	4.78	20	0.50–2.18	1.1% of diet 2.6% of CP	0.02–0.08	Weight gain, protein gain, threonine accretion	Broken-line regression	Bodin et al. (2008)
Channel catfish (*Ictalurus punctatus*)	195–205	24	2.29	9.6	0.30–1.25	0.5% of diet 2.2% of CP	0.08–0.16	Weight gain	Duncan's multiple range test	Wilson et al. (1978)
Chum salmon (*Oncorhynchus keta*)	1.5	40	3.85	16.1	0–2.56	1.2% of diet 3.0% of CP	0.03–0.09	Weight gain	Duncan's multiple range test	Akiyama et al. (1985)
Common carp (*Cyprinus carpio*)	1.9	48	3.51	14.7	0–2.00	1.5% of diet 3.9% of CP	–0.01–0.04	Weight gain	Broken-line regression	Nose (1979)
European sea bass (*Dicentrarchus labrax*)	7.5	49	4.25	17.8	0.76–2.46	1.1–1.3% of diet 2.3–2.6% of CP	0.05–0.06	Weight gain	Broken-line, quadratic regressions and sigmoidal model	Tibaldi and Tulli (1999)
Hybrid striped bass (*Morone chrysops* × *Morone saxatilis*)	3.0	35	3.25	13.6	0.49–1.25	0.9% of diet 2.6% of CP	0.06–0.08	Weight gain, feed efficiency	Broken-line regression	Keembiyehetty and Gatlin (1997b)
	9.8	35	3.25	13.6	0.49–1.75	0.8% of diet 1.8% of CP	0.07–0.10	Weight gain, feed efficiency	Broken-line regression	Keembiyehetty and Gatlin (1997b)
Indian catfish (*Heteropneustes fossilis*)	3.6	40	ND	ND	0.50–1.75	1.3% of diet 3.2% of CP	0.02–0.05	Weight gain	Polynomial regression	Ahmed (2007)
Mrigal carp (*Cirrhinus mrigala*)	0.5	40	3.51	14.7	1.00–2.25	1.8% of diet 4.5% of CP	0.02–0.03	Weight gain	Polynomial regression	Ahmed et al. (2004)
Rohu carp (*Labeo rohita*)	0.6	40	3.51	14.7	0.75–2.00	1.5–1.7% of diet 3.8–4.2% of CP	0.01–0.03	Weight gain	Polynomial regression	Abidi and Khan (2008)
Nile tilapia (*Oreochromis niloticus*)	0.05	28	2.01	8.4	0.20–1.60	1.1% of diet 3.8% of CP	0.04–0.05	Weight gain	Broken-line regression	Santiago and Lovell (1988)
Rainbow trout (*Oncorhynchus mykiss*)	1.8	40	4.78	20.0	0.50–2.18	1.1% of diet 2.6% of CP	0.07–0.16	Weight gain, protein gain, threonine accretion	Broken-line regression	Bodin et al. (2008)
Red drum (*Sciaenops ocellatus*)	2.8	35	3.20	13.4	0.49–1.25%	0.8% of diet 2.3% of CP	–0.04–0.14	Weight gain, feed efficiency, PER	Broken-line regression	Boren and Gatlin (1995)

[a]ND = not determined.

of the first limiting dietary EAA decreases with increasing protein level. At higher levels of protein, or higher digestible protein to energy ratio (DP:DE), more amino acids are catabolized to supply energy, and the first-limiting dietary EAA may not be spared at the expense of NEAAs or other nonlimiting EAA. This assumption implies that feeds formulated to high protein levels would still need to be formulated to a "balanced" amino acid profile.

Recent studies reported that lysine requirement (percent diet) of Atlantic salmon fry (Abboudi et al., 2006), rainbow trout (Nang Thu et al., 2009), and red drum (Webb and Gatlin, 2003) was independent of dietary protein level. These

TABLE 5-14 Tryptophan

Species	IBW (g/fish)	Diet CP (%)	Diet DE (kcal/g)	Diet DE (kJ/g)	Trp Levels (% diet)	Estimated Trp Requirement	TGC (%)	Response Variable	Model	Reference
Channel catfish (*Ictalurus punctatus*)	195–205	24	2.29	9.6	0.05–0.34	0.1% of diet 0.5% of CP	0.04–0.16	Weight gain	Duncan's multiple range test	Wilson et al. (1978)
Common carp (*Cyprinus carpio*)	1.8	48	3.51	14.7	0–0.50	0.3% of diet 0.8% of CP	–0.01–0.04	Weight gain	Broken-line regression	Nose (1979)
Hybrid striped bass (*Morone chrysops* × *Morone saxatilis*)	42.4	35	3.61	15.1	0.10–1.40	0.2–0.3% of diet 0.6–0.7% of CP	0.02–0.05	Weight gain, protein retention efficiency	Nonlinear regression	Gaylord et al. (2005)
Mrigal carp (*Cirrhinus mrigala*)	0.6	40	4.09	17.1	0.06–0.56	0.4% of diet 1.0% of CP	0.01–0.03	Weight gain	Polynomial regression	Ahmed and Khan (2005b)
Nile tilapia (*Oreochromis niloticus*)	0.06	28	2.01	8.4	0.05–0.45	0.28% of diet 1.0% of CP	0.04–0.05	Weight gain	Broken-line regression	Santiago and Lovell (1988)
Rainbow trout (*Oncorhynchus mykiss*)	1.6	42	3.92	16.4	0.00–0.38	0.2% of diet 0.6% of CP	0.01–0.12	Weight gain	Duncan's multiple range test	Poston and Rumsey (1983)
	7.0	45	3.90	16.3	0.08–2.60	0.3% of diet 0.4% of CP	0.02–0.13	Weight gain	Break-point plot	Walton et al. (1986)
	14.0	55	4.37	18.3	0.08–0.60	0.3% of diet 0.9% of CP	0.02–0.11	Weight gain	Duncan's multiple range test	Walton et al. (1984a)
	50.0	33	4.80	20.1	0.13–0.56	0.1–0.2% of diet 0.3–0.4% of CP	0.07–0.14	Weight gain, protein deposition	Exponential function	Rodehutscord et al. (1997)

TABLE 5-15 Valine[a]

Species	IBW (g/fish)	Diet CP (%)	Diet DE (kcal/g)	Diet DE (kJ/g)	Val Levels (% diet)	Estimated Val Requirement	TGC (%)	Response Variable	Model	Reference
Channel catfish (*Ictalurus punctatus*)	200	24	2.87	12.0	0.40–1.60	0.7% of diet 3.0% of CP	0.07–0.16	Weight gain	Broken-line regression	Wilson et al. (1980)
Common carp (*Cyprinus carpio*)	1.5	48	3.51	14.7	0–1.50	1.40% of diet 3.60% of CP	0.00–0.03	Weight gain	Broken-line regression	Nose (1979)
Mrigal carp (*Cirrhinus mrigala*)	0.6	40	3.51	14.7	0.75–2.00	1.52% of diet 3.80% of CP	0.02–0.03	Weight gain	Polynomial regression	Ahmed and Khan (2006)
Rohu carp (*Labeo rohita*)	0.16	40	ND	ND	0.75–2.00	1.50% of diet 3.75% of CP	–0.01	Weight gain	Polynomial regression	Abidi and Khan (2004)
Lake trout (*Salvelinus namaycush*)	7.0	35	3.87	16.2	0.78–1.82	0.8–1.0% of diet 1.8–2.2% of CP	0.03–0.06	Weight gain	Duncan's multiple range test	Hughes et al. (1983)
Nile tilapia (*Oreochromis niloticus*)	0.09	28	2.01	8.4	0.04–1.40	1.6% of diet 2.8% of CP	0.04–0.05	Weight gain	Broken-line regression	Santiago and Lovell (1988)
Rainbow trout (*Oncorhynchus mykiss*)	49.0	34	4.80	20.1	0.62–3.42	0.8–1.6% of diet 1.7–3.4% of CP	0.03–0.14	Weight gain, protein deposition	Exponential function	Rodehutscord et al. (1997)

[a]ND = not determined.

TABLE 5-16 Summary of Studies on Essential Amino Acid Requirements of Shrimp[a]

Species	EAA Studied	IBW (g/shrimp)	Diet CP (%)	Diet DE (kcal/g)	Diet DE (kJ/g)	EAA Level (% diet)	Estimated EAA Requirement	Response variable	Model	Reference
Atlantic ditch shrimp (*Palaemonetes varians*)	Arg	0.02	45	3.51	14.7	1.10–2.70	1.9–2.1% of diet 4.2–4.7% CP	Weight gain, feed efficiency	Broken-line regression, exponential model	Palma et al. (2009)
	Lys	0.02	45	3.51	14.7	1.10–2.80	1.9–2.1% of diet 4.2–4.7% CP	Weight gain, feed efficiency	Broken-line regression, exponential model	Palma et al. (2009)
	Met	0.02	45	3.51	14.7	0.50–1.30	0.9–1.1% of diet (with 0.8% Cys) 2.0–2.4% CP	Weight gain, feed efficiency	Broken-line regression, exponential model	Palma et al. (2009)
Kuruma shrimp (*Marsupenaeus japonicus*)	Arg	0.25	50	3.37	14.1	1.20–3.19	2.7% of diet 5.3% of CP	Weight gain	Broken-line regression	Alam et al. (2004)
	Arg	0.79	50	3.37	14.1	ND	1.6% of diet 3.2% of CP	Weight gain	Factorial model based on EAA profiles of whole-body protein	Teshima et al. (2002)
	His	0.79	50	3.37	14.1	ND	0.6% of diet 1.2% of CP	Weight gain	Factorial model based on EAA profiles of whole-body protein	Teshima et al. (2002)
	Ile	0.79	50	3.37	14.1	ND	1.3% of diet 2.6% of CP	Weight gain	Factorial model based on EAA profiles of whole-body protein	Teshima et al. (2002)
	Leu	0.79	50	3.37	14.1	ND	1.9% of diet 3.8% of CP	Weight gain	Factorial model based on EAA profiles of whole-body protein	Teshima et al. (2002)
	Lys	0.79	50	3.37	14.1	ND	1.9% of diet 3.8% of CP	Weight gain	Factorial model based on EAA profiles of whole-body protein	Teshima et al. (2002)
	Met	0.79	50	3.37	14.1	ND	0.7% of diet 1.4% of CP	Weight gain	Factorial model based on EAA profiles of whole-body protein	Teshima et al. (2002)
	Phe	0.79	50	3.37	14.1	ND	1.5% of diet 3.0% of CP	Weight gain	Factorial model based on EAA profiles of whole-body protein	Teshima et al. (2002)
	Thr	0.79	50	3.37	14.1	ND	1.3% of diet 2.6% of CP	Weight gain	Factorial model based on EAA profiles of whole-body protein	Teshima et al. (2002)
	Trp	0.79	50	3.37	14.1	ND	0.4% of diet 0.8% of CP	Weight gain	Factorial model based on EAA profiles of whole-body protein	Teshima et al. (2002)
	Val	0.79	50	3.37	14.1	ND	1.4% of diet 2.8% of CP	Weight gain	Factorial model based on EAA profiles of whole-body protein	Teshima et al. (2002)

continued

TABLE 5-16 Continued

Species	EAA Studied	IBW (g/shrimp)	Diet CP (%)	Diet DE (kcal/g)	Diet DE (kJ/g)	EAA Level (% diet)	Estimated EAA Requirement	Response variable	Model	Reference
Black tiger shrimp (*Penaeus monodon*)	Arg	0.02	40	ND	ND	0.60–3.00	1.9% of diet 5.3% of CP	Weight gain	Broken-line regression	Millamena et al. (1998)
	Arg	0.32	45	ND	ND	1.31–3.61	2.5% of diet 5.5% of CP	Weight gain	Broken-line regression	Chen et al. (1992)
	His	0.02	35–40	ND	ND	0.42–1.17	0.8% of diet 2.2% of CP	Weight gain	Quadratic regression	Millamena et al. (1999)
	Ile	0.02	35–40	ND	ND	0.52–2.02	1.0% of diet 2.7% of CP	Weight gain	Broken-line regression	Millamena et al. (1999)
	Leu	0.02	35–40	ND	ND	0.95–2.95	1.7% of diet 4.3% of CP	Weight gain	Quadratic regression	Millamena et al. (1999)
	Lys	0.02	40	ND	ND	1.18–3.28	2.1% of diet 5.2% of CP	Weight gain	Broken-line regression	Millamena et al. (1998)
		2.4	34	ND	ND	0.6–4.5	2.0% of diet (5.8% of CP)	Weight gain	Dilution technique/ factorial model	Richard et al. (2010)
	Met	0.02	37	3.61	15.1	0.72–1.12	0.9% of diet (with 0.4% Cys) 2.4% of CP	Weight gain	Broken-line regression	Millamena et al. (1996)
		2.4	34	ND	ND	0.3–1.6%	0.9% of diet (0.1–0.3% Cys) 2.9% of CP	Weight gain	Dilution technique/ factorial model	Richard et al. (2010)
	Phe	0.02	35–40	ND	ND	0.62–2.12	1.4% of diet 3.7% of CP	Weight gain	Quadratic regression	Millamena et al. (1999)
	Thr	0.05	40	ND	ND	0.72–2.12	1.4% of diet 3.5% of CP	Weight gain	Quadratic regression	Millamena et al. (1997)
	Trp	0.02	35–40	ND	ND	0.04–0.36	0.2% of diet 0.5% of CP	Weight gain	Quadratic regression	Millamena et al. (1999)
Pacific white shrimp (*Litopenaeus vannamei*)	Lys	0.10	35	3.13	13.1	1.2–2.3	1.6% of diet 4.5% of CP	Weight gain	Broken-line regression	Fox et al. (1995)
		0.10	45	3.39	14.2	1.5–3.0	2.1% of diet 4.7% of diet	Weight gain	Broken-line regression	Fox et al. (1995)

[a]ND = not determined.

results also suggest that expressing EAA requirements as "percentage of protein" is not totally appropriate. In the study of Webb and Gatlin (2003), similar estimates of lysine requirement for red drum (1.5% diet) were obtained using diets with different dietary protein levels (35% and 45% CP). Evidence generated by Encarnação et al. (2006) corroborated these findings. Requirement studies with rainbow trout, using similar dietary models but differing in the dietary protein level (32% CP; Rodehutscord et al., [1997], and 40% CP; Encarnação et al. [2004]), achieved similar growth rates and similar estimates of lysine requirement, expressed as dietary concentration (2.3% of diet). Dramatically different estimates of requirements (7.3 g/16 g N; Rodehutscord et al. [1997] vs. 5.8 g/16 g N) are obtained when lysine requirements are expressed in relation to protein content of the diet, thus suggesting that expressing lysine requirements as a percentage of protein is generally not appropriate.

In most EAA requirement studies, to avoid deficiencies and to ensure that the tested EAA is the first limiting at all tested levels, the remaining EAA (and NEAA) are supplied in excess of their requirements. This relative oversupply of dietary amino acids should result in an underestimation of EAA requirements when expressed as percentage of dietary protein. Moreover, the use of diets with excessive protein for a particular life stage of a fish species is likely to underestimate requirements expressed as a percentage of dietary protein (Hauler and Carter, 2001b). Finally, as described above, crude protein concentration is rarely a true reflection of the true protein and amino acid composition of the diet (Mariotti et al., 2008). Crude protein content of the diet represent mixture of different N containing compounds in different proportions and does not offer a very stable and rational basis for the expression of EAA requirements.

It is clear that all current modes of expression of EAA

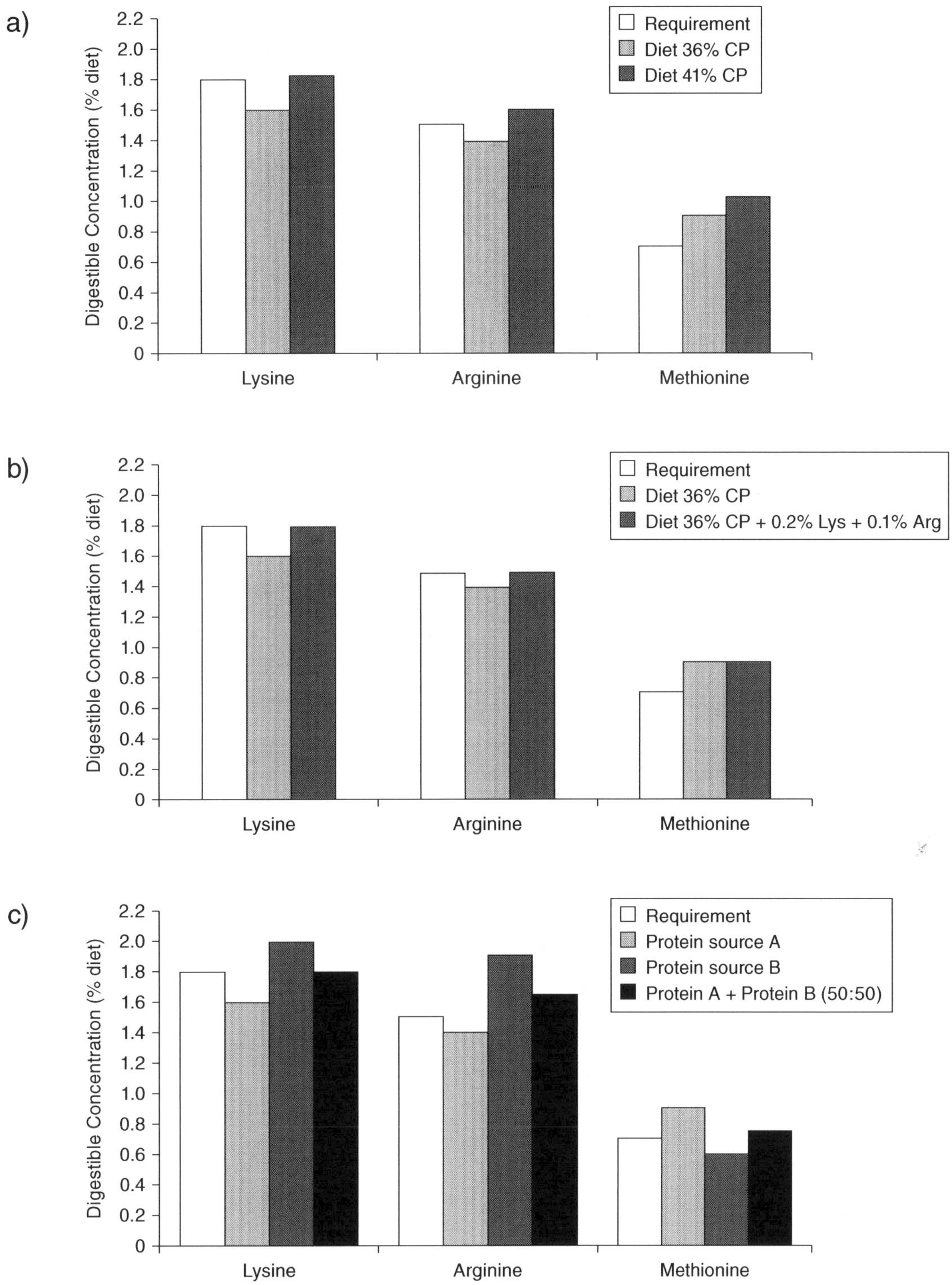

FIGURE 5-5 Meeting essential amino acid requirements of rainbow trout (*Oncorhynchus mykiss*) using three different approaches: (a) increasing dietary protein levels, (b) use of crystalline amino acid supplements, and (c) the use of combinations of proteins with different (complementary) amino acid profiles.

TABLE 5-17 Arginine Requirement of Rainbow Trout (*Oncorhynchus mykiss*) According to Different Modes of Expression

Reference	Requirement
NRC (1993)	1.5% of diet
Rodehutscord et al. (1997)[a]	4.184 g/Mcal (1.0 g/MJ) digestible energy
Mambrini and Guillaume (1999)	4.4% of protein (g/16 g N)

[a]Interpretation of NRC (1993) requirement based on the recommendation of Rodehutscord et al. (1997) to express requirement in relation to digestible energy content of the diet.

requirements have significant limitations, and more work is needed to develop more rational and robust modes of expressing or estimating EAA requirements of fish and shrimp. Given the limitations of expressing EAA in relation to protein content or DE of the diet, the committee decided to express requirements of essential amino acids as percentage diet dry matter as was the case in the previous NRC report (1993) (Table 18-1).

Ideal Protein Concept

An "ideal protein" has been defined as the amino acid profile that meets exactly the requirement of the animal with no excess or deficit (Wang and Fuller, 1989; Emmert and Baker, 1997). All EAAs are balanced to be equally limiting. The ideal protein concept is based on the famous barrel analogy put forward by von Liebig around 1840 and adapted to animal nutrition by Mitchell and Scott in the 1950s. The ideal protein concept has been used by swine and poultry nutritionists for several years, and the applicability of this concept to fish has been the focus of a few studies (Green and Hardy, 2002; Rollin et al., 2003; Furuya et al., 2004a,b).

TABLE 5-18 Dietary Arginine Level Expected to Meet Requirement of Rainbow Trout (*Oncorhynchus mykiss*) Fed Different Feed Formulae According to Different Modes of Expressing Requirement

Item	Starter	Grower	High Energy
Composition			
Crude Protein, %	50	42	45
Lipids, %	12	18	25
Digestible energy, Mcal/kg (MJ/kg)	4.06 (17)	4.30 (18)	4.78 (20)
School of Thought	**Arginine level deemed adequate (g/kg feed)**		
1. % diet	15	15	15
2. Mcal/kg (MJ/kg)	4.06 (17)	4.30 (18)	4.78 (20)
3. % Protein	22	18	20
High-Low, % difference[a]	36	18	29

[a](Highest value – lowest value)/lowest concentration.

According to the ideal protein concept, optimum dietary EAA quantities are considered as proportions relative to total EAAs, rather than as proportions relative to the whole diet (Wang and Fuller, 1989). Ideal amino acid patterns are usually stated as the ratio of each of the nine other EAAs to lysine, which is given the arbitrary value of 100. The use of this approach implies the need to first define the lysine requirement of the fish at different sizes and under various conditions and as such cannot substitute for traditional EAA requirement studies. The ideal protein concept may be especially useful as a guideline to establish an EAA profile to formulate practical or experimental diets for fish species whose EAA requirements are partially unknown.

The ideal protein concept assumes that efficiency of utilization of the first limiting dietary EAA decreases with increasing protein level. At higher levels of protein, or higher protein:energy ratios, more amino acids are catabolized to supply energy, and the first limiting dietary EAA is assumed to not be spared at the expense of NEAAs or other nonlimiting EAAs. Expression of EAA requirements in relation to other EAAs assumes a need to "balance" amino acid levels in diets formulated to high protein levels with "imbalanced" protein sources. Whether amino acids catabolized for energy need to be in an "ideal" balance is very much a matter of debate. Studies have suggested that an "imbalance" in the amino acid profile of the diet may depress feed intake and growth of fish (e.g., Yamamoto et al., 2004). An "imbalanced amino acid profile" is also something that is not easy to define. Many different amino acids can be manipulated and different amino acids (and combinations thereof) may have different effects. Understanding of the effect of the "amino acid balance" of the diet, in its multitude of definitions and forms, is not very advanced. Excess histidine, arginine, and leucine had no effect in rainbow trout fed a diet with "balanced amino acid profile" according to the ideal protein concept (Green and Hardy, 2008). Richard et al. (2010) also observed no effect of excess methionine and lysine in shrimp diet formulated according to the ideal protein concept. These studies indicate that excess amino acids (amino acid catabolized for energy) do not need to be "ideally" balanced.

The main application for the ideal protein concept lies in the formulation of low-protein diets resulting in optimal efficiency of protein utilization and minimal nitrogenous wastes. Defining the ideal dietary amino acid profile has been the focus of a significant number of studies. Most studies have based the ideal amino acid profile on EAA profile of the whole body or egg of the species of interest (Wilson and Cowey, 1985), and very few studies have attempted to defined the ideal amino acid profile of fish diet through feeding trials (Alam et al., 2002b; Green and Hardy, 2002; Rollin et al., 2003; Peres and Oliva-Teles, 2009). Variation in the estimates of ideal amino acid profiles across species is no greater than variability of estimates for one given species. Relative contributions of the growth and maintenance amino acid requirements may result in significant changes in the ideal protein profile for fish. The very limited amount of information on the maintenance requirement of different amino acids prevents defining more precisely how the ideal amino acid profile may vary with growth rate or other factors. Table 5-19 proposes an "average" ideal protein profile for teleost fish and one for penaeid shrimp based on compilation of about 15 studies carried out with approximately 10 fish and shrimp species. These "ideal" profiles are most probably imperfect but provide a reasonable starting point for formulating diets on an ideal protein basis.

Estimating Essential Amino Acid Requirements of Fish across Life Stages using a Factorial Model

Factorial models have been used to estimate EAA requirements for poultry (Baker et al., 1996), swine (Moughan, 1989; Susenbeth, 1995), and, more recently, fish (Hauler and Carter, 2001b; Abboudi et al., 2006; Bodin et al., 2008). In the most commonly used framework (reviewed earlier in this chapter), EAA utilization is explicitly represented in terms of maintenance requirement, catabolism of EAAs supplied in excess of requirement, inevitable catabolism, preferential catabolism of EAAs for energy use, and deposition as body protein (Moughan, 2003). This type of approach has a number of limitations when applied to fish (reviewed by Bureau and Encarnação, 2006), but it provides a relatively straightforward means of estimating EAA requirements of fish.

The EAAs are individually represented, while NEAA are not differentiated and converted to equivalents of nonspecific nitrogen. The simplest factorial approach assumes individual EAA retention per unit of live weight gain or protein deposition, efficiency of utilization of each EAA, and the obligatory loss (maintenance) for each EAA. The requirement (grams per animal per day) can then be calculated for each EAA by multiplying protein retention by EAA concentration in the protein retained and dividing by a constant efficiency of utilization and accounting for obligatory losses (Susenbeth, 1995). Target dietary concentration of each EAA is calculated on the basis of the predicted feed intake and weight gain (or feed conversion ratio) of the animal. Predicted feed intake is obtained from past performance record or calculated using feed requirement models.

Tables 5-20 and 5-21 present estimates of EAA requirements of rainbow trout and Atlantic salmon of different weight classes. The estimates of EAA requirements were based on a simulation using the "hybrid nutrient-flow bioenergetics model" (Bureau and Hua, 2006), which combines the classical concepts of bioenergetics and advanced nutrient partitioning scheme for fish and estimates of maintenance requirements and inevitable catabolism of amino acids from the study of Hua et al. (2010). The model is based on the

TABLE 5-19 Ideal Amino Acid Profile for Teleost Fish and Penaeid Shrimp Derived from a Review of the Literature

Amino Acid	Teleost Fish	Penaeid Shrimp
Lysine (Reference)	100	100
Arginine	82	95
Histidine	35	38
Isoleucine	54	48
Leucine	70	81
Methionine	38	48
Met + Cys	54	65
Phenylalanine	55	55
Phe + Tyr	90	100
Threonine	56	67
Tryptophan	14	10
Valine	61	65

TABLE 5-20 Digestible Essential Amino Acid Requirements (% Diet Dry Matter) Estimated Using a Factorial Model for Rainbow Trout (*Oncorhynchus mykiss*) of Different Weights Fed Diets with 4.78 Mcal DE (20 MJ DE)

	Weight Class		
	0.2–20 g	20–500 g	500–1,500 g
Amino Acid	% diet DM		
Arginine	1.91	1.77	1.62
Histidine	0.83	0.77	0.69
Isoleucine	1.27	1.19	0.98
Leucine	2.26	2.11	1.78
Lysine	2.47	2.31	1.92
Met + Cys	1.32	1.23	1.10
Phe + Tyr	2.49	2.33	1.82
Threonine	1.77	1.63	1.60
Tryptohan	0.43	0.40	0.42
Valine	1.90	1.76	1.64

TABLE 5-21 Digestible Essential Amino Acid Requirements (% Diet Dry Matter) Estimated Using a Factorial Model for Atlantic Salmon (*Salmo salar*) of Different Weights Fed Diets with 4.78 Mcal DE (20 MJ DE)

	Weight Class			
	0.2–20 g	20–500 g	500–1,500 g	> 1,500 g
Amino Acid	% diet DM			
Arginine	1.79	1.82	1.70	1.46
Histidine	0.80[a]	0.80[a]	0.75[a]	0.64[a]
Isoleucine	1.32	1.32	1.22	1.04
Leucine	2.31	2.31	2.14	1.82
Lysine	2.55	2.54	2.35	2.00
Met + Cys	1.28	1.30	1.21	1.03
Phe + Tyr	2.71	2.68	2.46	2.09
Threonine	1.55	1.60	1.51	1.30
Tryptohan	0.35	0.37	0.35	0.30
Valine	1.75	1.79	1.67	1.44

[a]Histidine level adequate to support optimal growth but not optimal ocular health.

determining effect of protein deposition on weight gain, which is driven by digestible protein and amino acids and digestible energy supply. Maintenance EAA requirements were those reported in Rodehutscord et al. (1997) but converted to mg/kg MBM per degree day to account for the effects of metabolic body weight (MBM, kg $BW^{0.8}$) and water temperature. A rate of inevitable catabolism of 30% of digestible amino acid intakes was derived from Rodehutscord et al. (2000).

The model allowed estimates of EAA requirements with increases in live weight of the animal. The estimated EAA requirement generally progressively decreased when the live weight of the fish increased in the two species. These estimates should be viewed with some degree of skepticism because they are derived from a first application (or iteration) of novel model. The reliability of these estimates remains to be tested. More work is needed to develop rational and accurate factorial EAA requirement models for different fish species of commercial importance.

REFERENCES

Abboudi, T., M. Mambrini, W. Ooghe, Y. Larondelle, and X. Rollin. 2006. Protein and lysine requirements for maintenance and protein tissue accretion in Atlantic salmon (*Salmo salar*) fry. Aquaculture 261:383.

Abboudi, T., W. Ooghe, Y. Larondelle, and X. Rollin. 2007. Determination of the threonine requirement for maintenance in Atlantic salmon (*Salmo salar*) fry with the diet dilution procedure. Aquacult. Nutr. 13:281–290.

Abboudi, T., M. Mambrini, Y. Larondelle, and X. Rollin. 2009. The effect of dispensable amino acids on nitrogen and amino acid losses in Atlantic salmon (*Salmo salar* L.) fry fed a protein-free diet. Aquaculture 289:327–333.

Abidi, S. F., and M. A. Khan. 2004. Dietary valine requirement of Indian major carp, *Labeo rohita* (Hamilton) fry. J. Appl. Ichthyol. 20:118–122.

Abidi, S. F., and M. A. Khan. 2007. Dietary leucine requirement of fingerling Indian major carp, *Labeo rohita* (Hamilton). Aquac. Res. 38:478–486.

Abidi, S. F., and M. A. Khan. 2008. Dietary threonine requirement of fingerling Indian major carp, *Labeo rohita* (Hamilton). Aquac. Res. 39:1498–1505.

Adeola, O. 1995. Dietary lysine and threonine utilization by young pigs: Efficiency for carcass growth. Can. J. Anim. Sci. 75:445–452.

Ahmed, I. 2007. Dietary amino acid L-threonine requirement of fingerling Indian catfish, *Heteropneustes fossilis* (Bloch) estimated by growth and biochemical parameters. Aquacult. Int. 15:337–350.

Ahmed, I. 2009. Dietary total aromatic amino acid requirement and tyrosine replacement value for phenylalanine in Indian major carp: *Cirrhinus mrigala* (Hamilton) fingerlings. J. Appl. Ichthyol. 25:719–727.

Ahmed, I., and M. A. Khan. 2004a. Dietary arginine requirement of fingerling Indian major carp, *Cirrhinus mrigala* (Hamilton). Aquacult. Nutr. 10:217–225.

Ahmed, I., and M. A. Khan. 2004b. Dietary lysine requirement of fingerling Indian major carp, *Cirrhinus mrigala* (Hamilton). Aquaculture 235:499–511.

Ahmed, I., and M. A. Khan. 2005a. Dietary histidine requirements of fingerling Indian major carp, *Cirrhinus mrigala* (Hamilton). Aquacult. Nutr. 11:359–366.

Ahmed, I., and M. A. Khan. 2005b. Dietary tryptophan requirements of fingerling Indian major carp, *Cirrhinus mrigala* (Hamilton). Aquac. Res. 36:687–695.

Ahmed, I., and M. A. Khan. 2006. Dietary branched-chain amino acid valine, isoleucine and leucine requirements of fingerling Indian major carp, *Cirrhinus mrigala* (Hamilton). Brit. J. Nutr. 96:450–460.

Ahmed, I., M. A. Khan, and A. K. Jafri. 2003. Dietary methionine requirement of fingerling Indian major carp, *Cirrhinus mrigala* (Hamilton). Aquacult. Int. 11:449–462.

Ahmed, I., M. A. Khan, and A. K. Jafri. 2004. Dietary threonine requirement of fingerling Indian major carp, *Cirrhinus mrigala* (Hamilton). Aquac. Res. 35:162–170.

Akiyama, T., S. Arai, T. Murai, and T. Nose. 1985. Threonine, histidine and lysine requirements of chum salmon fry. B. Jpn. Soc. Sci. Fish. 51:635–639.

Akiyama, T., T. Murai, and K. Mori. 1986. Role of tryptophan metabolites in inhibition of spinal deformity of chum salmon fry caused by tryptophan deficiency. B. Jpn. Soc. Sci. Fish. 52:1255–1259.

Alam, M. S., S. Teshima, M. Ishikawa, and S. Koshio. 2000. Methionine requirement of juvenile Japanese flounder *Paralichthys olivaceus*. J. World Aquacult. Soc. 31:618–626.

Alam, M. S, S. I. Teshima, M. Ishikawa, and S. Koshio. 2002a. Effects of dietary arginine and lysine levels on growth performance and biochemical parameters of juvenile Japanese flounder *Paralichthys olivaceus*. Fish. Sci. 68:509–516.

Alam, M. S., S. Teshima, S. Koshio, and M. Ishikawa. 2002b. Arginine requirement of juvenile Japanese flounder *Paralichthys olivaceus* estimated by growth and biochemical parameters. Aquaculture 205:127–140.

Alam, M. S., S. Teshima, M. Ishikawa, D. Hasegawa, and S. Koshio. 2004. Dietary arginine requirement of juvenile kuruma shrimp *Marsupenaeus japonicus* (Bate). Aquac. Res. 35:842–849.

Allmang, C., and A. Krol. 2006. Selenoprotein synthesis: UGA does not end the story. Biochimie 88:1561–1571.

Anderson, J. S., S. P. Lall, D. M. Anderson, and M. A. McNiven. 1993. Quantitative dietary lysine requirement of Atlantic salmon (*Salmo salar*) fingerlings. Can. J. Fish. Aquat. Sci. 50:316–322.

Arai, S. 1981. A purified test diet for coho salmon, *Oncorhynchus kisutch*, fry. B. Jpn. Soc. Sci. Fish. 47:547–550.

Austic R. E., and R. L. Scott. 1975. Involvement of food intake in the lysine-arginine antagonism in chick. J. Nutr. 105:1122–1131.

Azevedo, P. A., C. Y. Cho, S. Leeson, and D. P. Bureau. 1998. Effects of feeding level and water temperature on growth, nutrient and energy utilization and waste outputs of rainbow trout (*Oncorhynchus mykiss*). Aquat. Living Resour. 11:227–238.

Azevedo, P. A., S. Leeson, C. Y. Cho, and D. P. Bureau. 2004a. Growth and feed utilization of large size rainbow trout (*Oncorhynchus mykiss*) and Atlantic salmon (*Salmo salar*) reared in freshwater: Diet and species effects, and responses over time. Aquacult. Nutr. 10:401–411.

Azevedo, P. A., S. Leeson, C. Y. Cho, and D. P. Bureau. 2004b. Growth, nitrogen and energy utilization of juveniles from four salmonid species: Diet, species and size effects. Aquaculture 234:393–414.

Azevedo, P. A., J. van Milgen, S. Leeson, and D. P. Bureau. 2005. Comparing efficiency of metabolizable energy utilization by rainbow trout (*Oncorhynchus mykiss*) and Atlantic salmon (*Salmo salar*) using factorial and multivariate approaches. J. Anim. Sci. 83:842–851.

Baker, D. H. 2006. Comparative species utilization and toxicity of sulfur amino acids. J. Nutr. 136:1670S–1675S.

Baker, D. H. 2007. Lysine, arginine, and related amino acids: An introduction to the 6th Amino Acid Assessment Workshop. J. Nutr. 137:1599S–1601S.

Baker, D. H., S. R. Fernandez, C. M. Parsons, H. M. Edwards, III, J. L. Emmert, and D. M. Webel. 1996. Maintenance requirement for valine and efficiency of its use above maintenance for accretion of whole body valine and protein in young chicks. J. Nutr. 126:1844–1851.

Baldwin, R. L., and A. C. Bywater 1984. Nutritional energetics of animals. Annu. Rev. Nutr. 4:101–114.

Batterham, E. S. 1984. Utilization of free lysine by pigs. Pig News Info 5:85–86.

Batterham, E. S., and R. D. Morrison. 1981. Utilization of free lysine by growing pigs. J. Nutr. Res. Rev. 5:1–18.

Behne, D., and A. Kyriakopoulos. 2001. Mammalian selenium-containing proteins. Annu. Rev. Nutr. 21:453–473.

Berge, G. E., E. Lied, and H. Sveier. 1997. Nutrition of Atlantic salmon (*Salmo salar*): The requirement and metabolism of arginine. Comp. Biochem. Phys. A 117:501–509.

Bermudes, M., B. D. Glencross, K. Austen, and W. Hawkins. 2010. Effect of high water temperatures on nutrient and energy retention in barramundi (*Lates calcarifer*). Aquaculture 306:160–166.

Bjerkas, E., and H. Sveier. 2004. The influence of nutritional and environmental factors on osmoregulation and cataracts in Atlantic salmon (*Salmo salar* L). Aquaculture 235:101–122.

Block, K. P. and A. E. Harper. 1991. High levels of dietary amino and branched-chain α-keto acids alter plasma and brain amino acid concentrations in rats. J. Nutr. 121:663–671.

Bodin, N., M. Mambrini, J. P. Wauters, T. Abboudi, W. Ooghe, E. Le Boulenge, Y. Larondelle, and X. Rollin. 2008. Threonine requirements in rainbow trout (*Oncorhynchus mykiss*) and Atlantic salmon (*Salmo salar*) at the fry stage are similar. Aquaculture 274:353–365.

Bodin, N, B. Govaerts, T. Abboudi, C. Detavernier, S. De Saeger, Y. Larondelle, and X. Rollin. 2009. Protein level affects the relative lysine requirement of growing rainbow trout (*Oncorhynchus mykiss*) fry. Brit. J. Nutr. 102:37–53.

Boren, R. S., and D. M. Gatlin, III. 1995. Dietary threonine requirement of juvenile red drum *Sciaenops ocellatus*. J. World Aquacult. Soc. 26:279–283.

Boujard, T., and F. Médale. 1994. Regulation of voluntary feed intake in juvenile rainbow trout fed by hand and by self-feeders with diets containing different protein/energy ratios. Aquat. Living Resour. 7:211–215.

Brachet, P., and A. Puigserver. 1987. Transport of methionine hydroxy analog across the brush border membrane of rat jejunum. J. Nutr. 117:1241–1246.

Breck, O., E. Bjerkas, P. Campbell, P. Arnesen, P. Haldorsen, and R. Waagbo. 2003. Cataract preventative role of mammalian blood meal, histidine, iron and zinc in diets for Atlantic salmon (*Salmo salar* L.) of different strains. Aquacult. Nutr. 9:341–350.

Breck, O., E. Bjerkas, P. Campbell, J. D. Rhodes, J. Sanderson, and R. Waagbo. 2005a. Histidine nutrition and genotype affect cataract development in Atlantic salmon, *Salmo salar* L. J. Fish. Dis. 28:357–371.

Breck, O., E. Bjerkas, J. Sanderson, R. Waagbo, and P. Campbell. 2005b. Dietary histidine affects lens protein turnover and synthesis of N-acetylhistidine in Atlantic salmon (*Salmo salar* L.) undergoing parr-smolt transformation. Aquacult. Nutr. 11:321–332.

Brody, T. 1999. Classification of biological structure. Pp. 1–56 in Nutritional Biochemistry, Second Edition. San Diego, CA: Academic Press.

Brosnan, J. T., and M. E. Brosnan. 2006a. Branched-chain amino acids: Enzyme and substrate regulation. J. Nutr. 136:207S–211S.

Brosnan, J. T., and M. E. Brosnan. 2006b. The sulfur-containing amino acids: An overview. J. Nutr. 136:1636S–1640S.

Brown, P. B. 1995. Using whole body amino acid patterns and quantitative requirements to rapidly develop diets for new species such as striped bass (*Morone saxatilis*). J. Appl. Ichthyol. 11:342–346.

Bureau, D. P. 1997. The Partitioning of Energy from Digestible Carbohydrates by Rainbow Trout (*Oncorhynchus mykiss*). Ph.D. Thesis, University of Guelph, Guelph, Ontario, Canada.

Bureau, D. P., and P. M. Encarnação. 2006. Adequately determining the amino acid requirements of fish: The case example of lysine. Pp. 29–54 in Avances en Nutricion Acuicola, VIII Simposium Internacional de Nutricion Acuicola, L. E. Cruz-Suarez, D. Ricque-Marie, M. Tapia-Salazar, M. G. Nieto-Lopez, D. A. Villareal Cavazos, A. C. Puello Cruz, and A. Garcia-Ortega, eds. Mazatlan, Sinaloa, Mexico, 15–17 November, 2006.

Bureau, D. P., and K. Hua. 2006. Predicting feed efficiency of rainbow trout: Transitioning from bioenergetics models to approaches based on protein accretion. XII International Symposium on Fish Nutrition and Feeding, 28 May–1 June 2006, Biarritz, France.

Bureau, D. P., S. J. Kaushik, and C. Y. Cho. 2002. Bioenergetics. Pp. 1–56 in Fish Nutrition, J. E. Halver and R. W. Hardy, eds. San Diego, CA: Academic Press.

Buxbaum, E. 2007. Fundamentals of Protein Structure and Function. New York: Springer.

Carpenter, K. J. 1960. The estimation of the available lysine in animal-protein foods. Biochem. J. 77:604–610.

Carpenter, K. J. 2003. A short history of nutritional science: Part 3 (1912–1944). J. Nutr. 133:3023–3032.

Castellano S., A. M. Andrés, E. Bosch, M. Bayes, R. Guigó, A. G. Clark. 2009. Low exchangeability of selenocysteine, the 21st amino acid, in vertebrate proteins. Mol. Biol. Evol. 26:2031–2040.

Catacutan, M. R., and R. M. Coloso. 1995. Effect of dietary protein to energy ratios on growth, survival, and body composition of juvenile Asian seabass, *Lates calcarifer*. Aquaculture 131:125–133.

Chadwick, T. D., and P. A. Wright. 1999. Nitrogen excretion and expression of urea cycle enzymes in the Atlantic cod (*Gadus morhua* L.): A comparison of early life stages with adults. J. Exp. Biol. 202:2653–2662.

Chance, R. E., E. T. Mertz, and J. E. Halver. 1964. Nutrition of salmonoid fishes: XII. Isoleucine, leucine, valine and phenylalanine requirements of Chinook salmon and interrelations between isoleucine and leucine for growth. J. Nutr. 83:177–185.

Chaplin, M. 2006. Do we underestimate the importance of water in cell biology? Nat. Rev. Mol. Cell. Bio. 7:861–866.

Chen, H. Y., Y. T. Len, and I. Roelants. 1992. Quantification of arginine requirements of juvenile marine shrimp *Penaeus monodon* using microencapsulated arginine. Mar. Biol. 114:229–233.

Cheng, Z. J., R. W. Hardy, and M. Blair. 2003a. Effects of supplementing methionine hydroxy analogue in soybean meal and distiller's dried grain-based diets on the performance and nutrient retention of rainbow trout *Oncorhynchus mykiss* (Walbaum). Aquac. Res. 34:1303–1310.

Cheng, Z. J., R. W. Hardy, and J. L. Usry. 2003b. Effects of lysine supplementation in plant protein-based diets on the performance of rainbow trout (*Oncorhynchus mykiss*) and apparent digestibility coefficients of nutrients. Aquaculture 215:255–265.

Chiu, Y. N., R. E. Austic, and G. L. Rumsey. 1986. Urea cycle activity and arginine formation in rainbow trout (*Salmo gairdneri*). J. Nutr. 116:1640–1650.

Chiu, Y. N., R. E. Austic, and G. L. Rumsey. 1988. Effect of feeding level and dietary electrolytes on the arginine requirement of rainbow trout (*Salmo gairdneri*). Aquaculture 69:79–91.

Cho, C. Y., and D. P. Bureau. 1997. Reduction of waste output from salmonid aquaculture through feeds and feedings. Prog. Fish Cult. 59:155–160.

Cho, C. Y., and S. J. Kaushik. 1985. Effects of protein intake on metabolizable and net energy values of fish diets. Pp. 95–177 in Nutrition and Feeding in Fish in Nutrition and Feeding in Fish, C. B. Cowey, A. M. Mackie, and J. G. Bell, eds. London, UK: Academic Press, Proc. of the Int. Symp. on Fish Feeding and Nutrition, C. B. Cowey, A. M. Mackie, and J. G. Bell, eds. Aberdeen, UK.

Cho C. Y. and W. D. Woodward. 1989. Studies on the protein-to-energy ratio in diets for rainbow trout (*Salmo gairdneri*). Pp. 37–40 in Energy Metabolism of Farm Animals, Y. Van der Honing and W. H. Close, eds. Proceedings of the 11th Symposium, 18–24 September 1988, Lunteren, Netherlands. Wageningen, The Netherlands: Pudoc.

Cho, C. Y., S. J. Kaushik, and B. Woodward. 1992. Dietary arginine requirement of young rainbow trout (*Oncorhynchus mykiss*). Comp. Biochem. Phys. A 102:221–216.

Choo, P.-S., T. K. Smith, C. Y. Cho, and H. W. Ferguson. 1991. Dietary excesses of leucine influence growth and body composition of rainbow trout. J. Nutr. 121:1932–1939.

Coloso, R. M., and L. J. Cruz. 1980. Preliminary studies in some aspects of amino acid biosynthesis in juveniles of *Penaeus monodon Fabricius*. I. Incorporation of ^{14}C (U-^{14}C) acetate into amino acids of precipitable proteins. B. Phil. Biochem. Soc. 3:12–22.

Coloso, R. M., D. P. Murillo-Gurrea, I. G. Borlongan, and M. R. Catacutan. 1999. Sulphur amino acid requirement of juvenile Asian sea bass *Lates calcarifer*. J. Appl. Ichthyol. 15:54–58.

Cowey, C. B. 1994. Amino acid requirements of fish—a critical appraisal of present values. Aquaculture 124:1–11.

Cowey, C. B., and C. Y. Cho. 1993. Nutritional requirements of fish. Proc. Nutr. Soc. 52:417–426.

Cowey, C. B., and M. J. Walton. 1988. Studies on the uptake of (14C) amino acids derived from both dietary (14C) protein and dietary (14C) amino acids by rainbow trout, *Salmo gairdneri* Richardson. J. Fish Biol. 33:293–305.

Cowey, C. B., and M. J. Walton. 1989. Intermediary metabolism. Pp. 259–329 in Fish Nutrition, Second Edition, J. E. Halver, ed. New York: Academic Press.

Cowey, C. B., C. Y. Cho, J. G. Sivak, J. A. Weerheim, and D. D. Stuart. 1992. Methionine intake in rainbow trout (*Oncorhynchus mykiss*), relationship to cataract formation and the metabolism of methionine. J. Nutr.122:1154–1163.

Dabrowski, K., K. J. Lee, and J. Rinchard. 2003. The smallest vertebrate, teleost fish, can utilize synthetic dipeptide based diets. J. Nutr. 133:4225–4229.

Dabrowski, K., Y. F. Zhang, K. Kwasek, P. Hliwa, and T. Ostaszewska. 2010. Effects of protein-, peptide- and free amino acid-based diets in fish nutrition. Aquac. Res. 41:668–683.

de Lange, C. F. M., A. M. Gillis, and G. J. Simpson. 2001. Influence of threonine intake on whole-body protein deposition and threonine utilization in growing pigs fed purified diets. J. Anim. Sci. 79(12):3087–3095.

de la Higuera, M., A. Garzón, M. C. Hidalgo, J. Peragón, G. Cardenete, and J. A. Lupiáñez. 1998. Influence of temperature and dietary-protein supplementation either with free or coated lysine on the fractional protein-turnover rates in the white muscle of carp. Fish Physiol. Biochem. 18:85–95.

Deng, J., X. Zhang, L. Tao, H. Bi, L. Kong, and X. Lei. 2010. D-lysine can be effectively utilized for growth by common carp (*Cyprinus carpio*). Aquacult. Nutr. doi:10.1111/j.1365-2095.2010.00783.x

Dibner, J. J. 2003. Review of the metabolism of 2-hydroxy-4-(methylthio) butanoic acid. World Poultry Sci. J. 59:99–110.

D'Mello, J. P. F. 1994. Amino acid imbalances, antagonisms and toxicities. Pp. 63–97 in Amino Acids in Farm Animal Nutrition, J. P. F. D'Mello, ed. Wallingford, UK: CAB International.

D'Mello, J. P. F. 2003. Amino Acids in Animal Nutrition. Edinburgh, UK: CABI Publishing.

D'Mello, J. P. F., and D. Lewis. 1970. Amino acid interactions in chick nutrition. 3. Interdependence in Amino Acid Requirements. Braz. J. Poult. Sci. 11:367–385.

Deshimaru, O. 1976. Studies on a purified diet for prawn. VI. Absorption rate of amino acids in amino acid test diet. B. Jpn. Soc. Sci. Fish. 40:421–424.

Drew, M. D., A. G. Van Kessel, and D. D. Maenz. 2003. Absorption of methionine and 2-hydroxy-4-methylthiobutanoic acid in conventional and germ-free chickens. Poultry Sci. 82:1149–1153.

Dumas, A., C. F. M. de Lange, J. France, and D. P. Bureau. 2007. Quantitative description of body composition and rates of nutrient deposition in rainbow trout (*Oncorhynchus mykiss*). Aquaculture 263:165–181.

Eddy, F. B., and J. E. Frasert. 1982. Sialic acid and mucus production in rainbow trout (Salmo gairdnei Richardson) in response to zinc and seawater. Comp. Biochem. Physiol. 73C:357–359.

Edmonds, M. S., and D. H. Baker. 1987. Amino acid excesses for young pigs: Effects of excess methionine, tryptophan, threonine or leucine. J. Anim. Sci. 64:1664–1671.

Einen, O., and A. J. Roem. 1997. Dietary protein/energy ratios for Atlantic salmon in relation to fish size: Growth, feed utilization and slaughter quality. Aquacult. Nutr. 3:115–126.

El Haroun, E. R., and D. P. Bureau. 2006. Effect of increasing digestible energy of feeds by increasing digestible protein using protein sources with different amino acid profiles on growth and feed efficiency of rainbow trout (*Oncorhynchus mykiss*). Abstract in XII International Symposium on Fish Nutrition and Feeding, May 28–June 1, 2006, Biarritz, France.

Emmert, J. L., and D. H. Baker. 1997. Use of the ideal protein concept for precision formulation of amino acid levels in broiler diets. J. Appl. Poultry Res. 6:462–470.

Encarnação, P. M. 2005. Effect of Diet Composition on Lysine Requirement and Utilization in Rainbow Trout (*Oncorhynchus mykiss*). Ph.D. Dissertation, University of Guelph, Guelph, Ontario, Canada.

Encarnação, P., C. de Lange, M. Rodehutscord, D. Hoehler, W. Bureau, and D. P. Bureau. 2004. Diet digestible energy content affects lysine utilization, but not dietary lysine requirements of rainbow trout (*Oncorhynchus mykiss*) for maximum growth. Aquaculture 235:569–586.

Encarnação, P., C. F. M. de Lange, and D. P. Bureau. 2006. Diet energy source affects lysine utilization for protein deposition in rainbow trout (*Oncorhynchus mykiss*). Aquaculture 261:1371–1381.

Espe, M., and E. Lied. 1994. Do Atlantic salmon (*Salmo salar*) utilize mixtures of free amino acids to the same extent as intact protein sources for muscle protein synthesis? Comp. Biochem. Phys. 107A:249–254.

Espe, M., and L. R. Njaa. 1991. Growth and chemical composition of Atlantic salmon (*Salmo salar*) given a fish meal diet or a corresponding free amino acid diet. Fisk. Dir. Skr. Ser. Ernæring, 6:260–266.

Espe, M., A. Lemme, A. Petri, and A. El-Mowafi. 2006. Can Atlantic salmon (*Salmo salar*) grow on diets devoid of fish meal? Aquaculture 255:255–262.

Espe, M., A. Lemme, A. Petri, and A. El-Mowafi. 2007. Assessment of lysine requirement for maximal protein accretion in Atlantic salmon using plant protein diets. Aquaculture 263:168–178.

Espe, M., E. M. Hevroy, B. Liaset, A. Lemme, and A. El-Mowafi. 2008. Methionine intake affect hepatic sulphur metabolism in Atlantic salmon, *Salmo salar*. Aquaculture 274:132–141.

Fauconneau, B., A. Basseres, and S. J. Kaushik. 1992. Oxidation of phenylalanine and threonine in response to dietary arginine supply in rainbow trout (*Salmo gairdneri* R.). Comp. Biochem. Phys. A 101:395–401.

Ferrer, J. V., E. Gasco, J. Sastre, F. V. Pallardo, M. Asensi, and J. Vina. 1990. Age-related changes in glutathione synthesis in the eye lens. Biochem. J. 269:531–534.

Fico, M. E., A. S. Hassan, and J. A. Mimer. 1982. The influence of excess lysine on urea cycle operation and pyrimidine biosynthesis. J. Nutr. 112:1854–1861.

Finn, R. N., and H. J. Fyhn. 2010. Requirement for amino acids in ontogeny of fish. Aquac. Res. 41:684–716

Finn, R. N., I. Ronnestad, T. van der Meeren, and H. J. Fyhn. 2002. Fuel and metabolic scaling during the early life stages of Atlantic cod *Gadus morhua*. Mar. Ecol-Prog. Ser. 243:217–234.

Forster, I., and H. Y. Ogata. 1998. Lysine requirement of juvenile Japanese flounder *Paralichthys olivaceus* and juvenile red sea bream *Pagrus major*. Aquaculture 161:131–142.

Fournier, V., M. F. Gouillou-Coustans, R. Metailler, C. Vachot, J. Moriceau, H. Le Delliou, C. Huelvan, E. Desbruyeres, and S. J. Kaushik. 2003. Excess dietary arginine affects urea excretion but does not improve N utilisation in rainbow trout *Oncorhynchus mykiss* and turbot *Psetta maxima*. Aquaculture 217:559–576.

Fox, J. M., A. L. Lawrance, and E. Li-Chan. 1995. Dietary requirements for lysine by juvenile *Penaeus vannamei* using intact and free amino acid sources. Aquaculture 131:279–290.

Fox, J. M., D. A. Davis, M. Wilson, and A. L. Lawrence. 2006. Current status of amino acid requirement research with marine penaeid shrimp. Pp. 182–196 in Avances en Nutricion Acuicola, VIII Simposium Internacional de Nutricion Acuicola, L. E. Cruz-Suarez, D. Ricque-Marie, M. Tapia-Salazar, M. G. Nieto-Lopez, D. A. Villareal Cavazos, A. C. Puello Cruz, and A. Garcia-Ortega, eds. 15–17 November, 2006, Mazatlan, Sinaloa, Mexico.

Fuller, M. F. 1994. Amino acid requirements of maintenance, body protein accretion and reproduction in pigs. Pp. 155–184 in Amino Acids in Farm Animal Nutrition, J. P. F. D'Mello, ed. Wallingford, UK: CAB International.

Furuya, W. M., C. Hayashi, V. R. B. Furuya, D. Botaro, L. C. da Silva, and P. R. Neves. 2001. Exigências de metionina + cistina para alevinos revertidos de tilápia do Nilo (*Oreochromis niloticus*), baseadas no conceito de proteína ideal. Maringa 23:885–889.

Furuya, W. M., D. Botaro, P. R. Neves, L. C. R. Silva, and C. Hayashi. 2004a. Lysine requirement of Nile tilapia, (*Oreochromis niloticus*), for grow-out phase. Cienc. Rural 34:1571–1577.

Furuya, W. M., L. E. Pezzato, M. M. Barros, A. C. Pezzato, V. R. B. Furuya, and E. C. Miranda. 2004b. Use of ideal protein concept for precision formulation of amino acid levels in fish-meal-free diets for juvenile Nile tilapia (*Oreochromis niloticus* L.). Aquac. Res. 35:1110–1116.

Furuya, W. M., V. G. dos Santos, L. C. R. Silva, V. R. B. Furuya, and E. S. Sakaguti. 2006. Digestible lysine requirements of Nile tilapia juveniles. Rev. Bras. Zootecn. 35:937–942.

Gahl, M. J., T. D. Crenshaw, and N. J. Benevenga. 1995. Diminishing returns in weight, nitrogen, and lysine gain of pigs fed six levels of lysine from three supplemental sources. J. Anim. Sci. 73:3177–3187.

Gahl, M. J., M. D. Finke, T. D. Crenshaw, and N. J. Benevenga. 1996. Efficiency of lysine or threonine retention in growing rats fed diets limiting in either lysine or threonine. J. Nutr. 126:3090–3099.

Ganichkin, O. M., X. M. Xu, B. A. Carlson, H. Mix, D. L. Hatfield, V. N. Gladyshev, and M. C. Wahl. 2008. Structure and catalytic mechanism of eukaryotic selenocysteine synthase. J. Biol. Chem 283:5849–5865.

Gatlin, D. M., W. E. Poe, and R. P. Wilson. 1986. Protein and energy requirements of fingerling channel catfish for maintenance and maximum growth. J. Nutr. 116:2121–2131.

Gaull, G. E., and C. E Wright. 1986. Proteins and growth modulators in human milk. Pp. 87–95 in Energy and Protein Needs during Infancy, S. J. Fomon and W. C. Heird, eds. Orlando, FL: Academic Press.

Gaylord, T. G., S. D. Rawles, and K. B. David. 2005. Dietary tryptophan requirements of hybrid striped bass (*Morone chrysops* × *M. saxatilis*). Aquacult. Nutr. 11:367–374.

Gaylord, T. G., A. M. Teague, and F. T. Barrows. 2006. Taurine supplementation of all-plant protein diets for rainbow trout (*Oncorhynchus mykiss*). J. World Aquacult. Soc. 37:509–517.

Glencross, B. D. 2008. A factorial growth and feed utilization model for barramundi, *Lates calcarifer*, based on Australian production conditions. Aquacult. Nutr. 14:360–373.

Glencross, B. D., T. T. T. Hien, N. T. Phuong, and T. L. Cam Tu. 2010. A factorial approach to defining the energy and protein requirements of Tra catfish, *Pangasianodon hypothalamus*. Aquacult. Nutr. In press DOI 10.111/j.1365-2095.2010.00774.x.

Goff, J. B., and D. M. Gatlin. 2004. Evaluation of different sulfur amino acid compounds in the diet of red drum, *Sciaenops ocellatus*, and sparing value of cystine for methionine. Aquaculture 241:465–477.

Goto, T., S. Takagi, T. Ichiki, T. Sakai, M. Endo, T. Yoshida, M. Ukawa, and H. Murata. 2001. Studies on the green liver in cultured red sea bream fed low level and non-fish meal diets: Relationship between hepatic taurine and billiverdin levels. Fish. Sci. 67:58–63.

Goto, T., A. Mochizuki, and F. Hasumi. 2002. Distribution and activities of enzymes involved in taurine biosynthesis in the liver of fish. Suisan Zoshoku 50:443–449.

Gous, R. M., and T. R. Morris. 1985. Evaluation of a diet dilution technique for measuring the response of broiler chickens to increasing concentrations of lysine. Braz. J. Poult. Sci. 26:147–161.

Green, J. A., and R. W. Hardy. 2002. The optimum dietary essential amino acid pattern for rainbow trout (*Oncorhynchus mykiss*), to maximize nitrogen retention and minimize nitrogen excretion. Fish Physiol. Biochem. 27:97–108.

Green, J. A., and R. W. Hardy. 2008. The effects of dietary protein:energy ratio and amino acid pattern on nitrogen utilization and excretion of rainbow trout *Oncorhynchus mykiss* (Walbaum). J. Fish. Biol. 73:663–682.

Griffin, M. E., P. B. Brown, and A. L. Grant. 1992. The dietary lysine requirement of juvenile hybrid striped bass. J. Nutr. 122:1332–1337.

Grisdale-Helland, B. and S. J. Helland. 1997. Replacement of protein by fat and carbohydrate in diets for Atlantic salmon (*Salmo salar*) at the end of the freshwater stage. Aquaculture 152:167–180.

Guillaume, J., S. Kaushik, P. Bergot, and R. Métailler. 1999. Nutrition et Alimentation des Poissons et Crustacés. Paris: INRA Edition.

Gurure, R. M. 1997. Defining the protein and amino acid requirements of Arctic charr using in vivo and modelling techniques. Ph.D. Dissertation, University of Guelph, Guelph, Ontario, Canada.

Halver, J. E. 1957. Nutrition of salmonoid fishes: III. Water soluble vitamin requirements of Chinook salmon. J. Nutr. 62:225–243.

Halver, J. E., and W. E. Shanks. 1960. Nutrition of salmonoid fishes:VIII. Indispensable amino acids for sockeye salmon. J. Nutr. 72:340–346.

Harding, D. E., O. W. Allen, Jr., and R. P. Wilson. 1977. Sulfur amino acid requirement of channel catfish: L-Methionine and L-Cystine. J. Nutr. 107:2031–2035.

Hauler, R. C., and C. G. Carter. 2001a. Lysine deposition responds linearly to marginal lysine intake in Atlantic salmon (*Salmo salar* L.) parr. Aquac. Res. 32(Suppl. 1):147–156.

Hauler, R. C., and C. G. Carter. 2001b. Re-evaluation of the quantitative dietary lysine requirements of fish (review paper). Rev. Fish. Sci. 9:133–163.

Hauler, R. C., C. G. Carter, and S. J. Edwards. 2007. Feeding regime does not influence lysine utilization by Atlantic salmon, *Salmo salar* L., parr. Aquaculture 273:545–555.

Helland, S. J., and B. Grisdale-Helland. 1998. The influence of replacing fish meal in the diet with fish oil on growth, feed utilization and body composition of Atlantic salmon (*Salmo salar*) during the smoltification period. Aquaculture 162:1–10.

Helland, S. J., B. Hatlen, and B. Grisdale-Helland. 2010. Energy, protein and amino acid requirements for maintenance and efficiency of utilization for growth of Atlantic salmon post-smolts determined using increasing ration levels. Aquaculture (in press).

Hillestad, M., and F. Johnsen. 1994. High energy/low protein diets for Atlantic salmon: Effects on growth, nutrient retention and slaughter quality. Aquaculture 124:109–116.

Hillestad, M., F. Johnsen, E. Austreng, and T. Asgard. 1998. Long-term effects of dietary fat-level and feeding rate on growth, feed utilization and carcass quality of Atlantic salmon. Aquacult. Nutr. 4:89–97.

Hoglund, E., M. J. Bakke, Ø. Øverli, S. Winberg, and G. E. Nilsson. 2005. Suppression of aggressive behaviour in juvenile Atlantic cod (*Gadus morhua*), by L-tryptophan supplementation. Aquaculture 249:525–531.

Hoglund, E, C. Sørensen, M. Jørgensen Bakke, G. E. Nilsson and Ø. Øverli. 2007. Attenuation of stress-induced anorexia in brown trout (*Salmo trutta*) by pre-treatment with dietary L-tryptophan. Brit. J. Nut. 97:786–789.

Hseu, J. R., F. I. Lu, H. M. Su, L. S. Wang, C. L. Tsai, and P. P. Hwang. 2003. Effect of exogenous tryptophan on cannibalism, survival and growth in juvenile grouper, *Epinephelus coioides*. Aquaculture 218:251–263.

Hua, K., and D. P. Bureau. 2010. Quantification of differences in digestibility of phosphorus among cyprinids, cichlids, and salmonids through a mathematical modeling approach. Aquaculture 308(3-4):152–158.

Hua, K., C. F. M. de Lange, S. Birkett, and D. P. Bureau. 2010. Adaptation of a non-ruminant nutrient-based growth model to rainbow trout (*Oncorhynchus mykiss* Walbaum). J. Agr. Sci. 148:17–29.

Huggins, A. K., G. Skutch, and E. Baldwin. 1969. Ornithine-urea cycle enzymes in teleost fish. Comp. Biochem. Physiol. 28:587–602.

Hughes, S. G., G. L. Rumsey, and M. C. Nesheim. 1983. Dietary requirements for essential branched-chain amino acids by lake trout. T. Am. Fish. Soc. 112:812–817.

Hughes, S. G., G. L. Rumsey, and M. C. Nesheim. 1984. Effects of dietary excesses of branched-cahin amino acids on the metabolism and tissue composition of lake trout (Salvelinus namaycush). Comp. Biochem. Phys. A 78A:413–418.

Hung, L. T., N. Suhenda, J. Slembrouck, J. Lazard, and Y. Moreau. 2004. Comparison of dietary protein and energy utilization in three Asian catfishes (Pangasius bocourti, P. hypophthalmus and P. djambal). Aquacult. Nutr. 10(5):317–326.

Huxtable, R. J. 1992. Physiological actions of taurine. Physiol. Rev. 72(1):101–164.

IUPAC-IUB Joint Commission on Biochemical Nomenclature (JCBN). 1984. Nomenclature and symbolism for amino acids and peptides. Eur. J. Biochem. 138:9–37.

Jackson, A. J., and B. S. Capper. 1982. Investigations into the requirements of the tilapia *Sarotherodon mossambicus* for dietary methionine, lysine and arginine in semi-synthetic diets. Aquaculture 29:289–297.

Jansman, A. J. M., C. A. Kan, and J. Wiebenga. 2003. Comparison of the biological efficacy of DL-methionine and hydroxy-4-methylthiobutanoic acid (HMB) in pigs and poultry. CVB, Central Bureau for Livestock Feeding: Documentation Rep. No. 29. Lelystad, the Netherlands: Centraal Veevoederbureau.

Jobling, M., and A. Wandsvik. 1983. An investigation of factors contributing to the control of food intake in Arctic charr, *Salvelinus alpines*. J. Fish. Biol. 43:487–489.

Jones J. D., S. J. Petersburg, and P. C. Burnett. 1967. The mechanism of the lysine-arginine antagonism in the chick: effect of lysine on digestion, kidney arginase liver transaminase. J. Nutr. 93:103–116.

Kanazawa, A., and S. Teshima. 1981. Essential amino acids of the prawn. B. Jpn. Soc. Sci. Fish. 47:1375–1377.

Kaushik, S. J. 1998. Nutritional bioenergetics and estimation of waste production in non-salmonids. Aquat. Living Resour. 11:211–217.

Kaushik, S. J., and C. B. Cowey. 1991. Dietary factors affecting nitrogen excretion by fish. Pp. 3–20 in Nutritional Strategies and Aquaculture Waste, C. B. Cowey and C. Y. Cho, eds. Guelph, Ontario: University of Guelph.

Kaushik, S. J., and K. Dabrowski. 1983. Nitrogen and energy utilization in juvenile carp fed casein, amino acids or a protein-free diet. Reprod. Nutr. Dev. 23:741–754.

Kaushik, S. J., and B. Fauconneau. 1984. Effects of lysine administration on plasma arginine and on some nitrogenous catabolites in rainbow trout. Comp. Biochem. Phys. A 79A:459–462.

Kaushik, S. J., and P. Luquet. 1979. Influence of dietary amino acid patterns on the free amino acid contents of blood and muscle of rainbow trout *Salmo gairdneri*. Comp. Biochem. Phys. B 64:175–180.

Kaushik, S. J., and F. Médale. 1994. Energy requirements, utilization and dietary supply to salmonids. Aquaculture 124:81–97.

Kaushik, S. J., and A. Oliva-Teles. 1985. Effect of digestible energy on nitrogen and energy balance in rainbow trout. Aquaculture 50:89–101.

Kaushik, S. J., and I. Seiliez. 2010. Protein and amino acid nutrition and metabolism in fish. Aquac. Res. 41:322–332.

Keembiyehetty, C. N., and D. M. Gatlin. 1992. Dietary lysine requirement of juvenile hybrid striped bass (*Morone chrysops* × *M. saxatilis*). Aquaculture 104:271–277.

Keembiyehetty, C. N., and D. M. Gatlin. 1993. Total sulfur amino acid requirement of juvenile hybrid striped bass (*Morone chrysops* × *M. saxatilis*). Aquaculture 110:331–339.

Keembiyehetty, C. N., and D. M. Gatlin. 1995. Evaluation of different sulfur compounds in the diet of juvenile sunshine bass (*Morone chrysops* (female) × *M. saxatilis* (male)). Comp. Biochem. Phys. A 112A:155–159.

Keembiyehetty, C. N., and D. M. Gatlin. 1997a. Performance of sunshine bass fed soybeanmeal-based diets supplemented with different methionine compounds. Prog. Fish Cult. 59:25–30.

Keembiyehetty, C. N., and D. M. Gatlin, III. 1997b. Dietary threonine requirement of juvenile hybrid striped bass (*Morone chrysops* × *M. saxatilis*). Aquacult. Nutr. 3:217–221.

Kelly, M., B. Grisdale-Helland, S. J. Helland, and D. M. Gatlin. 2006. Refined understanding of sulphur amino acid nutrition in hybrid striped bass, *Morone chrysops* × *M. saxatilis*. Aquac. Res. 37:1546–1555.

Ketola, H. G. 1982. Amino acid nutrition of fishes: Requirements and supplementation of diets. Comp. Biochem. Phys. B 73:1–17.

Ketola, H. G. 1983. Requirement for dietary lysine and arginine by fry of rainbow trout. J. Anim. Sci. 56:101–107.

Khan, M. A., and S. F. Abidi. 2007a. Dietary isoleucine requirement of fingerling Indian major carp, *Labeo rohita* (Hamilton). Aquacult. Nutr. 13:424–430.

Khan, M. A., and S. F. Abidi. 2007b. Total aromatic amino acid requirement of Indian major carp *Labeo rohita* (Hamilton) fry. Aquaculture 267:111–118.

Kim, K.–I. 1993. Requirement for phenylalanine and replacement value of tyrosine for phenylalanine in rainbow trout (*Oncorhynchus mykiss*). Aquaculture 113:243–250.

Kim, K. I., I. McMillan, and H. S. Bayley. 1983. Determination of amino acid requirements of young pigs using an indicator amino acid. Brit. J. Nutr. 50:369–382.

Kim, K.–I., T. B. Kayes, and C. H. Amundson. 1991. Purified diet development and re-evaluation of the dietary protein requirement of fingerling rainbow trout (*Oncorhynchus mykiss*). Aquaculture 96:57–67.

Kim, K.-I., T. B. Kayes, and C. H. Amundson. 1992a. Requirements for sulfur amino acids and utilization of D-methionine by rainbow trout (*Oncorhynchus mykiss*). Aquaculture 101:95–103.

Kim, K.-I., T. B. Kayes, and C. H. Amundson. 1992b. Requirements for lysine and arginine by rainbow trout (*Oncorhynchus mykiss*). Aquaculture 106:333–344.

Kim, K.-I, T. W. Grimshaw, T. B. Kayes, and C. H. Amundson. 1992c. Effect of fasting or feeding diets containing different levels of protein or amino acids on the activities of the liver amino acid-degrading enzymes and amino acid oxidation in rainbow trout (*Oncorhynchus mykiss*). Aquaculture 107(1):89–105.

Kim, K.-W., X. Wang, S. M. Choi, G. J. Park, and S. C. Bai. 2004. Evaluation of optimum dietary protein-to-energy ratio in juvenile olive flounder *Paralichthys olivaceus* (Temminck et Schlegel). Aquac. Res. 35:250–255.

Kim, S.-K., T. Takeuchi, M. Yokoyama, and Y. Murata. 2003. Effect of dietary supplementation with taurine, β-alanine and GABA on the growth of juvenile and fingerling Japanese flounder *Paralichthys olivaceus*. Fish. Sci. 69:242–248.

Kim, S.-K., T. Takeuchi, A. Akimoto, H. Furuita, T. Yamamoto, M. Yokoyama, and Y. Murata. 2005a. Effect of taurine supplemented practical diet on growth performance and taurine contents in whole body and tissues of juvenile Japanese flounder *Paralichthys olivaceus*. Fish. Sci. 71:627–632.

Kim, S.-K., T. Takeuchi, M. Yokoyama, Y. Murata, M. Kaneniwa, and Y. Sakakura. 2005b. Effect of dietary taurine levels on growth and feeding behavior of juvenile Japanese flounder *Paralichthys olivaceus*. Aquaculture 250:765–774.

Kim, S.-K., H. Matsunari, T. Takeuchi, M. Yokoyama, H. Furuita, Y. Murata, and T. Goto. 2008. Comparison of taurine biosynthesis ability between juveniles of Japanese flounder and common carp. Amino Acids 35:161–168.

Klein, R. G., and J. E. Halver. 1970. Nutrition of salmonoid fishes: Arginine and histidine requirements of Chinook and coho salmon. J. Nutr. 100:1105–1109.

Lall, S. P., S. J. Kaushik, P. Y. Le Bail, R. Keith, J. S. Anderson, and E. Plisetskaya. 1994. Quantitative arginine requirement of Atlantic salmon (*Salmo salar*) reared in sea water. Aquaculture 124:13–25.

Langer, S., P. W. D. Scislowski, D. S. Brown, P. Dewey, and M. F. Fuller. 2000. Interactions among the branched-chain amino acids and their effects on methionine utilization in growing pigs: Effects on plasma amino- and keto-acid concentrations and branched-chain keto-acid dehydrogenase activity. Brit. J. Nutr. 83:49–58.

Lee, D. J., and G. B. Putnam. 1973. The response of rainbow trout to varying protein/energy ratios in a test diet. J. Nutr. 103:916–922.

Lemme, A. 2007. How to adjust dietary amino acids to energy in broiler diets? AminoNews 8:1–13.

Lepage, O., O. Tottmar, and S. Winberg. 2002. Elevated dietary intake of L-tryptophan counteracts the stress-induced elevation of plasma cortisol in rainbow trout (*Oncorhynchus mykiss*). J. Exp. Biol. 205:3679–3687.

Liebert, F., and K. Benkendorff. 2007. Modeling lysine requirements of *Oreochromis niloticus* due to principles of the diet dilution technique. Aquaculture 267:100–110.

Lim, C. 1993. Effect of dietary pH on amino acid utilization by shrimp (Penaeus vannamei). Aquaculture 114:293–303.

Liou, C. H. 1989. Lysine and Sulfur Amino Acid Requirements of Juvenile Blue Tilapia (*Oreochromis aureus*). Ph.D. Dissertation, Texas A&M University, College Station, TX.

Liu, Y. J., L. X. Tian, and D. H. Liu. 2002. Influence of practical diet supplemented with free or coated lysine on the growth, plasma free amino acids and protein synthesis rates in the muscle of Ctenopharyngodon idellus (*in Chinese with English Abstract*). J. Fish. China 26:252–258.

Lombardini, J. B., P. K. T. Pang, and R. W. Griffith. 1979. Amino acids and taurine in intracellular osmoregulation in marine animals. Occas. Pap. Calif. Acad. Sci. 134:160–171.

Lunger, A. N., E. McLean, T. G. Gaylord, D. Kuhn, and S. R. Craig. 2007. Taurine supplementation to alternative dietary proteins used in fish meal replacement enhances growth of juvenile cobia (*Rachycentron canadum*). Aquaculture 27:401–410.

Luo, Z., Y.-J. Liu, K.-S. Mai, L.-X. Tian, X.-Y. Tan, H.-J. Yang, G.-Y. Liang, and D.-H. Liu. 2006. Quantitative L-lysine requirement of juvenile grouper *Epinephelus coioides*. Aquacult. Nutr. 12(3):165–172.

Lupatsch, I., and G. W. Kissil. 2005. Feed formulations based on energy and protein demands in white grouper (Epinephelus aeneus). Aquaculture 248:83–95.

Lupatsch, I., G. W. Kissil, D. Sklan, and E. Pfeffer. 1998. Energy and protein requirements for maintenance and growth in gilthead seabream (*Sparus aurata* L.). Aquacult. Nutr. 4:165–173.

Lupatsch, I., G. W. Kissil, and D. Sklan. 2001a. Optimization of feeding regimes for European sea bass *Dicentrarchus labrax*: A factorial approach. Aquaculture 202:289–302.

Lupatsch, I., G. W. Kissil, D. Sklan, and E. Pfeffer. 2001b. Effects of varying dietary protein and energy supply on growth, body composition and protein utilization in gilthead seabream (*Sparus aurata* L.). Aquacult. Nutr. 7:71–80.

Lupatsch, I., G. W. Kissil, and D. Sklan. 2003. Comparison of energy and protein efficiency among three fish species, gilthead sea bream (*Sparus aurata*), European sea bass (*Dicentrarchus labrax*) and white grouper (Epinephelus aeneus): Energy expenditure for protein and lipid deposition. Aquaculture 225:175–189.

Lupatsch, I., L. Cuthbertson, S. Davies, and R. J. Shields. 2008. Studies on energy and protein requirements to improve feed management of the Pacific white shrimp, *Litopenaeus vannamei*. Pp. 281–295 in Avances en Nutricion Acuicola IX, L. E. Cruz Suarez, D. Ricque Marie, M. Tapia Salazar, M. G. Nieto Lopez, D. A. Villarreal Cavazos, J. Pablo Lazo, and M. Teresa Viana, eds. IX Symposio Internacional de Nutricion Acuicola, Universidad Autonoma de Nuevo Leon, Monterrey, Nuevo Leon, Mexico, 24–27 November, 2008.

Lupatsch, I., R. Deshev, and I. Magen. 2010. Energy and protein demands for optimal egg production including maintenance requirements of female tilapia *Oreochromis niloticus*. Aquac. Res. 41:763–769.

Luzzana, U., R. W. Hardy, and J. E. Halver. 1998. Dietary arginine requirement of fingerling coho salmon (*Oncorhynchus kisutch*). Aquaculture 163:137–150.

Magendie, P. F. 1817. Précis Élémentaire de Physiologie, Tome Second. Paris: Chez Méquignon-Marvis.

Mai, K., L. Zhang, Q. Ai, Q. Duan, C. Zhang, H. Li, J. Wan, and Z. Liufu. 2006. Dietary lysine requirement of Japanese seabass, *Lateolabrax japonicus*. Aquaculture 258:535–542.

Mambrini, M., and J. Guillaume. 1999. Nutrition proteique. Pp. 113–146 in Nutrition et Alimentation des Poissons et des Crustaces, J. Guillaume, S. J. Kaushik, P. Bergot, and R. Metailler, eds. Paris: Collection du Labo au terrain, Editions INRA.

Mambrini, M., and S. J. Kaushik. 1995. Indispensable amino acid requirements of fish: Correspondence between quantitative data and amino acid profiles of tissue proteins. J. Appl. Ichthyol. 11:240–247.

Mariotti, F., D. Tome, and P. P. Mirand. 2008. Converting nitrogen into protein—beyond 6.25 and Jones' Factors. Crit. Rev. Food Sci. Nutr. 48:177–184.

Martín-Venegas, R., M. J. Rodríguez-Lagunas, P. A. Geraert, and R. Ferrer. 2007. Monocarboxylate Transporter 1 mediates DL-2-hydroxy-(4-methylthio)butanoic acid transport across the apical membrane of Caco-2 cell monolayers. J. Nutr. 137:49–54.

Martinez, J. B., S. Chatzifotis, P. Divanach, and T. Takeuchi. 2004. Effect of dietary taurine supplementation on growth performance and feed selection of sea bass *Dicentrarchus labrax* fry fed with demand-feeders. Fish. Sci. 70:74–79.

Matsunari, H., T. Takeuchi, M. Takahashi, and K. Mushiake. 2005. Effect of dietary taurine supplementation on growth performance of yellowtail juveniles Seriola quinqueradiata. Fish. Sci. 71:1131–1135.

McLean, E., B. Ronsholdt, C. Sten, and J. F. Najamuddin. 1999. Gastrointestinal delivery of peptide and protein drugs to aquacultured teleosts. Aquaculture 177:231–247.

Millamena, O. M., M. N. Bautista, and A. Kanazawa. 1996. Methionine requirement of juvenile tiger shrimp *Penaeus monodon* Fabricius. Aquaculture 143:403–410.

Millamena, O. M., M. N. Bautista, O. S. Reyes, and A. Kanazawa. 1997. Threonine requirement of juvenile marine shrimp *Penaeus monodon*. Aquaculture 151:9–14.

Millamena, O. M., M. N. Bautista-Teruel, O. S. Reyes, and A. Kanazawa. 1998. Requirements of juvenile marine shrimp, *Penaeus monodon* (Fabricius) for lysine and arginine. Aquaculture 164:95–104.

Millamena, O. M., M. B. Teruel, A. Kanawaza, and S. Teshima. 1999. Quantitative dietary requirements of postlarval tiger shrimp, *Penaeus monodon*, for histidine, isoleucine, leucine, phenylalanine and tryptophan. Aquaculture 179:169–179.

Millward, D. J. 1989. The nutritional regulation of muscle growth and protein turnover. Aquaculture 79:1–28.

Miranda, S. G., Y. J. Wang, N. G. Purdie, V. R. Osborne, B. L. Coomber, and J. P. Cant. 2009. Selenomethionine stimulates expression of glutathione peroxidase 1 and 3 and growth of bovine mammary epithelial cells in primary culture. J. Dairy Sci. 92:2670–2683.

Moehn, S., R. O. Ball, M. F. Fuller, A. M. Gillis, and C. F. M. de Lange. 2004. Growth potential, but not body weight or moderate limitation of lysine intake, affects inevitable lysine catabolism in growing pigs. J. Nutr. 134:2287–2292.

Mohn, S., and C. F. M. de Lange. 1998. The effect of body weight on the upper protein limit to protein deposition in a defined population of growing gilts. J. Anim. Sci. 76:124–133.

Montes-Girao, P. J., and D. M. Fracalossi. 2006. Dietary lysine requirement as basis to estimate the essential dietary amino acid profile for jundia, *Rhamdia quelen*. J. World Aquacult. Soc. 37:388–396.

Moon, H. Y., and D. M. Gatlin, III. 1991. Total sulfur amino acid requirement of juvenile red drum, *Sciaenops ocellatus*. Aquaculture 95:97–106.

Moughan, P. J. 1989. Simulation of the daily partitioning of lysine in the 50kg liveweight pig—A factorial approach to estimating amino acid requirements for growth and maintenance. Res. Dev. Agric. 6:7–14.

Moughan, P. J. 1995. Modelling protein metabolism in the pig—Critical evaluation of a simple reference model. Pp. 103–112 in Modelling Growth in the Pig, P. J. Moughan, M. W. A. Verstegen, and M. I. Visser-Reyneveld, eds. Wageningen Pers, The Netherlands: EAAP Publishing.

Moughan, P. J. 1999. Protein metabolism in the growing pig. Pp. 299–331 in A Quantitative Biology of the Pig, I. Kyriazakis, ed. Wallingford, UK: CAB International.

Moughan, P. J. 2003. Simulating the partitioning of dietary amino acids: New directions. J Anim. Sci. 81:60–67.

Moughan, P. J., and S. M. Rutherfurd. 1996. A new method for determining digestible reactive lysine in foods. J. Agric. Food Chem 44:2202–2209.

Murai, T., T. Akiyama, and T. Nose. 1981. Use of crystalline amino acids coated with casein in diets for carp. B. Jpn. Soc. Sci. Fish. 47:523–527.

Murai, T., H. Ogata, Y. Hirasawa, T. Akiyama, and T. Nose. 1987. Portal absorption and hepatic uptake of amino acids in rainbow trout force-fed complete diets containing casein or crystalline amino acids. Nippon Suisan Gakk. 53:1847–1859.

Murillo-Gurrea, D. P., R. M. Coloso, I. G. Borlongan, and A. E. Serrano, Jr. 2001. Lysine and arginine requirements of juvenile Asian sea bass (*Lates calcarifer*). J. Appl. Ichthyol. 17:49–53.

Nang Thu, T. T., C. Parkouda, S. de Saeger, Y. Larondelle, and X. Rollin. 2007. Comparison of the lysine utilization efficiency in different plant protein sources supplemented with L-lysine•HCl in rainbow trout (*Oncorhynchus mykiss*) fry. Aquaculture 272:477–488.

Nang Thu, T. T., C. Parkouda, S. de Saeger, Y. Larondelle, and X. Rollin. 2009. Protein level does not affect lysine utilization efficiency at marginal lysine intake in growing rainbow trout (*Oncorhynchus mykiss*) fry. Aquaculture 288:312–320.

Ng, W.-K., G. Serrini, Z. Zhang, and R. P. Wilson. 1997. Niacin requirement and inability of tryptophan to act as a precursor of NAD^+ in channel catfish, *Ictalurus punctatus*. Aquaculture 152:273–285.

Nguyen, T. N., and D. A. Davis. 2009. Methionine requirement in practical diets of juvenile Nile tilapia, *Oreochromis niloticus*. J. World Aquacult. Soc. 40:410–416.

Nichols, N. L., and R. F. Bertolo. 2008. Luminal threonine concentration acutely affects intestinal mucosal protein and mucin synthesis in piglets. J. Nutr. 138:1298–1303.

Nose, T. 1979. Summary report on the requirements of essential amino acids for carp. Pp. 145–156 in Finfish Nutrition and Fishfeed Technology, J. E. Halver and K. Tiews, eds. Berlin, Germany: Heenemann GmbH.

Nose, T., S. Arai, D.-L. Lee, and Y. Hashimoto. 1974. A note on amino acids essential for growth of young carp. B. Jpn. Soc. Sci. Fish. 40:903–908.

NRC (National Research Council). 1981. Nutrient Requirements of Coldwater Fishes. Washington, DC: The National Academy Press.

NRC. 1993. Nutrient Requirements of Fish. Washington, DC: National Academy Press.

Ogino, C. 1980. Requirements of carp and rainbow trout for essential amino acids. B. Jpn. Soc. Sci. Fish. 46:171–174.

Pack, M., M. Rodehutscord, S. Jacobs, and E. Pfeiffer. 1995. Amino acid requirement of rainbow trout (*Oncorhynchus mykiss*) II. Protein deposition as function of dietary methionine, threonine and arginine. J. Appl. Ichthyol. 11:390–393.

Page, J. W., and J. W. Andrews. 1973. Interactions of dietary levels of protein and energy on channel catfish (*Ictalurus punctatus*). J. Nutr. 103:1339–1346.

Palma, J., D. P. Bureau, A. Lemme, M. Correia, and J. P. Andrade. 2009. Quantitative dietary requirement of juvenile Atlantic ditch shrimp *Palaemonetes varians* (Leach) for lysine, methionine and arginine. Unpublished. (Abstract ISFNF 2006).

Park, G.-S., T. Takeuchi, M. Yokoyama, and T. Seikai. 2002. Optimal dietary taurine level for growth of juvenile Japanese flounder *Paralichthys olivaceus*. Fish. Sci. 68:824–829.

Pei, Z., S. Xie, W. Lei, X. Zhu, and Y. Yang. 2004. Comparative study on the effect of dietary lipid level on growth and feed utilization for gibel carp (*Carassius auratus gibelio*) and Chinese longsnout catfish (*Leiocassis longirostris* Gunther). Aquacult. Nutr. 10:209–216.

Pellett, P. L., and V. R. Young. 1984. Evaluation of the use of amino acid composition data in assessing the protein quality of meat and poultry products. Am. J. Clin. Nutr. 40:718–736.

Peña-Ortega, L. O., and D. P. Bureau. 2004. Investigation into the differences in nitrogen metabolism between rainbow trout and Atlantic salmon of different sizes/ages. International Congress of Fish Biology, August 1–6, 2004, Manaus, Brazil.

Peres, H., and A. Oliva-Teles. 2005. The effect of dietary protein replacement by crystalline amino acid on growth and nitrogen utilization of turbot *Scophthalmus maximus* juveniles. Aquaculture 250:755–764.

Peres, H., and A. Oliva-Teles. 2008. Lysine requirement and efficiency of lysine utilization in turbot (*Scophthalmus maximus*) juveniles. Aquaculture 275:283–290.

Peres, H., and A. Oliva-Teles. 2009. The optimum dietary essential amino acid profile for gilthead seabream (*Sparus aurata*) juveniles. Aquaculture 296(1–2):81–86.

Pfeffer, E., H. Al-Sabty, and R. Haverkamp. 1992. Studies on lysine requirements of rainbow trout (*Oncorhynchus mykiss*) fed wheat gluten as only source of dietary protein. J. Anim. Physiol. An. N. 67:74–82.

Poppi, D. A., V. M. Quinton, K. Hua, and D. P. Bureau. 2010. Development of a protocol for assessing the bioavailability of arginine in feather meal fed to rainbow trout (*Oncorhynchus mykiss*). Aquaculture (accepted for publication August 2010). doi:10.1016/j.aquaculture.2011.01.06.

Portz, L., and J. E. P. Cyrino. 2003. Comparison of the amino acid contents of roe, whole body and muscle tissue and their A/E ratios for largemouth bass *Micropterus salmoides* (Lacepéde, 1802). Aquac. Res. 34:585–592.

Poston, H. A., and G. L. Rumsey. 1983. Factors affecting dietary requirement and deficiency signs of L-tryptophan in rainbow trout. J. Nutr. 113:2568–2577.

Poston, H. A., R. C. Riis, G. L. Rumsey, and H. G. Ketola. 1977. The effect of supplemental dietary amino acids, minerals and vitamins on salmonids fed cataractogenic diets. Cornell Vet. 67:472–509.

Randall, D. J., C. M. Wood, S. F. Perry, H. Bergman, G. M. O. Maloiy, T. P. Mommsen, and P. A. Wright. 1989. Urea excretion as a strategy for survival in a fish living in a very alkaline environment. Nature 337:165–166.

Refstie, S., T. Storebakken, G. Baeverfjord, and A. J. Roem. 2001. Long-term protein and lipid growth of Atlantic salmon (*Salmo salar*) fed diets with partial replacement of fish meal by soy protein products at medium or high lipid level. Aquaculture 193:91–106.

Regina, M., V.-P. Korhonen, T. K. Smith, L. Alakuijala, and T. O. Eloranta. 1993. Methionine toxicity in the rat in relation to hepatic accumulation of S-adenosylmethionine: Prevention by dietary stimulation of the hepatic transsulfuration pathway. Arch. Biochem. Biophys. 300:598–607.

Richard, L., P. P. Blanc, V. Rigolet, S. J. Kaushik, and I. Geurden. 2010. Maintenance and growth requirements for nitrogen, lysine and methionine and their utilisation efficiencies in juvenile black tiger shrimp, Penaeus monodon, using a factorial approach. Brit. J. Nutr. 103:984–995.

Richards, J. D., C. A. Atwell, M. Vazquez-Anon, and J. J. Dibner. 2005. Comparative in vitro and in vivo absorption of 2-hydroxy-4-(methylthio) butanoic acid and methionine in the broiler chicken. Poultry Sci. 84:1397–1405.

Robinson, E. H., O. W. Allen, Jr., W. E. Poe, and R. P. Wilson. 1978. Utilization of dietary sulfur compounds by fingerling channel catfish: L-methionine, DL-methionine, methionine hydroxy analogue, taurine and inorganic sulfate. J. Nutr. 108:1932–1936.

Robinson, E. H., R. P. Wilson, and W. E. Poe. 1980a. Re-evaluation of the lysine requirement and lysine utilization by fingerling channel catfish. J. Nutr. 110:2313–2316.

Robinson, E. H., R. P. Wilson, and W. E. Poe. 1980b. Total aromatic amino acid requirement, phenylalanine requirement and tyrosine replacement value for fingerling channel catfish. J. Nutr. 110:1805–1812.

Robinson, E. H., R. P. Wilson, and W. E. Poe. 1981. Arginine requirement and apparent absence of a lysine-arginine antagonist in fingerling channel catfish. J. Nutr. 111:46–52.

Robinson, E. H., W. E. Poe, and R. P. Wilson. 1984. Effects of feeding diets containing an imbalance of branched-chain amino acids on fingerling channel catfish. Aquaculture 37:51–62.

Rodehutscord, M., and M. Pack. 1999. Estimates of essential amino acid requirements from dose-response studies with rainbow trout and broiler chicken: Effect of mathematical model. Arch. Anim. Nutr. 52:223–244.

Rodehutscord, M., and E. Pfeffer. 1999. Maintenance requirement of digestible energy and efficiency of utilization of digestible energy for retention in rainbow trout, (*Oncorhynchus mykiss*). Aquaculture 179:95–107.

Rodehutscord, M., S. Jacobs, M. Pack, and E. Pfeffer. 1995a. Response of rainbow trout (*Oncorhynchus mykiss*) growing from 50 to 150 g to supplements of DL-methionine in a semipurified diet containing low or high levels of cystine. J. Nutr. 125:964–969.

Rodehutscord, M., S. Jacobs, M. Pack, and E. Pfeffer. 1995b. Response of rainbow trout (*Oncorhynchus mykiss*) growing from 50 to 170 g to supplements of either l-arginine or l-threonine in a semipurified diet. J. Nutr. 125:970–975.

Rodehutscord, M., A. Becker, M. Pack, and E. Pfeffer. 1997. Response of rainbow trout (*Oncorhynchus mykiss*) to supplements of individual essential amino acids in a semipurified diet, including an estimate of the maintenance requirement for essential amino acids. J. Nutr. 127:1166–1175.

Rodehutscord, M., F. Borchert, Z. Gregus, M. Pack, and E. Pfeffer. 2000. Availability and utilization of free lysine in rainbow trout (*Oncorhynchus mykiss*): 1. Effect of dietary crude protein level. Aquaculture 187:163–176.

Rollin, X. 1999. Critical Study of the Indispensable Amino Acids Requirements in Atlantic Salmon (*Salmo salar* L.) Fry. Ph.D. Dissertation, Université Catholique de Louvain, Louvain-la-Neuve, Belgium.

Rollin, X., M. Mambrini, T. Abboudi, Y. Larondelle, and S. J. Kaushik. 2003. The optimum dietary indispensable amino acid pattern for growing Atlantic salmon (*Salmo salar* L.) fry. Brit. J. Nutr. 90:865–876.

Rollin, X., J. B. Wauters, N. Bodin, Y. Larondelle, W. Ooghe, B. Wathelet, and T. Abboudi. 2006. Maintenance threonine requirement and efficiency of its use for accretion of whole-body threonine and protein in Atlantic salmon (*Salmo salar* L.) fry. Brit. J. Nutr. 95:234–245.

Ronnestad, I., A. Thorsen, and R. N. Finn. 1999. Fish larval nutrition: recent advances in amino acid metabolism. Aquaculture 177:201–216.

Ronnestad, A., S. K. Tonheim, H. J. Fyhn, C. R. Rojas-Garcia, Y. Kamisaka, W. Koven, R. N. Finn, B. F. Terjesen, Y. Barr, and L. E. C. Conceicao. 2003. The supply of amino acids during early feeding stages of marine fish larvae: A review of recent findings. Aquaculture 227:147–164.

Rumsey, G. L., J. W. Page, and M. L. Scott. 1983. Methionine and cystine requirements of rainbow trout. Prog. Fish Cult. 45:139–143.

Santiago, C. B., and R. T. Lovell. 1988. Amino acid requirements for growth of Nile tilapia. J. Nutr. 118:1540–1546.

Sara, J. R. 2007. Establishing the Protein Requirement in the Marine Prawn *Fenneropenaeus indicus*. Ph.D. Dissertation, School of Biological and Conservation Sciences, University of KwaZulu-Natal, Durban, South Africa.

Sassi, M. J. 2001. Carboxyterminal Degradation Products of Type I Collagen. Academic Dissertation, Oulu University Press, Oulun, Finland.

Satoh, S., M. A. Alam, K. I. Satoh, and V. Kiron. 2004. Effects of dietary lipid and phosphorus levels on nitrogen and phosphorus excretion in young yellowtail Seriola quinqueradiata: A preliminary observation. Fisheries Sci. 70:1082–1088.

Sauer, N., K. Emrich, H.-P. Piepho, A. Lemme, M. S. Redshaw, and R. Mosenthin. 2008. Meta-analysis of the relative efficiency of methionine-hydroxy-analogue-free-acid compared with dl-methionine in broilers using nonlinear mixed models. Poultry Sci. 87:2023–2031.

Schrauzer, G. N. 2000. Selenomethionine: A review of its nutritional significance, metabolism and toxicity. J. Nutr. 130:1653–1656.

Schuhmacher, A., C. Wax, and J. M. Gropp. 1997. Plasma amino acids in rainbow trout (*Oncorhynchus mykiss*) fed intact protein or a crystalline amino acid diet. Aquaculture 151:15–28.

Segovia-Quintero, M. A., and R. C. Reigh. 2004. Coating crystalline methionine with tripalmitin-polyvinyl alcohol slows its absorption in the intestine of Nile tilapia, *Oreochromis niloticus*. Aquaculture 238:355–367.

Shearer, K. D. 1994. Factors affecting the proximate composition of cultured fishes with emphasis on salmonids. Aquaculture 119:63–88.

Shiau, S.-Y., and B.-S. Chou. 1994. Grass shrimp, *Penaeus monodon*, growth as influenced by dietary taurine supplementation. Comp. Biochem. Biophys. A 108(1):137–142.

Simmons, L., R. D. Moccia, and D. P. Bureau. 1999. Dietary methionine requirement of juvenile Arctic charr *Salvelinus alpinus* (L.). Aquacult. Nutr. 5:93–100.

Singh, S., and M. A. Khan. 2007. Dietary arginine requirement of fingerling hybrid Clarias (*Clarias gariepinus* × *Clarias macrocephalus*). Aquac. Res. 38:17–25.

Statter, M., A. Russell, and S. Rotaro. 1978. Competitive interrelationships between lysine and arginine in rat under normal conditions and in experimental hyperammonia. Life Sci. 22:2097–2102.

Steffens, W., B. Rennert, M. Wirth, and R. Krüger. 1999. Effect of two lipid levels on growth, feed utilization, body composition and some biochemical parameters of rainbow trout, *Oncorhynchus mykiss* (Walbaum 1792). J. Appl. Ichthyol. 15:159–164.

Susenbeth, A. 1995. Factors affecting lysine utilization in growing pigs: An analysis of literature data. Livest. Prod. Sci. 43:193–204.

Suzuki, K. T., C. Doi, and N. Suzuki. 2006. Metabolism of ^{76}Se-methylselenocysteine compared with that of ^{77}Se-selenomethionine and ^{82}Se-selenite. Toxicol. Appl. Pharmacol. 217:185–195.

Sveier, H., H. Nordås, G. E. Berge, and E. Lied. 2001. Dietary inclusion of crystalline D- and L-methionine: Effects on growth, feed and protein utilization, and digestibility in small and large Atlantic salmon (*Salmon salar* L.). Aquacult. Nutr. 7(3):169–181.

Tacon, A. G. J., and C. B. Cowey. 1985. Protein and amino acid requirements. Pp. 155–183 in Fish Energetics: New Perspectives, P. Tyler and P. Calow, eds. London, UK: Croom Helm.

Tagaki, S., H. Murata, T. Goto, T. Ichiki, D. Munasinghe, M. Endo, T. Matsumoto, A. Sakurai, H. Hatate, T. Yoshida, T. Sakai, H. Yamashita, M. Ukawa, and T. Kuramoto. 2005. The green liver syndrome is caused by taurine deficiency in yellowtail, *Seriola quinqueradiata*, fed diets without fishmeal. Suisan Zoshoku 53:279–290.

Tantikitti, C., and B. E. March. 1995. Dynamics of plasma free amino acids in rainbow trout (*Oncorhynchus mykiss*) under variety of dietary conditions. Fish Physiol. Biochem. 14(3):179–194.

Tantikitti, C., and N. Chimsung. 2001. Dietary lysine requirement of freshwater catfish (*Mystus nemurus* Cuv. and Val). Aquac. Res. 32:135–141.

Teshima, S., A. Kanazawa, and S. Koshio. 1990. Effects of methionine enriched plastein supplemented to soybean-protein based diets on common carp and tilapia. Pp. 279–282 in The Second Asian Fisheries Forum, R. Hirano and I. Hanyu, eds. Manilla, the Philipines: Asian Fisheries Society.

Teshima, S., A. Kanazawa, and S. Koshio. 1992. Supplemental effects of methionine-enriched plastein in *Penaeus japonicus* diets. Aquaculture 101:85–93.

Teshima, S., M. S. Alam, S. Koshio, M. Ishikawa, and A. Kanazawa. 2002. Assessment of requirement values for essential amino acids in the prawn, *Marsupenaeus japonicus* (Bate). Aquac. Res. 33:395–402.

Tibaldi, E., and D. Lanari. 1991. Optimal dietary lysine levels for growth and protein utilisation of fingerling sea bass (*Dicentrarchus labrax* L.) fed semipurified diets. Aquaculture 95:297–304.

Tibaldi, E., and F. Tulli. 1999. Dietary threonine requirement of juvenile European sea bass (*Dicentrarchus labrax*). Aquaculture 175:155–166.

Tibaldi, E., F. Tulli, and D. Lanari. 1994. Arginine requirement and effect of different dietary arginine and lysine levels for fingerling sea bass (*Dicentrarchus labrus*). Aquaculture 127:207–218.

Tibbetts, S. M., S. P. Lall, and J. E. Milley. 2005. Effects of dietary protein and lipid levels and DP DE) 1 ratio on growth, feed utilization and hepatosomatic index of juvenile haddock, *Melanogrammus aeglefinus* L. Aquacult. Nutr. 11:67–75.

Tröße, C., R. Waagbo, O. Breck, A.-K. Stavrum, K. Petersen, and P. A. Olsvik. 2009. Genome-wide transcription analysis of histidine-related cataract in Atlantic salmon (*Salmo salar* L). Mol. Vis. 15:1332–1350.

Tulli, F., M. Messina, M. Calligaris, and E. Tibaldi. 2010. Response of European sea bass (*Dicentrarchus labrax*) to graded levels of methionine (total sulfur amino acids) in soya protein-based semi-purified diets. Brit. J. Nutr. doi:10.1017/S0007114510001029.

Twibell, R. G., and P. B. Brown. 1997. Dietary arginine requirement of juvenile yellow perch. J. Nutr. 127:1838–1841.

Twibell, R. G., K. A. Wilson, and P. B. Brown. 2000. Dietary sulfur amino acid requirement of juvenile yellow perch fed the maximum cystine replacement value for methionine. J. Nutr. 130:612–616.

Vergara, A. M., L. Robaina L., M. Izquierdo, and M. de la Higuera. 1996. Protein sparing effect of lipids in diets for fingerlings of gilthead sea bream. Fish. Sci. 62:624–628.

Waagbø, R., C. Trosse, W. Koppe, R. Fontanillas and O. Breck. 2010. Dietary histidine supplementation prevents cataract development in adult Atlantic salmon, *Salmo salar* L., in seawater. Brit. J. Nutr. doi:10.1017/S0007114510002485.

Walton, M. J., C. B. Cowey, and J. W. Adron. 1982. Methionine metabolism in rainbow trout fed diets of differing methionine and cystine content. J. Nutr. 112:1525–1535.

Walton, M. J., R. M. Coloso, C. B. Cowey, J. W. Adron, and D. Knox. 1984a. The effects of dietary tryptophan levels on growth and metabolism of rainbow trout (*Salmo gairdneri*). Brit. J. Nutr. 51:279–287.

Walton, M. J., C. B. Cowey, and J. W. Adron. 1984b. The effect of dietary lysine levels on growth and metabolism of rainbow trout (*Salmo gairdneri*). Brit. J. Nutr. 52:115–122.

Walton, M., R. M. Coloso, C. B. Cowey, and J. W. Adron. 1986. Dietary requirements of rainbow trout for tryptophan, lysine and arginine determined by growth and biochemical measurements. Fish Physiol. Biochem. 2:161–169.

Wan, J., K. Mai, and Q. Ai. 2006. The recent advance on arginine nutritional physiology in fish. J. Fish. Sci. 13:679–685.

Wang, S., Y. J. Liu, L. X. Tian, M. Q. Xie, H. J. Yang, Y. Wang, and G. Y. Liang. 2005. Quantitative dietary lysine requirement of juvenile grass carp *Ctenopharyngodon idella*. Aquaculture 249:419–429.

Wang, S., P. M. Encarnacao, R. L. Payne, and D. P. Bureau. 2010. Estimating dietary lysine requirements for live weight gain and protein deposition in juvenile rainbow trout (*Oncorhynchus mykiss*). International Symposium on Fish Nutrition and Feeding, 31 May–4 June 2010, Qingdao, China.

Wang, T. C., and M. F. Fuller. 1989. The optimum dietary amino acid pattern for growing pigs. 1. Experiments by amino acid deletion. Brit. J. Nutr. 62:77–89.

Wang, Y., J. Guo, K. Li, and D. P. Bureau. 2006. Effects of dietary protein and energy levels on growth, feed utilization and body composition of cuneate drum (Nibea miichthioides). Aquaculture 252:421–428.

Watanabe, K., H. Aoki, Y. Yamagata, V. Kiron, S. Satoh, and T. Watanabe. 2000. Energy and protein requirements of yellowtail during winter season. Fish. Sci. 66:521–527.

Webb, K. A., Jr., and D. M. Gatlin, III. 2003. Effects of dietary protein level and form on production characteristics and ammonia excretion of red drum *Sciaenops ocellatus*. Aquaculture 225:17–26.

Webb, K. A., L. T. Rawlinson, and, G. J. Holt. 2010. Effects of dietary starches and the protein to energy ratio on growth and feed efficiency of juvenile cobia (*Rachycentron canadum*). Aquacult. Nutr. 16:447–456.

Weijs, P. J. M. 1993. Dietary Protein, Physiological Condition and Metabolic Amino Acid Utilization. Ph.D. Dissertation, Proefschrift Wageningen.

Were, K.E. 1989. Determination of the Tryptophan Requirement of the Juvenile Rainbow Trout Using the Indicator Oxidation Technique. M.Sc. Thesis, University of Guelph, Guelph, Ontario, Canada.

Williams, K., C. Barlow, and L. Rodgers. 2001. Efficacy of crystalline and protein-bound amino acid enrichment of diets for barramundi/Asian seabass (*Lates calcarifer* Bloch). Aquac. Res. 32:415–429.

Wilson, R. P. 1989. Amino acids and proteins. Pp. 111–151 in Fish Nutrition, Second Edition, J. E. Halver, ed. New York: Academic Press.

Wilson, R. P. 1994. Amino acid requirements of fish. Pp. 377–399 in Amino Acids in Farm Animal Nutrition, J. P. F. D'Mello, ed. Wallingford, UK: CAB International.

Wilson, R. P. 2002. Amino acids and proteins. Pp. 143–179 in Fish Nutrition, Third Edition, J. E. Halver and R. W. Hardy, eds. New York: Academic Press.

Wilson, R. P., and C. B. Cowey. 1985. Amino acid composition of whole body tissue of rainbow trout and Atlantic salmon. Aquaculture 48:373–376.

Wilson, R. P., and J. E. Halver. 1986. Protein and amino acid requirements of fish. Annu. Rev. Nutr. 6:225–244.

Wilson, R. P., and W. E. Poe. 1987. Apparent inability of channel catfish to utilize dietary mono- and disaccharides as energy sources. J. Nutr. 117:280–285.

Wilson, R. P., D. E. Harding, and D. L. Garling, Jr. 1977. Effect of dietary pH on amino acid utilization and the lysine requirement of fingerling channel catfish. J. Nutr. 107:166–170.

Wilson, R. P., O. W. Allen, Jr., E. H. Robinson, and W. E. Poe. 1978. Tryptophan and threonine requirements of fingerling channel catfish. J. Nutr. 108:1595–1599.

Wilson, R. P., W. E. Poe, and E. H. Robinson. 1980. Leucine, isoleucine, valine and histidine requirements of fingerling channel catfish. J. Nutr. 110:627–633.

Winberg, S., Ø. Øverli, and O. Lepage. 2001. Suppression of aggression in rainbow trout (*Oncorhynchus mykiss*) by dietary L-tryptophan. J. Exp. Biol. 204:3867–3876.

Winfree, R. A., and R. R. Stickney. 1981. Effects of dietary protein and energy on growth, feed conversion efficiency and body composition of Tilapia aurea. J. Nutr. 111:1001–1012.

Yamada, S., K. L. Simpson, and Y. Tanaka, 1981a. Plasma amino acid changes in rainbow trout *Salmo Gairdneri* force-fed casein and a corresponding amino acid mixture. B. Jpn. Soc. Sci. Fish. 47:1035–1040.

Yamada, S., Y. Tanaka, and T. Katayama. 1981b. Feeding experiments with carp fry. B. Jpn. Soc. Sci. Fish. 47:1247.

Yamamoto, T., T. Shima, H. Furuita, M. Shiraishi, F. J. Sánchez-Vázquez, and M. Tabata. 2000. Self-selection of diets with different amino acid profiles by rainbow trout (*Oncorhynchus mykiss*). Aquaculture 187:375–386.

Yamamoto, T., T. Shima, H. Furuita, and N. Suzuki. 2002. Influence of dietary fat level and whole-body adiposity on voluntary energy intake by juvenile rainbow trout *Oncorhynchus mykiss* (Walbaum) under self-feeding conditions. Aquac. Res. 33:715–723.

Yamamoto, T., T. Shima, and H. Furuita. 2004. Antagonistic effects of branched-chain amino acids induced by excess protein-bound leucine in diets for rainbow trout (*Oncorhynchus mykiss*). Aquaculture 232:539–550.

Yamamoto, T., T. Sugita, and H. Furuita. 2005. Essential amino acid supplementation to fish meal-based diets with low protein to energy ratios improves the protein utilization in juvenile rainbow trout *Oncorhynchus mykiss*. Aquaculture 246:379–391.

Yang, X., C. Yang, A. Farberman, T. C. Rideout, C. F. M. de Lange, J. France, and M. Z. Fan. 2008. The mammalian target of rapamycin-signaling pathway in regulating metabolism and growth. J. Anim. Sci. 86(E Suppl):E36–E50.

Yokoyama, M., and J.–I. Nakazoe. 1992. Accumulation and excretion of taurine in rainbow trout (*Oncorhynchus mykiss*) fed diets supplemented with methionine, cystine and taurine. Comp. Biochem. Phys. A 102:565–568.

Zarate, D. D., and R. T. Lovell. 1997. Free lysine (L-lysine.HCl) is utilized for growth less efficiently than protein-bound lysine (soybean meal) in practical diets by young channel catfish (*Ictalurus punctatus*). Aquaculture 159:87–100.

Zarate, D. D., R. T. Lovell, and M. Payne. 1999. Effects of feeding frequency and rate of stomach evacuation on utilization of dietary free and protein-bound lysine for growth by channel catfish *Ictalurus punctatus*. Aquacult. Nutr. 5:17–22.

Zhang, Y., K. Dabrowski, P. Hilwa, and P. Gomulka. 2006. Indispensable amino acid concentrations decrease in tissues of stomachless fish, common carp in response to free amino acid- or peptide-based diets. Amino Acids 31:165–172.

Zhou, F., W. Xiong, J. X. Xiao, Q. J. Shao, O. N. Bergo,Y. Hua, and X. Chai. 2010a. Optimum arginine requirement of juvenile black sea bream, *Sparus macrocephalus*. Aquac. Res. doi:10.1111/j.1365-2109.2009.02474.x.

Zhou, F., J. Shao, R. Xu, J. Ma, and Z. Xu. 2010b. Quantitative L-lysine requirement of juvenile black sea bream (*Sparus macrocephalus*). Aquacult. Nutr. 16(2):194–204.

Zhou, Q.-C., Z.-H. Wu, B.-P. Tan, S.-Y. Chi, Q.-H. Yang. 2006. Optimal dietary methionine requirement for Juvenile Cobia (*Rachycentron canadum*). Aquaculture 258:551–557.

Zhou, Q.-C., Z.-H. Wu, S.-Y. Chi, Q.-H. Yang 2007. Dietary lysine requirement of juvenile cobia (*Rachycentron canadum*). Aquaculture 273:634–640.

Zhou, X. Q., C. R. Zhao, and Y. Lin. 2007. Compare the effect of diet supplementation with uncoated or coated lysine on juvenile Jian carp (*Cyprinus carpio* Var. Jian). Aquacult. Nutr. 13:457–461.

Zubay, G. 1993. Biochemistry, Third Edition, W. C. Brown, ed. Dubuque, IA: Brown Publishers.

6

Lipids

Defining dietary lipid requirements is complicated by the varied chemical nature and functional roles of lipids. Unlike proteins, lipids are a diverse range of very different compounds that are grouped together simply on the basis of their solubility in organic solvents. Most lipids are "complex," meaning that they contain fatty acids usually esterified to alcohol groups in the case of acylglycerols (glycerides) and to amino groups in the case of sphingolipids. In terms of function, lipids can be divided into two groups: (1) polar lipids, including phospholipids that play predominantly structural roles, and (2) neutral lipids, including triacylglycerols (TAG or triglycerides), whose role is storage, primarily of energy but also cellular components. Simple lipids do not contain fatty acids, with the most important one in animals, including fish, being cholesterol that can be unesterified as a functional component of cell membranes or in a storage form esterified to a fatty acid.

As most lipids are complex, fatty acids usually comprise the bulk of dietary lipid intake. The requirement for specific fatty acids depends upon their differing functional roles and whether they can be synthesized endogenously. All fatty acids can serve as energy sources, but some specific long-chain polyunsaturated fatty acids (LC-PUFA) also have a number of essential roles in metabolism. Lipid requirements therefore encompass a gross requirement for energy and more specific requirements for functional lipid classes such as cholesterol, intact phospholipid, and essential fatty acids. In addition, individual fatty acids can be delivered in a variety of chemical forms including TAG, phospholipids, steryl and wax esters, free fatty acids, or synthetic concentrates such as methyl- or ethyl-esters.

FATTY ACID STRUCTURE AND NOMENCLATURE

In the n- or "omega" nomenclature, fatty acids are described by the general formula, X:Yn-z, where X is the chain length, Y is the number of ethylenic/double bonds, and n-z (or ωz) denotes the position of the first double bond relative to the methyl end of the aliphatic chain. Thus, 16:0 denotes a saturated fatty acid containing 16 carbons and no double bonds (all carbons saturated with hydrogen), and 18:1n-9 (18:1ω9) designates a monounsaturated fatty acid with 18 carbon atoms with a single, normally *cis*, double bond 9 carbon atoms from the methyl end (Figure 6-1). Polyunsaturated fatty acids (PUFA) contain two or more double bonds, generally separated by a single methylene (CH_2) group. Thus, 20:4n-6 (20:4ω6) is a 20-carbon chain with four methylene-interrupted double bonds with the first double bond situated 6 carbon atoms from the methyl end of the molecule. Similarly, 22:6n-3 (22:6ω3) is a 22-carbon chain with 6 double bonds with the first double bond situated 3 carbon atoms from the methyl end (Figure 6-2).

In the alternative Δ (delta) nomenclature, 22:6n-3 is written as 22:6Δ4,7,10,13,16,19 with Δ signifying the positions of the double bonds from the carboxyl end of the molecule. The n-nomenclature is more convenient and commonly used, although the Δ nomenclature is often used for specifying fatty acyl desaturase (Fad) activities. Thus, a desaturase that introduces an ethylenic (double) bond 6 carbons from the carboxyl end of the molecule is termed Δ6 Fad. Fatty acids also have trivial names often reflecting their sources such as palmitic acid (16:0) from palm oil, oleic acid (18:1n-9) from olive oil, and α-linolenic acid (18:3n-3) from linseed oil. Slightly more useful are their chemical names, such as eicosapentaenoic acid (EPA; 20:5n-3) and docosahexaenoic acid (DHA; 22:6n-3), that reflect both the numbers of carbon atoms and double bonds they contain.

LIPID CLASS STRUCTURES

Sterols, tetracyclic hydrocarbon alcohol compounds are the most important simple lipids (i.e., not containing fatty acids) with cholesterol being the predominant sterol in animals including fish (Figure 6-3). Unesterified cholesterol is an essential component of all cell membranes, but can also be found in steroidogenic tissues esterified to fatty acid in the form of neutral lipid droplets as a store of hormone precursor.

The major neutral lipid is triacylglycerol (TAG), which

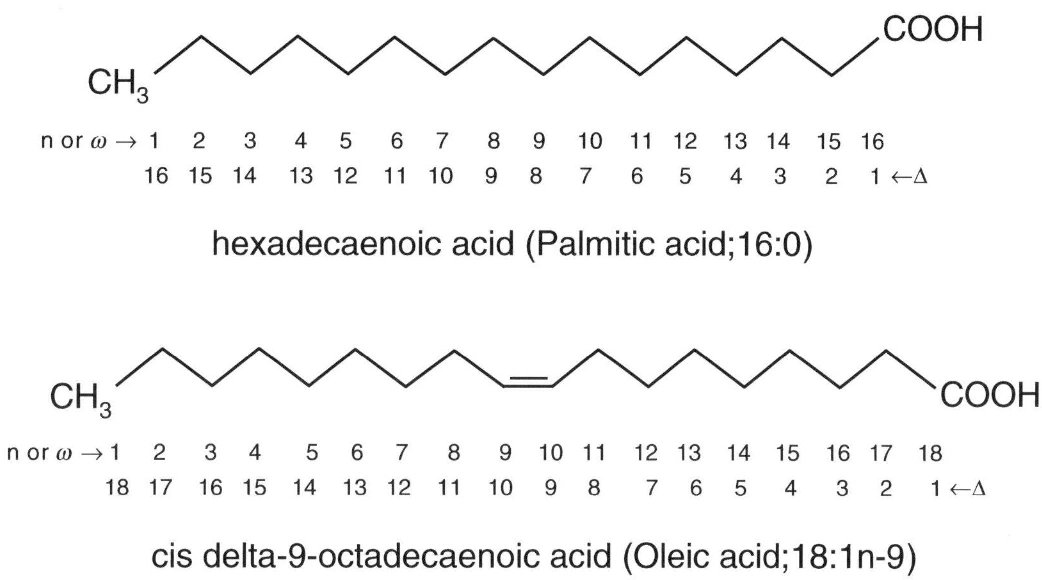

FIGURE 6-1 Palmitic (16:0) and oleic (18:1n-9) acids showing the *n* carbon numbering system.

consists of three fatty acids esterified to the alcohol groups of glycerol (Figure 6-3). Generally saturated and monounsaturated fatty acids are preferentially located in the *sn*1 and *sn*3 positions, whereas PUFA are preferentially located in the *sn*2 (middle) position, although, as with phospholipids, there are many exceptions (Tocher, 2003). The main role of TAG is for energy storage, and another form of lipid store is wax ester that consists of a fatty acid esterified to a fatty alcohol. Wax ester is abundant in marine zooplankton, particularly calanoid copepods and euphausiids, which are natural foods for many marine fish. The fatty alcohols are generally saturated or monounsaturated and, in the case of high-latitude marine zooplankton, can be rich in $C_{20/22}$ monounsaturated alcohols. The large amounts of 20:1n-9 and 22:1n-11 fatty acids in northern hemisphere fish oils are derived from the oxidation of the corresponding fatty alcohols during digestion and absorption of wax esters in zooplanktonivorous fish (Tocher, 2003).

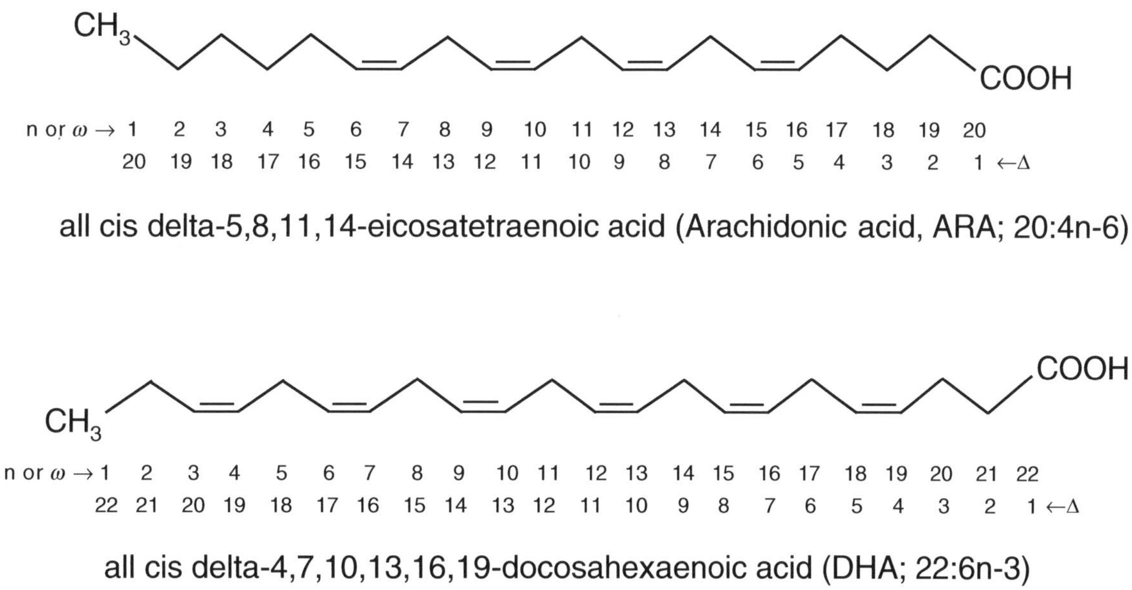

FIGURE 6-2 Arachidonic (20:4n-6) and docosahexaenoic (22:6n-3) acids showing the *n* and Δ carbon numbering systems.

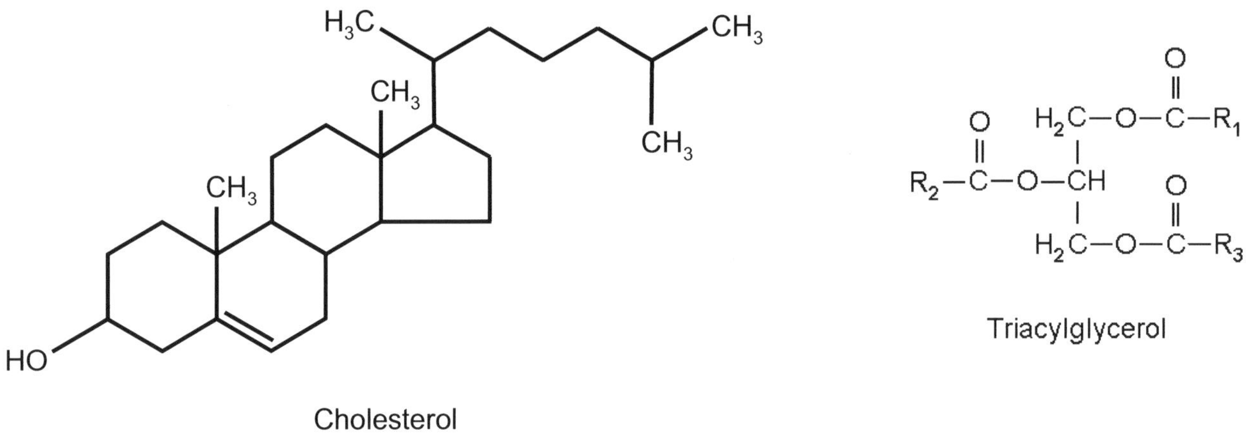

Cholesterol

Triacylglycerol

FIGURE 6-3 The structures of cholesterol and triacylglycerol.

Phospholipid is a general term comprising all lipids containing phosphorus including sphingomyelin, although the term is commonly used to describe phosphoglycerides, which are the predominant polar lipids. Phosphoglycerides are all derived from phosphatidic acid (PA), which is L-glycerol 3-phosphate esterified with two fatty acids. Esterification of the "bases" choline, ethanolamine, serine, and inositol to the phosphate group of PA results in the major phosphoglycerides, phosphatidylcholine (PC), phosphatidylethanolamine (PE), phosphatidylserine (PS), and phosphatidylinositol (PI) (Figure 6-4). Generally, saturated and monounsaturated fatty acids are preferentially esterified at *sn*-1 with PUFA preferentially esterified on position *sn*-2, although there are many exceptions (Tocher, 1995, 2003). Sphingolipids are a group of complex polar lipids based on the long-chain amino alcohol sphingosine, or a related base, with a long-chain saturated or monounsaturated fatty acid linked to the amino group to form a ceramide, and different polar head groups are attached to the primary alcohol group. Further groups can be esterified to the alcohol group of sphingosine such as phosphocholine to form sphingomyelin (Figure 6-4) or sugar moieties to form the cerebrosides.

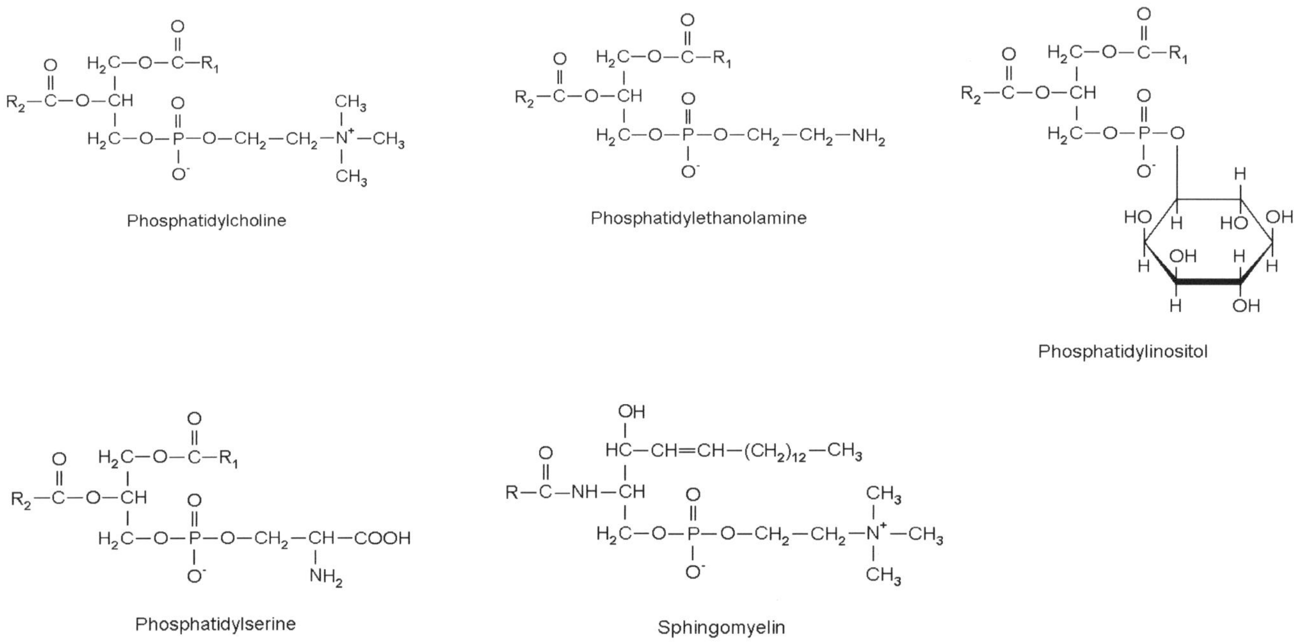

Phosphatidylcholine

Phosphatidylethanolamine

Phosphatidylinositol

Phosphatidylserine

Sphingomyelin

FIGURE 6-4 The structures of the main phospholipid classes.

GENERAL LIPID METABOLISM

Most of the basic pathways of lipid metabolism including digestion and absorption, lipid transport, lipogenesis, and β-oxidation are essentially the same in fish as they are in mammals. The following summary is based on several reviews where more detailed accounts can be found (Sargent et al., 1989, 2002; Olsen and Ringø, 1997; Tocher, 2003; Mourente et al., 2007; Tocher et al., 2008).

In general, dietary lipids are digested mainly in the proximal intestine and pyloric caeca, if present, with the pancreas or hepatopancreas generally assumed to be the major source of digestive enzymes, TAG lipases, and phospholipases. The main products of digestion of the major classes of lipids are free fatty acids along with partial acylglycerols, predominantly 2-monoacylglycerols, lyso-phospholipids, cholesterol, and fatty alcohols. The basic physical processes of digestion and absorption, including bile-enhanced emulsification and transport of the hydrolysed products, are assumed to be generally similar to that in mammals. Thus, the main hydrolytic products are solubilized or emulsified in bile salt micelles, followed by diffusion to the intestinal mucosa where uptake into the enterocytes occurs, probably mainly by passive diffusion. In the intestinal mucosal cells, the predominant fate of the absorbed free fatty acids is reesterification with glycerol, partial acylglycerols, and lyso-phospholipids to reform TAG and phospholipids. Steryl and wax esters may also reform, although free cholesterol is easily transported from the mucosal cells, and fatty alcohols are oxidized to the corresponding fatty acid in the epithelial cells.

Although some free fatty acids can be transported bound to albumin-like proteins, the majority of lipids are transported in the blood in the form of lipoproteins that are qualitatively similar to those found in mammals. The relative proportions of lipid and protein in these lipoproteins vary from chylomicrons (which transport the majority of absorbed dietary lipid away from the intestine) and very low density lipoproteins (VLDL) that have a high lipid:protein ratio, through low density lipoprotein (LDL), to high density lipoproteins (HDL) that have a low lipid:protein ratio. The proportions of the different lipoprotein classes vary with species. Female fish also produce vitellogenin, another lipoprotein, specifically for transporting lipid to the developing oocytes during the process of vitellogenesis. Intracellular transport of fatty acids is facilitated by specific low molecular weight, highly conserved tissue-specific cytoplasmic fatty acid binding proteins (FABP) that bind both long-chain fatty acids and other hydrophobic ligands. Excess dietary lipid is deposited in adipose cells although the precise tissue location can vary between fish species. Most species have intraperitoneal (intestinal) adipose tissue, and some deposit lipid as a layer between the skin and flesh. However, the so-called oily fish, such as herring and salmon, also deposit lipid in the flesh, and others, such as cod and halibut store lipid predominantly in the liver.

Lipogenesis describes the biosynthetic reactions for the endogenous formation of new lipid. The carbon source for the biosynthesis of new lipids is acetyl-CoA formed in mitochondria from the oxidative decarboxylation of pyruvate (carbohydrate source) or the oxidative degradation of some amino acids (protein source). The key lipogenesis pathway is catalyzed by the cytosolic fatty acid synthetase (FAS) multienzyme complex, which produces the saturated fatty acids 16:0 and 18:0. Monounsaturated fatty acids are produced by microsomal stearoyl CoA desaturase (SCD) or Δ9 desaturase, producing 18:1n-9 and, to a lesser extent, 16:1n-7. Fatty acid elongases produce longer chain saturated and monounsaturated fatty acids, such as 20:0 and 20:1n-9. However, PUFA cannot be synthesized de novo by any vertebrate and must be obtained in the diet. Fatty acids are esterified into complex lipids, including membrane phospholipids and TAG, by essentially the same pathways as in mammals. Fish lipids, and specifically membrane phospholipids, are rich in LC-PUFA that must be protected from peroxidation by endogenous systems including a suite of enzymes such as catalase, glutathione peroxidase, glutathione S-transferase and glutathione reductase, and antioxidant compounds including glutathione, vitamin E, and vitamin C.

Fatty acid catabolism is the major source of energy in many species of fish. Whereas the biosynthesis of fatty acids occurs in the cytosol, the catabolism of fatty acids occurs in the cellular organelles, mitochondria, and peroxisomes. The process is termed β-oxidation and involves the sequential cleavage of two-carbon units, released as acetyl-CoA, through a cyclic series of reactions catalyzed by several distinct enzyme activities rather than a multienzyme complex. Briefly, activated fatty acids are transported into the mitochondrion in the form of fatty acylcarnitine esters formed through the action of carnitine acyltransferase, converted back into fatty acyl-CoA derivatives that then undergo a round of dehydrogenation, hydration, second hydrogenation, and cleavage steps to produce acetyl-CoA and NADH. Acetyl-CoA can then be metabolized via the tricarboxylic cycle to produce more NADH, which can then provide metabolic energy in the form of ATP through the process of oxidative phosphorylation. In vitro studies comparing relative oxidation rates of fatty acids have generally shown the following orders of preference: saturated/monounsaturated > PUFA > LC-PUFA, with shorter chain > longer chain and n-6 > n-3. However, in vivo studies investigating fatty acid deposition show that, generally, the higher the concentration of a fatty acid in the diet, the lower its relative deposition (retention), implying increased concentration leads to increased oxidation. Therefore, oxidation of a fatty acid is a balance between enzyme specificities and substrate fatty acid concentrations (competition). One possible exception to this is DHA that is resistant to β-oxidation, as the Δ4 double bond requires peroxisomal oxidation to be removed, and so is poorly oxidized in mitochondria resulting in DHA appearing to be retained in tissues, independent of dietary concentration.

DIETARY LIPID LEVEL

A true lipid requirement for any species of fish or shrimp cannot be specifically defined because it is influenced by a variety of nutritional factors. As a macronutrient, lipid is principally a source of energy.The amount of dietary lipid required is influenced by the contents of dietary protein and carbohydrate, which can also serve as sources of energy. As previously described, protein and carbohydrate can also be sources of lipid through lipogenesis with amino acids and pyruvate serving as the main carbon sources. As protein sources are the most costly ingredients in diets formulated for commercial use, the goal is to minimize dietary protein that might be used as a source of energy. Therefore, with an appropriate amount of energy supplied by lipid, protein requirements can be reduced or "spared." In turn, the level of lipid required to satisfy the energy requirement could be reduced through the provision of sources of carbohydrates in species that can effectively utilize these nutrients. However, carbohydrates are more efficiently digested by some species (often herbivorous/omnivorous) than others. Thus, some species may have limited capabilities of digesting carbohydrate, thereby restricting its use as an effective source of energy. It may be that species that have evolved on high-lipid food sources are more likely to have a poor utilization of dietary carbohydrate. Environmental temperature may be another factor for the difference as most investigated carnivores are coldwater species and most herbivorous species are warmwater fish. The amount of dietary lipid is also affected by its source relative to the satisfaction of requirements for essential fatty acids (EFA). The relative amount of lipid to satisfy the EFA requirements is dependent upon lipid source(s) and corresponding fatty acid profiles. For commercial diet formulation, generally TAG-rich oils/fats are provided as ingredients of diets to ensure that specific requirements for PUFA and/or LC-PUFA are effectively satisfied.

Although an "optimum" level of dietary lipid cannot be truly defined for any species, there is a range within which dietary lipid should be supplied. The lower limit will be defined as the amount of lipid required to supply the requirements for EFA (and cholesterol and phospholipid in some species at specific life stages), which will depend upon the precise lipid source(s) and their corresponding fatty acid profiles. However, higher dietary levels may be necessary to satisfy obligatory lipid deposition required to successfully fulfill or realize certain physiological stages often associated with reproduction (migration/spawning). Increasing dietary lipid above the minimum level will support higher growth rates, possibly partly based on protein sparing, toward an upper limit where excess lipid leads to unwanted deposition of lipid in the peritoneal cavity, liver, or other tissues (Company et al., 1999; Craig et al., 1999; Gaylord and Gatlin, 2000). This represents wasted energy as there is little point in supplying an energy-yielding nutrient that is simply deposited unused in tissue stores. Of course, deposited lipid contributes to increased weight, but as it is not flesh (muscle), it is not contributing to yield. This is highlighted in species such as Atlantic cod (*Gadus morhua*) that deposit lipid in the liver or other species with large perivisceral storage. So-called "oily" fish such as Atlantic salmon (*Salmo salar*), which deposit significant amounts of lipid in the flesh, are able to tolerate and utilize higher dietary lipid levels.

Fish

Not withstanding the above caveats, various studies have investigated the relationships between dietary lipid contents, growth, and lipid deposition in fish. Weight gain was increased in rainbow trout (*Oncorhynchus mykiss*) in fish fed dietary lipid at 21% compared to 8–11% (Luzzana et al., 1994), and growth was higher in brown trout (*Salmo trutta*) fed dietary lipid at 29% compared to 21% (Arzel et al., 1993). Furthermore, weight gain in Atlantic salmon was higher in fish fed diets containing 38–47% lipid compared to fish fed 31% lipid (Hemre and Sandnes, 1999). However, high dietary lipid increases flesh lipid levels in freshwater fish and salmonids including rainbow trout (Dias et al., 1999) and Atlantic salmon (Bell et al., 1998; Hemre and Sandnes, 1999). Despite this, the upper level for dietary lipid in salmon diets doubled between the 1970s and late 1990s, when an optimal dietary lipid level of 35% was suggested (Einen and Roem, 1997). However, deposition of excess dietary lipid in the flesh can impact carcass and product quality, causing problems of oily texture and pigmentation that lead to consumer and processor resistance (Bell et al., 1998; Hillestad et al., 1998) and may influence early sexual maturation in males (Shearer and Swanson, 2000). Some problems may be alleviated by feeding a low-fat "finishing" diet prior to slaughter (Rasmussen et al., 2000).

However, in contrast to the above situation with salmonids, it should be noted that > 85% of all farmed finfish production is of freshwater, predominantly low trophic level fish species including carps and tilapia (Tacon et al., 2010), which generally cannot tolerate such high levels of dietary lipid (often < 10%). This may be associated with these species having natural diets that generally contain lower levels of lipid and, perhaps, higher levels of carbohydrate that they are thus adapted to utilize more effectively and efficiently (see Chapter 7). As a result these species seem to have a lower ability to utilize high dietary lipid and so commercial feeds.

Weight gain of European sea bass (*Dicentrarchus labrax*) was increased in fish fed diets containing lipid at 15% compared to 9% lipid (Manuel Vergara et al., 1996), and 19% compared to 11 and 15% (Lanari et al., 1999), but a lower limit to the growth-promoting effect of high-fat diets in marine fish was indicated because growth rate was higher in sea bass fed 24% lipid compared to fish fed 30% lipid (Peres and Oliva-Teles, 1999). Flesh, organ, and visceral lipid increases as dietary lipid increases in marine fish including turbot (*Psetta maximus*) (Saether and Jobling, 2001) and sea bass (Catacutan and Coloso, 1995). High-fat diets may also promote the development of fatty liver pathology (Caballero et al., 1999).

Shrimp

In shrimp and other crustaceans, weight gain responses to different levels of dietary oils, either alone or in combination, indicate that highest gains are generally achieved at dietary levels of 5–6% inclusion. Higher levels (> 10%) often retard growth (Kanazawa et al., 1977a; Davis and Robinson, 1986; Sheen and D'Abramo, 1991), most probably due to a reduction in consumption caused by high caloric content and/or an inability to metabolize high levels efficiently (reduced digestibility). Reduced growth has been shown to be associated with accumulation of lipid in tissue (Castell and Covey, 1976; Ponat and Adelung, 1983; González-Félix et al., 2002a). These conclusions on dietary lipid levels were drawn from experiments in which marine-derived sources containing good profiles of n-3 LC-PUFA, including cod liver oil, menhaden fish oil, pollock liver oil, and short-neck clam oil, were used. Some studies have included plant oils that are good sources of n-6 PUFA.

In summary, because of the complex metabolic interactions between protein, lipid, and carbohydrate mentioned at the beginning of the section, definition of precise dietary lipid requirements in fish and shrimp are not particularly useful or meaningful. Although lipid up to 20% of the dry weight of the diet allows protein to be effectively utilized for growth in many fish species without depositing excessive lipid in the tissues (Sargent et al., 2002), lipid can have a protein-sparing effect in many species that has driven the use of so-called "high-energy" (high-lipid) diets to become increasingly widespread in aquaculture. High-energy diets can have consequences by altering lipid and fatty acid metabolism with health and welfare implications for the fish and product quality for the consumer (Sargent and Tacon, 1999). More detailed accounts of nutritional energetics and the role of lipid as an energy source and its interaction with other dietary components, including protein and carbohydrate, are provided in Chapter 4.

SPECIFIC REQUIREMENTS

Essential Fatty Acids

As vertebrate and crustacean species cannot synthesize any PUFA from monounsaturated fatty acids de novo (see Figure 6-5), they therefore have an absolute dietary requirement for certain specific n-3 and n-6 PUFA. Dietary deficiency of these "essential fatty acids" results in various pathologies, the animal stops growing and reproducing, and eventually dies (Das, 2006). The biologically active PUFA required for many essential metabolic and physiological processes are the LC-PUFA, 20:4n-6 (ARA, arachidonic acid), 20:5n-3 (EPA) and 22:6n-3 (DHA) (Das, 2006). In contrast, the shorter chain C_{18} PUFA, typified by linoleic acid 18:2n-6 and α-linolenic acid 18:3n-3, have no specific metabolic roles in themselves, although they can serve as precursors for the corresponding n-6 and n-3 LC-PUFA (Sargent et al., 1995a). Note that vertebrates and crustaceans are unable to interconvert the n-6 and n-3 PUFA families (Figure 6-5). Species vary in their capacity to convert C_{18} PUFA to LC-PUFA. In species that cannot perform these conversions, dietary C_{20} and C_{22} LC-PUFA are essential, and their C_{18} homologues do not satisfy EFA requirements. In species that can perform the conversions, C_{18} PUFA, and C_{20-22} LC-PUFA can all be termed EFA with the LC-PUFA often being more effective nutritionally than their C_{18} counterparts. Definition of the optimal amounts of EFA to satisfy the requirements for normal growth and development has been a well-studied area of lipid metabolism in fish, driven by the needs of the aquaculture industry. Of particular importance, the requirements can vary quantitatively during ontogenesis and, therefore, accurate definition of EFA requirements for a given species involves determining not only the absolute requirements of specific PUFA and the optimal balance between different PUFA, but also how these requirements vary at different life stages (Tocher, 2010).

Methodological Challenges

Appreciation of quantitative EFA data requires some considerations of the methodology used for determining these requirements. The methodology is difficult because an EFA-deficient feed has to be produced, which requires an essentially lipid-free diet. This is hard to achieve without affecting other important aspects of the diet, such as attraction and palatability. The consequence of these difficulties is that EFA requirements were generally measured in small fish fed diets with much lower lipid levels than are commonly used today, with consequently lower growth rates. Therefore, the quoted estimates of EFA requirements probably represent the levels that were sufficient to (1) prevent appearance of deficiency signs and (2) to maintain growth at that particular, albeit low, dietary lipid level. The EFA requirements can be expressed as a percentage of the total lipid, percentage of diet, or percentage of total fatty acids. It is apparent that the quantitative requirement for EFA may vary with the total dietary lipid level, and this may also vary with the stage of development (Izquierdo, 1996). For instance, the requirement for n-3 LC-PUFA appeared to increase as the level of lipid in the diet increased in red sea bream (*Pagrus major*) fingerlings (Takeuchi et al., 1992a), yellowtail (*Seriola quinqueradiata*) fingerlings (Takeuchi et al., 1992b), and *Penaeus monodon* (Glencross et al., 2002a), although there was no apparent variation in the requirement for n-3 LC-PUFA as

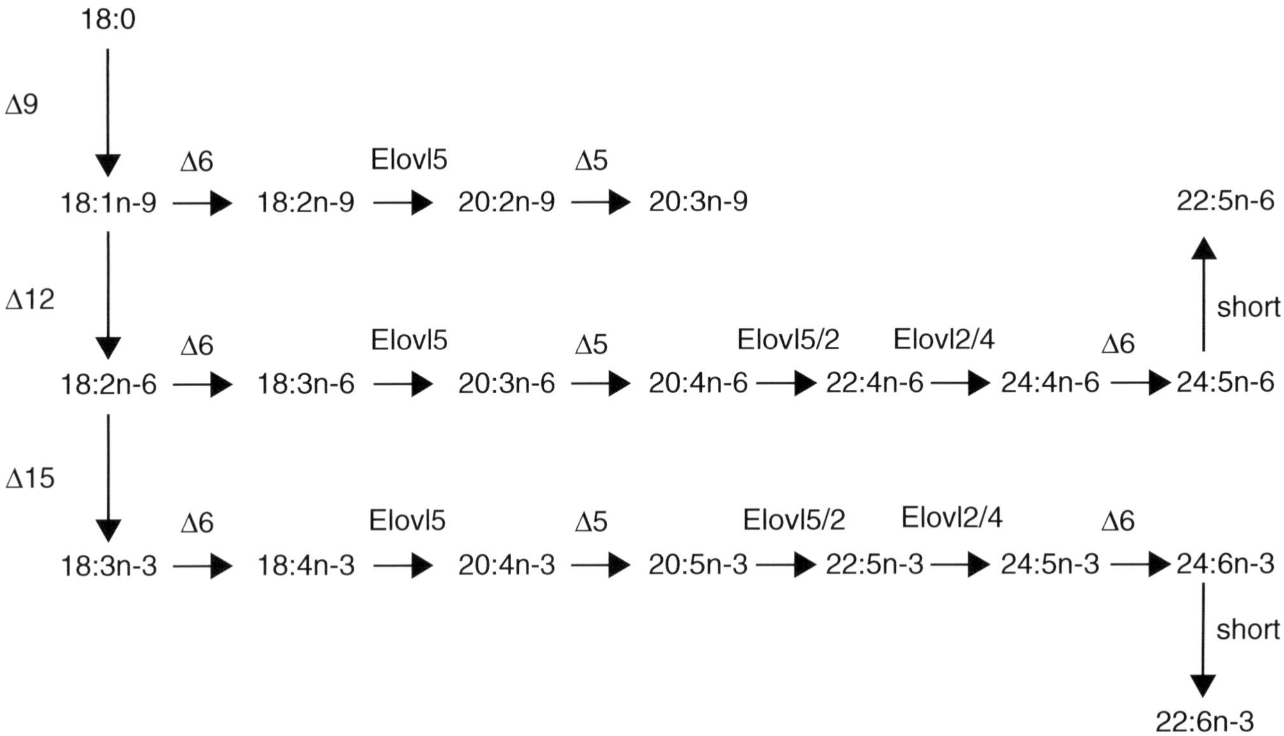

FIGURE 6-5 Pathways of biosynthesis of C20 and C22 long-chain polyunsaturated fatty acids (LC-PUFA) from n-3, n-6, and n-9 C_{18} PUFA. Δ9, stearoyl CoA desaturase (SCD); Δ5 and Δ6, front-end fatty acyl desaturases (Fad). Evidence suggests that the same Δ6 Fad operates on both C_{18} and C_{24} fatty acyl substrates; Δ12 and Δ15, Fads found only in plants and some invertebrates, and hence 18:2n-6 and 18:3n-3 cannot be formed in any vertebrate; Elovl2, Elovl4, and Elovl5, PUFA elongases; short, peroxisomal chain shortening.

the dietary lipid level increased in larval gilthead sea bream (*Sparus aurata*) (Salhi et al., 1994) or *Litopenaeus vannamei* (González-Félix et al., 2002a, 2003). However, in later studies, growth retardation has been observed in sea bream and turbot fed diets with high levels of dietary fish oil substituted with vegetable oils (devoid of LC-PUFA), despite the diets being formulated to supply EPA and DHA above the estimated EFA requirements (Caballero et al., 2003; Regost et al., 2003). These later studies used higher lipid levels (16–20%) supporting higher growth rates than in previous EFA requirement studies (8–12%) (Kalegeropoulos et al., 1992; Ibeas et al., 1994, 1997). One explanation may be that the higher growth rates supported by diets with increased lipid can only be achieved with similarly increased EFA. Therefore when EFA levels are reduced, albeit still above the "EFA requirement," by substituting vegetable oil, decreased growth can be observed (Caballero et al., 2003; Regost et al., 2003). Growth retardation was not apparent in salmonids fed similar vegetable oil diets, suggesting that endogenous production of EPA and DHA from dietary 18:3n-3 may be sufficient to maintain the physiological requirements for these fatty acids and prevent growth suppression (Bell et al., 2004; Torstensen et al., 2005). Therefore, increments in dietary EFA level above the reported "requirement" may improve growth and survival, suggesting that there may also be "optimal" EFA levels. Despite the suggestion that EFA requirements can vary based on diet formulation, it is likely, for the reasons argued at the beginning of this section, that the quoted EFA requirement levels are good indicators of minimum levels that should be provided to prevent pathology.

Fish

The quantitative and semiquantitative requirements for EFA have been reported for around 30 species of fish (Tables 6-1 through 6-3). In the past 10 years studies focused more on larval marine fish and the relative requirements of ARA, EPA, and DHA rather than defining absolute EFA requirements in juveniles and subadults of more species (Lund et al., 2007, 2008; Hamre and Harboe, 2008a,b). This is probably because the experiments are difficult and expensive, because, in addition to the problems of diet formulation discussed above, a regression protocol should be used requiring significant numbers of experimental units. However, there is probably sufficient information on a wide enough range of species to predict qualitative and semiquantitative EFA requirements for new species of interest (Tocher, 2003, 2010). Requirements for EFA also vary with developmental and possibly physiological stage, further complicating the definition of absolute quantitative requirements (Sargent et al., 2002).

TABLE 6-1 Reported Quantitative Essential Fatty Acid (EFA) Requirements of Juvenile and Subadult Freshwater and Diadromous Species of Finfish[a]

Species	Scientific Name	EFA	Requirement (% Dry Diet)	Reference
Arctic charr	*Salvelinus alpinus*	18:3n-3	1.0–2.0	Yang et al. (1994)
Atlantic salmon	*Salmo salar*	18:3n-3	1.0	Ruyter et al. (2000a)
		n-3 LC-PUFA	0.5–1.0	Ruyter et al. (2000b)
Ayu	*Plecoglossus altivelus*	18:3n-3 or EPA	1.0	Kanazawa et al. (1982)
Channel catfish	*Ictalurus punctatus*	18:3n-3	1.0–2.0	Satoh et al. (1989)
Cherry salmon	*Oncorhynchus masou*	18:3n-3 or n-3 LC-PUFA	1.0	Thongrod et al. (1990)
Chum salmon	*Oncorhynchus keta*	18:2n-6 and 18:3n-3	1.0 of each	Takeuchi et al. (1979)
Coho salmon	*Oncorhynchus kisutch*	18:2n-6 and 18:3n-3	1.0 of each	Yu and Sinnhuber (1979)
Common carp	*Cyprinus carpio*	18:2n-6	1.0	Takeuchi and Watanabe (1977)
		18:3n-3	0.5–1.0	Takeuchi and Watanabe (1977)
Grass carp	*Ctenopharyngodon idella*	18:2n-6	1.0	Takeuchi et al. (1991)
		18:3n-3	0.5	Takeuchi et al. (1991)
Japanese eel	*Anguilla japonicus*	18:2n-6 and 18:3n-3	0.5 of each	Takeuchi et al. (1980)
Milkfish	*Chanos chanos*	18:2n-6 and 18:3n-3	0.5 of each	Bautista and de la Cruz (1988)
Rainbow trout	*Oncorhynchus mykiss*	18:3n-3	0.7–1.0	Castell et al. (1972)
		n-3 LC-PUFA	0.4–0.5	Takeuchi and Watanabe (1976)
Sheatfish	*Silurus glanis*	18:3n-3	1.0	Borgut et al. (1998)
Striped bass	*Morone chrysop* × *Morone saxatilis*	n-3 LC-PUFA	1.0	Gatlin et al. (1994)
Tilapia	*Tilapia zilli*	18:2n-6	1.0	Kanazawa et al. (1980)
	Oreochromis nilotica	18:2n-6	0.5	Takeuchi et al. (1983)
	Oreochromis niloticus × *Oreochromis aureus*	n-3 required	?	Chou and Shiau (1999)
Whitefish	*Coregonus laveratus*	18:3n-3	> 1.0	Thongrod et al. (1989)
		n-3 LC-PUFA	0.5–1.0	Watanabe et al. (1989)

[a]Based on Tocher (2010).

Freshwater and Diadromous Species

Reported estimates for juveniles and subadults of freshwater and diadromous fish species indicate that EFA requirements can be satisfied by the C_{18} PUFA, 18:3n-3, and 18:2n-6, at around 1% of the diet dry weight (Table 6-1). In terms of EFA, freshwater/diadromous species were traditionally subdivided into three groups: coldwater species including salmonids that have a higher requirement for 18:3n-3 (compared to 18:2n-6), warmwater species such as tilapia (*Oreochromis* spp.) that have a higher requirement for 18:2n-6, and species that require significant amounts of both such as common carp (*Cyprinus carpio*). However, growth of hybrid tilapia (*O. niloticus* × *O. aureus*) was significantly improved by feeding cod liver oil compared to corn oil and so, although not quantitatively defined, tilapia also require n-3 fatty acids, or at least n-3 LC-PUFA, for maximal growth (Chou and Shiau, 1999). Therefore, it is likely that all freshwater/diadromous fish require both n-3 and n-6 PUFA, with coldwater fish possibly having a requirement for higher levels of n-3 compared to n-6. Although the C_{18} PUFA are usually effective in satisfying the EFA requirements of freshwater fish, for some species, including salmonids, n-3 LC-PUFA can satisfy the EFA requirements at lower levels than 18:3n-3 and increase growth over that obtained with 18:3n-3 alone (Ruyter et al., 2000b). Similarly, growth was significantly improved in channel catfish (*Ictalurus punctatus*) by inclusion of dietary n-3 LC-PUFA (Santha and Gatlin, 1991). There are few data on the requirements of freshwater fish for the main n-6 LC-PUFA, ARA (Bell and Sargent, 2003).

The early life stages of freshwater fish species have received little attention, and so there are few data reports providing estimated EFA requirements (Table 6-2). Newly hatched larvae or fry of many freshwater fish are large enough to accept formulated feeds whose composition can be defined to ensure maximal growth and survival such that feeds are not a problem in rearing high-quality fry. However, there is evidence that n-3 LC-PUFA and DHA may be more important and, possibly, essential in larvae of some species of freshwater fish compared to adults or juveniles (Webster and Lovell, 1990; Wirth et al., 1997). Broodstock nutrition is also critical to produce high-quality eggs and larvae with

TABLE 6-2 Reported Quantitative Essential Fatty Acid (EFA) Requirements of Larvae and Early Juveniles of Finfish[a]

Species	Scientific Name	EFA	Requirement (% Dry Diet)	Reference
Freshwater				
Common carp	*Cyprinus carpio*	n-6 PUFA	1.0	Radunzneto et al. (1996)
		n-3 PUFA	~ 0.05	Radunzneto et al. (1996)
Rainbow trout	*Oncorhynchus mykiss*	DHA essential	?	Wirth et al. (1997)
Marine				
Atlantic cod	*Gadus morhua*	EPA required	?	Zheng et al. (1996)
		DHA	~ 1.0	Takeuchi et al. (1994)
Gilthead sea bream	*Sparus aurata*	n-3 LC-PUFA	5.5 (DHA:EPA = 0.3)	Rodriguez et al. (1994a)
		n-3 LC-PUFA	1.5 (DHA:EPA = 2.0)	Rodriguez et al. (1998a)
		n-3 LC-PUFA	1.5 (in phospholipid)	Salhi et al. (1999)
		DHA:EPA	~ 2	Rodriguez et al. (1994b)
Mahi mahi	*Coryphaena hippurus*	n-3 LC-PUFA	0.6–1.0	Ostrowski and Kim (1993)
Red sea bream	*Pagrus major*	n-3 LC-PUFA	2.1 (with 1.0 DHA)	Furuita et al. (1996a)
		DHA	1.0–1.6	Furuita et al. (1996a)
		EPA	2.3	Furuita et al. (1996a)
Striped bass	*Morone chrysops* × *Morone saxatilis*	18:3n-3	?	Webster and Lovell (1990)
		n-3 LC-PUFA	< 0.5%	Webster and Lovell (1990)
Striped jack	*Pseudocaranx dentex*	DHA	1.6–2.2	Takeuchi et al. (1996)
		EPA	< 3.1	Takeuchi et al. (1996)
Turbot	*Psetta maxima*	DHA required	?	Reitan et al. (1994)
Yellowtail	*Seriola quinqueradiata*	n-3 LC-PUFA	3.9 (DHA:EPA = 0.5)	Furuita et al. (1996b)
		DHA	1.4–2.6	Furuita et al. (1996b)
		EPA	3.7	Furuita et al. (1996b)

[a]Based on Tocher (2010).

EFA contents optimized for the specific requirements of the developing embryos and larvae (Tandler et al., 1995; Izquierdo et al., 2001; Quintero et al., 2010), and broodstock diets have been shown to affect egg fatty acid compositions such as in Eurasian perch (*Perca fluviatilis*) (Abi-ayad et al., 1997) and Nile tilapia (*Oreochromis niloticus*) (Santiago and Reyes, 1993).

Marine Species

The reported EFA requirements of juvenile and subadult marine fish suggest that C_{18} PUFA cannot satisfy the requirement, and so the n-3 LC-PUFA, EPA, and DHA are required (Table 6-3). Levels of n-3 LC-PUFA of less than or up to 1% of diet dry weight can meet the requirements for juvenile turbot, red sea bream, European sea bass, red drum (*Sciaenops ocellatus*), and Korean rockfish (*Sebastes schlegeli*), whereas levels above 1% appear to be required by silver bream (*Rhabdosargus sarba*), striped jack (*Pseudocaranx dentex*), and yellowtail flounder (*Pleuronectes ferrugineus*). The quantitative requirement for n-6 LC-PUFA has not been fully determined (Bell and Sargent, 2003), but studies suggested ARA was essential in turbot with a requirement of around 0.3% of diet (dry weight) estimated in weaned fish (Castell et al., 1994; Bell et al., 1995a). Quantitative EFA requirements of juvenile sea bream varied with the dietary DHA:EPA ratio, with requirements for total n-3 LC-PUFA of around 1.9 and 0.9% of diet with dietary ratios of 0.5 and 1.0% of DHA:EPA, respectively, consistent with DHA generally having a higher EFA value for fish than EPA (Kalogeropoulos et al., 1992; Watanabe, 1993; Ibeas et al., 1994; Brinkmeyer and Holt, 1998; Wu et al., 2002).

The definition of precise EFA requirements of larval marine fish is complicated by their small size and generally poorly developed digestive system, difficulties of preparing microdiets, and the use of live feeds (Izquierdo et al., 2000; Cahu and Zambonino-Infante, 2001; Koven et al., 2001a; Robin and Vincent, 2003; Kvale et al., 2006; Conceição et al.,

TABLE 6-3 Reported Quantitative Essential Fatty Acid (EFA) Requirements of Juvenile and Subadult Marine Species of Finfish[a]

Species	Scientific Name	EFA	Requirement (% Dry Diet)	Reference
European sea bass	*Dicentrarchus labrax*	n-3 LC-PUFA	1.0	Coutteau et al. (1996a)
Gilthead sea bream	*Sparus aurata*	n-3 LC-PUFA	0.9 (DHA:EPA = 1)	Kalegeropoulos et al. (1992)
		n-3 LC-PUFA	1.9 (DHA:EPA = 0.5)	Ibeas et al. (1994)
		DHA:EPA	0.5	Ibeas et al. (1997)
Grouper	*Epinephelus malabaricus*	n-3 LC-PUFA, DHA > EPA	1.0	Wu et al. (2002)
Japanese flounder	*Paralicthys olivaceus*	n-3 LC-PUFA	1.4	Takeuchi (1997)
Korean rockfish	*Sebastes schlegeli*	n-3 LC-PUFA	0.9	Lee et al. (1993)
		EPA or DHA	1.0	Lee et al. (1994)
Red drum	*Sciaenops ocellatus*	n-3 LC-PUFA	0.5–1.0	Lochman and Gatlin (1993)
		EPA + DHA	0.3–0.6	Lochman and Gatlin (1993)
Red sea bream	*Pagrus major*	n-3 LC-PUFA or EPA	0.5	Yone (1978)
		EPA	1	Takeuchi et al. (1990)
		DHA	0.5	Takeuchi et al. (1990)
Silver bream	*Rhabdosargus sarba*	n-3 LC-PUFA	1.3	Leu et al. (1994)
Starry flounder	*Paralichthys stellatus*	n-3 LC-PUFA	0.9	Lee et al. (2003)
Striped bass	*Morone chrysop* × *Morone saxatilis*	n-3 LC-PUFA	1.0	Gatlin et al. (1994)
Striped jack	*Pseudocaranx dentex*	DHA	1.7	Takeuchi et al. (1992c)
Turbot	*Psetta maxima*	n-3 LC-PUFA	0.8	Gatesoupe et al. (1977)
		ARA	~ 0.3	Castell et al. (1994)
Yellowtail flounder	*Pleuronectes ferrugineus*	n-3 LC-PUFA	2.5	Whalen et al. (1999)
Yellowtail/Kingfish	*Seriola* spp.	n-3 LC-PUFA	2.0–2.4	Deshimaru et al. (1982)

[a]Based on Tocher (2010).

2007, 2010; Yufera and Darias, 2007). However, the quantitative and semiquantitative EFA requirements of larvae of various marine species have been estimated using a combination of enriched live feeds and fabricated microdiets (Table 6-2). The reported values can vary dependent upon the criteria measured, such as survival, growth, and vitality, as well as dietary lipid level (Salhi et al., 1994; Furuita et al., 1996b). Although there are few species where the requirements at larval and juvenile stages can be directly compared, larvae are generally characterized by having a higher requirement than juveniles and preadult fish for n-3 LC-PUFA (Tables 6-2 and 6-3). As with juveniles, EFA requirements in larval marine fish can often be satisfied by a lower level of DHA than can be achieved with EPA (Watanabe, 1993), with the higher efficacy of DHA related to its role in the rapidly developing visual and neural tissues, which account for a relatively greater proportion of total body mass in larval stages (Sargent et al., 2002). Thus, the relative proportions of the different EFA are important in larval marine fish with the absolute requirement for n-3 LC-PUFA decreasing with increasing DHA:EPA ratio (Rodriguez et al., 1994a, 1998a). Growth in larval gilthead sea bream was influenced by ARA (Rodriguez et al., 1994a), and, at a fixed level of dietary n-3 LC-PUFA and DHA:EPA ratio, ARA up to 1.5% and 1% of diet dry weight improved growth in larval sea bream (Bessonart et al., 1999) and Japanese flounder (*Paralichthys olivaceus*), respectively (Estevez et al., 1997). Dietary ARA also improved survival after handling stress in sea bream larvae, particularly when fed prior to the stress (Koven et al., 2001b), whereas high dietary ARA inhibited growth, increased mortality, and had negative effects on pigmentation in yellowtail flounder larvae (Ishizaki et al., 1998).

In recent years, increasing attention has been paid to the role of EFA, particularly ARA, in metamorphosis of marine flatfish including pigmentation and eye migration (Lund et al., 2007, 2008). Decreased n-3 LC-PUFA and increased ARA and ARA:EPA were associated with malpigmentation and impaired eye migration, increasing the focus on dietary DHA:EPA:ARA ratios (Villalta et al., 2005; Hamre and Harboe, 2008a,b). During the premetamorphic stages, there are critical periods when the absolute and relative amounts of EFA and the duration of feeding are particularly important, although these vary among species. In turbot, the early supply of DHA was essential for correct pigmentation (Reitan et al., 1994), and ARA levels in neural tissues were negatively correlated with pigmentation, with the optimum dietary EPA level being more dependent on dietary ARA than DHA level, emphasizing the importance of dietary DHA:EPA:ARA ratios (Estevez et al., 1999). Pigmentation success was related to dietary levels of ARA and LC-PUFA in neural tissues of Japanese flounder (Estevez and Kanazawa, 1996; Estevez et al., 1997) and dietary ARA in common sole (*Solea solea*) (Lund et al., 2008). Therefore, although there is increasing evidence for the essentiality of dietary ARA for optimal growth and development of marine fish larvae and although precise requirements are not defined, excess can cause problems at metamorphosis in flatfish (Rodriguez et al., 1994a; Ishizaki et al., 1998; Bessonart et al., 1999; Estevez et al., 1999; Hamre et al., 2007; Lund et al., 2007, 2008).

As with freshwater fish, broodstock nutrition is vital in marine fish to produce high-quality eggs and larvae with EFA contents optimized to give the developing embryos and larvae the best chance of success at a time of increased EFA requirement (Tandler et al., 1995; Izquierdo et al., 2001). Many studies have demonstrated that egg fatty acid compositions are affected by broodstock diets in various species including sea bream (Fernandez-Palacios et al., 1995; Almansa et al., 1999), sea bass (Bell et al., 1997), striped jack (Vassallo Agius et al., 1998), Atlantic cod (Silversand et al., 1995), and yellowtail (Verakunpiriya et al., 1996). Egg quality criteria, such as hatching, fertilization rates, and early survival, were positively correlated with increased levels of n-3 LC-PUFA and ARA in eggs of sea bream (Harel et al., 1992; Fernandez-Palacios et al., 1995; Rodriguez et al., 1998b), Atlantic cod (Pickova et al., 1997; Salze et al., 2005), and sea bass (Bruce et al., 1999), and with DHA:EPA ratio in cod (Pickova et al., 1997).

Shrimp

Kanazawa and Teshima (1977) conducted studies with the Kuruma prawn/shrimp *Marsupenaeus japonicus* and demonstrated the inability to synthesize n-3 and n-6 PUFA and LC-PUFA. Over the past 30 years, all investigations with shrimp and other crustaceans have supported these initial studies. Therefore, all crustaceans are reported to have an absolute requirement for specific PUFA and/or LC-PUFA (Table 6-4). Different studies have expressed the EFA requirements of shrimp in different ways, including as a percentage of diet weight, total dietary lipid, or total dietary fatty acids. Some dietary experiments have used different sources of oils with various fatty acid profiles to study responses to various fatty acids in the diet. Other approaches to understanding nutrition and nutritional requirements were approached through the use of pure TAG, or fatty acid methyl or ethyl ester concentrates. For most of the studies conducted with species of juvenile shrimp, TAG sources have been used and the dietary lipid content has commonly ranged between 30 and 75 g/kg diet (3.0 and 7.5%).

Early investigations by Kanazawa et al. (1979a,b) were based upon the use of pure fatty acids in the form of methyl esters in diets containing 18:1n-9 (40 g/kg) and 10 g/kg of either 18:2n-6, 18:3n-3, EPA, and DHA, or 18:1n-9 (50 g/kg) and fed to *M. japonicus*. Weight gains of prawns fed the diets containing either PUFA or LC-PUFA were higher than those fed 18:1n-9 alone. A hierarchy of effectiveness of fatty acids relative to LC-PUFA and PUFA was shown according to the following order: EPA > DHA > 18:3n-3 > 18:2n-6. This work has been supported by other investigations with marine shrimp demonstrating that LC-PUFA, particularly EPA, were more biologically active and elicited significantly higher growth rates than PUFA. Merican and Shim (1997) found that DHA had the highest EFA activity measured as weight gain in the marine tiger shrimp *P. monodon*.

An array of studies by Glencross and coworkers with *P. monodon* supported EFA requirements for n-3 and n-6 PUFA and LC-PUFA (Glencross and Smith, 1999, 2001a,b; Glencross et al., 2002a,b). Glencross and Smith (1999) found that the addition of either 18:2n-6 or 18:3n-3 yielded maximum growth when included at a concentration of 12 g/kg with the overall lipid level being 75 g/kg. They also found that the requirements differed when both EFA were included. Single additions of either EPA or DHA at about 9 g/kg also enhanced weight gain (Glencross and Smith, 2001a). Additional studies confirmed the interactive effects of EFA with requirements for both fatty acids being about 1/3 of what was observed when they were added as exclusive sources. Additional investigations led to an estimate of an ideal n-3 to n-6 ratio of 2.5:1 (Glencross et al., 2002a). This observation supported the results of early experiments that found that best growth responses were elicited by a combination of marine and plant oils, sources of n-3 and n-6 fatty acids, respectively (Deshimaru et al., 1979). Glencross et al. (2002b) also demonstrated that requirements for EFA are based upon the total amount of dietary lipid. Therefore, the proportion of the EFA in the lipid is key to the satisfaction of EFA requirements rather than the absolute level. However, other studies with *L. vannamei* did not indicate a change in the absolute requirement of LC-PUFA with increasing levels of dietary lipid (González-Félix et al., 2002a, 2003). It was also shown that addition of dietary PUFA and LC-PUFA increased weight gain and that the requirement for 18:3n-3 was between 7 and 10 g/kg, and for DHA it was 10 g/kg (Xu et al., 1993, 1994). Kanazawa et al. (1979b,c) also observed the best growth response using a combination of dietary PUFA and LC-PUFA, with the best growth in juvenile *M. japonicus* achieved with either 18:2n-6 or 18:3n-3 added at 10 g/kg in combination with n-3 LC-PUFA derived from pollock residual liver oil and short-necked clam lipids.

D'Abramo and Sheen (1993) examined the qualitative EFA requirements of the caridean shrimp *Macrobrachium rosenbergii*, which spend most of their life cycle in freshwater, by feeding juveniles diets containing pure sources of 18:2n-6, 18:3n-3, ARA, and DHA. Relative to the control diet that contained 60 g/kg of lipid composed of a mixture of

TABLE 6-4 Reported Quantitative Essential Fatty Acid (EFA) Requirements of Shrimp

Species	Requirement Level	Reference
18:3n-3 or 18:2n-6		
Brown shrimp (*Penaeus aztecus*)	1–2% (18:3n-3)	Shewbart and Mies (1973)
Tiger shrimp (*Penaeus monodon*)	1.2% (18:3n-3)	Glencross and Smith (1999)
P. monodon	1.2% (18:2n-6)	Glencross and Smith (1999)
Fleshy prawn (*Fenneropenaeus chinensis*)	0.7–1.0% (18:3n-3)	Xu et al. (1993)
18:3n-3 > 18:2n-6		
Kuruma prawn (*Marsupenaeus japonicus*)	—	Kanazawa et al. (1977b)
F. chinensis	—	Xu et al. (1994)
20:5n-3 or 22:6n-3		
M. japonicus	1.1% (20:5n-3)	Kanazawa et al. (1978)
M. japonicus	1.1% (22:6n-3)	Kanazawa et al. (1979b)
P. monodon	0.9% (22:6n-3)	Glencross and Smith (2001a)
P. monodon	0.9% (20:5n-3)	Glencross and Smith (2001a)
F. chinensis	1.0% (22:6n-3)	Xu et al. (1994)
Giant river prawn (*Macrobrachium rosenbergii*)	0.075% (22:6n-3)	D'Abramo and Sheen (1993)
Pacific white shrimp (*Litopenaeus vanamei*)	0.50% (20:5n-3; 22:6n-3)	Gonzalez-Félix et al. (2003)
20:4n-6		
M. rosenbergii	0.08%	D'Abramo and Sheen (1993)
P. monodon	No requirement identified	Glencross and Smith (2001b)
L. vanamei	0.50%	Gonzalez-Félix et al. (2003)
n-3 to n-6 ratio		
Blue shrimp (*Penaeus stylirostris*)	1.18:1 (n-3:18:2n-6)	Fenucci et al. (1981)
Common prawn (*Palaemon serratus*)	0.45 (18:3n-3:18:2n-6)	Martin (1980)
M. rosenbergii	0.083 (18:3n-3:18:2n-6)	Teshima et al. (1994)
P. monodon	2.5:1 (n-3:n-6)	Glencross et al. (2002a)

pure TAG of saturated and monounsaturated fatty acids, significant increases in weight gains were observed for dietary additions of either DHA or ARA. Also, a mixture of EPA or DHA, provided by extracts of cuttlefish liver oil and prepared for inclusion as methyl esters, significantly increased growth. No growth-enhancing effects were found for either 18:2n-6 or 18:3n-3, but a possible higher biological activity for 18:2n-6 was suggested. An improved growth response may be possible if both of these fatty acids are available in the diet, probably with an n-3 to n-6 PUFA ratio lower than 1. Evidence that 18:3n-3 and 18:2n-6 can be metabolized to their corresponding LC-PUFA forms was found, but apparently at a rate that was insufficient to sustain the highest growth rates. The n-6 LC-PUFA may have greater EFA activity than n-3 LC-PUFA.

For marine (penaeid) shrimp, requirement levels of about 10 g/kg of 18:2n-6 and about 15 g/kg for 18:3n-3, at a ratio of 0.7, are suggested. At these dietary levels, about equal amounts of EPA and DHA (3 g/kg) are additionally required to achieve the highest growth rates. These requirements are generally based upon a total dietary lipid level of 50 g/kg, about 10–15% and 3% of the total lipid for C_{18} PUFA and LC-PUFA, respectively. Martin (1980) found that a dietary ratio of 18:2n-6 to 18:3n-3 for the freshwater crustacean *Palaemon serratus* should be 2.2. Therefore a high-performance formulated diet should contain a total of approximately 30 g/kg of PUFA and LC-PUFA at the appropriate levels provided through TAG sources. The amount of dietary lipid needed to satisfy these recommended requirements would undoubtedly exceed 30 g/kg, and choice of fatty acid sources to achieve the requirements would ultimately be an important consideration in deciding what the total amount of dietary lipid (TAG plus phospholipid) should be. For freshwater crustaceans, the satisfaction of the EFA requirement seems to lie in a sufficient supply of LC-PUFA, either as ARA, EPA, DHA, or their combinations, with equivalent responses achieved at levels within a range of 0.8 to 6.0 g/kg. Therefore, LC-PUFA requirements are an order of magnitude lower than those of marine shrimp. Thus, both marine and freshwater shrimp require dietary LC-PUFA, but levels required to meet requirements differ dramatically, most probably reflective of the fatty acid composition of the food in their natural environment.

The amount of lipid that achieves the best growth and survival in a shrimp diet is based on a combination of satisfaction of EFA requirements and lipid-derived energy requirements, the ability of the species to use dietary carbohydrate, and how much protein may still be used for energy. The best results are obtained with a combination of marine- and plant-derived oils to provide ideal amounts and ratios of n-3 and n-6 PUFA and LC-PUFA. For marine shrimp, total dietary lipid levels between 50 and 60 g/kg levels seem sufficient with levels of 18:3n-3 and 18:2n-6 being about 15 g/kg and 10g/kg, and EPA and DHA being included at about 3 g/kg. For freshwater crustaceans, requirements for LC-PUFA appear lower, and n-6 EFA may be sufficient to achieve desired growth. Application of information concerning EFA requirements of shrimp species to the manufacture of practical feeds for culture may not always be possible because these ideal proportions cannot be achieved and the amount of dietary lipid in TAG may be restricted by the manufacturing process.

Digestibility is also a factor influencing the amount of dietary lipid. Most studies suggest that the digestibility of lipid, expressed as apparent lipid digestibility (ALD), in experimental and practical formulated diets that contain marine-derived triacylglyerols as the primary lipid source, is generally very high. However, dietary levels < 4.5 and > 10% were associated with reduced digestibility in adults of the shrimp *P. monodon* (Glencross et al., 2002b). Merican and Shim (1994, 1995) evaluated the comparative digestibility of a variety of terrestrial and marine oils, fatty acids, and lipids in meals used as dietary sources of lipid in formulated diets for the shrimp *P. monodon*. They found that marine-derived oils, such as cod liver, sardine, and refined squid, were more digestible than terrestrial derived oils such as soy and palm. Among classes of lipids, triacylglycerols were found to be most digestible followed by phospholipids, steryl esters, free sterols, and free fatty acids. Digestibility of PUFA was significantly higher than that of saturated fatty acids. Digestibility of saturated fatty acids decreased as the chain length increased, whereas the digestibility of monounsaturated fatty acids (monoenes) increased as chain length increased. The ALD of different meal ingredients (full fat soy, liver and whole meals from squid and fish) were lower (62–81%) than the respective apparent dry matter digestibility (ADMD) (78–95%). Liver-derived marine meals were more digestible than their whole meal counterparts.

The system in which the shrimp are grown may also be a consideration in meeting the EFA requirements in commercial diets. In semi-intensive pond culture systems, where there is considerable input from the consumption of natural, particularly secondary, productivity, the amounts of EFA required in formulations could be reduced considerably. This is an important consideration because the sources of EPA and DHA from marine-derived oils are in short supply and may soon become the limiting factor in the manufacture of shrimp feeds (Tacon and Metian, 2008). Under these conditions, the practicality of high-density culture systems where a nutritionally complete and balanced diet is essential for successful culture becomes questionable. The advent of other systems may allow a major decrease in the amount of EFA provided by diets. With the goal to minimize fish meal in crustacean diets, this source of n-3 LC-PUFA is lost and compensation will need to be addressed. A goal to reduce levels of dietary LC-PUFA to only satisfy EFA requirements may reduce the content of n-3 LC-PUFA in tissues such that consumer acceptance may be affected (see Chapter 16).

Biochemical/Molecular Basis of Requirements

The qualitative EFA requirements of both fish and crustaceans appear to vary with environment and/or feeding habit (Sargent et al., 2002). This may be an evolutionary adaptation, as the primary producers, the phytoplankton, produce high levels of the n-3 LC-PUFA, EPA, and DHA in the marine environment (Sargent et al., 1995b), and so marine fish and crustaceans have had less evolutionary pressure to retain the ability to endogenously produce LC-PUFA. In contrast, freshwater food webs are generally characterized by lower levels of EPA and, especially, DHA (Sargent et al., 1995b), and so evolutionary pressure for endogenous production of LC-PUFA has been maintained in freshwater species. Although the data, to date, generally support this hypothesis, there are confounding factors including precise feeding habit of different species (herbivorous vs. carnivorous/piscivorous) as well as phylogenetic issues. Indeed, defining fish species as marine and freshwater is often not ideal especially considering diadromous and euryhaline species. Furthermore, a similar generalization as above can be made for feeding habit with the ability to endogenously produce LC-PUFA being retained in herbivorous fish, but not in omnivorous, carnivorous, or piscivorous fish. Almost all the fish species studied so far would fit either of these generalizations (environment or feeding habit). However, recent data have helped to clarify this situation. A feeding study suggested that *Siganus canaliculatus* or rabbitfish, which consumes benthic algae and seagrasses and is thus a rare example of an exclusively herbivorous marine species, was able to biosynthesize EPA and DHA (Li et al., 2008). Very recently, it was shown that rabbitfish possesses all the potential fatty acyl desaturase activities required for endogenous synthesis of LC-PUFA (Li et al., 2010). These data suggest that trophic level and/or feeding habit are important factors associated with or determining a species' ability for endogenous LC-PUFA synthesis.

Synthesis of EPA in vertebrates is achieved by Δ6 desaturation of 18:3n-3 to produce 18:4n-3 that is elongated to 20:4n-3 followed by Δ5 desaturation (Cook and McMaster, 2004), with DHA synthesis from EPA requiring two further elongation steps, a second Δ6 desaturation, and a chain shortening step, at least in rats and rainbow trout (Sprecher, 2000; Tocher, 2003) (Figure 6-5). Synthesis of ARA from 18:2n-6 uses the same enzymes and pathway as for EPA. The ability of any species to convert C_{18} PUFA to LC-PUFA is thus associated with their complement of Fad and elongase (Elovl, *El*ongation of *v*ery *l*ong-chain fatty acids) enzymes (Bell and Tocher, 2009). The EFA requirements described above suggest that most freshwater fish are capable of producing the biologically active LC-PUFA from C_{18} PUFA, and must express all the biosynthetic activities necessary. Most marine fish and crustaceans, however, cannot or have only limited ability (Tocher, 2003). Biochemical studies using radioactive fatty acids supported the hypothesis that deficiencies in the LC-PUFA synthesis pathway account for the differences in EFA requirements between different fish species. In vivo injection studies suggested that EPA and ARA were produced from 18:3n-3 and 18:2n-6, respectively, in rainbow trout, but not in turbot (Owen et al., 1975), and in vitro studies indicated that cell lines from marine fish had very low Δ5 Fad or fatty acyl elongase activities compared to salmonid lines (Ghioni et al., 1999; Tocher and Ghioni, 1999). There is evidence that C_{18} PUFA can be metabolized to their corresponding LC-PUFA forms in the freshwater shrimp *M. rosenbergii*, but apparently at a rate that was insufficient to sustain the highest growth rates (D'Abramo and Sheen, 1993). Experiments with muscle and midgut gland tissue suggest that marine crustaceans have little or no capability to biosynthesize C_{18} PUFA into LC-PUFA (Kanazawa et al., 1979d).

The molecular basis of these enzyme deficiencies is being elucidated. The cDNAs for Δ6 Fad have been cloned and characterized from all fish so far investigated including freshwater (common carp), salmonid (rainbow trout and Atlantic salmon), and marine species (turbot, cod, sea bream, cobia *Rathycentron canadum*, European and Asian sea bass, *Lates calcarifer*) (Hastings et al., 2001; Seiliez et al., 2001, 2003; Zheng et al., 2004a, 2005a; Tocher et al., 2006; González-Rovira et al., 2009; Zheng et al., 2009a; Mohd-Yusof et al., 2010; Monroig et al., 2010). In contrast, until recently, cDNAs for Δ5 Fad have only been cloned from Atlantic salmon and zebrafish (*Danio rerio*) (Hastings et al., 2001, 2005). Zebrafish actually expressed a bifunctional Δ6/Δ5 Fad that also showed low activity toward C_{24} PUFA, indicating that it could function at two steps in the LC-PUFA synthesis pathway (Figure 6-5), consistent with it being the only PUFA Fad represented in the zebrafish genome (Hastings et al., 2001, 2005). A similar bifunctional Δ6/Δ5 Fad has very recently been isolated from the marine herbivore, rabbitfish, which also expresses a Δ4 Fad that offers an alternative pathway for DHA synthesis by direct desaturation of 22:5n-3 (Li et al., 2010). However, despite significant efforts, no Δ5 Fad has been found in any other marine fish, and searches of the three sequenced Acanthopterygii genomes showed that the medaka (*Oryzias latipes*) possess a single LC-PUFA Fad and the stickleback (*Gasterosteus aculeatus*) two very closely related genes, and no Δ5 or Δ6 homologues were identified in the pufferfish (both *Tetraodon nigroviridis* and *Takafugu rubripes*) (Leaver et al., 2008a). In mammals, several Elovl genes are known and at least two, Elovl2 and Elovl5, participate in LC-PUFA biosynthesis (Jakobsson et al., 2006). The cDNAs for Elovl5 have been cloned from a number of fish, including freshwater (zebrafish, common carp, and tilapia), salmonid (rainbow trout and Atlantic salmon), and marine (cod, turbot, gilthead sea bream, cobia, and Asian sea bass) species (Agaba et al., 2004, 2005; Hastings et al., 2005; Zheng et al., 2009a; Mohd-Yusof et al., 2010). In contrast, Elovl2 has only been cloned from Atlantic salmon and zebrafish (Monroig et al., 2009; Morais et al., 2009). Functional characterization of the fish Elovls showed that Elovl5 had activity predominantly toward C_{18} and C_{20} PUFA, whereas Elovl2 had activity predominantly toward C_{20} and C_{22} PUFA. Searches of the sequenced genomes showed that pufferfish, stickleback, and medaka, and possibly all other Acanthopterygii, do not possess Elovl2 homologues, and so it is likely that the characterized Elovl5 cDNAs of sea bream and turbot are the sole PUFA Elovl genes in these species (Leaver et al., 2008a). Therefore, the varying competences of different species to biosynthesize LC-PUFA probably depend on their genome complement of both desaturase and elongase genes, with many, predominantly marine species appearing to lack Δ5 Fad and Elovl2 elongase.

The Δ6 Fad cDNAs cloned from all fish species studied so far showed significant activity in heterologous yeast expression systems. In contrast, Δ6 desaturation activity and expression of Δ6 Fad are very low in cod liver and intestine compared to the activity and expression of Δ6 Fad in salmon tissues (Tocher et al., 2006). Furthermore, Δ6 Fad expression and activity are under nutritional regulation in freshwater and salmonid fish. The activity of the LC-PUFA pathway in carp cells was increased by EFA-deficiency (Tocher and Dick, 1999) and modulated by different C_{18} PUFA (Tocher and Dick, 2000). In vivo dietary trials have shown that activity of the LC-PUFA biosynthetic pathway is increased in freshwater and salmonid fish fed vegetable oils rich in C_{18} PUFA compared to fish fed fish oil, rich in EPA and DHA (Tocher et al., 1997, 2002, 2003). Expression of Δ6 Fad mRNA was reduced in salmon fed diets containing fish oil, in comparison to fish fed diets containing vegetable oils lacking LC-PUFA (Zheng et al., 2004b, 2005a,b; Leaver et al., 2008b; Taggart et al., 2008). Again, in contrast, there was little difference in Δ6 Fad expression and activity in cod fed diets containing either vegetable or fish oil (Tocher et al., 2006). Thus many marine fish, like cod, may be deficient both in critical LC-PUFA biosynthetic genes and in the expression levels and regulation of Δ6 Fad compared to freshwater or salmonid fish. Recent work suggested that differences in the *fad* gene promoters in cod and salmon may influence gene expression (Zheng et al., 2009b). The highest level of expression of Δ6 *fad* in the marine fish cod, cobia, and Asian sea bass was found in the brain, whereas in salmon, liver and intestine were the tissues of highest expression (Zheng et al., 2005a, 2009a; Tocher et al., 2006; Mohd-Yusof et al., 2010). This may suggest that the retention of Δ6 Fad in marine fish may be related to a requirement to maintain membrane DHA levels, particularly in neural tissues at times of high demand such as embryonic and larval development.

Differences among species that have the genes/enzymes required for EPA/DHA synthesis and those that do not have the genes/enzymes affect the biochemical markers of EFA deficiency in fish. In mammals, 20:3n-9 (Mead acid) is the biochemical marker, with a ratio of 20:3n-9:ARA in tissue phospholipids of 0.4 or higher indicating a state of EFA deficiency. The biochemical mechanism of 20:3n-9 production is based on the expression and fatty acid specificity of Fads that are in the rank order 18:3n-3 > 18:2n-6 > 18:1n-9 (Tocher et al., 1998). In the absence of dietary PUFA, 18:1n-9 can serve as a substrate for the LC-PUFA synthesis pathway and 20:3n-9 is produced (Figure 6-5). In freshwater fish and salmonids, 20:3n-9 is also a marker of EFA deficiency as they contain all the activities necessary for its production from 18:1n-9. However, Castell et al. (1972) suggested that 20:3n-9:DHA in tissue phospholipids would be a better indicator ratio, with 0.4 still being the value indicating EFA deficiency, and this was supported in subsequent studies (Watanabe et al., 1974, 1989). In many marine fish, production of 20:3n-9 is not possible, but 18:2n-9 and 20:2n-9 were reported in sea bream fed diets with low levels of EFA (Kalogeropoulos et al., 1992), and 18:2n-9 accumulated in a turbot cell line grown in the absence of PUFA (Tocher et al., 1988).

Functions of Fatty Acids

All fatty acids are important sources of cellular energy, independent of any role that some PUFA may have as EFA (Tocher, 2003). The extent to which any fatty acid is utilized for energy is largely dependent upon its dietary concentration, and so in high concentration all fatty acids will be oxidized. Possible exceptions include 22:1n-11, which tends to be highly oxidized irrespective of concentration, and DHA, which tends to be conserved due to being a relatively poor substrate for mitochondrial β-oxidation (Sargent et al., 2002). However, even DHA can have relatively low retention when fed in high concentrations (Stubhaug et al., 2007). Long-term absence of EFA from the diet leads to deficiency signs that most often include reduced growth and increased mortality (Glencross, 2009), although a range of other pathologies have been reported including myocarditis, pale/swollen (fatty) liver, intestinal steatosis, fin erosion, bleeding from gills, lordosis, reduced reproductive potential, and shock syndrome (Tacon, 1996). Clearly, these deficiency signs point to at least some fatty acids having critical roles other than simply energy production.

Large amounts of PUFA are also required for cellular membrane structure and function, as they are integral elements of phospholipids that are the fundamental components of lipid bilayers (Gylfason et al., 2010). Changes in the fatty acid composition of membrane lipids are important in homeostatic adaptation to environmental change (Farkas et al., 2001). For instance, adaptation to lower water temperature and increased hydrostatic pressure results in increased proportions of PUFA and monoenes, and reduced saturated fatty acids, in membrane phospholipids (Tocher, 2003). Phospholipid molecular species compositions also affect the physical properties of membranes, and so redistribution of fatty acids between phospholipid classes and within *sn* positions can also have effects without gross changes in fatty acid composition (Farkas et al., 1994; Farkas and Halver, 1996). There is evidence that the specificities of the acyltransferase enzymes are altered by temperature and so contributing to restructuring (Tocher, 1995). DHA in particular has important structural and functional roles in all membranes, but especially neural membranes (Feller, 2008; Wasall and Stillwell, 2008), and its importance for the proper development of neural tissues in larval fish has been demonstrated. Dietary deficiency of DHA resulted in impaired ability to capture prey at natural light intensities in larval Atlantic herring (*Clupea harengus*) (Bell et al., 1995b), delayed response to visual stimuli in larval sea bream (Benitez-Santana et al., 2007), and impaired schooling behavior in yellowtail (*S. quinqueradiata*) (Masuda et al., 1998; Ishizaki et al., 2001) and Pacific threadfin (*Polydactylus sexfilis*) (Masuda et al., 2001).

Small amounts of PUFA, particularly specific C_{20} PUFA, have unique and functionally important roles in controlling and regulating cellular metabolism and animal physiology. Central to this role is the regulated, dioxygenase-catalyzed oxidation of ARA and EPA to produce highly bioactive eicosanoids, including prostaglandins (PG), leukotrienes (LT), and lipoxins, autocrine hormones with a short half-life produced by cells to act in their immediate vicinity (Schmitz and Ecker, 2008). Almost all tissues produce eicosanoids, and they have a wide range of physiological actions in blood clotting, immune and inflammatory responses, cardiovascular tone, renal and neural functions, and reproduction (Schmitz and Ecker, 2008). Of particular interest is their role in the inflammatory response. Competition exists between ARA and EPA for the eicosanoid-producing enzymes, with ARA producing proinflammatory 2-series PG and 4-series LT, whereas EPA produces less potent 3-series PG and 5-series LT (Tocher, 1995). Despite the excess of EPA in their tissue lipids, ARA is still the primary eicosanoid precursor in fish (Tocher, 1995). Although inflammation is a protective response to injury and infection, excessive or inappropriate inflammation, driven by ARA-derived eicosanoids, contributes to various acute and chronic pathologies, which n-3 LC-PUFA, EPA, and DHA can mediate. The mechanism of EPA and DHA action is not fully defined, but recently n-3 LC-PUFA-derived mediators including resolvins, such as resolvin E1 from EPA, and D-series resolvins and protectin D1 from DHA, with potent antiinflammatory and proresolving properties, have been implicated (Seki et al., 2009). Resolvins and protectins from DHA have been identified in brain cells from rainbow trout (Hong et al., 2005).

In recent years it has also become apparent that fatty acids, particularly PUFA, play key roles in lipid homeostasis through influencing the regulation of gene transcription (Jump, 2002). Although PUFA can potentially affect gene transcription by a number of mechanisms, including changes in membrane composition and eicosanoid production, specific PUFA themselves are now known to influence the activities of a variety of transcription factors including peroxisome proliferator-activated receptors (PPAR), liver X receptors (LXR), and sterol regulatory element binding proteins (SREBP) that are critical regulators of genes involved in lipid homeostatic processes (Sampath and Ntambi, 2005). These transcription factors are all present in fish and assumed to have similar roles in the regulation of lipid and fatty acid metabolism (Leaver et al., 2008a; Cruz-Garcia et al., 2009). In mammals, some monounsaturated fatty acids have been shown to have quite specific metabolic roles including 18:1n-9, the content of which in mammalian biomembranes was shown to affect intracellular signaling via G-protein associated cascades involving adenyl cyclase and phospholipase C (Teres et al., 2008) and 16:1n-7, which has recently been identified as an adipose tissue-derived hormone (lipokine) that serves as a lipid signal that mediates communication between that tissue and other tissues such as liver and muscle (Cao et al., 2008).

Phospholipids

Shrimp

Phospholipid provided in different dietary forms has been demonstrated to be required for growth and survival of juvenile and larval forms of shrimp (Table 6-5). In most investigations, the source of phospholipids has been a form of lecithin derived from soybeans. The apparent absence of such a requirement in adult forms suggests that the requirement is age-specific and is founded on an insufficient rate of synthesis to meet the demands of the comparatively rapid growth rates characteristic of early life stages.

A phospholipid requirement was first demonstrated for a species of shrimp by Kanazawa et al. (1979c) in *M. japonicus* through the dietary addition of phospholipids derived from the short-necked clam. The study of Kanazawa et al. (1985) with larvae of *M. japonicus* suggests a comparatively noteworthy growth and survival response to the addition of 3.5–6% soybean lecithin, containing 23.6% PC. Paibulkichakul et al. (1998) found that the addition of dietary phospholipids at 1.0–1.5% of diet (10 to 15 g/kg) significantly increased the growth of juvenile *P. monodon*. Thongrod and Boonyaratpalin (1998) observed significant increases in body weight gain of juvenile banana shrimp when 2.5% of 60% pure soybean lecithin was included in the diet.

Not all phospholipids have equivalent effect. For juvenile lobsters (*Homarus americanus*), PC was found to be the active compound that significantly reduced mortality when compared to PE (ovine source) and PI (soybean source) (D'Abramo et al., 1981). A level of 1.0% of pure PC or PI, extracted from bonito eggs and soybean, respectively, was found to be most effective in growth and survival of larval *M. japonicus* (Kanazawa et al., 1985). Pure PC derived from chicken egg and pure PE derived from bonito eggs and ovine brain did not have an equivalent effect.

For species of juvenile penaeid shrimp, reported dietary phospholipid requirements, expressed as PC content, commonly fall within the range of 1.2 to 1.5% (Chen and Jenn, 1991; Chen, 1993; Kanazawa, 1993; Coutteau et al., 1996b). Some of these estimates are based on investigations in which highly purified sources of phospholipids have been added to diets. Chen and Jenn (1991) and Chen (1993) used an 80% pure soy PC. Coutteau et al. (1996b) showed that an addition of 1.5% PC (95% pure) from either a soybean source or 6.5% deoiled soybean lecithin (23% PC) significantly increased growth of *L. vannamei* relative to a PC-deficient diet. Addition of 1.5% PC (94% pure) derived from chicken egg yielded growth similar to that observed with the 95% pure PC from the soybean source.

Despite the above, some shortcomings still exist in stating accurately a PC or PC/PI requirement because dietary sources of lecithin reported in investigations often are not based on a consistent composition. The experimental sources of dietary phospholipid differ in quantitative and qualitative composition. For example, soybean lecithin, a known effective dietary ingredient as a source of phospholipid, is available in different forms of purity relative to the amount of phospholipid that is in the form of PC. In addition, the amount of phospholipid varies considerably when comparing a deoiled versus a raw form of soy lecithin. Therefore, the

TABLE 6-5 Reported Phospholipid Requirements in Juvenile and Larval Shrimp Species

Species	Requirement	Reference
Tiger shrimp (*Penaeus monodon [juvenile]*)	1.0–1.5%	Paibulkichakui et al. (1998)
P. monodon	80% pure soybean PC	Chen (1993)
Marine shrimp (*Penaeus penicillatus*)	80% pure soybean PC	Chen and Jenn (1991)
Pacific white shrimp (*Litopenaeus vannamei*)	1.5% PC (from soybean) 6.5% deoiled soybean lecithin	Coutteau et al. (1996b)
Kuruma prawn (*Marsupenaeus japonicus [juvenile]*)	1.0% (PC + PE)	Kanazawa et al. (1979c)
M. japonicus (juvenile)	3.0% soybean (lecithin) PE and PI	Teshima et al. (1986a,b)
M. japonicus (larvae)	3.0% soybean lecithin	Kanazawa (1983)
M. japonicus (larvae)	0.5 to 1.0% (PC and PI)	Kanazawa et al. (1985)
Banana shrimp (*Fenneropenaeus merguiensis*)	2.5% pure soybean lecithin	Thongrod and Boonyaratpalin (1998)

required quantitative and qualitative amount of dietary phospholipid is difficult to state. However, a relatively confident and conservative estimate based upon the collective results of all the studies is the provision of PC at a level between 0.5 to 1.5%. For commercial diets, the level of a dietary lecithin ingredient will vary based upon source and proportional composition of the phospholipid classes, but "feed grade" types generally should be added within a range of 2.5 to 3.5%. There appears to be no differences in requirements between larval and juvenile forms within a particular species. However, noteworthy is the lack of a specific requirement for dietary phospholipids for juveniles of freshwater species *M. rosenbergii* and *Cherax quadricarinatus*, suggesting that the requirement may be inexplicably unique to marine shrimp/crustaceans (Table 6-5). When an array of studies are collectively examined, a possible role of dietary phospholipid in sparing (reducing) the dietary cholesterol requirement is suggested. However, evidence remains inconclusive.

Fish

Table 6-6 summarizes the data obtained from studies investigating the qualitative and quantitative phospholipid requirements in finfish. The inclusion of dietary intact phospholipids improves growth in both larvae and early juveniles, but also increases survival rates and decreases incidence of malformation in larvae, and perhaps increases stress resistance of various freshwater and marine fish species (Tocher et al., 2008). Defining absolute dietary phospholipid requirements is complicated by the use of a variety of phospholipid preparations that vary both in phospholipid content and class composition. Furthermore, larval studies also have often been compromised by the need to supply phospholipid through enrichment of live feeds with possible remodeling of the phospholipid and fatty acid composition in the feed organisms. The levels of phospholipid requirement can be as high as 8–12% of diet dry matter for larval fish (Cahu et al., 2009), rather less (around 2–4%) for juvenile fish, and a requirement for dietary phospholipids has not been established for adult fish (Tocher et al., 2008). The majority of studies have used crude mixed phospholipid preparations, including soybean and other plant lecithins, and egg yolk lecithin that are enriched in several phospholipid classes making identification of which specific phospholipid class imparts beneficial effects more difficult. However, based on the few studies where single pure phospholipid species have been used, the order of efficacy appears to be PC > PI > PE > PS, with PC possibly more important for growth and PI being more important for survival and preventing deformities (Geurden et al., 1998a; Tocher et al., 2008). The efficacy of other phospholipid classes or sphingolipids is not known.

Biochemical Basis of Requirements

The mechanism underpinning the role of the phospholipids in larval and early juvenile shrimp and fish must also explain their lack of effect in adults. Relatively little work has been carried out in shrimp (crustaceans), but the physiological role of phospholipids is principally attributed to their role as components of lipoprotein molecules, specifically high-density lipoproteins that serve as transport molecules for cholesterol and TAG. Teshima and Kanazawa (1980) found that the high-density lipoproteins in the serum of *M. japonicus* contained 65–85% polar lipids, with 50% of the fatty acids being DHA. Lipoproteins transport lipids from the epithelial cells of the gut to the hemolymph, which transfers them to tissues. Removal of dietary soy lecithin in semipurified diets fed to juvenile lobsters resulted in lower levels of both PC and cholesterol (D'Abramo et al., 1982) and reduced rates of cholesterol transport from the midgut

TABLE 6-6 Reported Quantitative and Qualitative Phospholipid Requirements of Finfish[a]

Species	Developmental Stage	Phospholipid Supplement[b] and Levels Studied[c]	Optimal Requirement and Criteria Used[d]	Feeding Period	Reference
Atlantic salmon (*Salmo salar*)	Juvenile (180 mg)	0, 2, 4, 6, and 8% SL/CPL	6% (G)	14 weeks	Poston (1991)
	Juvenile (180 mg)	0 and 4% SL	4% (G)	16 weeks	Poston (1990b)
	Juvenile (1.0 g)	0 and 4% SL	4% (G)	12 weeks	Poston (1990b)
	Juvenile (1.7 g)	0 and 4% SL	4% (G)	12 weeks	Poston (1990b)
	Juvenile (7.5 g)	0 and 4% SL	0% (no requirement)	12 weeks	Poston (1990b)
Ayu (*Plecoglossus altivelus*)	Larvae	0 and 3% SL or EL	3% (G,S,M)	20 days	Kanazawa et al. (1981)
	Larvae	0, 1, 3, and 5% SL	3% (M), 5% (G,S)	50 days	Kanazawa et al. (1983a)
	Larvae	0 and 3% EL or BPL	3% (G,S,M)	50 days	Kanazawa et al. (1983a)
	Juvenile	0 and 3% SL or BPL	3% (G)	33 days	Kanazawa et al. (1981)
	Juvenile	0, 1, 3, and 5% EL	3% (G)	33 days	Kanazawa et al. (1981)
Common carp (*Cyprinus carpio*)	Larvae	0 and 2% EL	2% (G,S)	25 days	Geurden et al. (1995a)
	Larvae	0 and 2% PL	2% (G,S)	21 days	Geurden et al. (1995a)
	Larvae	0 and 2% SPC, SPI, or EL	2% (G,S,M except EL)	25 days	Geurden et al. (1997a)
European sea bass (*Dicentrarchus labrax*)	Larvae	3, 6, 9, and 12% SL	12% (G,S,M)	40 days	Cahu et al. (2003)
	Juvenile	0 and 3% SL	3% (G)	40 days	Geurden et al. (1995b)
	Juvenile	0 and 2% EPC or SPC	2% (G)	40 days	Geurden et al. (1995b)
Gilthead sea bream (*Sparus aurata*)	Larvae	9, 11, and 15% SL	> 9% (G,S)	23 days	Seiliez et al. (2006)
Japanese flounder (*Paralichthys olivaceus*)	Larvae	0, 3, 5, and 7% SL	7% (G,S)	30 days	Kanazawa (1993)
	Juvenile	0, 3, 5, and 7% SL	7% (G)	30 days	Kanazawa (1993)
Knife jaw (*Oplegnathus fasciatus*)	Larvae	0, 2.5, 5 ,and 7.4% SL	7.4% (G,S)	22 days	Kanazawa et al. (1983b)
	Larvae	0, 3, 5, and 7% SL	5% (G,S,R)	28 days	Kanazawa (1993)
	Juvenile	0, 3, 5, and 7% SL	3% (G)	60 days	Kanazawa (1993)
Pikeperch (*Sander lucioperca*)	Larvae	1, 5, and 9% SL	9% (G)	24 days	Hamza et al. (2008)
Rainbow trout (*Oncorhynchus mykiss*)	Juvenile	0, 2, 4, and 8% SL	4% (G)	20 weeks	Poston (1990a)
	Juvenile	0 and 14%	14% (G)	8 weeks	Rinchard et al. (2007)
Red sea bream (*Pagrus major*)	Larvae	0 and 5% SL	5% (G,S)	20 days	Kanazawa et al. (1983b)
Striped jack (*Pseudocaranx dentex*)	Juvenile	0, 0.5, 1, 1.5, and 2% SPC	1.5% (G,S,R)	6 weeks	Takeuchi et al. (1992c)
	Juvenile	0 and 1.5% SPE	1.5% (G)	6 weeks	Takeuchi et al. (1992c)
Turbot (*Psetta maximus*)	Juvenile	0 and 2% EL	2% (G)	—	Geurden et al. (1997b)
White sturgeon (*Acipenser transmontanus*)	Juvenile (5–10 g)	0 and 8% SL	0% (no requirement)	6 weeks	Hung and Lutes (1988)

[a]Based on Tocher et al. (2008).

[b]BPL, bonito egg polar lipid; CPL, corn polar lipid; EL, chicken egg lecithin; EPC, purified egg PC; PL, various phospholipid sources supplemented to supply 2% dietary phospholipids including EL, SL, sunflower, rapeseed, and marine phospholipids; SL, soybean lecithin; SPC, purified soybean PC; SPE, purified soybean PE; SPI, purified soybean PI.

[c]Percentage of diet weight.

[d]G, growth; S, survival; M, malformations; R, stress resistance.

gland to the hemolymph (D'Abramo et al., 1985a). The results of the later investigation were supported by the studies of Teshima et al. (1986c,d) working with juvenile *M. japonicus*, i.e., the enhanced transport of cholesterol from the hepatopancreas (midgut gland) to the hemolymph and the eventual deposition in the muscle tissue. The aforementioned studies that identified the role of PC in lipid transport did not examine the possible relationship between a requirement for cholesterol and the level of dietary phospholipid.

Gong et al. (2000) reported an interaction between the requirements of dietary cholesterol and the presence of deoiled soybean lecithin in the diet. In the absence of the lecithin ingredient, the dietary requirement of *L. vannamei* for cholesterol was reported to be 0.35%. When dietary levels of the lecithin ingredient increased to 1.5% and 3.0%, the cholesterol requirement correspondingly decreased to 0.14% and 0.13%, respectively. However, Chen (1993) and Chen and Jenn (1991) did not observe any interaction between the presence of PC and the requirement for cholesterol as determined by a weight gain response of *P. monodon*.

The comparatively higher nutritional value of PC compared to other phospholipids may also be attributed to the need for PC as the phospholipid component of cellular membranes. Within the structure of the PC molecule, EFA are preferentially esterified to the *sn*2 position. These fatty acids may be assimilated intact with the PC molecule or separately. Some evidence of this possible role was found when Chen and Jenn (1991) fed a diet to *Penaeus penicillatus* that contained soy lecithin, which principally contains 18:2n-6 at the *sn*2 position. They found that levels of n-3 LC-PUFA in the polar lipid fraction of the muscle tissue decreased as levels of 18:2n-6 increased. The reciprocal interaction was reported by González Félix et al. (2002b) for 18:2n-6 when DHA or a mixture of n-3 LC-PUFA was included in the diet.

The possibility of another active component of the PC molecule, choline, was investigated by Kanazawa et al. (1985), but no response was observed. Kanazawa et al. (1979c) also demonstrated that the delivery of EFA as part of the phospholipid molecule was not the reason for the essentiality of phospholipid. Phospholipids provided in the diet in a form such as a soy lecithin ingredient may increase the physical stability of experimental diets when introduced into water and thereby reduce the rate of loss of water-soluble nutrients. The possible beneficial effects of lecithin as an emulsifier to enhance the rate of lipid digestion have been postulated, but D'Abramo et al. (1981), Kanazawa et al. (1985), and Teshima et al. (1986c,d) did not observe a possible role of dietary phospholipid in the enhancement of digestion.

More studies have investigated the mechanism of phospholipid requirement in fish. Although dietary phospholipids may be a more effective source of EFA to larval fish than neutral lipids (TAG) due to higher proportions of EFA and increased digestibility (Tocher, 1995; Sargent et al., 1997, 1999a,b), the role of dietary phospholipid in growth promotion and increasing survival in larvae appears to be independent of EFA requirements (Geurden et al., 1995a). Similarly, although dietary phospholipids increased digestibility in juvenile fish (Craig and Gatlin, 1997; Kasper and Brown, 2003), the growth-promoting effect of phospholipids was not due to generally enhanced emulsification and digestion of lipids (Geurden et al., 1997c, 1998b). Furthermore, the phospholipid requirement was not related to the delivery of other essential dietary components such as the bases choline and inositol (Geurden et al., 1995a). Intestinal steatosis in larvae fed phospholipid-deficient diets led to the hypothesis that early developing stages of fish had impaired ability to transport dietary lipids away from the intestine, possibly through limitations in lipoprotein synthesis (Fontagné et al., 1998; Geurden et al., 1998b; Olsen et al., 1999; Salhi et al., 1999). More specifically, the stimulating effects of phospholipids on larval fish growth were due to larvae having a limited ability to biosynthesize phospholipids de novo (Geurden et al., 1995a; Coutteau et al., 1997; Fontagné et al., 1998; Geurden et al., 1999). Sargent et al. (2002) hypothesized that the enzymatic location of the limitation was possibly in the production of the glycerophosphobase backbone. Tocher et al. (2008) speculated that the CDP-choline and CDP-ethanolamine phosphotransferases (CPT and EPT) involved in the conversion of diacylglycerol (DAG) to PC and PE, respectively, were possible candidates for the deficiency. However, limitations in CPT and EPT cannot account for the observed effects of PI, as it is not formed via the CPT/EPT pathways, suggesting that the limiting step would have to be further back in the phospholipid synthesis pathway; however, there is no single enzyme that could be responsible for limiting both PC and PI synthesis (Tocher et al., 2008). In summary, phospholipid synthesis is limited in larval fish, and so intact dietary phospholipids are required for the efficient transport of dietary fatty acids and lipids from the gut to the rest of the body, possibly by enhancing lipoprotein synthesis.

Functions of Phospholipids

The amphipathic nature of phospholipids is key to their major structural roles as components of cell biomembranes and lipoproteins. In mammalian cell membranes, phospholipids are asymmetrically distributed, with choline-containing phospholipids, PC, and sphingomyelin concentrated in the outer leaflet and amine-containing PE and PS concentrated in the inner leaflet, and this appears to be also the case in fish cell membranes (Kagan et al., 1984). The composition and metabolism of membrane phospholipids respond dynamically to environmental factors as part of a homeostatic mechanism regulating fluidity (Hochachka and Mommsen, 1995). For instance, reduced temperature and cold acclimation result in increased PE and decreased PC (Tocher, 1995). Biomembranes also serve as the source of phospholipid for several metabolic pathways. In lipoproteins, phospholipids enable hydrophobic lipids such as TAG and steryl esters to be transported in aqueous environments by forming the lipid/water interfaces along with cholesterol and proteins (Tocher, 1995). However, as alluded to above, phospholipids or molecules derived from phospholipids, also have a number of specific metabolic roles as intra- and inter-cellular lipid mediators involved in many important signaling mechanisms. In general, phospholipid metabolism is poorly studied in fish, but evidence suggests that most of these pathways occur in fish and that phospholipid-derived mediators play similar roles in fish as they do in mammals (Tocher, 1995, 2003).

Phosphoinositides and Protein Kinase C

Phosphoinositides are intracellular mediators that include phosphorylated derivatives of PI, such as PIP_2, which have important metabolic roles including golgi/lysosome/endosome trafficking, cell proliferation, survival, and migration (Hirsch et al., 2007), often mediated through the regulation of ion channel and transport proteins (Gamper and Shapiro,

2007). Further intracellular second messengers are produced by the cleavage of phosphorylated PI derivatives (Michell, 2007). Thus, the cleavage of PIP_2 by phospholipase C-β in response to various stimuli produces DAG and the inositol phosphate, IP_3, which stimulates calcium mobilization from the endoplasmic reticulum (Berridge, 2005). The subsequently increased intracellular Ca^{2+} and DAG are activators of protein kinase C (PKC), a threonine/serine kinase and important regulator of metabolism (Gomez-Fernandez and Corbalan-Garcia, 2007). However, the activity of DAG as a second messenger is not limited to interaction with PKC, and so DAG can trigger a range of biological responses through other proteins, including phospholipases, and also alter the biophysical properties of membranes and hence modulate membrane associated-proteins and processes (Gomez-Fernandez and Corbalan-Garcia, 2007). Other phospholipids, particularly PS, also have a metabolic role as activators of protein kinase C, along with DAG and Ca^{2+} ions (Newton, 2009). In fish, inositol phospholipid metabolism has been studied in metabolically active electrocytes from the electric ray (*Discopyge tschudii*), protein kinase C and PS-activation have been reported in rainbow trout and dogfish (*Scyliorhinus canicula*) tissues, and protein kinase C was implicated in the stimulation of steroidogenesis in goldfish (*Carassius auratus*) (see Tocher et al., 2008).

Eicosanoids

Phospholipids are the source of LC-PUFA substrates for synthesis of these intercellular fatty acid mediators. The first step in the eicosanoid biosynthesis pathway is the activation of phospholipase A_2 resulting in the release of fatty acid from the *sn*2 position of membrane phospholipids. In mammals, the key enzyme is type IV cytosolic phospholipase A_2 ($cPLA_2$) that is specific for phospholipids that contain ARA (or EPA) at the *sn*2 position, but little is known about the specificity of phospholipases in fish (Tocher, 1995, 2003). The roles of eicosanoids were described above.

Platelet-Activating Factor

Platelet-activating factor (PAF) or 1-*O*-alkyl-2-acetyl-*sn*-glycero-3-phosphocholine, an ether analog of PC, is synthesized by inflammatory cells and is a potent intercellular mediator of many leukocyte functions, including platelet aggregation, inflammation, and anaphylaxis (Snyder, 1990). The main pathway for PAF biosynthesis is through acetylation of lyso-PAF by lyso-PAF acetyltransferase and acetyl-CoA, and degradation (terminating activity) is via PAF acetylhydrolases that are related to phospholipase A_2. In mammals, reacylation of lyso-PAF is highly specific for ARA and, in turn, 1-alky-2-arachidonyl-glycerophosphocholine is substrate for the synthesis of PAF via an ARA-specific phospholipase A2 that also produces ARA for eicosanoid synthesis (Snyder, 1990). Both PAF biosynthesis and degradation have been demonstrated in fish (Tocher et al., 2008).

Other Intercellular Lipid Mediators and G-Protein Coupled Receptors

Phosphatidic acid (PA), the simplest diacyl phospholipid, is well established as an important intracellular signaling molecule involved in the regulation of many cell processes including proliferation and differentiation, transformation, tumor progression, and survival signaling (Wang et al., 2006). The anionic phosphate head group linked as a phosphomonoester sets PA apart from other phospholipids and is key to the specificity of PA–protein interaction and its functional roles (Kooijman and Burger, 2009). In contrast, many intercellular lipid mediators are now known to function by interacting with G-protein-coupled receptors (GPCR) in the plasma membrane and thereby modulate functions of the target cells (Im, 2009). For instance, both eicosanoids and PAF act via GPCR. Others that act this way include lysophospholipid mediators such as lyso-PA and sphingosine-1-phosphate that have well-known essential roles in many cell processes in vivo (Makide et al., 2009). Other lysophospholipids, including lyso-PS, lyso-PE, lyso-PI, lysophosphatidylglycerol, and lyso-phoshatidylthreonine have been shown to have lipid mediator-like responses, although their roles are not well understood (Makide et al., 2009). Other novel lipids now recognized to have intercellular signaling mediators acting on GPCR include resolvin E1, acylethanolamides (e.g., the endogenous cannabinoid, arachidonylethanolamine or anandamide), bile acids, and fatty acids (Im, 2009). Although it is likely that most or all of these lipid mediators will also be important in biochemical and physiological regulation in all vertebrates, nothing is currently known about these mechanisms in fish.

Cholesterol

Shrimp

A peculiarity of crustaceans is that they are unable to biosynthesize ex novo sterols (i.e., cholesterol) (Teshima, 1997). All studies that have investigated cholesterol requirements of shrimp have demonstrated a true requirement that has been generally established through observation of a reduced growth response or comparatively higher mortality. The reported dietary requirement, expressed as percentage of diet, commonly ranges from 0.2% to 1.0% (Table 6-7). Evidence suggests that higher levels of dietary cholesterol may exert an adverse affect on growth (Thongrod and Boonyaratpalin, 1998). Duerr and Walsh (1996) suggested that part or the entire sterol requirement may be satisfied through the ingredients in the feed without the need for supplementation. The cholesterol requirement in some species of shrimp may be somewhat spared (reduced) through

TABLE 6-7 Reported Cholesterol/Sterol Requirements of Shrimp and Other Crustaceans

Species	Requirement	Reference
Cholesterol alone		
Pacific white shrimp (*Litopenaeus vannamei*)	0.35%	Gong et al. (2000)
Kuruma prawn (*Marsupenaeus japonicus*)	0.50%	Kanazawa et al. (1971)
M. japonicus (larval)	1.00%	Teshima et al. (1983)
M. japonicus	0.20%	Shudo et al. (1971)
M. japonicus	2.00%	Deshimaru and Kuroki (1974)
American lobster (*Homarus americanus*)	0.50%	Kean et al. (1985)
H. americanus	0.50%	Castell et al. (1975)
Signal crayfish (*Pacifastacus leniusculus*)	0.40%	D'Abramo et al. (1985b)
Cholesterol + phospholipid		
L. vannamei	0.14% (1.5% deoiled soybean lecithin)	Gong et al. (2000)
L. vannamei	0.13% (3.0% deoiled soybean lecithin)	Gong et al. (2000)
H. americanus	(8% soybean lecithin)	D'Abramo et al. (1985a)

the ability of some species of shrimp to effectively utilize dietary plant sterols, phytosterols. D'Abramo et al. (1985b) found that a combination of a phytosterol mix (1.39% of the diet) and cholesterol (0.22% of the diet) could partially spare the cholesterol requirement (0.4% of the diet) of the freshwater crayfish *Pascifastacus leniusculus*. Sitosterol was the primary proportional component (~ 63%) in the mixture. However, a similar effect was not observed with juvenile lobsters (*H. americanus*) (D'Abramo et al., 1984). No sparing of dietary cholesterol for larval (1.0% level) or juvenile (0.5% level) *M. japonicus* with dietary sitosterol at ratios of 1:1 to 1:100 (cholesterol:sitosterol) was found by Teshima et al. (1989). Teshima and Kanazawa (1986) found that dietary sitosterol was somewhat effective, but could not completely replace dietary cholesterol (0.5%), or partially spare the cholesterol requirement when a mixture of 0.45% sitosterol and 0.05% cholesterol was added to the diet of *M. japonicus*. None of an array of sterol compounds (cholesterol precursors) served as total replacements for dietary cholesterol for larval *M. japonicus* (Teshima et al., 1983). Effective application of this knowledge via supplements of phytosterol or combination of phytosterols in commercial diets appears impractical due to the limited ability to convert these precursors to cholesterol, which is the exclusive sterol found in crustacean tissue.

Fish

Although there is no known requirement for cholesterol in finfish, this may be an area requiring more consideration in the future as dietary cholesterol concentrations decline and phytosterols increase with increasing inclusion levels of plant meals and oils in dietary formulations (see Chapter 16). There are few reports on the effects of dietary cholesterol on growth or metabolism in fish. In one study, dietary cholesterol level had no significant effect on specific growth rate (SGR), mortality, apparent digestibility coefficients of macronutrients, and total lipid content in Atlantic salmon (Bjerkeng et al., 1999). However, both hepatic cholesterol concentration and hepatosomatic index were increased by the dietary cholesterol supplement. Recently, it was shown that genes of the cholesterol biosynthesis pathway in liver were upregulated in salmon fed high levels of vegetable oil compared to fish fed diets containing fish oil (Leaver et al., 2008b; Taggart et al., 2008). Tissue levels of cholesterol were unaffected, suggesting that the reduced dietary cholesterol intake in fish fed vegetable oil was sufficiently compensated by increased synthesis (Leaver et al., 2008b).

Biochemistry of Requirement in Shrimp

The requirement for cholesterol in shrimp is founded upon an inability to synthesize this nutrient de novo, but the basis for the deficiency in biochemical or molecular terms, such as which genes are absent or nonfunctional, or which enzyme(s) or activities are limited, has not been determined.

Functions of Cholesterol

Cholesterol is an essential component of all animal cell membranes with important structural roles, reducing both fluidity and the permeability of the plasma membrane to protons and sodium ions (Lange and Steck, 2008). The distribution of membrane cholesterol between compartments is partly regulated by sphingolipids, toward which it has high affinity (Ikonen, 2008). Membrane cholesterol also has functional roles in intracellular transport as an important component in caveola-dependent and clathrin-dependent endocytosis, cell signalling through the formation of lipid rafts, and nerve conduction as an important component of the insulating myelin sheath (Simons and Ikonen, 2000).

Cholesterol is also the precursor of several other functional molecules in different tissues. In liver, it is converted by cholesterol-7α hydroxylase via cytochrome P450-mediated oxidation to bile acids, including cholate, that are then conjugated, for instance with taurine, before storage as bile in the gallbladder (Moschetta et al., 2005). Bile is secreted into the intestine where bile acids/salts aid emulsification of dietary lipids and fat-soluble vitamins (Olsen and Ringø, 1997). In skin, the cholesterol precursor (7-dehydrocholesterol) is converted by UV radiation into cholecalciferol (vitamin D_3), a secosteroid (with a broken ring structure) that is the precursor of the active form of vitamin D (Halver, 2002). In steroidogenic tissues, cholesterol is the precursor for the synthesis of steroids, classic nuclear hormones having genomic effects (i.e., affecting gene transcription) but also more rapid nongenomic effects (Wehling, 1997). Adrenal steroid hormones include mineralocorticoids (e.g., aldosterone) that have important roles in electrolyte and water balance, and the corticosteroids that have key roles in protein and carbohydrate metabolism and, in the case of cortisol, important functions in stress responses (Jobling, 1994). The gonadal sex hormones including estrogens (e.g., estradiol and progesterone), androgens (e.g., testosterone), and their derivatives have crucial roles in sex differentiation, gametogenesis, and the control of reproduction including vitellogenesis (Jobling, 1994).

OTHER ISSUES IN LIPID NUTRITION

Lipid Digestibility

The efficacy of a dietary lipid will depend upon the proportion of ingested nutrients that pass from the luminal side to the serosal side of the intestine, and thus its efficient digestion and absorption, which are measured and defined as digestibility (Bell and Koppe, 2010). Several factors affect lipid and fatty acid digestibility in fish including species, dietary lipid content and composition, other dietary constituents, rearing conditions such as water temperature, and the adaptation of the animal to the diet (Hua and Bureau, 2009). The proportions of different fatty acid groups are also important as fatty acids of shorter chain length and higher degree of unsaturation have higher digestibility, and saturated fatty acids have lower digestibility than unsaturated fatty acids at lower temperature. Dietary nonlipid components, such as nonstarch polysaccharides commonly found in some plant feedstuffs, and chitin/chitosan found in crustacean meals can have potential negative effects on lipid digestibility. Nutrient digestibility in general, including lipids and fatty acids, is covered in more detail in Chapter 12.

Larval Feeds

The small size and often poorly developed digestive system of marine fish larvae have major consequences for larval lipid nutrition (Izquierdo et al., 2000; Conceição et al., 2007; Yufera and Darias, 2007). Formulated first feeds such as defined microdiets have been difficult and slow to develop, necessitating the use of live feeds (Cahu and Zambonino-Infante, 2001; Koven et al., 2001a; Robin and Vincent, 2003; Kvale et al., 2006). Natural live feeds such as copepods have been shown to be nutritionally superior to the alternatives, but their use is still constrained by technical problems associated with both extensive and intensive culture systems that limit the consistent and reliable provision of adequate amounts of copepods at a financially viable cost (Stottrup, 2000). In contrast, the preferred live feeds such as rotifers and *Artemia*, although convenient, are nutritionally inadequate for marine fish, being relatively poor in LC-PUFA and so enrichment processes are required. Despite considerable work, it is still difficult to enrich live feeds to provide adequately balanced levels of LC-PUFA and, in particular, sufficient DHA (Conceição et al., 2010). Various fish oils, including tuna orbital oil, algal extracts, and ethyl-ester LC-PUFA concentrates, have all been investigated as means for more precisely controlling enrichment of LC-PUFA ratios (Sargent et al., 1999b; Izquierdo et al., 2000). However, problems include potential oxidation during enrichment and endogenous metabolism of the LC-PUFA by the feed organism, including fatty acid oxidation and retroconversion (Sargent et al., 1997, 1999b). Marine larval fish nutrition, live feeds, and enrichments are covered in more in Chapter 14.

Chemical Forms of Lipid Sources

The specific requirement for intact phospholipid indicates that the chemical form of dietary lipid or fatty acids can be important. Studies for determining EFA requirements of fish and crustaceans often employed fatty acid concentrates, usually as methyl- or ethyl-esters, as their composition can be more easily controlled (Castell et al., 1972; Kanazawa et al., 1979b; Lochman and Gatlin, 1993), but few studies have investigated the efficacy of these fatty acid esters. Growth suppression has been reported in rainbow trout and red drum fed n-3 LC-PUFA as ethyl esters (Castell et al., 1972; Lochman and Gatlin, 1993), and sea bream larvae showed high mortality and poor growth when fed rotifers enriched with n-3 LC-PUFA methyl esters compared to rotifers enriched with n-3 LC-PUFA as TAG (Izquierdo et al., 1989; Rodriguez, 1994). However, more recently, methyl esters appeared to be equally effective in supplying n-3 LC-PUFA as TAG in juvenile gilthead sea bream (Ibeas et al., 2000).

The majority of global oils and fats are TAG-rich and contain fatty acids with chain lengths from C_{14} to C_{22} (Gunstone, 2010), and so this is the normal nature of dietary lipids and fatty acids (Sargent et al., 2002). However, TAG containing high levels of medium-chain fatty acids (6:0, 8:0, 10:0, and 12:0) have been investigated as alternative energy sources in fish. Growth and survival was reduced by feeding 8:0 (tricaprylin) in comparison with 12:0 (coconut oil) and triolein in carp larvae (Fontagné et al., 1999). Although both 6:0 (tricaproin) and 8:0 stimulated growth in the first week of feeding, 8:0 later decreased growth and survival compared to triolein in carp larvae (Fontagné et al., 2000a,b). In contrast, sea bass larvae fed 6:0 or 8:0 showed better growth, and fish fed 8:0 better survival, than fish fed 10:0 (tricaprin) or triolein (Fontagné et al., 2000c). Feeding medium-chain TAG reduced body neutral lipids in carp (Fontagné et al., 2000a) and reduced intraperitoneal fat deposition in juvenile red drum (Craig and Gatlin, 1997; Davis et al., 1999). Similarly, although 8:0 and 10:0 were highly digestible in Atlantic salmon, they appeared to reduce absorption of other fatty acids and decreased muscle fat content (Roesjoe et al., 2000). Small amounts of medium-chain fatty acids can be deposited in fish tissues, with 1–3% observed in sea bass larvae, whereas in carp larvae there was deposition of both 8:0 and 10:0 in neutral lipids, with 8:0 also being significantly elongated to 10:0 (Fontagné et al., 1999, 2000a,b,c). Medium-chain TAG may therefore be potentially useful as energy sources and may have beneficial effects in lowering body fat levels. In contrast, inclusion of short-chain fatty acids (C_2, C_3, and C_4) up to 2% of total diet dry weight had no effect on growth, mortality, lipid levels, or fatty acid composition in Atlantic salmon (Bjerkeng et al., 1999).

Wax ester appears to be effectively utilized by Atlantic salmon. In salmon fed for 140 days on a diet formulated with copepod oil containing 38% of lipid as wax ester, weight gain, SGR, and FCR were not significantly different to fish fed a diet containing fish oil supplying all lipid as TAG (Olsen et al., 2004).

Dietary Lipid, Fatty Acid, and Antioxidant Requirements

As dietary lipid content and PUFA, especially LC-PUFA levels, increase, the resulting unsaturation index of the diet potentially increases prooxidant stress in the fish and so has to be balanced by increasing dietary antioxidant content, especially vitamin E (tocopherol). In tilapia, the vitamin E requirement increased as the level of dietary lipid increased and optimal levels in juvenile tilapia (*O. niloticus* × *O. aureus*) were reported as 42–44 and 60–66 mg/kg in diets containing 5% and 12% lipid, respectively (Shiau and Shiau, 2001). Tocopherol levels in the flesh decreased as dietary lipid, supplied as fish oil, increased in Atlantic salmon (Hemre and Sandnes, 1999), and muscle homogenates of rainbow trout and sea bass were more susceptible to lipid peroxidation from fish fed high-fat compared to low-fat diets (Dias et al., 1999). However, this is dependent upon lipid source as increased flesh tocopherol levels were observed with increasing dietary lipid when supplied as tocopherol-rich crude palm oil (Lim et al., 2001).

Bioactive Lipids and Related Compounds

Various potential modulators of lipid and fatty acid metabolism, including bioactive fatty acids, fibrates, and sesamin, have been investigated in fish with a view to altering tissue lipid content or distribution, and LC-PUFA synthesis.

In mammals, conjugated linoleic acid (CLA), a collective term for various positional and geometric isomers of 18:2n-6, have beneficial physiological effects, including anticarcinogenic and immune-enhancing properties (Belury, 2002), and effects on lipid metabolism such as increasing lean body mass (DeLany and West, 2000; Wang and Jones, 2004). In fish, dietary CLA up to 5% (usually 1–2%) of diet has had no major effects on growth rate or feed efficiency in a number of species, including hybrid striped bass (*Morone saxatilis* × *M. chrysops*), tilapia, yellow perch (*Perca flavescens*), channel catfish, rainbow trout, and Atlantic salmon (Twibell et al., 2000, 2001; Twibell and Wilson, 2003; Berge et al., 2004; Yasmin et al., 2004; Figueiredo-Silva et al., 2005; Kennedy et al., 2005, 2007a,b; Valente et al., 2007a). Furthermore, CLA had no effects on whole-body proximate composition or flesh lipid levels in salmon or trout (Berge et al., 2004; Kennedy et al., 2005; Bandarra et al., 2006), carcass, intraperitoneal or liver fat contents in channel catfish (Twibell and Wilson, 2003), or tissue lipid contents in tilapia (Yasmin et al., 2004). However, intraperitoneal fat and liver lipid content were decreased by dietary CLA in striped bass although hepatosomatic index was increased (Twibell et al., 2000), as it was in yellow perch and tilapia (Twibell et al., 2001; Yasmin et al., 2004). Therefore, there is little evidence that CLA will have any beneficial effects on growth, body composition, or fat content in fish. However, CLA was accumulated in flesh of fish without any detrimental effects on n-3 LC-PUFA levels (Twibell et al., 2000; Berge et al., 2004; Kennedy et al., 2005, 2007a,b; Valente et al., 2007b). Indeed, there was some evidence that CLA may increase DHA in salmon and trout (Berge et al., 2004; Kennedy et al., 2007b), but these effects of CLA on LC-PUFA have not been consistently observed. In summary, it appears the only beneficial effect of dietary CLA on the nutritional quality of fish is through the presence of the bioactive fatty acids in the flesh.

The 3-thia fatty acids, including tetradecylthioacetic acid (TTA), a 16-carbon sulfur-containing saturated fatty acid, have also been studied as dietary supplements in mammals, where they have been shown to be peroxisome proliferators (Berge et al., 2002) and to reduce body fat (Madsen et al., 2002). However, dietary TTA (0.5–0.6%) significantly depressed growth and increased mortality in salmon smolts (Moya-Falcon et al., 2004) and increased mortality in cod (Kennedy et al., 2007a). Dietary TTA was shown to alter the expression of a number of genes involved in lipid metabolism in salmon and trout liver (Kleveland et al., 2006; Kennedy et al., 2007b) and reduced carcass fat in salmon when fed at 0.6% (Moya-Falcon et al., 2004). In contrast, 0.5% TTA had no effect on growth, proximate composition, or liver and flesh fat levels in cod and trout (Kennedy et al., 2007a,b). Dietary TTA significantly increased the percentage of DHA, and decreased the proportion of EPA, in flesh of cod (Kennedy et al., 2007a), which appeared to be due to enhanced production of hexaenes from EPA rather than an overall stimulation of the LC-PUFA synthesis pathway (Kennedy et al., 2007b). As with CLA, TTA is deposited in the tissues, including flesh, but there is little to suggest that dietary supplementation with 3-thia fatty acids will be viable in fish nutrition.

Little is known regarding the effects of hypolipidemic drugs such as fibrates (e.g., ciprofibrate, clofibrate, gemfibrozil, and fenofibrate) in fish. The results from the few studies available suggest that fibrates have only very weak peroxisomal proliferation activity and only mild effects on fatty acid oxidation in trout, salmon, and medaka (Donohue et al., 1993; Scarano et al., 1994; Ruyter et al., 1997). An early study using primary cell cultures reported an increase in LC-PUFA synthesis in rainbow trout brain astroglial cells treated with clofibrate (Tocher and Sargent, 1993), but liver EPA and DHA levels were decreased in liver of fenofibrate-treated rainbow trout (Du et al., 2004). The effects of dietary supplementation with fibrates in fish require further investigation.

Sesamin is the main lignan in sesame seed and has been shown to increase β-oxidation, lower serum TAG and cholesterol levels, and affect elongation and desaturation of fatty acids in rats (Fujiyama-Fujiwara et al., 1995; Jeng and Hou, 2005). In rainbow trout, dietary sesamin (0.58%) increased the percentages of DHA in white muscle lipid by up to 37% (Trattner et al., 2008a). In vitro, sesamin (0.05 mM) lowered TAG secretion from salmon hepatocytes, and increased the β-oxidation of [1-^{14}C]18:3n-3 associated with increased expression of CPT1, and also its desaturation and elongation to DHA although, paradoxically, Δ5 and Δ6 Fad gene expression was down-regulated (Trattner et al., 2008b).

REFERENCES

Abi-ayad, S.-M. E.-A., C. Melard, and P. Kestemont. 1997. Effects of n-3 fatty acids in Eurasian perch broodstock diet on egg fatty acid composition and larvae stress resistance. Aquacult. Int. 5:161–168.

Agaba, M., D. R. Tocher, C. A. Dickson, J. R. Dick, and A. J. Teale. 2004. Zebrafish cDNA encoding multifunctional fatty acid elongase involved in production of eicosapentaenoic (20:5n-3) and docosahexaenoic (22:6n-3) acids. Mar. Biotechnol. 6:251–261.

Agaba, M. K., D. R. Tocher, X. Zheng, C. A. Dickson, J. R. Dick, and A. J. Teale. 2005. Cloning and functional characterisation of polyunsaturated fatty acid elongases of marine and freshwater teleost fish. Comp. Biochem. Phys. B 142:342–352.

Almansa, E., M. J. Perez, J. R. Cejas, P. Badia, J. E. Villamandos, and A. Lorenzo. 1999. Influence of broodstock gilthead seabream (*Sparus aurata* L.) dietary fatty acids on egg quality and egg fatty acid composition throughout the spawning season. Aquaculture 170:323–336.

Arzel, J., M. Cardinal, J. Cornet, R. Metailler, and J. C. Guillaume. 1993. Nutrition of brown trout (*Salmo trutta*) reared in seawater, effect of dietary lipid on growth performances, body composition andfillet quality. P. 19 in From Discovery to Commercialization, Special Publication no.19. Oostende, Belgium: European Aquaculture Society.

Bandarra, N. M., M. L. Nunes, A. M. Andrade, J. A. M. Prates, S. Pereira, M. Monteiro, P. Rema, and L. M. P. Valente. 2006. Effect of conjugated linoleic acid on muscle, liver and visceral lipid deposition in rainbow trout juveniles (*Oncorhynchus mykiss*). Aquaculture 254:496–505

Bautista, M. N., and M. C. de la Cruz. 1988. Linoleic and linolenic acids in the diet of fingerling milkfish (*Chanos chanos* Forsskal). Aquaculture 71:347–359.

Bell, J. G., and W. Koppe. 2010. Lipids in aquafeeds. Pp. 21–59 in Fish Oil Replacement and Alternative Lipid Sources in Aquaculture Feeds, G. M. Turchini, W. K. Ng, and D. R. Tocher, eds. Boca Raton, FL: Taylor & Francis, CRC Press.

Bell, J. G., and J. R. Sargent. 2003. Arachidonic acid in aquaculture feeds: Current status and future opportunities. Aquaculture 218:491–499.

Bell, M. V., and D. R. Tocher. 2009. Biosynthesis of fatty acids; general principles and new directions. Pp. 211–236 in Lipids in Aquatic Ecosystems, M. T. Arts, M. Brett, and M. Kainz, eds. New York: Springer-Verlag.

Bell, J. G., J. D. Castell, D. R. Tocher, F. M. MacDonald, and J. R. Sargent. 1995a. Effects of different dietary arachidonic acid:docosahexaenoic acid ratios on phospholipid fatty acid compositions and prostaglandin production in juvenile turbot (*Scophthalmus maximus*). Fish Physiol. Biochem. 14:139–151.

Bell, M. V., R. S. Batty, J. R. Dick, K. Fretwell, J. C. Navarro, and J. R. Sargent. 1995b. Dietary deficiency of docosahexaenoic acid impairs vision at low light intensities in juvenile herring (*Clupea harengus* L.). Lipids 30:443–449.

Bell, J. G., B. M. Farndale, M. P. Bruce, J. M. Navas, and M. Carillo. 1997. Effects of broodstock dietary lipid on fatty acid compositions of eggs from sea bass (*Dicentrarchus labrax*). Aquaculture 149:107–119.

Bell, J. G., J. McEvoy, J. L. Webster, F. McGhee, R. M. Millar, and J. R. Sargent. 1998. Flesh lipid and carotenoid composition of Scottish farmed Atlantic salmon (*Salmo salar*). J. Agr. Food Chem. 46:119–127.

Bell, J. G., R. J. Henderson, D. R. Tocher, and J. R. Sargent. 2004. Replacement of dietary fish oil with increasing levels of linseed oil; modification of flesh fatty acid compositions in Atlantic salmon (*Salmo salar*) using a fish oil finishing diet. Lipids 39:223–232.

Belury, M. A. 2002. Dietary conjugated linoleic acid in health: Physiological effects and mechanisms of action. Annu. Rev. Nutr. 22:505–531.

Benitez-Santana, T., R. Masuda, E. J. Carrillo, E. Ganuza, A. Valencia, C. M. Hernandez-Cruz, and M. S. Izquierdo. 2007. Dietary n-3 HUFA deficiency induces a reduced visual response in gilthead seabream *Sparus aurata* larvae. Aquaculture 264:408–417.

Berge, G. M., B. Ruyter, and T. Asgard. 2004. Conjugated linoleic acid in diets for juvenile Atlantic salmon (*Salmo salar*); effects on fish performance, proximate composition, fatty acid and mineral content. Aquaculture 237:365–380.

Berge, R. K., J. Skorve, K. J. Tronstad, K. Berge, O. A. Gudbrandsen, and H. Grav. 2002. Metabolic effects of thia fatty acids. Curr. Opin. Lipidol. 13:295–304.

Berridge, M. J. 2005. Unlocking the secrets of cell signaling. Annu. Rev. Physiol. 67:1–21.

Bessonart, M., M. S. Izquierdo, M. Salhi, C. M. Hernandez-Cruz, M. M. Gonzalez, and H. Fernandez-Palacios. 1999. Effect of dietary arachidonic acid levels on growth and survival of gilthead sea bream (*Sparus aurata* L.) larvae. Aquaculture 179:265–275.

Bjerkeng, B., T. Storebakken, and E. Wathne. 1999. Cholesterol and short-chain fatty acids in diets for Atlantic salmon *Salmo salar* (L.): Effects on growth, organ indices, macronutrient digestibility, and fatty acid composition. Aquacult. Nutr. 5:181–191.

Borgut, I., Z. Bukvic, Z. Steiner, Z. Milakovic, and I. Stevic. 1998. Influence of linolenic fatty acid (18:3ω3) additive on European sheat fish (*Silurus glanis*) growth bred in cages. Czech. J. Anim. Sci. 43:133–137.

Brinkmeyer, R. L., and G. J. Holt. 1998. Highly unsaturated fatty acids in diets for red drum (*Sciaenops ocellatus*) larvae. Aquaculture 161:253–268.

Bruce, M., F. Oyen, G. Bell, J. F. Asturiano, B. Farndale, M. Carrillo, S. Zanuy, J. Ramos, and N. Bromage. 1999. Development of broodstock diets for the European sea bass (*Dicentrarchus labrax*) with special emphasis on the importance of n-3 and n-6 highly unsaturated fatty acid to reproductive performance. Aquaculture 177:85–97.

Caballero, M. J., G. Lopez-Calero, J. Socorro, F. J. Roo, M. S. Izquierdo, and A. J. Fernandez. 1999. Combined effect of lipid level and fish meal quality on liver histology of gilthead seabream (*Sparus aurata*). Aquaculture 179:277–290.

Caballero, M. J., M. S. Izquierdo, E. Kjørsvik, D. Montero, J. Socorro, A. J. Fernández, and G. Rosenlund. 2003. Morphological aspects of intestinal cells from gilthead seabream (*Sparus aurata*) fed diets containing different lipid sources. Aquaculture 225:325–340.

Cahu, C., and J. Zambonino-Infante. 2001. Substitution of live food by formulated diets in marine fish larvae. Aquaculture 200:161–180.

Cahu, C., J. Zambonino Infante, and V. Barbosa. 2003. Effect of dietary phospholipid level and phospholipid: Neutral lipid value on the development of sea bass (*Dicentrarchus labrax*) larvae fed a compound diet. Brit. J. Nutr. 90:21–28.

Cahu, C. L., E. Gilbert, L. A. N. Villeneuve, S. Morais, N. Hamza, P.-A. Wold, J. L. Zamonino Infante. 2009. Influence of dietary phospholipids on early ontogenesis of fish. Aquac. Res. 40:989–999.

Cao, H., K. Gerhold, J. R. Mayers, M. M. Wiest, S. M. Watkins, and G. S. Hotamisligil. 2008. Identification of a lipokine, a lipid hormone linking adipose tissue to systemic metabolism. Cell 134:933–944.

Castell, J. D., and J. F. Covey. 1976. Dietary lipid requirements of adult lobsters, homarus americanus (M.E.). J. Nutr. 106:1159–1165.

Castell, J. D., R. O. Sinnhuber, J. H. Wales, and J. D. Lee. 1972. Essential fatty acids in the diet of rainbow trout: Lipid metabolism and fatty acid composition. J. Nutr. 102:77–86.

Castell, J. D., E. G. Mason, and J. F. Covey. 1975. Cholesterol requirements of juvenile American lobster (*Homarus americanus*). J. Fish. Res. Board Can. 32:1432–1435.

Castell J. D., J. G. Bell, D. R. Tocher, and J. R. Sargent. 1994. Effects of purified diets containing different combinations arachidonic and docosahexaenoic acid on survival, growth and fatty acid composition of juvenile turbot (*Scophthalmus maximus*). Aquaculture 128:315–333.

Catacutan, M. R., and R. M. Coloso. 1995. Effect of dietary protein to energy ratios on growth, survival, and body composition of juvenile Asian seabass, *Lates calcarifer*. Aquaculture 131:125–133.

Chen, H. Y. 1993. Requirements of marine shrimp, *Penaeus monodon*, juveniles for phosphatidylcholine and cholesterol. Aquaculture 109:165–176.

Chen, H. Y., and J. S. Jenn. 1991. Combined effects of dietary phosphatidylcholine and cholesterol on the growth, survival, and body composition of marine shrimp, *Penaeus penicillatus*. Aquaculture 96:167–178.

Chou, B.-S., and S.-Y. Shiau. 1999. Both n-6 and n-3 fatty acids are required for maximal growth of juvenile hybrid tilapia. N. Am. J. Aquacult. 61:13–20.

Company, R., J. A. Calduch-Giner, S. Kaushik, and J. Perez-Sanchez. 1999. Growth performance and adiposity in gilthead sea bream (*Sparus aurata*): Risks and benefits of high energy diets. Aquaculture 171:279–292.

Conceição, L. E. C., S. Morais, and I. Rønnestad. 2007. Tracers in fish larvae nutrition: A review of methods and applications. Aquaculture 267:62–75.

Conceição, L. E. C., M. Yufera, P. Makridis, S. Morais, and M. T. Dinis. 2010. Live feeds for early stages of fish rearing. Aquac. Res. 41:613–640.

Cook, H. W., and R. C. R. McMaster. 2004. Fatty acid desaturation and chain elongation in eukaryotes. Pp. 181–204 in Biochemistry of Lipids, Lipoproteins and Membranes, 4th Edition, D. E. Vance, and J. E. Vance, eds. Amsterdam, the Netherlands: Elsevier.

Coutteau, P., G. VanStappen, and P. Sorgeloos. 1996a. A standard experimental diet for the study of fatty acid requirements of weaning and first ongrowing stages of the European sea bass *Dicentrarchus labrax* L: Comparison of extruded and extruded/coated diets. Arch. Anim. Nutr. 49:49–59.

Coutteau, P., M. R. Camara, and P. Sorgeloos. 1996b. The effect of different levels and sources of dietary phosphatidylcholine on the growth, survival, stress resistance, and fatty acid composition of *Penaeus vannamei*. Aquaculture 147:261–273.

Coutteau, P., I. Geurden, M. R. Camara, P. Bergot, and P. Sorgeloos. 1997. Review on the dietary effects of phospholipids in fish and crustacean larviculture. Aquaculture 155:149–164.

Craig, S. R., and D. M. Gatlin. 1997. Growth and body composition of juvenile red drum (*Sciaenops ocellatus*) fed diets containing lecithin and supplemental choline. Aquaculture 151:259–267.

Craig, S. R., B. S. Washburn, and D. M. Gatlin. 1999. Effects of dietary lipids on body composition and liver function in juvenile red drum, *Sciaenops ocellatus*. Fish Physiol. Biochem. 21:249–255.

Cruz-Garcia, L., M. Minghetti, I. Navarro, and D. R. Tocher. 2009. Molecular cloning, tissue expression and regulation of Liver X Receptor (LXR) transcription factors of Atlantic salmon (*Salmo salar*) and rainbow trout (*Oncorhynchus mykiss*). Comp. Biochem. Phys. B 153:81–88.

D'Abramo, L. R., and S. S. Sheen. 1993. Polyunsaturated fatty acid nutrition in juvenile freshwater prawn *Macrobrachium rosenbergii*. Aquaculture 115:63–86.

D'Abramo, L. R., C. E. Bordner, D. E. Conklin, and N. A. Baum. 1981. Essentiality of dietary phosphatidylcholine for the survival of juvenile lobsters. J. Nutr. 111:425–431.

D'Abramo, L. R., C. E. Bordner, and D. E. Conklin. 1982. Relationship between dietary phosphatidylcholine and serum cholesterol in the American lobster. Mar. Biol. 67:231–235.

D'Abramo, L. R., C. E. Bordner, D. E. Conklin, and N. A. Baum. 1984. Sterol requirements of juvenile lobsters *Homarus* sp. Aquaculture 42:13–25.

D'Abramo, L. R., N. A. Baum, C. E. Bordner, D. E. Conklin, and E. S. Chang. 1985a. Diet-dependent cholesterol transport in the American lobster. J. Exp. Mar. Biol. Ecol. 49:245–255.

D'Abramo, L. R., J. S. Wright, K. H. Wright, C. E. Bordner, and D. E. Conklin. 1985b. Sterol requirement of cultured juvenile crayfish *Pascifastacus leniusculus*. Aquaculture 49:245–255.

Das, U. N. 2006. Essential fatty acids—a review. Curr. Pharm. Biotechno. 7:467–482.

Davis, D. A., and E. H. Robinson. 1986. Estimation of the dietary lipid requirement level of the white crayfish *Procambarus acutus acutus*. J. World Aquacult. Soc. 17:37–43.

Davis, D. A., J. P. Lazo, and C. R. Arnold. 1999. Response of juvenile red drum (*Sciaenops ocellatus*) to practical diets supplemented with medium chain triglycerides. Fish Physiol. Biochem. 21:235–247.

DeLany, J. P., and D. B. West. 2000. Changes in body composition with conjugated linoleic acid. J. Am. Coll. Nutr. 19:487S–493S.

Deshimaru, O., and K. Kuroki. 1974. Studies on a purified diet for prawn-II. Optimum contents of cholesterol and glucosamine in the diet. B. Jpn. Soc. Sci. Fish. 40:421–424.

Deshimaru O., K. Kuroki, and Y. Yone. 1979. The composition and level of dietary lipid appropriate for the growth of prawn. B. Jpn. Soc. Sci. Fish. 45:591–594.

Deshimaru, O., K. Kuroki, and Y. Yone. 1982. Nutritive value of various oils for yellowtail. B. Jpn. Soc. Sci. Fish. 48:1155–1157.

Dias, J., G. Corraze, J. Arzel, M. J. Alvarez, J. M. Bautista, C. Lopez-Bote, and S. J. Kaushik. 1999. Nutritional control of lipid deposition in rainbow trout and European seabass: Effect of dietary protein/energy ratio. Cybium 23(suppl.):127–137.

Donohue, M., L. A. Baldwin, D. A. Leonard, P. T. Kostecki, and E. J. Calabrease. 1993. Effect of hypolipidemic drugs gemfibrozil, ciprofibrate, and clofibric acid on peroxisomal β-oxidation in primary cultures of rainbow trout hepatocytes. Ecotox. Environ. Safe. 26:127–132.

Du, Z., L. Demizieuz, P. Degrace, J. Gresti, B. Moindrot, Y. Liu, L. Tian, J. Cao, and P. Clouet. 2004. Alteration of 20:5n-3 and 22:6n-3 fat contents and liver peroxisomal activities in fenofibrate-treated trout. Lipids 39:849–855.

Duerr, E. O., and W. A. Walsh. 1996. Evaluation of cholesterol additions to a soybean meal based diet for juvenile Pacific white shrimp *Penaeus vannamei* (Boone) in an outdoor growth trial. Aquacult. Nutr. 2:111–116.

Einen, O., and A. J. Roem. 1997. Dietary protein/energy ratios for Atlantic salmon in relation to fish size: Growth, feed utilisation and slaughter quality. Aquacult. Nutr. 3:115–126.

Estevez, A., and A. Kanazawa. 1996. Fatty acid composition of neural tissues of normally pigmented and unpigmented juveniles of Japanese flounder using rotifer and *Artemia* enriched in n-3 HUFA. Fish. Sci. 62:88–93.

Estevez, A., M. Ishikawa, and A. Kanazawa. 1997. Effects of arachidonic acid on pigmentation and fatty acid composition of Japanese flounder, *Paralichthys olivaceus* (Temminck & Schlegel). Aquac. Res. 28:279–289.

Estevez, A., L. A. McEvoy, J. G. Bell, and J. R. Sargent. 1999. Growth, survival, lipid composition and pigmentation of turbot (*Scophthalmus maximus*) larvae fed live-prey enriched in arachidonic and eicosapentaenoic acids. Aquaculture 180:321–343.

Farkas, T., and J. E. Halver. 1996. Involvement of phospholipid molecular species in controlling structural order of vertebrate brain synaptic membranes during thermal evolution. Lipids 31:1045–1050.

Farkas, T., I. Dey, C. Buda, and J. E. Halver. 1994. Role of phospholipid molecular species in maintaining lipid membrane structure in response to temperature. Biophys. Chem. 50:147–155.

Farkas, T., E. Fodor, K. Kitajka, and J. E. Halver. 2001. Response of fish membranes to environmental temperature. Aquac. Res. 32:645–655.

Feller, S. E. 2008. Acyl chain conformations in phospholipid bilayers: a comparative study of docosahexaenoic acid and saturated fatty acids. Chem. Phys. Lipids 153:76–80.

Fenucci, J. L., A. L. Lawrence, and Z. P. Zein-Eldin. 1981. The effects of fatty acid and shrimp meal composition of prepared diets on growth of juvenile shrimp, *Penaeus stylirostris*. J. World Maricult. Soc. 12:315–324.

Fernandez-Palacios, H., M. S. Izquierdo, L. Robaina, A. Valencia, M. Salhi, and J. M. Vergara. 1995. Effect of n-3HUFA level in broodstock diets on egg quality of gilthead sea bream (*Sparus aurata* L.). Aquaculture 132:325–337.

Figueiredo-Silva, A. C., P. Rema, N. M. Bandarra, M. L. Nunes, and L. M. P. Valente. 2005. Effects of dietary conjugated linoleic acid on growth, nutrient utilization, body composition, and hepatic lipogenesis in rainbow trout juveniles (*Oncorhynchus mykiss*). Aquaculture 248:163–172.

Fontagné, S., I. Geurden, A.-M. Escaffre, and P. Bergot. 1998. Histological changes induced by dietary phospholipids in intestine and liver of common carp (*Cyprinus carpio* L.) larvae. Aquaculture 161:213–223.

Fontagné, S., T. Pruszynski, G. Corraze, and P. Bergot. 1999. Effect of coconut oil and tricaprylin vs. triolein on survival, growth and fatty acid composition of common carp (*Cyprinus carpio* L.) larvae. Aquaculture 179:241–251.

Fontagné, S., L. Burtaire, G. Corraze, and P. Bergot. 2000a. Effects of dietary medium-chain triacylglycerols (tricaprylin and tricaproin) and phospholipid supply on survival, growth and lipid metabolism in common carp (*Cyprinus carpio* L.) larvae. Aquaculture 190:289–303.

Fontagné, S., G. Corraze, and P. Bergot. 2000b. Tricaproin, tricaprin and trilaurin are utilized more efficiently than tricaprylin by carp (*Cyprinus carpio* L.) larvae. J. Nutr. 130:2009–2015.

Fontagné, S., J. Robin, G. Corraze, and P. Bergot. 2000c. Growth and survival of European sea bass (*Dicentrarchus labrax*) larvae fed from first feeding on compound diets containing medium-chain triacylglycerols. Aquaculture 190:261–271.

Fujiyama-Fujiwara, Y., R. Umeda-Sawada, M. Kuzuyama, and O. Igarashi. 1995. Effects of sesamin on the fatty acid composition of the liver of rats fed n-6 and n-3 fatty acids-rich diet. J. Nutr. Sci. Vitaminol. 41:217–225.

Furuita, H., T. Takeuchi, M. Toyota, and T. Watanabe. 1996a. EPA and DHA requirements in early juvenile red sea bream using HUFA enriched *Artemia* Nauplii. Fish. Sci. 62:246–251.

Furuita, H., T. Takeuchi, T. Watanabe, H. Fujimoto, S. Sekiya, and K. Imaizumi. 1996b. Requirements of larval yellowtail for eicosapentaenoic acid, docosahexaenoic acid, and n-3 highly unsaturated fatty acid. Fish. Sci. 62:372–379.

Gamper, N., and M. S. Shapiro. 2007. Regulation of ion transport proteins by membrane phosphoinositides. Nat. Rev. Neurosci. 8:921–934.

Gatesoupe, F. J., C. Leger, R. Metailler, and P. Luquet. 1977. Alimentation lipidique du turbot (*Scophthalmus maximus* L.) I. Influence de la langeur de chaire de acides gras de la serie ω3. Ann. Hydrobiol. 8:89–97.

Gatlin, D. M., M. L. Brown, C. N. Keembiyehetty, F. Jaramillo, and G. R. Nematipour. 1994. Nutritional requirements of hybrid striped bass (*Morone chrysops* × *M. saxatilis*). Aquaculture 124:127.

Gaylord, T. G., and D. M. Gatlin, III. 2000. Dietary lipid level but not L-carnitine affects growth performance of hybrid striped bass (*Morone chrysops* × *M. saxatilis*). Aquaculture 190:237–246.

Geurden, I., J. Radünz-Neto, and P. Bergot. 1995a. Essentiality of dietary phospholipids for carp (*Cyprinus carpio* L.) larvae. Aquaculture 131:303–314.

Geurden, I., P. Coutteau, and P. Sorgeloos. 1995b. Dietary phospholipids for European sea bass (*Dicentrarchus labrax* L.) during first ongrowing. Pp. 175–178 in Larvi '95—Fish and Shellfish Symposium, September 3–7, 1995, EAS Spec. Publ. 24, P. Lavens, R. Jaspers, and I. Roelants, eds. Ghent, Belgium: European Aquaculture Society.

Geurden, I., N. Charlon, D. Marion, and P. Bergot. 1997a. Influence of purified soybean phospholipids on early development of common carp. Aquacult. Int. 5:137–149.

Geurden, I., P. Coutteau, and P. Sorgeloos. 1997b. Effect of dietary phospholipid supplementation on growth and fatty acid composition of European sea bass (*Dicentrarchus labrax* L.) and turbot (*Scophthalmus maximus* L.) juveniles from weaning onwards. Fish Physiol. Biochem. 16:259–272.

Geurden, I., P. Coutteau, and P. Sorgeloos.1997c. Increased docosahexaenoic acid levels in total and polar lipid of European sea bass (*Dicentrarchus labrax* L.) postlarvae fed vegetable or animal phospholipids. Mar. Biol. 129:689–698.

Geurden, I., D. Marion, N. Charlon, P. Coutteau, and P. Bergot. 1998a. Comparison of different soybean phospholipidic fractions as dietary supplements for common carp, *Cyprinus carpio*, larvae. Aquaculture 161:225–235.

Geurden, I., P. Bergot, I. Schwarz, and P. Sorgeloos. 1998b. Relationship between dietary phospholipids class composition and neutral lipid absorption in postlarval turbot. Fish Physiol. Biochem. 19:217–228.

Geurden, I., P. Bergot, K. Van Ryckeghem, and P. Sorgeloos. 1999. Phospholipid composition of common carp (*Cyprinus carpio*) larvae starved or fed different phospholipid classes. Aquaculture 171:93–107.

Ghioni, C., D. R. Tocher, M. V. Bell, J. R. Dick, and J. R. Sargent. 1999. Low C_{18} to C_{20} fatty acid elongase activity and limited conversion of stearidonic acid, 18:4n-3, to eicosapentaenoic acid, 20:5n-3, in a cell line from the turbot, *Scophthalmus maximus*. Biochim. Biophys. Acta 1437:179–181.

Glencross, B. D. 2009. Exploring the nutritional demand for essential fatty acids by aquaculture species. Rev. Aquacult. 1:71–124.

Glencross, B. D., and D. M. Smith. 1999. The linoleic and linolenic acids requirements of the prawn, *Penaeus monodon*. Aquacult. Nutr. 5:53–64.

Glencross, B. D., and D. M. Smith. 2001a. Optimising the essential fatty acids, eicosapentaenoic and docosahexaenoic acid in the diet of the prawn, *Penaeus monodon*. Aquacult. Nutr. 7:101–112.

Glencross, B. D., and D. M. Smith. 2001b. A study of the arachidonic acid requirements of the giant tiger prawn *Penaeus monodon*. Aquacult. Nutr. 7:59–70.

Glencross, B. D., D. M. Smith, M. R. Thomas, and K. C. Williams. 2002a. The effect of dietary n-3 to n-6 fatty acid balance on the growth of the prawn, *Penaeus monodon*. Aquacult. Nutr. 8:43–52.

Glencross, B. D., D. M. Smith, M. R. Thomas, and K. C. Williams. 2002b. Optimising the essential fatty acid and total neutral lipid requirements for weight gain of the prawn, *Penaeus monodon*. Aquaculture 204:85–59.

Gomez-Fernandez, J. C., and S. Corbalan-Garcia. 2007. Diacylglycerols, multivalent membrane modulators. Chem. Phys. Lipids148:1–25.

Gong, H., A. L. Lawrence, J. Dong-Huo, F. L. Castille, and D. M. Gatlin. 2000. Lipid nutrition of juvenile *Litopenaeus vanamei*. I. Dietary cholesterol and de-oil lecithin requirements and their interaction. Aquaculture 190:305–324.

González-Félix, M. L., D. M. Gatlin, III, A. L. Lawrence, and M. Perez-Velazquez. 2002a. Effect of various dietary lipid levels on quantitative essential fatty acid requirements of juvenile Pacific white shrimp *Litopenaeus vannamei*. J. World Aquacult. Soc. 33:330–340.

González-Félix, M. L., D. M. Gatlin, III, A. L. Lawrence, and M. Perez-Velazquez. 2002b. Effect of dietary phospholipids on essential fatty acid requirements and tissue composition of *Litopenaeus vannamei* juveniles. Aquaculture 207:151–167.

González-Félix, M. L., D. M. Gatlin, III, A. L. Lawrence, and M. Perez-Velazquez. 2003. Nutritional evaluation of fatty acids for the open thelycum shrimp, *Litopenaeus vannamei*: II. Effect of dietary n-3 and n-6 polyunsaturated and highly unsaturated fatty acids on juvenile shrimp growth, survival, and fatty acid composition. Aquacult. Nutr. 9:115–122.

González-Rovira, A., G. Mourente, X. Zheng, D. R. Tocher, and C. Pendon. 2009. Cloning, molecular and functional characterization and expression analysis of a fatty acyl Δ6 desaturase cDNA of European sea bass (*Dicentrarchus labrax* L.). Aquaculture 298:90–100.

Gunstone, F. D. 2010. The world's oils and fats. Pp. 61–98 in Fish Oil Replacement and Alternative Lipid Sources in Aquaculture Feeds, G. M. Turchini, W.-K. Ng, and D. R. Tocher, eds. Boca Raton, FL: Taylor & Francis, CRC Press.

Gylfason, G. A., E. Knútsdóttir, and B. Ásgeirsson. 2010. Isolation and biochemical characterisation of lipid rafts from Atlantic cod (*Gadus morhua*) intestinal enterocytes. Comp. Biochem. Phys. B 155:86–95.

Halver, J. E. 2002. The vitamins. Pp. 61–141 in Fish Nutrition, 3rd Edition, J. E. Halver and R. W. Hardy, eds. San Diego, CA: Academic Press.

Hamre, K., and T. Harboe. 2008a. Critical levels of essential fatty acids for normal pigmentation in Atlantic halibut (*Hippoglossus hippoglossus* L.) larvae. Aquaculture 277:101–108.

Hamre, K., and T. Harboe. 2008b. Artemia enriched with n-3 HUFA may give a large improvement in performance of Atlantic halibut (*Hippoglossus hippoglossus* L.) larvae. Aquaculture 277:239–243.

Hamre, K., E. Holen, and M. Moren. 2007. Pigmentation and eye-migration in Atlantic halibut (*Hippoglossus hippoglossus* L.) larvae: New findings and hypotheses. Aquacult. Nutr. 13:65–80.

Hamza, N., M. Mhetli, I. B. Khemis, C. Cahu, and P. Kestemont. 2008. Effect of dietary phospholipids levels on performance, enzyme activities and fatty acid composition of pikeperch (*Sander lucioperca*) larvae. Aquaculture 275:274–282.

Harel, M., A. Tandler, G. W. Kissil, and S. Applebaum. 1992. The kinetics of nutrient incorporation into body tissues of gilthead seabream *S. aurata* females and the subsequent effects on egg composition and egg quality. Isr. J. Aquacult. Bamidgeh 44:127.

Hastings, N., M. Agaba, D. R. Tocher, M. J. Leaver, J. R. Dick, J. R. Sargent, and A. J. Teale. 2001. A vertebrate fatty acid desaturase with Δ6 and Δ5 activities. P. Natl. Acad. Sci. USA 98:14304–14309.

Hastings, N., M. K. Agaba, D. R. Tocher, X. Zheng, C. A. Dickson, J. R. Dick, and A. J. Teale. 2005. Molecular cloning and functional characterization of fatty acyl desaturase and elongase cDNAs involved in the production of eicosapentaenoic and docosahexaenoic acids from α-linolenic acid in Atlantic salmon (*Salmo salar*). Mar. Biotechnol. 6:463–474.

Hemre, G.-I., and K. Sandnes. 1999. Effect of dietary lipid level on muscle composition in Atlantic salmon *Salmo salar*. Aquacult. Nutr. 5:9–16.

Hillestad, M., F. Johnsen, E. Austreng, and T. Asgard. 1998. Long-term effects of dietary fat level and feeding rate on growth, feed utilization and carcass quality of Atlantic salmon. Aquacult. Nutr. 4:89–97.

Hirsch, E., C. Costa, and E. Ciraolo. 2007. Phosphoinositide 3-kinase as a common platform for multi-hormone signaling. J. Endocrinol. 194:243–256.

Hochachka, P. W., and T. P. Mommsen, eds. 1995. Biochemistry and Molecular Biology of Fishes, Environmental and Ecological Biochemistry, Vol. 5. Amsterdam, the Netherlands: Elsevier Press.

Hong, S., E. Tjonahen, E. L. Morgan, Y. Lu, C. N. Serhan, and A. F. Rowley. 2005. Rainbow trout (*Oncorhynchus mykiss*) brain cells biosynthesize novel docosahexaenoic acid-derived resolvins and protectins—Mediator lipidomic analysis. Prostag. Leukotr. Ess. 78:107–116.

Hua, K., and D. P. Bureau. 2009. Development of a model to estimate digestible lipid content of salmonid fish feeds. Aquaculture 286:271–276.

Hung, S. S., and P. B. Lutes. 1988. A preliminary study on the non-essentiality of lecithin for hatchery-produced juvenile white sturgeon (*Acipenser transmontanus*). Aquaculture 68:353–360.

Ibeas, C., M. S. Izquierdo, and A. Lorenzo. 1994. Effect of different levels of n-3 highly unsaturated fatty acids on growth and fatty acid composition of juvenile gilthead seabream (*Sparus aurata*). Aquaculture 127:177–188.

Ibeas, C., J. R. Cejas, R. Fores, P. Badia, T. Gomez, A. Lorenzo, and A. Hernandez. 1997. Influence of eicosapentaenoic to docosahexaenoic acid ratio (EPA/DHA) of dietary lipids on growth and fatty acid composition of gilthead seabream (*Sparus aurata*) juveniles. Aquaculture 150:91–102.

Ibeas, C., C. Rodriguez, P. Badia, J. R. Cejas, F. J. Santamaria, and A. Lorenzo. 2000. Efficacy of dietary methyl esters of n-3 HUFA vs. triacylglycerols of n-3 HUFA by gilthead seabream (*Sparus aurata* L.) juveniles. Aquaculture 190:273–287.

Ikonen, E. 2008. Cellular cholesterol trafficking and compartmentalization. Nat. Rev. Mol. Cell. Bio. 9:125–138.

Im, D. S. 2009. New intercellular lipid mediators and their GPCRs: An update. Prostag. Oth. Lipid M. 89:53–56.

Ishizaki, Y., T. Takeuchi, T. Watanabe, M. Arimoto, and K. Shimizu. 1998. A preliminary experiment on the effect of *Artemia* enriched with arachidonic acid on survival and growth of yellowtail. Fish. Sci. 64:295–299.

Ishizaki, Y., R. Masuda, K. Uematsu, K. Shimizu, M. Arimoto, and T. Takeuchi. 2001. The effect of dietary docosahexaenoic acid on schooling behaviour and brain development in larval yellowtail. J. Fish Biol. 58:1691–1703.

Izquierdo, M. S. 1996. Essential fatty acid requirements of cultured marine fish larvae. Aquacult. Nutr. 2:183–191.

Izquierdo, M. S., T. Watanabe, T. Takeuchi, T. Arakawa, and C. Kitajima. 1989. Requirement of larval red seabream *Pagrus major* for essential fatty acids. Nippon Suisan Gakk. 55:859–867.

Izquierdo, M. S., J. Socorro, L. Arantzamendi, and C. M. Hernandez-Cruz. 2000. Recent advances in lipid nutrition in fish larvae. Fish Physiol. Biochem. 22:97–107.

Izquierdo, M. S., H. Fernandez-Palacios, and A. G. J. Tacon. 2001. Effect of broodstock nutrition on reproductive performance of fish. Aquaculture 197:25–42.

Jakobsson, A., R. Westerberg, and A. Jacobsson. 2006. Fatty acid elongases in mammals: Their regulation and roles in metabolism. Prog. Lipid Res. 45:237–249.

Jeng, K. C. G., and R. C. W. Hou. 2005. Sesamin and sesamolin: Nature's therapeutic lignans. Curr. Enzym. Inhib. 1:11–20.

Jobling, M. 1994. Fish Bioenergetics–Fish and Fisheries Series, Vol. 13. London, UK: Chapman & Hall.

Jump, D. B. 2002. Dietary polyunsaturated fatty acids and regulation of gene transcription. Curr. Opin. Lipidol. 13:155–164.

Kagan, V. E., V. A. Tyurin, N. V. Gorbunov, L. L. Prilipko, and V. P. Chelomin. 1984. Are changes in the microviscosity and an asymmetrical distribution of phospholipids in the membrane necessary conditions for signal transmission? A comparison of the mechanisms of signal transmission in plasma membranes of brain synaptosomes and photoreceptor membranes of the retina. J. Evol. Biochem. Phys. 20:6–11.

Kalogeropulos, N., M. N. Alexis, and R. J. Henderson. 1992. Effects of dietary soybean and cod-liver oil levels on growth and body composition of gilthead bream (*Sparus aurata*). Aquaculture 104:293–308.

Kanazawa, A. 1983. Penaeid nutrition. In Proceedings of the Second International Conference on Aquaculture Nutrition: Biochemical and Physiological Approaches to Shellfish Nutrition, G. D. Pruder, D. E. Conklin, and C. Langdon, eds. Baton Rouge, LA: Louisiana State University, Division of Continuing Education.

Kanazawa, A. 1993. Essential phospholipids of fish and crustaceans. Pp. 519–530 in Fish Nutrition in Practice, IV International Symposium on Fish Nutrition and Feeding, S. J. Kaushik and P. Luquet, eds. Paris, France: National Institute for Agricultural Research.

Kanazawa, A., and S. Teshima. 1977. Biosynthesis of fatty acids from acetate in the prawn, *Penaeus japonicus*. Mem. Fac. Fish. Kagoshima University 26:49–53.

Kanazawa, A., N. Tanaka, and S. Teshima. 1971. Nutritional requirements of prawn-II. Requirement for sterols. B. Jpn. Soc. Sci. Fish. 37:211–215.

Kanazawa, A., S. Teshima, and S. Tokiwa. 1977a. Nutritional requrements of prawn-VII. Effect of dietary lipids on growth. B. Jpn. Soc. Sci. Fish. 43:849–856.

Kanazawa A., S. Tokiwa, M. Kayama, and M. Hirata. 1977b. Essential fatty acids in the diet of prawn. I. Effects of linoleic and linolenic acids on growth. B. Jpn. Soc. Sci. Fish. 43:1111–1114.

Kanazawa, A., S. Teshima, M. Endo, and M. Kayama. 1978. Effects of eicosapentaenoic acid on growth and fatty acid composition of the prawn. Mem. Fac. Fish. Kagoshima University 27:35–40.

Kanazawa, A., S. Teshima, and M. Endo. 1979a. Requirements of the prawn, *Penaeus japonicus* for essential fatty acids. Mem. Fac. Fish. Kagoshima University 28:27–33.

Kanazawa, A., S. Teshima, S. Tokiwa, M. Kayama, and M. Hirata. 1979b. Essential fatty acids in the diet of the prawn. II. Effect of docosahexenoic acid on growth. B. Jpn. Soc. Sci. Fish. 45:1151–1153.

Kanazawa, A., S. Teshima, S. Tokiwa, M. Endo, and F.A. Abdel Razek. 1979c. Effects of short-necked clam phospholipids on the growth of the prawn. B. Jpn. Soc. Sci. Fish. 45:961–965.

Kanazawa A., S. Teshima, and S. Ono. 1979d. Relationship between essential fatty acid requirements of aquatic animals and the capacity for bioconversion of linolenic acid to highly unsaturated fatty acids. Comp. Biochem. Physiol. 63:295–298.

Kanazawa, A., S. Teshima, M. Sakamoto, and M. A. Awal. 1980. Requirement of *Tilapia zilli* for essential fatty acids. B. Jpn. Soc. Sci. Fish. 46:1353–1356.

Kanazawa, A., S. Teshima, S. Inamori, T. Iwashita, and A. Nagao. 1981. Effects of phospholipids on survival rate and incidence of malformation in the larval ayu. Mem. Fac. Fish. Kagoshima University 30:301–309.

Kanazawa, A., S. Teshima, and M. Sakamoto. 1982. Requirements of essential fatty acids for the larval ayu. B. Jpn. Soc. Sci. Fish. 48:586–590.

Kanazawa, A., S. Teshima, T. Kobayashi, M. Takae, T. Iwashita, and R. Uehara. 1983a. Necessity of phospholipids for growth of the larval ayu. Mem. Fac. Fish. Kagoshima University 32:115–120.

Kanazawa, A., S. Teshima, S. Inamori, and H. Matsubara. 1983b. Effects of dietary phospholipids on growth of the larval red sea bream and knife jaw. Mem. Fac. Fish. Kagoshima University 32:109–114.

Kanazawa, A., S. Teshima, and M. Sakamoto 1985. Effects of dietary lipids, fatty acids and phospholipids on growth and survival of the prawn (*Penaeus japonicus*) larvae. Aquaculture 50:39–49.

Kasper, C. S., and P. B. Brown. 2003. Growth improved in juvenile Nile tilapia fed phosphatidylcholine. N. Am. J. Aquacult. 65:39–43.

Kean, J. C., J. D. Castell, A. G. Boghen, L. R. D'Abramo, and D. E. Conklin. 1985. A re-evaluation of the lecithin and cholesterol requirements of juvenile lobster (*Homarus americanus*) using crab protein based diets. Aquaculture 47:143–149.

Kennedy, S. R., P. J. Campbell, A. Porter, and D. R. Tocher. 2005. Influence of dietary conjugated linoleic acid (CLA) on lipid and fatty acid composition in liver and flesh of Atlantic salmon (*Salmo salar*). Comp. Biochem. Phys. B 141:168–178.

Kennedy, S. R., R. Bickerdike, R. K. Berge, A. R. Porter, and D. R. Tocher. 2007a. Influence of dietary conjugated linoleic acid (CLA) and tetradecylthioacetic acid (TTA) on growth, lipid composition and key enzymes of fatty acid oxidation in liver and muscle of Atlantic cod (*Gadus morhua* L.). Aquaculture 264:372–382.

Kennedy, S. R., R. Bickerdike, R. K. Berge, J. R. Dick, and D. R. Tocher. 2007b. Effects of dietary conjugated linoleic acid (CLA) and tetradecylthioacetic acid (TTA) on growth, lipid composition, fatty acid metabolism and lipid gene expression of rainbow trout (*Oncorhynchus mykiss* L.). Aquaculture 272:489–501.

Kleveland, E. J., B. Ruyter, A. Vegusdal, H. Sundvold, R. K. Berge, and T. Gjøen. 2006. Effects of 3-thia fatty acids on expression of some lipid related genes in Atlantic salmon (*Salmo salar* L.). Comp. Biochem. Phys. B 145:239–248.

Kooijman, E. E., and K. N. J. Burger. 2009. Biophysics and function of phosphatidic acid: A molecular perspective. Biochim. Biophys. Acta 1791:881–888.

Koven, W. M., S. Kolkovski, E. Hadas, K. Gamsiz, and A. Tandler. 2001a. Advances in the development of microdiets for gilthead seabream, *Sparus aurata*: A review. Aquaculture 194:107–121.

Koven, W., Y. Barr, S. Lutzky, I. Ben-Atia, R. Weiss, M. Harel, P. Behrens, and A. Tandler. 2001b. The effect of dietary arachidonic acid (20:4n-6) on growth, survival and resistance to handling stress in gilthead seabream (*Sparus aurata*) larvae. Aquaculture 193:107–122.

Kvale, A., M. Yufera, E. Nygard, K. Aursland, T. Harboe, and K. Hamre. 2006. Leaching properties of three different microparticulate diets and preference of the diets in cod (*Gadus morhua* L.) larvae. Aquaculture 251:402–415.

Lanari, D., B. M. Poli, R. Ballestrazzi, P. Lupi, E. D'Agaro, and M. Mecatti. 1999. The effects of dietary fat and NFE levels on growing European sea bass (*Dicentrarchus labrax* L.). Growth rate, body and fillet composition, carcass traits and nutrient retention efficiency. Aquaculture 179:351–364.

Lange, Y., and T. L. Steck. 2008. Cholesterol homeostasis and the escape tendency (activity) of plasma membrane cholesterol. Prog. Lipid Res. 47:319–332.

Leaver, M. J., J. M. Bautista, T. Björnsson, E. Jönsson, G. Krey, D. R. Tocher, and B. E. Torstensen. 2008a. Towards fish lipid nutrigenomics: Current state and prospects for fin-fish aquaculture. Rev. Fish. Sci. 16:71–92.

Leaver, M. J., L. A. N. Villeneuve, A. Obach, L. Jensen, J. E. Bron, D. R. Tocher, and J. B. Taggart. 2008b. Functional genomics reveals increased cholesterol and highly unsaturated fatty acid biosynthesis after dietary substitution of fish oil with vegetable oils in Atlantic salmon (*Salmo salar*). BMC Genomics 9:299. doi:10.1186/1471-2164-9-299

Lee, S.-M., J.-Y. Lee, Y. J. Kang, H.-D. Yoon, and S. B. Hur. 1993. n-3 highly unsaturated fatty acid requirement of the Korean rockfish *Sebastes schlegeli*. B. Kor. Fish. Soc. 26:477–492.

Lee, S.–M., J.–Y. Lee, and S.–B. Hur. 1994. Essentiality of dietary eicosapentaenoic acid and docosahexaenoic acid in Korean rockfish, *Sebastes schlegeli*. B. Kor. Fish. Soc. 27:712–726.

Lee, S. M., J. H. Lee, and K. D. Kim. 2003. Effect of dietary essential fatty acids on growth, body composition and blood chemistry of juvenile starry flounder (*Platichthys stellatus*). Aquaculture 225:269–281.

Leu, M.-Y., S.-D. Yang, C.-H. Wu, and C.-H. Liou. 1994. Effect of dietary n-3 highly unsaturated fatty acids on growth, feed efficiency and fatty acid composition of juvenile silver bream *Rhabdosargus sarba* (Sparidae). Asian Fish. Sci. 7:233–240.

Li, Y., C. Hu, Y. Zheng, X. Xia, W. Xu, S. Wang, W. Chen, Z. Sun, and J. Huang. 2008. The effects of dietary fatty acids on liver fatty acid composition and delta 6-desaturase expression differ with ambient salinities in *Siganus canaliculatus*. Comp. Biochem. Phys. B 151:183–190.

Li, Y., Ó. Monroig, L. Zhang, S. Wang, X. Zheng, J. R. Dick, C. You, and D. R. Tocher. 2010. A vertebrate fatty acyl desaturase with Δ4 activity. Natl. Acad. Sci. USA 107:16840–16845.

Lim, P.–K., P.–L. Boey, and W.–K. Ng. 2001. Dietary palm oil level affects growth performance, protein retention and tissue vitamin E concentration of African catfish, *Clarias gariepinus*. Aquaculture 202:101–112.

Lochmann, R. T., and D. M. Gatlin. 1993. Essential fatty acid requirement of juvenile red drum (*Sciaenops ocellatus*). Fish Physiol. Biochem. 12:221–235.

Lund, I., S. J. Steenfeldt, and B. W. Hansen. 2007. Effect of dietary arachidonic acid, eicosapentaenoic acid and docosahexaenoic acid on survival, growth and pigmentation in larvae of common sole (*Solea solea* L.). Aquaculture 276:143–153.

Lund, I., S. J. Steenfeldt, G. Banta, and B. W. Hansen. 2008. The influence of dietary concentrations of arachidonic acid and eicosapentaenoic acid at various stages of larval ontogeny on eye migration, pigmentation and prostaglandin content of common sole larvae (*Solea solea* L.). Aquaculture 276:143–153.

Luzzana, U., G. Serrini, V. M. Moretti, C. Gianesini, and F. Valfre. 1994. Effect of expanded feed with high fish oil content on growth and fatty acid composition of rainbow trout. Aquacult. Int. 2:239–248.

Madsen, L., M. Guerre-Millo, E. N. Flindt, K. Berge, K. J. Tronstad, E. Bergene, E. Sebokova, A. C. Rustan, J. Jensen, S. Mandrup, K. Kristiansen, I. Klimes, B. Staels, and R. K. Berge. 2002. Tetradecylthioacetic acid prevents high fat diet induced adiposity and insulin resistance. J. Lipid Res. 43:742–750.

Makide, K., H. Kitamura, Y. Sato, M. Okutani, and J. Aoki. 2009. Emerging lysophospholipid mediators, lysophosphatidylserine, lysophosphatidylthreonine, lysophosphatidylethanolamine and lysophosphatidylglycerol. Prostag. Oth. Lipid M. 89:135–139.

Manuel Vergara, A., L. Robaina, M. Izquierdo, and M. Higuera. 1996. Protein sparing effect of lipids in diets for fingerlings of gilthead sea bream. Fish. Sci. 62:624–628.

Martin, B. J. 1980. Croissance et acides gras de la crevette *Paleamon serratus* (Crustacea, decapods) nourrie avec des aliments composes contenant differentes proportions d'acide linoleique et linolenique. Aquaculture 19:325–337.

Masuda, R., T. Takeuchi, T. Tsukamoto, Y. Ishizaki, M. Kanematsu, and K. Imaizumi. 1998. Critical involvement of dietary docosahexaenoic acid in the ontogeny of schooling behaviour in the yellowtail. J. Fish Biol. 53:471–484.

Masuda, R., D. A. Ziemann, and A. C. Ostrowski. 2001. Patchiness formation and development of schooling behavior in pacific threadfin *Polydactylus sexfilis* reared with different dietary highly unsaturated fatty acid contents. J. World Aquacult. Soc. 32:309–316.

Merican, Z. O., and K. F. Shim. 1994. Lipid and fatty acid utilization in adult *Penaeus monodon* fed diets supplemented with various oils. Aqaculture 123:335–347.

Merican, Z. O., and K. K. Shim. 1995. Apparent digestibility of lipid and fatty acids in residual lipids of meals by adult *Penaeus monodon*. Aquaculture 133:275–286.

Merican, Z. O., and K. F. Shim. 1997. Quantitative requirements of essential fatty acids for juvenile Penaeus monodon. Aquaculture 157:277–295.

Michell, R. H. 2007. Evolution of the diverse biological roles of inositols. Biochem. Soc. Symp. 74:223–246.

Mohd-Yusof, N. Y., Ó. Monroig, A. Mohd-Adnan, K.-L. Wan, and D. R. Tocher. 2010. Investigation of highly unsatiurated fatty acid metabolism in barramundi, the Asian sea bass *Lates calcarifer*. Fish Physiol. Biochem. doi: 10.1007/s10695-010-9409-4.

Monroig, Ó., J. Rotllant, E. Sánchez, J. M. Cerdá-Reverter, and D. R. Tocher. 2009. Expression patterns of genes of long-chain polyunsaturated fatty acid (LC-PUFA) biosynthesis during embryonic development of zebrafish *Danio rerio*. Biochim. Biophys. Acta 1791:1093–1101.

Monroig, Ó., X. Zheng, S. Morais, M. J. Leaver, J. B. Taggart, and D. R. Tocher. 2010. Multiple fatty acyl desaturase (FAD) genes in Atlantic salmon: Cloning and functional expression of cDNAs confirm presence of three Δ6 FADs. Biochim. Biophys. Acta 1801:1072–1081.

Morais, S., O. Monroig, X. Zheng, M. J. Leaver, and D. R. Tocher. 2009. Highly unsaturated fatty acid synthesis in Atlantic salmon: Isolation of genes of fatty acyl elongases and characterisation of ELOVL5- and ELOVL2-like elongase cDNAs. Mar. Biotechnol. 11:627–639.

Moschetta, A., F. Xu, L. R. Hagey, G. P. van Berge-Henegouwen, K. J. van Erpecum, J. F. Browers, J. C. Cohen, M. Bierman, H. H. Hobbs, J. H. Steinbach, and A. F. Hofmann. 2005. A phylogenetic survey of biliary lipids in vertebrates. J. Lipid Res. 46:2221–2232.

Mourente, G., J. G. Bell, and D. R. Tocher. 2007. Does dietary tocopherol level affect fatty acid metabolism in fish? Fish Physiol. Biochem. 33:269–280.

Moya-Falcon, C., E. Hvattum, E. Dyrøy, J. Skorve, S. O. Stefansson, M. Thomassen, J. V. Jakobsen, R. K. Berge, and B. Ruyter, B., 2004. Effects of 3-thia fatty acids on feed intake, growth, tissue fatty acid composition, β-oxidation and Na^+,K^+-ATPase activity in Atlantic salmon. Comp. Biochem. Phys. B 139:657–668.

Newton, A. C. 2009. Lipid activation of protein kinases. J. Lipid Res. 50:S266–S271.

Olsen, R. E., and E. Ringø. 1997. Lipid digestibility in fish: A review. Rec. Res. Dev. Lipids Res. 1:199–265.

Olsen, R. E., R. Myklebust, T. Kaino, and E. Ringø. 1999. Lipid digestibility and ultrastructural changes in the enterocytes of Arctic char (*Salvelinus alpinus* L.) fed linseed oil and soybean lecithin. Fish Physiol. Biochem. 21:35–44.

Olsen, R. E., R. J. Henderson, J. Sountama, G.-I. Hemre, E. Ringø, W. Melle, and D. R. Tocher. 2004. Atlantic salmon effectively utilizes wax ester rich oil from *Calanus finmarchicus*. Aquaculture 240:433–449.

Ostrowski, A. C., and B. G. Kim. 1993. Responses of larval and juvenile mahimahi (*Coryphaena hippurus*) to various dietary lipid sources and n-3 HUFA contents. Pp. 424 in From Discovery to Commercialization, Special Publication No. 19. Oostende, Belgium: European Aquaculture Society.

Owen, J. M., J. A. Adron, C. Middleton, and C. B. Cowey. 1975. Elongation and desaturation of dietary fatty acids in turbot (*Scophthalmus maximus*) and rainbow trout (*Salmo gairdneri*). Lipids 10:528–531.

Paibulkichakui, C., S. Piyatiratitivorakul, P. Kittakpp, V. Voranop, A. W. Fast, and P. Menasveta. 1998. Optimal dietary levels of lecithin and cholesterol for black tiger prawn *Penaeus monodon* larvae and postlarvae. Aquaculture 167:273–281.

Peres, H., and A. Oliva-Teles. 1999. Effect of dietary lipid level on growth performance and feed utilization by European sea bass juveniles (*Dicentrarchus labrax*). Aquaculture 179:325–334.

Pickova, J., P. C. Dutta, P.-O. Larsson, and A. Kiessling. 1997. Early embryonic cleavage pattern, hatching success, and egg-lipid fatty acid composition: Comparison between two cod (*Gadus morhua*) stocks. Can. J. Fish. Aquat. Sci. 54:2410–2416.

Ponat, A., and D. Adelung. 1983. Studies to establish an optimal diet for *Carcinus maenus*: III. Vitamin and quantitative lipid requirements. Mar. Biol. 74:275–279.

Poston, H. A. 1990a. Performance of rainbow trout fed supplemental soybean lecithin and choline. Prog. Fish Cult. 52:218–225.

Poston, H. A. 1990b. Effect of body size on growth, survival and chemical composition of Atlantic salmon fed soy lecithin and choline. Prog. Fish Cult. 52:226–230.

Poston, H. A. 1991. Response of Atlantic salmon fry to feed-grade lecithin and choline. Prog. Fish Cult. 53:224–228.

Quintero, H., E. Durland, D. A. Davis, and R. Dunham. 2010. Effects of lipid supplementation on reproductive performance of female channel catfish *Ictalurus punctatus*, induced and strip-spawned for hybridization. Aquacult. Nutr. In press.

Radunzneto, J., G. Corraze, P. Bergot, and S. J. Kaushik. 1996. Estimation of essential fatty acid requirements of common carp larvae using semipurified diets. Arch. Anim. Nutr. 49:41–48.

Rasmussen, R. S., T. H. Ostenfeld, B. Roensholdt, and E. McLean. 2000. Manipulation of end-product quality of rainbow trout with finishing diets. Aquacult. Nutr. 6:17–23.

Regost, C., J. Arzel, J. Robin, G. Rosenlund, and S. J. Kaushik. 2003. Total replacement of fish oil by soybean or linseed oil with a return to fish oil in turbot (*Psetta maxima*) 1. Growth performance, flesh fatty acid profile, and lipid metabolism. Aquaculture 217:465–482.

Reitan, K. I., J. R. Rainuzzo, and Y. Olsen. 1994. Influence of lipid composition of live feed on growth, survival and pigmentation of turbot larvae. Aquacult. Int. 2:33–48.

Rinchard, J., S. Czesny, and K. Dabrowski. 2007. Influence of lipid class and fatty acid deficiency on survival, growth, and fatty acid composition in rainbow trout juveniles. Aquaculture 264:363–371.

Robin, J. H., and B. Vincent. 2003. Microparticulate diets as first food for gilthead sea bream larvae (*Sparus aurata*): Study of fatty acid incorporation. Aquaculture 225:463–474.

Rodriguez, C. 1994. Estudio de los requerimientos de acidos grasos esenciales de la dorada europea *Sparus aurata* L. durante las dos primeras semanas de alimentacion. Ph.D. Dissertation, Universidad de La Laguna, Tenerife, Spain.

Rodriguez, C., J. A. Perez, M. S. Izquierdo, J. Mora, A. Lorenzo, and H. Fernandez-Palacios. 1994a. Essential fatty acid requirements of larval gilthead sea bream *Sparus aurata* (L). Aquacult. Fish. Manage. 25:295–304.

Rodriguez, C., J. A. Perez, M. S. Izquierdo, A. Lorenzo, and H. Fernandez-Palacios. 1994b. The effect of n-3 HUFA proportions in diets for gilthead sea bream (*Sparus aurata*) larval culture. Aquaculture 124:284.

Rodriguez, C., J. A. Perez, P. Badia, M. S. Izquierdo, H. Fernandez-Palacios, and A. L. Hernandez. 1998a. The n-3 highly unsaturated fatty acids requirements of gilthead seabream (*Sparus aurata* L.) larvae when using an appropriate DHA/EPA ratio in the diet. Aquaculture 169:9–23.

Rodriguez, C., J. R. Cejas, M. V. Martin, P. Badia, M. Samper, and A. Lorenzo. 1998b. Influence of n-3 highly unsaturated fatty acid deficiency on the lipid composition of broodstock gilthead seabream (*Sparus aurata* L.) and on egg quality. Fish Physiol. Biochem. 18:177–187.

Roesjoe, C., S. Nordrum, J. J. Olli, A. Krogdahl, B. Ruyter, and H. Holm. 2000. Lipid digestibility and metabolism in Atlantic salmon (*Salmo salar*) fed medium-chain triglycerides. Aquaculture 190:65–76.

Ruyter, B., Ø. Anderson, A. Dehli, A. Östlund Farrants, T. Gjøen, and M. S. Thomassen. 1997. Peroxisome proliferator activated receptors in Atlantic salmon (*Salmo salar*): Effects on PPAR transcription and acyl-CoA oxidase activity in hepatocytes by peroxisome proliferators and fatty acids. Biochim. Biophys. Acta 1348:331–338.

Ruyter, B., C. Rosjo, O. Einen, and M. S. Thomassen. 2000a. Essential fatty acids in Atlantic salmon: Time course of changes in fatty acid composition of liver, blood and carcass induced by a diet deficient in n-3 and n-6 fatty acids. Aquacult. Nutr. 6:109–118.

Ruyter, B., C. Rosjo, O. Einen, and M. A. Thomassen. 2000b. Essential fatty acids in Atlantic salmon: Effects of increasing dietary doses of n-3 and n-6 fatty acids on growth, survival and fatty acid composition of liver, blood and carcass. Aquacult. Nutr. 6:119–127.

Saether, B. S., and M. Jobling. 2001. Fat content in turbot feed: Influence on feed intake, growth and body composition. Aquac. Res. 32:451–458.

Salhi, M., M. S. Izquierdo, C. M. Hernandez-Cruz, M. Gonzalez, and H. Fernandez-Palacios. 1994. Effect of lipid and n-3 HUFA levels in microdiets on growth, survival and fatty acid composition of larval gilthead sea bream (*Sparus aurata*). Aquaculture 124:275–282.

Salhi, M., C. M. Hernandez-Cruz, M. Bessonart, M. S. Izquierdo, and H. Fernandez-Palacios. 1999. Effect of different dietary polar lipid levels and different n-3 HUFA content in polar lipids on gut and liver histological structure of gilthead seabream (*Sparus aurata*) larvae. Aquaculture 179:253–263.

Salze, G., D. R. Tocher, W. J. Roy, and D. A. Robertson. 2005. Egg quality determinants in cod (*Gadus morhua* L.): Egg performance and lipids in eggs from farmed and wild broodstock. Aquac. Res. 36:1488–1499.

Sampath, H., and J. M. Ntambi. 2005. Polyunsaturated fatty acid regulation of genes of lipid metabolism. Annu. Rev. Nutr. 25:317–340.

Santha, C. R., and D. M. Gatlin. 1991. Growth response and fatty acid composition of channel catfish fry fed practical diets supplemented with menhaden fish oil. Prog. Fish Cult. 53:135–140.

Santiago, C. B., and O. S. Reyes. 1993. Effects of dietary lipid source on reproductive performance and tissue lipid levels of Nile tilapia *Oreochromis niloticus* (Linnaeus) broodstock. J. Appl. Ichthyol. 9:33–40.

Sargent, J. R., and A. G. J. Tacon. 1999. Development of farmed fish: A nutritionally necessary alternative to meat. Proc. Nutr. Soc. 58:377–383.

Sargent, J. R., R. J. Henderson, and D. R. Tocher. 1989. The lipids. Pp. 154–219 in Fish Nutrition, 2nd Edition, J. E. Halver, ed. San Diego, CA: Academic Press.

Sargent, J. R., J. G. Bell, M. V. Bell, R. J. Henderson, and D. R. Tocher. 1995a. Requirement criteria for essential fatty acids. J. Appl. Ichthyol. 11:183–198.

Sargent, J. R., J. G. Bell, M. V. Bell, R. J. Henderson, and D. R. Tocher. 1995b. Dietary origins and functions of long-chain (n-3) polyunsaturated fatty acids in marine fish. J. Mar. Biotechnol. 3:26–28.

Sargent, J. R., L.A. McEvoy, and J. G. Bell. 1997. Requirements, presentation and sources of polyunsaturated fatty acids in marine fish larval feeds. Aquaculture 155:119–129.

Sargent, J., G. Bell, L. McEvoy, D. R. Tocher, and A. Estevez. 1999a. Recent developments in the essential fatty acid nutrition of fish. Aquaculture 177:191–199.

Sargent, J., L. McEvoy, A. Estevez, G. Bell, M. Bell, J. Henderson, and D. R. Tocher. 1999b. Lipid nutrition of marine fish during early development: Current status and future directions. Aquaculture 179:217–229.

Sargent, J. R., D. R. Tocher, and J. G. Bell. 2002. The lipids. Pp. 181–257 in Fish Nutrition, 3rd Edition, J. E. Halver and R. W. Hardy, eds. San Diego, CA: Academic Press.

Satoh, S., W. E. Poe, and R. P. Wilson. 1989. Studies on the essential fatty acid requirement of channel catfish, *Ictalurus punctatus*. Aquaculture 79:121–128.

Scarano, L. J., E. J. Calabrese, P. T. Kostecki, L. A. Baldwin, and D.A Leonard. 1994. Evaluation of rodent peroxisome proliferator in two species of freshwater fish: Rainbow trout (*Oncorhynchus mykiss*) and Japanese medaka (*Oryzias latipes*). Ecotox. Environ. Safe. 29:13–19.

Schmitz, G., and J. Ecker. 2008. The opposing effects of *n-3* and *n-6* fatty acids. Prog. Lipid Res. 47:147–155.

Seiliez, I., S. Panserat, S. Kaushik, and P. Bergot. 2001. Cloning, tissue distribution and nutritional regulation of a delta 6-desaturase-like enzyme in rainbow trout. Comp. Biochem. Phys. B 130:83–93.

Seiliez, I., S. Panserat, G. Corraze, S. Kaushik, and P. Bergot. 2003. Cloning and nutritional regulation of a delta 6-desaturase-like enzyme in the marine teleost gilthead seabream (*Sparus aurata*). Comp. Biochem. Phys. B 135:449–460.

Seiliez, I., J. S. Bruant, J. Zambonino-Infante, S. Kaushik, and P. Bergot. 2006. Effect of dietary phospholipid level on the development of gilthead sea bream (*Sparus aurata*) larvae fed a compound diet. Aquacult. Nutr. 12:372–378.

Seki, H., Y. Tani, and M. Arita. 2009. Omega-3 PUFA derived anti-inflammatory lipid mediator resolving E1. Prostag. Oth. Lipid M. 89:126–130.

Shearer, K. D., and P. Swanson. 2000. The effect of whole body lipid on early sexual maturation of 1+ age Chinook salmon (*Oncorhynchus tshawytscha*). Aquaculture 190:343–367.

Sheen, S.-S., and L. R. D'Abramo. 1991. Response of juvenile freshwater prawn, *Macrobrachium rosenbergii*, to different levels of a cod liver oil/corn oil mixture in a semi-purified diet. Aquaculture 93:121–134.

Shewbart, K. L., and W. L. Mies. 1973. Studies on the nutritional requirements of brown shrimp: The effect of linolenic acid on growth of *Penaeus aztecus*. P. World Maricult. Soc. 4:277–287.

Shiau, S. Y., and L. F. Shiau 2001. Re-evaluation of the vitamin E requirements of juvenile tilapia (*Oreochromis niloticus* × *O. aureus*). Anim. Sci. 72:529–534.

Shudo, K., K. Nakamura, S. Ishikawa, and K. Kitabayshi. 1971. Studies on the growth-promoting effects of both squid liver oil and cholesterol. B. Tokai Regional Fish. Res. Lab. 65:129–137.

Silversand, C., B. Norberg, J. C. Holm, Ø. Lie, and C. Haux. 1995. Dietary influence on the fatty acid composition of vitellogenin and the subsequent effect on the egg composition in cod (*Gadus morhua*). P. 375 in Proceedings of the 5th International Symposium on the Reproductive Physiology of Fish, July 2–8, 1995, in Austin, TX. Austin: University of Texas at Austin.

Simons, K., and E. Ikonen. 2000. How cells handle cholesterol. Science 290:1721–1726.

Snyder, F. 1990. Platelet-activating factor and related acetylated lipids as potent biologically active cellular mediators. Am. J. Physiol.-Cell Physiol. 259:C697–C708.

Sprecher, H. 2000. Metabolism of highly unsaturated n-3 and n-6 fatty acids. Biochim. Biophys. Acta 1486:219–231.

Stottrup, J. G. 2000. The elusive copepods: Their production and suitability in marine aquaculture. Aquac. Res. 31:703–711.

Stubhaug, I., Ø. Lie, and B. E. Torstensen. 2007. Fatty acid productive value and β-oxidation capacity in Atlantic salmon tissues (*Salmo salar* L.) fed on different lipid sources along the whole growth period. Aquacult. Nutr. 13:145–155.

Tacon, A. G. J. 1996. Lipid nutritional pathology in farmed fish. Arch. Anim. Nutr. 49:33–39.

Tacon, A. G. J., and M. Metian. 2008. Global overview on the use of fish meal and fish oil in industrially compounded aquafeeds: Trends and future prospects. Aquaculture 285:146–158.

Tacon, A. G. J., M. Metian, G. M. Turchini, and S. S. DeSilva. 2010. Responsible aquaculture and trophic level implications to global fish supply. Rev. Fish. Sci. 18:94–105.

Taggart, J. B., J. E. Bron, S. A. M. Martin, P. J. Seear, B. Hoyheim, R. Talbot, L. Villeneuve, G. E. Sweeney, D. F. Houlihan, C. J. Secombes, D. R. Tocher, and A. J. Teale. 2008. A description of the origins, design and performance of the TRAITS/SGP Atlantic salmon (*Salmo salar* L.) cDNA microarray. J. Fish Biol. 72:2071–2094.

Takeuchi, T. 1997. Essential fatty acids requirements of aquatic animals with emphasis on fish larvae and fingerlings. Rev. Fish. Sci. 5:1–25.

Takeuchi, T., and T. Watanabe. 1976. Nutritive value of ω3 highly unsaturated fatty acids in pollock liver oil for rainbow trout. B. Jpn. Soc. Sci. Fish. 42:907 919.

Takeuchi, T., and T. Watanabe. 1977. Requirement of carp for essential fatty acids. B. Jpn. Soc. Sci. Fish. 43:541–551.

Takeuchi, T., T. Watanabe, and T. Nose. 1979. Requirement for essential fatty acids of chum salmon (*Oncorhyncus keta*) in freshwater environment. B. Jpn. Soc. Sci. Fish. 45:1319–1323.

Takeuchi, T., S. Arais, T. Watanabe, and Y. Shimma. 1980. Requirements of the eel *Anguilla japonica* for essential fatty acids. B. Jpn. Soc. Sci. Fish. 46:345–353.

Takeuchi, T., S. Satoh, and T. Watanabe. 1983. Requirement of *Tilapia nilotica* for essential fatty acids. B. Jpn. Soc. Sci. Fish. 49:1127–1134.

Takeuchi, T., M. Toyota, S. Satoh, and T. Watanabe. 1990. Requirement of juvenile red sea bream (*Pagrus major*) for eicosapentaenoic and docosahexaenoic acids. Nippon Suisan Gakk. 56:1263–1269.

Takeuchi, T., K. Watanabe, W.-Y. Yong, and T. Watanabe. 1991. Essential fatty acids of grass carp (*Ctenopharyngodon idella*). Nippon Suisan Gakk. 57:467–473.

Takeuchi, T., Y. Shiina, and T. Watanabe. 1992a. Suitable levels of n-3 highly unsaturated fatty acids in diet for fingerlings of red sea bream. Nippon Suisan Gakk. 58:509–514.

Takeuchi, T., Y. Shiina, T. Watanabe, S. Sekiya, and K. Imaizumi. 1992b. Suitable levels of n-3 highly unsaturated fatty acids in diet for fingerlings of yellowtail. Nippon Suisan Gakk. 58:1341–1346.

Takeuchi, T., T. Arakawa, S. Satoh, and T. Watanabe. 1992c. Supplemented effect of phospholipids and requirement of eicosapentaenoic acid and docosahexaenoic acid of juvenile striped jack. Nippon Suisan Gakk. 58:707–713.

Takeuchi, T., Z. Feng, K. Yoseda, J. Hirokawa, and T. Watanabe. 1994. Nutritive value of DHA-enriched rotifer for larval cod. Nippon Suisan Gakk. 60:641–652.

Takeuchi, T., R. Masuda, Y. Ishizaki, T. Watanabe, M. Kanematsu, K. Imaizumi, and K. Tsukamoto. 1996. Determination of the requirement of larval striped jack for eicosapentaenoic acid and docosahexaenoic acid using enriched *Artemia* nauplii. Fish. Sci. 62:760–765.

Tandler, A., M. Harel, W. M. Koven, and S. Kolkovski. 1995. Broodstock and larvae nutrition in gilthead seabream *Sparus aurata*—new findings on its mode involvement in improving growth, survival and swimbladder inflation. Isr. J. Aquacult. Bamidgeh 47:95–111.

Teres, S., G. Barcelo-Coblijn, M. Benet, R. Alvarez, R. Bressani, J. E. Halver, and P. V. Escriba. 2008. Oleic acid content is responsible for the reduction in blood pressure induced by olive oil. P. Natl. Acad. Sci. USA 105:13811–13816.

Teshima, S. I. 1997. Phospholipids and sterols. Pp. 85–107 in Crustacean Nutrition, L. R. D'Abramo, D. E. Conklin, and D. M. Akiyama, eds. Baton Rouge, LA:World Aquaculture Society.

Teshima, S., and A. Kanazawa. 1980. Lipid constituents of serum lipoproteins in the prawn. B. Jpn. Soc. Sci. Fish. 46:57–62.

Teshima, S., and A. Kanazawa. 1986. Nutritive value of sterols for the juvenile prawn. B. Jpn. Soc. Sci. Fish. 52:1417–1422.

Teshima, S., A. Kanazawa, and H. Sasada. 1983. Nutritional value of dietary cholesterol and other sterols to larval prawn, *Penaeus japonicus* Bate. Aquaculture 31:159–167.

Teshima, S., A. Kanazawa, and Y. Kakuta 1986a. Effects of dietary phospholipids on growth and body composition of the juvenile prawn. B. Jpn. Soc. Sci. Fish. 52:155–158.

Teshima, S., A. Kanazawa, and Y. Kakuta. 1986b. Growth, survival, and body composition of the prawn larvae receiving several dietary phospholipids. Mem. Fac. Fish. Kagoshima University 35:17–27.

Teshima, S., A. Kanazawa, and Y. Kakuta. 1986c. Role of dietary phospholipids in the transport of ^{14}C tripalmitin in the prawn. B. Jpn. Soc. Sci. Fish. 52:519–524

Teshima, S., A. Kanazawa, and A. Kakuta. 1986d. Role of dietary phospholipids in the transport of ^{14}C cholesterolin the prawn. B. Jpn. Soc. Sci. Fish. 52:719–723.

Teshima, S., A. Kanazawa, S. Koshio, and N. Kondo. 1989. Nutritive value of sitosterol for prawn, *Penaeus japonicus*. Nippon Suisan Gakk. 55:153 157.

Teshima, S., S. Koshio, A. Kanazawa and K. Oshida. 1994. Essential fatty acids of the prawn Macrobrachium rosenbergii. Pp. 705–708 in Proceedings of the 3rd Asian Fisheries Society Forum, L. M. Chou, A. D. Munro, T. J. Lam, T. W. Chen, L. K. K. Cheong, J. K. Ding, K. K. Hooi, H. W. Khoo, V. P. E. Phang, K. F. Shim, and C. H. Tan, eds., October 26–30, 1992, Singapore. Manila, the Philipines: Asian Fisheries Society.

Thongrod, S., and M. Boonyaratpalin. 1998. Cholesterol and lecithin requirement of juvenile banana shrimp *Penaeus merguiensis*. Aquaculture 161:315–321.

Thongrod, S., T. Takeuchi, S. Satoh, and T. Watanabe. 1989. Requirement of fingerling white fish *Coregonus lavaretus maraena* for dietary n-3 fatty acids. B. Jpn. Soc. Sci. Fish. 55:1983–1987.

Thongrod, S., T. Takeuchi, S. Satoh, and T. Watanabe. 1990. Requirement of Yamane (*Oncorhynchus masou*) for essential fatty acids. Nippon Suisan Gakk. 56:1255–1262.

Tocher, D. R. 1995. Glycerophospholipid metabolism. Pp. 119–157 in Biochemistry and Molecular Biology of Fishes, Volume 4: Metabolic and Adaptational Biochemistry, P. W. Hochachka and T. P. Mommsen, eds. Amsterdam, the Netherlands: Elsevier Press.

Tocher, D. R. 2003. Metabolism and functions of lipids and fatty acids in teleost fish. Rev. Fish. Sci. 11:107–184.

Tocher, D. R. 2010. Fatty acid requirements in ontogeny of marine and freshwater fish. Aquac. Res. 41:717–732.

Tocher, D. R., and J. R. Dick. 1999. Polyunsaturated fatty acid metabolism in a cell culture model of essential fatty acid deficiency in a freshwater fish, carp (*Cyprinus carpio*). Fish. Physiol. Biochem. 21:257–267.

Tocher, D. R., and J. R. Dick. 2000. Essential fatty acid deficiency in freshwater fish: The effects of linoleic, α-linolenic, γ-linolenic and stearidonic acids on the metabolism of [1-^{14}C]18:3n-3 in a carp cell culture model. Fish Physiol. Biochem. 22:67–75.

Tocher, D. R., and C. Ghioni. 1999. Fatty acid metabolism in marine fish: Low activity of fatty acyl Δ5 desaturation in gilthead sea bream (*Sparus aurata*) cells. Lipids 34:433–440.

Tocher, D. R., and J. R. Sargent. 1993. No relationship between morphology and metabolism of α-linolenate and eicosapentaenoate in rainbow trout (*Oncorhynchus mykiss*) astroglial cells in primary culture. Comp. Biochem. Phys. B 106:211–219.

Tocher, D. R., J. R. Sargent, and G. N. Frerichs. 1988. The fatty acid compositions of established fish cell lines after long-term culture in mammalian sera. Fish Physiol. Biochem. 5:219–227.

Tocher, D. R., J. G. Bell, J. R. Dick, and J. R. Sargent. 1997. Fatty acyl desaturation in isolated hepatocytes from Atlantic salmon (*Salmo salar*): Stimulation by dietary borage oil containing γ-linolenic acid. Lipids 32:1237–1247.

Tocher, D. R., M. J. Leaver, and P. A. Hodgson. 1998. Recent advances in the biochemistry and molecular biology of fatty acyl desaturases. Prog. Lipid Res. 37:73–117.

Tocher, D. R., M. Agaba, N. Hastings, J. G. Bell, J. R. Dick, and A. J. Teale. 2002. Nutritional regulation of hepatocyte fatty acid desaturation and polyunsaturated fatty acid composition in zebrafish (*Danio rerio*) and tilapia (*Oreochromis nilotica*). Fish Physiol. Biochem. 24:309–320.

Tocher, D. R., J. G. Bell, J. R. Dick, and V. O. Crampton. 2003. Effects of vegetable oil diets on Atlantic salmon hepatocyte desaturase activities and liver fatty acid compositions. Lipids 38:723–732.

Tocher, D. R., X. Zheng, C. Schlechtriem, N. Hastings, J. R. Dick, A. J. Teale. 2006. Highly unsaturated fatty acid synthesis in marine fish; cloning, functional characterization and nutritional regulation of fatty acid Δ6 desaturase of Atlantic cod (*Gadus morhua* L.). Lipids 41:1003–1016.

Tocher, D. R., E. Å. Bendiksen, P. J. Campbell, and J. G. Bell. 2008. The role of phospholipids in nutrition and metabolism of teleost fish. Aquaculture 280:21–34.

Torstensen, B. E., J. G. Bell, G. Rosenlund, R. J. Henderson, I. E. Graff, D. R. Tocher, Ø. Lie, and J. R. Sargent. 2005. Tailoring of Atlantic salmon (*Salmo salar* L.) flesh lipid composition and sensory quality by replacing fish oil with a vegetable oil blend. J. Agr. Food Chem. 53:10166–10178.

Trattner, S., A. Kamal-Eldin, E. Brännäs, A. Moazzami, V. Zlabek, P. Larsson, B. Ruyter, T. Gjøen, and J. Pickova. 2008a. Sesamin supplementation increases white muscle docosahexaenoic acid (DHA) levels in rainbow trout (*Oncorhynchus mykiss*) fed high alpha-linolenic acid (ALA) containing vegetable oil: Metabolic actions. Lipids 43:989–997.

Trattner, S., B. Ruyter, T. K. Østbye, T. Gjøen, V. Zlabek, A. Kamal-Eldin, and J. Pickova. 2008b. Sesamin increases alpha-linolenic acid conversion to docosahexaenoic acid in Atlantic salmon (*Salmo salar* L.) hepatocytes: Role of altered gene expression. Lipids 43:999–1008.

Twibell, R. G., and R. P. Wilson. 2003. Effects of conjugated linoleic acids and total dietary lipid concentrations on growth responses of juvenile channel catfish, *Ictalurus punctatus*. Aquaculture 221:621–628.

Twibell, R. G., B. A. Watkins, L. Rogers, and P. B. Brown. 2000. Effects of dietary conjugated linoleic acids on hepatic and muscle lipids in hybrid striped bass. Lipids 35:155–161.

Twibell, R. G., B. A. Watkins, and P. B. Brown. 2001. Dietary conjugated linoleic acids and lipid source alter fatty acid composition of juvenile yellow perch, *Perca flavescens*. J. Nutr. 131:2322–2328.

Valente, L. M. P., N. M. Bandarra, A. C. Figueiredo-Silva, P. Rema, P. Vaz-Pires, S. Martins, J. A. M. Prates, and M. L. Nunes. 2007a. Conjugated linoleic acid in diets for large-size rainbow trout (*Oncorhynchus mykiss*): Effects on growth, chemical composition, and sensory attributes. Brit. J. Nutr. 97:289–297.

Valente, L. M. P., N. M. Bandarra, A. C. Figueiredo-Silva, A. R. Cordeiro, R. M. Simoes, and M. L. Nunes. 2007b. Influence of conjugated linoleic acid on growth, lipid composition, and hepatic lipogenesis in juvenile European sea bass (*Dicentrarchus labrax*). Aquaculture 267:225–235.

Vassallo Agius, R., T. Watanabe, K. Mushiake, K. Kawano, and S. Satoh. 1998. Chemical components of eggs and yolksac larvae obtained from striped jack broodstock fed on a raw fish mix or dry pellets. Fish. Sci. 64:759–765.

Verakunpiriya, V., T. Watanabe, K. Mushiake, V. Kiron, S. Satoh, and T. Takeuchi. 1996. Effect of broodstock diets on the chemical components of milt and eggs produced by yellowtail. Fish. Sci. 62:610–619.

Villalta, M., A. Estevez, and M. P. Bransden. 2005. Arachidonic acid enriched live prey induces albinism in Senegalese sole (*Solea senegalensis*) larvae. Aquaculture 245:193–209.

Wang, X., S. P. Devaiah, W. Zhang, and R. Welti. 2006. Signaling functions of phosphatidic acid. Prog. Lipid Res. 45:250–278.

Wang, Y., and P. J. Jones. 2004. Dietary conjugated linoleic acid and body composition. Am. J. Clin. Nutr. 79:1153S–1158S.

Wasall, S. R., and W. Stillwell. 2008. Docosahexaenoic acid domains: The ultimate non-raft membrane domain. Chem. Phys. Lipids 153:57–63.

Watanabe, T. 1993. Importance of docosahexaenoic acid in marine fish larvae. J. World Aquacult. Soc. 24:152–161.

Watanabe, T., C. Ogino, Y. Kashiishi, and T. Matsunaga. 1974. Requirement of rainbow trout for essential fatty acids. B. Jpn. Soc. Sci. Fish. 40:493–497.

Watanabe, T., S. Thongrod, T. Takeuchi, S. Satoh, S. S. Kubota, Y. Fujimaki, and C. Y. Cho. 1989. Effect of dietary n-6 and n-3 fatty acids on growth, fatty acid composition and histological changes of white fish *Coregonus lavaretus maraena*. B. Jpn. Soc. Sci. Fish. 55:1977–1982.

Webster, C. D., and R. T. Lovell. 1990. Response of striped bass larvae fed brine shrimp from different sources containing different fatty acid compositions. Aquaculture 90:49–61.

Wehling, M. 1997. Specific, nongenomic actions of steroid hormones. Annu. Rev. Physiol. 59:365–393.

Whalen, K. S., J. A. Brown, C. C. Parrish, S. P. Lall, and J. S. Goddard. 1999. Effect of dietary n-3 HUFA on growth and body composition of juvenile yellowtail flounder (*Pleuronectes ferrugineus*). Bull. Aqua. Assoc. Can. 98:21–22.

Wirth, M., W. Steffens, T. Meinelt, and C. Steinberg. 1997. Significance of docosahexaenoic acid for rainbow trout (*Oncorhynchus mykiss*) larvae. Fett 99:251–253.

Wu, P. C., Y. Y. Ting, and H. Y. Chen. 2002. Docosahexaenoic acid is superior to eicosapentaenoic acid as the essential fatty acid for growth of grouper, *Epinephelus malabancus*. J. Nutr. 132:72–79.

Xu, X. L., W. J. Ji, J. D. Castell, and R. K. O'Dor. 1993. The nutritional value of dietary n-3 and n-6 fatty acids for the Chinese prawn (*Penaeus chinensis*). Aquaculture 118:277–285.

Xu, X. L., W. J. Ji, J. D. Castell, and R. K. O'Dor. 1994. Essential fatty acid requirements of the Chinese prawn, *Penaeus chinensis*. Aquaculture 127:29–40.

Yang, X., J. L. Tabachek, and T. A. Dick. 1994. Effects of dietary n-3 polyunsaturated fatty acids on lipid and fatty acid composition and haematology of juvenile Arctic charr *Salvelinus alpinus* (L.). Fish Physiol. Biochem. 12:409–420.

Yasmin, A., T. Takeuchi, M. Hayashi, T. Hirota, W. Ishizuka, and S. Ishida. 2004. Effect of conjugated linoleic and docosahexaenoic acids on growth of juvenile tilapia *Oreochromis niloticus*. Fish. Sci. 70:473–481.

Yone, Y. 1978. Essential fatty acids and lipid requirements of marine fish. Pp. 43–59 in Dietary Lipids in Aquaculture, Japanese Society of Scientific Fisheries. Tokyo, Japan: Koseisha-Koseik-Abu.

Yu, T. C., and R. O. Sinnhuber. 1979. Effect of dietary ω3 and ω6 fatty acids on growth and feed conversion efficiencies of coho salmon (*Oncorhyncus kisutch*). Aquaculture 16:31–38.

Yufera, M., and M. J. Darias. 2007. The onset of exogenous feeding in marine fish larvae. Aquaculture 268:53–63.

Zheng, F., T. Takeuchi, K. Yoseda, M. Kobayashi, J. Hirokawa, and T. Watanabe. 1996. Requirement of larval cod for arachidonic acid, eicosapentaenoic acid, and docosahexaenoic acid using by their enriched *Artemia* nauplii. Nippon Suisan Gakk. 62:669–676.

Zheng, X., I. Seiliez, N. Hastings, D. R. Tocher, S. Panserat, C. A. Dickson, P. Bergot, and A. J. Teale. 2004a. Characterization and comparison of fatty acyl Δ6 desaturase cDNAs from freshwater and marine teleost fish species. Comp. Biochem. Phys. B 139:269–279.

Zheng, X., D. R. Tocher, C. A. Dickson, J. G. Bell, and A. J. Teale. 2004b. Effects of diets containing vegetable oil on expression of genes involved in highly unsaturated fatty acid biosynthesis in liver of Atlantic salmon (*Salmo salar*). Aquaculture 236:467–483.

Zheng, X., D. R. Tocher, C. A. Dickson, J. G. Bell, and A. J. Teale. 2005a. Highly unsaturated fatty acid synthesis in vertebrates: New insights with the cloning and characterization of a Δ6 desaturase of Atlantic salmon. Lipids 40:13–24.

Zheng, X., B. E. Torstensen, D. R. Tocher, J. R. Dick, R. J. Henderson, and J. G. Bell. 2005b. Environmental and dietary influences on highly unsaturated fatty acid biosynthesis and expression of fatty acyl desaturase and elongase genes in liver of Atlantic salmon (*Salmo salar*). Biochim. Biophys. Acta 1734:13–24.

Zheng, X., Z. Ding, Y. Xu, O. Monroig, S. Morais, and D. R. Tocher. 2009a. Physiological roles of fatty acyl desaturase and elongase in marine fish: Characterisation of cDNAs of fatty acyl Δ6 desaturase and Elovl5 elongase of cobia (*Rachycentron canadum*). Aquaculture 290:122–131.

Zheng, X., M. J. Leaver, and D. R. Tocher. 2009b. Regulation of fatty acyl desaturase (FAD) gene transcription: Isolation and characterisation of Δ6 FAD gene promoters of Atlantic salmon (*Salmo salar* L.) and Atlantic cod (*Gadus morhua* L.). Comp. Biochem. Phys. B 154:255–263.

7

Carbohydrates and Fiber

It is generally accepted that fish and shrimp do not have a specific requirement for dietary carbohydrates. Regardless of species, they are able to survive and grow when fed diets without carbohydrates. This is probably because glucose is synthesized efficiently from nonglucose precursors through gluconeogenesis, especially amino acids, which are the major substrates for gluconeogenesis (Walton and Cowey, 1982; Suarez and Mommsen, 1987; Cowey and Walton, 1989). Glucose may also be derived from the mobilization of liver and muscle glycogen stores through glycogenolysis (Suarez and Mommsen, 1987).

Carbohydrates, which include low-molecular-weight sugars, starch, and various cell wall and storage nonstarch polysaccharides (NSP), provide the major part of energy in the diets for human and most farmed animals. They are the least expensive dietary energy sources. The oxidative pathways use the glucose arising from carbohydrate digestion for the production of ATP (adenosine triphosphate) required to meet cellular energy requirements. Glucose is also a substrate for the pentose pathway and the production of NADPH (nicotinamide adenine dinucleotide phosphate) required for lipid biosynthesis and ribose 5-phosphate required for nucleotide synthesis. Glucose can be used for glycogenesis that supplies glycogen stores, and for lipogenesis.

Usually, inclusion of dietary carbohydrates in feed for aquatic farmed species is limited compared to feed for poultry and mammals. The digestibility of carbohydrates greatly depends on the sources composition, processing, and level of inclusion, and it is often lower than that of protein and fat sources. In addition, a glucose load (following ingestion of digestible carbohydrates or injection of glucose) results in persistent hyperglycemia. Numerous studies have evaluated the maximum levels of carbohydrates that fish and shrimp can tolerate without physiological disorders and growth impairment, rather than the optimal level for growth. However, when carbohydrates are not provided in the diet, other energy-yielding nutrients, such as protein and lipids, are catabolized to produce energy and the biological compounds usually derived from carbohydrates. Thus, the provision of an appropriate amount of digestible carbohydrates in diets formulated for farmed aquatic species is important to spare the use of lipids and protein as sources of energy. In addition, dietary carbohydrate utilization by fish deserves attention because of efforts to effectively replace fish meal with plant protein sources, which contain significant quantities of different carbohydrates components. The ability of fish to use dietary carbohydrates for growth varies greatly among species and basically corresponds to the feeding habits of the species. The different feeding habits result in differences in activities of digestive carbohydrases and endogenous metabolic enzymes, glucose clearance from the bloodstream and uptake by tissues, and, finally, efficiency of utilization of digestible carbohydrates for growth and protein sparing. Within species, the main factors that affect carbohydrates' utilization as glucose and energy sources are their composition, molecular complexity, physical state, and the amount included in the diet.

This chapter provides an overview of the different categories of carbohydrates in feed ingredients for aquatic species and the structure that determines their physicochemical properties, which in turn affects digestibility in fish and shrimp. The metabolic routes of glucose and their nutritional regulation in aquatic species are described. Finally, the main conclusions drawn from feeding trials with different types of carbohydrates in fish and shrimp species are summarized.

TYPES OF CARBOHYDRATES

The term "carbohydrates" refers to a large number of molecules composed primarily of carbon, oxygen, and hydrogen that are polyalcohols with one or several aldehyde or ketone functions. This group of molecules is quite complex in terms of sugar units, linkage between units, and branching. The composition and molecular structure of carbohydrates determine their susceptibility to enzymatic degradation or bacterial fermentation, and their effect on animal physiology.

Carbohydrates can be classified according to their degree of polymerization (DP) as: sugars (mono- and disaccharides containing 1 or 2 molecules), oligosaccharides (3 to 9 monosaccharide units), and polysaccharides (≥10 or more monosaccharide units) (Table 7-1).

The monosaccharides contain a single unit of hydroxyaldehyde or hydroxyketone. They are characterized according to their length of carbon chain as a triose, tetrose, pentose, or hexose when they contain three, four, five, or six carbon atoms. The crystalline forms of pentoses and hexoses are cyclic compounds (Figure 7-1) that exist as both α- and β-anomers. In the α-anomer, one of the -OH groups is perpendicular to the plane of the ring (Figure 7-2). As a result, the α-anomer is less stable than the β-anomer.

Disaccharides are formed by condensing two monosaccharides, and oligosaccharides are formed by the condensation of three or more monosaccharides. Oligosaccharides are subclassified into trisaccharides, tetrasaccharides, pentasaccharides, and so on (Table 7-1), according to the number of molecules they contain. Oligosaccharides are not widely represented, except for a series of galactosylsucroses (often designated as α-galactosides) and fructo-oligosaccharides. The galactosylsucrose family of oligosaccharides includes raffinose (a trisaccharide), stachyose (a tetrasaccharide),

TABLE 7-1 Carbohydrate Categories

Classes	Examples
Monosaccharides	
Trioses	Glyceraldehyde, dihydroxyacetone
Tetroses	Erythrose
Pentoses	Ribose, arabinose, xylose
Hexoses	Glucose, galactose, mannose, fructose
Disaccharides	
Lactose	glucose + galactose
Maltose	glucose + glucose
Saccharose or sucrose	glucose + fructose
Oligosaccharides	
Trisaccharides	Raffinose (galactose + glucose + fructose)
Tetrasaccharides	Stachyose (galactose + galactose + glucose + fructose)
Pentasaccharides	Verbascose (galactose + galactose + galactose + glucose + fructose)
Polysaccharides	
Homopolysaccharides	Hexosans: starch, cellulose, β-glucans, laminarin
	Fructans
	Galactans
	Mannans
Heteropolysaccharides	Hemicelluloses, pectins, guar-gums
	Chitin (polymer of N-acetyl-glucosamine)

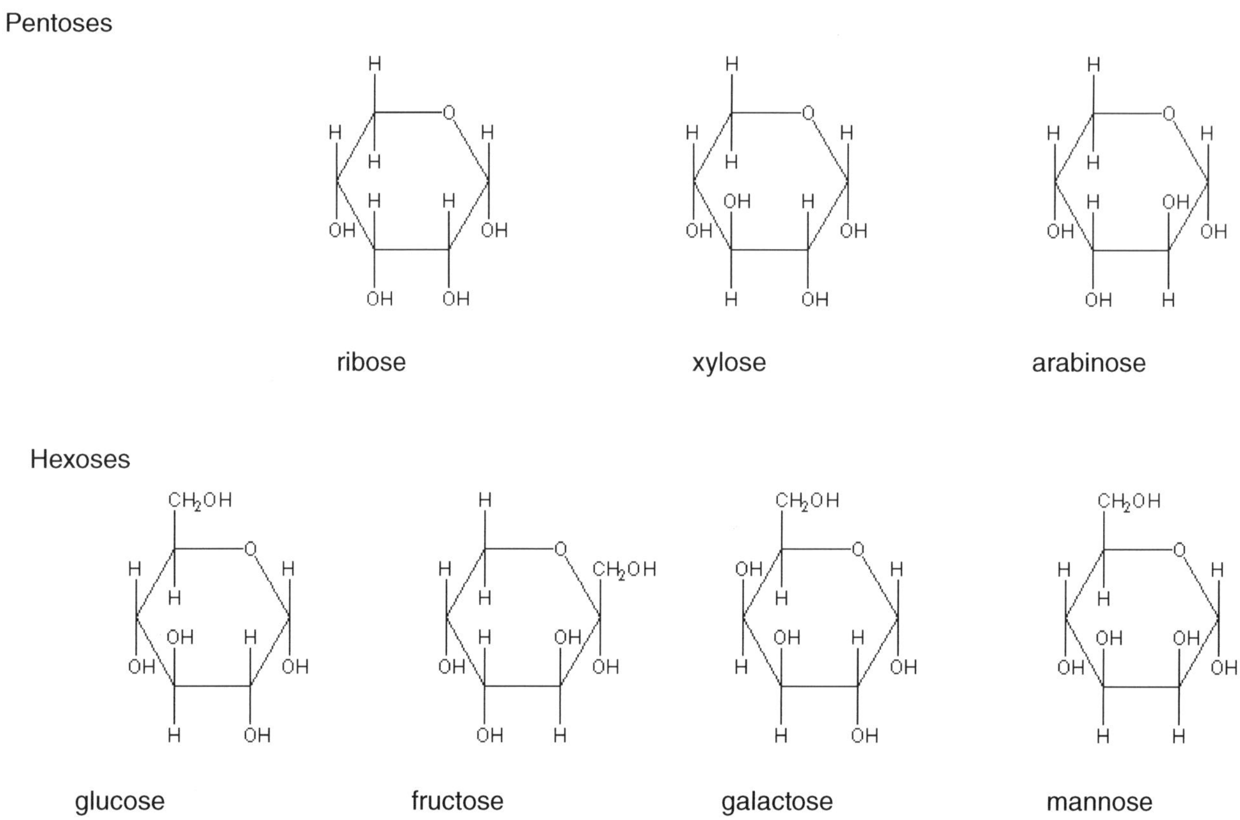

FIGURE 7-1 Structure of the main pentoses and hexoses.

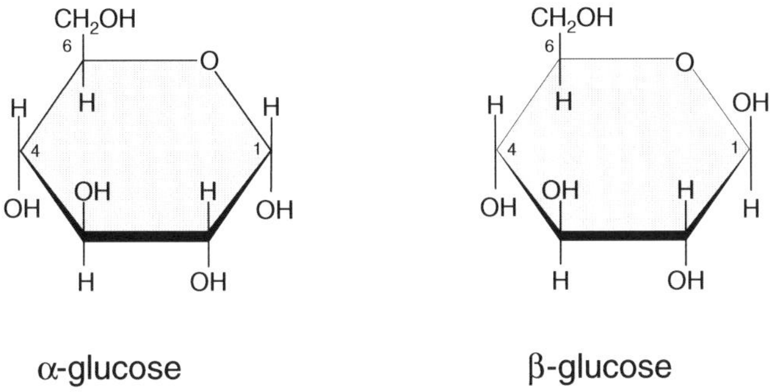

FIGURE 7-2 The α- and β-anomers of glucose.

and verbascose (a pentasaccharide) that are not digested by gastrointestinal enzymes. In pulses, such as peas, beans, and lupins, the content of these oligosaccharides can range from 5–8% on a dry matter basis (Asp, 1995). Fructo-oligosaccharides (FOS), composed of fructose residues linked by β(2→1) bonds, are present in wheat, rye, triticale, and a number of other plants. Because of the configuration of their osidic bonds, FOS are resistant to hydrolysis by intestinal digestive enzymes. They are sometimes included in fish and shrimp diets as prebiotics (see Chapter 10, "Feed Additives").

Polysaccharides comprise a heterogeneous group of compounds made up of large aggregates of monosaccharide units, joined through glycosidic bonds. When all the monosaccharide units are the same type, the polysaccharide is called a homopolysaccharide; when more than one type of monosaccharide is present, they are called heteropolysaccharides (Table 7-1). The polysaccharides are the predominant carbohydrates present in nature and serve two principal functions: they store saccharides as sources of future energy and they provide some of the physical structure of cells.

Starch is the major polysaccharide stored as an energy reserve in many plants. Starch accumulates in the endosperm of cereals (e.g., wheat, corn, and barley) and in tubers such as potatoes, whereas starch levels are very low in other land plants such as oilseeds and pulses and in seaweeds (Table 7-2).

Starch is an α-glycan, a homopolysaccharide composed exclusively of glucose units linked by α-glycosidic bonds that can be hydrolyzed by α-amylase. Starch molecules are arranged in the form of semicrystalline granules, the size and composition of which determine the susceptibility of starch to enzymatic digestion (Buleon et al., 1998).

Starch granule size varies according to its botanical origin (Table 7-3) and is approximately 3–10 μm for oats starch, 8 μm for rice starch, 22 μm for wheat starch, 35 μm for corn (maize) starch, and 40–100 μm for potato starch. In rye, wheat, maize, and triticale, a bimodal distribution in granule size has been reported with the coexistence of large and small granules. Size of the starch granule affects digestibility because surface area, and thus contact between substrate and enzyme, decreases as size of the granule increases (Zobel, 1988).

Starch granules consist of a mixture of two polymers: amylose, the water-soluble inner portion of the granule, and amylopectin, the nearly insoluble substance of the outer part of starch granules. Amylose (Figure 7-3) is a linear chain of glucose units linked by α(1→4) glycosidic bonds, whereas amylopectin (Figure 7-4) is a highly branched polymer of glucose. The chains of glucose units that are linked in a linear way by α(1→4) glycosidic bonds are branched with α(1→6) bonds every 24 to 30 glucose units. The proportions of amylose and amylopectin in starches of different origin influence the digestion rate, with amylopectin being more susceptible to enzymatic cleavage than amylose (Zhou and Kaplan, 1997; Svihus et al., 2005). The enzyme α-amylase attacks one end of the polymer chain and cleaves glucose molecules. The more branches there are, the more points at which the enzyme can attack the polysaccharide. Thus, a highly branched polysaccharide is better suited for the rapid release of glucose than a linear polymer. The relative proportions of amylose and amylopectin vary with plant species and cultivar. Most cereal starches contain 720–790 g/kg of amylopectin and 210–280 g/kg of amylose (Table 7-3). However, genotypes exist with amylose contents < 10 g/kg (waxy types) as well as with amylose content as high as 700 g/kg (amylose-rich types) (Cummings and Englyst, 1995; Gaylord et al., 2009).

Starch is insoluble in cold water but swells on heating in excess water. Water penetrates the outer layers of the granules and the granules swell until they disrupt. Amylose and portions of the amylopectin leach from the granule, producing a viscous suspension. The whole process of disruption and swelling is known as gelatinization (Zeng et al., 1997). The gelatinization temperature ranges for some selected starches sources are presented in Table 7-3, together with the starch content, mean granule size, and amylose proportion.

Starch content can be measured by acid or enzyme hydrolysis on residues that have been previously extracted with ethanol to remove soluble sugars (Smith, 1981). Enzymatic hydrolysis is preferred because of its specificity. The enzyme α-amylase hydrolyzes the starches to glucose. Glucose concentrations are then converted into equivalent starch concentrations. Determination of starch in plants is described by the Method 948.02 (AOAC, 2002b) and in animal feeds by the Method 920.40 (AOAC, 2002a).

The other polysaccharides present in plants (NSP) are predominantly structural components of cell walls. The NSP comprise cellulose and β-glucans (polymers of glucose with β-linkages), hemicellulose (various sugars linked by β-bonds), and pectins and gums (α- and β-galactosides respectively). The NSP form the major part of the "dietary fiber" group (Figure 7-5), which is defined as "the dietary components resistant to degradation by mammalian enzymes" (Bach Knudsen, 2001). They can be classified according to their water-binding capacity into two categories: viscous or water-soluble polysaccharides, and nonviscous

TABLE 7-2 Carbohydrates (g/kg dry matter) in Selected Ingredients Used for Fish and Shrimp Diets

Carbohydrate	Maize (Whole grain)	Maize (Feed meal)	Maize (Flour)	Maize (Bran)	Wheat (Whole grain)	Wheat (Flour)	Wheat (Bran)	Rye (Whole grain)	Rye (Bran)	Barley (Hulled)	Barley (Dehulled)	Oats (Hulled)	Oats (Feed meal)
Total sugars	20	40	10	32	19	17	53	32	75	21	15	17	17
Monosaccharides	4	6	4	5	3	1	7	6	6	4	2	2	2
Sucrose	1.3	29	6	21	11	8	30	19	39	12	7	11	12
Raffinose	2	4	1	4	4	3	12	4	24	5	4	3	2
Stachyose	1	1	0	1	2	3	4	3	6	1	1	2	2
Verbascose	0	0	0	0	0	0	0	0	0	0	0	0	0
Starch	690	566	902	376	651	820	222	613	87	587	654	468	623
Fructan	6	2	2	4	15	16	20	31	23	4	5	3	2
Total dietary fiber	108	174	25	379	138	35	449	174	490	221	146	298	108
Total NSP[a]	97	156	21	354	119	35	374	152	422	186	127	232	89
β-Glucan	1	1	1	2	8	4	24	16	45	42	44	28	42
Soluble NCP[b]	9	10	8	32	25	16	29	42	63	56	50	40	42
Rhamnose	0	0	0	0	0	0	0	0	0	0	0	0	0
Arabinose	3	1	3	6	7	8	7	12	11	6	4	3	2
Xylose	2	2	3	5	9	7	10	20	33	6	7	2	1
Mannose	2	1	1	1	2	0	1	2	1	2	1	2	2
Galactose	1	1	1	2	2	2	2	1	2	1	2	2	2
Glucose	1	0	0	6	4	2	8	6	13	39	34	28	33
Uronic acids	1	4	1	12	1	2	2	1	2	2	1	3	2
Insoluble NCP	66	114	13	240	74	17	273	94	321	88	58	110	39
Rhamnose	0	0	0	0	0	0	0	0	0	0	0	0	0
Arabinose	19	32	3	66	22	6	83	24	67	22	17	15	8
Xylose	28	46	3	111	38	8	138	41	180	50	29	78	15
Mannose	1	1	0	3	1	1	4	3	2	2	2	1	1
Galactose	4	8	0	18	2	0	7	4	10	2	0	5	1
Glucose	9	10	5	10	7	4	27	20	53	8	8	5	10
Uronic acids[c]	6	16	2	32	4	0	13	3	8	4	2	7	3
Cellulose	22	33	0	83	20	3	72	16	39	43	19	82	8

Carbohydrate	Cottonseed (Cake)	Cottonseed (Meal)	Linseed (Meal)	Coconut (Cake)	Palm (Cake)	Sunflower (Cake)	Sunflower (Cake, partly dehulled)	Soybean (Meal)	Rapeseed (Meal)	Rapeseed (Cake)	Peas	White Lupin
Total sugars	66	69		124	24	58	56	137	82	90	88	104
Monosaccharides	2	3	3	7	5	5	5	7	8	4	9	3
Sucrose	10	16	28	113	17	36	33	70	58	68	30	24
Raffinose	39	35	10	2	2	14	14	10	4	3	5	10
Stachyose	14	13	2	2	0	3	5	47	12	13	23	53
Verbascose	1	2	0	0	0	0	0	3	0	0	22	14
Starch	18	19	27	10	11	10	17	27	18	15	454	14
Fructan	0	0	0	0	0	0	0	—	—	—	—	—
Total dietary fiber	340	375	423	488	602	448	326	233	354	295	192	416
Total NSP[a]	257	283	303	422	466	315	240	217	220	205	180	405
β-Glucan	0	0	0	0	0	0	0	—	—	—	—	—
Soluble NCP[b]	61	66	138	32	32	57	52	63	55	43	52	144
Rhamnose	1	0	6	1	0	2	1	1	1	1	1	2
Arabinose	16	11	17	5	3	8	9	9	12	13	19	19
Xylose	6	17	38	0	0	4	4	2	4	2	1	0
Mannose	1	1	1	12	16	1	1	5	1	1	1	4
Galactose	7	6	21	7	3	5	5	16	6	5	4	80
Glucose	6	14	10	3	3	5	5	6	9	3	5	1
Uronic acids	23	18	45	4	1	34	27	23	22	18	20	27
Insoluble NCP	103	127	112	336	361	136	99	92	123	103	76	139
Rhamnose	1	1	1	0	0	2	1	2	2	2	0	1
Arabinose	18	23	19	9	9	23	17	17	31	31	17	24
Xylose	54	68	28	8	31	55	38	17	13	15	12	36
Mannose	3	3	3	294	293	11	9	8	5	4	1	8
Galactose	5	6	12	18	12	8	7	25	13	15	3	64
Glucose	2	0	27	1	4	12	5	1	12	5	31	1
Uronic acids[c]	22	27	23	6	12	33	26	23	39	32	12	12
Cellulose	92	90	53	54	73	123	89	62	52	59	53	131

[a]NSP: nonstarch polysaccharides.
[b]NCP: noncellulosic polysaccharides.
[c]Uronic acids = pectins.
SOURCE: Adapted from Bach Knudsen (1997).

continued

TABLE 7-2 Continued

Carbohydrate	Norwegian Kelp *Ascophyllum nodosum*	Oarweed *Laminaria digitata*	Giant Kelp *Macrocystis pyrifera*	Gulfweeds *Sargassum* spp.
Total carbohydrates	500–760	500–650	500–750	500–680
β-Glucan	0–10	0–180	—	—
Laminaran	—	—	—	—
Mannitol	50–100	40–160	70–80	—
Alginic acid	150–300	200–450	200–400	—
Fucoidan	40–100	20–40	—	200
Uronic acid (pectin)	—	—	—	460
Cellulose	80	80	ND	ND

ND = not determined.
SOURCE: Adapted from Guiry and Blunden (1991) and Ortiz et al. (2009).

TABLE 7-3 Starch Content[a] (% Dry Matter) and Characteristics of Starches of Some Selected Ingredients for Fish and Shrimp Diets

Source	Starch Content (% dry matter)	Gelatinization Temperature (°C)	Granule Size (μm)	Amylose Content (%)
Barley	58.7	51–60	20–25	22
Maize	69	63–72	35–40	21–28
Waxy maize	—	63–72	20	1
High amylose maize	—	67–80	—	70
Potato	73	59–68	40 (15–100)	20–23
Rice	88	—	—	17–22
Waxy rice	—	68–78	8	0
Rye	61	57–70	28 (12–40)	27
Sorghum	69.5	68–78	25 (15–35)	23–28
Waxy sorghum	—	—	—	0
Wheat	65–68	58–64	22 (2–26)	26 (23–27)

[a]Mean value, concentration is dependant on culture location and varies from year to year—cereals as whole grain. Mean values from Zobel (1988), Buleon et al. (1998), Svihus et al. (2005), Zeng et al. (2008), and Hirose et al. (2010).

or water-insoluble polysaccharides. Soluble NSP (pectins, gums, and some hemicelluloses) disperse when mixed with water and have the ability to increase the viscosity of digesta. Insoluble NSP (cellulose and most of the hemicelluloses) increase transit rate and are characterized by their fecal bulking capacity (Davidson and McDonald, 1998).

Cellulose is the most abundant cell wall structural polysaccharide in plants. It is a linear polymer of glucose. The glucose units are linked by β-linkages (Figure 7-6), unlike in starch where the linkage is of α-type. Many animals (including humans and fish) lack the enzymes to break the β-linkages, so they are unable to digest cellulose. Ruminant animals can digest cellulose, albeit to a limited extent, because of the presence in their gut of bacteria possessing the adequate enzyme. Cellulose can be used as a bulking agent in feed for aquatic species due to its water-absorbing ability and very low solubility.

The other polysaccharides made of glucose units linked by different β-bonds (1→3 and 1→6 in addition to 1→4 bonds) are referred to as β-glucans or mixed linked β-glucans. Unlike cellulose, they are water-soluble. Some of them have immunostimulant properties (see Chapter 10). The β-glucans

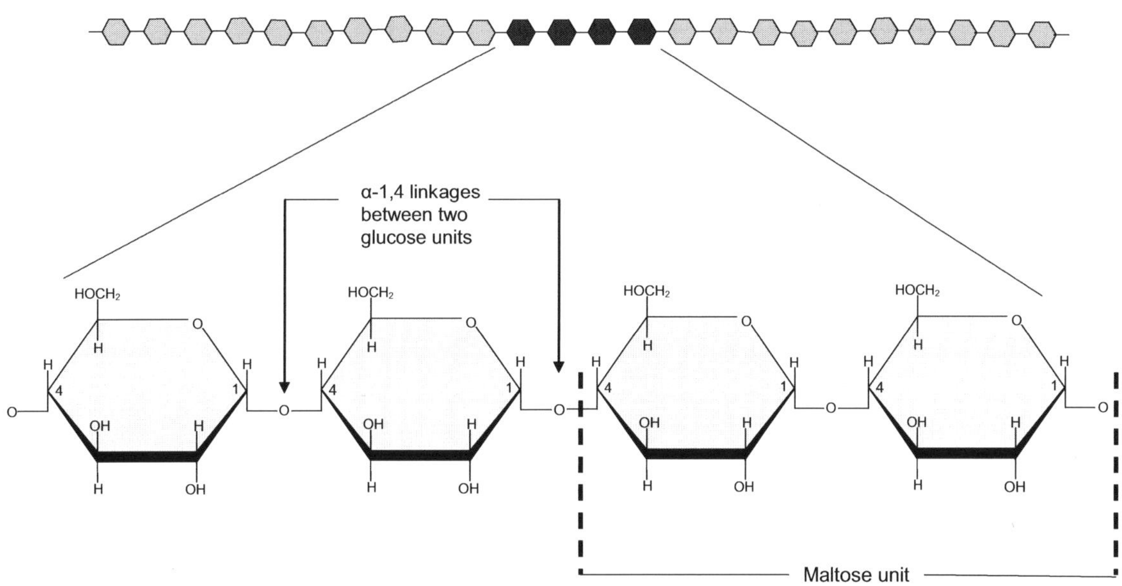

FIGURE 7-3 Structure of amylose.

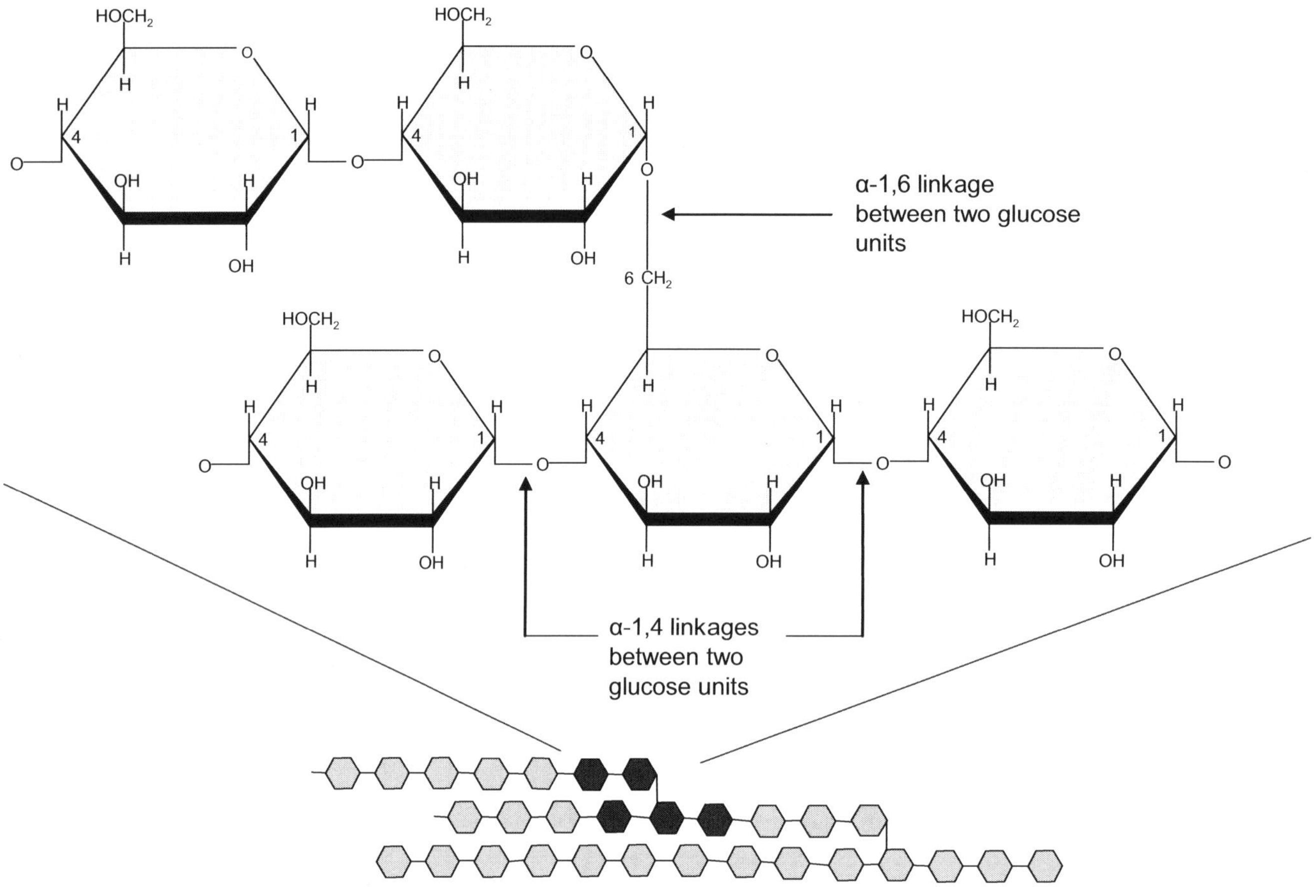

FIGURE 7-4 Structure of amylopectin.

are the major constituents of the endosperm cell walls of oats and barley. Laminarins, low molecular β(1→3), β(1→6) glucans are reserve materials of brown seaweeds (Table 7-2).

Other cell wall polysaccharides present in almost all plants along with cellulose belong to the family of hemicellulose. In contrast to cellulose that contains only glucose, hemicelluloses contain many different pentoses and hexoses linked by β-bonds. Sugar monomers in hemicellulose can include glucose, xylose, mannose, galactose, rhamnose, and arabinose, with xylose always the sugar monomer present in the largest amount. Unlike cellulose, hemicellulose consists of shorter chains (500–3,000 sugar units per polymer in hemicellulose vs. 7,000–15,000 glucose molecules per polymer in cellulose). In addition, hemicellulose is a branched polymer, whereas cellulose is unbranched. Hemicellulose exists in both soluble and insoluble forms. Arabinoxylans are the major NSP in maize, wheat, and rye. Arabinogalactans and galactomannans are storage polysaccharides in coconut and legume seeds such as soybean, rapeseed, and lupin, where they substitute for starch as the primary energy source for germination (Table 7-2).

Pectins (uronic acids) and guar gums contain galactose and some sugars other than glucose. They are water soluble. Guar gum is composed of galactose and mannose. The backbone is a linear chain of β(1→4)-linked mannose residues to which galactose residues are 1→6-linked at every second mannose, forming short side-branches. In the pectins family, galactose residues are linked by α(1→4) bonds. The most common pectins are homogalacturonans, linear chains of α(1→4)-linked galactose residues; substituted galacturonans that are characterized by the presence of saccharide-branched residues (such as xylose in the case of xylogalacturonan); and rhamnogalacturonans that contain a backbone of galactose and rhamnose. From many of the rhamnose residues, side-chains of various neutral sugars are branched. The types and proportions of the neutral sugars vary with the origin of pectin; they are mainly galactose, arabinose, and xylose (Figure 7-1). One of the most common gums found in seaweeds is the alginic acid, a linear polymer consisting mainly of β(1→4)-linked mannuronic acid and α(1→4)-linked glucuronic acid (Table 7-2).

Analytical methods of dietary fiber components have been primarily developed for ruminant animals and for humans. The conventional methods led to the categorization of carbohydrates on the basis of the fractions that they measured (Figure 7-6):

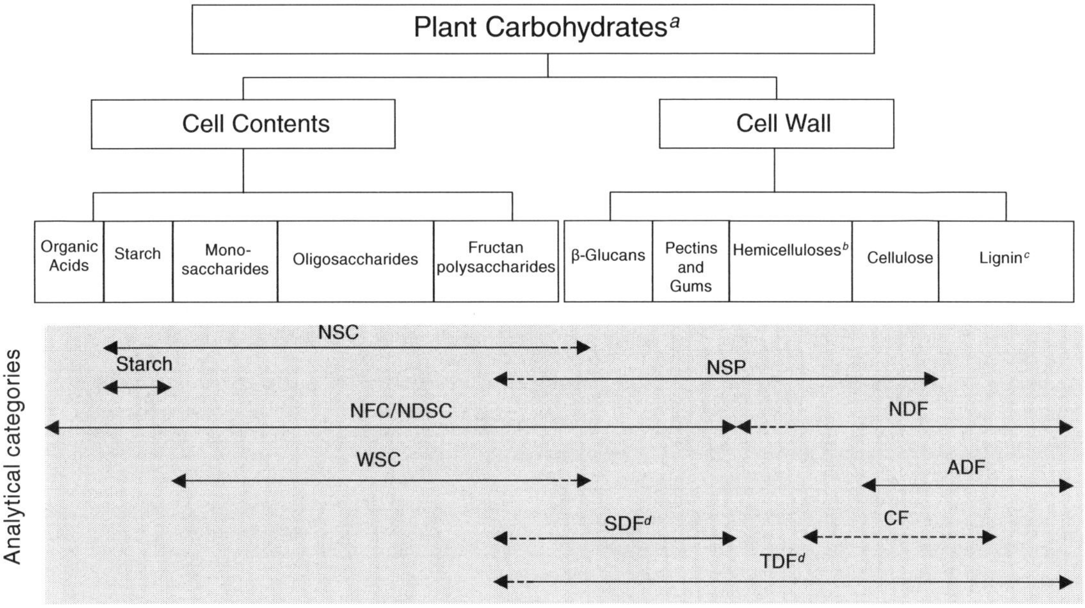

FIGURE 7-5 Categories of dietary carbohydrates based on current analytical methods.
ABBREVIATIONS: ADF = acid detergent fiber; CF = crude fiber; NDF = neutral detergent fiber; NDSC = neutral detergent soluble carbohydrates; NFC = nonfiber carbohydrates; NSC = nonstructural carbohydrates; NSP = non-starch polysaccharides; SDF = soluble dietary fiber; TDF = total dietary fiber; WSC = water-soluble carbohydrates. Dashed lines indicate that recovery of included compounds may be incomplete.

[a]These categories may not include all carbohydrates produced by plants.

[b]Some hemicellulose may be soluble in neutral detergent and thus recovered in the NFC/NDSC fraction, rather than the NDF fraction.

[c]Some noncarbohydrate components are included here because they are components of the specific analytical fractions.

[d]From a nutritional perspective, TDF includes all carbohydrates resistant to mammalian digestion. However, the analytical method for TDF (and SDF) does not recover oligosaccharides and may recover a variable amount of fructan polysaccharides.

- Crude fiber (CF) is the sum of cellulose, insoluble hemicellulose, and lignin (even though lignin is not actually a carbohydrate). Lignin is a heterogeneous complex that fills the spaces in the plant cell wall between cellulose, hemicellulose, and pectins and is crosslinked with the other plant polysaccharides.
- Neutral detergent fiber (NDF) is the sum of lignin, cellulose, and neutral detergent-insoluble hemicellulose.
- Acid detergent fiber (ADF) is the sum of lignin, cellulose, and acid-insoluble hemicellulose.

To analyze crude fiber, samples are treated with acid and then with alkali to mimic digestion by gastric secretions (Method 962.09; AOAC, 2002c). This procedure is still being used by regulatory agencies despite the errors in the determination of the structural and nonstructural carbohydrates. The

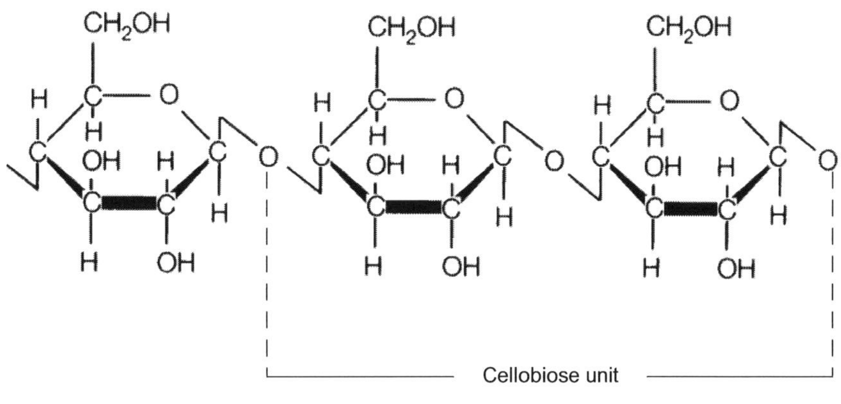

FIGURE 7-6 Structure of cellulose.

most significant error is because of the solubilization and loss of part of the lignin, cellulose, and hemicellulose (Van Soest, 1994). The Van Soest detergent system (Van Soest et al., 1991) is based on an aqueous treatment followed by recovery of an insoluble residue. The main disadvantage is the loss of water-soluble NSP fraction. The original NDF method has been modified: NDF is amylase-treated (aNDF) to minimize the contamination by starch, and sodium sulfite is used to remove contaminating nitrogen from NDF (Mertens, 2002). The ADF fraction is still analyzed according to the method proposed by Van Soest and Wine (1967). Hemicelluloses can be estimated by subtracting ADF from NDF (Robertson and Van Soest, 1981). To reduce errors, a sequential analysis for NDF and then ADF on the same sample is recommended.

Since the early 1990s, some enzymatic-gravimetric methods and enzymatic-chemical methods have been developed to measure both the soluble and insoluble fractions of fibers. Total dietary fibers (TDF; Figure 7-6) are measured by using three enzymes (heat stable α-amylase, protease, and amyloglucosidase) incubated in a phosphate buffer. The procedure, detailed in AOAC Method 991.43, section 32.1.17 (AOAC, 1995), involves extraction of low-molecular-weight sugars and lipids and enzymatic degradation of protein and starch. Soluble dietary fiber (SDF, the sum of β-glucans, pectins, gums, and the soluble forms of hemicellulose) can be estimated by subtracting NDF from TDF (Baer et al., 1997) or by the AOAC method (Method 991.43; AOAC, 2002d).

The NSP can also be measured by the enzymatic chemical method known as the Englyst method, which includes enzymatic removal of starch, precipitation of NSP, followed by acid hydrolysis and measurement of the released constituent sugars. In the initial protocol, the sugars were assayed by colorimetry; in the more recent procedures, they are quantified by gas-liquid chromatography or high-performance liquid chromatography (Englyst et al., 1994). Uronic acids (pectins) are analyzed by colorimetry. The NSP in the selected ingredients for feed for fish and shrimp shown in Table 7-2 have been quantified using the enzymatic chemical method (Bach Knudsen, 1997).

Besides plant carbohydrates, the diet of aquatic species may contain chitin, a long-chain polymer of N-acetylglucosamine linked by β(1→4) bonds (Figure 7-7), which is the main component of the exoskeletons of crustaceans and insects and the beaks of cephalopods such as squid and octopus. Chitin is also the main component of the cell wall of fungi. Its breakdown requires a series of three specific chitinolytic enzymes (chitinase, chitobiase, and lyzozyme).

NONSTARCH POLYSACCHARIDES IN FISH AND SHRIMP DIETS: PHYSIOLOGICAL CONSEQUENCES

Although some types of starch can be well digested (see below), there are few NSP that succumb to the digestion processes in fish and shrimp (Kraugerud et al., 2007; Ovrum-Hansen and Storebakken, 2007). Chemical composition and chemical bonds of carbohydrates explain the differences in digestibility among different types of polysaccharides. The α-glycosidic bonds of starch can be hydrolyzed by α-amylase that is present in the digestive tract of most fish species. In contrast, the presence of digestive enzymes that specifically hydrolyze the β-glycosidic bonds of NSP such as β-amylase, β-glucanases, β-mannases, and β-xylanases seems to be very low or nonexistent in fish, including herbivorous species (Rust, 2002). Fish also lack the enzyme α-galactosidase required to hydrolyze galactose polymers (Glencross et al., 2003). Cellulase activity and cellulolytic microflora have been detected in some warmwater fish species such as Indian major carps and some marine-species-eating invertebrates (Lindsay and Harris, 1980). Although Stickney and Shumway (1974) reported that cellulase activity of freshwater catfish resulted from intestinal microflora of the fish, it has been suggested that, in most fish species, cellulase is not endogenous. Endogenous cellulase activity has, however, been detected in the gastric fluid and the digestive tract of a range of crustaceans including prawn (*Macrobrachium rosenbergii*) and crayfish (*Cherax quadricarinatus*) (Crawford et al., 2005). However, increasing dietary levels of α-cellulase do not promote survival and growth of crayfish and prawns, suggesting that the supply of digestible energy from cellulose is limited despite the presence of cellulose hydrolyzing enzymes (Pavasovic et al., 2006). At the other end of the spectrum, essentially no cellulase activity and associated microflora have been found in salmonids.

Because of their limited digestibility, NSP play a minor role as carbohydrates or energy sources in fish and shrimp nutrition. Some of them are added into aquaculture feed in small quantities as prebiotics or as binders that stabilize food pellets or feces (see Chapter 10). At higher dietary levels, NSP may have detrimental effects resulting from their physicochemical properties that include viscosity, water-holding capacity, and capacity to bind organic and inorganic molecules. Rather than supplying nutrients, they may reduce the utilization of other nutrients by acting as antinutritional factors in carnivorous as well as herbivorous species. Because some NSP can be introduced into diets with plant products used as alternatives to fish meal (Francis et al., 2001; Gatlin et al., 2007), there have been increasing concerns in recent years about NSP in feed for fish and shrimp. The reports have mainly focused on the effects of soluble and insoluble NSP on nutrient digestibility and growth of fish and shrimp.

Shiau et al. (1989) studied the effects of four soluble fibers (guar gum, agar, carrageenan, and carboxymethylcellulose [CMC]) and one insoluble fiber (cellulose) on the utilization of dextrin by tilapia. The diets contained 30% dextrin as the carbohydrate source and 10% of one fiber type, except for the fiber-free diet. The weight gain was significantly lower for tilapia fed fiber-containing diets than for those fed the fiber-free diet. This is likely caused by the reduced digestible energy (DE) of diets containing fiber. The intestinal absorption of glucose released from dextrin digestion was lower in

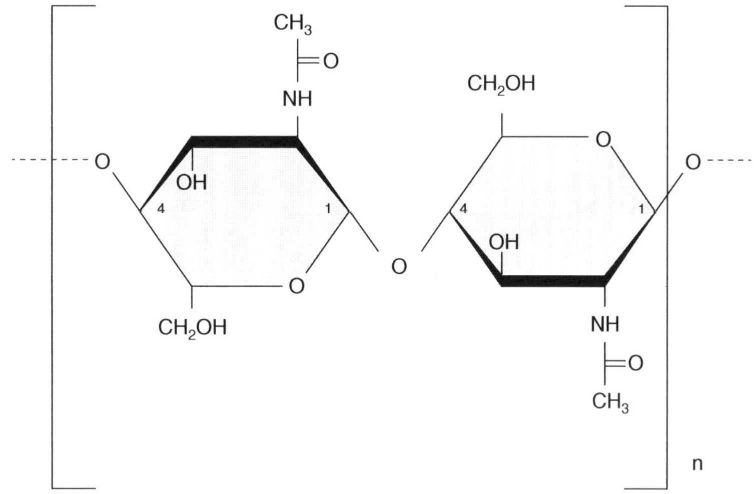

FIGURE 7-7 Structure of a chitosan unit that composes chitin.

fish fed diets that contained fiber, regardless of source. The lower rate of carbohydrate absorption in hybrid tilapia fed fiber-containing diets suggests that dextrin digestion or glucose absorption was delayed because of the presence of fiber. This effect of dietary fiber on the digestibility of the other dietary components was confirmed in several other studies with tilapia (Amirkolaie et al., 2005), common carp (Hossain et al., 2001), Atlantic salmon (Refstie et al., 1999), rainbow trout (Glencross et al., 2003; Ovrum-Hansen and Storebakken, 2007), European sea bass (Dias et al., 1998; Leenhouwers et al., 2004), African catfish (Leenhouwers et al., 2006), and crayfish (Pavasovic et al., 2006). Both soluble (guar gum, pectins, and oligosaccharides) and insoluble (cellulose) NSP, at levels higher than 10%, had adverse effects on digestibility of organic matter and energy of the diets. In almost all studies, there was little or no effect on the digestibility of protein by the inclusion of insoluble fiber, but protein digestibility was reduced by the inclusion of soluble fiber. Water-soluble dietary fibers such as guar gum and pectin have been reported to delay the rate of stomach-emptying in terrestrial animals. The delay has been attributed to an increased viscosity of the diet. Whether the adverse effects induced by soluble fibers in fish result from changes in the gastric emptying rate has not been demonstrated. Leenhouwers et al. (2006) found increased intestinal viscosity in African catfish when adding 4–8% guar gum to the diets. They suggested that the resulting reduced digestibility was related to changes in distribution of digestive enzymes in the viscous solution and a lowered intestinal flow rate. The adverse effects on lipid digestibility may also be related to the ability of fiber to bind dietary lipids, thereby reducing intestinal lipid absorption.

Because shrimp are known to sometimes consume their molt, it has been thought that addition of chitin, the major structural component of the exoskeleton of shrimp, to their diets could have beneficial effects. Different responses were observed among shrimp species, depending on the form of dietary supply. A dietary level of 5% chitin enhanced tiger shrimp (*P. monodon*) growth, whereas chitosan depressed shrimp growth, regardless of the level of supplementation (Shiau and Yu, 1998). Deshimaru and Kuroki (1974) observed that the presence of glucosamine in the diet inhibited the growth-promoting effect of dietary cholesterol in Kuruma shrimp (*M. japonicus*) juveniles. Some trials were also performed with fish. Addition of either chitin or chitosan to a diet fed to tilapia or trout depressed nutrient digestibility and growth, regardless of the level (Shiau, 1997; Shiau and Yu, 1999). On the other hand, presence of endogenous chitinolytic enzymes, i.e., active in presence of antibiotics, has been recently demonstrated in the digestive tract of juveniles cobia. ADC values of chitin ranged from 67–78% in cobia fed diets containing 3 or 10% chitin provided by shrimp meal or crab meal (Fines and Holt, 2010).

Overall, studies evaluating the use of NSP (oligosaccharides, cellulose, CMC, and guar gum) for fish and shrimp have reported a reduction in growth when the dietary NSP level exceeded 10% (Hilton et al., 1983; Shiau et al., 1988; Dioundick and Stom, 1990). Some attempts have been made to improve the nutritional value of NSP. Treatment of dietary material with carbohydrases specific for the NSP they contain may enhance energy digestibility by enabling the release of previously unavailable glucose, galactose, and xylose. However, total depolymerization of NSP requires several classes of enzymes, and no convincing results have been obtained so far. Glencross et al. (2003) reported an improvement in protein digestibility in rainbow trout when lupin was treated with α-galactosidase, but the digestibility of nitrogen-free extract was not significantly changed. In addition, investigations into carbohydrate tolerance have shown that fish are not able to tolerate simple sugars such as galactose and xylose that could be released by NSP digestion. Stone (2003) and Stone et al. (2003a) reported that barramundi and silver perch were unable to clear circulating xylose and galactose from the bloodstream 24 hours after an intraperitoneal injection (1 g/kg body weight). According to Shikata et al. (1994), the clearance rate of galactose was lower than that of glucose in carp after an oral administration of 0.5 g/kg body weight. The prolonged galactosemia resulting from feeding a diet containing 30% galactose was accompanied by reduced feed intake, reduced growth rate, and reduced activity of enzymes of intermediary metabolism. Poor utilization of galactose was also reported in salmonids (Stone, 2003). Fermentation of grains has also been performed in an attempt to alter carbohydrate content of NSP. Skrede et al. (2002) reported that lactic fermentation of barley and wheat using a *Lactobacillus* strain reduced by half the content of soluble mixed β-glucans and led to an increase in lipid, starch, and energy digestibility in Atlantic salmon. An improvement in growth and feed efficiency was observed in rohu fingerlings when duckweed leaf meal was prefermented by a cellulose-degrading bacillus. However, Ng et al. (2002) found that prefermentation of palm kernel meal with *Trichoderma koningii* reduced digestibility and growth rates in red hybrid tilapia compared to raw kernel meal.

These results emphasize the need to pay special attention to the NSP component of diets. As aquafeed formulations move toward increased use of plant-based protein sources and processing byproducts, the levels of NSP will increase. In diets containing low or moderate levels of protein, a considerable amount of carbohydrates is included. Hence, a better understanding of chemical characteristics and tolerable levels of NSP is becoming increasingly more important.

DIGESTIBILITY OF STARCH

Precise knowledge about digestibility of starch is required because it determines the amount of glucose provided to fish and shrimp. Digestion is the primary limiting step of efficient utilization of starch for growth. Starch components are hydrolyzed into maltose by α-amylase, and the residues (maltose and short-chain dextrins) are further hydrolyzed by maltase and isomaltase. In combination, these enzymes hydrolyze the α-glycosidic bonds of starch to yield glucose. These enzymes are present in the digestive tract of fish and shrimp, and their levels of activity generally correspond to the trophic types of the species. Low, medium, and high activities have been measured in carnivorous, omnivorous, and herbivorous species, respectively (Stone, 2003). For example, Hidalgo et al. (1999) measured higher amylase activities in carp and goldfish than in tench, seabream, rainbow trout, and eel (see Chapter 3, "Comparative Digestive Physiology").

Within species, four major factors affect starch digestibility: molecular complexity, starch origin, degree of gelatinization, and dietary level (Table 7-4; Krogdahl et al., 2005). Digestibility and absorption rate increase as the molecular complexity of carbohydrate sources decreases. Apparent digestibility coefficient (ADC) values are higher for maltose and dextrins than for starch even in carnivorous species with lower α-amylase activities (Table 7-4; Singh and Nose, 1967; Stone et al., 2003b). The differences in digestibility among starch sources are attributed to the size and amylose/amylopectin content of the granules (see Table 7-3) (i.e., factors that affect the ability to enzymatically break down the starch). Cereals with small to medium granule size such as wheat have greater starch digestibility in salmonids and marine fish and shrimp species than potato with larger starch granules (Table 7-4; Hemre et al., 1990; Bergot, 1993; Cousin et al., 1996). The proportions of amylose and amylopectin affect starch digestibility (Bergot, 1993; Cousin et al., 1996; Gaylord et al., 2009), because the branched amylopectin is more susceptible to enzymatic cleavage than the linear amylose. It is known that native granules are degraded very slowly by amylases and that starch gelatinization markedly increases susceptibility to amylolytic degradation because of loss of crystalline structure. An improvement in digestibility is obtained in many fish and shrimp species when processes that lead to gelatinization of starch, such as cooking, steaming, expansion, and extrusion, are applied to the feedstuffs (Table 7-4). This change is particularly helpful for feeding carnivorous species that have low amylase activities because these processes reduce the work required from the digestive enzymes. Partial gelatinization and subsequent increases in starch digestibility can also be achieved through extrusion of the whole diet.

When the dietary level of starch increases, a reduction in digestibility is observed in almost all fish and shrimp species, irrespective of the starch origin (Table 7-4). According to Stone (2003), the reduction in digestibility of carbohydrates when dietary levels increase may be the result of an overload on the digestive enzymes that become saturated with substrate. The threshold for enzyme saturation by substrate differs among species as suggested by the lack of a significant effect of dietary levels of carbohydrate on digestibility in tilapia (Wang et al., 1985), common carp (Appleford and Anderson, 1996), and Pacific white shrimp *Litopenaeus vannamei* (Cuzon et al., 2004). Because of great variability among fish and shrimp species and among carbohydrate sources, carbohydrate digestibility data are an absolute prerequisite when formulating diets. Hua and Bureau (2009) constructed a model to explain variation in digestibility of starch across studies and to estimate digestible starch content of salmonids feed by integrating data from 26 studies conducted with rainbow trout and Atlantic salmon. The model confirms that the digestibility of raw and gelatinized starches differs significantly and is negatively affected by dietary inclusion levels. The model also revealed a significant positive effect of water temperature on starch digestibility with the effect slightly higher on raw starch than on gelatinized starch. It seems that rainbow trout are able to digest starch more efficiently than are Atlantic salmon. Feces collection method was found to contribute significantly to the variability of starch digestibility reported across studies.

TABLE 7-4 Apparent Digestibility Coefficient of Starch According to the Sources and the Dietary Levels in Different Fish and Shrimp Species

Fish (*Genus species*)	Water Temperature[a]	Fish Weight (g)	Feces Collection Method	Carbohydrate Sources	Diet[b]	Level in the Diet (g/kg)	ADC (%)	Reference
Rainbow trout (*Oncorhynchus mykiss*)	FW 15°C	10–25	Stripping	Potato starch	Moist P	200	69	Singh and Nose (1967)
						300	65	
						400	53	
						500	38	
						600	26	
				Dextrin		200	77	
						300	74	
						400	60	
						500	50	
						600	46	
	FW 17°C	120	Filtration	Raw maize starch	P			Bergot and Breque (1983)
				50% feeding		310	55	
				100% feeding		310	38	
				Cooked maize starch	P			
				50% feeding		270	90	
				100% feeding		270	87	
	FW 17°C	100	Filtration	Potato starch	P	270	5	Bergot (1993)
				Manioc starch		270	19	
				Rice starch		260	43	
				Wheat starch		260	58	
				Maize starch		280	34	
				Amylomaize starch		260	24	
				Waxy maize starch		270	56	
				Gelatinized maize starch		280	96	
				Extruded maize starch		260	96	
	FW 17°C	230	Filtration	Raw wheat starch	P	271	35	Brauge et al. (1994)
				Wheat starch raw + gelatinized (1:1)		320	73	
				Gelatinized wheat starch		369	81	
	FW 8°C	80	Filtration	Raw maize starch	P	300	45	Aguirre et al. (1995)
	FW 18°C					300	63	
	FW 8°C			Gelatinized maize starch		300	79	
	FW 18°C					300	85	
	FW 10°C	100	Stripping	Whole oat	P	140	68	Arnesen and Krogdahl (1995)
				Flaked oat		140	91	
	FW 8–9°C	700	Stripping	Maize dextrin	P	100	38	Storebakken et al. (1998)
						200	48	
	SW 8–9°C					100	38	
						200	44	
	FW 17°C	38	Filtration	Wheat starch raw + gelatinized (3:1)	Ex	280	78	Yamamoto et al., (2001)

continued

TABLE 7-4 Continued

Fish (*Genus species*)	Water Temperature[a]	Fish Weight (g)	Feces Collection Method	Carbohydrate Sources	Diet[b]	Level in the Diet (g/kg)	ADC (%)	Reference
Atlantic salmon (*Salmo salar*)	SW 8°C	145	Stripping	Whole oat	P	70	61	Arnesen et al. (1995)
						90	50	
						120	43	
						140	30	
						160	33	
	SW 7.3°C	1900	Dissection	Precooked maize	Ex	60	80	Krogdahl et al. (1999)
						170	39	
	FW 9°C	200	Stripping	Gelatinized maize	Ex	70	90	Krogdahl et al. (2004)
						230	60	
	SW 9°C					70	83	
						230	56	
	SW 10.2°C	80	Stripping	Wheat starch	Ex	105	88	Grisdale-Helland and Helland (1997)
						145	82	
						161	78	
						213	75	
	SW 7–8°C	940	Stripping	Wheat starch	Ex	120	72	Storebakken et al., (2000)
	SW 10°C	275	Stripping	Potato starch	Ex	80	62	Hillestad et al. (2001)
						120	54	
						180	46	
	SW 8°C	500	Stripping	Rye flour	Ex	120	51	Thodesen and Storebakken (1998)
						15	53	
						130	53	
						160	54	
Cod (*Gadus morhua*)	SW	150	Dissection	Raw wheat meal	P	200	45	Hemre et al. (1990)
				Cooked wheat meal		200	59	
				Extruded wheat		200	62	
				Extruded potato		200	46	
	SW 8°C	200	Dissection	Extruded wheat	Ex	70	95	Hemre et al. (2003)
						180	92	
Atlantic halibut (*Hippoglossus hippoglossus*)	SW 10–12°C	1000	Stripping	Extruded wheat	Ex	90	83.0	Grisdale-Helland and Helland (1998)
						170	55.0	
Turbot (*Psetta maxima*)	SW 16°C	110	Settling	Gelatinized wheat	P	144	82	Burel et al. (2000)
European sea bass (*Dicentrarchus labrax*)	SW 25°C	28	Settling	Raw starch	P	250	66	Peres and Oliva-Teles (2002)
				Raw + gel starch		250	85	
				Gelatinized starch		250	98	
	SW 18°C	15	Settling	Gelatinized starch	P	100	97	Moreira et al. (2008)
						200	94	
						300	91	
	SW 25°C			Gelatinized starch	P	100	98	
						200	95	
						300	94	
Silver perch (*Bidyanus bidyanus*)	FW 24°C	2.7	Settling	Raw wheat starch	P	300	76	Stone et al. (2003a,b)
				Raw + Gelatinized 75:25		300	73	
				Raw + Gelatinized 50:50		300	79	
				Raw + Gelatinized 80:20		300	89	
				Raw wheat starch		600	41	
				Raw + Gelatinized 75:25		600	48	
				Raw + Gelatinized 50:50		600	58	
				Raw + Gelatinized 80:20		600	70	
		16	Settling	Raw pea starch	P	300	76	
				Raw wheat starch		300	90	
				Gelatinized wheat starch		300	98	
				Dextrin		300	99.7	
				Maltose		300	99.7	
				Glucose		300	99.8	

TABLE 7-4 Continued

Fish (*Genus species*)	Water Temperature[a]	Fish Weight (g)	Feces Collection Method	Carbohydrate Sources	Diet[b]	Level in the Diet (g/kg)	ADC (%)	Reference
Common carp (*Cyprinus carpio*)	FW 25°C	180	Filtration	Flaked wheat	P	300	99	Médale et al. (1999)
	FW 25°C	45	Filtration	Wheat flour + gelatinized starch (3:1)	Ex	280	89	Yamamoto et al. (2001)
Tilapia (*Oreochromis niloticus* × *O. aureus*)	FW 24°C	6	Siphoning	Raw corn starch	P	310	92	Shiau and Liang (1994)
						470	86	
	FW 28°C	8	Filtration	Raw wheat starch	P	280	96	Kaushik et al. (1995)
						420	92	
						550	82	
						660	71	
Pacific white shrimp (*Litopenaeus vannamei*)	SW 27°C	21	Siphoning	Raw maize starch	P	350	85	Cousin et al. (1996)
				High amylose maize starch		350	63	
				Waxy maize starch		350	85	
				Gelatinized maize starch		350	94	
				Gelatinized waxy maize starch		350	96	
				Raw potato starch		350	72	
				Gelatinized potato starch		350	93	
				Raw wheat starch		350	92	
	SW 27°C	15	Siphoning	Wheat flour	P	360	78	Rivas-Vega et al. (2006)
				Raw cowpea		150	77	
				Cooked cowpea		150	83	
				Extruded cowpea		150	82	

[a]Temp, water temperature; FW, freshwater; SW, Seawater.
[b]P, pelleted diet; Ex, extruded diet.

METABOLIC FATE OF GLUCOSE

Glucose is the major product of carbohydrate digestion although other monosaccharides, such as fructose, galactose, and xylose, may be present in small quantities.

Absorption of the glucose released by digestion is very efficient (> 92%, Singh and Nose, 1967; Furuichi and Yone, 1981) and leads to increased blood glucose levels in all fish and shrimp species (Table 7-5). The intensity of the blood glucose peak increases with the levels of digestible carbohydrates in the diet (Bergot, 1979; Brauge et al., 1995a; Stone et al., 2003a). The rate of glucose loading and clearance seems to be related to both the complexity and the inclusion levels of the carbohydrates. Complex carbohydrates, such as starch and dextrin, generally take longer to be digested and absorbed into the bloodstream than do maltose and glucose. The duration of the elevation of blood glucose is shorter in warmwater herbivorous and omnivorous fish than in coldwater carnivorous fish (Table 7-5). This has been confirmed by experiments of glucose loading by intravenous injection. For example, Legate et al. (2001) found that an intravenous glucose injection of 250 mg/kg body weight resulted in a similar increase in blood glucose (+ 250% of the basal level) in rainbow trout and black bullhead catfish, but the return to the basal level took 30 minutes in black bullhead catfish and 24 hours in rainbow trout. Based on the response to glucose loading (also called a "glucose tolerance test"), teleost fish are generally considered to be glucose intolerant compared to mammals, because the clearance rate of glucose from bloodstream is sluggish (Moon, 2001).

Different hypotheses have been offered to explain such differences in blood glucose regulation among aquatic species and the persistent hyperglycemia compared to mammals. They include inefficiency of glucose as a stimulator of insulin secretion; low number or low responsiveness of insulin receptors in muscle (Parrizas et al., 1994); absence of glucose transporters in muscle (Wright et al., 1998); a limited glucose phosphorylation capacity (Cowey and Walton, 1989), which is an indispensable prerequisite for any further metabolic utilization of glucose; and an inadequate regulation of endogenous production of glucose through gluconeogenesis (Panserat et al., 2000a, 2001).

Insulin: Circulating Levels and Receptors

The initial hypothesis for the persistent hyperglycemia was based on the belief that insulin levels in response to dietary carbohydrate intake were too low or inefficient at regulating blood glucose in fish (Palmer and Ryman, 1972). Radioimmunoassays developed specifically for fish insulin measurements demonstrated that insulin levels in fish (0.2 to 5 nmol/L) are in the same range or even higher than those found in mammals. In addition, digestible carbohydrate intake that induces hyperglycemia does also induce

TABLE 7-5 Effect of Oral (O), Intraperitoneal (IP), or Intravenous (IV) Administration of Different Carbohydrates Sources on Blood Glucose Levels and Return to the Basal Level

Fish	Carbohydrate Source	Administration	Dose (g/kg BW)	Basal Blood Glucose (mM)	Blood Glucose Peak (mM)	Time to Peak (hours after administration)	Return to Basal Level (hours after administration)	Reference
Atlantic salmon (*Salmo salar*)	Glucose	IP	1.0	5.0	15	3	> 72[a]	Hemre et al. (1995)
Rainbow trout (*Oncorhynchus mykiss*)	Glucose	IV	0.25	6.2	21.9	—	24	Legate et al. (2001)
American eel (*Anguilla rostrata*)	Glucose	IV	0.25	9.3	32.6	—	1	Legate et al. (2001)
Gilthead sea bream (*Sparus auratus*)		IP	1.0	3.8	20	3	24	Peres et al. (1999)
European sea bass (*Dicentrarchus labrax*)		IP	1.0	3.6	15.5	6	24	
Asian sea bass (*Lates calcarifer*)		IP	1.0	4.2	10.2	8	16	Stone (2003)
Red sea bream (*Pagellus bogaraveo*)		O	0.2	2.6	9.8	2	> 5[a]	Furuichi and Yone (1981)
Yellowtail (*Seriola lalandi*)		O	0.2	6.1	10.8	3	> 5[a]	
Turbot (*Scophthalmus maximus*)		IP	0.1	4.5	13.8	3	> 24[a]	Garcia Riera and Hemre (1995)
White sturgeon (*Acipenser transmontanus*)	None (sham)	O	0	3.9	4	—	—	Deng et al. (2001)
	Glucose	O	0.1	3.9	10.2	6	24	
	Maltose	O	0.1	3.9	8	6	24	
	Maize dextrin	O	0.1	3.9	6.7	6	15	
	Maize starch	O	0.1	3.9	5.3	6	15	
Hybrid tilapia (*Oreochromis niloticus* × *Oreochromis aureus*)	Maize starch	O	2.5	2.2	4.4	3	6	Shiau and Chuang (1995)
	Glucose	O	2.5	2.2	6.6	2	6	
	Maltose	O	2.5	3.1	4.9	3	6	
	Sucrose	O	2.5	3.0	4.6	4	6	
	Glucose	IP	1.0	4.3	15	2	4	Stone (2003)
Common carp (*Cyprinus carpio*)	Glucose	O	1.7	2.2	10	1	5	Furuichi and Yone (1981)
	Dextrin	O	0.2	2.6	9.4	1	5	Stone (2003)
Channel catfish (*Ictalurus punctatus*)	Glucose	O	1.7	3.7	9.8	3	> 6[a]	Wilson and Poe (1987)
	Dextrin	O	0.8	3.4	3.8	< 1	< 1	Ng and Wilson (1997)
	Glucose	O	0.8	3.3	4.8	1	> 6[a]	
Black bullhead catfish (*Ameiurus melas*)	Glucose	IV	0.25	3.4	12.3	—	0.5	Legate et al. (2001)
Silver perch (*Bidyanus bidyanus*)	Glucose	IP	1.0	3.4	22.2	1	12	Stone et al. (2003a)
	Glucose	IP	2.0	4.8	30.2	3	24	
	Glucose	IP	4.0	4.8	30.7	3	> 24[a]	

[a]Experiment ended before the blood glucose reached basal level.

an increase in circulating levels of insulin (Mommsen and Plisetskaya, 1991; Blasco et al., 1996; Capilla et al., 2003, 2004a). However, the magnitude of the insulin response to digestible carbohydrate intake is fish-species dependent (Krogdahl et al., 2004).

By combining in vivo and in vitro approaches, it was recently demonstrated that insulin in rainbow trout possesses the intrinsic ability to activate its signaling pathway and to regulate expression of hepatic target genes (Plagnes-Juan et al., 2008). A lower number of insulin receptors in extrahepatic tissues may contribute, at least in part, to the relative insulin resistance of fish compared with mammals (Navarro et al., 1999). Insulin receptors have been found in major insulin-responsive tissues of fish (i.e., white muscle, liver, and adipose tissues; Navarro et al., 1999). Upregulation of insulin binding and tyrosine kinase activity has been observed after insulin treatment and ingestion of a carbohydrate-rich diet, respectively (Parrizas et al., 1994; Baños et al., 1998; Capilla et al., 2003, 2004a). However, the number of receptors (semipurified by affinity chromatography) found in the skeletal muscle of rainbow and brown trout is lower than that found in mammals using an identical experimental procedure (Gutiérrez and Plisetskaya, 1991). The number of muscle receptors of tilapia and carp is higher than in salmonids, although always lower than values in mammals (Parrizas et al., 1994), which led Navarro et al. (1999) to conclude that the number of muscle insulin receptors is in agreement with the feeding preference of fish species. Insulin acts together with glucagon in the control of glucose homeostasis (Puviani et al., 1990). Glucagon, secreted by the α cells of the endocrine pancreas, is best known as a glycogenolytic and hyperglycemic hormone. Circulating levels of glucagon in fish are higher than those in rats and humans, compensating for the lower receptor affinity, which does not necessarily reflect a deficiency in the physiological effect of the hormone (Sundby et al., 1991; Navarro et al., 1999). The relationships between insulin, glucagon, and glucagon-like peptides remain to be defined in fish. It is likely that other hormones may also alter plasma glucose levels. The other hormones may include insulin-like growth factor (IGF), growth hormone, the somatostatins, and cortisol.

The presence of insulin/insulin-like growth factor peptides has been suggested in crustaceans. An insulin immunoreactive peptide has been identified in the hepatopancreas of the lobster *Homarus americanus* (Sanders, 1983a). It was shown to stimulate glycogen synthesis in muscle cells (Sanders, 1983b). Tyrosine kinase insulin-like receptors have been characterized in the hepatopancreas and the muscle of the shrimp *P. monodon* (Lin et al., 1993) and *M. japonicus* (Chuang and Wang, 1994). Because the primary structure of *Penaeus* insulin is yet unknown, recent investigations were conducted with bovine insulin (oral or intraabdominal injection). Gutierrez et al. (2007) and Gonzales et al. (2010) have found that bovine insulin altered glucose levels in hemolyph and increase glycogen content in hepotopancreas, gills, and muscle, suggesting the functionality of the insulin pathways in shrimp. According to Verri et al. (2001), the crustacean hyperglycemic hormone (CHH) plays an important role in the regulation of glucose homeostasis by the mobilization of glucose from the hepatopancreas and muscle glycogen stores when glucose level in the hemolymph is low.

Glucose Transport

Once in the bloodstream, glucose arising from digestion moves across cell membranes principally through facilitative glucose transporters (Pilkis and Granner, 1992). Glucose transport inside the cells is the first step in glucose utilization in all organisms. Molecular studies have proved the existence of glucose transporter 1 (GLUT1, an ubiquitous transporter), glucose transporter 2 (GLUT2, a transporter involved in the movement of high concentrations of glucose from intestine to blood and between blood and liver cells), and glucose transporter 4 (GLUT4, an insulin-sensitive transporter of glucose in muscle and adipose tissues) in fish (Wright et al., 1998; Planas et al., 2000; Krasnov et al., 2001; Capilla et al., 2002). Fish GLUT2 and GLUT4 are characterized by a lower affinity for glucose than the same transporters in mammals, which might at least partially explain the persistent postprandial hyperglycemia observed in fish that consume diets containing digestible carbohydrates (Krasnov et al., 2001; Capilla et al., 2002, 2004b; Diaz et al., 2007). Regulation of glucose transporters by dietary carbohydrates requires further investigations. To our knowledge, GLUT proteins of crustaceans have not been described yet.

Routes of Glucose Metabolism

Under aerobic conditions, glucose is catabolized through the glycolytic pathway, tricarboxylic cycle, and respiratory chain resulting in ATP genesis, or through the pentose phosphate pathway leading to the production of NADPH, which is needed for lipid biosynthesis, and ribose 5-phosphate, which is required for nucleotide synthesis. Excess glucose may be stored as glycogen through glycogenesis, converted to lipids, or excreted. Under anaerobic conditions, which prevail in fish muscle during exercise, pyruvate is converted into lactate. All the enzymes involved in these pathways have been identified in fish and shrimp (Knox et al., 1980; Cowey and Walton, 1989). A summary of the knowledge gained since the publication of the Nutrient Requirements of Fish (NRC, 1993) about the regulation of these routes is provided below.

Phosphorylation of Glucose into Glucose-6-Phosphate

The first step of glucose utilization in cells is the phosphorylation of glucose into glucose-6-phosphate mediated by hexokinases (HK). In mammals, four hexokinase isozymes have been described. Three of them (HK I-III), which differ in tissue distribution, have a relatively high affinity for glucose and are inhibited by high concentrations of glucose-6-phosphate. The function of hexokinases I to III is to ensure that the glycolytic pathway is provided with glucose even when blood glucose levels are low. The fourth hexokinase, known as glucokinase (GK; EC 2.7.1.2) or hexokinase IV, is characterized by a low affinity for glucose and a lack of inhibition by glucose-6-phosphate (Printz et al., 1993). In contrast to other hexokinases, the function of glucokinase is to remove glucose from the blood after a meal. It was suggested that the prolonged hyperglycemia observed in fish after glucose tolerance tests or after ingestion of high levels of digestible carbohydrates might result from a limited glucose phosphorylation by HK, or from the absence of an inducible hepatic GK expression (Nagayama et al., 1973; Nagayama and Ohshima, 1974; Cowey et al., 1977; Walton and Cowey, 1982; Shiau, 1997).

Since then, biochemical and molecular analyses have refuted this hypothesis. Expression and activity of hexokinases and glucokinase have been detected in all fish species examined so far (Borrebaek et al., 1993; Tranulis et al., 1996, 1997; Panserat et al., 2000b; Soengas et al., 2006). Although the activity of the low K_m HK (HK I–III) seems not to be nutritionally regulated (Kirchner et al., 2005; Enes et al., 2006, 2008a,b; Moreira et al., 2008), several studies reported that dietary carbohydrates promote changes in both GK activity and gene expression (Caseras et al., 2000; Borrebaek and Christophersen, 2001; Panserat et al., 2000c; Borrebaek et al., 2003; Capilla et al., 2003; Meton et al., 2003; Enes et al., 2006, 2008b). Hepatic GK activity increases with the presence of dietary digestible carbohydrates and is independent of the levels of plasma insulin. For instance, in European sea bass and gilthead sea bream a significant increase in hepatic GK activity was observed with the increase of dietary starch level from 10 to 20% (Enes et al., 2006, 2008a; Moreira et al., 2008). However, there was a lack of further increase of GK activity in the liver of European sea bass as the dietary carbohydrate level increased from 20 to 30%. Moreira et al. (2008) suggested that 20% digestible starch is probably near the tolerance threshold for metabolic utilization of glucose by this species.

Knowledge about glucose metabolism in shrimp is rather limited. According to Cuzon et al. (2004), uptake of glucose released by starch digestion resulted in an increased hemolymph glucose level up to 8–9 mM when *L. vannamei* is fed a diet containing 40% wheat starch. Glucose is then phosphorylated into glucose-6-phosphate and enters the same metabolic pathways as in fish. The response of hexokinases to dietary carbohydrates was found to be dependant upon molt stage (Gaxiola et al., 2005).

Glycolysis

Glycolysis is the major route of glucose metabolism in fish as in other animals and consists of a progressive oxidation of one molecule of glucose into two molecules of pyruvate (Figure 7-8). Besides hexokinases, two other enzymes control the regulation of the glycolytic pathway: 6-phosphofructo-1-kinase (PFK-1, EC 2.7.1.11), which catalyzes the phosphorylation of fructose-6-phosphate into fructose-1,6-bisphosphate, and pyruvate kinase (PK, EC 2.7.1.40), which catalyzes the last step of glycolysis, the conversion of phosphoenolpyruvate to pyruvate. Among fish tissues, the highest activity level of PFK-1 is recorded in skeletal muscle. As in mammals, higher hepatic PFK-1 activity has been found in fish fed high-carbohydrate/low-protein diets than in those given low-carbohydrate/high-protein diets (Fideu et al., 1983; Walton, 1986; Baanante et al., 1991; Meton et al., 1999, 2000). Meal frequency was reported to significantly affect hepatic PFK-1 activities in hybrid tilapia (*Oreochromis niloticus* × *O. aureus*) (Tung and Shiau, 1991). No differences in hepatic PFK-1 activities were observed in hybrid tilapia fed glucose or starch (Lin and Shiau, 1995). In common carp, supplementation of the diet with 30% galactose decreased PFK-1 activity (Shikata et al., 1994).

A recent review of the nutritional regulation of hepatic glucose metabolism in fish (Enes et al., 2009) identified the discrepancy of data regarding regulation of PK activity by the sources and levels of dietary carbohydrates. In European sea bass, glucose induced a higher liver PK activity than starch did, whereas in gibel carp *Carassius auratus* and Chinese longsnout catfish *Leiocassis longirostris*, PK activity was similar when fish were fed diets containing glucose, dextrin, or starch (Tan et al., 2006). In common carp, diet supplementation with 30% starch or glucose had no effect on hepatic PK activity (Shikata et al., 1994). Some studies found unchanged high PK activity regardless of the dietary carbohydrate level, whereas many other studies revealed an induction of PK activity by dietary carbohydrate levels in different fish species.

Overall, recent knowledge gained about the nutritional regulation of the glycolytic pathway strongly suggests that persistent hyperglycemia induced by dietary carbohydrate intake does not result from inefficient regulation of the key enzymes of this pathway. Experiments conducted with ^{14}C-glucose in fish demonstrated that the major part of the administered glucose was catabolized (Brauge et al., 1995b; Garcia-Riera and Hemre, 1996; Hemre and Kahrs, 1997; Hemre and Storebakken, 2000). Glucose oxidation rates within the period of experiments were found to be lower than in mammals (at least two orders of magnitude), which is probably due to the low metabolic rate related to the lower body temperature of fish compared to mammals. Other experiments have demonstrated that glucose was actually used as a fuel in muscle and brain (Johnston and Goldspink, 1973; Soengas and Aldegunde, 2002).

One hypothesis currently under investigation is related to the endogenous production of glucose through gluconeogenesis that could be responsible for persistent hyperglycemia.

Gluconeogenesis

Gluconeogenesis involves a series of metabolic process that use substrates such as α-ketoacids (derived from the catabolism of glucogenic amino acids), glycerol (derived from the hydrolysis of triacylglycerols), and lactate (Walton and Cowey, 1982; Suarez and Mommsen, 1987;

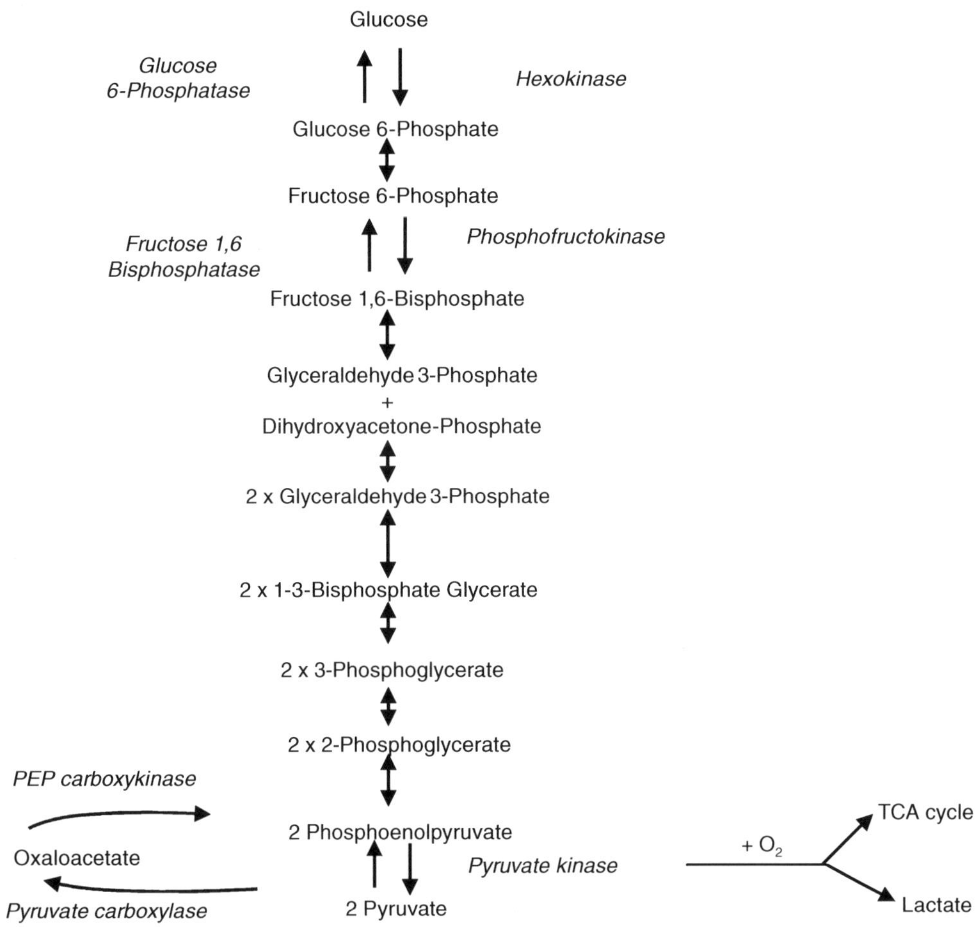

FIGURE 7-8 Scheme of glycolysis and gluconeogenesis.

Cowey and Walton, 1989). Phosphoenolpyruvate carboxykinase (PEPCK), fructose-1,6-bisphosphatase (FBPase), and glucose-6-phosphatase (G6Pase) are the key enzymes that control gluconeogenesis. In rats, dietary carbohydrates inhibit expression and activity of these enzymes (Pilkis and Granner, 1992). In rainbow trout hepatocytes primary culture, the presence of glucose (10 mmol/L) in the medium did not modify gluconeogenesis from alanine. Data regarding the nutritional and hormonal regulation of the gluconeogeneic enzymes in fish were recently examined by Enes et al. (2009). In common carp, ingestion of digestible carbohydrate resulted in down-regulation of gene expression or activities of PEPCK, FBPase, and G6Pase (Panserat et al., 2002). In contrast, no modification of these enzymes was observed after ingestion of digestible carbohydrates in the liver of Atlantic salmon, rainbow trout, gilthead sea bream, and European sea bass (Tranulis et al., 1996; Panserat et al., 2000a, 2001; Caseras et al., 2002; Enes et al., 2008c; Moreira et al., 2008). In addition, there was a high expression level of FBPase in extrahepatic tissues such as intestine and kidney, suggesting a persistent high level of endogenous glucose production in the fish because of a lack of regulation of gluconeogenesis by dietary carbohydrates. This lack of regulation could be related to the high levels of gluconeogenic amino acids and/or fatty acids in fish diets, particularly in the diets of salmonids and marine species. This hypothesis is further supported by the lower expression of PEPCK and G6Pase in the liver of rainbow trout fed low-protein diets (Kirchner et al., 2003).

Storage of Glucose as Glycogen and Mobilization

Glucose may be stored as glycogen through glycogenesis catalyzed by glycogen synthase (GSase). Studies with trout hepatocytes provided evidence that both the glycogen synthesis (glycogenesis) and breakdown (glycogenolysis catalyzed by glycogen phosphorylase [GPase]) are active concurrently in vitro (Pereira et al., 1995). Both GSase and GPase are regulated through phosphorylation and dephosphorylation reactions. The phosphorylation of GSase results in its inactivation and thus in the inactivation of glycogen synthesis, whereas the phosphorylation of GPase results in its activation and subsequent breakdown of glycogen. Phosphorylations result from activation of adenyl cyclase that increases the formation of intracellular cAMP, which in turn activates protein kinase by binding. This regulation is under hormonal control; the role of glucagon/insulin ratio, epinephrine, and norepinephrine has been demonstrated in fish (Moon et al., 1999). Data about the effect of dietary carbohydrates on expression and activity of GSase are lacking.

Glycogen reserves in white muscle are very low (0.4 to 2 mg/g) regardless of the species. The heart contains 3–9 mg/g and brain approximately 3.5 mg/g (Soengas et al., 1996; Kaushik, 1999). The liver is the major site of glycogen storage in fish. Hepatic glycogen concentrations can reach up to 200 mg/g and vary greatly both among and within species according to the feeding status (Tables 7-6 and 7-7) and many other factors such as water temperature (Hemre et al., 2002). Liver glycogen content increases when fish are refed after a period without feed that depletes hepatic glycogen stores (Table 7-6). Some studies reported an increase in glycogen hepatic concentrations with the level of digestible carbohydrate in the diet (Table 7-7; Moreira et al., 2008 in sea bass and Rosas et al., 2000, 2001 in shrimp as examples). Some others failed to find a relationship between liver glycogen content and dietary digestible carbohydrate or blood glucose levels (Table 7-7). According to Moon and Foster (1995) and Pereira et al. (1995), most glycogen in fish would be synthesized from glucose formed through gluconeogenesis (i.e., from noncarbohydrate substrates).

Although the glycogen stores are low in white muscle of fish, glycogen breakdown is used by the fast-twitch glycolytic fibers that compose the white muscle for powering intensive burst of exercise. During high-speed swimming, glycogenolysis is rapidly activated in white muscle and leads to an accumulation of lactate (Johnston and Goldspink, 1973; Kieffer, 2000). West et al. (1993) reported a 30-fold increase in glucose utilization during peak of swimming activity. The limited information available on fish brain metabolism suggests that oxidation of glucose provides most of the ATP required for brain function in teleosts as in other animals. When fish are fed, exogenous glucose is used. Fasting increases mobilization of the brain glycogen reserves. When they are depleted, the brain of teleosts utilizes other fuels such as lactate or ketones (Soengas and Aldegunde, 2002). In the liver, the rate of glycogen mobilization during fasting differs among species. In cod, carp, and tilapia, mobilization of the liver glycogen stores is moderate. In brown trout, sea bass, and sea bream, glycogen mobilization occurs within a few days of fasting until depletion (Table 7 7). Meton et al. (2003) and Perez-Jimenez et al. (2007) reported that refeeding rapidly replenished the liver reserves of gilthead sea bream and European sea bass, respectively. Concentrations close to those observed before fasting were reached after 2 or 3 days of refeeding.

TABLE 7-6 Changes in Liver Glycogen Content (mg/g liver) with Fasting and Feeding Status

Fish (*Genus species*)	Feeding Status	Duration	Glycogen	References
Atlantic salmon (*Salmo salar*)	Fed	7 weeks	20.2	Sundby et al. (1991)
	Fasted	7 weeks	14.8	
Brown trout (*Salmo trutta fario*)	Fed	30 days	39	Navarro et al. (1992)
	Fasted	8 days	9	
	Fasted	30 days	9	
	Fasted	50 days	6	
Common carp (*Cyprinus carpio*)	Fed	10 days	85	Nagai and Ikeda (1971)
	Fasted	2 days	122	
	Fasted	4 days	107	
	Fasted	10 days	88	
	Fasted	15 days	111	
	Fasted	22 days	107	
	Fasted	110 days	16	
Hybrid tilapia (*Oreochromis niloticus* × *Oreochromis aureus*)	Fed	8 weeks	102	Hsieh and Shiau (2000)
	Fasted	1 week	89	
	Fasted	2 weeks	86	
	Fasted	3 weeks	79	
Cod (*Gadus morhua*)	Fed	8 weeks	47	Sundby et al. (1991)
	Fasted	4 weeks	34	
European sea bass (*Dicentrarchus labrax*)	Fed	22 days	115	Perez-Jimenez et al. (2007)
	Fasted	1 day	94	
	Fasted	3 days	67	
	Fasted	9 days	17	
	Refed	1 day	4	
	Refed	3 days	127	
	Refed	9 days	101	
Gilthead sea bream (*Sparus aurata*)	Fed	17 days	180	Meton et al. (2003)
	Fasted	8 days	50	
	Fasted	18 days	0	
	Refed	8 hour	55	
	Refed	2 days	150	
	Refed	8 days	170	
	Refed	21 days	210	

Synthesis of Lipid

Several studies with different fish species have shown that body fat deposition increased with the levels of digestible carbohydrates in the diet (Kaushik et al., 1989; Shimeno et al. 1993; Brauge et al., 1994; Hemre et al., 2002). In mammals, glucose, as one of the precursors of acetyl-CoA, is the primary substrate for fatty acid synthesis in the liver. Hepatic enzymes that produce NADPH required for fatty acid synthesis (glucose-6-phosphate dehydrogenase, malic enzyme, and ATP-citrate lyase) and that are involved in the lipogenic pathway (acetyl-CoA-carboxylase and fatty acid synthase) are stimulated by feeding a carbohydrate-rich diet (Hellerstein et al., 1996). A stimulation of hepatic lipogenic enzymes by high dietary levels of digestible carbohydrate has also been reported in most fish species, including rainbow trout (Hilton and Atkinson, 1982; Rollin et al., 2003), sea bass (Dias et al., 1998, 2004), white sturgeon (Fynn-Aikins et al.,

TABLE 7-7 Effect of Carbohydrate Sources and Levels on Glycogen Content in Different Fish and Shrimp Species

Fish (*Genus species*) (reference)	Carbohydrate Source	Level in the Diet (g/kg DM)	Carbohydrate ADC[a] (%)	Liver Glycogen (mg/g)	Muscle Glycogen (mg/g)
European sea bass	Gelatinized starch	100	98	4.1	—
(*Dicentrarchus labrax*)		200	95	8.8	—
(Moreira et al., 2008)		300	94	10.5	—
White sturgeon	Carbohydrate-free	0	—	14	0.5
(*Acipenser transmontanus*)	Hydrolyzed potato starch	45	—	38	0.5
(Deng et al., 2005)		103	—	68	1.0
	Glucose	45	—	48	2.7
		103	—	76	2.0
Silver perch	None	0	—	55	0.5
(*Bidyanus bidyanus*)	Raw pea starch	300	76	63	0.9
(Stone et al., 2003)	Raw wheat starch	300	90	62	1.4
	Gel wheat starch	300	98	60	1.4
	Dextrin	300	99.7	59	1.1
	Maltose	300	99.7	63	0.8
	Glucose	300	99.8	60	1.1
Hybrid tilapia	Raw starch	400	89.0	102.0	—
(*Oreochromis niloticus* × *Oreochromis aureus*)	Glucose	400	99.0	82.0	—
(Hsieh and Shiau, 2000)					

Shrimp (*Genus species*) (reference)	Carbohydrate Source	Level in the Diet (g/kg DM)	Carbohydrate ADC (%)	Digestive Gland (mg/g)	Muscle Glycogen (mg/g)
Blue shrimp	Wheat flour	50	—	6.4	—
(*Litopenaeus stylirostris*)		150	—	9.3	—
(Rosas et al., 2000)		350	—	13.3	—
Pacific white shrimp	Wheat + starch	10	—	3.4	—
(*Litopenaeus vannamei*)		360	—	6.7	—
(Rosas et al., 2001)					

[a]Apparent digestiblity coefficient.

1992), channel catfish (Likimani and Wilson, 1982), tilapia (Lin and Shiau, 1995), and common carp (Shimeno et al., 1995). However, the few studies that examined the recovery of radio-labeled glucose into lipid fractions led to the conclusion that the amount of de novo synthesis of lipid from glucose is quite limited in fish (Brauge et al., 1995b; Hemre and Kahrs, 1997; Hemre and Storebakken, 2000). Hemre et al. (2002) suggested that the role of digestible carbohydrates in lipogenesis is more for the production of cytosolic reducing equivalents than for the delivery of carbon backbones for lipid synthesis per se. As an explanation for the enhanced fat deposition in fish fed carbohydrate-rich diets, they hypothesized that glucose could increase lipid deposition from dietary lipid by being oxidized instead of dietary lipid, thus reducing the contribution of lipid to oxidative metabolism. The contribution of glucose to lipid synthesis and its role in the regulation of lipid oxidation deserve further attention. Recent studies using metformin, a drug that reduces blood glucose level (Panserat et al., 2009) or rainbow trout strains with different glucose clearance rates (Skiba-Cassy et al., 2009) have provided evidence of a relationship between expression and activity of fatty acid synthase and the regulation of glucose homeostasis.

Glucose Excretion

Excess glucose may be excreted in the urine and/or through the gills when blood glucose levels are elevated. Glucosuria has been observed in Atlantic salmon, red sea bream, tilapia, yellowtail, and white sturgeon fed diets containing high levels of digestible carbohydrates (Furuichi and Yone, 1981; Lin et al., 2000; Deng et al., 2001; Hemre et al., 2002), whereas glucose excretion through gills has been reported in cod (Hemre and Kahrs, 1997). In red sea bream and yellowtail, glucosuria was dependant on the source of carbohydrates. Glucose excretion was low or zero in fish fed diets with starch and increased as molecular complexity of the carbohydrate sources decreased (starch < dextrin < glucose). This could explain that, in most fish species and shrimp, dietary digestible complex carbohydrates such as starch and dextrin are more efficient in promoting growth than is glucose.

NUTRITIONAL ROLE OF DIGESTIBLE CARBOHYDRATES IN FISH AND SHRIMP

Glucose can thus be used in fish and shrimp as a carbon substrate for the same purposes as in other animals, i.e., the production of ATP through oxidative pathways, the production of NADPH and ribose 5-phosphate through the pentose phosphate pathway, the synthesis of glycogen, and the de novo lipogenesis in some circumstances. All of these roles can be achieved either by exogenous glucose (supplied by the digestion of starch and starch-digestion products) or by glucose produced through gluconeogenesis. This is the reason why glucose is not an indispensable nutrient although it is used as fuel. Gluconeogenesis is described as very efficient in fish (Cowey and Walton, 1989). Besides lactate, amino acids are the major substrates for gluconeogenesis in fish liver and kidney. When they are used as substrates for glucose production, amino acids are deaminated and diverted away from the protein synthesis pathway. Such metabolic utilization lowers protein retention and increases nitrogen excretion. Addition of digestible starch in fish and shrimp feeds can thus help to meet environmental regulation objectives in addition to economical objectives, because starchy ingredients are cheaper than protein and lipid sources. Effectiveness of the protein-sparing effect of dietary carbohydrate varies greatly among species. A range of criteria have been examined to find an explanation for such variations and to establish maximum recommended levels. They include growth rate, digestive amylase level, blood glucose level, clearance rate of glucose from the bloodstream, and, more recently, the number of insulin receptors and regulation of gluconeogenesis by dietary carbohydrate. These criteria have led to distinctions between omnivorous and carnivorous species (NRC, 1993; Wilson, 1994; Hemre et al., 2002; Stone, 2003), which can be partly related to the thermal environment of species and the associated metabolic rate. Omnivorous fish generally digest raw starch more efficiently than do carnivorous fish. They exhibit limited glycemia levels and a faster glucose clearance rate, suggesting a more efficient delivery of glucose to tissues, which could be related to the higher number of insulin receptors in muscle. In addition, gluconeogenesis has been found to be inhibited by digestible carbohydrates in carp liver and not in salmonids and marine species (Panserat et al., 2002; Enes et al., 2009).

However, a protein-sparing effect of digestible starch has been described in many carnivorous species, and differences have been found within the categories (omnivorous and carnivorous) (Hemre et al., 2002; Stone, 2003). To define precisely a tolerable level for each species remains difficult because physical state, molecular complexity, and inclusion level of dietary starch are factors that influence digestibility, glucose tolerance, and, ultimately, efficient utilization of carbohydrates. In addition, starch sources interact with the other macronutrients (protein and fat) and likely with environmental factors, especially temperature, within a same species (Brauge et al., 1995a; Médale et al., 1999; Hemre et al., 2002). Differences in metabolic use of glucose have been also found among strains (Mazur et al., 1992; Kolditz et al., 2008; Skiba-Cassy et al., 2009).

Phillips et al. (1948) were the first to describe utilization of carbohydrates as a dietary energy source for salmonids and concluded that 12% starch was an optimum dietary level for rainbow trout. Higher levels resulted in reduced growth rate and increased mortality, probably linked to the low digestibility of the carbohydrate sources incorporated in the diet. Since then, numerous studies have demonstrated that carbohydrates must be supplied in available form in the diet, and the positive effect of starch gelatinization on digestion efficiency was demonstrated. Starch gelatinization can be achieved either by heat treatment of the feed ingredient or by extrusion of the whole feed. Addition of α-amylase into feed is sometimes applied to improve starch digestion. The amount of digestible carbohydrate currently included in fish feeds generally exceeds the level recommended by Phillips et al. (1948). Some studies evaluated the influence of dietary carbohydrate on several health traits of salmonids. Results did not reveal any adverse effect. Long-term feeding of a high-carbohydrate diet to rainbow trout did not significantly affect the nonspecific immunity, such as lysozyme activity or macrophage superoxide production (Page et al., 1999). In Atlantic salmon, humoral immune responses after vaccination with *Vibrio salmonicida* were similar in fish fed diets with carbohydrate levels ranging from 0.5–30% dry matter.

The rate of glucose delivery from digestion and glucose clearance from the bloodstream seem to be key to the efficient metabolic use of glucose. For most fish and shrimp species, digestible complex carbohydrates lead to better growth rates than do simple carbohydrates, as demonstrated by the results of an array of studies using fish and shrimp species with different feeding habits. The improved growth rate obtained with dietary gelatinized starch relative to glucose or other monosaccharides corresponded to higher activity of liver lipogenic enzymes and higher body lipid content in tilapia (Shiau and Liang, 1995), sturgeon (Hung et al., 1989), and Atlantic salmon (Arnesen et al., 1995). The grouper seems to respond differently than most other fish species because the utilization of starch and glucose was reported to be similar (Shiau and Lin, 2001).

Shrimp, regardless of the species, also utilize complex carbohydrates for growth more efficiently than simple sugar molecules, especially glucose. Deshimaru and Yone (1978) fed diets containing 10% of starch, dextrin, glucose, or sucrose to *M. japonicus* and found that feed efficiency was highest for the diet containing starch, followed by sucrose, and dextrin in decreasing order. This result was confirmed by Abdel-Rahman et al. (1979) in the same species *M. japonicus*, by Cuzon et al. (2004) in *L. vannamei*, and by Shiau and Peng (1992) in *Penaeus monodon*. In addition, Shiau and Peng (1992) demonstrated that corn starch produced a greater protein-sparing effect than glucose. Accordingly, the dietary protein requirement for *P. monodon* was found to be lower when starch, rather than glucose or dextrin, was included in the diet as the carbohydrate source.

Based on the reported findings for different fish and shrimp species, the maximum recommended levels of digestible starch inclusion in feed fall within 15–25% in salmonids and marine fish, within 20 40% in shrimp and up to 50% in omnivorous species. The levels of dietary fiber must be kept as low as possible and not exceed 10%. Further investigations are needed to improve the precision of these recommendations.

REFERENCES

Abdel-Rahman, S. H., A. Kanazawa, and S. I. Teshima. 1979. Effects of dietary carbohydrate on the growth and the levels of the hepatopancreatic glycogen and serum glucose of prawn. B. Jpn. Soc. Sci. Fish. 45:1491–1494.

Aguirre, P., F. Médale, and S. J. Kaushik. 1995. Influence of the type of dietary starch on the use of two protein sources by rainbow trout reared at 8° or 18°C. Reprod. Nutr. Dev. 35:129–136.

Amirkolaie, A. K., J. I. Leenhouwers, J. A. J. Verreth, and J. W. Schrama. 2005. Type of dietary fibre (soluble versus insoluble) influences digestion, faeces characteristics and faecal waste production in Nile tilapia (*Oreochromis niltoticus*). Aquac. Res. 36:1157–1166.

AOAC Method 991.43. 1995. Total, Soluble and Insoluble Dietary Fiber in Foods. Official Methods of Analysis, 16th ed. Gaithersburg, MD: Assoc. Official Anal. Chem.

AOAC Method 920.40. 2002a. Starch in Animal Feed. Official Methods of Analysis, 17th ed. Gaithersburg, MD: Assoc. Official Anal. Chem.

AOAC Method 948.02. 2002b. Starch in Plants. Official Methods of Analysis, 17th ed. Gaithersburg, MD: Assoc. Official Anal. Chem.

AOAC Method 962.09. 2002c. Fiber (Crude) in Animal Feed and Pet Food. Official Methods of Analysis, 17th ed. Gaithersburg, MD: Assoc. Official Anal. Chem.

AOAC Method 991.43. 2002d. Total Soluble, and Insoluble Dietary Fiber in Foods. Official Methods of Analysis, 17th ed. Gaithersburg, MD: Assoc. Official Anal. Chem.

Appleford, P., and T. A. Anderson. 1996. The effect of inclusion level and time on digestibility of starch for common carp (*Cyprinus carpio*). Asian Fish. Sci. 9:121–126.

Arnesen, P., and Å. Krogdahl. 1995. Nutrient digestibilities weight gain and plasma and liver levels of carbohydrate in rainbow trout, *Oncorhynchus mykiss* (Walbaum), fed diets with oats. Aquacult. Nutr. 1:201–211.

Arnesen P., Å. Krogdahl, and A. Sundby. 1995. Nutrient digestibilities, weight gain and plasma and liver levels of carbohydrate in Atlantic salmon (*Salmo salar*, L) fed diets containing oats and maize. Aquacult. Nutr. 1:151–158.

Asp, N-G. L. 1995. Classification and methodology of food carbohydrates as related to nutritional effects. Am. J. Clin. Nutr. 61(4 Suppl.):930S–937S.

Baanante, I. V., P. Garcia de Frutos, L. Bonamusa, and F. Fernandez. 1991. Regulation of fish glycolysis-gluconeogenesis: Role of fructose 2,6 P2 and PFK-2. Comp. Biochem. Phys. B 100(1):11–17.

Bach Knudsen, K. E. 1997. Carbohydrate and lignin contents of plant materials used in animal feeding. Anim. Feed Sci. Tech. 67:319–338.

Bach Knudsen, K. E. 2001. The nutritional significance of "dietary fibre" analysis. Anim. Feed Sci. Tech. 90:3–20.

Baer, D. J., W. V. Rumpler, C. W. Miles, and G. C. Fahey, Jr. 1997. Dietary fiber decreases the metabolizable energy content and nutrient digestibility of mixed diets fed to humans. J. Nutr. 127:579–586.

Baños, N., J. Baró, C. Castejón, I. Navarro, and J. Gutiérrez. 1998. Influence of high-carbohydrate enriched diets on plasma insulin levels and insulin and IGF-I receptors in trout. Regul. Pept. 77(1–3):55–62.

Bergot, F. 1979. Effects of dietary carbohydrates and of their mode of distribution on glycemia in rainbow trout (*Salmo gairdneri* Richardson). Comp. Biochem. Phys. A 64:543–547.

Bergot, F. 1993. Digestibility of native starches of various botanic origins by rainbow trout (*Oncorhynchus mykiss*). Pp. 857–865 in Fish Nutrition in Practice, S. J. Kaushik and P. Luquet, eds. Paris, France: INRA.

Bergot, F., and J. Breque. 1983. Digestibility of starch by rainbow trout: Effects of the physical state of starch and the intake level. Aquaculture 34:203–212.

Blasco, J., J. Fernandez-Borras, I. Marimon, and A. Requena. 1996. Plasma glucose kinetics and tissue uptake in brown trout in vivo: Effect of an intravascular glucose load. J. Comp. Physiol. B 165:534–541.

Borrebaek, B., and B. Christophersen. 2001. Activities of glucose phosphorylation, glucose-6-phosphatase and lipogenic enzymes in the liver of perch, *Perca fluviatilis*, after different dietary treatment. Aquac. Res 32:221–224.

Borrebaek, B., R. Waagbo, B. Christophersen, M. A. Tranulis, and G. I. Hemre. 1993. Adaptable hexokinase with low affinity for glucose in the liver of Atlantic salmon (*Salmo salar*). Comp. Biochem. Phys. B 106:833–836.

Borrebaek, B., B. Christophersen, and A. Sundby. 2003. Metabolic function of hepatic hexokinase in perch, Perca fluviatilis. Aquac. Res 34:235–239.

Brauge, C., F. Médale, and G. Corraze. 1994. Effect of dietary carbohydrate levels on growth, body composition and glycaemia in rainbow trout (*Oncorhynchus mykiss*) reared in seawater. Aquaculture 123:109–120.

Brauge, C., G. Corraze, and F. Médale. 1995a. Effect of dietary levels of lipid and carbohydrate on growth performance, body composition, nitrogen excretion and plasma glucose levels in rainbow trout reared at 8 or 18°C. Reprod. Nutr. Dev. 35:277–290.

Brauge, C., G. Corraze, and F. Médale. 1995b. Effects of dietary levels of carbohydrate and lipid on glucose oxidation and lipogenesis from glucose in rainbow trout, *Oncorhynchus mykiss* reared in freshwater or in seawater. Comp. Biochem. Phys. A 111:117–124.

Buléon, A., P. Colonna, V. Planchot, and S. Ball. 1998. Starch granules: Structure and biosynthesis. Int. J. Biol. Macromol. 23:85–112.

Burel, C., T. Boujard, F. Tulli, and S. J. Kaushik. 2000. Digestibility of extruded peas, extruded lupin, and rapeseed meal in rainbow trout (*Oncorhynchus mykiss*) and turbot (*Psetta maxima*). Aquaculture 188:285–298.

Capilla, E., M. Diaz, J. Gutierrez, and J. V. Planas. 2002. Physiological regulation of the expression of a GLUT4 homolog in fish skeletal muscle. Am. J. Physiol. Endoc. M. 283:44–49.

Capilla, E., F. Médale, I. Navarro, S. Panserat, C. Vachot, S. Kaushik, and J. Gutierrez. 2003. Muscle binding and plasma levels in relation to liver glucokinase activity, glucose metabolism and dietary carbohydrates in rainbow trout. Regul. Pept. 110(2):123–132.

Capilla, E., F. Médale, S. Panserat, C. Vachot, P. Rema, E. Gomes, S. Kaushik, I. Navarro, and J. Gutierrez. 2004a. Response of hexokinase enzymes and insulin system to dietary carbohydrates in common carp, *Cyprinus carpio*. Reprod. Nutr. Dev. 44(3):1–10.

Capilla, E., M. Diaz, A. Albalat, I. Navarro, J. E. Pessin, K. Keller, and J. V. Planas. 2004b. Functional characterization of an insulin-responsive glucose transporter (GLUT4) from fish adipose tissue. Am. J. Physiol. Endoc. M. 287:E348–357.

Caseras, A., I. Meton, F. Fernandez, and I. V. Baanante. 2000. Glucokinase gene expression is nutritionally regulated in liver of gilthead sea bream (*Sparus aurata*). Biochim. Biophys. Acta. 1493:135–141.

Caseras, A., I. Meton, C. Vives, M. Egea, F. Fernandez, and I. V. Baanante. 2002. Nutritional regulation of glucose-6-phosphatase gene expression in liver of the gilthead sea bream (*Sparus aurata*). Brit. J. Nutr. 88:607–614.

Chuang, N. N., and P. C. Wang. 1994. Characterization of insulin receptor from the muscle of the *Penaeus japonicus*. Comp. Biochem. Physiol. 108:289–297.

Cousin, M., G. Cuzon, and J. Guillaume. 1996. Digestibility of starch in *Penaeus vannamei:* In vivo and vitro study on eight samples of various origin. Aquaculture 140:361–372.

Cowey, C. B., and M. J. Walton. 1989. Intermediary metabolism. Pp. 260–329 in Fish Nutrition, J. E. Halver, ed. San Diego, CA: Academic Press.

Cowey, C. B., D. Knox, M. J. Walton, and J. W. Adron. 1977. The regulation of gluconeogenesis by diet and insulin in rainbow trout (*Salmo gairdneri*). Brit. J. Nutr. 38:463–470.

Crawford, A. C., N. R. Richardson, and P. B. Mather. 2005. A comparative study of cellulose and xylanase activity in freshwater crayfish and marine prawns. Aquac. Res. 36:586–592.

Cummings, J. H., and H. N. Englyst. 1995. Gastrointestinal effects of food carbohydrate. Am. J. Clin. Nutr. 61:938–945.

Cuzon, G., A. Lawrence, G. Gaxiola, C. Rosas, and J. Guillaume. 2004. Nutrition of *Litopenaeus vannamei* reared in tanks or in ponds. Aquaculture 235:513–551.

Davidson, M. H., and A. Mc Donald. 1998. Fiber: Forms and functions. Nutrition Res. 18:617–624.

Deng, D. F., S. Refstie, and S. S. O. Hung. 2001. Glycemic and glycosuric responses in white sturgeon (*Acipenser transmontanus*) after oral administration of simple and complex carbohydrates. Aquaculture 199:107–117.

Deng, D. F., G.-I. Hemre, T. Storebakken, S.-Y. Shiau, and S. S. O. Hung. 2005. Utilization of diets with hydrolyzed potato starch, or glucose by juvenile white sturgeon (*Acipenser transmontanus*), as affected by Maillard reaction during feed processing. Aquaculture 248:103–109.

Deshimaru, O., and K. Kuroki. 1974. Studies on purified diet for prawn: II. Optimum contents of cholesterol and glucosamine in the diet. B. Jpn. Soc. Sci. Fish. 40:421–424.

Deshimaru, O., and Y. Yone. 1978. Effect of dietary carbohydrate source on the growth and feed efficiency of prawn. B. Jpn. Soc. Sci. Fish. 44: 1161–1163.

Dias, J., C. Huelvan, M. T. Dinis, and R. Métailler. 1998. Influence of dietary bulk agents (silica, cellulose and a natural zeolite) on protein digestibility, growth, feed intake and feed transit time in European seabass (*Dicentrarchus labrax*) juveniles. Aquat. Living. Resour. 11:219–226.

Dias, J., R. Rueda Jasso, S. Panserat, L. E. C. da Conceicao, E. F. Gomes, and M. T. Dinis. 2004. Effect of dietary carbohydrate-to-lipid ratios on growth, lipid deposition and metabolic hepatic enzymes in juvenile Senegalese sole (*Solea senegalensis*, Kaup). Aquac. Res. 35:1122–1130.

Diaz, M., E. Capilla, and J. V. Planas. 2007. Physiological regulation of glucose transporter (GLUT4) protein content in brown trout (*Salmo trutta*) skeletal muscle. J. Exp. Biol. 210:2346–2351.

Dioundick, O. B., and D. I. Stom. 1990. Effects of dietary α-cellulose levels on the juvenile tilapia, *Oreochromis mossambicus* (Peters). Aquaculture 91:311–315.

Enes, P., S. Panserat, S. Kaushik, and A. Oliva-Teles. 2006. Effect of normal and waxy maize starch on growth, food utilization and hepatic glucose metabolism in European sea bass (*Dicentrarchus labrax*) juveniles. Comp. Biochem. Phys. A 143:89–96.

Enes, P., S. Panserat, S. Kaushik, and A. Oliva-Teles. 2008a. Growth performance and metabolic utilization of diets with native and waxy maize starch by gilthead seabream (*Sparus auratus*) juveniles. Aquaculture 274:101–108.

Enes, P., S. Panserat, S. Kaushik, and A. Oliva-Teles. 2008b. Hepatic glucokinase and glucose-6-phosphatase responses to dietary glucose and starch in gilthead sea bream (*Sparus aurata*) juveniles reared at two temperatures. Comp. Biochem. Phys. A. 149:80–86.

Enes, P., S. Panserat, S. Kaushik, and A. Oliva-Teles. 2008c. Rearing temperature enhances hepatic glucokinase but not glucose-6-phosphatase activities in European sea bass (*Dicentrarchus labrax*) and gilthead sea bream (*Sparus aurata*) juveniles fed with the same level of glucose. Comp. Biochem. Phys. A 150:355–358.

Enes, P., S. Panserat, S. Kaushik, and A. Oliva-Teles. 2009. Nutritional regulation of hepatic glucose metabolism in fish. Fish Physiol. Biochem. 35:519–539.

Englyst, H. N., M. E. Quigley, and G. J. Hudson. 1994. Determination of dietary fibre as non-starch polysaccharides with gas-liquid chromatographic, high-performance liquid chromatographic or spectrophotometric measurement of constituent sugars. Analyst 119:1497–1509.

Fideu, M. D., G. Soler, and M. Ruiz-Amil. 1983. Nutritional regulation of glycolysis in rainbow trout (*Salmo gairdneri* R.). Comp. Biochem. Phys. B 74:795–799.

Fines, B. C., and G. J. Holt. 2010. Chitinase and apparent digestibility of chitin in the digestive tract of juvenile cobia, *Rachycentron canadum*. Aquaculture 303:34–39.

Francis, G., H. P. S. Makkar, and K. Becker. 2001. Antinutritional factors present in plant-derived alternate fish feed ingredients and their effect in fish. Aquaculture 199:197–227.

Furuichi, M., and Y. Yone. 1981. Change of blood sugar and plasma insulin levels of fishes in glucose tolerance test. B. Jpn. Soc. Sci. Fish. 47:761–764.

Fynn-Aikins, K., S. S. O. Hung, W. Liu, and H. Li. 1992. Growth, lipogenesis and liver composition of juvenile white sturgeon fed different levels of D-glucose. Aquaculture 105:61–72.

Garcia-Riera, M. P., and G. I. Hemre. 1995. Glucose tolerance in turbot, *Scophthalmus maximus* (L.). Aquacult. Nutr. 2:117–120.

Garcia-Riera, M. P., and G.-I. Hemre. 1996. Effect of adaptation to three different levels of dietary carbohydrates on the incorporation of 14C-glucose in several organs of Atlantic halibut (*Hippoglossus hippoglossus*). Aquac. Res. 27:565–571.

Gatlin, D. M., F. T. Barrows, P. Brown, K. Dabrowski, T. G. Gaylord, R. W. Hardy, E. Herman, G. Hu, K. Krogdahl, R. Nelson, K. Overturf, M. Rust, W. Sealy, D. Skonberg, E. J. Souza, D. Stone, R. Wilson, and E. Wurtele. 2007. Expanding the utilisation of sustainable plant products in aquafeeds: A review. Aquac. Res. 38:551–579.

Gaxiola, G., G. Cuzon, T. García, G. Taboada, R. Brito, M. E. Chiml, A. Paredes, L. Soto, C. Rosas, and A. VanWormhoudt. 2005. Factorial effects of salinity, dietary carbohydrate and moult cycle on digestive carbohydrases and hexokinases in *Litopenaeus vannamei* (Boone, 1931). Comp. Biochem. Phys A. 140:29–39.

Gaylord, T. G., F. T. Barrows, S. D. Rawles, K. Liu, P. Bregitzer, A. Hang, D. E. Obert, and C. Morris. 2009. Apparent digestibility of nutrients and energy in extruded diets from cultivars of barley and wheat selected for nutritional quality in rainbow trout *Oncorhynchus mykiss*. Aquacult. Nutr. 15:306–312.

Glencross, B. D., T. B. Boujard, and S. J. Kaushik. 2003. Evaluation of the influence of oligosaccharides on the nutritional value of lupin meals when fed to rainbow trout, *Oncorhynchus mykiss*. Aquaculture 219:703–713.

Gonzalez, J. P., O. Carrillo Farnés, G. Gaxiola, C. Pascual, and C. Rosas. 2010. The effects of microencapsulated bovine insulin given to *Litopenaeus vannamei* juveniles as a feed additive on growth, metabolism, and digestive enzyme activities. Aquaculture 306:252–258.

Grisdale-Helland, B., and S. J. Helland. 1997. Replacement of protein by fat and carbohydrate in diet for Atlantic salmon (*Salmo salar*) at the end of the freshwater stage. Aquaculture 152:167–180.

Grisdale-Helland, B., and S. J. Helland. 1998. Macronutrient utilization by Atlantic halibut (*Hippoglossus hippoglossus*): Diet digestibility and growth of 1 kg fish. Aquaculture 166:57–65.

Guiry, M. D., and G. Blunden. 1991. Seaweeds resources in Europe: Uses and potential. Chichester, UK: John Wiley and Sons.

Gutiérrez, A., J. Nieto, F. Pozo, S. Stern, and L. Schoofs. 2007. Effect of insulin/IGF-I like peptides on glucose metabolism in the white shrimp *Penaeus vannamei*. Gen. Comp. Endocr. 153:170–175.

Gutiérrez, J., and E. M. Plisetskaya. 1991. Insulin binding to liver plasma membranes in salmonids with modified plasma insulin levels. Can. J. Zoolog. 69:2745–2750.

Hellerstein, M. K., J. M. Schwarz, and R. A. Neese. 1996. Regulation of hepatic de novo lipogenesis in humans. Annu. Rev. Nutr. 16:523–557.

Hemre, G.-I., and F. Kahrs. 1997. 14C-injection in Atlantic cod (*Gadus morhua*), metabolic responses and excretion via the gill membrane. Aquacult. Nutr. 2:3–8.

Hemre, G.-I., and T. Storebakken. 2000. Tissue and organ distribution of 14C-activity in dextrin-adapted Atlantic salmon after oral administration of radiolabelled 14C1-glucose. Aquacult. Nutr. 6:229–234.

Hemre, G.-I., O. Lie, and G. Lambertsen. 1990. Digestibility of different carbohydrates sources in cod (*Gadus morhua*) and its relation to glucose content in blood and urine. Fisk Dir. Skr. 3:3–9.

Hemre, G.-I., O. Torrissen, Å. Krogdahl, and O. Lie. 1995. Glucose tolerance in Atlantic salmon (*Salmo salar*), dependence on preadaptation to dietary starch and water temperature. Aquacult. Nutr. 2:69–75.

Hemre, G.-I., T. P. Mommsen, and Å. Krogdahl, 2002. Carbohydrates in fish nutrition: Effects on growth, glucose metabolism and hepatic enzymes. Aquacult. Nutr. 8:175–194.

Hemre, G.-I., Ø. Karlsen, A. Mangor-Jensen, and G. Rosenlund. 2003. Digestibility of dry matter, protein, starch and lipid by cod, *Gadus morhua*: Comparison of sampling methods. Aquaculture 225:225–232.

Hidalgo, M. C., E. Urea, and A. Sanz. 1999. Comparative study of digestive enzymes in fish with different nutritional habits. Proteolytic and amylase activities. Aquaculture 170:267–283.

Hillestad, M., F. Johnsen, and T. Asgard. 2001. Protein to carbohydrate ratio in high-energy diets for Atlantic salmon (*Salmo salar* L.). Aquac. Res. 32:517–529.

Hilton, J. W., and J. L. Atkinson. 1982. Response of rainbow trout (*Salmo gairdneri*) to increased levels of available carbohydrate in practical trout diets. Brit. J. Nutr. 47:597–607.

Hilton, J. W., J. L. Atkinson, and S. J. Slinger. 1983. Effect of dietary fiber on the growth of rainbow trout. Can. J. Fish. Aquat. Sci. 40:81–85.

Hirose, R., Y. Tezuka, T. Kondo, K. Hirao, T. Hatta, S. Nemoto, K. Saio, S. Takahashi, and K. Kainuma. 2010. Characteristic physico-chemical properties and potential uses of enset (*Ensete ventricosum*) starch: comparative studies with starches of potato, sago and corn. J. Appl. Glycosci. 57:185–192.

Hossain, M. A., U. Focken, and K. Becker. 2001. Evaluation of an unconventional legume seed, Sesbnia aculeate, as a dietary protein source for common carp, *Cyprinus carpio* L. Aquaculture 198:129–140.

Hsieh, S. L., and S.-Y. Shiau. 2000. Effects of diets containing different carbohydrates on starved condition in juvenile tilapia *Oreochromis niloticus* × *O. aureus*. Fisheries Sci. 66:32–37.

Hua, K., and D. P. Bureau. 2009. A mathematical model to explain variations in estimates of starch digestibility and predict digestible starch content of salmonid fish feeds. Aquaculture 294:282–287.

Hung, S. S. O., K. Fynn-Aikins, P. B. Lutes, and R. P. Xu. 1989. Ability of juvenile white sturgeon (*Acipenser transmontanus*) to utilize different carbohydrate sources. J. Nutr. 119:727–733.

Johnston, I. A., and G. Goldspink. 1973. A study of the swimming performance of the crucian carp *Carassius cavassius* L. in relation to the effects of exercise and recovery on biochemical changes in the myotomal muscles and liver. J. Fish Biol. 5:249–260.

Kaushik, S. 1999. Nutrition glucidique: Intérêt et limites des apports de glucides. Pp. 171–186 in Nutrition et Alimentation des Poissons et Crustacés, J. Guillaume, S. Kaushik, P. Bergot, and R. Métailler, eds. Paris, France: INRA.

Kaushik, S. J., F. Médale, B. Fauconneau, and D. Blanc. 1989. Effect of digestible carbohydrates on protein/energy utilization and on glucose metabolism in rainbow trout (*Salmo gairdneri* R.). Aquaculture 79:63–74.

Kaushik, S. J., T. Doudet, F. Médale, P. Aguirre, D. Blanc. 1995. Protein and energy needs for maintenance and growth'of Nile tilapia (*Oreochromis niloticus*) using different criteria. J. Appl. Ichthyol. 11:290–296.

Kieffer, J. D. 2000. Limits to exhaustive exercise in fish. Comp. Biochem. Physiol. 126:161–179.

Kirchner, S., S. Kaushik, and S. Panserat. 2003. Low protein intake is associated with reduced hepatic gluconeogenic enzyme expression in rainbow trout (*Oncorhynchus mykiss*). J. Nutr. 133:2561–2564.

Kirchner, S., P. Seixas, S. Kaushik, and S. Panserat. 2005. Effects of low protein intake on extra-hepatic gluconeogenic enzyme expression and peripheral glucose phosphorylation in rainbow trout (*Oncorhynchus mykiss*). Comp. Biochem. Phys. B. 140:333–340.

Knox, D., M. J. Walton, and C. B. Cowey. 1980. Distribution of enzymes of glycolysis and gluconeogenesis in fish tissues. Mar. Biol. 56:7–10.

Kolditz, C., M. Borthaire, N. Richard, G. Corraze, S. Panserat, C. Vachot, F. Lefevre, and F. Médale. 2008. Liver and muscle metabolic changes induced by dietary energy content and genetic selection in rainbow trout (*Oncorhynchus mykiss*). Am. J. Physiol. Reg. I. 294:R1154–R1164.

Krasnov, A., H. Teerijoki, and H. Molsa. 2001. Rainbow trout (*Oncho-rhynchus mykiss*) hepatic glucose transporter. Biochim Biophys. Acta 1520:174–178.

Kraugerud, O. F., M. Penn, T. Storebakken, S. Refstie, Å. Krogdahl, and B. Svilhus. 2007. Nutrient digestibilities and gut function in Atlantic salmon (*Salmo salar*) fed diets with cellulose or non-starch polysaccharides form soy. Aquaculture 273:96–107.

Krogdahl, Å., S. Nordrum, M. Sørensen, L. Brudeseth, and C. Røsjø. 1999. Effects of diet composition on apparent nutrient absorption along the intestinal tract and of subsequent fasting on mucosal disaccharidase activities and plasma nutrient concentration in Atlantic salmon, *Salmo salar* L. Aquacult. Nutr. 5:121–133.

Krogdahl, Å., A. Sundby, and J. J. Olli. 2004. Atlantic salmon (*Salmo salar*) and rainbow trout (*Oncorhynchus mykiss*) digest and metabolize nutrients differently depending on water salinity and dietary starch level. Aquaculture 229:335–360.

Krogdahl, Å., G.-I. Hemre, and T. P. Mommsen. 2005. Carbohydrates in fish nutrition: Digestion and absorption in postlarval stages. Aquacult. Nutr. 11:103–122.

Leenhouwers, J. I., G. A. Santos, J. W. Schrama, and J. A. J. Verreth. 2004. Digesta viscosity and nutrient digestibility in European sea bass in a fresh water and marine environment. Proceedings "Biotechnologies for Quality", October 20–23, 2004, Barcelona, Spain, pp. 475–476.

Leenhouwers, J. I., D. Adjei-Boateng, J. A. J. Verreth, and J. W. Schrama. 2006. Digesta viscosity, nutrient digestibility and organ weights in African catfish (Clarias gariepinus) fed diets supplemented with different levels of a soluble non-starch polysaccharide. Aquacult. Nutr. 12:111–116.

Legate, N. J., A. Bonen, and T. W. Moon. 2001. Glucose tolerance and peripheral glucose utilization in three fish species. Gen. Comp. Endocr. 122:48–59.

Likimani, T. A., and R. P. Wilson. 1982. Effect of diet on lipogenic enzyme activities in channel catfish hepatic and adipose tissue. J. Nutr. 112:112–117.

Lin, C. L., P. C. Wang, and N. N. Chuang. 1993. Specific phosphorylation of membrane proteins of Mr. 44,000 and Mr. 32,000 by the autophosphorylated insulin receptor from the hepatopancreas of the shrimp *Penaeus monodon* (crustacea: decapoda). J. Exp. Zool. 267:113–119.

Lin, J. H., and S.-Y. Shiau. 1995. Hepatic enzyme adaptation to different carbohydrates in juvenile tilapia (*Oreochromis niloticus* × *O. aureus*). Fish Physiol. Biochem. 14:165–170.

Lin, S. C., C. H. Liou, and S.-Y. Shiau. 2000. Renal threshold for urinary glucose excretion by tilapia in response to orally administered carbohydrate and injected glucose. Fish Physiol. Biochem. 23:127–132.

Lindsay, G. J. H, and J. E. Harris. 1980. Carboxymethylcellulase activity in the digestive tracts of fish. J. Fish Biol. 16:219–233.

Mazur, C. N., D. A. Higgs, E. Plisetskaya, and B. E. March. 1992. Utilization of dietary starch and glucose tolerance in juvenile Chinook salmon (*Oncorhynchus tshawytscha*) of different strains in seawater. Fish Physiol. Biochem. 10:303–313.

Médale, F., J. M. Poli, F. Vallée, and D. Blanc. 1999. Utilisation of a carbohydrate-rich diet by common carp reared at 18 and 25° C. Cybium 23:139–152.

Mertens, D. R. 2002. Gravimetric determination of amylase-treated neutral detergent fiber in feeds with refluxing in beakers or crucibles: Collaborative study. J. AOAC Int. 85:1217–1240.

Meton, I., A. Caseras, D. Mediavilla, F. Fernandez, and I. V. Baanante. 1999. Molecular cloning of a cDNA encoding 6-phosphofructo-2-kinase/fructose-2,6-bisphosphatase from liver of *Sparus aurata*: Nutritional regulation of enzyme expression. Biochim. Biophys. Acta 1444:153–165.

Meton, I., A. Caseras, F. Fernandez, I. V. Baanante. 2000. 6-Phosphofructo-2-kinase/fructose-2,6-bisphosphatase gene expression is regulated by diet composition and ration size in liver of gilthead sea bream, *Sparus aurata*. Biochim. Biophys. Acta 1491:220–228.

Meton, I., F. Fernandez, and I. V. Baanante. 2003. Short- and long-term effects of refeeding on key enzyme activities in glycolysis-gluconeogenesis in the liver of gilthead seabream (*Sparus aurata*). Aquaculture 225:99–107.

Mommsen, T. P., and E. M. Plisetskaya. 1991. Insulin in fishes and agnathans: History, structure and metabolic regulation. Rev. Aquat. Sci. 4:225–259.

Moon, T. W. 2001. Glucose intolerance in teleost fish: Fact or fiction? Comp. Biochem. Phys. B 129:243–249.

Moon, T. W., and G. D. Foster. 1995. Tissue carbohydrate metabolism, gluconeogenesis and hormonal and environmental influences. Pp. 65–100 in Biochemistry and Molecular Biology of Fishes, P. W. Hochachka and T. P. Mommsen, eds. Amsterdam, the Netherlands: Elsevier.

Moon, T. W., E. R. Busby, G. A. Cooper, and T. P. Mommsen. 1999. Fish hepatocyte glycogen phosphorylase—a sensitive indicator for hormonal modulation. Fish Physiol. Biochem. 21:15–24.

Moreira, I. S., H. Peres, A. Couto, P. Enes, and A. Oliva-Teles. 2008. Temperature and dietary carbohydrate levels effects on performance and metabolic utilisation of diets in European sea bass (*Dicentrarchus labrax*) juveniles. Aquaculture 274:153–160.

Nagai, M., and S. Ikeda. 1971. Carbohydrate metabolism in fish. I. Effects of starvation and dietary composition on the blood glucose level and the hepatopancreatic glycogen and lipid contents in carp. B. Jpn. Soc. Sci. Fish. 37:404–409.

Nagayama, F., and H. Ohshima. 1974. Studies on enzyme-system of carbohydrate-metabolism in fish. 1. Properties of liver hexokinase. B. Jpn. Soc. Sci. Fish. 40:285–290.

Nagayama, F., H. Ohshima, and T. Takeushi. 1973. Activities of hexokinase and glucose dehydrogenase in fish liver. B. Jpn. Soc. Sci. Fish. 39:1349.

Navarro, I., J. Gutierrez, and J. Planas. 1992. Changes in plasma glucagon, insulin and tissues metabolites associated with prolonged fasting in brown trout (*Salmo trutta* Fario) during 2 different seasons of the year. Comp. Biochem. Phys. A 102:401–407.

Navarro, I., B. Leibush, T. W. Moon, E. M. Plisetskaya, N. Baños, E. Méndez, V. Planas, and J. Gutiérrez. 1999. Insulin, insulin-like growth factor-I (IGF-I) and glucagon: The evolution of their receptors. Comp. Biochem. Phys. B 122:137–153.

Ng, W.-K., and R. P. Wilson. 1997. Chromic oxide inclusion in the diet does not affect glucose utilization or chromium retention by Channel catfish, *Ictalurus punctatus*. J. Nutr. 127:2357–2362.

Ng, W. K., H. A. Lim, S. L. Lim, and C. O. Ibrahim. 2002. Nutritive value of palm kernel meal pretreated with enzyme or fermented with *Trichoderma koningii* (Oudemans) as a dietary ingredient for red hybrid tilapia (*Oreochromis* sp.). Aquac. Res. 33:1199–1207.

NRC (National Research Council). 1993. Nutrient Requirements of Fish. Washington, DC: National Academy Press.

Ortiz, J., E. Uquiche, P. Robert, P. N. Romero, V. Quitral, and C. Llantén. 2009. Functional and nutritional value of the Chilean seaweeds *Codium fragile*, *Gracilaria chilensis* and *Macrocystis pyrifera*. Eur. J. Lipid Sci. Tech. 111:320–327.

Ovrum-Hansen, J., and T. Storebakken. 2007. Effects of dietary cellulose level on pellet quality and nutrient digestibilities in rainbow trout (*Oncorhynchus mykiss*). Aquaculture 272:458–465.

Page, G. I., K. M. Hayworth, R. R. Wade, A. M. Harris and, D. P. Bureau. 1999. Non-specific immunity parameters and the formation of advanced glycosylation end-products (AGE) in rainbow trout, *Oncorhynchus mykiss* (Walbaum), fed high levels of dietary carbohydrates. Aquac. Res. 30:287–297.

Palmer, T. N., and B. E. Ryman. 1972. Studies on glucose intolerance in fish. J. Fish Biol. 4:311–319.

Panserat, S., F. Médale, J. Brèque, E. Plagnes-Juan, and S. Kaushik. 2000a. Lack of significant long-term effect of dietary carbohydrates on hepatic glucose-6-phosphatase expression in rainbow trout (*Oncorhynchus mykiss*). J. Nutr. Biochem. 11:22–29.

Panserat, S., C. Blin, F. Médale, E. Plagnes-Juan, J. Brèque, J. Krishnamoorthy, and S. J. Kaushik. 2000b. Molecular cloning, tissue distribution and sequence analysis of complete glucokinase cDNAs from gilthead seabream (*Sparus aurata*), rainbow trout (*Oncorhynchus mykiss*) and common carp (*Cyprinus carpio*). Biochim. Biophys. Acta 1474:61–69.

Panserat, S., F. Médale, C. Blin, J. Brèque, C. Vachot, E. Plagnes-Juan, J. Krishnamoorthy, and S. J. Kaushik. 2000c. Hepatic glucokinase is induced by dietary carbohydrates in rainbow trout, gilthead seabream, and common carp. Am. J. Physiol. Reg. I 278:R1164–R1170.

Panserat, S., E. Plagnes-Juan, J. Brèque, and S. Kaushik. 2001. Hepatic phosphoenolpyruvate carboxykinase gene expression is not repressed by dietary carbohydrates in rainbow trout (*Oncorhynchus mykiss*). J. Exp. Biol. 204:359–365.

Panserat, S., E. Plagnes-Juan, and S. Kaushik. 2002. Gluconeogenic enzyme gene expression is decreased by dietary carbohydrates in common carp (*Cyprinus carpio*) and gilthead seabream (*Sparus aurata*). Biochim. Biophys. Acta 1579:35–42.

Panserat, S., S. Skiba-Cassy, I. Seiliez, M. Lansard, E. Plagnes-Juan, C. Vachot, P. Aguirre, L. Larroquet, G. Charvegnac, F. Médale, G. Corraze, S. Kaushik, and T. W. Moon. 2009. Metformin improves postprandial glucose homeostasis in rainbow trout fed dietary carbohydrates: A link with the induction of hepatic lipogenic capacities? Am. J. Physiol. Reg. I 297:707–715.

Parrizas, M., J. Planas, E. M. Plisetskaya, and J. Gutierrez. 1994. Insulin binding and receptor tyrosine kinase activity in skeletal muscle of carnivorous and omnivorous fish. Am. J. Physiol. Reg. I 266:R1944–R1950.

Pavasovic, A., N. R. Richardson, P. B. Mather, and A. J. Anderson. 2006. Influence of insoluble dietary cellulose on digestive enzyme activity, feed digestibility and survival in the red clay crayfish, *Cherax quadricarinatus* (von Martens). Aquac. Res. 37:25–32.

Pereira, C., M. M. Vijayan, K. B. Storey, R. A. Jones, and T. W. Moon. 1995. Role of glucose and insulin in regulating glycogen synthase and phosphorylase activities in rainbow trout hepatocytes. J. Comp. Physiol. B 165:62–70.

Peres, H., and E. Oliva-Teles. 2002. Utilization of raw and gelatinized starch by European sea bass (*Dicentrarchus labrax*) juveniles. Aquaculture 205:287–299.

Peres, H., P. Goncalves, and A. Oliva-Teles. 1999. Glucose tolerance in gilthead seabream (*Sparus aurata*) and European sea bass (*Dicentrarchus labrax*). Aquaculture 179:415–423.

Perez-Jimenez, A., M. J. Guedes, A. E. Morales, and A. Oliva-Teles. 2007. Metabolic responses to short starvation and refeeding in *Dicentrarchus labrax*. Effect of dietary composition. Aquaculture 265:325–335.

Phillips, S. N., Jr., A. V. Tunison, and D. R. Brockway. 1948. Utilization of carbohydrates by trout. Fish. Res. B. NY 11:1–44.

Pilkis, S., and D. K. Granner. 1992. Molecular physiology of the regulation of hepatic gluconeogenesis and glycolysis. Annu. Rev. Physiol. 54:885–909.

Plagnes-Juan, E., M. Lansard, I. Seiliez, F. Médale, G. Corraze, S. Kaushik, S. Panserat, and S. Skiba-Cassy. 2008. Insulin regulates the expression of several metabolism-related genes in the liver and primary hepatocytes of rainbow trout (*Oncorhynchus mykiss*). J. Exp. Biol. 211:2510–2518.

Planas, J. V., E. Capilla, and J. Gutierrez. 2000. Molecular identification of a glucose transporter from fish muscle. FEBS Lett. 481:266–270.

Printz, R. L., M. A. Magnuson, and D. K. Granner. 1993. Mammalian glucokinase. Annu. Rev. Nutr. 13:463–496.

Puviani, A. C., C. Ottolenghi, M. E. Gavioli, E. Fabbri, and L. Brighenti. 1990. Action of glucagon and glucagon-like peptide on glycogen metabolism of trout isolated hepatocytes. Comp. Biochem. Phys. B 96:387–391.

Refstie, S., B. Svihus, K. D. Shearer, and T. Storebakken. 1999. Nutrient digestibility in Atlantic salmon and broiler chickens related to viscosity and non-starch polysaccharide content in different soyabean products. Anim. Feed Sci. Tech. 79:331–345.

Rivas-Vega, M. E., E. Goytortua-Bores, J. M. Ezquerra-Brauer, M. G. Salazar-Garcia, L. E. Cruz-Suarez, H. Nolasco, and R. Civera-Cerecedo. 2006. Nutritional value of cowpea (*Vigna unguiculata L. Walp*) meals as ingredients in diets for Pacific white shrimp (*Litopenaeus vannamei* Boone). Food Chem. 97:41–49.

Robertson, J. B., and P. J. Van Soest. 1981. The detergent system of analysis and its application to human foods. Pp. 123–158 in The Analysis of Dietary Fiber in Food, W. P. T. James and O. Theander, eds. New York: Marcel Dekker.

Rollin, X., F. Médale, S. Gutières, D. Blanc, and S. J. Kaushik. 2003. Short and long term nutritional modulation of acetyl-coenzyme A carboxylase activity in selected tissues of rainbow trout (*Oncorhynchus mykiss*). Brit. J. Nut. 89:1–9.

Rosas, C., G. Cuzon, G. Gaxiola, L. Arena, P. Lemaire, C. Soyez, and A. Van Wormhoud. 2000. Influence of dietary carbohydrate on the metabolism of juvenile *Litopenaeus stylirostris*. J. Exp. Mar. Biol. Ecol. 249:181–198.

Rosas, C., G. Cuzon, G. Gaxiola, Y. Le Priol, C. Pascual, J. Rossignyol, F. Contreras, A. Sanchez, and A. Van Wormhoud. 2001. Metabolism and growth of juveniles of *Litopenaeus vannamei*: Effect of salinity and dietary carbohydrate levels. J. Exp. Mar. Biol. Ecol. 259:1–22.

Rust, M. B. 2002. Nutritional Physiology. Pp. 367–505 in Fish Nutrition, J. E. Halver and R. W. Hardy, eds. San Diego, CA: Academic Press.

Sanders, B. 1983a. Insulin-like peptides in the lobster *Homarus americanus*. II Insulin immunoreactivity. Gen. Comp. Endocr. 50:366–373.

Sanders, B. 1983b. Insulin-like peptides in the lobster *Homarus americanus*. I Insulin-like Biological activity. Gen. Comp. Endocr. 50:374–377.

Shiau, S-Y., 1997. Utilization of carbohydrates in warmwater fish—with particular reference to tilapia, *Oreochromis niloticus* × *O. aureus*. Aquaculture 151:79–96.

Shiau, S-Y., and J. C. Chuang. 1995. Utilization of disaccharides by juvenile tilapia, *Oreochromis niloticus* × *O. aureus*. Aquaculture 133:249–256.

Shiau, S. Y., and H. S. Liang. 1994. Nutrient digestibility and growth of hybrid tilapia, *Oreochromis niloticus* × *Oreochromis aureus*, as influenced by agar supplementation at 2 dietary protein levels. Aquaculture 127:41–48.

Shiau, S-Y., and H. S. Liang. 1995. Carbohydrate utilization and digestibility by tilapia, *Oreochromis niloticus* × *O. aureus*, are affected by chromic oxide inclusion in the diet. J. Nutr. 125:976–982.

Shiau, S. Y., and Y. H. Lin. 2001. Carbohydrate utilization and its protein-sparing effect in diets for grouper, *Epinephelus malabaricus*. Animal Sci. 73:299–304.

Shiau, S. Y., and C. Y. Peng. 1992. Utilization of different carbohydrates at different dietary protein levels in grass prawn, *Penaeus monodon*, reared in seawater. Aquaculture 101:241–250.

Shiau, S-Y., and Y. P. Yu. 1998. Chitin but not chitosan supplementation enhances grass shrimp, Penaeus monodon, growth. J. Nutr. 128:908–912.

Shiau, S-Y., and Y. P. Yu. 1999. Dietary supplementation of chitin and chitosan depresses growth in tilapia, *Oreochromis niloticus* × *O. aureus*. Aquaculture 179:439–446.

Shiau, S-Y., H. L. Yu, S. Hwa, S. Y. Chen, and S. I. Hsu. 1988. The influence of carboxymethylcellulose on growth, digestion, gastric emptying time and body composition of tilapia. Aquaculture 70:345–354.

Shiau, S-Y., C. C. Kwok, C. J. Chen, H. T. Hong, and H. B. Hsieh. 1989. Effects of dietary fibre on the intestinal absorption of dextrin, blood sugar level and growth of tilapia, *Oreochromis niloticus* × *O. aureus*. J. Fish Biol. 34:929–935.

Shikata, T., S. Iwanaga, and S. Shimeno. 1994. Effects of dietary glucose, fructose, and galactose on hepatopancreatic enzyme activities and body composition in carp. Fisheries Sci. 60:613–617.

Shimeno, S., D-C. Ming, and M. Takeda. 1993. Regulation of carbohydrate metabolism in fish XVI. Metabolic response to dietary carbohydrate to lipid ratios in *Oreochromis niloticus*. B. Jpn. Soc. Sci. Fish. 59:827–833.

Shimeno, S., D. Kheyyalid, and T. Shikata. 1995. Metabolic response to dietary carbohydrate to protein ratios in carp. Fisheries Sci. 61:277–281.

Singh, R. P., and T. Nose. 1967. Digestibility of carbohydrates in young rainbow trout. B. Fresh. Fish Res. Lab. 17:21–25.

Skiba-Cassy, S., M. Lansard, S. Panserat, and F. Médale. 2009. Rainbow trout genetically selected for greater muscle fat content display increased activation of liver TOR signaling and lipogenic gene expression. Am. J. Physiol. Reg. I 297:1421–1429.

Skrede, G., T. Storebakken, A. Skrede, S. Sahlstrom, M. Sorensen, K. D. Shearer, and E. Slind. 2002. Lactic acid fermentation of wheat and barley whole meal flours improves digestibility of nutrients and energy in Atlantic salmon (*Salmo salar*) diets. Aquaculture 210:305–321.

Smith, D. 1981. Removing and analyzing total non-structural carbohydrates from plant tissues. Wisconsin Ag. Exp. Stn. Rep. No R2107, Madison.

Soengas, J. L., and M. Aldegunde. 2002. Energy metabolism of fish brain. Comp. Biochem. Phys. B 131:271–296.

Soengas, J. L., E. F. Strong, J. Fuentes, J. A. R. Veira, M. D. Andres. 1996. Food deprivation and refeeding in Atlantic salmon, *Salmo Salar*—effects on brain and liver carbohydrate and ketone bodies metabolism. Fish Physiol. Biochem. 15:491–511.

Soengas, J. L., S. Polakof, X. Chen, S. Sangiao-Alvarellos, and T. W. Moon. 2006. Glucokinase and hexokinase expression and activities in rainbow trout tissues: Changes with food deprivation and refeeding. Am. J. Physiol. Reg. I 291: R810–R821

Stickney, R. R., and S. E. Shumway. 1974. Occurrence of cellulose activity in the stomachs of fishes. J. Fish Biol. 6:779–790.

Stone, D. A. J. 2003. Dietary carbohydrate utilization by fish. Rev. Fish. Sci. 11:337–369.

Stone, D. A. J., G. L. Allan, and A. J. Anderson. 2003a. Carbohydrate utilization by juvenile silver perch, *Bidyanus bidyanus* (Mitchell). I. Uptake and clearance of monosaccharides following intraperitoneal injection. Aquac. Res. 34:97–107.

Stone, D. A. J., G. L. Allan, and A. J. Anderson. 2003b. Carbohydrate utilization by juvenile silver perch, *Bidyanus bidyanus* (Mitchell). II. Digestibility and utilization of starch and its breakdown products. Aquac. Res. 34:109–121.

Storebakken, T., K. D. Shearer, S. Refstie, S. Lagocki, and J. McCool. 1998. Interactions between salinity, dietary carbohydrate source and carbohydrate concentration on the digestibility of macronutrients and energy in rainbow trout (*Oncorhynchus mykiss*). Aquaculture 163:347–359.

Storebakken, T., K. D. Shearer, G. Baeverfjord, B. G. Nielsen, T. Asgard, T. Scott, and A. DeLaporte. 2000. Digestibility of macronutrients, energy and amino acids, absorption of elements and absence of intestinal enteritis in Atlantic salmon, *Salmo salar*, fed diets with wheat gluten. Aquaculture 184:115–132.

Suarez, R. K., and T. P. Mommsen. 1987. Gluconeogenesis in teleost fishes. Can. J. Zoolog. 65:1869–1882.

Sundby, A., G. I. Hemre, B. Borrebaek, B. Christophersen, and A. K. Blom. 1991. Insulin and glucagon family peptides in relation to activities of hepatic hexokinase and other enzymes in fed and starved Atlantic salmon (*Salmo salar*) and cod (*Gadus morhua*). Comp. Biochem. Phys. B 100:467–470.

Svihus, B., A. K. Uhlen, and O. M. Harstad. 2005. Effect of starch granule structure, associated components and processing on nutritive value of cereal starch: A review. Anim. Feed Sci. Tech. 122:303–320.

Tan, Q., S. Xie, X. Zhu, W. Lei, and Y. Yang. 2006. Effect of dietary carbohydrate sources on growth performance and utilization for gibel carp (*Carassius auratus gibelio*) and Chinese longsnout catfish (*Leiocassis longirostris* Gunther). Aquacult. Nutr. 12:61–70.

Thodesen, J., and T. Storebakken. 1998. Digestibility of diets with precooked rye or wheat by Atlantic salmon, *Salmo salar* L. Aquacult. Nutr. 4:123–126.

Tranulis, M. A., O. Dregni, B. Christophersen, Å. Krogdahl, and B. Borrebaek. 1996. A glucokinase-like enzyme in the liver of Atlantic salmon (*Salmo salar*). Comp. Biochem. Phys. B. 114:35–39.

Tranulis, M. A., B. Christophersen, and B. Borrebaek. 1997. Glucokinase in Atlantic halibut (*Hippoglossus hippoglossus*) Brockmann bodies. Comp. Biochem. Phys. B. 116:367–370.

Tung, P. S., and S. Y. Shiau. 1991. Effects of meal frequency on growth performance of hybrid tilapia, *Oreochromis niloticus* × *O. aureus*, fed different carbohydrate diets. Aquaculture 92:343–350.

Van Soest, P. J. 1994. Nutritional Ecology of the Ruminant. Ithaca, NY: Cornell University Press.

Van Soest, P. J., and R. H. Wine. 1967. Use of detergents in the analysis of fibrous feeds. IV. J. Assoc. Off. Agric. Chem. 50:50–55.

Van Soest, P. J., J. B. Robertson, and B. A. Lewis. 1991. Methods for dietary fiber, neutral detergent fiber, and nonstarch polysaccharides in relation to animal nutrition. J. Dairy Sci. 74:3583–3597.

Verri, T., A. Mandal, L. Zilli, D. Bossa, P. K. Mandal, L. Ingrosso, V. Zonno, S. Vilella, G. A. Ahearn, and C. Storelli. 2001. Review: D-Glucose transport in decapod crustacean hepatopancreas. Comp. Biochem. Phys. A. 130:585–606.

Walton, M. J. 1986. Metabolic effects of feeding a high protein/low carbohydrate diet as compared to a low protein/high carbohydrate diet to rainbow trout *Salmo gairdneri*. Fish Physiol. Biochem. 1:7–15.

Walton, M. J., and C. B. Cowey. 1982. Aspects of intermediary metabolism in salmonid fish. Comp. Biochem. Phys. B 73:59–79.

Wang, K. W., T. Takeuchi, and T. Watanabe. 1985. Optimum protein and digestible energy levels in diets for tilapia nilotica. B. Jpn. Soc. Sci. Fish. 51:141–146.

West, T. G., P. G. Arthur, R. K. Suarez, C. J. Doll, and P. W. Hochachka. 1993. In vivo utilization of glucose by heart and locomotory muscles of exercising rainbow trout (*Oncorhynchus mykiss*). J. Exp. Biol. 177:63–79.

Wilson, R. P. 1994. Utilization of dietary carbohydrate by fish. Aquaculture 124:67–80.

Wilson, R. P., and W. E. Poe. 1987. Apparent inability of Channel catfish to utilize dietary monosaccharides and disaccharides as energy sources. J. Nutr. 117:280–285.

Wright, J. J. R., W. Ohali, H. Yang, X. Han, and A. Bonen. 1998. GLUT-4 Deficiency and severe peripheral resistance to insulin in the teleost fish tilapia. Gen. Comp. Endocr. 111:20–27.

Yamamoto, T., T. Shima, H. Furuita, N. Suzuki, and M. Shiraishi. 2001. Nutrient digestibility values of a test diet determined by manual feeding and self-feeding in rainbow trout and common carp. Fish Sci. 67:355–357.

Zeng, M., C. F. Morris, I. L. Batey, and C. W. Wrigley. 1997. Sources of variation for starch gelatinization, pasting, and gelation properties in wheat. Cereal Chem. 74:63–71.

Zhou, X., and M. L. Kaplan. 1997. Soluble amylose corn starch is more digestible than soluble amylopectin potato starch in rats. J. Nutr. 127:1349–1356.

Zobel, H. F. 1988. Molecules to granules: A comprehensive starch review. Starch 40:44–50.

8

Minerals

Information concerning mineral nutrition of fish and crustaceans is limited compared to most other nutrient groups. In addition, there is much less information on mineral requirements of aquatic species compared to terrestrial animals, in part because of complications in that fish can absorb some minerals from the aquatic medium in which they live, in addition to from their diet. Nevertheless, the basic metabolic functions of the various mineral elements are the same for aquatic and terrestrial animals with the exception of osmoregulation (Table 8-1). The large discrepancy in knowledge of mineral nutrition between terrestrial and aquatic species is due to the relative infancy of aquatic animal nutrition and the related difficulty of conducting research on mineral nutrition of aquatic species. Problems associated with the quantification of mineral requirements include identification of the potential contribution of minerals from the water, leaching of minerals from the diet prior to consumption, availability of suitable test diets that have a low concentration of the targeted mineral, and limited data on mineral bioavailability. Despite these problems, mineral requirement data for a variety of species originating from diverse environments are becoming established (Tables 8-2, 8-3, and 8-4).

The metabolism of various minerals by aquatic organisms is influenced not only by dietary concentrations but also by the concentration and relative composition of dissolved ions in the aquatic medium, because they may influence the organism's osmoregulation, ion regulation, and acid:base balance (Moyle and Cech, 2000). Numerous minerals can be absorbed by the gills and contribute to meeting metabolic requirements. The uptake of minerals from the diet or aquatic medium and excretion of minerals in the urine and feces are influenced by osmoregulatory processes in response to the salinity of the aquatic medium. Organisms in freshwater are characterized as being hyperosmotic to the environment. They continually lose small ions from the gills, and water is passively taken up such that they do not drink water but excrete large quantities of dilute urine. In contrast, organisms in seawater are hyposmotic to the environment and thus lose

TABLE 8-1 Minerals and Some of Their Prominent Functions and Deficiency Signs Observed in Fish and Shrimp

Mineral	Functions	Deficiency Signs
Macromineral		
Calcium	Skeletal tissues, membrane permeability	Impaired growth and hard tissue mineralization
Chloride	Osmotic balance	Impaired growth
Magnesium	Enzyme activator	Tetany, muscle flaccidity
Phosphorus	Skeletal tissue, phospholipids	Impaired growth, reduced hard tissue mineralization, skeletal deformities, fat accumulation
Potassium	Osmotic balance, acid-base equilibrium	Convulsions, tetany
Sodium	Osmotic balance, acid-base equilibrium	Impaired growth
Micromineral		
Copper	Metalloenzymes	Impaired growth and reduced activity of copper-containing enzymes
Cobalt	Vitamin B_{12}	Anemia
Chromium	Carbohydrate metabolism	Impaired glucose utilization
Iodine	Thyroid hormones	Thyroid hyperplasia
Iron	Hemoglobin	Impaired growth, anemia
Manganese	Organic matrix of bone	Impaired growth, skeletal abnormalities, cataracts
Molybdenum	Xanthine oxidase	Reduced enzyme activity
Selenium	Glutathione peroxidase	Impaired growth, anemia, exudative diathesis, reduced activity of glutathione peroxidase
Zinc	Metalloenzymes	Impaired growth, cataracts, skeletal abnormalities, reduced activity of various zinc metalloenzymes

TABLE 8-2 Macromineral Requirements of Fish

Mineral	Species	Recommended Supplement (g/100 g diet)	Rearing Condition[a]	Reference
Calcium	Channel catfish (*Ictalurus punctatus*)	1.5	FW Practical diet	Andrews et al. (1973)
		Dispensable	FW (14 mg Ca/L)	Lovell (1978)
		0.45	FW (Ca-free)	Robinson et al. (1986)
	Blue tilapia (*Oreochromis aureus*)	0.17–0.65	FW (Ca-free)	Robinson et al. (1984)
		0.7	FW (Ca-free)	Robinson et al. (1987)
	Red sea bream (*Chrysophrys major*)	Dispensable	SW	Sakamoto and Yone (1976a)
	Common carp (*Cyprinus carpio*)	Dispensable	FW (20 mg Ca/L)	Ogino and Takeda (1976)
	Rainbow trout (*Oncorhynchus mykiss*)	Dispensable	FW (20–23 mg Ca/L)	Ogino and Takeda (1978)
	Chum salmon (*Oncorhynchus keta*)	Dispensable	FW (20 mg Ca/L)	Watanabe et al. (1980)
	Hybrid tilapia (*Oreochromis niloticus* × *Oreochromis aureas*)	0.35–0.43	FW (27–33 mg Ca/L)	Shiau and Tseng (2007)
Phosphorus	*I. punctatus*	0.8	FW Practical diet	Andrews et al. (1973)
		0.45	FW (0.03 mg P/L)	Lovell (1978)
		0.33 (available P)	FW (0.04 mg P/L)	Wilson et al. (1982)
	Cyprinus carpio	0.6–0.7	FW (0.002 mg P/L)	Ogino and Takeda (1976)
	Oreochromis aureus	0.5	FW (Ca-free)	Robinson et al. (1987)
	Oncorhynchus mykiss	0.7–0.8	FW (0.002 mg P/L)	Ogino and Takeda (1978)
		0.54–0.61	FW	Ketola and Richmond (1994)
	Oncorhynchus keta	0.5–0.6	FW (0.002 mg P/L)	Watanabe et al. (1980)
	Atlantic salmon (*Salmo salar*)	0.6 (available P)	FW (< 0.5 mg P/L) 0.7% dietary P from plant sources	Ketola (1975)
		0.83–0.93	FW	Vielma and Lall (1998)
		1.0 (0.9 available P)	FW	Asgard and Shearer (1997); Lall and Bishop (1977)
	Hybrid striped bass (*Morone chrysops* × *Morone saxatili*)	0.5	FW (150 mg/L hardness as $CaCO_3$)	Brown et al. (1993)
	Chrysophrys major	0.68	SW	Sakamoto and Yone (1978a)
	Milkfish (*Chanos chanos*)	0.85	SW	Borlongan and Satoh (2001)
	Red drum (*Sciaenops ocellatus*)	0.86	BW (5–6‰)	Davis and Robinson (1987)
	Yellow croaker (*Pseudosciaena crocea*)	0.89–0.91 (available P)	SW	Ma et al. (2006)
	Haddock (*Melanogrammus aeglefinus*)	0.96 (0.72 available P)	SW	Roy and Lall (2003)
	Japanese flounder (*Paralichthys olivaceus*)	0.6–1.5 (total P)	SW	Choi et al. (2005); Wang et al. (2005); Uyan et al. (2007)
	Black sea bream (*Acanthopagrus schlegeli*)	0.55 (available P)	SW	Shao et al. (2008)
	Japanese sea bass (*Lateolabrax japonicus*)	0.86–0.90	SW	Zhang et al. (2006)
	Orange-spotted grouper (*Epinephelus coioides*)	1.09	SW	Ye et al. (2006)
	Gilthead sea bream (*Sparus auratus*)	0.75	SW	Pimentel-Rodrigues and Olivia-Teles (2001)
	European sea bass (*Dicentrarchus labrax*)	0.65	SW	Olivia-Teles and Pimentel-Rodrigues (2004)

TABLE 8-2 Continued

Mineral	Species	Recommended Supplement (g/100 g diet)	Rearing Condition[a]	Reference
Ca:P ratio	*I. punctatus*	1.5:0.8	FW Practical diet	Andrews et al. (1973)
		No relationship	FW (14 mg Ca/L)	Lovell (1978)
	Chrysophrys major	0.34:0.68	SW	Sakamoto and Yone (1973)
		No relationship	SW	Sakamoto and Yone (1976a)
	Cyprinus carpio	No relationship	FW (20 mg Ca/L)	Ogino and Takeda (1976)
	Oncorhynchus mykiss	No relationship	FW (20–23 mg Ca/L)	Ogino and Takeda (1978)
	Oncorhynchus keta	No relationship	FW (20 mg Ca/L)	Watanabe et al. (1980)
	Chinook salmon (*Oncorhynchus tshawytscha*)	0.6–1.2	FW	Shearer (1988)
Potassium	*I. punctatus*	0.26	FW (4 mg K/L)	Wilson and El Naggar (1992)
	Chrysophrys major	Dispensable	SW	Sakamoto and Yone (1978b)
	I. punctatus	Dispensable	FW Practical diet	Murray and Andrews (1979)
	Oreochromis niloticus × *Oreochromis aureas*	0.2–0.3	FW	Shiau and Hsieh (2001a)
Sodium/ chloride	*Oncorhynchus mykiss*	Dispensable	FW; Levels up to 11.6% produced no adverse effects	Salman and Eddy (1988)
	Red drum (*Sciaenops ocellatus*)	2	FW and BW (6‰)	
		Dispensable	35‰	Gatlin et al. (1992)
Sodium	*Chrysophrys major*	Dispensable	SW	Sakamoto and Yone (1978b)
	Oreochromis niloticus × *Oreochromis aureas*	0.15	FW	Shiau and Lu (2004)
	Oreochromis niloticus × *Oreochromis aureas*	Dispensable	SW	Shiau and Lu (2004)
Magnesium	*I. punctatus*	0.04	FW (1.6 mg Mg/L)	Gatlin et al. (1982)
	Oncorhynchus mykiss	0.06–0.07	FW (3.1 ppm Mg)	Ogino et al. (1978)
		0.05	FW (1.2 mg Mg/L)	Knox et al. (1981)
		0.06	FW (1.3 mg Mg/L)	Shearer (1989)
	Nile tilapia (*Oreochromis niloticus*)	0.059–0.077	FW (1.0 mg Mg/L)	DaBrowska et al. (1989)
	Mozambique tilapia (*Oreochromis mossambicus*)	Dispensable	FW	van der Velden et al. (1991)
	Oreochromis aureus	0.023	FW	Reigh et al. (1991)
	Cyprinus carpio	0.06	FW (9.4 mg Mg/L)	Dabrowska et al. (1991)
	Chrysophrys major	Dispensable	SW (0.012% basal diet)	Sakamoto and Yone (1979a)

[a]Information about diet and/or water used in the experiment. FW = freshwater, BW = brackish water, SW = seawater.

water (but gain monovalent ions from the environment) such that they must drink water. The excess salts ingested are primarily excreted by specialized chloride cells in the gills and opercular skin epithelia via active transport, while the kidney excretes primarily divalent ions in small volumes of urine and other salts are concentrated in feces. In addition to maintaining stable internal osmotic concentrations relative to the aquatic environment, organisms also exert energy to maintain appropriate ionic and acid:base balance via active and passive processes in various organs including the gills, kidney, and gastrointestinal tract (Moyle and Cech, 2000). As such, these various processes may directly affect the metabolism of certain minerals.

The functions of macrominerals, those required in the diet and body at relatively high concentrations, include the formation of skeletal structures and other hard tissues (e.g., fin rays, scales, teeth, and exoskeleton), electron transfer, regulation of acid:base equilibrium, the production of membrane potentials, and osmoregulation. Six minerals, including calcium, chlorine, magnesium, phosphorus, potassium, and sodium, are the most commonly recognized macrominerals. Specific information about each of the macrominerals will be detailed in individual sections later in this chapter. Trace minerals or microminerals, which are typically required in the diet and body at much lower concentrations than the macrominerals, are important components of hormones and enzymes, serve as cofactors and/or activators of a variety of enzymes, as well as participate in a wide variety of biochemical processes. The most commonly recognized trace minerals include chromium, copper, iodine, iron, manganese, selenium, and zinc. Detailed information about each of these microminerals also will be provided in individual sections later in this chapter. Among the most important dietary minerals, eight are cations: calcium (Ca^{++}), copper (Cu^{2+}), iron (Fe^{2+}), magnesium (Mg^{2+}), manganese (Mn^{2+}), potassium (K^{+}), sodium (Na^{+}), and zinc (Zn^{2+}). Five are anions or are

TABLE 8-3 Micromineral Requirements of Fish

Mineral	Species	Recommended Supplement (mg/kg diet)	Rearing Condition[a]	Reference
Copper	Channel catfish	1.5	FW	Murai et al. (1981)
	(Ictalurus punctatus)	5	FW	Gatlin and Wilson (1986a)
	Common carp *(Cyprinus carpio)*	3	FW	Ogino and Yang (1980)
	Rainbow trout	3	FW	Ogino and Yang (1980)
	(Oncorhynchus mykiss)	3.5	FW	Julshamn et al. (1988)
	Atlantic salmon	5–10 (as supplement)	FW; 3.5 mg Cu/kg	Lorentzen et al. (1998)
	(Salmo salar)	≥ 500 toxic	FW (0.6 µg/L)	Berntssen et al. (1999)
	Hybrid tilapia *(Oreochromis niloticus × Oreochromis aureas)*	4	FW	Shiau and Ning (2003)
	Malabar grouper *(Epinephelus malabaricus)*	4–6	SW	Lin et al. (2008a)
Iron	*I. punctatus*	30	FW (0.43 mg Fe/L)	Gatlin and Wilson (1986b)
	Cyprinus carpio	199	FW (dietary levels 1 and 199 mg Fe/kg)	Sakamoto and Yone (1978b)
	Red sea bream *(Chrysophrys major)*	199	SW (dietary levels 1 and 199 mg Fe/kg)	Sakamoto and Yone (1976b)
		150	SW	Sakamoto and Yone (1978c)
	Oreochromis niloticus × Oreochromis aureas	85	FW	Shiau and Su (2003)
Zinc	*I. punctatus*	20	FW (25 µg Zn/L)	Gatlin and Wilson (1983)
		150	FW with 1.1% phytate	Gatlin and Wilson (1984c)
	Cyprinus carpio	15–30	FW (10 µg Zn/L)	Ogino and Yang (1979)
	Oncorhynchus mykiss	15–30	FW (11 µg Zn/L)	Ogino and Yang (1978)
		20–40	FW	Satoh et al. (1987)
		40	4% Tricalcium phosphate	
		80	7% Tricalcium phosphate	
	Blue tilapia *(Oreochromis aureus)*	20	FW (4 µg Zn/L)	McClain and Gatlin (1988)
	Red drum *(Sciaenops ocellatus)*	20	BW (6‰)	Gatlin et al. (1991)
	Nile tilapia *(Oreochromis niloticus)*	30	FW	Eid and Ghonim (1994)
	Oreochromis niloticus × Oreochromis aureas	26–29	FW	Lin et al. (2008c)
Manganese	*I. punctatus*	2.4	FW (2 µg Mn/L)	Gatlin and Wilson (1984a)
	Cyprinus carpio	12–13	FW	Ogino and Yang (1980)
	Oncorhynchus mykiss	12–13	FW	Ogino and Yang (1980)
	Mossambique tilapia *(Oreochromis mossambica)*	1.7	FW	Ishac and Dollar (1968)
	Oreochromis niloticus × Oreochromis aureas	7	FW	Lin et al. (2008b)
Selenium	*I. punctatus*	0.25	FW; adequate vitamin E	Gatlin and Wilson (1984b)
	Oncorhynchus mykiss	0.15–0.38	FW (0.4 µg Se/L); adequate vitamin E	Hilton et al. (1980)
		0.07	Prevented overt deficiency	Poston et al. (1976)
		3	May produce toxicity	
		13	Toxic	
	Malabar grouper *(Epinephelus malabaricus)*	0.7	SW	Lin and Shiau (2005)
Iodine	Chinook salmon *(Oncorhynchus tshawytscha)*	0.6–1.1	FW (0.2 µg I/L)	Woodall and LaRoche (1964)

[a]Information about diet and/or water used in the experiment. FW = freshwater, BW = brackish water SW = seawater.

TABLE 8-4 Mineral Requirements of Crustaceans

Macromineral	Species	Dietary Protein	Requirement (g/100 g)	Reference
Calcium	Kuruma prawn *(Marsupenaeus japonicus)*	Casein—egg	Dispensable	Deshimaru and Yone (1978)
		Squid meal	1.2	Kitabayashi et al. (1971)
		Casein	1.0–2.0	Kanazawa et al. (1984)
	Pacific white shrimp *(Litopenaeus vannamei)*	Casein—gelatin	Dispensable	Davis et al. (1993a)
	Tiger shrimp *(Penaeus monodon)*	Casein—gelatin	Dispensable	Penaflorida (1999)
Phosphorus	*M. japonicus*	Casein—egg	2.0	Deshimaru and Yone (1978)
		Casein	1.0–2.0	Kanazawa et al. (1984)
		Squid meal	1.0	Kitabayashi et al. (1971)
	L. vannamei	Casein—gelatin	0.03% dietary Ca, ≤ 0.34% P 1% Ca, 0.5–1.0% P 2% Ca, 1.0–2.0% P	Davis et al. (1993a)
		Casein—gelatin	0.5% Ca, 0.93% total P (EAP[a] = 0.77% P) 1.5% Ca, 2% total P (EAP = 1.22% P)	Cheng et al. (2006)
		Fish meal, soybean meal	> 1.33%, practical diet	Pan et al. (2005)
	P. monodon	Casein—gelatin	0.5% (0.74 total) at low Ca	Penaflorida (1999)
Ca:P ratio	American lobster *(Homarus americanus)*	Casein, yeast	0.56:1.10	Gallagher et al. (1978)
	H. americanus (juvenile)	Casein—fish meal	1:1	Gallagher et al. (1982)
	H. americanus (adult)	Casein	1:1	Kanazawa et al. (1984)
	M. japonicus	Squid meal	1.24:1.04	Kitabayashi et al. (1971)
	Yellow leg shrimp *(Penaeus californiensis)*	Soybean meal, shrimp head meal, and fish meal	2.06:1, < 2.42:1	Huner and Colvin (1977)
Potassium	*L. vannamei*	Casein—gelatin	Unclear	Davis et al. (1993a)
	M. japonicus	Casein—egg	1	Deshimaru and Yone (1978)
		Casein	0.9	Kanazawa et al. (1984)
	P. monodon	Casein	1.20	Shiau and Hsieh (2001b)
		Casein	1.09	Zhu et al. (2006)
Magnesium	*M. japonicus*	Casein—egg	Dispensable	Deshimaru and Yone (1978)
		Casein	0.3	Kanazawa et al. (1984)
	L. vannamei	Casein—gelatin	0.26–0.35	Cheng et al. (2005)
Micromineral	**Species**	**Dietary Protein**	**Requirement (mg/kg diet)**	**Reference**
Copper	*M. japonicus*	Casein	Dispensable	Kanazawa et al. (1984)
	Fleshy prawn *(Penaeus orientalis)*	Fish meal, peanut meal	53	Liu et al. (1990)
	L. vannamei	Casein—gelatin	16–32	Davis et al. (1993b)
	P. monodon	Casein	10–30	Lee and Shiau (2002)
	Fleshy prawn *(Fenneropenaeus chinensis)*		25.3	Wang et al. (1997)
	M. japonicus	Casein—egg	Dispensable	Deshimaru and Yone (1978)
Iron	*M. japonicus*	Casein	Dispensable	Kanazawa et al. (1984)
	L. vannamei	Casein—gelatin	Dispensable	Davis et al. (1992b)
	M. japonicus	Casein	Dispensable	Kanazawa et al. (1984)
Manganese	*M. japonicus*	Casein	Dispensable	Kanazawa et al. (1984)
	L. vannamei	Casein—gelatin	Required	Davis et al. (1992a)
Selenium	*L. vannamei*	Casein—gelatin	0.2–0.4	Davis (1990)
Zinc	*P. monodon*	Casein	32–34 (growth) 35–48 (immunity)	Shiau and Jiang (2006)
	L. vannamei	Casein—gelatin	15 (32 total) 200 (218 total) in the presence of phytate	Davis et al. (1993c)

[a]Estimated available phosphorus.

usually found in anionic groupings: chloride (Cl^-), iodide (I^-), molybdate (MoO_4^{2-}), phosphate (PO_4^{3-}), and selenite (SeO_3^{2-}) (Scott et al., 1982). Other trace minerals such as aluminum, arsenic, cobalt, fluorine, molybdenum, nickel, silicon, tin, and vanadium are typically required in such small amounts that dietary supplementation is not required.

The remainder of this chapter will summarize pertinent information about mineral nutrition of fish and crustaceans, including general functions of the macro- and microminerals, dietary essentiality and/or quantitative requirements, bioavailability and dietary interactions, and general recommendations for dietary supplementation.

CALCIUM AND PHOSPHORUS

Calcium and phosphorus are two of the major constituents of the inorganic portion of diets. Quantitatively, calcium and phosphorus function primarily as structural components of hard tissues (e.g., bone, exoskeleton, scales, and teeth). In addition to its structural functions, calcium is essential for blood clotting (vertebrates), muscle function, proper nerve impulse transmission, osmoregulation, and as a cofactor for enzymatic processes (Lall, 2002). Phosphorus is a component of a variety of organic phosphates, such as nucleotides, phospholipids, coenzymes, deoxyribonucleic acid (DNA), and ribonucleic acid. Inorganic phosphates also serve as important buffers to maintain normal pH of intra- and extracellular fluids (Zubay, 1983).

Dietary deficiencies of most macrominerals such as calcium have been generally difficult to produce with fish species because of the presence of these ions in the water. However, supplementation of phosphorus in fish diets is usually most critical because its presence in the water and utilization by fish is limited. Dietary deficiency of phosphorus impairs intermediary metabolism, resulting in reduced growth and feed conversion. Various skeletal malformations associated with reduced mineralization of hard tissues also occur at suboptimal phosphorus intake (Sugiura et al., 2004).

The influence of excreted phosphorus on eutrophication of receiving waters has resulted in a considerable amount of research being focused on phosphorus nutrition in recent years with the aim of minimizing phosphorus excretion, especially for salmonid species cultured in flowing-water or net pen systems where effluents have direct effects on the surrounding waters. The experimental measurement of urinary phosphorus concentration has been a sensitive indicator of dietary phosphorus metabolism in fish (Sigiura et al., 2000). Various dietary manipulations, such as reducing total dietary phosphorus (e.g., Green et al., 2002a) or increasing the availability of phosphorus in the diet by adding the enzyme phytase (see Chapter 10; reviewed in Gatlin and Li, 2008) or other additives such as citric acid (Sugiura et al., 1998; also see Chapter 10), have been effective in reducing urinary and fecal phosphorus excretions. Such nutritional manipulations have been shown to reduce solid and dissolved phosphorus wastes produced by rainbow trout by more than 50% (e.g., Green et al., 2002a,b). Phase-feeding strategies also have been evaluated in conjunction with different dietary phosphorus levels in an effort to minimize dietary phosphorus inputs but maintain adequate production characteristics of the cultured organism (Lellis et al., 2004). The occurrence of spinal deformities in commercially produced Atlantic salmon, such as compression of the vertebral column, has been recognized for several years as a problem due to the reduction in processed fish quality (Sullivan et al., 2007a) and likely was caused by inadequate mineral nutrition. Extensive investigations have led to the conclusion that this condition does not have a genetic basis (Sullivan et al., 2007b), but is primarily caused by phosphorus deficiency in first-feeding fry (Sullivan et al., 2007c). In addition, elevated calcium and phosphorus in the diet of early seawater smolts decreased the incidence of vertebral deformities (Fjellhal et al., 2009). Another recent study on rainbow trout fry revealed that a phosphorus deficiency not only induced phosphorus depletion in whole-body tissues, but also lowered or delayed ossification of both the endochondrial and dermal skeleton (Fontagne et al., 2009). This study also reported that calcium deficiency delayed the ontogeny of skeletal development without affecting final bone mineralization, but did lead to modifications in the size and shape of vertebrae.

Models have been developed using multiple regression approaches to accurately predict the availability of phosphorus in diets formulated with a wide variety of feedstuffs and phosphorus supplements (Hua and Bureau, 2006). Factorial models also have been developed to predict phosphorus waste output in salmonid culture systems (Hua et al., 2008).

In contrast to calcium, concentrations of phosphorus in natural waters are generally very low (Boyd, 1981). Consequently, absorption of significant amounts of phosphorus from freshwater and saltwater is unlikely (Lall, 1991), making a dietary source of phosphorus potentially more critical for both fish and shrimp.

The dietary phosphorus requirements of fish species have been reported to range from 0.3 to 1.5% of diet (Lall, 2002). Some of the variability in these requirement values may be due to differences in phosphorus availability of diets used in quantifying requirements. Available phosphorus requirement values as low as 0.3 and 0.34% of diet have been reported for subadult (Eya and Lovell, 1997a) and fingerling (Wilson et al., 1982) channel catfish (*Ictalurus punctatus*). A higher value of 0.56% of diet was reported for rainbow trout (*Oncorhynchus mykiss*) (Rodehutscord et al., 1995). Values of 0.45% and 0.58% were reported for hybrid striped bass (*Morone chrysops* × *M. saxatilis*) (Brown et al., 1993) and striped bass juveniles (Dougall et al., 1996). The European whitefish (*Coregonus lavaretus*) was reported to have available phosphorus requirements of 0.62 to 0.65% of diet based on maximum growth and vertebral ash content, respectively (Vielma et al., 2002). The minimum available phosphorus requirement value for Japanese sea bass (*Lateolabrax japonicas*) was reported to be 0.68% of diet based on weight gain; higher values of 0.86 and 0.90% were estimated based on whole-body and vertebrae phosphorus (Zhang et al., 2006). An available phosphorus requirement of 0.7% of diet for maximum growth and minimum phosphorus excretion was reported for juvenile common carp (*Cyprinus carpio*) (Kim et al., 1998) as well as for silver perch (*Bidyanus bidyanus*) (Yang et al., 2006). The same requirement level was estimated for the marine yellow croaker (*Pseudosciaena crocea*) based on weight gain with a high value of 0.9% based on phosphorus deposition in vertebrae or whole body (Ma et al., 2006). A similar requirement value of 0.72% of diet was determined for haddock (*Melanogrammus aeglefinus*) (Roy and Lall, 2003) while a value of 0.86% of diet was reported for red drum (*Sciaenops ocellatus*) (Davis and Robinson, 1987). The phosphorus requirement of black sea bream (*Acanthopagrus schlegelii*) was reported to be 0.55% of diet based on weight gain but 0.81, 0.87, and 0.88% of diet based on phosphorus deposition in whole fish, vertebrae, and scales, respectively (Shao et al., 2008). A much lower requirement of 0.44% of diet was estimated for yellowtail (*Seriola quinqueradiata*) based on nonfecal phosphorus excretion (Sarker et al., 2009).

The effects of dietary phosphorus on immunity and disease resistance of fish have been investigated to only a limited extent to date. Eya and Lovell (1998) reported that phosphorus deficiency in channel catfish reduced their antibody production and resistance to *Edwardsiella ictaluri* infection. The minimum phosphorus requirement for maximum protection from *E. ictaluri* was 0.4% of diet, which is similar to that required for maximum weight gain, while phosphorus at 0.5% of diet maximized antibody production (Eya and Lovell, 1998).

Phosphorus requirement estimates for crustacean species in early studies were much higher than the values previously reported for various fish species. For example, phosphorus requirements of 1% (Kitabayashi et al., 1971), 1 to 2% (Kanazawa et al., 1984), and 2% (Deshimaru and Yone, 1978) of the diet were recommended for kuruma prawn (*Marsupenaeus japonicas*). Pan et al. (2005) supplemented graded amounts of calcium phosphate monobasic to practical diets for *Litopenaeus vannamei* and found that total phosphorus at 1.33% was required for optimal weight gain and feed efficiency. However, Davis et al. (1993a) and Cheng et al. (2006) demonstrated that the dietary phosphorus requirement of *L. vannamei* was dependent upon the calcium content of the diet. In the absence of calcium supplementation, Davis et al. (1993a) observed the basal diet (0.03% Ca, 0.34% P) contained adequate phosphorus for normal growth. Cheng et al. (2006) reported that *L. vannamei* weight gain was optimal with available phosphorus at 0.77% of diet in the absence of supplemental calcium, but an increase to 1.22% available phosphorus was needed with 1% supplemental calcium. Similarly, Penaflorida (1999) reported that based on weight gain, *Penaeus monodon* required 0.5% supplemental phosphorus (0.74% total phosphorus) when the diet was not supplemented with calcium, and higher phosphorus supplementation was required with calcium supplementation. Ambasankar and Ali (2002) fed semipurified diets containing graded levels of phosphorus with 1.25% calcium to *Penaeus indicus* and reported that weight gain and body composition was optimal for shrimp fed the diet containing 1% total phosphorus, and shrimp maintained their body phosphorus to calcium ratio at 1:2 regardless of dietary phosphorus level.

Gallagher et al. (1978) examined the effects of varying dietary calcium-to-phosphorus (Ca:P) ratio on juvenile lobsters (*Homarus americanus*). Based on growth and histological evaluation of the endocuticle, a Ca:P of 0.51 (0.56:1.10) was found to be best for lobster juveniles, with Ca:P of 1.55 and greater resulting in abnormalities of the endocuticle. In *Marsupenaeus japonicus*, a dietary Ca:P of 1:1 was recommended (Kitabayashi et al., 1971; Kanazawa et al., 1984). Davis et al. (1993a) reported that supplementation of calcium to the basal diet appeared to inhibit phosphorus bioavailability, and the Ca:P did not totally explain the inhibitory effects of calcium. Based on these studies, dietary calcium may affect phosphorus availability such that calcium levels in excess of 2.5% should be avoided. Although there does not appear to be a fixed Ca:P ratio that will produce optimal results, it appears that a ratio < 2:1 (Ca:P) provides good results in commercial formulations for shrimp as well as for fish such as Atlantic salmon (Vielma and Lall, 1998) and grouper (*Epinephelus coioides*) (Ye et al., 2006).

In contrast to phosphorus, calcium deficiency has been difficult to establish in fish species such as common carp (Ogino and Takeda, 1976) and channel catfish (Andrews et al., 1973; Lovell, 1978) cultured in freshwater with normal calcium hardness because of their ability to absorb waterborne calcium via the gills; however, overt deficiency signs such as reduced weight gain, feed efficiency, and bone mineralization were produced in low-calcium water (Robinson et al., 1984, 1986, 1987). Uptake of waterborne calcium occurs principally from the gills, although other tissues such as fins and oral epithelium also have been associated with this function. Adequate calcium hardness is an important characteristic of freshwater for fish culture, and calcium ions are abundant in seawater; therefore, metabolic requirements for calcium are commonly met by uptake from the aquatic medium. Studies conducted in low-calcium freshwater have established dietary calcium requirements of channel catfish and blue tilapia (*Oreochromis aureus*) of 0.45 and 0.7%, respectively (Robinson et al., 1986, 1987). Shiau and Tseng (2007) reported a dietary calcium requirement ranging from 0.35 to 0.43% for hybrid tilapia (*Oreochromis niloticus* × *O. aureus*) reared in water with 27–33 mg Ca/L based on weight gain, bone calcium, and scale calcium. Calcium concentrations as low as 0.34% of diet were reported as required for common carp and Japanese eel (*Anguilla japonicas*) (Ogino and Takeda, 1976). Atlantic salmon (*Salmo salar*) in seawater utilized waterborne calcium such that dietary supplementation was unnecessary (Lall and Bishop, 1977). In contrast, red sea bream (*Chrysophrys major*) were unable to obtain enough calcium from seawater and required calcium at 0.34% of diet (Sakamoto and Yone, 1973, 1976a). More recently, a series of experiments was conducted on different marine species by Hossain and Furuichi (1998, 1999, 2000) in which a dietary supply of calcium from calcium lactate was needed to support normal growth of tiger puffer (*Takifugu rubripes*) and scorpion fish (*Sebastiscus marmoratus*), but not required for black sea bream (*A. schlegelii*).

Calcium requirements of various crustacean species have been reported (Kitabayashi et al., 1971; Gallagher et al., 1978; Kanazawa et al., 1984); however, *L. vannamei* raised in seawater did not require supplemental calcium (Davis et al., 1993a). Even in low-salinity (2‰ [parts per thousand]) water, a dietary calcium requirement of *L. vannamei* was not apparent, but the amount of calcium in the diet affected the dietary phosphorus requirement (Cheng et al., 2006) as previously observed with *P. monodon* (Penaflorida, 1999). Although the diet plays an important role, it appears that crustaceans also can absorb some minerals from the water via drinking and by direct absorption via the gills, exoskeleton, or both (Deshimaru et al., 1978; NRC, 1993).

Many factors influence the absorption, distribution, and excretion of calcium and phosphorus. Dietary calcium is primarily absorbed from the intestine by active transport. In vertebrates, the vitamin 1,25-dihydroxycholecalciferol functions in the maintenance of serum calcium and phosphorus levels by altering the rate of intestinal absorption (via Ca^{2+}-binding protein), renal resorption, and bone mobilization (Zubay, 1983). There is some evidence that vitamin D and its metabolites may affect calcium homeostasis in teleosts. However, O'Connell and Gatlin (1994) observed that dietary supplementation of cholecalciferol (vitamin D_3) was not required for blue tilapia to utilize dietary calcium for growth and tissue mineralization in low-calcium water. In rainbow trout the major regulator of phosphorus metabolism was dietary phosphorus as compared to vitamin D_3 and its metabolites or the calcium phosphate cotransporter (Coloso et al., 2003). The role of vitamin D_3 in intestinal calcium uptake of fish, therefore, is not well established as it is in terrestrial vertebrates. The control of extracellular calcium concentrations in marine invertebrates appears to be rather limited (Cameron, 1990).

MAGNESIUM

In vertebrates, approximately 60% of total body magnesium is located in bone, of which about one-third is combined with phosphate and the remainder is adsorbed loosely on the surface of the mineral structure (Pikę and Brown, 1975). In soft tissues, magnesium occurs both intra- and extracellularly. Magnesium is essential for maintenance of intra- and extracellular homeostasis in fish (Moyle and Cech, 2000) and crustaceans (Mantel and Farmer, 1983). In addition, magnesium is essential for cellular respiration and phosphate transfer reactions involving adenosine tri-, di-, and monophosphate. It is an activator for all thiamine pyrophosphate reactions and is involved in the metabolism of fats, carbohydrates, and proteins.

Dietary magnesium deficiencies have been documented for a variety of freshwater fish and include poor growth, anorexia, lethargy, muscle flaccidity, convulsions, vertebral curvature, high mortality, and depressed magnesium levels in the whole-body, blood serum, and bone (Lall, 2002). Fish in freshwater, which contains 1 to 3 mg Mg/L, have been shown to require 0.025 to 0.07% magnesium in the diet (Lall, 2002). Rainbow trout were able to uptake waterborne magnesium (1.3 mg Mg/L) to meet a portion of their metabolic requirement (Shearer, 1989). Seawater typically contains high levels of magnesium (1,350 mg/L), and magnesium is excreted by marine crustaceans and fish, resulting in blood levels lower than that of the external medium. Thus, marine species may not require a dietary source of magnesium (Dall and Moriarty, 1983). Atlantic salmon reared in brackish water (54 mg Mg/L) needed at least 0.01% magnesium in the diet to maintain normal magnesium concentrations in whole-body and serum as well as for proper bone mineralization (El-Mowafi and Maage, 1998). In contrast, red sea bream reared in seawater showed no signs of deficiency when fed diets containing as little as 0.012% magnesium (Sakamoto and Yone, 1979a).

The effects of dietary magnesium on immune responses of fish have received limited consideration to date. Atlantic salmon vaccinated against *Vibrio anguillarum* and fed graded levels of magnesium had similar antibody titers, lysozyme, and serum complement hemolytic activity regardless of dietary magnesium level (El-Mowafi et al., 1997). However, both lysozyme and serum hemolytic activity were elevated in vaccinated fish compared with unvaccinated fish.

In crustaceans, early research with *M. japonicus* indicated that supplementation of 0.3% magnesium did not improve the nutritive value of a semipurified diet (Deshimaru and Yone, 1978). Kanazawa et al. (1984) reevaluated the magnesium requirement of *M. japonicus* and reported that dietary supplementation of 0.1 to 0.5% magnesium improved growth responses. However, in this series of studies, weight gain was very low (< 100%) and a dietary essentiality was not established. Davis et al. (1992a) reported a depression of hepatopancreas magnesium levels in *L. vannamei* in response to the deletion of magnesium from a semipurified diet; however, weight gain and magnesium levels of the carapace were unaffected. Based on the high levels of magnesium in seawater and the magnesium requirements of freshwater fish, a dietary magnesium requirement for marine species would not be expected. More recently, the magnesium requirement of *L. vannamei* cultured in low-salinity (2‰) water was estimated to be 0.26 to 0.35% of diet based on weight gain (Cheng et al., 2005). Another study in artificial low-salinity (4‰) water did not observe an improvement in growth of *L. vannamei* when magnesium chloride was supplemented at 150 and 300 mg/kg (Roy et al., 2007).

In terms of practical diet formulations, most feed ingredients, especially those of plant origin, are high in magnesium, and supplementation of magnesium to practical diets is generally not necessary. However, because of the low bioavailability of magnesium from white fish meals (Watanabe et al., 1988), some diets containing this ingredient may require magnesium supplementation.

SODIUM, POTASSIUM, AND CHLORIDE

Sodium, potassium, and chloride are recognized as being essential for a number of physiological processes including acid:base balance and osmoregulation (Lall, 2002). Dietary deficiencies of sodium and chloride have been difficult to demonstrate in fish (NRC, 1993). These minerals are typically abundant in water and feedstuffs, and thus metabolic deficiencies have rarely been observed.

The supplementation of high levels (4.5 to 11.6% of the diet) of sodium chloride (NaCl) to the diet has been reported to inhibit feed efficiency of rainbow trout raised in freshwater, presumably due to nutrient dilution (Salman and Eddy, 1988). Additionally, there were no positive or negative effects of sodium chloride supplementation on channel catfish raised in freshwater (Murray and Andrews, 1979) or on Atlantic salmon cultured in freshwater or seawater (Shaw et al., 1975). However, hybrid tilapia reared in freshwater required approximately 0.15% Na in purified diets for optimal weight gain, gill Na^+-K^+ ATPase activity, and whole-body sodium retention, whereas no dietary requirement was apparent in seawater (Shiau and Lu, 2004).

Research with the euryhaline red drum demonstrated that at low salinities, the supplementation of sodium chloride at 2 to 10% of diet resulted in increased growth (Gatlin et al., 1992). Similar improvements in growth and feed efficiency were reported for European sea bass (*Dicentrarchus labrax*) (Eroldogan et al., 2005) and Asian sea bass (*Lates calcarifer*) (Harpaz et al., 2005) in freshwater with the addition of sodium chloride to the diet at 3 and 4%, respectively. One possible explanation for this positive response is an increase in amino acid absorption. Boge et al. (1983) demonstrated that, in addition to being energized by a Na^+-gradient, amino acid transport by brush-border membrane vesicles prepared from enterocytes of the European sea bass is dependent upon the presence of Cl^- ions. The addition of dietary sodium chloride at low salinities may increase the absorption of amino acids and/or satisfy other metabolic requirements, thus resulting in a physiological advantage for some species. Harpaz et al. (2005) reported that addition of dietary salt in freshwater increased the activity of several brush-border enzymes including alkaline phosphatase, lactase, and leucine amino peptidase, with the effect most pronounced in the pyloric caeca.

The supplementation of sodium chloride to practical diet formulations at 7 to 10% also has been found to reduce osmoregulatory stress and increase survival of fish being transferred from freshwater to saltwater (Zaugg et al., 1983; Al-Amoudi, 1987; Duston, 1993). This was presumably through the stimulation of osmoregulatory function and gill sodium and potassium ATPase activity (Al-Amoudi, 1987).

Due to the recent interest in culturing shrimp such as *L. vannamei* in low-salinity ground waters, the effects of dietary supplementation of sodium chloride and other minerals have been evaluated with this species in natural and artificial low-salinity water. Sodium chloride at 1 and 2% of diet did not improve weight gain of *L. vannamei* in artificial water of 4‰ salinity (Roy et al., 2007).

Dietary potassium requirements have been identified for channel catfish (Wilson and El Naggar, 1992) and Chinook salmon (*O. tshawytscha*) in freshwater (Shearer, 1988), but not for red sea bream in seawater (Sakamoto and Yone, 1978b), possibly indicating that marine fish can obtain adequate levels of potassium from the water. Channel catfish and Chinook salmon both absorbed waterborne potassium but could not meet their metabolic requirements without a dietary supply. Dietary potassium requirement estimates for channel catfish and Chinook salmon were 0.26 and 0.8%, respectively. Hybrid tilapia reared in freshwater required potassium at 0.2 to 0.3% of diet for optimal weight gain, gill Na^+-K^+ ATPase activity, and whole-body potassium retention (Shiau and Hsieh, 2001a). More recently, the Asian sea bass or barramundi was reported to require more dietary potassium at hypoosmotic salinity (5‰) compared to iso- or hyper-osmotic environments (Partridge and Lymbery, 2008). One study with African catfish (*Clarius gariepinus*) indicated that a dietary sodium to potassium ratio of 1.5:2.5 produced the best growth and nutrient utilization, while excess dietary potassium in relation to sodium reduced growth.

Potassium nutrition of crustaceans also has been studied to an appreciable extent with marine shrimp. Kanazawa et al. (1984) reported that diets containing 0.9% potassium improved growth of *M. japonicus* as compared to diets containing 1.8% potassium. Deshimaru and Yone (1978) recommended dietary supplementation of 1% potassium based on the comparative growth of shrimp fed a diet without magnesium and potassium supplementation; however, potential interactions between potassium and magnesium were not evaluated. Davis et al. (1992a) reported that the individual deletion of potassium from a semipurified diet did not result in a significant depression in tissue potassium or growth of *L. vannamei*; however, tissue levels of magnesium were affected, indicating a potential interaction. In another study, *P. monodon* reared in water of 19 to 21‰ salinity (360 mg K/L) showed a minimum potassium requirement of 1.2% of diet based on weight gain and protein efficiency ratio (Shiau and Hsieh, 2001b). *L. vannamei* responses to graded levels of dietary potassium in conjunction with different concentrations of potassium in seawater at 30‰ were monitored in a recent study (Zhu, 2006). Dietary potassium had limited influences on shrimp responses compared to potassium levels in the water. No interaction between dietary and waterborne potassium was observed, possibly due to limited assimilation of dietary potassium at 30‰ salinity. As mentioned earlier, culture of marine shrimp in low-salinity environments has increased in recent years and prompted more research investigating dietary and environmental mineral requirements of these organisms. Dietary supplementation of chelated potassium at 1% improved weight gain of *L. vannamei* in low-salinity (4‰) water in at least three experiments, but addition of potassium to the water to correct imbalances with sodium and chloride appeared to be more effective in improving growth and survival (Roy et al., 2007; Saoud et al., 2007). Thus, with the exception of culturing marine shrimp in low-salinity water, most freshwater and all seawater probably contains sufficient amounts of sodium, potassium, and chloride ions to satisfy the physiological needs of various cultured species (Lovell, 1989). Those ions are also found in substantial amounts in most feedstuffs, making the necessity of dietary supplementation unlikely. However, commercial diets may be supplemented with sodium chloride at 1 to 2% because of its relatively low cost.

CHROMIUM

Chromium, which is found predominantly in trivalent form, has been recognized since the late 1950s as an essential nutrient for humans and animals. One of its major metabolic roles is in association with the glucose tolerance factor, an organometallic molecule that potentiates the action of insulin in carbohydrate metabolism (NRC, 1997). In addition, chromium has been shown to be integral in activating various enzymes, influencing lipid metabolism, as well as maintaining the stability of proteins and nucleic acids (NRC, 1997).

To date, research on the influences of dietary chromium on fish has been rather limited with even less effort directed toward crustaceans. An early study with rainbow trout in which chromium chloride was supplemented to chemically defined diets at 0, 1, 3, or 6 mg/kg did not produce differences in weight gain or changes in tissue chromium distribution (Tacon and Beveridge, 1982). In another study with rainbow trout, supplementation of a practical diet with chelated chromium at 0.5 mg/kg increased blood glucose clearance but did not affect weight gain or protein and energy retention (Bureau et al., 1995). Gatta et al. (2001a) fed semipurifed diets supplemented with up to 4.1 mg Cr/kg from chromium yeast to rainbow trout and reported increased serum lysozyme as well as increased phagocytosis and respiratory burst of head-kidney macrophages after 6 weeks. However, organic chromium supplementation did not affect growth performance, carcass and fillet composition, or hepatic xenobiotic-metabolizing enzymes of gilthead sea bream (*Sparus aurata*) (Gatta et al., 2001b).

Several other studies with hybrid tilapia (*O. niloticus* × *O. aureas*) have shown that supplementation of chromium, especially in diets containing glucose compared to corn starch, caused significant increases in weight gain, energy deposition, and liver glycogen, as well as altered postprandial plasma glucose concentrations (Shiau and Chen, 1993; Shiau and Lin, 1993; Shiau and Liang, 1995). Only one subsequent study reported no significant effect of chromium picolinate supplementation at 2 mg/kg on growth or carbohydrate utilization of hybrid tilapia (Pan et al., 2003). In the other studies, chromium supplementation was effective at doses as low as 2 mg/kg from chromium chloride (Shiau and Lin, 1993) and as high as 0.5% of chromium oxide (Shiau and Liang, 1995), with chromium oxide being most efficacious relative to chromium chloride and sodium dichromate (Shiau and Chen, 1993). A minimum dietary chromic oxide level of 204 mg/kg was estimated based on maximum weight gain of hybrid tilapia fed diets containing glucose (Shiau and Shy, 1998). In contrast to the studies with hybrid tilapia, chromic oxide supplementation of diets containing glucose did not alter weight gain, feed efficiency, protein efficiency, or glucose utilization of channel catfish (Ng and Wilson, 1997). In addition, the dietary chromium did not influence whole-body chromium concentrations, indicating the dietary chromium from chromic oxide was not retained by the channel catfish. Hertz et al. (1989) reported improved glucose utilization of common carp fed practical diets supplemented with chromium. Increased intestinal absorption of glucose also was observed in snakehead (*Channa punctatus*) exposed to 1 mmol/L chromium (Sastry and Sunita, 1982). Specific mechanisms by which chromium influences carbohydrate utilization of fish have not been elucidated.

COPPER

Copper functions in hematopoiesis and in numerous copper-dependent enzymes (O'Dell, 1976) including lysyl oxidase, cytochrome *C* oxidase (CCO), ferroxidase, tyrosinase, and superoxide dismutase (SOD). Lysyl oxidase functions in the formation of crosslinks during the synthesis of collagen and elastin. The failure of collagen maturation (crosslinking) in the organic matrix of bone accounts for increased fragility of bones and the associated abnormalities of copper deficiencies (O'Dell, 1976). Failure of collagen and elastin crosslinking and an undefined muscular defect result in enlargement of the heart and cardiac failure in copper-deficient animals.

Copper also is involved in the absorption and metabolism of iron and functions in the formation of hemoglobin in vertebrates. Crustaceans utilize hemocyanin as the oxygen-carrying pigment. This copper-containing pigment has an analogous role to hemoglobin in red-blooded animals (Lovell, 1989). Depledge (1989) estimated that, on a fresh-weight basis, 40% of the whole-body copper load in shrimp is found in hemocyanin. This suggests a considerable increase in the physiological demand for copper by crustaceans above that required by vertebrates.

In addition to the physiological functions of copper, high levels of dietary copper (≥ 500 mg/kg) may be toxic (Julshamn et al., 1988; Berntssen et al., 1999), as is excessive environmental copper (Bryan, 1976). Exposure to waterborne copper does not increase tissue copper accumulation as much as feeding elevated dietary copper (Clearwater et al., 2002; Bielmyer et al., 2005; Hoyle et al., 2007). Copper contamination in the marine environment is generally due to an increase in anthropogenic input and occurs primarily in coastal and estuarine areas, where copper concentrations (up to 0.6 mg Cu/L) far exceed the background copper level in seawater (0.5 μg Cu/L) (Bjerregaard and Vislie, 1986). Due to toxic effects of dissolved copper and other heavy metals, shrimp hatcheries routinely use water that has been treated with ethylenediaminetetraacetic acid to chelate free copper (Lawrence et al., 1981). Although a great deal of research has been conducted on the toxicity of dissolved copper, the dietary essentiality of this nutrient for marine species has received limited attention.

Dietary deficiencies of copper have been documented for several freshwater fish (Ogino and Yang, 1980; Murai et al., 1981; Gatlin and Wilson, 1986a; Julshamn et al., 1988), but fewer marine species (Lin et al., 2008a). Dietary requirements for copper have been established to range from 1.5 to 5 mg Cu/kg diet in species such as common carp, rainbow trout (Ogino and Yang, 1980), channel catfish (Gatlin and Wilson, 1986a), and hybrid tilapia (Shiau and Ning, 2003). Similarly, the copper requirement of grouper was determined to be 4–6 mg Cu/kg diet (Lin et al., 2008a). In addition to growth and feed efficiency, copper-dependent enzymes such as ceruloplasmin, copper- and zinc-dependent SOD, and CCO have been shown to be excellent indicators of copper nutriture (Gatlin and Wilson, 1986a).

Kanazawa et al. (1984) found that the dual deletion of iron and copper had no significant effect on growth and survival of *M. japonicus*. However, in this series of experiments, percentage weight gain was low (40%) and survival was poor (57%); hence, the nutritional stress or the quality of the diet may not have been adequate to induce a dietary deficiency. Davis et al. (1993b) demonstrated a dietary copper deficiency in *L. vannamei* fed semipurified diets containing < 34 mg Cu/kg diet. Deficiency signs included poor growth; reduced copper levels in the carapace, hepatopancreas, and hemolymph; and enlargement of the heart. This response is similar to that of W. Wang et al. (1997), who reported a dietary copper requirement of 25.3 mg Cu/kg diet for *Fennerpenaeus chinensis* based on growth and tissue mineralization. Liu et al. (1990) reported a dietary copper requirement of 53 mg Cu/kg diet for *P. orientalis* based on growth, survival, CCO activity, and tissue mineralization. The dietary copper requirement of *P. monodon* was quantified at 15–21 mg Cu/kg based on weight gain, feed efficiency, and whole-body copper retention (Lee and Shiau, 2002). These results indicate that shrimp cannot meet their physiological needs for copper from seawater and that a dietary source is required for maximum growth and tissue mineralization. It also appears that invertebrate species utilizing copper as a component of their respiratory pigment have an increased copper requirement over vertebrates utilizing iron-based respiratory pigments.

An interaction between dietary copper and zinc has been observed in some terrestrial animals whereby elevated levels of one reduced the bioavailability of the other (NRC, 2005); however, such interactions have not been well established in fish (Knox et al., 1982, 1984; Gatlin et al., 1989). An interaction between copper and selenium was reported in Atlantic salmon in which liver selenium was inversely related to dietary copper (Lorentzen et al., 1998). In addition, elevated dietary copper (20 mg Cu/kg) resulted in increased oxidative stress in grouper, which was relieved by increased dietary selenium (Lin and Shiau, 2007).

IODINE

Iodine is an essential element for a variety of animals. Nearly every cell in the body contains iodine; however, in vertebrates the thyroid gland is the main location of iodine reserves. Thyroid hormones, which contain iodine, are known to have roles in thermoregulation, intermediary metabolism, reproduction, growth and development, hematopoiesis and circulation, as well as neuromuscular functioning (NRC, 2005). Thyroid hyperplasia (goiter) is the most overt sign of iodine deficiency in vertebrates and has been occasionally observed in fish, primarily when goitrogenic compounds such as glucosinolates are present in the diet (Lall, 2002).

One of the first mineral studies with fish evaluated the Chinook salmon's requirement for iodine (Woodall and LaRoche, 1964). Although weight gain of fish was not affected by feeding graded levels of sodium iodide for 24 weeks and an additional 9 months, requirement values of 0.6 mg I/kg diet for fingerlings and 1.1 mg I/kg diet for advanced parr were estimated based on maximal thyroid iodine storage. Levels of 1 to 5 mg I/kg diet have been estimated as adequate for most species (Lall, 2002). Iodine supplementation of a commercial diet was reported to enhance weight gain and reduce stress responses of steelhead trout (Gensic et al., 2004). Another study indicated 4.5 mg I/kg diet protected Atlantic salmon from bacterial kidney disease (Lall, 2002). Iodine supplementation of moist pellets up to 80 mg I/kg increased fillet iodine without impacting health, growth, or plasma thyroid hormone status (Julshamn et al., 2006). The physiological essentiality of iodine has not been evaluated in shrimp.

IRON

Iron is a trace element that is essential for the production and normal functioning of hemoglobin, myoglobin, cytochromes, and many other enzyme systems. In vertebrates, the principal role of iron is as a component of hemoglobin. Red blood cells are regenerated periodically, and most of the iron is recycled. That which is not recycled is excreted via the bile into the intestine. Like other elements of low solubility, such as zinc and copper, iron is absorbed and transported in the body in a protein-bound form (Lovell, 1989). In vertebrates, mucosal transferrin binds Fe^{2+} in the intestinal lumen and transports it across the mucosal brush border. Within the cell, Fe^{3+} is bound to apoferritin forming ferritin. The amount of apoferritin in the mucosa is regulated by physiological needs for iron. The iron-bound transferrin is then transported in the blood, where the iron is again released at target sites (liver and hematopoietic tissue). Iron and other minerals of low solubility are not easily excreted; thus, mineral excesses are deposited in cells of the digestive system, which are sloughed into the digestive tract for elimination.

In crustaceans, the hepatopancreas has been found to be the organ richest in iron. Storage cells containing iron have been reported in crayfish, *Procambarus clarkii* (Ogura 1959; Miyawaki et al., 1961), and the crab *Cancer irroratus* (Martin, 1973). Iron-transporting proteins have been found in the hemolymph of two species of crabs (Ghidalia et al., 1972; Depledge et al., 1986). These observations indicate the presence of a regulatory mechanism similar to that of vertebrates. In addition to the digestive system, gills appear to play an active role in iron metabolism. In *C. irroratus* iron accumulates by forming a coating around the branchial lamellae during the intermolt cycle, which is then rejected at ecdysis along with the integument (Martin, 1973). Absorption from the water through the gills could provide an additional source of iron.

Iron deficiencies have been documented for several species of fish; however, dietary deficiencies for shrimp have not been observed (Deshimaru and Yone, 1978; Kanazawa et al., 1984; Davis et al., 1992b). Iron deficiency causing hypochromic microcytic anemia has been reported for freshwater fish such as the brook trout (*Salvelinus fontinalis*) (Kawatsu, 1972) and common carp (Sakamoto and Yone, 1978c) and in marine fish such as red sea bream (Sakamoto and Yone, 1976b) and yellowtail (Ikeda et al., 1973). However, growth depression was not observed in these iron-deficient fish. Gatlin and Wilson (1986b) characterized iron deficiency signs of channel catfish and found that fish fed the basal diet (9.6 mg Fe/kg) exhibited suppressed growth and feed efficiency, as well as reduced hemoglobin, hematocrit, plasma iron, transferrin saturation, and erythrocyte-count values. Similar deficiency signs were reported in channel catfish by Lim and Klesius (1997). Additionally, catfish fed an iron-deficient diet experienced earlier mortality when exposed to the bacterial pathogen *Edwardsiella ictaluri* (Lim et al., 2000).

Gatlin and Wilson (1986b) concluded that a minimum of 20 mg supplemental Fe/kg diet (30 mg total Fe/kg) was required by channel catfish for best growth and hematological values. A much higher requirement value of 150 mg Fe/kg from iron chloride was recommended for red sea bream to maintain maximum hematological values (Sakamoto and Yone, 1978d), while 100 mg Fe/kg was reported to maximize weight gain, feed efficiency, and hepatic iron storage in grouper (Ye et al., 2007). Additionally, the dietary iron requirement of hybrid tilapia was estimated to be 85 mg Fe/kg from ferrous sulfate based on weight gain, hepatic iron, and hemoglobin concentrations, whereas approximately twice as much iron from ferric citrate was required due to reduced bioavailability (Shiau and Su, 2003).

Iron has received particular attention with regard to disease resistance because its availability to microorganisms affects their ability to cause infection. A study by Ravndal et al. (1994) indicated that Atlantic salmon families with high levels of serum iron were more susceptible to *Vibrio* infection; however, no relationship was apparent with regard to furunculosis or bacterial kidney disease. Another study with Atlantic salmon reported that supplementation of 400 mg Fe/kg diet to a basal diet containing 160 mg Fe/kg did not alter serum total protein, serum total antibody, hemolytic complement activity, or lysozyme activity in serum, head kidney, or spleen but did increase catalase activity in the head kidney (Andersen et al., 1998). Iron deficiency in channel catfish resulted in increased mortality due to *E. ictaluri* and reduced chemotactic migration of peritoneal macrophages (Sealey et al., 1997; Lim and Klesius, 1997) but antibody production was not affected by dietary iron level (Sealey et al., 1997). Diets containing 60 mg Fe/kg from either iron methionine or iron sulfate provided the highest chemotactic index (Sealey et al., 1997).

In addition to physiological problems caused by iron deficiency, excessive levels of iron may be toxic as well. Excessive iron supplementation appears to have potentially adverse effects on growth of *M. japonicus* (Deshimaru and Yone, 1978; Kanazawa et al., 1984). Additionally, iron-catalyzed lipid oxidation increases with iron supplementation, which in turn may adversely affect feed stability (Desjardins et al., 1987; Sutton et al., 2006). However, Andersen et al. (1998) supplemented extruded fish meal-based diets with 400 mg Fe/kg from iron sulfate and observed no effects on growth, hematology, antioxidant status, or health of Atlantic salmon smolts.

Iron is one of the primary metals involved in lipid oxidation, and ferrous iron is a more potent catalyst of lipid peroxidation than ferric iron (Chvapil et al., 1974; Lee et al., 1981). Ferrous iron catalyzes the formation of hydroperoxides and free radical peroxides by providing a free radical initiator in the presence of unsaturated fatty acids and oxygen. Thus, excessive iron supplementation of marine fish and crustacean diets should be avoided due to the presence of polyunsaturated fats that make these diets particularly susceptible to oxidation. Such oxidation also may reduce the stability of ascorbic acid (Hilton, 1989). Although many practical diets may contain considerable levels of endogenous iron, little is known about its form and availability (Lall, 1989). Hence, a low level of supplementation ($\cong$ 10% of dietary requirement) of an available source is often recommended to ensure adequacy of the diet.

MANGANESE

Manganese functions as a cofactor in several enzyme systems, including those involved in urea synthesis from ammonia, amino acid metabolism, fatty acid metabolism, and glucose oxidation (Lall, 2002). Principal signs of manganese deficiency in terrestrial species include reduced growth rate, skeletal abnormalities, convulsions, reduced righting ability, abnormal reproductive function in males and females, and ataxia in the newborn (Lall, 1991).

Dietary deficiencies in fish have resulted in poor growth, skeletal abnormalities, high embryo mortalities, and poor hatch rates (Lall, 2002). A total dietary manganese content of 12 to 13 mg/kg has been recommended for the common carp and rainbow trout (Ogino and Yang, 1980); however, Gatlin and Wilson (1984a) found that 2.4 mg Mn/kg diet was sufficient for normal growth and health of channel catfish. A requirement of 7 mg Mn/kg was estimated for hybrid tilapia based on hepatic Mn superoxide dismutase and whole-body manganese retention (Lin et al., 2008b). Pan et al. (2008) estimated a higher requirement of approximately 14 mg Mn/kg for gibel carp (*Carassius auratus*).

In terms of immunity and disease resistance, deficiencies of manganese along with zinc were reported to decrease leukocyte natural killer cell activity of rainbow trout; however, that activity was restored with supplementation of those minerals (Inoue et al., 1998). However, supplementation of both manganese and zinc at more than 100 mg/kg diet did not enhance resistance of sockeye salmon to bacterial kidney disease nor did it have a significant effect on their production of serum antibodies (Bell et al., 1984).

Even less information on manganese requirements of crustaceans is currently available. Kanazawa et al. (1984) found that supplementation of 10 and 100 mg Mn/kg diet did not improve the growth of *M. japonicus*. However, it should be noted that percentage weight gain during the study was < 70% and the nutritional stress placed on the shrimp would not be considered adequate to reduce body stores enough to induce a deficiency. Because the manganese content of seawater is very low (0.01 mg/L), significant absorption from the water is unlikely. Thus, a dietary source of manganese could be necessary for marine shrimp and fish.

SELENIUM

Selenium is a trace element that functions as a component of the enzyme family called glutathione peroxidase, which converts hydrogen peroxide and lipid hydroperoxides into water and lipid alcohols, respectively. Thus, this enzyme group functions in protecting the cell from deleterious effects of peroxides (Little et al., 1970). Glutathione peroxidase acts along with vitamin E to function as a biological antioxidant to protect polyunsaturated phospholipids in cellular and subcellular membranes from peroxidative damage (Lovell, 1989). The function of this enzyme is complementary to that of vitamin E, which is a lipid-soluble antioxidant. A dietary deficiency of selenium has been generally reported to result in reduced activity of glutathione peroxidase as well as growth reduction. However, a combined deficiency of selenium and the fat-soluble antioxidant vitamin E was required to produce more overt deficiency signs such as nutritional muscular dystrophy and exudative diathesis in channel catfish (Gatlin and Wilson, 1984b; Lall, 2002).

The toxicity of selenium to various animals was established well before its dietary essentiality. Toxicity of dietary selenium has been shown to vary with several factors, including source and duration of exposure. Chronic selenium toxicity has been demonstrated in several fish species at dietary levels of 13 to 15 mg Se/kg from sodium selenite resulting in reduced growth and elevated mortality (Hilton et al., 1980; Gatlin and Wilson, 1984b; NRC, 2005) as well as renal calcinosis (Hilton and Hodson, 1983). More recently, the threshold of selenium toxicity for white sturgeon (*Acipenser transmontanus*) fed graded levels of selenomethionine was estimated to be 10 to 20 mg Se/kg (Tashjian et al., 2006). Cutthroat trout (*Oncorhynchus clarki bouvieri*) were fed diets containing up to 10 mg Se/kg from selenomethionine without showing signs of toxicity (Hardy et al., 2010). A dietary selenium level of 8 mg Se/kg was required for hatchery-reared coho salmon (*Oncorhynchus kisutch*) in freshwater to have eviscerated body selenium concentrations similar to that of their wild counterparts, and this dietary level provided similar seawater survival of the cultured fish as observed in the wild fish (Felton et al., 1996). A subsequent investigation (Halver et al., 2004) reported that the stress of confinement for 30 h during barge transport reduced carcass selenium of Chinook salmon by 20% while liver glutathione peroxidase activity was increased. These studies indicated elevated dietary selenium beneficially affected salmon against the stressful conditions associated with confinement and saltwater transfer by increasing glutathione peroxidase activity.

Fish mortality in natural settings related to waterborne selenium toxicity also has been studied extensively over the past two decades (Hamilton, 2004). Based on these laboratory and field studies, threshold levels for adverse effects of selenium on fish have been estimated at 3–4 mg Se/kg diet and 2–5 μg/L in water (NRC, 2005).

Selenium and vitamin E interrelationships have been investigated in several animal species, and a variety of common and unique deficiency signs have been described (NRC, 1993). Differing responses, especially with respect to gross deficiency signs, were observed when Atlantic salmon (Poston et al., 1976), rainbow trout (Bell et al., 1985), and channel catfish (Gatlin et al., 1986) were fed diets without supplemental selenium, vitamin E, or both nutrients. Mutual sparing of metabolic requirements for either selenium or vitamin E by the other nutrient, as measured by weight gain and liver lipid oxidation, was recently demonstrated in grouper (Lin and Shiau, 2009). A dietary interaction between dietary selenium and copper also was reported for grouper in that elevated dietary copper induced oxidative stress and reduced the fish's immune response, but selenium at twice the minimum requirement level (1.6 mg Se/kg) reduced the oxidative stress and improved immune responses (Lin and Shiau, 2007). Dietary selenium concentration and supplemental form were reported to influence various immunological responses and resistance of channel catfish to *E. ictaluri* with fish fed 0.20 mg Se/kg from selenomethionine and 0.40 mg Se/kg from selenoyeast or sodium selenite having the greatest resistance to *E. ictaluri* (C. Wang et al., 1997). Antibody production was generally increased with dietary selenium supplementation but was greatest in catfish fed selenoyeast. Macrophage chemotactic response also was enhanced in catfish fed selenoyeast and selenomethionine (C. Wang et al., 1997). Dietary supplementation of organic selenium also has been reported to reduce the harmful effects of waterborne copper toxicity to African catfish (Abdel-Tawwab et al., 2007) and cadmium toxicity to Nile tilapia (Abdel-Tawwab and Wafeek, 2010).

Levels of 0.15 to 0.38 mg Se/kg diet (Hilton et al., 1980) and 0.25 mg Se/kg diet (Gatlin and Wilson, 1984b) were required to provide maximum growth and glutathione peroxidase activity in rainbow trout and channel catfish, respectively. More recently, Lin and Shiau (2005) estimated the minimum selenium requirement of grouper to be 0.7 mg Se/kg based on weight gain and whole-body selenium retention. As indicated previously, elevated dietary selenium from 5 to 8 mg Se/kg increased liver glutathione peroxidase activity in salmon smolts and improved their responses to the stressful conditions associated with confinement and saltwater transfer. The U.S. Food and Drug Administration currently allows selenium supplementation of up to 0.3 mg Se/kg from sodium selenate or sodium selenite in feeds for all animals including aquatic species.[1]

Information on selenium nutrition of crustaceans is rather limited at this time. Davis (1990) found that juvenile *L. vannamei* grew best when fed semipurified diets supplemented with 0.2 to 0.4 mg Se/kg diet. Although a specific dietary level was not quantified, that study indicated shrimp have a dietary requirement for selenium.

ZINC

Zinc is required for normal growth, development, and function in all animal species that have been studied (NRC, 1980). Zinc functions as a cofactor in several enzyme systems and is a component of a large number of metalloenzymes, which include carbonic anhydrase, carboxypeptidases A and B, alcohol dehydrogenase, glutamic dehydrogenase, D-glyceraldehyde-3-phosphate dehydrogenase, lactate dehydrogenase, malic dehydrogenase, alkaline phosphatase, aldolase, SOD, ribonuclease, and DNA polymerase (NRC, 1980; Lall, 2002).

A dietary requirement for zinc has been quantified for a variety of freshwater fish fed semipurified diets based on weight gain, tissue zinc saturation, and/or whole-body zinc retention. These requirement estimates include: 20 mg Zn/kg diet for channel catfish (Gatlin and Wilson, 1983; Scarpa and Gatlin, 1992) and blue tilapia (McClain and Gatlin, 1988), 15–30 mg Zn/kg diet for common carp (Ogino and Yang, 1979), 15–30 mg Zn/kg diet for rainbow trout (Ogino and Yang, 1978), and 26–29 mg Zn/kg for hybrid tilapia (Lin et al., 2008c). The zinc requirement of red drum also has been determined to be 20 mg Zn/kg diet (Gatlin et al., 1991).

The effects of zinc on immunity and disease resistance of fish also have been investigated. One study determined that supplementation of zinc at 200 mg/kg diet did not enhance resistance of nonimmunized channel catfish juveniles to *Aeromonas hydrophila* (Scarpa and Gatlin, 1992). In a more recent study, zinc deficiency caused 100% mortality in channel catfish challenged with *E. ictaluri*, and maximum survival after challenge was achieved with 5 mg Zn/kg provided by zinc methionine or 30 mg Zn/kg provided by zinc sulfate (Paripatananont and Lovell, 1995a). Maximum antibody production in that study was achieved with 15 mg Zn/kg from zinc methionine or 30 mg Zn/kg or greater from zinc sulfate. In a similar study by Lim et al. (1996), channel catfish fed zinc methionine at 20 and 60 mg/kg diet and zinc sulfate at 60 mg/kg diet had higher chemotactic responses of macrophages; however, dietary zinc did not influence phagocytic activity of macrophages for *E. ictaluri*. Additionally, the source or level of dietary zinc did not provide protection against *E. ictaluri* infection.

Zinc nutrition of various crustaceans also has been investigated. Davis et al. (1993c) reported that *L. vannamei* required 33 mg Zn/kg diet to maintain normal tissue mineralization although growth was not affected by graded levels of dietary zinc. A similar requirement value of 32–34 mg Zn/kg was quantified for *P. monodon* by Shiau and Jiang (2006) based on weight gain and whole-body zinc retention.

As previously mentioned in the section on copper, elevated levels of dietary zinc have been shown to reduce the bioavailability of copper in terrestrial animals (NRC, 2005). However, such interactions have not been well established in fish (Knox et al., 1982, 1984; Gatlin et al., 1989).

OTHER MINERALS

There are several other trace minerals established as being essential in terrestrial animals and humans, but for which there is much less information available on aquatic species. Cobalt is one such mineral included in the group that has been studied to only a limited extent in aquatic species. Cobalt is of nutritional significance because it is a component of vitamin B_{12}. Synthesis of vitamin B_{12} by intestinal bacteria

[1]Title 21, Code of Federal Regulations, Part 573.920.

of channel catfish was reduced by removal of cobalt from the diet (Limsuwan and Lovell, 1981). A series of studies conducted in Russian ponds (reviewed by Castell et al., 1986) reported the provision of cobalt in the diet or water increased growth and hemoglobin formation in common carp. Enhanced growth and survival of mullet (*Mugil parsia*) fry was reported by Ghosh (1975) when cobalt chloride at 0.6 to 1 ppm was dissolved in water and dispensed in powdered feed consisting of each portions of rice bran, mustard oil cake and fish meal. Another study (Hertz et al., 1989) investigated the effects of both cobalt and chromium on blood glucose regulation of common carp, and noted improved glucose utilization and increased amino acid incorporation into protein with cobalt supplementation. Molybdenum has been implicated in enhancing growth and survival of carp (George, 1970 cited by Lall, 2002). However, similar studies with other fish species have not been conducted. Intake of dietary fluoride was shown to increase fluoride content of bones in rainbow trout but did not result in any overt responses (Tiews et al., 1982). Other inorganic elements such as arsenic, barium, bromine, cadmium, and strontium are potentially required by the body but their essentiality has been difficult to establish except under highly controlled conditions. Dietary requirements for these elements have not been specifically studied in fish. However, some of these minerals, such as arsenic and mercury, may accumulate in fish tissues and potentially have adverse effects on human consumers. Thus, research on these minerals has primarily focused on bioaccumulation in fish and nutritional quality of seafood products, which is addressed in Chapter 16.

SOURCES AND FORMS

Of the feed ingredients used in practical animal diets, fish meal is the richest source of endogenous minerals. Research on the bioavailability of minerals contained in fish meals has demonstrated that there is considerable variation among fish species (perhaps due to luminal pH) and that the bioavailability of minerals is affected by meal type and ash content. Fish meals, as well as other feedstuffs of animal origin such as meat and bone meal, poultry byproduct meal, and feather meal, are relatively rich in minerals. However, due to low mineral availability and potential inhibitory interactions, prepared diets are routinely supplemented with available sources of copper, phosphorus, manganese, and zinc to prevent dietary deficiencies and maximize fish growth (Watanabe et al., 1988). As the aquatic animal feed industry increases its use of plant feedstuffs, which are generally poor sources of minerals and may contain factors that reduce the bioavailability of minerals, the need for mineral supplements should increase.

Regardless of the form in which calcium and phosphorus are ingested, their absorption is dependent upon solubility at the point of contact with the absorbing membranes; hence, mineral sources that are soluble at luminal pH are potentially more available. The NRC (1983) summarized the relative availability or apparent absorption of various sources of phosphorus for four species of fish. In general, bioavailability of phosphorus (and minerals in general) has been found to be positively correlated with the solubility of the mineral in water. Monobasic phosphates of sodium and potassium are highly available (90 to 95%) sources of phosphorus for channel catfish (Lovell, 1978), common carp (Ogino et al., 1979), red sea bream (Sakamoto and Yone, 1979b), and rainbow trout (Ogino et al., 1979). Di- and tribasic phosphates are highly available to red sea bream (Sakamoto and Yone, 1979b). Dibasic calcium phosphate has an availability of 65% for channel catfish (Lovell, 1989). More recently, phosphorus availability values for monosodium phosphate, monoammonium phosphate, defluorinated rock phosphate, and mono-calcium phosphate to channel catfish were determined to be 88.8, 85.4, 81.7, and 81.2%, respectively (Eya and Lovell, 1997b). Li et al. (1996) also confirmed that dicalcium phosphate and defluorinated phosphates were equally efficacious as phosphorus supplements for channel catfish. Di- and tribasic calcium phosphates have availability values of 71 and 64%, respectively, for rainbow trout (Ogino et al., 1979). Phosphorus availability of inorganic sources to European sea bass decreased in the order of dicalcium phosphate (68%), monocalcium phosphate (56%), and tricalcium phosphate (50%) (Pimentel-Rodrigues and Oliva-Teles, 2007). In yellowtail, apparent phosphorus availability values greater than 92% were reported for monobasic forms of sodium, potassium, and calcium phosphate compared to calcium phosphate di-basic (52.9%) and tribasic (48.8%) (Sarker et al., 2009). For the agastric common carp, the bioavailability of phosphorus from different sources was reported to be more variable with bioavailability values of only 13% for tribasic calcium phosphate, 46% for dibasic calcium phosphate, and 94% for monobasic calcium phosphate (Ogino et al., 1979). In the presence of adequate phosphorus, Nakamuara and Yamada (1980) determined the availability of calcium for common carp to be 58, 37, and 27% from calcium lactate, tri-basic calcium phosphate, and calcium carbonate, respectively. The differences in calcium availability were influenced by the lack of acidic digestion in common carp.

The bioavailability of phosphorus to marine shrimp appears to be intermediate to that of fish with and without true stomachs. Apparent phosphorus availability values for *L. vannamei* include calcium phosphate monobasic, 46%; calcium phosphate dibasic, 19%; calcium phosphate tribasic, 10%; potassium phosphate monobasic, 68%; and sodium phosphate monobasic, 68% (Davis and Arnold, 1994).

In addition to chemical form (solubility) affecting mineral availability, other nutrients affect the availability of certain minerals. For example, lactose may interact with absorptive cells to increase their permeability to calcium ions; large intakes of iron, aluminum, and magnesium may interfere with the absorption of phosphorus by forming insoluble phosphates; and fats may interact with calcium to form insoluble soaps (Maynard et al., 1979). Hence, availability of calcium and phosphorus may be dependent upon the mineral form (solubility at intestinal pH), and dietary levels of calcium, phosphorus, vitamin D, iron, aluminum, manganese, potassium, magnesium, and fat.

Phosphorus in fish meal exists mainly in the form of insoluble hydroxyapatite originating from hard tissues such as bones. The availability of phosphorus contained in fish meal is fairly low for carp (10–33%) as compared with rainbow trout (60–81%) (Watanabe et al., 1988). Based on unpublished data, Akiyama and Dominy (1989) reported the apparent absorption of phosphorus from fish meal to be 46.5% for *L. vannamei*. Phytate phosphorus, which constitutes approximately 67% of the phosphorus in plant feedstuffs, also is poorly available to fish (Lovell, 1989) and shrimp (Civera et al., 1990; Davis et al., 1993c). Civera et al. (1990) found that the presence of phytate inhibited the availability of dietary calcium and phosphorus to *L. vannamei* and *M. japonicus*, presumably because of the formation of insoluble complexes in the digestive system. The apparent availability of phytate phosphorus was determined to be 47.3% for *M. japonicus* and 8.4% for *L. vannamei* (Civera et al., 1990). Similarly, Davis et al. (1993c) reported that the presence of 1.5% phytate inhibited the availability of dietary phosphorus and zinc to *L. vannamei*. It should be noted that phytate phosphorus can account for a considerable portion of the phosphorus in practical diet formulations. The treatment of plant protein feedstuffs with phytase has been demonstrated to increase the availability of phosphorus to a variety of fish species including rainbow trout (Ketola, 1994; Riche and Brown, 1996; Dalsgaard et al., 2009), Atlantic salmon (Storebakken et al., 1998; Sajjadi and Carter, 2004), channel catfish (Eya and Lovell, 1997b), *Pangasius* spp. catfish (Debnath et al., 2005), striped bass (Hughes and Soares, 1998), and Korean rockfish (*Sebastes schlegeli*) (Yoo et al., 2005). Consequently, phytase supplementation of plant feedstuffs or complete diets may be utilized to increase the availability of phosphorus from this component of the diet. Advancements in plant breeding also have resulted in low-phytic acid varieties of crops such as barley (Overturf et al., 2003; Buentello et al., 2010).

The bioavailability of minerals chelated to organic molecules has generally been reported to be higher than inorganic forms. If an element is chelated by a compound that will release it in ionic form at the site of absorption or will be readily absorbed as the intact chelate, this form may greatly enhance the absorption of the element by preventing its conversion to insoluble chemical compounds in the intestine or by preventing its strong adsorption on insoluble colloids (Scott et al., 1982). Compared with inorganic sources, chelated minerals also are generally less sensitive to the inhibitory action of other compounds (i.e., phytate and fiber) and may have a higher bioavailability in practical diets. Another potential benefit of mineral chelates for aquatic species is reduced solubility in water. However, the higher cost of most mineral chelates relative to inorganic sources has generally limited their use in aquaculture to date.

Several different forms of organic or chelated minerals have been developed for various elements. Amino acid chelates of copper, manganese, and zinc were determined to be more readily available than their inorganic salts based on increased bone and liver deposition in rainbow trout (Apines et al., 2001; Apines-Amar et al., 2004). Copper exchanged montmorillonite is another relatively uncommon form of supplemental copper. When it was included at 1.5 g/kg diet to provide 30 mg Cu/kg, it increased growth of Nile tilapia and protected the intestinal mucosa from invasion by pathogenic bacteria (Hu et al., 2007).

Selenium content and availability have been shown to vary considerably among different feedstuffs. Diets formulated with predominantly plant ingredients may require a selenium supplement. Selenium availability from various inorganic and organic sources has been evaluated in aquatic species. Bell and Cowey (1989) reported that selenium from selenomethionine was most available to Atlantic salmon, and selenium from fish meal was least available. However, based on glutathionine peroxidase activity, selenite and selenocystine were better sources than selenomethionine and fish meal. Lorentzen et al. (1994) determined that dietary selenite yielded higher levels of hepatic selenium in Atlantic salmon compared to selenomethionine, but the latter provided higher muscle and whole-body selenium concentrations. Due to their very toxic nature, sodium selenite or sodium selenate should not be handled in pure form. Selenomethionine was recently determined to be approximately three times more available to hybrid striped bass than sodium selenite (Jaramillo et al., 2009). Likewise, selenomethionine appears to be the most toxic form of selenium at elevated levels (NRC, 2005). A tentative maximum tolerable selenium level for fish in which health or performance would not be impaired was suggested as 2 mg Se/kg diet, although more research to establish appropriate safety factors was recommended (NRC, 2005).

Several different sources of zinc have been evaluated with various fish species. Zinc gluconate was found to be equivalent to zinc sulfate for Atlantic salmon (Maage et al., 2001). The organic chelate zinc methionine was compared to zinc sulfate in both purified and practical diets with channel catfish (Paripatananont and Lovell, 1995b). Zinc methionine was estimated to be three times as potent as zinc sulfate in purified diet, and four to five times more available in soybean-meal-based diets. Zinc picolinate was evaluated by supplementing the diet to provide 30 and 60 mg Zn/kg. This supplement linearly increased serum and whole-body zinc, as well as reduced oxidative stress of rainbow trout (Kucukbay et al., 2006).

INTERACTIONS WITH OTHER DIETARY COMPONENTS

In order to meet an animal's physiological requirements for various minerals, dietary sources must be available. Numerous factors affect the availability of minerals. The most soluble, and consequently the most readily absorbed, forms are the simple state of the element or ionic group of atoms (e.g., Ca^{2+}, Mg^{2+}, Mn^{2+}, PO_4^{3-}). However, in nature, compounds differing in electric charge may bind with minerals forming stable compounds that are less soluble in water. Although such compounds have low solubility in water, the acidic condition of the gastric stomach allows dissociation of the compounds into salts that can be easily absorbed in the intestine. Consequently, in animals without acidic digestive systems, the bioavailability of minerals is generally reduced.

Although the gastric stomach generally increases the availability of minerals, some minerals may interact after being released into the alkaline intestine by forming insoluble precipitates. Excessive levels of calcium and phosphorus react with magnesium and zinc to form insoluble precipitates. Additionally, colloids such as particles of clay, insoluble salts of aluminum, magnesium, iron, and other elements strongly absorb cations. This absorption occurs both through chemical union with highly electronegative areas of the colloidal surface and through attraction of the cation by physical forces (Scott et al., 1982). Consequently, the bioavailability of a given mineral will be dependent upon its dissociation as well as interactions with other dietary components.

The bioavailability of zinc in feedstuffs is generally low, making supplementation essential (Watanabe et al., 1988). The bioavailability of zinc in various fish meals has been found to be inversely related to the tri-calcium phosphate content of the meal. Thus, it is generally lowest in white fish meals, which contain the highest level of tri-calcium phosphate, and slightly higher in brown fish meals (Watanabe et al., 1988). Reduced bioavailability of zinc in response to calcium phosphate supplementation also has been observed in rainbow trout (Ketola, 1979; Hardy and Shearer, 1985; Satoh et al., 1987). Bioavailability of manganese also is reduced when high levels of calcium and phosphorus are in the diet (Watanabe et al., 1997).

Practical diets often contain feedstuffs that are relatively high in phytate, which may also reduce the bioavailability of zinc. The effect of phytate on zinc bioavailability is well established in a variety of terrestrial animals (Oberleas et al., 1962; O'Dell et al., 1964; Savage et al., 1964; Lo et al., 1981), fish (Gatlin and Wilson, 1984c; Richardson et al., 1985; McClain and Gatlin, 1988; Gatlin and Phillips, 1989; Satoh et al., 1989), and shrimp (Davis et al., 1993c). Consequently, practical diets are often supplemented at levels in excess (100 to 150 mg Zn/kg diet) of the established minimum dietary requirement to overcome the effects of inhibitory agents such as phytate. For example, Eid and Ghonim (1994) determined the minimum zinc requirement of Nile tilapia fed purified diets was 30 mg Zn/kg based on weight gain, feed efficiency, serum zinc, and bone zinc. However, Do Carmo e Sa et al. (2004) determined that plant-based diets had to be supplemented with zinc sulfate at 79.5 mg Zn/kg to maximize bone zinc storage of Nile tilapia.

Of all the microminerals, copper, iron, manganese, selenium, and zinc have been demonstrated in some fish species to be most important to supplement in diets due to low levels in practical feedstuffs and/or interactions with other dietary components that may reduce bioavailability (Watanabe et al., 1997). Although supplementation of practical diets with other microminerals has not been shown to be essential in most instances, an inexpensive trace mineral premix containing inorganic forms may be added to most nutritionally complete diets in order to ensure adequacy.

REFERENCES

Abdel-Tawwab, M., and M. Wafeek. 2010. Response of Nile tilapia, *Oreochromis niloticus* (L.) to environmental cadmium toxicity during organic selenium supplementation. J. World Aquacult. Soc. 41:106–114.

Abdel-Tawwab, M., M. A. A. Mousa, and F. E. Abbass. 2007. Growth performance and physiological response of African catfish, *Clarias gariepinus* (B.) fed organic selenium prior to the exposure to environmental copper toxicity. Aquaculture 272:335–345.

Akiyama, D. M., and W. G. Dominy. 1989. Penaeid Shrimp Nutrition for the Commercial Feed Industry. St. Louis, MO: American Soybean Association.

Al-Amoudi, M. M. 1987. The effect of high salt diet on the direct transfer of *Oreochromis mossambicus*, *O. spilurus* and *O. aureus/O. niloticus* hybrids to sea water. Aquaculture 64:333–338.

Ambasankar, K., and S. A. Ali. 2002. Effects of dietary phosphorus on growth and phosphorus excretion in Indian white shrimp. J. Aquacult. Trop. 17:119–126.

Andersen, F., B. Lygren, and A. Maage. 1998. Interaction between two dietary levels of iron and two forms of ascorbic acid and the effect on growth, antioxidant status and some non-specific immune parameters in Atlantic salmon (*Salmo salar*) smolts. Aquaculture 161:437–451.

Andrews, J. W., T. Murai, and C. Campbell. 1973. Effects of dietary calcium and phosphorus on growth, food conversion, bone ash and hematocrit levels of catfish. J. Nutr. 103:766–771.

Apines, M. J., S. Satoh, V. Kiron, T. Watanabe, N. Nasu, and S. Fujita. 2001. Bioavailability of amino acids chelated and glass embedded zinc to rainbow trout, *Oncorhynchus mykissi* fingerlings. Aquacult. Nutr. 7:221–228.

Apines-Amar, M. J. S., S. Satoh, C. M. A. Caipang, V. Kiron, T. Watanabe, and T. Aoki. 2004. Amino acid-chelate: A better source of Zn, Mn and Cu for rainbow trout, *Oncorhynchus mykissi*. Aquaculture 240:345–358.

Åsgård, T., and K. Shearer. 1997. Dietary phosphorus requirement of juvenile Atlantic salmon, *Salmo salar* L. Aquacult. Nutr. 3:17–23.

Bell, J. G., and C. B. Cowey. 1989. Digestibility and bioavailability of dietary selenium from fishmeal, selenite, selenomethionine and selenocystine in Atlantic salmon (*Salmo salar*). Aquaculture 81:61–68.

Bell, G. R., D. A. Higgs, and G. S. Traxler. 1984. The effect of dietary ascorbate, zinc and manganese on the development of experimentally induced bacterial kidney disease in sockeye salmon (*Oncorhynchus nerka*). Aquaculture 36:293–311.

Bell, J. G., C. B. Cowey, J. W. Adron, and A. M. Shanks. 1985. Some effects of vitamin E and selenium deprivation on tissue enzyme levels and indices of tissue peroxidation in rainbow trout (*Salmo gairdneri*). Brit. J. Nutr. 53:149–157.

Berntssen, M. H. G., A.-K. Lundebye, and A. Maage. 1999. Effects of elevated dietary copper concentrations on growth, feed utilization and nutritional status of Atlantic salmon (*Salmo salar* L.) fry. Aquaculture 174:167–181.

Bielmyer, G. K., D. M. Gatlin, III, J. J. Isely, J. Tomasso, and S. J. Klaine. 2005. Responses of hybrid striped bass to waterborne and dietary copper in freshwater and saltwater. Comp. Biochem. Phys. C 140:131–137.

Bjerregaard, P., and T. Vislie. 1986. Effect of copper on ion- and osmoregulation in the shore crab *Carcinus maenas*. Mar. Biol. 91:69–76.

Boge, G., A. Rigal, and G. Peres. 1983. Analysis of the chloride requirements for sodium-dependent amino acid and glucose transport by intestinal brush-border membrane vesicles of fish. Biochim. Biophys. Acta 729:209–218.

Borlongan, I. G., and S. Satoh. 2001. Dietary phosphorus requirement of juvenile milkfish, *Chanos chanos* (Forsskal). Aquac. Res. 32(Suppl. 1):26–32.

Boyd, C. E. 1981. Water Quality in Warmwater Fish Ponds. Opelika, AL: Craftmaster Printers Inc.

Brown, M. L., F. Jaramillo, Jr., and D. M. Gatlin, III. 1993. Dietary phosphorus requirement of juvenile sunshine bass, *Morone chrysops* × *M. saxatilis*. Aquaculture 119:359–264.

Bryan, G. W. 1976. Heavy metal contamination in the sea. Pp. 185–302 in Marine Pollution, R. Johnston, ed. New York: Academic Press, Inc.

Buentello, A. J., and D. M. Gatlin, III. 2010. Phosphorus availability in low-phytate varieties of barley: Digestibility studies with channel catfish and red drum. N. Am. J. Aquacult. 72:188–194.

Bureau, D. P., J. B. Kirkland, and C. Y. Cho. 1995. The effects of dietary chromium supplementation on performance, carcass yield and blood glucose of rainbow trout (*Oncorhynchus mykiss*) fed two practical diets. J. Anim. Sci. 73(Suppl 1):194.

Cameron, J. N. 1990. Unusual aspects of calcium metabolism in aquatic animals. Annu. Rev. Physiol. 52:77–95.

Castell, J. D., D. E. Conklin, J. S. Craigie, S. P. Lall, and K. Norman-Boudreau. 1986. Aquaculture nutrition. Pp. 251–308 in Realism in Aquaculture: Achievements, Constraints, Perspectives, M. Bilio, H. Rosenthal, and C. Sindermann, eds. Bredene, Belgium: European Aquaculture Society.

Cheng, K.-M., C.-Q. Hu, Y.-N. Liu, S.-X. Zheng, and X.-J. Qi. 2005. Dietary magnesium requirement and physiological responses of marine shrimp *Litopenaeus vannamei* reared in low salinity water. Aquacult. Nutr. 11:385–393.

Cheng, K.-M., C.-Q. Hu, Y.-N. Liu, S.-X. Zheng, and X.-J. Qi. 2006. Effects of dietary calcium, phosphorus and calcium/phosphorus ratio on the growth and tissue mineralization of *Litopenaeus vannamei* reared in low-salinity water. Aquaculture 251:472–483.

Choi, S.-M., K.-W. Kim, Y.-J. Kang, X. Wang, J.-W. Kim, G.-Y. Yoo, and S. C. Bai. 2005. Reevaluation of the phosphorus requirement of juvenile olive flounder *Paralichthys olivaceus* and the bioavailability of various inorganic phosphorus sources. J. World Aquacult. Soc. 36:217–212.

Chvapil, M., A. L. Aronson, and Y. M. Peng. 1974. Relation between zinc and iron and peroxidation of lipids in liver homogenate in Cu EDTA-treated rats. Exp. Mol. Pathol. 20:216–227.

Civera, R., M. Fakhfakh, and J. C. Guillaume. 1990. Digestive utilization of phytate phosphorus in two species of shrimp: *Penaeus japonicus* and *Penaeus vannamei* (Abstract.). World Aquaculture Meeting, 1990, Halifax, Canada.

Clearwater, S. J., A. M. Farag, and J. S. Meyer. 2002. Bioavailability and toxicity of dietborne copper and zinc to fish. Comp. Biochem. Phys. C 132:269–313.

Coloso, R. M., K. King, J. W. Fletcher, P. Weis, A. Werner, and R. P. Ferraris. 2003. Dietary P regulates phosphate transporter expression, phosphatase activity, and effluent P partitioning in trout culture. J. Comp. Physiol. 173:519–530.

Dabrowska, H., K. Meyer-Burgdorff, and K. D. Gunther. 1989. Interaction between dietary protein and magnesium level in tilapia (*Oreochromis niloticus*). Aquaculture 76:277–291.

Dabrowska, H., K. Meyer-Burgdorff, and K. D. Gunther. 1991. Magnesium status in freshwater fish, common carp (*Cyprinus carpio*) and the dietary protein-magnesium interaction. Fish Physiol. Biochem. 9:165–172.

Dall, W., and J. W. Moriarty. 1983. Functional aspects of nutrition and digestion. Pp. 215–216 in The Biology of the Crustacea, Vol. 5, D. E. Bliss, ed. New York: Academic Press, Inc.

Dalsgaard, J., K. S. Ekmann, P. B. Pedersen, and V. Verlhac. 2009. Effect of supplemented fungal phytase on performance and phosphorus availability by phosphorus-depleted juvenile rainbow trout (*Oncorhynchus mykiss*), and on the magnitude and composition of phosphorus waste output. Aquaculture 286:105–112.

Davis, D. A. 1990. Dietary Mineral Requirements of *Penaeus vannamei*: Evaluation of the Essentiality for Thirteen Minerals and the Requirements for Calcium, Phosphorus, Copper, Iron, Zinc and Selenium. Ph.D. Dissertation, Texas A&M University, College Station, TX.

Davis, D. A., and C. R. Arnold. 1994. Estimation of apparent phosphorus availability from inorganic phosphorus sources for *Penaeus vannamei*. Aquaculture 127:245–254.

Davis, D. A., and E. H. Robinson. 1987. Dietary phosphorus requirement of juvenile red drum *Sciaenops ocellatus*. J. World Aquacult. Soc. 18:129–136.

Davis, D. A., A. L. Lawrence, and D. M. Gatlin, III. 1992a. Mineral requirements of *Penaeus vannamei*: A preliminary examination of the dietary essentiality of thirteen minerals. J. World Aquacult Soc. 23:8–14.

Davis, D. A., A. L. Lawrence, and D. M. Gatlin, III. 1992b. Evaluation of the dietary iron requirement of *Penaeus vannamei*. J. World Aquacult. Soc. 23:15–22.

Davis, D. A., A. L. Lawrence, and D. M. Gatlin, III. 1993a. Response of *Penaeus vannamei* to dietary calcium, phosphorus, and calcium:phosphorus ratio. J. World Aquacult. Soc. 24:504–515.

Davis, D. A., A. L. Lawrence, and D. M. Gatlin, III. 1993b. Dietary copper requirement of *Penaeus vannamei*. Nippon Suisan Gakk. 59:117–122.

Davis, D. A., A. L. Lawrence, and D. M. Gatlin, III. 1993c. Dietary zinc requirement of *Penaeus vannamei* and the effects of phytic acid on zinc and phosphorus bioavailability. J. World Aquacult. Soc. 24:40–47.

Debnath, D., N. P. Sahu, A. K. Pal, K. K. Jain, S. Yengkokpam, and S. C. Mukherjee. 2005. Mineral status of *Pangasius pangasius* (Hamilton) fingerlings in relation to supplemental phytase: Absorption, whole-body and bone mineral content. Aquac. Res. 36:326–335.

Depledge, M. H. 1989. Re-evaluation of metabolic requirements for copper and zinc in decapod crustaceans. Mar. Environ. Res. 27:115–126.

Depledge, M. H., R. Chan, and T. T. Loh. 1986. Iron distribution and transport in *Scylla serrata* (Forskal). Asian Mar. Biol. 3:101–110.

Deshimaru, O., and Y. Yone. 1978. Requirement of prawn for dietary minerals. B. Jpn. Soc. Sci. Fish. 44:907–910.

Deshimaru, O., K. Kuroki, S. Sakamoto, and Y. Yone. 1978. Absorption of labeled-^{45}Ca by prawn from sea water. B. Jpn. Soc. Sci. Fish. 44:975–977.

Desjardins, L. M., B. D. Hicks, and J. W. Hilton. 1987. Iron catalyzed oxidation of trout diets and its effect on the growth and physiological response of rainbow trout. Fish Physiol. Biochem. 3:173–182.

Do Carmo e Sa, M. V., L. E. Pezzato, M. M. B. F. Lima, and P. De Magalhaes Padilha. 2004. Optimum zinc supplementation level in Nile tilapia *Oreochromis niloticus* juveniles diets. Aquaculture 238:385–401.

Dougall, D. S., L. C. Woods, III, L. W. Douglass, and J. H. Soares. 1996. Dietary phosphorus requirement of juvenile striped bass *Morone saxatilis*. J. World Aquacult. Soc. 27(1):82–91.

Duston, J. 1993. Effects of dietary betaine and sodium chloride on seawater adaptation in Atlantic salmon parr (*Salmo salar* L.). Comp. Biochem. Phys. A 105:673–677.

Eid, A. E., and S. I. Ghonim. 1994. Dietary zinc requirement of fingerling *Oreochromis niloticus*. Aquaculture 119:259–264.

El-Mowafi, A. R. A., and A. Maage. 1998. Magnesium requirement of Atlantic salmon (*Salmo salar* L.) parr in seawater-treated fresh water. Aquacult. Nutr. 4:31–38.

El-Mowafi, A. R. A., R. Waagbo and A. Maage. 1997. Effect of low dietary magnesium on immune response and osmoregulation of Atlantic salmon. J. Aquat. Anim. Health 9:8–17.

Eroldogan, O. T., M. Kumlu, M. Kir, and G.A. Kiris. 2005. Enhancement of growth and feed utilization of the European sea bass (*Dicentrarchus labrax*) fed supplementary dietary salt in freshwater. Aquac. Res. 36:361–369.

Eya, J. C., and R. T. Lovell. 1997a. Available phosphorus requirements of food-size channel catfish (*Ictalurus punctatus*) fed practical diets in ponds. Aquaculture 154(3–4):283–291.

Eya, J. C., and R. T. Lovell. 1997b. Net absorption of dietary phosphorus from various inorganic sources and effect of fungal phytase on net absorption of plant phosphorus by channel catfish *Ictalurus punctatus*. J. World Aquacult. Soc. 28:386–391.

Eya, J. C., and R. T. Lovell. 1998. Effects of dietary phosphorus on resistance of channel catfish to *Edwardsiella ictaluri* challenge. J. Aquat. Anim. Health 10:28–34.

Felton, S. P., M. L. Landolt, R. Grace, and A. Palmisano. 1996. Effects of selenium dietary enhancement on hatchery-reared coho salmon, *Oncorhynchus kisutch* (Walbaum), enzymes and seawater adaptation evaluated. Aquac. Res. 27:135–142.

Fjelldal, P. G., T. Hansen, O. Breck, R. Sandvik, R. Waagbo, A. Berg, and R. Ornsrud. 2009. Supplementation of dietary minerals during the early seawater phase increase vertebral strength and reduce the prevalence of vertebral deformities in fast-growing under-yearling Atlantic salmon (*Salmo salar* L.) smolt. Aquacult. Nutr. 15:366–378.

Fontagne, S., N. Silva, D. Bazin, A. Ramos, P. Aguirre, A. Surget, A. Abrantes, S. J. Kaushik, and D. M. Power. 2009. Effects of dietary phosphorus and calcium level on growth and skeletal development in rainbow trout (*Oncorhynchus mykiss*) fry. Aquaculture 297:141–150.

Gallagher, M. L., W. D. Brown, D. E. Conklin, and M. Sifri. 1978. Effects of varying calcium/phosphorus ratios in diets fed to juvenile lobsters (*Homarus americanus*). Comp. Biochem. Phys. A 60A:467–471.

Gallagher, M. L., R. C. Bayer, J. H. Rittenburg, and D. F. Leavitt. 1982. Studies on the mineral requirements of the adult American lobster. Prog. Fish Cult. 44:210–212.

Gatlin, D. M., III, and P. Li. 2008. Use of diet additives to improve nutritional value of alternative protein sources. Pp. 501–522 in Alternative Protein Sources in Aquaculture Diets, C. Lim, C. D. Webster, and C.-S. Lee, eds. New York: Haworth Press.

Gatlin, D. M., III, and H. F. Phillips. 1989. Dietary calcium, phytate and zinc interactions in channel catfish. Aquaculture 79:259–266.

Gatlin, D. M., III, and R. P. Wilson. 1983. Dietary zinc requirement of fingerling channel catfish. J. Nutr. 113:630–635.

Gatlin, D. M., III, and R. P. Wilson. 1984a. Studies on the manganese requirement of fingerling channel catfish. Aquaculture 41:85–92.

Gatlin, D. M., III, and R. P. Wilson. 1984b. Dietary selenium requirement of fingerling channel catfish. J. Nutr. 114:627–633.

Gatlin, D. M., III, and R. P. Wilson. 1984c. Zinc supplementation of practical channel catfish diets. Aquaculture 41:31–36.

Gatlin, D. M., III, and R. P. Wilson. 1986a. Dietary copper requirement of fingerling channel catfish. Aquaculture 54:277–285.

Gatlin, D. M., III, and R. P. Wilson. 1986b. Characterization of iron deficiency and the dietary iron requirement of fingerling channel catfish. Aquaculture 52:191–198.

Gatlin, D. M., III, E. H. Robinson, W. E. Poe, and R. P. Wilson. 1982. Magnesium requirement of fingerling channel catfish and signs of magnesium deficiency. J. Nutr. 112:1182–1187.

Gatlin, D. M., III, W. E. Poe, and R. P. Wilson. 1986. Effects of singular and combined dietary deficiencies of selenium and vitamin E on fingerling channel catfish (*Ictalurus punctatus*). J. Nutr. 116:1061–1067.

Gatlin, D. M., III, H. F. Phillips, and E. L. Torrans. 1989. Effects of various levels of dietary copper and zinc on channel catfish. Aquaculture 76:127–134.

Gatlin, D. M., III, J. P. O'Connell, and J. Scarpa. 1991. Dietary zinc requirement of the red drum, *Sciaenops ocellatus*. Aquaculture 92:259–265.

Gatlin, D. M., III, D. S. MacKenzie, S. R. Craig, and W. H. Neill. 1992. Effects of dietary sodium chloride on red drum juveniles in waters of various salinities. Prog. Fish-Cult. 54:220–227.

Gatta, P. P., K. D. Thompson, R. Smullen, A. Piva, S. Testi, and A. Adams. 2001a. Dietary organic chromium supplementation and its effect on the immune response of rainbow trout (*Oncorhynchus mykiss*). Fish Shellfish Immun. 11:371–382.

Gatta, P. P., A. Piva, M. Paolini, S. Testi, A. Bonaldo, A. Antelli, and A. Mordenti. 2001b. Effects of dietary organic chromium on gilthead seabream (*Sparus aurata* L.) performances and liver microsomal metabolism. Aquac. Res. 32:60–69.

Gensic, M., P. J. Wissing, T. R. Keefe, and A. Mustafa. 2004. Effects of iodized feed on stress modulation in steelhead trout, *Oncorhynchus mykiss* (Walbaum). Aquac. Res. 35:1117–1121.

Ghidalia, W., J. M. Fine, and M. Marneux. 1972. On the presence of an iron-binding protein in the serum of a decapod crustacean [*Macropipus puber* (Lnne)]. Comp. Biochem. Phys. B 41B:349–354.

Ghosh, S. R. 1975. Preliminary observation on the effect of cobalt on the survival and growth of *Mugil parsia*. Isr. J. Aquacult. Bamidgeh 27:110–111.

Green, J. A., R. W. Hardy, and E. L. Brannon. 2002a. Effects of dietary phosphorus and lipid levels on utilization and excretion of phosphorus and nitrogen by rainbow trout (*Oncorhynchus mykiss*). 1. Laboratory-scale study. Aquacult. Nutr. 8:279–290.

Green, J. A., E. L. Brannon, and R. W. Hardy. 2002b. Effects of dietary phosphorus and lipid levels on utilization and excretion of phosphorus and nitrogen by rainbow trout (*Oncorhynchus mykiss*). 2. Production-scale study. Aquacult. Nutr. 8:291–298.

Halver, J. E., S. P. Felton, and R. Zbanyszek. 2004. Carcass selenium loss as an indicator of stress in barge transported salmon (*Oncorhynchus tshawytscha* Walbaum) smolts. Aquac. Res. 35:1099–1103.

Hamilton, S. J. 2004. Review of selenium toxicity in the aquatic food chain. Sci. Total Environ. 326:1–31.

Hardy, R. W., and K. D. Shearer. 1985. Effect of dietary calcium phosphate and zinc supplementation on whole body zinc concentration of rainbow trout (*Salmo gairdneri*). Can. J. Fish. Aquat. Sci. 42:181–184.

Hardy, R. W., L. L. Oram and G. Moller. 2010. Effects of dietary selenomethionine on cutthroat trout (*Oncorhynchus clarki bouvieri*) growth and reproductive performance over a life cycle. Arch. Environ. Con. Tox. 58:237–245.

Harpaz, S., Y. Hakim, T. Slosman, and O. T. Eroldogan. 2005. Effects of adding salt to the diet of Asian sea bass *Lates calcarifer* reared in fresh or salt water recirculating tanks, on growth and brush border enzyme activity. Aquaculture 248:315–324.

Hertz, Y., Z. Madar, B. Hepher, and A. Gertler. 1989. Glucose metabolism in the common carp (*Cyrpinus carpio* l.): The effects of cobalt and chromium. Aquaculture 76:255–267.

Hilton, J. W. 1989. The interaction of vitamins, minerals and diet composition in the diet of fish. Aquaculture 79:223–244.

Hilton, J. W., and P. V. Hodson. 1983. Effect of increased dietary carbohydrate on selenium metabolism and toxicity in rainbow trout (*Salmo gairdneri*). J. Nutr. 113(6):1241–1248.

Hilton, J. W., P. V. Hodson, and S. J. Slinger. 1980. The requirement and toxicity of selenium in rainbow trout (*Salmo gairdneri*). J. Nutr. 110:2527–2535.

Hossain, M. A., and M. Furuichi. 1998. Availability of environmental and dietary calcium in tiger puffer. Aquacult. Int. 6:121–132.

Hossain, M. A., and M. Furuichi. 1999. Necessity of dietary calcium supplement in black sea bream. Fish. Sci. 65:893–897.

Hossain, M. A., and M. Furuichi. 2000. Essentiality of dietary calcium supplement in fingerling scorpion fish (*Sebastiscus marmoratus*). Aquaculture 189:155–163.

Hoyle, I., B. J. Shaw, and R. D. Handy. 2007. Dietary copper exposure in the African walking catfish, *Clarias gariepinus*: Transient osmoregulatory disturbances and oxidative stress. Aquat. Toxicol. 83:62–72.

Hu, C. H., Y. Xu, M. S. Xia, L. Xiong, and Z. R. Xu. 2007. Effects of Cu^{2+}-exchanged montmorillonite on growth performance, microbial ecology and intestinal morphology of Nile tilapia (*Oreochromis niloticus*). Aquaculture 270:200–206.

Hua, K., and D. P. Bureau. 2006. Modelling digestible phosphorus content of salmonid fish feeds. Aquaculture 254(1–4):455–465.

Hua, K., C. F. M. De Lange, A. J. Niimi, G. Cole, R. D. Moccia, M. Z. Fan, and D. P. Bureau. 2008. A factorial model to predict phosphorus waste output of rainbow trout (*Oncorhynchus mykiss*). Aquac. Res. 39(10):1059–1068.

Hughes, K. P., and J. H. Soares, Jr. 1998. Efficacy of phytase on phosphorus utilization in practical diets fed to striped bass *Morone saxatilis*. Aquacult. Nutr. 4:133–140.

Huner, J. V., and L. B. Colvin. 1977. A short-term study on the effects of diets with varied calcium:phosphorus ratios on the growth of juvenile shrimp, *Penaeus californiensis*. A short communication. Proc. 8th Ann. Mtg. World Maricult. Soc. 8:775–778.

Ikeda, Y., H. Ozaki, and K. Uematasu. 1973. Effect of enriched diet with iron in culture of yellowtail. J. Tokyo Univ. Fish. 59:91–99.

Inoue, M., S. Satoh, M. Maiat, V. Kiron, and N. Okamoto. 1998. Recovery from derangement of natural killer-like activity of leucocytes due to Zn and Mn deficiency in rainbow trout, *Oncorynchus mykiss* (Walbaum), by the oral administration of these elements. J. Fish Dis. 21:233–236.

Ishac, M. M., and A. M. Dollar. 1968. Studies on manganese uptake in *Tilapia mossambica* and *Salmo gairdneri*. 1. Growth of *Tilapia mossambica* in response to manganese. Hydrobiologia 31:572–584.

Jaramillo, F., Jr., P. Li, and D. M. Gatlin, III. 2009. Selenium nutrition of hybrid striped bass (*Morone chrysops* × *M. saxatilis*) bioavailability, toxicity and interaction with vitamin E. Aquacult. Nutr. 15:160–165.

Julshamn, K., K.-J. Andersen, O. Ringdal, and J. Brenna. 1988. Effect of dietary copper on the hepatic concentration and subcellular distribution of copper and zinc in the rainbow trout (*Salmo gairdneri*). Aquaculture 73:143–155.

Julshamn, K., A. Maage, R. Waagbo, and A. K. Lundebye. 2006. A preliminary study on tailoring of fillet iodine concentrations in adult Atlantic salmon (*Salmo salar* L.) through dietary supplementation. Aquacult. Nutr. 12:45 51.

Kanazawa, A., S. Teshima, and M. Sasaki. 1984. Requirements of the juvenile prawn for calcium, phosphorus, magnesium, potassium, copper, manganese and iron. Mem. Fac. Fish. Kagoshima University 33:63–71.

Kawatsu, H. 1972. Studies on the anemia of fish. V. Dietary iron deficient anemia in brook trout, *Salvelinus fontinalis*. B. Freshwater Fish. Res. Lab. 22:59–67.

Ketola, H. G. 1975. Requirement of Atlantic salmon for dietary phosphorus. T. Am. Fish. Soc. 104:548–551.

Ketola, H. G. 1979. Influence of dietary zinc on cataracts in rainbow trout (*Salmo gairdneri*). J. Nutr. 109:965–969.

Ketola, H. G. 1994. Use of enzymes in diets of trout to reduce environmental discharges of phosphorus. P. 211 in World Aquaculture Society Book of Abstracts. Baton Rouge, LA: World Aquaculture Society.

Ketola, H. G., and M. E. Richmond. 1994. Requirement of rainbow trout for dietary phosphorus and its relationship to the amount discharged in hatchery effluent. T. Am. Fish. Soc. 123:587–594.

Kim, J. D., K. S. Kim, J. S. Song, J. Y. Lee, and K. S. Jeong. 1998. Optimum level of dietary monocalcium phosphate based on growth and phosphorus excretion of mirror carp, *Cyprinus carpio*. Aquaculture 161(1–4):337–344.

Kitabayashi, K., H. Kurata, K. Shudo, K. Nakamura, and S. Ishikawa. 1971. Studies on formula feed for kuruma prawn—I. On the relationship among glucosamine, phosphorus and calcium. B. Tokai Regional Fish. Res. Lab. Tokyo 65:91–108.

Knox, D., C. B. Cowey, and J. W. Adron. 1981. Studies on the nutrition of salmonid fish. The magnesium requirement of rainbow trout (*Salmo guairdneri*). Br. J. Nutr. 45:137–148.

Knox, D., C. B. Cowey, and J. W. Adron. 1982. Effects of dietary copper and copper:zinc ratio on rainbow trout *Salmo gairdneri*. Aquaculture 27:111–119.

Knox, D., C. B. Cowey, and J. W. Adron. 1984. Effects of dietary zinc intake upon copper metabolism in rainbow trout (*Salmo gairdneri*). Aquaculture 40:199–207.

Kucukbay, Z., H. Yazlak, N. Sahin, M. Tuzcu, M. N. Cakmak, F. Gurdogan, V. Juturu, and K. Sahin. 2006. Zinc picolinate supplementation decreases oxidative stress in rainbow trout (*Oncorhynchus mykiss*). Aquaculture 257:465–469.

Lall, S. P. 1989. The minerals. Pp. 219–257 in Fish Nutrition, Second Edition, J. E. Halver, ed. New York: Academic Press.

Lall, S. P. 1991. Digestibility, metabolism and excretion of dietary phosphorus. Pp. 77–90 in Nutritional Strategies and Aquaculture Waste. Proceedings of the First International Symposium on Nutritional Strategies on Management of Aquaculture Waste, 1990, C. B. Cowey and C. Y. Cho, eds. Guelph, Ontario, Canada: University of Guelph.

Lall, S. P. 2002. The Minerals. Pp. 259–308 in Fish Nutrition, Third Edition, J. E. Halver and R. W. Hardy, eds. London, UK: Academic Press.

Lall, S. P., and F. J. Bishop. 1977. Studies on mineral and protein utilization by Atlantic salmon (*Salmo salar*) grown in sea water. Fish. Mar. Serv. Tech. Rep. 688:1–16.

Lawrence, A. L., J. Fox, and F. L. Castille, Jr. 1981. Decreased toxicity of copper and manganese ions to shrimp nauplii (*Penaeus stylirostris* Stimpson) in the presence of EDTA. J. World Maricult. Soc. 12:271–280.

Lee, M.-H., and S.-Y. Shiau. 2002. Dietary copper requirement of juvenile grass shrimp, *Penaeus monodon*, and effects on non-specific immune responses. Fish Shellfish Immun. 13:259–270.

Lee, Y. H., D. K. Layman, R. R. Bell, and H. W. Norton. 1981. Response of glutathione peroxidase and catalase to excess iron in rats. J. Nutr. 111:2195–2202.

Lellis, W. A., F. T. Barrows, and R. W. Hardy. 2004. Effects of phase-feeding dietary phosphorus on survival, growth, and processing characteristics of rainbow trout (*Oncorhynchus mykiss*). Aquaculture 242(1–4):607–616.

Li, M. H., H. R. Robinette, and E. H. Robinson. 1996. Efficacy of dicalcium and defluorinated rock phosphates as dietary phosphorus sources for channel catfish (*Ictalurus punctatus*). Aquaculture 147(1–2):107–114.

Lim, C., and P. H. Klesius. 1997. Responses of channel catfish (*Ictalurus punctatus*) fed iron-deficient and replete diets to *Edwardsiella ictaluri* challenge. Aquaculture 157:83–93.

Lim, C., P. H. Klesius, and P. L. Duncan. 1996. Immune response and resistance of channel catfish to *Edwardsiella ictaluri* challenge when fed various dietary levels of zinc methionine and zinc sulfate. J. Aquat. Anim. Health 8:302–307.

Lim, C., P. H. Klesius, and M. H. Li. 2000. Interaction between dietary levels of iron and vitamin C on growth, hermatology, immune response and resistance of channel catfish (*Ictalurus punctatus*) to *Edwardsiella ictaluri* challenge. Aquaculture 185:313–327.

Limsuwan, T., and R. T. Lovell. 1981. Intestinal synthesis and absorption of vitamin B-12 in channel catfish. J. Nutr. 111:2125–2132.

Lin, Y.-H., and S.-Y. Shiau. 2005. Dietary selenium requirements of juvenile grouper, *Epinephelus malabaricus*. Aquaculture 250:356–363.

Lin, Y.-H., and S.-Y. Shiau. 2007. The effects of dietary selenium on the oxidative stress of grouper, *Epinephelus malabaricus*, fed high copper. Aquaculture 267:38–43.

Lin, Y.-H., and S.-Y. Shiau. 2009. Mutual sparing of dietary requirements for alpha-tocoperol and selenium in grouper, *Epinephelus malabaricus*. Aquaculture 294:242–245.

Lin, Y.-H., Y.-Y. Shie, and S.-Y. Shiau. 2008a. Dietary copper requirements of juvenile grouper, *Epinephelus malabaricus*. Aquaculture 274:161–165.

Lin, Y.-H., S.-M. Lin, and S.-Y. Shiau. 2008b. Dietary manganese requirements of juvenile tilapia, *Oreochromis niloticus* × *O. aureus*. Aquaculture 284:207–210.

Lin, Y.-H., L.-C. Jiang, and S.-Y. Shiau. 2008c. Dietary zinc requirements of juvenile tilapia, *Oreochromis niloticus* × *O. aureus*. J. Fish. Soc. Taiwan 35:117–125.

Little, C., R. Olinescu, K. G. Reid, and P. J. O'Brien. 1970. Properties and regulation of glutathione peroxidase. J. Biol. Chem. 245:3632–3636.

Liu, F., D. Liang, F. Sun, H. Li, and X. Lan. 1990. Effects of copper on the prawn *Penaeus orientalis*. Oceanol. Limnol. Sinica. 21:404–410.

Lo, G. S., S. L. Settle, F. H. Steinke, and D. T. Hopkins. 1981. Effect of phytate: Zinc molar ratio and isolated soybean protein on zinc bioavailability. J. Nutr. 111:2223–2235.

Lorentzen, M., A. Maage, and K. Julshamn. 1994. Effects of dietary selenite or selenomethionine on tissue selenium levels of Atlantic salmon (*Salmo salar*). Aquaculture 121:359–367.

Lorentzen, M., A. Maage, and K. Julshamn. 1998. Supplementing copper to a fish meal based diet fed to Atlantic salmon parr affects liver copper and selenium concentrations. Aquacult. Nutr. 4:67–72.

Lovell, T. 1978. Dietary phosphorus requirement of channel catfish (*Ictalurus punctatus*). T. Am. Fish. Soc. 107(4):617–621.

Lovell, T. 1989. Nutrition and Feeding of Fish. New York: Van Nostrand Reinhold.

Ma, K., C. Zhang, Q. Ai, Q. Duan, W. Xu, L. Zhang, Z. Luifu, and B. Tan. 2006. Dietary phosphorus requirement of large yellow croaker, *Pseudosciaena crocea* R. Aquaculture 251:346–353.

Maage, A., K. Julshamn, and G. E. Berge. 2001. Zinc gluconate and zinc sulphate as dietary zinc sources for Atlantic salmon. Aquacult. Nutr. 7:183–187.

Mantel, L. H., and L. L. Farmer. 1983. Osmotic and ionic regulation. Pp. 215–261 in The Biology of the Crustacea, Volume 5, D. E. Bliss, ed. New York: Academic Press.

Martin, J.-L. M. 1973. Iron metabolism in *Cancer irroratus* (Crustacean Decapoda) during the intermoult cycle, with special reference to iron in the gills. Comp. Biochem. Phys. A 46A:123–129.

Maynard, L. A., J. K. Loosli, H. F. Hintz, and R. G. Warner. 1979. Animal Nutrition. New York: McGraw-Hill Book Co.

McClain, W. R., and D. M. Gatlin, III. 1988. Dietary zinc requirement of *Oreochromis aureus* and effects of dietary calcium and phytate on zinc bioavailability. J. World Aquacult. Soc. 19:103–108.

Miyawaki, M., M. Matsuzaki, and N. Sasaki. 1961. Histochemical studies on the hepatopancreas of the crayfish, *Procambarus clarkii*. Kumamato J. Sci. 5:170–172.

Moyle, P. B., and J. J. Cech, Jr. 2000. Hydromineral balance. Pp. 79–96 in Fishes: An Introduction to Ichthyology. Upper Saddle River, NJ: Prentice-Hall, Inc.

Murai, T., J. W. Andrews, and R. G. Smith Jr. 1981. Effects of dietary copper on channel catfish. Aquaculture 22:353–357.

Murray, M. W., and J. W. Andrews. 1979. Channel catfish: The absence of an effect on dietary salt on growth. Prog. Fish-Cult. 41:151–152.

Nakamuara, Y., and J. Yamada. 1980. Effects of dietary calcium levels, Ca/P ratios, and calcium components on the calcium absorption rate in carp. Bull. Fac. Fish. Hokkaido University 31:277–282.

Ng, W.-K., and R. P. Wilson. 1997. Chromic oxide inclusion in the diet does not affect glucose utilization or chromium retention by channel catfish, *Ictalurus punctatus*. J. Nutr. 127:2357–2362.

NRC (National Research Council). 1980. Mineral Tolerance of Domestic Animals. Washington, DC: National Academy Press.

NRC. 1983. Nutrient Requirements of Warmwater Fishes and Shellfishes. Washington, DC: National Academy Press.

NRC. 1993. Nutrient Requirements of Fish. Washington, DC: National Academy Press.

NRC. 1997. The Role of Chromium in Animal Nutrition. Washington, DC: National Academy Press.

NRC. 2005. Mineral Tolerance of Domestic Animals, 2nd revised edition. Washington, DC: National Academies Press.

Oberleas, D., M. E. Muhrer, and B. L. O'Dell. 1962. Effects of phytic acid on zinc availability and parakeratosis in swine. J. Anim. Sci. 21:57–61.

O'Connell, J. P., and D. M. Gatlin, III. 1994. Effects of dietary calcium and vitamin D_3 on weight gain and mineral composition of the blue tilapia (*Oreochromis aureus*) in low-calcium water. Aquaculture 125:107–117.

O'Dell, B. L. 1976. Biochemistry of copper. Med. Clin. N. Am. 60:697–703.

O'Dell, B. L., J. M. Yohe, and J. E. Savage. 1964. Zinc availability in the chicken as affected by phytate, calcium and ethylenediaminetetraacetate. Poultry Sci. 43:45–419.

Ogino, C., and H. Takeda. 1976. Mineral requirements in fish–III calcium and phosphorus requirements in carp. B. Jpn. Soc. Sci. Fish. 42:793–799.

Ogino, C., and H. Takeda. 1978. Requirement of rainbow trout for dietary calcium and phosphorus. B. Jpn. Soc. Sci. Fish. 44:1019–1022.

Ogino, C., and G.-Y. Yang. 1978. Requirement of rainbow trout for dietary zinc. B. Jpn. Soc. Sci. Fish. 44:1015–1018.

Ogino, C., and G.-Y. Yang. 1979. Requirement of carp for dietary zinc. B. Jpn. Soc. Sci. Fish. 45:967–969.

Ogino, C., and G.-Y. Yang. 1980. Requirement of carp and rainbow trout for dietary manganese and copper. B. Jpn. Soc. Sci. Fish. 46:455–458.

Ogino, C., F. Takashima, and J. Y. Chiou. 1978. Requirement of rainbow trout for dietary magnesium. B. Jpn. Soc. Sci. Fish. 44:1105–1108.

Ogino, C., L. Takeuchi, H. Takeda, and T. Watanabe. 1979. Availability of dietary phosphorus in carp and rainbow trout. B. Jpn. Soc. Sci. Fish. 45:1527–1532.

Ogura, K. 1959. Midgut gland cells accumulating iron or copper in the crayfish, *Procambarus clarkii*. Annot. Zool. Japon. 32:133–142.

Oliva-Teles, A., and A. Pimentel-Rodrigues. 2004. Phosphorus requirement of European sea bass (*Dicentrarchus labrax* L.) juveniles. Aquac. Res. 35:636–642.

Overturf, K., V. Raboy, Z. Cheng, and R. Hardy. 2003. Mineral availability from barley low phytic acid grains in rainbow trout (*Oncorhynchus mykiss*) diets. Aquacult. Nutr. 9:239–246.

Pan, Q., S. Liu, Y.-G. Tan, and Y.-Z. Bi. 2003. The effect of chromium picolinate on growth and carbohydrate utilization in tilapia, *Oreochromis niloticus* × *Oreochromis aureus*. Aquaculture 225:421–429.

Pan, Q., X.-Y. Chen, F. Li, Y.-Z. Bi, and S.-X. Zheng. 2005. Response of juvenile *Litopenaeus vannamei* to varying levels of calcium phosphate monobasic supplemented to a practical diet. Aquaculture 248:97–102.

Pan, L., X. Zhu, S. Xie, W. Lei, D. Han, and Y. Yang. 2008. Effects of dietary manganese on growth and tissue manganese concentrations of juvenile gibel carp, *Carassius auratus gibelio*. Aquacult. Nutr. 14:459–463.

Paripatananont, T., and R. T. Lovell. 1995a. Response of channel catfish fed organic and inorganic sources of zinc to *Edwardsiella ictaluri* challenge. J. Aquat. Anim. Health 7:147–154.

Paripatananont, T., and R. T. Lovell. 1995b. Chelated zinc reduces the dietary zinc requirement of channel catfish, *Ictalurus punctatus*. Aquaculture 133:73–82.

Partridge, G. J., and A. J. Lymbery. 2008. The effect of salinity on the requirement for potassium by barramundi (*Lates calcarifer*) in saline groundwater. Aquaculture 278:164–170.

Penaflorida, V. D. 1999. Interaction between dietary levels of calcium and phosphorus on growth of juvenile shrimp, *Penaeus monodon*. Aquaculture 172:281–289.

Pike, R. L., and M. L. Brown. 1975. Nutrition: An Integrated Approach, 2nd Edition. New York: John Wiley & Sons.

Pimentel-Rodrigues, A. M., and A. Oliva-Teles. 2001. Phosphorus requirements of gilthead sea bream (*Sparus aurata* L.) juveniles. Aquac. Res. 32(Suppl. 1):157–161.

Pimentel-Rodrigues, A., and A. Oliva-Teles. 2007. Phosphorus availability of inorganic phosphates and fish meals in European sea bass (*Dicentrarchus labrax* L.) juveniles. Aquaculture 267(1–4):300–307.

Poston, H. A., G. F. Combs, Jr., and L. Leibovitz. 1976. Vitamin E and selenium interactions in the diet of Atlantic salmon (*Salmo salar*): Gross, histological and biochemical deficiency signs. J. Nutr. 106:892–904.

Ravndal, J., T. Lovold, H. B. Bentsen, K. H. Roed, T. Gjedrem, and K. A. Rorvik. 1994. Serum iron levels in farmed atlantic salmon: Family variation and associations with disease resistance: Aquaculture 125(1–2):37–45.

Reigh, R., E. H. Robinson, and P. B. Brown. 1991. Effects of dietary magnesium on growth and tissue magnesium content of blue tilapia *Oreochromis aureus*. J. World Aquacult. Soc. 22:192–200.

Richardson, N. L., D. A. Higgs, R. M. Beames, and J. R. McBride. 1985. Influence of dietary calcium, phosphorus, zinc and sodium phytate level on cataract incidence, growth and histopathology in juvenile Chinook salmon (*Oncorhynchus tshawytscha*). J. Nutr. 115:553–567.

Riche, M., and P. B. Brown. 1996. Availability of phosphorus from feedstuffs fed to rainbow trout, *Oncorhynchus mykiss*. Aquaculture 142:269–282.

Robinson, E. H., S. D. Rawles, H. E. Yette, and L. W. Greene. 1984. An estimate of the dietary calcium requirement of fingerling *Tilapia aurea* reared in calcium-free water. Aquaculture 41:389–393.

Robinson, E. H., S. D. Rawles, P. B. Brown, H. E. Yette, and L. W. Greene. 1986. Dietary calcium requirement of channel catfish *Ictalurus punctatus*, reared in calcium-free water. Aquaculture 53:263–270.

Robinson, E. H., D. LaBomascus, P. B. Brown, and T. L. Linton. 1987. Dietary calcium and phosphorus requirements of *Oreochromis aureus* reared in calcium-free water. Aquaculture 64:267–276.

Rodehutscord, M., E. Pfeffer, and C. B. Cowey. 1995. Requirement for phosphorus in rainbow trout (*Oncorhynchus mykiss*) growing from 50 to 200 g. Water Sci. Technol. 31:137–141.

Roy, L. A., D. A. Davis, I. P. Saoud, and R. P. Henry. 2007. Supplementation of potassium, magnesium and sodium chloride in practical diets for the Pacific white shrimp, *Litopenaeus vannamei*, reared in low salinity waters. Aquacult. Nutr. 13:104–113.

Roy, P. K., and S. P. Lall. 2003. Dietary phosphorus requirement of juvenile haddock (*Melanogrammus aeglefinus* L.). Aquaculture 221(1–4):451–468.

Sajjadi, M., and C. G. Carter. 2004. Dietary phytase supplementation and the utilization of phosphorus by Atlantic salmon (*Salmo salar* L.) fed a canola-meal-based diet. Aquaculture 240:417–431.

Sakamoto, S., and Y. Yone. 1973. Effect of dietary calcium/phosphorus ratio upon growth, feed efficiency and blood serum Ca and P level in red sea bream. B. Jpn. Soc. Sci. Fish. 39:343–348.

Sakamoto, S., and Y. Yone. 1976a. Requirement of red sea bream for dietary Ca. Rep. Fish. Res. Lab. Kyushu Univ. (Jpn) 3:59–64.

Sakamoto, S., and Y. Yone. 1976b. Requirement of red sea bream for dietary Fe-I. Rep. Fish. Res. Lab. Kyushu Univ. (Jpn) 3:53–58.

Sakamoto, S., and Y. Yone. 1978a. Effect of dietary phosphorus level on chemical composition of red sea bream. B. Jpn. Soc. Sci. Fish. 44:227–229.

Sakamoto, S., and Y. Yone. 1978b. Requirement of red sea bream for dietary Na and K. J. Fac. Agric. Kyushu Univ. (Jpn) 23:79–84.

Sakamoto, S., and Y. Yone. 1978c. Iron deficiency symptoms of carp. B. Jpn. Soc. Sci. Fish. 44:1157–1160.

Sakamoto, S., and Y. Yone. 1978d. Requirement of red sea bream for dietary iron–II. B. Jpn. Soc. Sci. Fish. 44:223–225.

Sakamoto, S., and Y. Yone. 1979a. Requirement of red sea bream for dietary Mg. B. Jpn. Soc. Sci. Fish. 45:57–60.

Sakamoto, S., and Y. Yone. 1979b. Availabilities of phosphorus compounds as dietary phosphorus sources for red sea bream. J. Fac. Agric. Kyushu Univ. (Jpn) 23:177–184.

Salman, N. A., and F. B. Eddy. 1988. Effect of dietary sodium chloride on growth, food intake and conversion efficiency in rainbow trout (*Salmo gairdneri* Richardson). Aquaculture 70:131–144.

Saoud, I. P., L. A. Roy, and D. A. Davis. 2007. Chelated potassium and arginine supplementation in diets of Pacific white shrimp reared in low-salinity waters of west Alabama. N. Am. J. Aquacult. 69:265–270.

Sarker, P. K., S. Satoh, H. Fukada, and T. Masumoto. 2009. Effects of dietary phosphorus level on non-faecal phosphorus excretion from yellowtail (*Seriola quinqueradiata* Temminck & Schlegel) fed purified and practical diets. Aquac. Res. 40:225–232.

Sastry, K. V., and K. Sunita. 1982. Effect of cadmium and chromium on the intestinal absorption of glucose in snakehead (*Channa punctatus*). Toxicol. Lett. 10:293–296.

Satoh, S., K. Tabata, K. Izume, T. Takeuchi, and T. Watanabe. 1987. Effect of dietary tricalcium phosphate on availability of zinc to rainbow trout. Nippon Suisan Gakk. 53:1199–1205.

Satoh, S., W. E. Poe, and R. P. Wilson. 1989. Effect of supplemental phytate and/or tricalcium phosphate on weight gain, feed efficiency and zinc content in vertebrae of channel catfish. Aquaculture 80:155–161.

Savage, J. E., J. M. Yohe, E. E. Picket, and B. L. O'Dell. 1964. Zinc metabolism in the growing chick. Tissue concentration and effect of phytate on absorption. Poultry Sci. 43:420–426.

Scarpa, J., and D. M. Gatlin, III. 1992. Dietary zinc requirements of channel catfish, *Ictalurus punctatus*, swim-up fry in soft and hard water. Aquaculture 106:311–322.

Scott, M. L., M. C. Nesheim, and R. J. Young. 1982. Nutrition of the Chicken. New York: M. L. Scott and Associates.

Sealey, W. M., C. Lim, and P. H. Klesius. 1997. Influence of the dietary level of iron from iron methionine and iron sulfate on immune response and resistance of channel catfish to *Edwardsiella ictaluri*. J. World Aquacult. Soc. 42:142–149.

Shao, Q., J. Ma, Z. Xu, W. Hu, J. Xu, and S. Xie. 2008. Dietary phosphorus requirement of juvenile black seabream, *Sparus macrocephalus*. Aquaculture 277:92–100.

Shaw, H. M., R. L. Saunders, H. C. Hall, and E. B. Henderson. 1975. Effect of dietary sodium chloride on growth of Atlantic salmon (*Salmo salar*). J. Fish. Res. Board Can. 32:1813–1819.

Shearer, K. D. 1988. Dietary potassium requirement of juvenile Chinook salmon. Aquaculture 73:119–129.

Shearer, K. D. 1989. Whole body magnesium concentration as an indicator of magnesium status in rainbow trout (*Salmo gairdneri*). Aquaculture 77:201–210.

Shiau, S.-Y., and M.-J. Chen. 1993. Carbohydrate utilization by tilapia, *Oreochromis niloticus* × *O. aureus* as influenced by different chromium sources. J. Nutr. 123:1747–1753.

Shiau, S.-Y., and J.-F. Hsieh. 2001a. Quantifying the dietary potassium requirement of juvenile hybrid tilapia (*Oreochromis niloticus* × *O. aureus*). Brit. J. Nutr. 85:213–218.

Shiau, S.-Y., and J.-F. Hsieh. 2001b. Dietary potassium requirement of juvenile grass shrimp *Penaeus monodon*. Fish. Res. 67:592–595.

Shiau, S.-Y., and L. C. Jiang. 2006. Dietary zinc requirements of grass shrimp, *Penaeus monodon*, and effects on immune responses. Aquaculture 254:476–482.

Shiau, S.-Y., and H.-S. Liang. 1995. Carbohydrate utilization and digestibility by tilapia, *Oreochromis niloticus* × *O. aureus* are affected by chromium oxide inclusion in the diet. J. Nutr. 125:976–982.

Shiau, S.-Y., and S.-F. Lin. 1993. Effects of supplemental dietary chromium and vanadium on the utilization of different carbohydrates in tilapia, *Oreochromis niloticus* × *O. aureus*. Aquaculture 110:321–330.

Shiau, S.-Y., and L.-S. Lu. 2004. Dietary sodium requirement determined for juvenile hybrid tilapia (*Oreochromis niloticus* × *O. aureus*) reared in fresh water and seawater. Brit. J. Nutr. 91:585–590.

Shiau, S.-Y., and Y. C. Ning. 2003. Estimation of dietary copper requirements of juvenile tilapia, *Oerochromis niloticus* × *O. aureus*. Anim. Sci. 77:287–292.

Shiau, S.-Y., and S.-M. Shy. 1998. Dietary chromic oxide inclusion level required to maximize glucose utilization in hybrid tilapia, *Oreochromis niloticus* × *O. aureus*. Aquaculture 161:357–362.

Shiau, S.-Y., and L.-W. Su. 2003. Ferric citrate is half as effective as ferrous sulfate in meeting the requirement of juvenile tilapia, *Oerochromis niloticus* × *O. aureus*. J. Nutr. 133:483–488.

Shiau, S.-Y., and H. C. Tseng. 2007. Dietary calcium requirements of juvenile tilapia, *Oerochromis niloticus* × *O. aureus*, reared in fresh water. Aquacult. Nutr. 13:298–303.

Storebakken, T., K. D. Shearer, and A. J. Roem. 1998. Availability of protein, phosphorus and other elements in fish meal, soy-protein concentrate and phytase-treated soy-protein-concentrate-based diets to Atlantic salmon, *Salmo salar*. Aquaculture 161:365–377.

Sugiura, S. H., F. M. Dong, and R. W. Hardy. 1998. Effects of dietary supplements on the availability of minerals in fish meal: Preliminary observations. Aquaculture 160:283–303.

Sugiura, S. H., F. M. Dong, and R. W. Hardy. 2000. Primary responses of rainbow trout to dietary phosphorus concentrations. Aquacult. Nutr. 6:235–245.

Sugiura, S. H., R. W. Hardy, and R. J. Roberts. 2004. The pathology of phosphorus deficiency in fish—A review. J. Fish Dis. 27:255–265.

Sullivan, M., G. Hammond, R. J. Roberts, and N. J. Manchester. 2007a. Spinal deformation in commercially cultured Atlantic salmon, *Salmo salar* L.: A clinical and radiological study. J. Fish Dis. 30:745–752.

Sullivan, M., D. R. Guy, R. J. Roberts, and N. J. Manchester. 2007b. The aetiology of spinal deformity in Atlantic salmon, *Salmo salar* L.: Influence of genetic factors on the frequency and severity in freshwater stages. J. Fish Dis. 30:753–758.

Sullivan, M., S. W. J. Reid, H. Ternent, N. J. Manchester, R. J. Roberts, D. A. J. Stone, and R. W. Hardy. 2007c. The aetiology of spinal deformity in Atlantic salmon, *Salmo salar* L.: Influence of different commercial diets on the incidence and severity of the preclinical condition in salmon parr under two contrasting husbandry regimes. J. Fish Dis. 30:759–767.

Sutton, J., S. Balfry, D. Higgs, C. H. Huang, and B. Skura. 2006. Impact of iron-catalyzed dietary lipid peroxidation on growth performance, general health and flesh proximate and fatty acid composition of Atlantic salmon (*Salmo salar* L.) reared in seawater. Aquaculture 257:534–557.

Tacon, A. J., and M. M. Beveridge. 1982. Effects of trivalent chromium on rainbow trout, *Salmo gairdneri*. Nutr. Rep. Int. 25:49–56.

Tashjian, D. H., S. J. The, A. Sogomonyan, and S. S. O. Hung. 2006. Bioaccumulation and chronic toxicity of dietary l-selenomethionine in juvenile white sturgeon (*Acipenser transmontanus*). Aquat. Toxicol. 79:401–409.

Tiews, K., M. Manthey, and M. Koops. 1982. The carry-over of fluoride from krill meal pellets into rainbow trout (*Salmo gairdneri*). Arch. Fischereiwiss 32:39–42.

Uyan, O., S. Koshio, M. Ishikawa, S. Uyan, T. Ren, S. Yokoyama, C. F. Komilus, F. R. Michael. 2007. Effects of dietary phosphorus and phospholipid level on growth, and phosphorus deficiency signs in juvenile Japanese flounder, *Paralichthys olivaceus*. Aquaculture 267:44–54.

van der Velden, J. A., F. A. T. Spanings, G. Flik, C. Zegers, Z. I. Kolar, and S. E. Wendelaar Bonga. 1991. Growth rate and tissue magnesium concentration in adult freshwater tilapia, *Oreochromis mossambicus* (Peters), fed diets differing in magnesium content. J. Fish Biol. 39:83–91.

Vielma, J., and S. P. Lall. 1998. Phosphorus utilization by Atlantic salmon (*Salmo salar*) reared in freshwater is not influenced by higher dietary calcium intake. Aquaculture 160:117–128.

Vielma, J., J. Koskela, and K. Ruohonen. 2002. Growth, bone mineralization, and heat and low oxygen tolerance in European whitefish (*Coregonus lavaretus* L.) fed with graded levels of phosphorus. Aquaculture 212:321–333.

Wang, C., R. T. Lovell, and P. H. Klesius. 1997. Response to *Edwardsiella ictaluri* challenge by channel catfish fed organic and inorganic sources of selenium. J. Aquat. Anim. Health 9:172–179.

Wang, W., A. Wang, C. Liu, S. Wang, R. Wang and Z. Ma. 1997. Effect of copper concentrations in diets on the growth and copper, zinc and iron contents of *Penaeus chinensis*. J. Fish. China 21:259–262.

Wang, X., S. Choi, S. Park, G. Yoo, K. Kim, J.-C. Kang, and S. C. Bai. 2005. Optimum dietary phosphorus level of juvenile Japanese flounder *Paralichthys olivaceus* reared in the recirculating system. Fish. Sci. 71:168–173.

Watanabe, T., A. Murakami, L. Takeuchi, T. Nose, and C. Ogino. 1980. Requirement of chum salmon held in freshwater for dietary phosphorus. B. Jpn. Soc. Sci. Fish. 46:361–367.

Watanabe, T., S. Satoh, and T. Takeuchi. 1988. Availability of minerals in fish meal to fish. Asian Fish. Soc. 1:175–195.

Watanabe. T., V. Kiron, and S. Satoh. 1997. Trace minerals in fish nutrition. Aquaculture 151:185–207.

Wilson, R. P., and G. El Naggar. 1992. Potassium requirement of fingerling channel catfish, *Ictalurus puntatus*. Aquaculture 108:169–175.

Wilson, R. P., E. H. Robinson, D. M. Gatlin, III, and W. E. Poe. 1982. Dietary phosphorus requirement of channel catfish. J. Nutr. 112:1197–1202.

Woodall, A. N., and G. LaRoche. 1964. Nutrition of salmonoid fishes. XI. Iodide requirements of Chinook salmon. J. Nutr. 82:475–482.

Yang, S.-D., T.-S. Lin, F.-G. Liu, and C.-H. Liou. 2006. Influence of dietary phosphorus levels on growth, metabolic response and body composition of juvenile silver perch (*Bidyanus bidyanus*). Aquaculture 253:592–601.

Ye, C. X., Y. J. Liu, L. X. Tian, K. S. Mai, Z. Y. Du, H. J. Yang, and J. Niu. 2006. Effects of dietary calcium and phosphorus on growth, feed efficiency, mineral content and body composition of juvenile grouper, *Epinephelus coioides*. Aquaculture 255:264–271.

Ye, C. X., Y. J. Liu, K. S. Mai, T. Li-Xia, H. J. Rang, J. Niu, and J. W. Huang. 2007. Effect of dietary iron supplement on growth, haematology and microelements of juvenile grouper, *Epinephelus coioides*. Aquacult. Nutr. 13:471–477.

Yoo, G.-Y., X. Wang, S. Choi, K. Han, J.-C. Kang, and S. C. Bai. 2005. Dietary microbial phytase increased the phosphorus digestibility in juvenile Korean rockfish *Sebastes schlegeli* fed diets containing soybean meal. Aquaculture 243:315–322.

Zaugg, W. S., D. D. Roley, E. F. Prentice, K. X. Gores, and F. W. Waknitz. 1983. Increased seawater survival and contribution to the fishery of chinook salmon (*Oncorhynchus tshawytscha*) by supplemental dietary salt. Aquaculture 32:183–188.

Zhang, C., K. Mai, Q. Ai, W. Zhang, Q. Duan, B. Tan, H. Ma, W. Xu, Z. Liufu, and X. Wang. 2006. Dietary phosphorus requirement of juvenile Japanese seabass, *Lateolabrax japonicus*. Aquaculture 255:201–209.

Zhu, C.-B., S.-L. Dong, F. Wang, H.-H. Zhang. 2006. Effects of seawater potassium concentration on the dietary potassium requirement of *Litopenaeus vannamei*. Aquaculture 258:543–550.

Zubay, G. L. 1983. Biochemistry. Reading, MA: Addison-Wesley Publishing Company, Inc.

9

Vitamins

Vitamins are organic compounds, distinct from amino acids, carbohydrates, and lipids, that are required in trace amounts from an exogenous source (usually the diet) for normal growth, reproduction, and health. Vitamins are classified as water-soluble and fat-soluble. Eight of the water-soluble vitamins are required in relatively small amounts, have primarily coenzyme functions, and are known as the vitamin B complex. Three of the water-soluble vitamins, choline, inositol, and vitamin C, are required in larger quantities and have functions other than coenzymes. Vitamins A, D, E, and K are the fat-soluble vitamins that function independently of enzymes or, in some cases such as vitamin K, may have coenzyme roles. In mammals, the absence of vitamins leads to characteristic deficiency diseases, but in aquatic species such diseases are less specifically identified. Vitamin deficiency signs reported in fish and shrimp are listed in Table 9-1.

Some vitamins may be synthesized from other essential nutrients to spare a portion of the dietary requirement. For example, channel catfish (*Ictalurus punctatus*) appear to synthesize choline if adequate methyl donors such as methionine are present in the diet; however, if the concentration of dietary methionine is limiting, a choline requirement can be demonstrated (Wilson and Poe, 1988). An exogenous source of some water-soluble vitamins for certain warmwater fish has been shown to be derived from microorganisms in the gastrointestinal tract (Limsuwan and Lovell, 1981; Lovell and Limsuwan, 1982; Burtle and Lovell, 1989; Shiau and Lung, 1993a). In coldwater carnivorous fish, microorganisms are not a significant source of vitamins (Hepher, 1988).

Both qualitative and quantitative vitamin requirements of fish and shrimp have been determined by feeding chemically defined diets deficient in a specific vitamin. The quantitative requirements for most of the vitamins have been established for Chinook salmon (*Oncorhynchus tshawytscha*), rainbow trout (*Oncorhynchus mykiss*), common carp (*Cyprinus carpio*), channel catfish, hybrid tilapia (*Oreochromis niloticus* × *O. aureus*), and yellowtail (*Seriola lalandi*). Qualitative requirements have been identified in several other species. Some shrimp, including tiger shrimp (*Penaeus monodon*), Kuruma shrimp (*Marsupenaeus japonicus*), fleshy prawn (*Fenneropenaeus chinensis*), Pacific white shrimp (*Litopenaeus vannamei*), *Farfantepenaeus californiensis*, and Indian white prawn (*Fenneropenaeus indicus*), have been studied to some extent regarding their vitamin requirements, of which tiger shrimp is the species that all the 15 required vitamins have been established. The requirements are affected by size, age, and growth rates as well as by various environmental factors and nutrient interrelationships. Thus, different researchers have reported fairly wide ranges in requirement values for growth in the same species. Growth performance may not be the only parameter for determining vitamin requirements in fish and shrimp. Other parameters, such as lipid accumulation, skeleton deformities, activity of specific enzymes, tissue vitamin storage, absence of deficiency sign, liver lipid content, hepatosomatic index, lipid oxidation degree, heat shock protein, and immune responses, have also been used for quantifying vitamin requirements of aquatic animals (Tables 9-2, 9-3, and 9-4).

FAT-SOLUBLE VITAMINS

The fat-soluble vitamins, A, D, E, and K, are absorbed in the intestine along with dietary fats; therefore, conditions favorable for fat absorption also enhance the absorption of fat-soluble vitamins. As fish seem to lack a lymphatic system as found in mammals, lipid and fat-soluble vitamins most likely are transported to the peripheral tissues via the portal vein and the liver. Animals store fat-soluble vitamins, either actively in specific cell compartments or by simple accumulation in the lipid compartment, if dietary intake exceeds metabolic needs. Thus, animals can accumulate enough fat-soluble vitamins in their tissues to produce a toxic condition (hypervitaminosis). This has been demonstrated in the laboratory with trout for vitamins A, D, and E, but it is unlikely to occur under practical conditions (Poston et al., 1966; Poston, 1969a; Poston and Livingston, 1969).

TABLE 9-1 Historical Vitamin Diagnostic Signs Reported in Fish and Shrimp (in alphabetical order)

Deficiency Signs	Vitamins	Species
Abnormal swimming (ataxia)	Riboflavin	Salmonids, channel catfish
	Pantothenate	Japanese eel
	Niacin	Japanese eel
	Biotin	Japanese eel
	C	Channel catfish, Korean rockfish, *Clarias* hybrid catfish
Ascites	A	Salmonids
	E	Salmonids
	C	Salmonids
Black death syndrome	C	Tiger shrimp, Pacific white shrimp, fleshy prawn, Kuruma shrimp, giant river prawn, blue shrimp
Congested fin, skin, or bronchial mantles	E	Yellowtail
	Thiamin	Yellowtail, Japanese eel, red sea bream
	Pantothenate	Japanese eel
	B_{12}	Yellowtail
	Folate	Yellowtail
Dark skin coloration	A	Yellowtail
	E	Yellowtail, hybrid striped bass
	Thiamin	Channel catfish, yellowtail
	Riboflavin	Salmonids, yellowtail, hybrid striped bass
	Biotin	Japanese eel, Indian catfish, African catfish
	Inositol	Salmonids, yellowtail
	Folate	Salmonids, Japanese eel
	Choline	Yellowtail
	C	Yellowtail, Korean rockfish, Mexican cichlid
Decreased liver lipid	Choline	Hybrid tilapia, red drum
Deformities (spinal, jaws, or snout)	A	Jian carp
	Riboflavin	Salmonids
	Niacin	Channel catfish, hybrid tilapia
	C	Channel catfish, yellowtail, olive flounder, Korean rockfish, Indian catfish, parrotfish, Asian sea bass, *Clarias* hybrid catfish, Mexican cichlid
Dermatitis	Riboflavin	Japanese eel
	Pantothenate	Japanese eel
Edema	A	Salmonids, channel catfish
	E	Salmonids
	Niacin	Salmonids, hybrid tilapia
Emaciation	Riboflavin	Common carp
	Pantothenate	Channel catfish
Erosion	A	Grouper
	Riboflavin	Yellowtail, blue tilapia, rainbow trout
	Pantothanate	Channel catfish
	Inositol	Salmonids
	C	Channel catfish, Japanese eel, hybrid tilapia, Indian catfish, Asian sea bass, *Clarias* hybrid catfish Japanese sea bass, Mexican cichlid
Exudative diathesis	E	Channel catfish
Eye pathological changes (iritis, cataract)	A	Salmonids
	Riboflavin	Yellowtail, salmonids, channel catfish, hybrid striped bass
	C	Olive flounder, Mexican chichlid
Gill pathological changes (clubbed gill; hyperplasia of the epithelial cells of gill lamellae; degenerative gill lamellae; distorted gill filaments)	Pantothenate	Salmonids, channel catfish, yellowtail, blue tilapia
	Biotin	Salmonids
	C	Salmonids

continued

TABLE 9-1 Continued

Deficiency Signs	Vitamins	Species
Hematological index changes (anemia; abnormal erythrocyte; reduced erythrocyte fragility value; reduced hematocrit value; megaloblastosis; misshapen nuclei in a small proportion of erythrocyte; low thrombocyte counts; higher hemocytoblast and neutrophil count; fragmented erythrocytes; prolonged blood clotting)	A	Yellowtail, rainbow trout
	E	Salmonids, Indian major carp, hybrid striped bass
	K	Salmonids, fleshy prawn
	Pyridoxine	Salmonids, common carp, Indian catfish
	Pantothenate	Salmonids, channel catfish, common carp, yellowtail, blue tilapia
	Niacin	Salmonids, channel catfish, common carp, Japanese eel, Indian catfish
	Folate	Salmonids, channel catfish, rainbow trout
	B_{12}	Salmonids, yellowtail, Indian major carp
	Inositol	Salmonids
Hemorrhage	A	Channel catfish, common carp, yellowtail, hybrid striped bass, Atlantic halibut, grouper, Jian carp, rainbow trout
	E	Channel catfish
	K	Salmonids, channel catfish
	Thiamin	Common carp, Japanese eel, red sea bream
	Riboflavin	Salmonids, common carp, Japanese eel
	Pantothenate	Common carp, Japanese eel, Jian carp
	Niacin	Common carp, yellowtail, Japanese eel, hybrid tilapia, Indian catfish
	Biotin	Yellowtail
	Choline	Salmonids, channel catfish
	C	Salmonids, yellowtail, Japanese eel, hybrid tilapia, olive flounder, Korean rockfish, *Clarias* hybrid catfish, Mexican cichlid
Hyperirritability (irritability)	Thiamin	Salmonids
	Riboflavin	Tiger shrimp
	Pyridoxine	Salmonids, channel catfish
	Pantothenate	Tiger shrimp
	Biotin	Channel catfish
Increased liver lipid (fatty liver; lipid infiltration of liver)	D	Salmonids, rainbow trout
	E	Channel catfish
	Biotin	Salmonids
	Choline	Salmonids, channel catfish, common carp, hybrid striped bass
	Inositol	Grouper, hybrid tilapia, tiger shrimp
Lesion	A	Salmonids
	Pyridoxine	Red hybrid tilapia
	Niacin	Salmonids, channel catfish, yellowtail, hybrid tilapia, common carp, rainbow trout
	Biotin	Salmonids
Lethargy (sluggishness; listlessness; sluggish movement; motionless)	A	Jian carp
	Riboflavin	Salmonids, Japanese eel, blue tilapia, parrotfish
	Pyridoxine	Yellowtail, red hybrid tilapia, Indian catfish
	Pantothenate	Common carp, blue tilapia
	Niacin	Channel catfish, Indian catfish
	Biotin	Common carp, yellowtail, Indian catfish
	Folate	Salmonids, channel carp
	Inositol	Yellowtail
	C	Salmonids, Pacific white shrimp
Lordosis	E	Common carp, grass carp
	C	Channel catfish, yellowtail, Indian catfish, red drum, Japanese sea bass
Loss of equilibrium	Thiamin	Salmonids, channel catfish
	C	Red drum
Loss of scale	C	Mexican cichlid
Mucosa (increase number of dermal mucous cells; loss of skin mucosa)	Biotin	Common carp
	Inositol	Common carp
Muscle dystrophy (muscular weakness)	Vitamin E	Salmonids, channel catfish, common carp, Korean rockfish, grass carp
	Niacin	Salmonids

TABLE 9-1 Continued

Deficiency Signs	Vitamins	Species
Organ pathological changes (pancreatic atrophy; kidney and pancreatic degeneration; anterior kidney necrosis; histopathological changes in liver, kidney, and intestine; atrophied pancreatic acinar cell; vacuolization of hepatic cell; white-grey intestine; severe atrophy of liver)	E	Channel catfish, common carp
	Riboflavin	Common carp
	Pyridoxine	Indian major carp, rainbow trout, gilthead sea bream
	Pantothenate	Salmonids
	Biotin	Salmonids
	Choline	Common carp
	Inositol	Japanese eel
	C	Parrotfish, Pacific white shrimp
Poor appetite (anorexia)	A	Hybrid striped bass, Jian carp, common carp
	Riboflavin	Channel catfish, hybrid striped bass, blue tilapia
	Pyridoxine	Hybrid tilapia, red hybrid tilapia, Indian catfish
	Pantothenate	Jian carp, parrotfish
	Niacin	Channel catfish, Indian catfish
	Biotin	African catfish
	B_{12}	Japanese eel
	C	Olive flounder, Korean rockfish, parrotfish, yellowtail
Reduced bone collagen	C	Channel catfish
Scoliosis	C	Salmonids, channel catfish, yellowtail, olive flounder, Korean rockfish, red drum, Japanese sea bass, Mexican cichlid
Short-body dwarfism (higher condition factor)	Riboflavin	Channel catfish, red hybrid tilapia, blue tilapia
	C	Indian catfish
Skin depigmentation	A	Salmonids, channel catfish, common carp, Jian carp
	E	Salmonids, channel catfish
	Thiamin	Common carp
	Riboflavin	Blue tilapia, rainbow trout, tiger shrimp
	Biotin	Channel catfish
	C	Pacific white shrimp
Swollen abdomen (distended abdomen)	Inositol	Salmonids
	C	Mexican cichlid
Tetany (nervous disorders; nervousness; spastic atrophy; spasms; convulsion)	D	Salmonids, rainbow trout
	Thiamin	Salmonids, channel catfish, common carp
	Pyridoxine	Salmonids, channel catfish, common carp, yellowtail, Japanese eel, red hybrid tilapia
	Pantothenate	Gilthead sea bream, Indian catfish
	Biotin	Parrotfish
	E	Salmonids, Indian catfish, African catfish, yellowtail
Twisted gill opercula	A	Common carp, yellowtail, rainbow trout
	E	Korean rockfish
	Pantothenate	Salmonids
	C	Mexican cichlid

Because fat-soluble vitamins can be stored in the body, the nutritional history of experimental fish prior to their use in requirement studies becomes critical. The time required to deplete fish of their stored fat-soluble vitamins is highly variable. Differences in vitamin intake prior to an experiment may be responsible for some of the conflicting findings on the induction and severity of deficiency signs.

Vitamin A

Vitamin A is crucial in a number of physiological processes necessary for optimal function of an animal. It is involved in cell differentiation, and hence is vital for reproduction as a key factor in embryo development; for development of epithelial cells from stem cells to fully functional layers, including mucus-producing cells; and proper differentiation of immune cells in response to exposure to pathogens or foreign proteins. Its function in vision is well established. The mechanisms of action of vitamin A are not equally well established for other functions. However, retionoic acid, bound to retionoic acid receptors in the nucleus, is a key factor in expression of genes involved in cell differenciation. Exsessive intake of vitamin A, in the range of 5–10 times the requirement, may disturb the same functions as for which it

TABLE 9-2 Historical Vitamin Requirements[a] Estimates for Growing Fish Determined with Chemically Defined Diets[b] in a Controlled Environment

Vitamin and Fish	Requirement (units/kg diet)	Response Criteria	Reference
Vitamin A[c]			
Pacific salmon (*Oncorhynchus* spp.)	R		Halver (1972)
Rainbow trout (*Oncorhynchus mykiss*)	0.75 mg	WG, ADS	Kitamura et al. (1967a)
Channel catfish (*Ictalurus punctatus*)	0.3–0.6 mg	WG	Dupree (1970)
Common carp (*Cyprinus carpio*)	1.2–6 mg	WG, MLS	Aoe et al. (1968)
Yellowtail (*Seriola lalandi*)	5.68 mg	WG, MLS	Shimeno (1991)
Grouper (*Epinephelus* spp.)	0.93 mg	WG	Shaik Mohamed et al. (2003)
Atlantic halibut (*Hippoglossus hippoglossus*)	2.5 mg	MLS, ADS	Moren et al. (2004)
Hybrid striped bass (*Morone chrysops* female × *Morone saxatilis*)	0.51–40.52 mg	WG	Hemre et al. (2004)
Japanese flounder (*Paralichthys olivaceus*)	2.7 mg	WG	Hernandez et al. (2005)
Hybrid tilapia (*Oreochromis niloticus* × *Oreochromis aureus*)	1.76–2.09 mg	WG, MLS	Hu et al. (2006)
European sea bass (*Dicentrarchus labrax*)	31 mg	WG	Villeneuve et al. (2005a,b)
Vitamin D[d]			
Pacific salmon (*Oncorhynchus* spp.)	NR		Halver (1972)
Rainbow trout (*Oncorhynchus mykiss*)	40–60 μg	WG, FE	Barnett et al. (1982a)
Channel catfish (*I. punctatus*)	12.5 μg	WG	Lovell and Li (1978)
	25 μg	WG	Andrews et al. (1980)
	6.25 μg	WG	Brown (1988)
Yellowtail (*Seriola lalandi*)	NR		Shimeno (1991)
Hybrid tilapia (*Oreochromis niloticus* × *Oreochromis aureus*)	9.35 μg	WG	Shiau and Hwang (1993)
Vitamin E			
Atlantic salmon (*Salmo salar*)	35 mg	WG, ADS	Lall et al. (1988)
	60 mg	WG	Hamre and Lie (1995)
Pacific salmon (*Oncorhynchus* spp.)	30 mg	WG, ADS	Woodall et al. (1964)
	40–50 mg	WG, MLS	Halver (1972)
Rainbow trout (*Oncorhynchus mykiss*)	30 mg	WG, ADS	Woodall et al. (1964)
	25 mg	WG, ADS	Hung et al. (1980)
	100 mg	MLS	Watanabe et al. (1981a)
	50 mg	AASLP	Cowey et al. (1983)
Channel catfish (*I. punctatus*)	25 mg	WG, ADS	Murai and Andrews (1974)
	50 mg	AASLP	Wilson et al. (1984)
Common carp (*Cyprinus carpio*)	100 mg	WG, ADS	Watanabe et al. (1970a)
Grass carp (*Ctenopharyngodon idella*)	200 mg	MLS	Takeuchi et al. (1992)
	62.9 mg	WG, FE	Wu et al. (1990)
Yellowtail (*Seriola lalandi*)	119 mg	MLS	Shimeno (1991)
Blue tilapia (*Oreochromis aureus*)	25 mg	WG	Roem et al. (1990)
Nile tilapia (*Oreochromis niloticus*)	50–100 mg	WG, ADS	Satoh et al. (1987)
Hybrid tilapia (*Oreochromis niloticus* × *Oreochromis aureus*)	60–67 mg	WG, AASLP	Shiau and Shiau (2001)
Hybrid striped bass (*M. chrysops* female × *M. saxatilis*)	28 mg	WG	Kocabas and Gatlin (1999)
Korean rockfish (*Sebastes schlegeli*)	45 mg	WG	Bai and Lee (1998)
Mrigal (*Cirrhinus mrigala*)	99 mg	WG	Paul et al. (2004)
Rohu (*Labeo rohita*)	131.91 mg	WG	Sau et al. (2004)
Grouper (*Epinephelus* spp.)	104–115 mg	WG, AASLP	Lin and Shiau (2005a)
Red drum (*Sciaenops ocellatus*)	31 mg	WG	Peng and Gatlin (2009)
Vitamin K			
Pacific salmon (*Oncorhynchus* spp.)	R		Halver (1972)
Lake trout (*Salvelinus namaycush*)	0.5–1 mg	NHV	Poston (1976a)
Channel catfish (*I. punctatus*)	R		Dupree (1966)
	NR		Murai and Andrews (1977)
Yellowtail (*Seriola lalandi*)	NR		Shimeno (1991)
Grass carp (*Ctenopharyngodon idella*)	1.9 mg	BCT	Jiang et al. (2007)
Atlantic cod (*Gadus morhua*)	0.2 mg	WG	Grahl-Madsen and Lie (1997)
Atlantic salmon (*Salmo salar*)	< 10 mg	WG	Krossøy et al. (2009)

TABLE 9-2 Continued

Vitamin and Fish	Requirement (units/kg diet)	Response Criteria	Reference
Thiamin			
Pacific salmon (*Oncorhynchus* spp.)	10–15 mg	MLS	Halver (1972)
Rainbow trout (*Oncorhynchus mykiss*)	1–10 mg	WG, ADS	McLaren et al. (1947)
	1 mg	WG, ED	Morito et al. (1986)
Channel catfish (*I. punctatus*)	1 mg	WG, ADS	Murai and Andrews (1978a)
Common carp (*Cyprinus carpio*)	0.5 mg	WG, ADS	Aoe et al. (1969)
Yellowtail (*Seriola lalandi*)	11.2 mg	MLS	Shimeno (1991)
Riboflavin			
Pacific salmon (*Oncorhynchus* spp.)	20–25 mg	MLS	Halver (1972)
	7 mg	WG, ADS	Leith et al. (1990)
Rainbow trout (*Oncorhynchus mykiss*)	5–15 mg	WG, ADS	McLaren et al. (1947)
	6 mg	MLS	Takeuchi et al. (1980)
	3 mg	ED	Hughes et al. (1981a)
	2.7 mg	MLS, ED	Amezaga and Knox (1990)
Channel catfish (*I. punctatus*)	9 mg	WG, ADS	Murai and Andrews (1978b)
	6 mg	ED	Serrini et al. (1996)
Common carp (*Cyprinus carpio*)	4 mg	WG, ADS	Aoe et al. (1967a)
	6.2 mg	MLS	Aoe et al. (1967a)
	7 mg	MLS	Takeuchi et al. (1980)
Yellowtail (*Seriola lalandi*)	11 mg	MLS	Shimeno (1991)
Blue tilapia (*Oreochromis aureus*)	6 mg	WG, ADS	Soliman and Wilson (1992a)
Red hybrid tilapia (*Oreochromis mossambicus* × *Oreochromis niloticus*)	5 mg	WG	Lim et al. (1993)
Hybrid striped bass (*M. chrysops* female × *M. saxatilis*)	4.1–5.0 mg	WG, MLS	Deng and Wilson (2003)
Jian carp (*Cyprinus carpio* var. Jian)	5.0 mg	WG	W. Li et al. (2010)
Vitamin B_6			
Atlantic salmon (*Salmo salar*)	5 mg	WG, ADS	Lall and Weerakoon (1990)
Pacific salmon (*Oncorhynchus* spp.)	10–20 mg	MLS	Halver (1972)
	6 mg	WG, ADS	Leith et al. (1990)
Rainbow trout (*Oncorhynchus mykiss*)	1–10 mg	WG, ADS	McLaren et al. (1947)
	2 mg	WG, ADS	Woodward (1990)
	3–6 mg	ED	Woodward (1990)
Channel catfish (*I. punctatus*)	3 mg	WG, ADS	Andrews and Murai (1979)
Common carp (*Cyprinus carpio*)	5–6 mg	WG, ADS	Ogino (1965)
Yellowtail (*Seriola lalandi*)	11.7 mg	MLS	Shimeno (1991)
Hybrid tilapia (*Oreochromis niloticus* × *Oreochromis aureus*)	15–16.5 mg	WG, ED	Shiau and Hsieh (1997)
Red hybrid tilapia (*Oreochromis mossambicus* × *Oreochromis niloticus*)	3 mg	WG	Lim et al. (1995)
Indian catfish (*Heteropneustes fossilis*)	3.21 mg	WG	Shaik Mohamed (2001a)
Jian carp (*C. carpio* var. Jian)	6.07 mg	WG	He et al. (2009)
Gibel carp (*Carassius auratus gibelio*)	7.26–11.36 mg	ED	Wang et al. (2009)
Pantothenic acid			
Pacific salmon (*Oncorhynchus* spp.)	40–50 mg	MLS	Halver (1972)
	17 mg	WG, ADS	Leith et al. (1990)
Rainbow trout (*Oncorhynchus mykiss*)	10–20 mg	WG, ADS	McLaren et al. (1947)
	20 mg	WG, ADS	Cho and Woodward (1990)
Channel catfish (*I. punctatus*)	10 mg	WG, ADS	Murai and Andrews (1979)
	15 mg	WG, ADS	Wilson et al. (1983)
Common carp (*Cyprinus carpio*)	30–50 mg	WG, ADS	Ogino (1967)
Yellowtail (*Seriola lalandi*)	35.9 mg	MLS	Shimeno (1991)
Blue tilapia (*Oreochromis aureus*)	10 mg	WG, ADS	Soliman and Wilson (1992b)
Grass carp (*Ctenopharyngodon idella*)	25 mg	WG	A. Liu et al. (2007)
Jian carp (*C. carpio* var. Jian)	23 mg	WG	Wen et al. (2009)

continued

TABLE 9-2 Continued

Vitamin and Fish	Requirement (units/kg diet)	Response Criteria	Reference
Niacin			
Pacific salmon (*Oncorhynchus* spp.)	150–200 mg	MLS	Halver (1972)
Rainbow trout (*Oncorhynchus mykiss*)	1–5 mg	WG, ADS	McLaren et al. (1947)
	10 mg	WG, ADS	Poston and Wolfe (1985)
Channel catfish (*I. punctatus*)	14 mg	WG, ADS	Andrews and Murai (1978)
	7.4 mg	WG, ED	Ng et al. (1997)
Common carp (*Cyprinus carpio*)	28 mg	WG, ADS	Aoe et al. (1967b)
Yellowtail (*Seriola lalandi*)	12 mg	MLS	Shimeno (1991)
Hybrid tilapia (*Oreochromis niloticus* × *Oreochromis aureus*)	26 mg	WG	Shiau and Suen (1992)
Grass carp (*Ctenopharyngodon idella*)	25.5 mg	SGR	Wu et al. (2007a)
Mrigal (*Cirrhinus mrigala*)	20 mg	WG	Shaik Mohamed and Ibrahim (2001)
African catfish (*Heterobranchus longifilis*)	33.1 mg	FE	Morris et al. (1998)
Biotin			
Pacific salmon (*Oncorhynchus* spp.)	1–1.5 mg	MLS	Halver (1972)
Rainbow trout (*Oncorhynchus mykiss*)	0.05–0.25 mg	WG, ADS	McLaren et al. (1947)
	0.08 mg	WG, ADS	Woodward and Frigg (1989)
	0.14 mg	ED	Woodward and Frigg (1989)
Lake trout (*Salvelinus namaycush*)	0.1 mg	WG, ADS	Poston (1976b)
	0.5–1 mg	OSS	Poston (1976b)
Channel catfish (*I. punctatus*)	R		Robinson and Lovell (1978)
Common carp (*Cyprinus carpio*)	1 mg	WG, ADS	Ogino et al. (1970a)
Yellowtail (*Seriola lalandi*)	0.67 mg	MLS	Shimeno (1991)
Hybrid tilapia (*Oreochromis niloticus* × *Oreochromis aureus*)	0.06 mg	WG, ED	Shiau and Chin (1999)
Asian catfish (*Clarias batrachus*)	2.49 mg	WG	Shaik Mohamed et al. (2000)
Indian catfish (*H. fossilis*)	0.25 mg	WG	Shaik Mohamed (2001b)
Goldspot mullet (*Liza parsia*)	1.6–3.2 mg	WG	Chavan et al. (2003)
Japanese sea bass (*Lateolabrax japonicus*)	0.046 mg	WG	J. Li et al. (2010)
Vitamin B_{12}			
Pacific salmon (*Oncorhynchus* spp.)	0.015–0.02 mg	MLS	Halver (1972)
Rainbow trout (*Oncorhynchus mykiss*)	R		Phillips et al. (1964)
Channel catfish (*I. punctatus*)	R		Limsuwan and Lovell (1981)
Common carp (*Cyprinus carpio*)	NR		Kashiwada et al. (1970)
Yellowtail (*Seriola lalandi*)	0.053 mg	MLS	Shimeno (1991)
Nile tilapia (*Oreochromis niloticus*)	NR		Lovell and Limsuwan (1982)
Hybrid tilapia (*Oreochromis niloticus* × *Oreochromis aureus*)	NR		Shiau and Lung (1993a)
Grass carp (*Ctenopharyngodon idella*)	0.094 mg	SGR	Wu et al. (2007b)
Folate			
Pacific salmon (*Oncorhynchus* spp.)	6–10 mg	MLS	Halver (1972)
	2 mg	WG, ADS	Leith et al. (1990)
Rainbow trout (*Oncorhynchus mykiss*)	1.0 mg	WG, ADS	Cowey and Woodward (1993)
Channel catfish (*I. punctatus*)	1.5 mg	WG, NHV	Duncan and Lovell (1991)
	1 mg	WG, NHV	Duncan et al. (1993)
Common carp (*Cyprinus carpio*)	NR		Aoe et al. (1967c)
Yellowtail (*Seriola lalandi*)	1.2 mg	MLS	Shimeno (1991)
Hybrid tilapia (*Oreochromis niloticus* × *Oreochromis aureus*)	0.82	WG, MLS, HSI	Shiau and Huang (2001a)
Grass carp (*Ctenopharyngodon idella*)	3.6–4.3 mg	WG, STG, STC, NHV	Zhao et al. (2008)

TABLE 9-2 Continued

Vitamin and Fish	Requirement (units/kg diet)	Response Criteria	Reference
Choline			
Pacific salmon (*Oncorhynchus* spp.)	600–800 mg	MLS	Halver (1972)
Rainbow trout (*Oncorhynchus mykiss*)	50–100 mg	WG, ADS	McLaren et al. (1947)
	714–813 mg	WG, LLC	Rumsey (1991)
Lake trout (*Salvelinus namaycush*)	1,000 mg	WG	Ketola (1976)
Channel catfish (*I. punctatus*)	400 mg	WG, LLC	Wilson and Poe (1988)
Common carp (*Cyprinus carpio*)	1,500 mg	WG, LLC	Ogino et al. (1970b)
Yellowtail (*Seriola lalandi*)	2,920 mg	MLS	Shimeno (1991)
Hybrid tilapia (*Oreochromis niloticus* × *Oreochromis aureus*)	1,000 mg	WG, MBS	Shiau and Lo (2000)
Hybrid striped bass (*M. chrysops* female × *M. saxatilis*)	500 mg	WG	Griffin et al. (1994)
Cobia (*Rachycentron canadum*)	696 mg	WG	Mai et al. (2009)
Yellow perch (*Perca flavescens*)	598–634 mg	WG	Twibell and Brown (2000)
Red drum (*Sciaenops ocellatus*)	588 mg	WG	Craig and Gatlin (1996)
Grass carp (*Ctenopharyngodon idella*)	3,000 mg	WG, FE, LLC	Wang et al. (1995)
Myoinositol			
Pacific salmon (*Oncorhynchus* spp.)	300–400 mg	MLS	Halver (1972)
Rainbow trout (*Oncorhynchus mykiss*)	250–500 mg	WG, ADS	McLaren et al. (1947)
Channel catfish (*I. punctatus*)	NR		Burtle and Lovell (1989)
Common carp (*Cyprinus carpio*)	440 mg	WG, ADS	Aoe and Masuda (1967)
Yellowtail (*Seriola lalandi*)	423 mg	MLS	Shimeno (1991)
Hybrid striped bass (*M. chrysops* female × *M. saxatilis*)	NR		Deng et al. (2002)
Nile tilapia (*Oreochromis niloticus*)	NR		Peres et al. (2004)
Hybrid tilapia (*Oreochromis niloticus* × *Oreochromis aureus*)	400 mg	WG, MLS	Shiau and Su (2005)
Grouper (*E.* spp.)	335–365 mg	WG, MLS	Su and Shiau (2004)
Olive flounder (*Paralichthys olivaceus*)	617 mg	WG	Lee et al. (2009)
Grass carp (*Ctenopharyngodon idella*)	166–214 mg	WG, FE	Wen et al. (2007)

NOTE: AAASLP, ascorbic acid stimulated lipid peroxidation; ADS, absence of deficiency signs; ED, enzyme data; FE, feed efficiency; HSI, hepatosomatic index; LLC, liver lipid content; MBS, maximum body storage; MLS, maximum liver storage; MKS, maximum kidney storage; NHV, normal hematocrit values; NR, no requirement determined; OSS, optimum swimming stamina; R, required but no value determined; VC, vertebral collagen content; and WG, weight gain.
[a]Without vitamin C.
[b]Refers to type and amount of each ingredient in the diet.
[c]10,000 IU ≈ 3,000 μg vitamin A (retinol).
[d]1 IU = 0.025 μg cholecalciferol.

is vital. Symptoms such as visual impairment, dry epithelia, reproduction failure, skin disorders, and bone and muscle pain may occur.

Vitamin A occurs in three forms: as an alcohol (retinol), an aldehyde (retinal), and an acid (retinoic acid). Vitamin A_1 (retinol) is found in mammals and marine fish, whereas both vitamin A_1 and vitamin A_2 (3-dehydroretinol) are found in freshwater fish (Braekkan et al., 1969; Lee, 1987). In freshwater fish, the oxidative conversion of retinol to 3-dehydroretinol occurs (Goswami, 1984), as well as the reversible oxidation and reduction reactions of retinol to retinal and of 3-dehydroretinol to 3-dehydroretinal (Wald, 1945–1946). For example, Nile tilapia (*Oreochromis niloticus*) have been shown to convert dietary retinol into 3-dehydroretinol and retinal into 3-dehydroretinal (Katsuyama and Matsuno, 1988).

Coldwater fish can use β-carotene as a vitamin A precursor (Poston et al., 1977). Dupree (1970) found that channel catfish could use β-carotene as a vitamin A source only if the dietary concentration exceeded 2,000 international units per kilogram (IU/kg). It has been shown that β-carotene and canthaxanthin can be biotransformed in the liver of Nile tilapia into vitamin A_1 and that dihydroxycarotenoids such as astaxanthin, zeaxanthin, lutein, and tunaxanthin were directly bioconverted into vitamin A_2 (Katsuyama and Matsuno, 1988). In mammals, carotenoids have been found to fulfill various biological functions independent of vitamin A (Olson, 1989). Recently, the conversion ratio of β-carotene to vitamin A has been established to be 19:1 in hybrid tilapia (Hu et al., 2006). More studies are needed on the conversion rate of carotenoids to vitamin A in different fish and shrimp species.

Vitamin A deficiency in rainbow trout causes anemia, twisted gill opercula, and hemorrhages in the eyes and base of fins (Kitamura et al., 1967a). Brook trout (*Salvelinus fontinalis*) exhibited poor growth, high mortality, and eye lesions, such as edematous eyes, displaced lens, and degeneration of the retina, when fed a vitamin A-deficient, purified diet from

TABLE 9-3 Historical Vitamin Requirements[a] Estimates for Growing Shrimp Determined with Chemically Defined Diets[b] in a Controlled Environment

Vitamin and Fish	Requirement (units/kg diet)	Response Criteria	Reference
Vitamin A[c]			
Tiger shrimp (*Penaeus monodon*)	2.51 mg	WG	Shiau and Chen (2000)
Pacific white shrimp (*Litopenaeus vannamei*)	1.44 mg	WG	He et al. (1992)
Fleshy prawn (*Fenneropenaeus chinensis*)	36–54 mg	WG	Chen and Li (1994)
Vitamin D[d]			
Tiger shrimp (*P. monodon*)	100 µg	WG, ED	Shiau and Hwang (1994)
Vitamin E			
Tiger shrimp (*P. monodon*)	85–89 mg	WG, THC	Lee and Shiau (2004)
Pacific white shrimp (*L. vannamei*)	99 mg	WG	He and Lawrence (1993a)
Vitamin K			
Tiger shrimp (*P. monodon*)	30–40 mg	WG	Shiau and Liu (1994a)
Fleshy prawn (*F. chinensis*)	185 mg	WG	Shiau and Liu (1994b)
Thiamin			
Tiger shrimp (*P. monodon*)	14 mg	WG	Chen et al. (1991)
Kuruma shrimp (*Marsupenaeus japonicus*)	60–120 mg	WG	Deshimaru and Kuroki (1979)
Indian white prawn (*Fenneropenaeus indicus*)	100 mg	WG	Boonyaratpalin (1998)
Riboflavin			
Tiger shrimp (*P. monodon*)	22.5 mg	MLS	Chen and Hwang (1992)
Kuruma shrimp (*M. japonicus*)	80 mg		NRC (1983)
Vitamin B_6			
Tiger shrimp (*P. monodon*)	72–89 mg	WG, ED, MLS	Shiau and Wu (2003)
Kuruma shrimp (*M. japonicus*)	120 mg	WG	Deshimaru and Kuroki (1979)
Pacific white shrimp (*L. vannamei*)	80–100 mg	WG	He and Lawrence (1991)
Indian white prawn (*F. indicus*)	100–200 mg	WG, SUR	Boonyaratpalin (1998)
Pantothenic acid			
Tiger shrimp (*P. monodon*)	101–139 mg	WG, ED, MLS	Shiau and C. W. Hsu (1999)
Fleshy prawn (*F. chinensis*)	100 mg	WG	Liu et al. (1995)
Indian white prawn (*F. indicus*)	750 mg	WG, SUR, FE	Boonyaratpalin (1998)
Niacin			
Tiger shrimp (*P. monodon*)	7.2 mg	WG	Shiau and Suen (1994)
Kuruma shrimp (*M. japonicus*)	400 mg		NRC (1983)
Indian white prawn (*F. indicus*)	250 mg	WG, SUR	Boonyaratpalin (1998)
Biotin			
Tiger shrimp (*P. monodon*)	2.0–2.4 mg	WG	Shiau and Chin (1998)
Fleshy prawn (*F. chinensis*)	0.4 mg	WG	Liu et al. (1995)
Vitamin B_{12}			
Tiger shrimp (*P. monodon*)	0.2 mg	WG	Shiau and Lung (1993b)
Fleshy prawn (*F. chinensis*)	0.01 mg	WG	Liu et al. (1995)
Folate			
Tiger shrimp (*P. monodon*)	1.9–2.1 mg	WG, MLS, HSI	Shiau and Huang (2001b)
Fleshy prawn (*F. chinensis*)	5 mg	WG	Liu et al. (1995)
Choline			
Tiger shrimp (*P. monodon*)	6,200 mg	WG	Shiau and Lo (2001)
Kuruma shrimp (*M. japonicus*)	600 mg	WG	Kanazawa et al. (1976)
	NR		Deshimaru and Kuroki (1979)
Myoinositol			
Tiger shrimp (*P. monodon*)	3,400 mg	WG, MLS, HSI	Shiau and Su (2004)
Kuruma shrimp (*M. japonicus*)	2,000 mg	WG	Kanazawa et al. (1976)
Fleshy prawn (*F. chinensis*)	4,000 mg	WG	Liu et al. (1993)

NOTE: ED, enzyme data; FE, feed efficiency; HSI, hepatosomatic index; MLS, maximum liver storage; NR, no requirement determined; R, required but no value determined; SUR, survival; THC, total hemocyte count; and WG, weight gain.

[a]Without vitamin C.

[b]Refers to type and amount of each ingredient in the diet.

[c]10,000 IU ≈ 3,000 µg vitamin A (retinol).

[d]1 IU = 0.025 µg cholecalciferol.

TABLE 9-4 Historical Vitamin C Requirements Estimates for Growing Fish and Shrimp with Chemically Defined Diets[a] in a Controlled Environment

Species	Vitamin C Source	Requirement (mg/kg diet)	Response Criteria	Reference
Rainbow trout	AA	250–500	WG, ADS	McLaren et al. (1947)
(*Oncorhynchus mykiss*)	AA	100	WG, ADS	Halver et al. (1969)
	AA	40	WG, ADS	Hilton et al. (1978)
	AA	20	WG, ADS,	Sato et al. (1982)
		5–100	C/H	Sato et al. (1982)
		500	MLS	Sato et al. (1982)
	C2PP	20	WG, ADS	Grant et al. (1989)
Coho salmon (*Oncorhynchus kisutch*)	AA	50–100	WG, ADS	Halver et al. (1969)
Atlantic salmon	AA	50	WG, ADS	Lall et al. (1990)
(*Salmo salar*)	C2MP-Ca	10	WG, ADS	Sandnes et al. (1992)
		20	C/H	Sandnes et al. (1992)
Channel catfish	AA	50	WG, FE	Andrews and Murai (1975)
(*Ictalurus punctatus*)	AA	60	WG, FE, C/H	Lim and Lovell (1978)
	AA	45	WG, ADS	Robinson (1990)
	AA	25	ADS	Murai et al. (1978)
		50	WG, FE	Murai et al. (1978)
	C2S	2,000	WG, FE	Murai et al. (1978)
	AA	30	WG, ADS, C/H	Durve and Lovell (1982)
	C2MP	11	WG, ADS, C/H	El Naggar and Lovell (1991)
	C2MP-Na	15	WG, ADS	Mustin and Lovell (1992)
	C2MP	50	WG, FE	Li et al. (1998)
		150	MLS	Li et al. (1998)
Blue tilapia (*Oreochromis aureus*)	AA	50	WG, ADS, FE	Stickney et al. (1984)
Hybrid tilapia	AA	79	WG, FE, MLS	Shiau and Jan (1992a)
(*Oreochromis niloticus* × *Oreochromis aureus*)	C2MP	17–20	WG, C/H	Shiau and Hsu (1995)
	C2S	19–23	WG, C/H	Shiau and Hsu (1995)
	C2MP-Mg	18.82	WG	Shiau and Hsu (1999)
	C2MP-Na	15.98	WG	Shiau and Hsu (1999)
Nile tilapia	AA	420	WG, FE	Soliman et al. (1994)
(*Oreochromis niloticus*)	C2PP	50	WG	Abdelghany (1996)
	C2S	50	WG	Abdelghany (1996)
Sabaki tilapia	C2S	75	ADS	Al-Amoudi et al. (1992)
(*Oreochromis spilurus*)		400	MMS, MLS	Al-Amoudi et al. (1992)
Rohu carp	AA	670–750	WG, ADS	Mahajan and Agrawal (1980)
(*Labeo rohita*)	AA	200	WG	Misra et al. (2007)
Hybrid striped bass (*Morone chrysops* female × *Morone saxatilis*)	C2PP	22	WG, ADS	Sealey and Gatlin (1999)
Mexican cichlid (*Cichlasoma urophthalmus*)	AA	40	WG	Chavez de Martinez (1990)
African catfish (*Clarius gariepinus*)	AA	46	ADS, C/H	Eya et al. (1996)
Indian catfish	AA	69	WG	Mishra and Mukhopadhyay (1996)
(*Heteropneustes fossilis*)	AA	82.2	WG	Ibiyo et al. (2007)
Common carp	C2PP	45	WG	Gouillou-Coustans et al. (1998)
(*Cyprinus carpio*)		354	MBS	Gouillou-Coustans et al. (1998)
Ayu	C2PP	116	WG	Xie and Niu (2006)
(*Plecoglossus altivelis*)		47	ADS, C/H	Xie and Niu (2006)
Parrot fish	AA	250	WG, ADS	Ishibashi et al. (1992)
(*Oplegnathus fasciatus*)		500	MLS, MKS	Ishibashi et al. (1992)
	C2MP	118	WG	Wang et al. (2003a)

continued

TABLE 9-4 Continued

Species	Vitamin C Source	Requirement (mg/kg diet)	Response Criteria	Reference
Plaice (*Pleuronectes platessa*)	AA	200	WG, ADS, C/H	Rosenlund et al. (1990)
Gilthead sea bream (*Sparus auratus*)	AA	63	ADS	Alexis et al. (1997)
Korean rockfish (*Sebastes schlegeli*)	AA	144	WG	Lee et al. (1998)
	AA	100–102	WG, SGR, PER, FE	Bai (2001)
	C2MP-Ca	112	WG	Wang et al. (2003b)
	C2MP-Na/Ca	101	WG	Wang et al. (2003b)
	C2D	50	WG, ADS	Wang et al. (2003c)
Yellowtail (*Seriola lalandi*)	AA	122	WG, ADS	Shimeno (1991)
	C2MP-Mg	14–28	WG, ADS, C/H	Kanazawa et al. (1992)
	C2MP-Mg	52	WG	Ren et al. (2008)
	C2MP-Na/Ca	43	WG	Ren et al. (2008)
Olive flounder (*Paralichthys olivaceus*)	C2MP-Mg	28–47	WG	Teshima et al. (1991)
	C2PP	91–93	WG, PER	Wang et al. (2002)
Turbot (*Scophthalmus maximus*)	C2PP	20	WG	Merchie et al. (1996)
Red drum (*Sciaenops ocellatus*)	C2PP	15	WG, ADS, FE	Aguirre and Gatlin (1999)
Asian sea bass (*Lates calcarifer*)	C2MP-Mg	30	WG, ADS, FE	Phromkunthong et al. (1997)
European sea bass (*Dicentrarchus labrax*)	AA	200	WG, ADS, MLS	Saroglia and Scarano (1992)
	C2PP	20	WG	Merchie et al. (1996)
	C2PP	5	WG, ADS, FE	Fournier et al. (2000)
		5–31	C/H	Fournier et al. (2000)
		121	MLS	Fournier et al. (2000)
Yellow croaker (*Pseudosciaena crocea*)	C2PP	28.2	SUR	Ai et al. (2006)
		87	MLS	Ai et al. (2006)
Angelfish (*Pterophylum scalare*)	C2MP-Mg	360	MLS	Blom et al. (2000)
Japanese sea bass (*Lateolabrax japonicus*)	C2PP	53.5	WG	Ai et al. (2004)
		93.4	MLS	Ai et al. (2004)
		207.2	MMS	Ai et al. (2004)
Grouper (*Epinephelus* spp.)	AA	45.3	WG	Lin and Shiau (2005b)
	C2MP-Mg	17.9	WG	Lin and Shiau (2004)
	C2MP-Na	8.3	WG	Lin and Shiau (2004)
	C2PP	17.8	WG	Lin and Shiau (2005c)
	C2S	46.2	WG	Lin and Shiau (2005c)
Clarias hybrid catfish (*Clarias gariepinus* × *Clarias macrocephalus*)	C2D	42	WG, FE, ADS	Khajarern and Khajarern (1997)
	C2MP-Ca	12.6	WG, ADS	Boonyaratpalin and Phromkunthong (2001)
Tiger puffer (*Takifugu rubripes*)	C2MP	29	WG, SGR	Eo and Lee (2008)
Pacu (*Piaractus mesopotamicus*)	Ascorbyl-6-palmitate	139	WG, ADS	Martins (1995)
Cobia (*Rachycentron canadum*)	C2PP	44.7–53.9	WG, MLS	Xiao et al. (2009)
Tiger shrimp (*Penaeus monodon*)	AA	2,000	WG	Shiau and Jan (1992b)
	C2PP	210	WG, MLS, MMS	Chen and Chang (1994)
	C2MP-Mg	100–200	ADS, FE, SUR	Catacutan and Lavilla-Pitogo (1994)
	C2MP-Mg	40	WG	Shiau and Hsu (1994)
	C2S	157	WG	Hsu and Shiau (1997)
	C2MP-Na	106	WG	Hsu and Shiau (1998)

TABLE 9-4 Continued

Species	Vitamin C Source	Requirement (mg/kg diet)	Response Criteria	Reference
Kuruma shrimp (*Marsupenaeus japonicus*)	AA	3,000	WG, FE, ADS	Deshimaru and Kuroki (1976)
	AA	5,000–10,000	WG, SUR	Guary et al. (1976)
	C2MP-Mg	215–430	ADS	Shigueno (1988)
	C2MP-Mg	71	WG	Moe et al. (2004)
	C2MP-Na/Ca	43	WG	Moe et al. (2004)
	C2MP-Na/Ca	91.8	WG	Moe et al. (2005)
Pacific white shrimp (*Litopenaeus vannamei*)	C2PP	90–120	MBS	He and Lawrence (1993b)
	C2MP	130	WG	Lavens et al. (1999)
	C2PP	150	WG	Zhou et al. (2004)
	C2PP	191	WG	Niu et al. (2009)
Fleshy prawn (*Fenneropenaeus califoriensis*)	C2PP	600	WG	Qin et al. (2007)
	AA	2,000	ADS	Lightner et al. (1979)
Indian white prawn (*Fenneropenaeus indicus*)	AA	4,000–8,000	WG, SUR, FE	Boonyaratpalin (1998)
Giant river prawn (*Macrobrachium rosenbergii*)	C2MP-Ca	104	SUR	D'Abramo et al. (1994)
	Ascorbyl-6-palmitate	104	SUR	D'Abramo et al. (1994)
	C2PP	135	SUR	Hari and Kurup (2002)

NOTE: AA, L-ascorbic acid; ADS, absence of deficiency signs; C2D, L-ascorbyl-2-glucose; C2MP, ascorbyl-2-monophosphate; C2PP, L-ascorbyl-2-polyphosphate; C2S, L-ascorbyl-2-sulfate; C/H, collagen or hydroxyproline concentration; FE, feed efficiency; MBS, maximum body storage; MKS, maximum kidney storage; MLS, maximum liver storage; MMS, maximum muscle storage; SGR, specific growth rate; SUR, survival; and WG, weight gain.

[a]Refers to type and amount of each ingredient in the diet.

first feeding (Poston et al., 1977). Channel catfish fed 0.4 mg of β-carotene/kg of diet for 3 years developed exophthalmia, edema, and hemorrhagic kidney (Dupree, 1966). Anorexia, pale body color, hemorrhagic skin and fins, exophthalmia, and twisted gill opercula occurred in common carp fed a vitamin A-deficient diet after 8 to 11 weeks (Aoe et al., 1968). Rapidly growing yellowtail fingerlings fed a vitamin A-deficient diet developed deficiency signs in 20 days, including arrested growth of gill opercula, dark pigmentation, anemia, and hemorrhages in the eyes and liver, accompanied by high mortality (Hosokawa, 1989). Vitamin A deficiency caused hemorrhages in the fins and the area surrounding the eyes in Atlantic halibut (*Hippoglossus hippoglossus*) (Moren et al., 2004) and hemorrhages on the skin overlaying the base of the fins and erosion on the caudal peduncle in grouper (*Epinephelus tauvina*) (Shaik Mohamed et al., 2003). Furuita et al. (2003) reported that low vitamin A content in the diet caused negative effect on reproduction such as buoyant egg rate and percentage of normal larvae of Japanese flounder (*Paralichthys olivaceus*). Mengqing et al. (2004) reported that high vitamin A (60 mg/kg diet) supplemented in diet improved fecundity and larval quality in fleshy prawn.

High dietary intake (660 mg/kg diet, or 2.2 million IU vitamin A) of retinyl palmitate caused slow growth, anemia, and severe necrosis of the caudal fin of brook trout at 8.3°C (Poston et al., 1966). Feeding up to 750 mg retinyl palmitate to trout at 12.4°C also reduced body fat and liver size (Poston, 1971a). A high intake of dietary protein (Poston and Livingston, 1971) or methionine (Eckhert and Kemmerer, 1974) by young trout reduced the toxicity of excess dietary vitamin A observed in fish fed a low-protein diet. Takeuchi et al. (1998) reported Japanese flounder larvae fed Atemia enriched by 30 mg/kg diet of retinol, retinyl palmitate, or retinyl acetate or 100 mg/kg diet of all-trans-retinoic acid compressed the fish vertebrae. Signs of vitamin A toxicity, such as increased mortality, abnormal vertebral growth, and reduced growth, were found in Atlantic salmon (*Salmo salar*) receiving 938 mg retinol/kg diet (Ørnsrud et al., 2002). Villeneuve et al. (2005a) reported that 31 mg vitamin A/kg diet is required for European sea bass (*Dicentrarchus labrax*) larvae, and excess levels of dietary retinyl acetate fed to European sea bass resulted in an alternation of head organization characterized by the abnormal development of the splanchnocranium and neurocranium, and scoliotic fish. Of the larvae fed 1,000 mg retinyl acetate/kg diet, 78.8% exhibited skeletal abnormalities. The authors indicated that a linear correlation between vitamin A level in larvae and malformation percentage was linked to a modification in the relative retinoic acid receptor gene expression. This demonstrates the influence of nutrition on the retinoid pathway that plays an important role in body patterning and the induction of skeletal malformations during post-hatching development in European sea bass larvae. Retinoid pathway can also be influenced by dietary lipid, leading to skeletal malformation during sea bass larvae development (Villeneuve et al., 2005b).

Vitamin A was reported to have only limited potential as an immunostimulatory agent in practical rainbow trout diet (Thompson et al., 1995), but it significantly improved antibacterial activity of Japanese flounder (Hernandez et al., 2007) and immune response and disease resistance of Jian carp (*Cyprinus carpio* var. Jian) (Yang et al., 2008).

Vitamin A is a chemically synthesized product that is enclosed in a beadlet to protect against oxidation. Indeed, vitamin A is one of the vitamins most affected by the aggression of mill machinery. Beadlets typically contain a matrix such as cross-linked gelatin, with vitamin A dispersed throughout the matrix. Within the matrix the vitamin can be additionally protected by an antioxidant. Beadlets are then coated with a protective layer, such as corn starch, which improves handling. The form of vitamin A in such beadlets is vitamin A acetate. Several manufacturers produce beadlets containing both vitamin A and vitamin D. Stability of vitamin A beadlets through extrusion and 3 months of room temperature storage is approximately 80% at best and 40% at worst for different vitamin A beadlet products, the latter having no crosslinked matrix (Gabaudan and Hardy, 2000).

Vitamin D

Vitamin D functions to maintain calcium homeostasis together with two peptide hormones, calcitonin and parathyroid hormone. It is critically important for the development, growth, and maintenance of a healthy skeleton from birth until death. It is also involved in alkaline phosphatase activity, promotes intestinal absorption of calcium, and influences the action of parathyroid hormone on bone. Besides these functions vitamin D has been found to play important roles in differentiation of bone, skin, and blood cells; in secretion of insulin and prolactin; muscle function; immune and stress responses; and melanin synthesis. Excessive intake of vitamin D may cause irreversible calcification of the heart, kidneys, and other soft tissue.

The two major natural sources of vitamin D are ergocalciferol (vitamin D_2, which occurs predominantly in plants) and cholecalciferol (vitamin D_3, which occurs in animals). Both forms of vitamin D are hydroxylated in the liver to the 25-hydroxy forms. The 25-hydroxy-D_3 is further hydroxylated in the kidney to 1,25-dihydroxyvitamin D_3, which is the biologically active form of vitamin D responsible for facilitating mobilization, transport, absorption, and use of calcium and phosphorus in concert with the actions of parathyroid hormone and calcitonin.

Cholecalciferol has been shown to be at least three times more effective than ergocalciferol in meeting the vitamin D requirement of rainbow trout (Barnett et al., 1982a). Andrews et al. (1980) found that vitamin D_3 was used more effectively by catfish than vitamin D_2 at dietary concentrations of 50 μg/kg diet (1 IU = 0.025 μg cholecalciferol) and that high concentrations of vitamin D_3 (500 to 1,250 μg/kg diet) reduced weight gain. Brown (1988), however, found that vitamin D_2 was utilized as well as vitamin D_3 up to 37.5 μg/kg of diet, but higher concentrations of vitamin D_2 depressed weight gain and feed efficiency in channel catfish reared in calcium-free water. In tiger shrimp, vitamin D_2 has been reported to have approximate 2% of the vitamin D potency of vitamin D_3 for growth (Shiau and Hwang, 1994).

Rainbow trout fed a vitamin D-deficient diet exhibited poor growth, elevated liver lipid content, impaired calcium homeostasis manifested by tetany of white skeletal muscles, and ultrastructural changes in the white muscle fibers of the epaxial musculature (George et al., 1981). However, in a similar study also with rainbow trout, no hypocalcemia or changes in bone ash were observed (Barnett et al., 1982a). A lordosis-like droopy tail syndrome observed in vitamin D-deficient trout (Barnett et al., 1982b) was suggested to be related to an epaxial muscle weakness. Channel catfish fed a vitamin D-deficient diet for 16 weeks showed poor growth, lowered body calcium and phosphorus levels, and lowered total body ash (Lovell and Li, 1978). Andrews et al. (1980) reported that vertebral ash level in channel catfish was not significantly affected by vitamin D deficiency. Cerezuela et al. (2009) reported that dietary vitamin D_3 administration in diet enhanced innate immune parameters of gilthead sea bream (*Sparus aurata*).

Fingerling brook trout fed 93.75 mg vitamin D_3/kg diet for 40 weeks had hypercalcemia and increased hematocrit levels but no difference in rates of growth and survival (Poston, 1969a). However, Hilton and Ferguson (1982) did not detect any incidence of renal calcinosis in rainbow trout fed a diet containing up to 25 mg vitamin D_3/kg diet. Supplementation of 1.25 mg vitamin D_3/kg diet significantly depressed the growth rate of channel catfish (Andrews et al., 1980). By contrast, a diet of 25 mg vitamin D_3/kg has been reported to show no toxic effects in channel catfish reared in calcium-free water for 14 weeks (Brown, 1988). Atlantic salmon fry seemed to be highly tolerant of megadoses of vitamin D_3 over a period of time. When the fry were fed diet supplemented with three levels of vitamin D_3 (0.2, 5, and 57 mg/kg diet) for 14 weeks, no differences in weight, length, specific growth rate, mortality, or kidney calcium concentration nor skeleton malformation or histopathological changes were observed among the three dietary groups (Graff et al., 2002).

Vitamin D is added to feeds as a beadlet enclosing cholecalciferol (D_3), or as a spray-dried product. The other form of vitamin D, ergocalciferol (D_2), is not used as a vitamin D supplement in animal or fish feeds. Antioxidant addition plus encapsulation within a beadlet provide protection of vitamin D_3 against oxidation. Typical stability ranges between 75–100% after extrusion and 3 months of room temperature storage (Gabaudan and Hardy, 2000).

Vitamin E

Vitamin E is a generic descriptor for all the molecules that possess the biological activity of α-tocopherol. Natural forms of vitamin E are all d-stereoisomers and consist of a substituted aromatic ring and a long isoprenoid side chain. There are eight naturally occurring compounds with vitamin E activity: d-α-; d-β-; d-γ-; d-δ-tocopherols, which differ in the number and position of the methyl groups in the aromatic ring; and their corresponding tocotrienols. The compound with the highest biopotency is d-α-tocopherol. The other tocopherol and tocotrienol isomers have some, but very low biological activity. When expressed as d-α-tocopherol equivalents, the following values were assigned to the various isoforms: α-tocopherol, 1.0; β-tocopherol, 0.5; γ-tocopherol, 0.1; δ-tocopherol, 0.03; α-tocotrienol, 0.3; β-tocotrienol, 0.05; γ-tocotrienol and δ-tocotrienol, unknown (Ng et al., 2004). More recent research has indicated that some of the vitamin E isomers have other biological functions not covered by α-tocopherol or have been shown to be more "potent" than α-tocopherol in some situations (e.g., antioxidant activity) in laboratory animals (i.e., rats). However, very limited information is currently available on the utilization and metabolism of the other vitamin E isomers by fish and shrimp. Li et al. (2008) reported α-tocopheryl acetate is a higher bioavailable source of vitamin E than α-tocopheryl succinate for red drum (*Sciaenops ocellatus*). No interconversion between α-tocopherol and the other tocopherol forms has been detected in liver or muscle tissue of rainbow trout (Watanabe et al., 1981b). Hsu and Shiau (1999a) reported dl-α-tocopherol is the form of vitamin E in liver, whereas dl-α-tocopheryl acetate is the form of vitamin E in blood and muscle of hybrid tilapia fed a diet supplemented with dl-α-tocopheryl acetate as the vitamin E source. However, dl-α-tocopheryl acetate was found as the only storage form of vitamin E in the hepatopancreas and muscle of tiger shrimp when fed the diet supplemented with dl-α-tocopheryl acetate (Hsu and Shiau, 1999b). The free tocopherol form of vitamin E is unstable to oxidizing conditions, whereas the acetate and succinate esters are quite stable. These ester forms possess no antioxidant activity, but they are readily hydrolyzed in the digestive tract to the biologically active free tocopherol. One IU of vitamin E is defined as the biological activity of 1 mg of dl-α-tocopheryl.

Vitamin E functions in vitro as a very good antioxidant in a manner similar to several synthetic antioxidants. In vivo, vitamin E and selenium (via glutathione peroxidase) function as parts of a multicomponent antioxidant defense system. This system protects the cell against the adverse effects of reactive oxygen and other free radical initiators of the oxidation of polyunsaturated membrane phospholipids, critical proteins, or both. Mutual sparing of dietary requirements for vitamin E and Se in grouper (*Epinephelus malabaricus*) has been reported at low ingestion levels of either vitamin E or Se, and higher dietary supplementation of one of these nutrients spares the metabolic requirements of the other (Lin and Shiau, 2009).

Gilthead sea bream fed diets deficient in vitamin E decreased the percentage of fertilized egg (Fernández-Palacios et al., 2005). This may have been related to the decrease in the number and motility of the spermatozoids, as has been described for fish such as ayu (Hsiao and Mak, 1978). Lee and Dabrowski (2004) found that the level of sperm plasma tocopherol decreased significantly and sperm viability was seriously compromised in yellow perch (*Perca flavescens*) broodstock fed with diets deficient in vitamin E. The vitamin E content is generally high in fish eggs and low in broodstock tissues after the spawning period (Mukhopadhyay et al., 2003). This may be a result of mobilization of vitamin E from peripheral tissues to the ovary during vitellogenesis as it has been shown in turbot and Atlantic salmon (Hemre et al., 1994; Lie et al., 1994). Furuita et al. (2008) found that vitamin E injection into broodstock during artificial maturation increase the vitamin content of eggs and can improve egg quality. It has been reported that 100 mg/kg supplemental vitamin E caused significantly bigger and more pleopodal eggs and more stage-1 juveniles in freshwater crayfish, *Astacus leptodactylus* (Harlıoğlu and Barum, 2004).

Vitamin E deficiency signs have been described for Chinook salmon (Woodall et al., 1964), Atlantic salmon (Poston et al., 1976), channel catfish (Dupree, 1968; Murai and Andrews, 1974; Lovell et al., 1984; Wilson et al., 1984), common carp (Watanabe et al., 1970a,b, 1981c), rainbow trout (Cowey et al., 1981, 1983; Hung et al., 1981; Watanabe et al., 1981a; Moccia et al., 1984), Nile tilapia (Roem et al., 1990), grass carp (*Ctenopharyngodon idella*) (Takeuchi et al., 1992), Korean rockfish (*Sebastes schlegeli*) (Bai and Lee, 1998), and hybrid striped bass (*Morone chrysops* female × *M. saxatilis*) (Kocabas and Gatlin, 1999). The deficiency signs of vitamin E in various fish are similar and include muscular dystrophy involving atrophy and necrosis of white muscle fibers; edema of heart, muscle, and other tissues because of increased capillary permeability allowing exudates to escape and accumulate, which are often green in color as a result of hemoglobin breakdown; anemia and impaired erythropoiesis; depigmentation; and ceroid pigment in the liver. The incidence and severity of these deficiency signs have been shown to be enhanced when diets deficient in both vitamin E and selenium were fed to Atlantic salmon (Poston et al., 1976), rainbow trout (Bell et al., 1985), and channel catfish (Gatlin et al., 1986a). These latter observations demonstrated a significant interaction between selenium and vitamin E in the nutrition of fish.

Erythrocyte fragility has been used as an indicator of vitamin E status in some animals (Draper and Csallany, 1969). Peroxide hemolysis of red blood cells has been used to determine vitamin E deficiency in rainbow trout (Hung et al., 1981), mrigal (*Cirrhinus mrigala*) (Paul et al., 2004), and rohu (*Labeo rohita*) (Sau et al., 2004); however, this procedure was not sensitive enough to aid in determining the vitamin E requirement in rainbow trout (Cowey et al., 1981) and channel catfish (Wilson et al., 1984). Cowey et al. (1981) found that in vitro ascorbic acid stimulated lipid peroxidation, i.e., thiobarbituric acid reactive substance (TBARS) value in liver microsomes of rainbow trout accurately reflected α-tocopherol status. This latter procedure has also been used to assess vitamin E status in channel catfish (Wilson et al., 1984; Gatlin et al., 1986a), hybrid tilapia (Shiau and Shiau, 2001), and grouper (Y. H. Lin and Shiau, 2005).

Dietary vitamin E requirements increased as the lipid level increased in diet of rainbow trout (Cowey et al., 1983), hybrid tilapia (Shiau and Shiau, 2001), and grouper (Y. H. Lin and Shiau, 2005). When high concentrations of dietary polyunsaturated fatty acids are included in the diets of common carp (Watanabe et al., 1981c) and rainbow trout (Watanabe et al., 1981a; Cowey et al., 1983), the requirement for vitamin E is increased.

Vitamin E-deficient rainbow trout have been reported to have significantly reduced immune and nonspecific responses to infection (Blazer and Wolke, 1984a). Low vitamin E in the diet reduced stress resistance of gilthead sea bream (Montero et al., 2001), and it led to higher levels of lipid peroxides in turbot, halibut, and gilthead sea bream (Tocher et al., 2002). Increased vitamin E supplementation in diet improved immune responses of Atlantic salmon (Hardie et al., 1990), turbot (*Scophthalmus maximus*) (Stéphan et al., 1995), gilthead sea bream (Ortuño et al., 2000), rainbow trout (Clerton et al., 2001; Puangkaew et al., 2004), and grouper (Y. H. Lin and Shiau, 2005) and enhanced natural cytotoxic activity of gilthead sea bream (Cuesta et al., 2001). Vitamin E and n-3 polyunsaturated fatty acids (PUFA) had a synergistic effect on the nonspecific immune responses and disease resistance in Japanese flounder (Z. Wang et al., 2006). However, Salte et al. (1988) could show no beneficial effect of dietary vitamin E supplementation alone or in combination with selenium as a prophylaxis for Hitra disease in Atlantic salmon. Also, increased dietary vitamin E did not enhance immune responses of rainbow trout (Kiron et al., 2004). Vitamin E exhibits an effective antioxidant role by regulating osmotic balance and resistance to salinity changes in Pacific white shrimp (Y. Liu et al., 2007).

High dietary concentrations of vitamin E (5,000 mg of DL-α-tocopherol/kg of diet) have been shown to cause reduced concentrations of erythrocytes in trout blood (Poston and Livingston, 1969). Kaewsrithong et al. (2001) also reported high dietary vitamin E (10,000 mg/kg diet) supplemented in diet promoted lipid peroxidation, increased accumulation of hydroperoxide in blood, and reduced erythrocyte osmotic fragility in sweet smelt.

Vitamin E is supplied to feeds as dl-α-tocopheryl acetate, which is an acetate ester of α-tocopherol. The acetate moiety is attached at the active site on the tocopherol molecule, thus preventing any other reactions from occurring that might result in loss of tocopherol activity. The most worrisome reaction in feeds is associated with oxidizing lipids, in which tocopherol donates a hydrogen atom, thus becoming a sacrificial antioxidant. The presence of the acetate moiety prevents oxidation, but also renders dl-α-tocopheryl acetate inactive with respect to antioxidant function in feeds. Once in the gut, the acetate moiety is enzymatically removed, restoring the antioxidant property to the tocopherol molecule. Vitamin E is relatively stable in extruded feeds when supplemented in the protected form, with no more than 10% loss after pelleting and extrusion (Gabaudan and Hardy, 2000).

Vitamin K

Vitamin K is typically associated with its role in coagulation of blood. It also has an important role in calcium transport. In vertebrates, the action of osteocalcin, the major bone matrix protein, is vitamin K-dependent. Vitamin K is required for the posttranslational carboxylation of specific glutamate residues to gamma-carboxyglutamate residues. These residues interact with calcium allowing osteocalcin to regulate the incorporation of calcium phosphates into bone tissue.

Vitamin K is required for stimulation of prothrombin activity in plasma and synthesis of blood clotting factors VII, IX, and X. The metabolic role of vitamin K involves the vitamin K-dependent carboxylase, which carries out the post-translational conversion of specific glutamyl residues in the vitamin K-dependent plasma proteins to γ-carboxy-glutamyl residues. These residues are essential for the normal, Ca^{2+}-dependent interaction of the vitamin K-dependent clotting factors with phospholipid surfaces (Suttie, 1985).

The term "vitamin K" is used as a generic descriptor for both 2-methyl-1,4-naphthoquinone and all 3-substituted derivatives of this compound, which exhibit an antihemorrhagic activity in animals fed a vitamin K-deficient diet. The three major forms of vitamin K are vitamin K_1 or phylloquinone, which can be isolated from plants; vitamin K_2 or the menaquinones, which are synthesized by bacteria; and vitamin K_3 or menadione, which is a synthetic product. Menadione is a yellow powder, technically easier to use than the natural oily compounds, and therefore commonly added to fish and shrimp feeds.

Many animals do not require vitamin K in the diet because of bacterial synthesis in the intestinal tract, but intestinal vitamin K-synthesizing microflora have not been described in fish (Margolis, 1953). Supplementation of sulfaguanidine to a vitamin K-deficient diet and low water temperature caused prolonged blood coagulation time and low hematocrit values without affecting growth performance of trout (Poston, 1964). Addition of antibiotic (nifurazolidon) to a vitamin K-deficient diet resulted in reduced growth, but did not provoke external signs of vitamin K deficiency in Atlantic cod (Grahl-Madsen and Lie, 1997). Dupree (1966) reported hemorrhages in channel catfish fed a vitamin K-deficient diet. However, Murai and Andrews (1977) failed to detect any deficiency signs in channel catfish fed a diet devoid of vitamin K and supplemented with sulfaguanidine. The addition of dicumarol, a vitamin K antagonist, did not increase prothrombin time in catfish. The addition of pivalyl, a stronger (20 times) vitamin K antagonist than dicumarol, completely blocked the blood coagulation of channel catfish (Murai and Andrews, 1977). High-dietary concentrations of menadione sodium bisulfite (2,400 mg/kg of diet) had no adverse affect on growth, survival, blood coagulation, or the number of erythrocytes of young trout (Poston, 1971b). The total plasma prothrombin concentration was low in fleshy prawn fed the vitamin K-unsupplemented diet (Shiau and Liu, 1994b). The amount of vitamin K found naturally in the presently used feed ingredients (approximately 0.1 mg/kg diet) may be enough to maintain optimal growth, health, and bone strength in Atlantic salmon fry from start feeding (Krossøy et al., 2009).

Vitamin K is supplied as a menadione (K_3) salt. There are four forms of menadione salts used in feeds: menadione sodium bisulfite (MSB, 50% active K_3); menadione nicotinamide bisulfite (MNB, 43% active K_3); menadione sodium bisulfite complex (MSBC, 33% active K_3); and menadione dimethlypyrimidinol bisulfite (MPB, 45.4% active K_3). All are affected by heat, moisture, and the presence of trace minerals. After extrusion pelleting and 3 months of room temperature storage, between 20–50% of vitamin K activity remains (Gabaudan and Hardy, 2000).

WATER-SOLUBLE VITAMINS

The water-soluble vitamins, with the exception of two water-soluble growth factors (choline and myoinositol) and ascorbic acid, have unique coenzyme functions in cellular metabolism. Yet, it is not always possible to correlate a sign of deficiency with a diminished function of an enzyme system for which that vitamin is essential. For some warmwater fish, intestinal synthesis by microorganisms supplies the requirement for certain vitamins. Thus, deficiency signs result only in those cases when antibiotics are fed along with a deficient diet. A constant supply of essential water-soluble vitamins is required to prevent deficiency signs in fish, because these vitamins are not stored in body tissues.

Thiamin

Thiamin was the first vitamin to be recognized. In animal tissue, thiamin occurs predominantly in a di-phosphate form known as thiamin pyrophosphate (TPP). TPP is an essential cofactor for a number of important enzymatic steps in energy production, including both decarboxylations and transketolase reactions.

The coenzyme form of thiamin is thiamin pyrophosphate. Thiamin pyrophosphate functions in the oxidative decarboxylation of α-keto acids, such as pyruvate and α-ketoglutarate, and in the transketolase reaction in the pentose shunt.

Dietary thiamin deficiency has been shown to result in neurological disorders such as hyperirritability in salmonids (Halver, 1957; Coates and Halver, 1958; Kitamura et al., 1967b; Lehmitz and Spannhof, 1977), channel catfish (Dupree, 1966; Comacho, 1978), Japanese eel (*Anguilla japonica*) (Hashimoto et al., 1970), and Japanese parrotfish (*Scarus coeruleus*) (Ikeda et al., 1988). However, Murai and Andrews (1978b) did not observe neurological disorders in thiamin-deficient channel catfish. Arai et al. (1972) found only subcutaneous hemorrhages and congested fins in subadult Japanese eels, and Hashimoto et al. (1970) observed neurological disorders in small Japanese eels. Similar deficiency signs with varying degrees of mortality have been reported in common carp (Aoe et al., 1969), red sea bream (Yone and Fujii, 1974), turbot (Cowey et al., 1975), and yellowtail (Hosokawa, 1989). Tiger shrimp fed thiamin-free diet did not exhibit specific deficiency signs, except universal signs such as retarded growth, poor feed conversion, and low survival (Chen et al., 1991). Unlike fish, aquatic crustaceans are slow feeders, and food particles usually remain suspended in water for extended periods before being consumed. Manipulation by shrimp of food particles during feeding may further increase leaching of water-soluble vitamins. Therefore, one would expect a higher thiamin requirement for crustaceans than for fish (Chen et al., 1991). Thiamin deficiency is observed in wild stocks of fish both in fresh- and saltwaters. The deficiency is induced in fish eating food with high thiaminase activity originating from certain algae.

Erythrocyte transketolase activity has been used as a specific indicator of thiamin status in the turbot (Cowey et al., 1975). Kidney or liver transketolase activity in rainbow trout (Lehmitz and Spannhof, 1977; Masumoto et al., 1987) and grouper (Huang et al., 2007) and thiamin content in the blood of yellowtail (Hosokawa, 1989) also have been shown to decrease much earlier than the appearance of external deficiency signs. Measurement of the thiamin pyrophosphate in the tissue of tiger shrimp is more sensitive of thiamin status than measurement of transketolase activity (Chen et al., 1994).

Thiamin (vitamin B_1) is commercially available as crystalline mononitrate or hydrochloride salts. Thiamin mononitrate (1 g thiamin = 1.088 g thiamin mononitrate) is typically used in animal feeds, while thiamin hydrochloride is typically used in liquid parenteral or oral vitamin products because it is more water soluble. Between 60–80% retention of thiamin activity is typically observed after feeds are extruded and stored at room temperature for 3 months (Gabaudan and Hardy, 2000).

Riboflavin

Riboflavin functions in the intermediary transfer of electrons in metabolic oxidation-reduction reactions as a component of two coenzymes, flavin mononucleotide (FMN) and flavin adenine dinucleotide (FAD). These coenzymes serve as prosthetic groups of oxidation-reduction enzymes involved in the metabolism of keto-acids, fatty acids, and amino acids in the mitochondrial electron transport system.

Species-specific deficiency signs are found in fish. The only common signs are anorexia and poor growth. The first sign of riboflavin deficiency observed in salmonids (McLaren et al., 1947; Halver, 1957; Steffens, 1970; Takeuchi et al.,

1980; Hughes et al., 1981a,b) appeared in the eyes and included photophobia, cataracts, corneal vascularization, and hemorrhages. Lack of coordinated swimming and dark skin coloration have also been reported for riboflavin-deficient Chinook salmon (Halver, 1957) and rainbow trout (Kitamura et al., 1967b; Steffens, 1970). In contrast, Woodward (1984) did not observe cataracts or corneal occlusion in riboflavin-deficient rainbow trout fry and fingerlings; however, severe fin erosion and light skin coloration accompanied by high mortality were observed. Monolateral or bilateral cataracts have been reported in riboflavin-deficient channel catfish (Dupree, 1966), but Murai and Andrews (1978a) found only poor growth and short-body dwarfism in two independent feeding trials with channel catfish. Riboflavin deficiency caused lethargy and high mortality in Japanese parrotfish (Ikeda et al., 1988); eye lesions, dark skin coloration, and high mortality in yellowtail (Hosokawa, 1989); hemorrhages in various parts of the body, nervousness, and photophobia in common carp (Aoe et al., 1967a; Ogino, 1967; Takeuchi et al., 1980) and Japanese eel (Arai et al., 1972); lethargy, fin erosion, anorexia, loss of normal body color, short-body dwarfism, and cataracts in blue tilapia (*Oreochromis aureus*) (Soliman and Wilson, 1992a); short-body dwarfism and cataracts in red hybrid tilapia (*Oreochromis mossambicus* × *O. niloticus*) (Lim et al., 1993); short-body dwarfism, anorexia, and poor growth in channel catfish (Serrini et al., 1996); anorexia, dark body color, and cataracts in hybrid striped bass (Deng and Wilson, 2003); and light coloration, irritability, protuberant cuticle at intersomites, and short-head dwarfism in tiger shrimp (Chen and Hwang, 1992). Huang et al. (2010) reported that riboflavin-deficient grouper showed high oxidative stress and low antioxiative enzyme activity.

Hughes et al. (1981a) used the activation coefficient (ratio of activity following preincubation with FAD:basal activity) of erythrocyte glutathione reductase to measure the riboflavin status of rainbow trout. However, Woodward (1983) found the activity of D-amino acid oxidase to be a more sensitive indicator of the riboflavin status in rainbow trout, because the low activity of erythrocyte glutathione reductase made its quantification difficult. Hepatic D-amino acid oxidase was also found as a reliable indicator of riboflavin status in rainbow trout (Amezaga and Knox, 1990), channel catfish (Serrini et al., 1996), and hybrid striped bass (Deng and Wilson, 2003). Amezaga and Knox (1990) pointed out, however, that an assay for glutathione reductase activity in erythrocytes would be advantageous because it could be used on live fish. Chen and Hwang (1992) reported hemolymph glutathione redutase activity is not a sensitive and specific indicator of riboflavin status in tiger shrimp. Woodward (1985) reported that the riboflavin requirement was not affected by temperature or by genetic differences in growth rate. This might be one reason why the riboflavin requirement values shown in Table 9-1 agree fairly well even among different species. Because weight gain of tiger shrimp did not vary with dietary riboflavin supplementation level, the requirement was estimated by body riboflavin concentration (Chen and Hwang, 1992).

Hughes (1984) found that feeding high concentrations of riboflavin (up to 600 mg/kg diet) had no adverse effects on growth of rainbow trout. These results were expected because riboflavin has not been shown to cause hypervitaminosis in other animals. However, two previous studies (McLaren et al., 1947; Woodward, 1982) had reported depressed growth in rainbow trout fed moderate concentrations of riboflavin. It was concluded that the growth depression observed in the earlier studies resulted from some factor other than riboflavin.

Riboflavin is produced as a crystalline compound or as a product of fermentation. The crystalline product is electrostatic and hygroscopic and does not distribute well when blended into a feed mixture. Its handling properties are significantly improved when it is formulated into a spray-dried powder. Riboflavin is relatively unaffected by extrusion pelleting and storage, with no more than 10% loss occurring after 3 months of storage of extruded pellets (Gabaudan and Hardy, 2000).

Vitamin B_6 (Pyridoxine)

The term "vitamin B_6" is the generic descriptor for the 2-methylpyridine derivatives that have the biological activity of pyridoxine. Pyridoxine is the main form found in plant products, whereas pyridoxal and pyridoxamine are the principal forms found in animal tissue. All three forms are readily converted in animal tissue to the coenzyme forms, pyridoxal phosphate and pyridoxamine phosphate. Pyridoxal phosphate is required for many enzymatic reactions involving amino acids such as transamination, decarboxylation, and dehydration. Pyridoxal phosphate also functions in the biosynthesis of porphyrins and in the catabolism of glycogen.

Pyridoxal phosphate is required for the synthesis of the neurotransmitters—5-hydroxytryptamine and serotonin—from tryptophan. Consequently, signs of pyridoxine deficiency include nervous disorders—erratic swimming, hyperirritability, and convulsions—that have been observed in salmonids (Halver, 1957; Coates and Halver, 1958), gilthead sea bream (Kissil et al., 1981), channel catfish (Andrews and Murai, 1979), common carp (Ogino, 1965), yellowtail (Sakaguchi et al., 1969), Japanese eel (Arai et al., 1972), and red hybrid tilapia (Lim et al., 1995). Other deficiency signs such as anorexia and poor growth were reported in hybrid tilapia (Shiau and Hsieh, 1997), and they usually appear in the fish within 3 to 6 weeks after being fed a pyridoxine-deficient diet. Pyridoxine deficiency has been reported to cause various histopathological changes in rainbow trout liver (Jurss and Jonas, 1981) and kidney (Smith et al., 1974), in the intestinal tissue of both rainbow trout (Smith et al., 1974) and gilthead sea bream (Kissil et al., 1981), and in the intestinal tissue and kidney of Indian carp (*Heteropneustes fossilis*) (Shaik Mohamed, 2001a).

The activity of certain aminotransferase enzymes that require pyridoxal phosphate as a coenzyme has been used as an index of pyridoxine status in fish. Serum or tissue alanine and/or aspartate aminotransferase activities have been used to evaluate pyridoxine status in common carp (Ogino, 1965), rainbow trout (Smith et al., 1974; Jurss, 1978), Chinook salmon (Hardy et al., 1979), turbot (Adron et al., 1978), gilthead sea bream (Kissil et al., 1981), hybrid tilapia (Shiau and Hsieh, 1997), and tiger shrimp (Shiau and Wu, 2003).

Vitamin B_6 requirements of hybrid tilapia has been reported to vary with dietary protein level, with 1.7–9.5, and 15–16.5 mg B_6/kg diet required in diets with 28% and 36% protein, respectively (Shiau and Hsieh, 1997). Vitamin B_6 supplementation increased the docosahexaenoic acid concentration of muscle lipids of rainbow trout (Maranesi et al., 2005).

Albrektsen et al. (1995) indicated that further increased supplementation level of vitamin B_6 beyond its minimal requirements did not enhance immune functions and disease resistance in Atlantic salmon. Feng et al. (2009) reported that with increasing dietary vitamin B_6 concentration up to 5 mg/kg diet, survival after pathogen challenge and phagocytic activity of leukocyte for Jian carp were improved and plateaued thereafter.

Vitamin B_6 is typically added to feeds as crystalline pyridoxine hydrochloride, which is 82.3% active. Pyridoxine is relatively unstable, especially in premixes exposed to moisture, e.g., high humidity, and containing trace minerals. Up to 50% of pyridoxine activity in premixes exposed to abusive conditions can be lost after 3 months of storage. The stability of pyridoxine partly depends on the size of its crystal particles. Therefore, fine granular crystals have improved stability during feed processing compared to very fine crystals. Properly formulated pyridoxine is relatively stable during pelleting, with typical extrusion and storage losses of 10–20% (Gabaudan and Hardy, 2000).

Pantothenic Acid

Pantothenic acid is a component of coenzyme A (CoA), acyl CoA synthetase, and acyl carrier protein. The coenzyme form of the vitamin is therefore responsible for acyl group transfer reactions. Coenzyme A is required in reactions in which the carbon skeletons of glucose, fatty acids, and amino acids enter into the energy-yielding tricarboxylic acid cycle. Acyl carrier protein is required for fatty acid synthesis.

A deficiency of this vitamin impairs the metabolism of mitochondria-rich cells that undergo rapid mitosis and high-energy expenditure. Thus, deficiency signs have been found to appear within 10 to 14 days in rapidly growing fish such as fingerling yellowtail (Hosokawa, 1989). Gill lamellar hyperplasia or clubbed gills is a characteristic sign of pantothenic acid deficiency in most fish. In addition to clubbed gills, anemia and high mortality have been observed in pantothenic acid-deficient salmonids (Phillips et al., 1945; McLaren et al., 1947; Coates and Halver, 1958; Kitamura et al., 1967b; Poston and Page, 1982; Karges and Woodward, 1984), channel catfish (Dupree, 1966; Murai and Andrews, 1979; Brunson et al., 1983; Wilson et al., 1983), and yellowtail (Hosokawa, 1989). Pantothenic acid-deficient Japanese parrotfish exhibited anorexia, convulsions, and cessation of growth followed by high mortality (Ikeda et al., 1988). Similar deficiency signs were observed in red sea bream (Yone and Fujii, 1974). Slow growth, anorexia, lethargy, and anemia were observed in common carp (Ogino, 1967). Poor growth, hemorrhage, skin lesions, and abnormal swimming were found in Japanese eel (Arai et al., 1972) fed pantothenic acid-deficient diets. Pantothenic acid deficiency causes poor growth, hemorrhage, sluggishness, high mortality, anemia and severe hyperplasia of the epithelial cells of gill lamellae in blue tilapia (Soliman and Wilson, 1992b), and it causes anorexia, poor growth, exophthalmus, and hemorrhage of body surface and fin in Jian carp (Wen et al., 2009). Irritability, light coloration, thin and soft shell, poor growth, and high mortality were observed in tiger shrimp fed a pantothenic acid-deficient diet (Shiau and Hsu, 1999).

Pantothenic acid is normally added to feeds as calcium d-pantothenate, which contains 92% of d-pantothenic acid. Calcium dl-pantothenate also exists but has half of this activity because the l forms of pantothenate are not biologically active. Calcium d-pantothenate is relatively stable during pelleting and feed storage, with losses after extrusion and storage of no more than 20% (Gabaudan and Hardy, 2000).

Niacin

Niacin is used as the generic descriptor of pyridine 3-carboxylic acids and their derivatives that exhibit the biological activity of nicotinamide (the amide of nicotinic acid). Of the compounds with niacin activity, nicotinic acid and nicotinamide have the greatest biological activity. Niacin is widely distributed in both plant and animal tissue. It is generally considered that much of the niacin in plant material is present in bound forms that have limited availability to fish. Bioavailability of niacin from feed ingredients commonly used in feeds for channel catfish (i.e., Mehaden fish meal, meat and bone/blood meal, wheat middlings, cooked corn, uncooked corn, cottonseed meal, and soybean meal) were estimated to be 100, 100, 60, 44, 28, 58, and 57%, respectively (Ng et al., 1998). These authors concluded that endogenous total niacin present in the ingredients used in commercial channel catfish diets was more than sufficient to meet the niacin requirements of this fish without any niacin supplementation.

Niacin is a component of two coenzymes: nicotinamide adenine dinucleotide (NAD) and nicotinamide adenine dinucleotide phosphate (NADP). These coenzymes are essential for several oxidation-reduction reactions involving the transfer of hydrogen and electrons in carbohydrate, lipid, and amino acid metabolism. They are also involved in various energy-yielding and biosynthetic pathways, including the mitochondrial electron transport system. Tryptophan can be metabolically converted to niacin in many animals, but not in certain salmonid fish (Poston and DiLorenzo, 1973; Poston and Combs, 1980). Ng et al. (1997) reported that excess tryptophan in niacin-deficient diets did not improve growth, hematocrits, or hepatic NAD concentrations in channel catfish. The fact that niacin deficiency can readily be induced in various fish indicates that most if not all fish lack the capacity for niacin synthesis.

Trout and salmon fed niacin-deficient diets exhibited anorexia, poor growth, poor feed conversion, photosensitivity or sunburn, intestinal lesions, abdominal edema, muscular weakness, spasms, and increased mortality (McLaren et al., 1947; Phillips and Brockway, 1947; Halver, 1957). Channel catfish (Andrews and Murai, 1978) and common carp (Aoe et al., 1967b) showed skin and fin lesions, high mortality, skin hemorrhages, anemia, and deformed jaws when fed niacin-deficient diets for 2 to 6 weeks. Skin hemorrhages, dermatitis, anemia, abnormal swimming, and ataxia were observed in Japanese eels fed a niacin-deficient diet for 14 weeks (Arai et al., 1972). Two weeks after exposure of niacin-deficient rainbow trout to ultraviolet radiation, a total loss of mucus-producing cells was observed in histopathological sections of the epidermis. Dermal lesions in rainbow trout (Poston and Wolfe, 1985); skin, fin, and mouth lesions and hemorrhages, as well as deformed snout and gill edema in hybrid tilapia (Shiau and Suen, 1992); and anemia, anorexia, lethargy, and skin hemorrhage in Indian catfish (*Heteropneustes fossilis*) (Shaik Mohamed and Ibrahim, 2001) were observed in niacin-deficient fish.

Complex carbohydrates in the diet may increase the requirement for niacin. Shiau and Suen (1992) reported that 26 and 121 mg niacin/kg diet were required for hybrid tilapia fed diets with glucose and dextrin as the carbohydrate source, respectively.

High dietary intake of niacin (10,000 mg/kg) increased liver fat, decreased body fat, and tended to reduce growth rate in fingerling brook trout (Poston, 1969b).

Niacin is added to aquatic feeds as either nicotinic acid or niacinamide; both have similar biological activity. Nicotinic acid or niacinamide is added to the multivitamin premix in a dry form and is quite stable during extrusion pelleting and storage, with losses generally 10% or less (Gabaudan and Hardy, 2000).

Biotin

Biotin acts in certain metabolic reactions as an intermediate carrier of carbon dioxide during carboxylation and decarboxylation reactions. Specific enzymes that require biotin include acetyl-CoA carboxylase, pyruvate carboxylase, and propionyl-CoA carboxylase. Metabolic pathways requiring biotin include the biosynthesis of long-chain fatty acids and the synthesis of purines.

In many animals, a biotin deficiency can only be induced by feeding avidin, a glycoprotein found in raw chicken egg white that binds biotin and prevents absorption of the vitamin from the intestine. Robinson and Lovell (1978) fed avidin in a biotin-free chemically defined diet to channel catfish and noted a growth suppression that led them to suggest some biotin synthesis by intestinal microflora in this species. However, in a later study by Lovell and Buston (1984), no synthesis of biotin by the intestinal microflora in channel catfish could be detected. Common carp required 8 to 12 weeks (Ogino et al., 1970a) and channel catfish took 11 weeks (Lovell and Buston, 1984) to show growth depression when fed biotin-deficient diets. A similar effect in rainbow trout took only 4 to 8 weeks in water temperatures of 15°C (Woodward and Frigg, 1989). Anorexia, reduced weight gain, and higher feed conversion were more noticeable in smaller than in larger rainbow trout fed biotin-deficient diets (Walton et al., 1984). Biotin-deficient channel catfish exhibited skin depigmentation (Robinson and Lovell, 1978), whereas biotin-deficient Japanese eels had darker skin coloration (Arai et al., 1972). Histological signs of biotin deficiency were not detected after 12 weeks in rainbow trout having an initial weight of 25 g (Walton et al., 1984). However, severe deficiency signs were produced in rainbow trout and lake trout having initial weights of 1.3 and 6.7 g, respectively (Poston and Page, 1982; Woodward and Frigg, 1989). Rainbow trout and lake trout (*Salvelinus namaycush*) developed biotin-related histopathological signs in the gills (Castledine et al., 1978; Poston and Page, 1982), liver (Poston, 1976b; Poston and Page, 1982), and kidney (Poston and Page, 1982). Biotin-deficient diets caused anorexia, dark skin color, convulsions, and high mortality in Asian catfish (*Clarias batrachus*) (Shaik Mohamed et al., 2000) and convulsions, heavy mortality, listlessness, poor feed conversion and feed intake, dark skin color, tetanus, and weight loss in Indian catfish (Shaik Mohamed, 2001b).

Hepatic pyruvate carboxylase activity in rainbow trout fed a lipid-free and biotin-deficient diet decreased to 3.3% of that in fish fed a diet sufficient in lipid and biotin, although the enzyme activity was restored to about 50% of normal levels following the addition of lipid to the diet (Walton et al., 1984). In contrast, lipid supplementation of biotin-deficient diets did not increase hepatic pyruvate carboxylase activity in channel catfish (Robinson and Lovell, 1978). Hepatic pyruvate carboxylase and acetyl-CoA carboxylase activities have been used to evaluate the biotin status in hybrid tilapia (Shiau and Chin, 1999), Asian catfish (Shaik Mohamed et al., 2000), Indian catfish (Shaik Mohamed, 2001b), and tiger shrimp (Shiau and Chin, 1998).

Signs of biotin deficiency were not detected in rainbow trout (Castledine et al., 1978) or channel catfish (Lovell and Buston, 1984) fed natural ingredient diets without supplemented biotin for 24 and 17 weeks, respectively. Mæland et al. (1998) reported no need for supplemental biotin in practical fish meal diets for Atlantic salmon fry to achieve optimum growth, survival, and maximal hepatic pyruvate carboxylase activity. These studies concluded that adequate biotin was available in the various feed ingredients in the natural ingredient diets used to meet the requirements of the fish.

Biotin is added to feeds as D-biotin, the biologically active form. The isomer L-biotin has no biological activity. The activity of D-biotin products is 2% on a weight basis, and the stability of biotin during extrusion pelleting and room temperature storage for 3 months ranges from 70 to 90% (Gabaudan and Hardy, 2000).

Folic Acid

The term "folate" is used as the generic descriptor for folic acid and related compounds exhibiting qualitatively the biological activity of folic acid. Folic acid is composed of a pteridine ring linked through a methylene bridge to p-aminobenzoic acid to form pteroic acid, which is in turn linked as an amide to glutamic acid. Folic acid undergoes enzymatic reduction in the tissues to its active coenzyme form, tetrahydrofolic acid. It functions as a pivoting intermediate carrier of one-carbon groups in a number of complex enzymatic reactions also involving vitamin B_{12}, biotin, niacin, methionine, choline, betaine, and homocysteine. In these reactions, methyl, methylene, and other one-carbon groups are transferred from one molecule to another. The C-3 of serine is the major source of one-carbon units for folate metabolism. Other sources include formate, much of which is derived from serine metabolism in the mitochondria, and the C-2 of histidine. The folate-dependent reactions are found in the metabolism of certain amino acids and the biosynthesis of purines and pyrimidines along with the nucleotides found in DNA and RNA.

Trout and salmon fed folate-deficient diets exhibited anorexia; reduced growth; poor feed conversion; and macrocytic normochromic, megaloblastic anemia (Smith, 1968; Smith and Halver, 1969) characterized by pale gills, anisocytosis, and poikilocytosis. The erythrocytes were large with abnormally segmented and constricted nuclei, and a large number of megaloblastic proerythrocytes were present in the erythropoietic tissue of the anterior kidney. Production of erythrocytes decreased with time in fish fed the folate-deficient diet. Some of these signs have also been observed in the rohu (John and Mahajan, 1979).

Poor growth and dark skin coloration were noted in Japanese eels fed a folate-deficient diet for 10 weeks (Arai et al., 1972). Folate-deficient yellowtail fingerlings also showed congestion in fins and bronchial mantle, dark skin coloration, and anemia (Hosokawa, 1989). Rainbow trout fed a folic acid-deficient diet showed a blood pathology (i.e., megaloblastosis; Cowey and Woodward, 1993). Folate deficiency signs in channel catfish included reduced growth, anemia, and increased sensitivity to bacterial infection (Duncan and Lovell, 1991). Deficiency signs were not observed in common carp (Aoe et al., 1967c), hybrid tilapia (Shiau and Huang, 2001b), and tiger shrimp (Shiau and Huang, 2001a) fed a folate-free diet, presumably due to bacterial synthesis of folate in the intestine (Kashiwada et al., 1971; Duncan et al., 1993).

The hepatosomatic index (HSI) has been reported to be a good parameter other than growth and tissue concentration for estimating folate requirements in hybrid tilapia (Shiau and Huang, 2001a) and tiger shrimp (Shiau and Huang, 2001b).

Folic acid is synthesized and added to vitamin premixes as a dry dilution either as a crystalline form or a spray-dried form. The crystalline form is electrostatic and tends to adhere to the machinery, while the spray-dried form does not and therefore contributes to a higher recovery of the vitamin in the feed. Stability of folic acid after extrusion and feed storage is relatively low, ranging from 50–65% (Gabaudan and Hardy, 2000).

Vitamin B_{12}

The term "vitamin B_{12}" should be used as the generic descriptor for all corrinoids exhibiting qualitatively the biological activity of cyanocobalamin. This vitamin was previously known as vitamin B_{12} or cyanocobalamin. Vitamin B_{12} is a large molecule (molecular weight 1,355) that contains a cobalt atom. Neither higher plants nor animals can synthesize vitamin B_{12}, but both depend on certain microorganisms for the trace amounts required. In animals, vitamin B_{12} is known to be involved in two separate enzyme systems: (1) methylmalonyl CoA mutase, by which propionic acid is converted to succinate, and (2) methyltetrahydrofolate-homocysteine methyltransferase, a catalyst in methionine, methane, and acetate synthetase. Vitamin B_{12} is required for normal maturation and development of erythrocytes, for the metabolism of fatty acids, in the methylation of homocysteine to methionine, and for the normal recycling of tetrahydrofolic acid. Thus, a deficiency of vitamin B_{12} can result in signs similar to folate deficiency.

Salmon (Halver, 1957) and trout (Phillips et al., 1964) fed low amounts of vitamin B_{12} showed a high variability in numbers of fragmented erythrocytes and in hemoglobin values, with a tendency for a microcytic, hypochromic anemia. Channel catfish fed a vitamin B_{12}-deficient diet for 36 weeks exhibited reduced growth but no other clinical deficiency signs (Dupree, 1966). John and Mahajan (1979) observed reduced growth and lower hematocrit in rohu fed a vitamin B_{12}-deficient diet. Japanese eel were found to require vitamin B_{12} for normal appetite and growth (Arai et al., 1972).

Intestinal microfloral synthesis appeared to satisfy the B_{12} requirement of common carp (Kashiwada et al., 1970), Nile tilapia (Lovell and Limsuwan, 1982), and hybrid tilapia (Shiau and Lung, 1993a), but channel catfish (Limsuwan and Lovell, 1981) and tiger shrimp (Shiau and Lung, 1993b) required dietary supplementation of B_{12} to prevent anemia. Intestinal microfloral synthesis of vitamin B_{12} has been demonstrated in common carp (Kashiwada et al., 1970; Sugita et al., 1991a), channel catfish (Limsuwan and Lovell, 1981; Sugita et al., 1990, 1991a), Nile tilapia (Lovell and Limsuwan, 1982; Sugita et al., 1990, 1991a), rainbow trout (Sugita et al., 1991b), ayu (*Plecoglossus altivelis*), and goldfish (*Carassius auratus*) (Sugita et al., 1991a). Sugita et al. (1991a) found a close relationship between the amount of vitamin B_{12} and the viable counts of *Bacteroides* type A in the intestinal contents of the various fish studied. They found that this bacterium was present in the intestinal contents of fish that do not require vitamin B_{12} and absent in those fish that do require vitamin B_{12}. Cobalt (Co) is recognized as the mineral in the large vitamin B_{12} molecule. Grouper has been reported to require 10 mg Co/kg diet for optimal growth. This amount of Co can promote gastrointestinal bacterial production of vitamin B_{12} in sufficient amounts to supply growth requirements of grouper, so no additional dietary vitamin B_{12} supplementation is needed (Lin et al., 2010).

Vitamin B_{12} is produced by fermentation and used in feeds as a dry dilution having 1% activity, on a weight basis. Vitamin B_{12} stability in extruded feeds after 3 months of room temperature storage ranges from 40–80% (Gabaudan and Hardy, 2000).

Choline

Unlike the other water-soluble vitamins, choline has no known coenzyme function. Choline has three major metabolic functions: as a component of phosphatidylcholine, which has structural functions in biological membranes and in tissue lipid utilization; as a precursor of the neurotransmitter acetylcholine; and as a precursor of betaine, which serves as a source of labile methyl groups for methylation reactions such as the formation of methionine from homocysteine and creatine from guanidoacetic acid.

Rainbow trout fed a choline-deficient diet developed light yellow-colored livers, protruded eyes, anemia, and extended abdomens (Kitamura et al., 1967b). Lake trout fed a choline-deficient diet for 12 weeks had depressed growth rate and increased liver fat content (Ketola, 1976). Depressed growth, loss of appetite, and white-gray colored intestines were observed in Japanese eels fed a choline-deficient diet (Arai et al., 1972). Increased liver lipid content has been observed in common carp (Ogino et al., 1970a), channel catfish (Wilson and Poe, 1988), and hybrid striped bass (Griffin et al., 1994) fed choline-deficient diets. However, decreased liver lipid content has been found in red drum (Craig and Gatlin, 1996) and hybrid tilapia (Shiau and Lo, 2000) fed choline-free diets. In addition, common carp developed vacuolization of hepatic cells after being on such a diet for 10 weeks (Ogino et al., 1970a). A thinning of the intestinal wall muscle and focal degeneration of the exocrine pancreas were observed in choline-deficient sturgeon (Hung, 1989).

Channel catfish fed casein-gelatin diets containing excess methionine did not develop signs of choline deficiency; however, catfish fed diets adequate but not excessive in methionine did develop deficiency signs (Wilson and Poe, 1988). Rumsey (1991) has suggested that 50% of the choline requirement of rainbow trout can be met from betaine. These observations indicate that certain fish can meet a part of their choline needs through the synthesis of choline by the methylation of ethanolamine, which uses methyl groups from S-adenosyl methionine.

A requirement of 7,500–8,500 mg choline/kg diet is reported for microcrustacean *Moina macrocopa* (D'Abramo and Baum, 1981). High-lipid diets may increase choline requirements. Shiau and Cho (2002) reported that 6,400 and 7,800 mg choline per kilogram of diet were required for tiger shrimp fed diets with 5 and 11% of lipid, respectively. No choline requirement was evident when additionally provided at 1.5% and 3% of phospholipid (lecithin) in a Pacific white shrimp diet (Gong et al., 2003).

Choline is produced for feed use as a chloride salt, which is available as a dry dilution product having 25, 50, or 60% activity on a weight basis, or as a liquid having 70% activity. On a molecular weight basis, choline chloride is 86.8% choline. Choline is completely stable during feed pelleting and storage, but it is a hygroscopic substance and a strong base. Its presence reduces the activity of other vitamins, such as vitamin E and vitamin K, when it is included in vitamin premixes. Thus, it should be added separately to feed mixtures (Gabaudan and Hardy, 2000).

Myoinositol

Inositol may exist in one of seven optically inactive forms and as one pair of optically active isomers. Only one of these forms, myoinositol, possesses biological activity. Inositol is a biologically active cyclohexitol and occurs as a structural component in biological membranes as phosphatidylinositol. Phosphatidylinositol has been shown to be involved in signal transduction of several metabolic processes (Mathews and van Holde, 1990). Although similar in many respects to the adenylate cyclase transduction system, the phosphoinositide system is distinctive in that the hormonal stimulus activates a reaction that generates two second messengers. Membrane bound phosphatidylinositol 4,5-bisphosphate is cleaved to release *sn*-1,2-diacylglycerol and inositol 1,4,5-triphosphate, following the interaction of a hormone or agonist with the receptor on the cell membrane. Inositol 1,4,5-triphosphate stimulates the release of calcium from its intracellular stores in the endoplasmic reticulum, and *sn*-1,2-diacylglycerol activates protein kinase C to phosphorylate specific target proteins. Examples of cellular processes controlled by the phosphoinositide second messenger system include amylase secretion, insulin release, smooth muscle contraction, liver glycogenolysis, platelet aggregation, histamine secretion, and DNA synthesis in fibroblasts and lymphoblasts.

Signs of inositol deficiency have been reported to include poor appetite, anemia, poor growth, fin erosion, dark skin coloration, slow gastric emptying, and decreased cholinesterase and certain aminotransferase activities in trout (McLaren et al., 1947; Kitamura et al., 1967b), red sea bream (Yone et al., 1971), Japanese eel (Arai et al., 1972), Japanese parrotfish (Ikeda et al., 1988), and yellowtail (Hosokawa, 1989). Rainbow trout fed a diet devoid of inositol had large accumulations of neutral lipids in the liver, increased levels of cholesterol and triglycerides, but decreased amounts of total phospholipid, phosphotidylcholine, phosphotidylethanolamine, and phosphotidylinositol (Holub et al., 1982). Total lipid content was increased in hybrid tilapia (Shiau and Su, 2005), grouper (Su and Shiau, 2004), and tiger shrimp (Shiau and Su, 2004) fed inositol-deficient diets. Lee et al. (2009) reported that olive flounder (*Paralichthys olivaceus*) fed inositol-free diet resulted in abnormal lipid metabolism and a decreased amount of polyunsaturated fatty acids.

Inositol appears to be synthesized in common carp intestine (Aoe and Masuda, 1967), but not in amounts sufficient to sustain normal growth of young fish without an exogenous source of this vitamin, because younger carp require a higher level of inositol than older fish. High concentrations of dietary glucose may increase the need for inositol in some fish (Yone et al., 1971). Burtle and Lovell (1989) demonstrated de novo synthesis of inositol in the liver of channel catfish, as well as intestinal synthesis. Deng et al. (2002) demonstrated that de novo synthesis of inositol was sufficient for normal growth and tissue storage of hybrid striped bass. Intestinal microbial synthesis was not a significant source of inositol for hybrid tilapia (Shiau and Su, 2005) and olive flounder (Lee et al., 2009), because addition of an antibiotic to a myoinositol-free diet had similar growth and tissue myoinositol levels with fish fed the myoinositol- and antibiotic-free diet. Peres et al. (2004) reported that Nile tilapia did not require an exogenous source of inositol for normal growth, feed utilization, and erythropoiesis. In addition, supplementation of dietary inositol did not improve disease resistance of the fish. There is no need for inositol supplementation to fish meal-based diets (containing 296 mg/kg diet of inositol) for young Atlantic salmon (Waagbø et al., 1998).

Myoinositol is added to aquatic feeds when necessary as a dry powder in a multivitamin premix.

Vitamin C

Most animals can synthesize vitamin C, or L-ascorbic acid (AA), from D-glucose, but many fish cannot (Kitamura et al., 1965; Poston, 1967; Halver et al., 1969; Wilson, 1973; Dabrowski, 1990). Shrimp has been suggested as having a limited ability to synthesize vitamin C (Lightner et al., 1979), although endogenous AA synthesis does not seem to meet the requirements in young crustaceans (He and Lawrence, 1993b). Ascorbic acid is a strong reducing agent and is readily oxidized to dehydroascorbic acid. Dehydroascorbic acid can be enzymatically reduced back to AA in animal tissue with glutathione or reduced NADP. Ascorbic acid is a cofactor in the hydroxylation of proline and lysine to hydroxyproline and hydroxylysine in procollagen, which is the precursor of collagen and thus is necessary for the formation of connective tissues, scar tissue in wound repair, and bone matrix (Sandel and Daniel, 1988). Ascorbic acid also facilitates the absorption of iron (Hsu and Shiau, 1999c), thus preventing the anemia often observed in AA-deficient fish. In addition, AA functions with vitamin E to minimize peroxidation of lipids in fish tissues (Heikkila and Manzino, 1987). Substantial experimental evidence indicates that vitamin C is involved in several physiological processes, including growth, reproduction, response to stressors, wound healing, and immune response (Gabaudan and Verlhac, 2001). It is also essential for the regeneration of vitamin E after oxidation.

Vitamin C-deficient salmon and trout exhibited structural deformities (scoliosis, lordosis, and abnormal support cartilage of the eye, gill, and fins) and internal hemorrhaging usually preceded by nonspecific signs such as anorexia and lethargy (Halver et al., 1969; Hilton et al., 1978; Tsujimura et al., 1978; Sato et al., 1983), ascites and hemorrhagic exophthalmia (Poston, 1967), and high level of plasma triglycerides and cholesterol (John et al., 1979). Similar structural deformities such as scoliosis and lordosis due to vitamin C deficiency have been observed in channel catfish (Wilson and Poe, 1973; Andrews and Murai, 1975; Lim and Lovell, 1978; Wilson et al., 1989), Indian major carp (*Cirrhina mrigala*) (Agrawal and Mahajan, 1980), common carp and roach (*Rutilus rutilus*) (Dabrowski et al., 1988, 1989), blue tilapia (Stickney et al., 1984), Nile tilapia (Soliman et al., 1986a,b), yellowtail (Sakaguchi et al., 1969), hybrid catfish (Khajarern and Khajarern, 1997), red drum (Aguirre and Gatlin, 1999), hybrid striped bass (Sealey and Gatlin, 1999), olive flounder (Wang et al., 2002), parrot fish (*Oplegnathus fasciatus*) (Wang et al., 2003a), Japanese sea bass (*Lateolabrax japonicus*) (Ai et al., 2004), and African catfish (*Heterobranchus longifilis*) (Ibiyo et al., 2007). Japanese eels fed a vitamin C-deficient diet showed reduced growth after 10 weeks and hemorrhage in the head and fins after 14 weeks (Arai et al., 1972). Opacity of the cornea and kidney granulomatosis associated with hypertyrosinemia have been described as signs of vitamin C deficiency in turbot (Messager, 1986; Messager et al., 1986). Anemia has been found in scorbutic hybrid tilapia (Shiau and Jan, 1992a).

Although some studies failed to demonstrate the essentiality of AA in crustacean species (Desjardins et al., 1985), others have shown that diet devoid of AA affects growth and collagen formation. Further, suboptimal AA status has been related to the "black death" syndrome in blue shrimp and *F. californiensis* (Lightner et al., 1977; Magarelli et al., 1979). In a study on Kuruma shrimp, Kanazawa (1985) found a clear dose-related preventive of AA on mortality. Similar lesions have been described in tiger shrimp (Catacutan and Lavilla-Pitogo, 1994) and Pacific white shrimp (Montoya and Molina, 1995). He and Lawrence (1993b) reported that besides "black death" syndrome, Pacific white shrimp fed vitamin C-deficient diet showed abnormal coloration, swollen hepatopancreata, motionlessness, and unresponsiveness to disturbances. Symptoms of vitamin C deficiency are somewhat different in Kuruma shrimp, being a decolorization and development of abnormal grayish-white color on the carapace margins, lower abdomen, and tips of walking legs (Deshimaru and Kuroki, 1976).

Reproduction appears to increase maternal demands for vitamin C. Female tilapia fed vitamin C-free diets for 21 weeks produced eggs and fry containing no detectable ascorbic acid (Soliman et al., 1986b). Reduced reproductive performance has also been reported in rainbow trout fed vitamin C-deficient diets (Sandnes et al., 1984). Ascorbic acid reserves are rapidly depleted during embryonic (Sato et al., 1987) and larval development of certain fish (Dabrowski et al., 1988, 1989; Dabrowski, 1990), suggesting that requirements during early life stages may be higher than for fingerlings or adults. In rainbow trout, sperm concentrations and motility were found to be correlated to seminal plasma concentrations of vitamin C (Ciereszko and Dabrowski, 1995). The dietary AA levels needed to raise seminal concentrations of this vitamin resulting in beneficial effects were 130 to 270 mg AA/kg. A correlation between hatching rate and the seminal plasma concentrations of AA showed that survival of embryos from rainbow trout males with less than 7.3 μg AA/ml of seminal plasma was reduced (Dabrowski and Ciereszko, 1996).

Liver (Hilton et al., 1977; Sato et al., 1983) and kidney (Halver et al., 1969) AA concentrations of less than 20 μg/g have been suggested as an indicator of vitamin C deficiency in salmonid fish. The values of less than 26 and 10 μg/g of liver in channel catfish (Lim and Lovell, 1978) and in Atlantic salmon (Sandnes et al., 1992), respectively, have been suggested as vitamin C deficiency indicators. A much higher value of 100 μg/g of kidney coincided with signs of vitamin C deficiency in snakehead (Mahajan and Agrawal, 1979).

Vertebral collagen levels have been shown to be a sensitive index of vitamin C status in channel catfish (Wilson and Poe, 1973; Lim and Lovell, 1978; El Naggar and Lovell, 1991) and rainbow trout (Sato et al., 1978).

L-ascorbic acid is relatively sensitive to oxidation at carbon 2 in the lactone ring in its molecular structure. Thus, moisture, heat, and exposure to prooxidants reduce vitamin C activity. Lovell and Lim (1978) and El Naggar and Lovell (1991) showed that approximately 50% of the vitamin potency of AA was lost during extrusion processing of fish feeds. Shiau and Hsu (1993) found about 75% of AA can be lost in shrimp feed during a mild processing and storage at ambient temperature for 1 hour. A great number of studies to establish dietary requirements for vitamin C in fish and shrimp have used AA as the dietary source. High requirements and high variability of vitamin C requirements were presented in the literature. A great part of the variability was due to the use of unstable form of dietary AA.

Various derivatives of AA, including L-ascorbyl-2-sulfate (C2S), L-ascorbyl-2-monophosphate-Mg (C2MP-Mg), L-ascorbyl-2-monophosphate-Na (C2MP-Na), L-ascorbyl-2-monophosphate-Ca (C2MP-Ca), L-ascorbyl-2-monophosphate-Na/Mg (C2MP-Na/Mg), L-ascorbyl-2-polyphosphate (C2PP), and ascorbate-2-glucose (C2D), have been demonstrated more stable than the parent compound and have also been shown to provide antiscorbutic activity in fish and shrimp. These include C2S in rainbow trout (Halver et al., 1975; Grant et al., 1989), channel catfish (Murai et al., 1978; Brandt et al., 1985; Wilson et al., 1989; El Naggar and Lovell, 1991), blue tilapia (Soliman et al., 1986a), hybrid tilapia (Shiau and Hsu, 1995), and grouper (M. F. Lin and Shiau, 2005b); C2MP-Mg in Asian sea bass (*Lates calcarifer*) (Phromkunthong et al., 1997), hybrid tilapia (Shiau and T. S. Hsu, 1999), angelfish (*Pterophyllum scalare*) (Blom et al., 2000), grouper (Lin and Shiau, 2004), tiger shrimp (Hsu and Shiau, 1998), and Kuruma shrimp (Moe et al., 2004); C2MP-Na in hybrid tilapia (Shiau and T. S. Hsu, 1999), grouper (Lin and Shiau, 2004), and tiger shrimp (Hsu and Shiau, 1998); C2PP in rainbow trout (Grant et al., 1989), channel catfish (Wilson et al., 1989), hybrid striped bass (Sealey and Gatlin, 1999), grouper (M. F. Lin and Shiau, 2005b), yellow croaker (Ai et al., 2006), Pacific white shrimp (He and Lawrence, 1993b), and tiger shrimp (Chen and Chang, 1994); C2MP-Ca in Atlantic salmon (Sandnes et al., 1992) and Korean rockfish (Wang et al., 2003b); C2MP-Na/Ca in Kuruma shrimp (Moe et al., 2004); and C2D in hybrid *Clarias* catfish (Khajaren and Khajaren, 1997) and Korean rockfish (Wang et al., 2003c). Ascorbate-2-sulfate does not appear to be used as well as other more stable forms of AA by certain fish (Murai et al., 1978; Soliman et al., 1986a; Dabrowski and Kock, 1989; Dabrowski et al., 1990), and in channel catfish it accounted for only 7% as much vitamin C activity as AA or C2MP (Lovell and El Naggar, 1990). El Naggar and Lovell (1991) reported that generally C2MP-Mg and AA had equal vitamin C potency, but that C2S had only 5.2% of the vitamin activity of AA. Shiau and Hsu (1995) indicated that hybrid tilapia could use C2S equally as well as C2MP for growth and collagen synthesis.

Strict comparison of the potency of each source of AA from the published results is perhaps misleading because the experimental condition and purity of these derivatives were not the same. Thus, a comparison between different AA derivatives within a study for an individual aquatic species is preferred to a cross-comparison of the potency of each AA derivative from different studies. Cross-comparison of C2S, C2PP, C2MP-Mg, and C2MP-Na has shown that the biopotencies of these compounds for tiger shrimp and grouper are C2MP-Mg (1) > C2MP-Na (84%) > C2PP (64%) > C2S (25%) (Shiau, 2001) and C2Mp-Na (1) > C2PP (84%) > C2Mp-Mg (46%) > C2S (32%) (Lin and Shiau, 2004, 2005b,c), respectively.

Phagocytic activity of cells of the immune system in fish produce reactive oxygen radicals that are potent microbicidal factors, but also autotoxic to fish macrophages (Secombes et al., 1988). Vitamin C appears to protect phagocytic cells and surrounding tissues from oxidative damage. An increased immune response due to high concentrations of vitamin C supplementation has been demonstrated in channel catfish (Durve and Lovell, 1982; Li and Lovell, 1985), rainbow trout (Blazer and Wolke, 1984b; Wahli et al., 1986; Navarre and Halver, 1989; Verlhac et al., 1996), turbot (Roberts et al., 1995), gilthead sea bream (Henrique et al., 1998; Ortuño et al., 1999), bagrid catfish (Anbarasu and Chandran, 2001), mrigal (Sobhana et al., 2002), Japanese sea bass (Ai et al., 2004), grouper (M. F. Lin and Shiau, 2005a), yellow croaker (*Pseudosciaena crocea*) (Ai et al., 2006), Indian major carp (Misra et al., 2007; Nayak et al., 2007), Japanese eel (Ren et al., 2007), and tiger shrimp (Lee and Shiau, 2002). Wang et al. (2006) indicated that vitamin C in enriched live food (*Artemia*) could enhance immune responses and resistance to environmental stress for Pacific white shrimp. It has been reported that an in vitro study showed a significant increase in the natural cytotoxic activity of gilthead sea bream head-kidney leucocytes treated with high dose (2 mg/ml) of vitamin C compared to the low-dose (0.002 mg/ml) treatment (Cuesta et al., 2002). However, Lall et al. (1990) observed no differences in humoral response and the complement system in Atlantic salmon fed diets containing 0 to 2,000 mg of vitamin C/kg after vaccination and subsequent live challenge with *Aeromonas salmonicida* and *Vibrio anguillarum*. Li et al. (1993) found that elevated dietary vitamin C concentrations did not improve resistance of channel catfish against the pathogen *Edwardsiella ictaluri*. Tiger puffer (*Takifugu rubripes*) fed diets with overdose of dietary AA (> 160 mg/kg) did not enhance their nonspecific immune responses (Eo and Lee, 2008).

Dietary and environmental contaminants, such as heavy metals (Yamamoto and Inoue, 1985) and chlorinated hydrocarbon pesticides (Mayer et al., 1978), increase the vitamin C requirements of fish. Derivatives of AA may interact with minerals differently. Tiger shrimp fed diets with C2MP-Mg and C2PP had higher hepatopancreatic Fe concentrations than shrimp fed diet with C2S. Increasing dietary ascorbate supplementation level could lower hepatopancreatic Cu concentration in tiger shrimp fed diets with C2MP-Mg or C2PP than that in shrimp fed diet with C2S (Hsu and Shiau, 1999c). Lee and Shiau (2003) found that an increase of dietary C2MP-Mg prevented tissue Cu accumulation in tiger shrimp fed with high dietary Cu.

Vitamin C may function in vivo to repair the membrane-bound oxidized vitamin E (Chan, 1993). Channel catfish (Gatlin et al., 1986b) and giant river prawn (*Macrobrachium rosenbergii*) (Cavalli et al., 2003) fed diet with adequate dietary vitamin C did not show a sparing effect on vitamin E. Increasing dietary ascorbate supplementation level, the vitamin C sparing on vitamin E was observed in hybrid tilapia (Shiau and Hsu, 2002), hybrid striped bass (Sealey and Gatlin, 2002a,b), yellow perch (Lee and Dabrowski, 2004), and channel catfish (Yildirim-Aksoy et al., 2008), but not in turbot (Ruff et al., 2003) and golden shiner (*Notemigonus crysoleucas*) (Chen et al., 2004). Both supplementation of high (900 mg/kg diet) vitamin C and E in giant river prawn diet enhanced larval quality and resistance to ammonia for the prawn (Cavalli et al., 2003).

Crystalline ascorbic acid (100% active on a weight basis) is extremely susceptible to oxidation, and early tests with the Oregon Moist Pellet indicated that within 3 days without frozen storage, all vitamin C activity was gone. Thus, feed was kept frozen until use, and coated forms (fat-coated, ethylcellulose-coated) more resistant to oxidation were used. In dry, pelleted fish feeds, approximately 20% of vitamin C activity remained after steam pelleting and storage, so feed formulators added five times more crystalline or coated ascorbic acid than they wanted to ensure that enough remained to meet the dietary requirements of fish at the time of feeding. The use of extrusion pelleting in fish feed production introduced additional heat and pressure to the pelleting process, sufficient to melt the fat-coating and otherwise accelerate the loss of ascorbic acid activity in fish feeds. Thus, conjugates were developed that added a functional group to the 2nd carbon position of ascorbic acid, thus protecting ascorbic acid from oxidation. The first such product was ascorbate-2-sulfate, which was very stable but had low biological activity. The second such product was ascorbate-2-polyphosphate, which had full biological activity but relatively low activity on a molecular weight basis, due to the relative weight of the polyphosphate moiety. More recently, ascorbate-2-monophosphate has been developed, which increases the ascorbic acid content on a molecular weight basis. This product, a Na/Ca salt of ascorbate-2-monophosphate, is in wide use in the fish feed industry today and exhibits less than 15% loss of activity in extrusion pelleting and 3 months of room temperature storage, compared to 70–90% loss of activity for ethylcellulose-coated or fat-coated ascorbic acid. For steam-pelleted feeds, losses of crystalline ascorbic acid range from 30–70%, depending upon pelleting, drying, and storage conditions, compared to less than 10% loss of ascorbate-2-phosphate. The relative ability to utilize the various protected forms of vitamin C is likely the result of differences in absorption. Accordingly, the biopotency of each ascorbate source is critical in determining the supplemental dietary level for fish and shrimp. Caution should be taken when comparing the requirement of each of the ascorbate derivatives from the published data. More information is needed on the biopotency of these derivatives in various aquatic species (Gabaudan and Hardy, 2000).

OTHER VITAMIN-LIKE COMPOUND

Astaxanthin

Many plants and animals contain a variety of natural pigments that impart yellow, orange, and red colors to the flesh, skin, and eggs of fish. One of the most important groups of natural pigments in the plant and animal kingdom is the carotenoids. In salmonids, two oxycarotenoids, astaxanthin (3,3'-dihydroxy-4,4'-diketo-β-carotene) and canthaxanthin (4-4'-diketo-β-carotene), are responsible for the red to orange coloring of the flesh, skin, and fins. Astaxanthin is the main carotenoid pigment of wild salmonids and is derived mainly from zooplankton. Astaxanthin was shown to be the principal pigment for rainbow trout (Storebakken and No, 1992), Atlantic salmon (Christiansen et al., 1995), Atlantic charr (Olsen and Mortensen, 1997), Australian snapper (Doolan et al., 2009), tiger shrimp (Okada et al., 1994), and Kuruma shrimp (Chien and Jeng, 1992).

Carotenoids have been shown to have several physiological functions, including (1) accessory pigment in photosynthesis, (2) protective pigment against photosensitization, (3) provitamin A source, and (4) communication in aquatic animals. Responses to physiological or pharmacological administration of carotenoids are normally classified as actions. Potential mechanisms associated with carotenoids actions include antioxidant and singlet oxygen quenching, provitamin A activity, upregulating of DNA expression, cooxidation, and enhancement of immune functions associated with increased tumor immunity and modulation of macrophage and lymphocyte activation (Bendich, 1993).

An improved growth of Atlantic salmon was found by supplementing diets with astaxanthin (Torrissen, 1984; Christiansen et al., 1994). Corresponding results were found for Nile tilapia (Boonyaratpalin and Unprasert, 1989) and Kuruma shrimp (Chien and Jeng, 1992). Consequently, astaxanthin has been suggested to be considered as a vitamin for fish and shrimp (Torrissen and Christiansen, 1995).

Astaxanthin supplemented in the diet showed improving egg quality and larval production in Atlantic cod (Sawanboonchun et al., 2008) and spawning and fecundity in tiger shrimp broodstock (Pangantihon-Kühlmann et al., 1998; Huang et al., 2008). The antioxidation capacity and survival after thermal and osmotic stress of tiger shrimp was enhanced while the shrimp were fed a diet with astaxanthin (Chien et al., 2003). Astaxanthin incorporation by feed (Kumar et al., 2009) or by injection (Angeles et al., 2009) improved immune response and disease resistance of giant river prawn. However, Christiansen and Torrissen (1997) reported that astaxanthin is not essential for fertilization and egg survival in Atlantic salmon. Choubert et al. (1998) found that frequency of maturing females or the date of maturation for rainbow trout was not improved by supplementing astaxanthin in diet.

More information on the essentiality of astaxanthin in diets for fish and shrimp, not only qualitatively but quantitatively (i.e., dose–response relationship) is needed. Evidence of deficiency symptoms associated with absence of the compound is also needed.

REFERENCES

Abdelghany, A. E. 1996. Growth response of Nile tilapia *Oreochromis niloticus* to dietary L-ascorbic acid, L-ascorbyl-2-sulfate, and L-ascorbyl-2-polyphosphate. J. World Aquacult. Soc. 27:449–455.

Adron, J. W., D. Knox, C. B. Cowey, and G. T. Ball. 1978. Studies on the nutrition of marine flatfish. The pyridoxine requirement of turbot (*Scophthalmus maximus*). Br. J. Nutr. 40:261–268.

Agrawal, N. K., and C. L. Mahajan. 1980. Nutritional deficiency disease in an Indian major carp, *Cirrhina mrigala* Hamilton, due to avitaminosis C during early growth. J. Fish Dis. 3:231–248.

Aguirre, P., and D. M. Gatlin, III. 1999. Dietary vitamin C requirement of red drum *Sciaenops ocellatus*. Aquacult. Nutr. 5:247–249.

Ai, Q., K. Mai, C. Zhang, W. Xu, Q. Duan, B. Tan, and Z. Liufu. 2004. Effects of dietary vitamin C on growth and immune response of Japanese seabass, *Lateolabrax japonicus*. Aquaculture 242:489–500.

Ai, Q., K. Mai, B. Tan, W. Xu, W. Zhang, H. Ma, and Z. Liufu. 2006. Effects of dietary vitamin C on survival, growth, and immunity of large yellow croaker, *Pseudosciaena crocea*. Aquaculture 261:327–336.

Albrektsen, S., K. Sandnes, J. Glette, and R. Waagbø. 1995. Influence of dietary vitamin B_6 on tissue vitamin B_6 contents and immunity in Atlantic salmon, *Salmo salar* L. Aquac. Res. 26:331–339.

Alexis, M. N., K. K. Karanikolas, and R. H. Richards. 1997. Pathological findings owing to the lack of ascorbic acid in cultured gilthead seabream (*Sparus aurata* L.). Aquaculture 151:209–218.

Al-Amoudi, M. M., A. M. N. El-Nakkadi, and B. M. El-Nouman. 1992. Evaluation of optimum dietary requirement of vitamin C for the growth of *Oreochromis spilurus* fingerlings in water from the Red Sea. Aquaculture 105:165–173.

Amezaga, M. R., and D. Knox. 1990. Riboflavin requirements in on-growing rainbow trout, *Oncorhynchus mykiss*. Aquaculture 88:87–98.

Anbarasu, K., and M. R. Chandran. 2001. Effect of ascorbic acid on the immune response of the catfish, *Mystus gulio* (Hamilton), to different bacterins of *Aeromonas hydrophila*. Fish Shellfish Immun. 11:347–355.

Andrews, J. W., and T. Murai. 1975. Studies on the vitamin C requirements of channel catfish (*Ictalurus punctatus*). J. Nutr. 105:557–561.

Andrews, J. W., and T. Murai. 1978. Dietary niacin requirements of channel catfish. J. Nutr. 108:1508–1511.

Andrews, J. W., and T. Murai. 1979. Pyridoxine requirements of channel catfish. J. Nutr. 109:533–537.

Andrews, J. W., T. Murai, and J. W. Page. 1980. Effects of dietary cholecalciferol and ergocalciferol on catfish. Aquaculture 19:49–54.

Angeles, I. P., Jr., Y. H. Chien, and M. M. Tayamen. 2009. Effects of different dosages of astaxanthin on giant freshwater prawn *Macrobrachium rosenbergii* (De Man) challenged with *Lactococcus garvieae*. Aquac. Res. 41:70–77.

Aoe, H., and I. Masuda. 1967. Water-soluble vitamin requirements of carp. 2. Requirements for *p*-aminobenzoic acid and inositol. B. Jpn. Soc. Sci. Fish. 33:674–680.

Aoe, H., I. Masuda, T. Saito, and A. Komo. 1967a. Water-soluble vitamin requirements of carp. 1. Requirement for vitamin B2. B. Jpn. Soc. Sci. Fish. 33:355–360.

Aoe, H., I. Masuda, and T. Takada. 1967b. Water-soluble vitamin requirements of carp. 3. Requirement for niacin. B. Jpn. Soc. Sci. Fish. 33:681–685.

Aoe, H., I. Masuda, T. Saito, and T. Takada. 1967c. Water-soluble vitamin requirements of carp. 5. Requirement for folic acid. B. Jpn. Soc. Sci. Fish. 33:1068–1071.

Aoe, H., I. Masuda, T. Mimura, T. Saito, and A. Komo. 1968. Requirement of young carp for vitamin A. B. Jpn. Soc. Sci. Fish. 34:959–964.

Aoe, H., I. Masuda, T. Mimura, T. Saito, A. Komo, and S. Kitamura. 1969. Water-soluble vitamin requirements for carp. 6. Requirements for thiamin and effects of antithiamins. B. Jpn. Soc. Sci. Fish. 35:459–465.

Arai, S., T. Nose, and Y. Hashimoto. 1972. Qualitative requirements of young eels, *Anguilla japonica*, for water-soluble vitamins and their deficiency symptoms. B. Freshwater Res. Lab. Tokyo 22:69–83.

Bai, S. C. 2001. Requirements of L-ascorbic acid in a viviparous marine teleost, Korean rockfish, *Sebastes schlegeli* (Hulgendorf). Pp. 69–85 in Ascorbic Acid in Aquatic Organisms, K. Dabrowski, ed. Boca Raton, FL: CRC Press.

Bai, S. C., and K. J. Lee. 1998. Different levels of dietary DL-α-tocopheryl acetate affect the vitamin E status of juvenile Korean rockfish, *Sebastes schlegeli*. Aquaculture 161:405–414.

Barnett, B. J., C. Y. Cho, and S. J. Slinger. 1982a. Relative biopotency of ergocalciferol and cholecalciferol and the role of and requirement for vitamin D in rainbow trout (*Salmo gairdneri*). J. Nutr. 112:2011–2019.

Barnett, B. J., G. Jones, C. Y. Cho, and S. J. Slinger. 1982b. The biological activity of 25-hydroxycholecalciferol and 1,25-dehydroxycholecalciferol for rainbow trout (*Salmo gairdneri*). J. Nutr. 112:2020–2026.

Bell, J. G., C. B. Cowey, J. W. Adron, and A. M. Shanks. 1985. Some effects of vitamin E and selenium deprivation on tissue enzyme levels and indices of tissue peroxidation in rainbow trout (*Salmo gairdneri*). Br. J. Nutr. 53:149–157.

Bendich, A. 1993. Biological functions of dietary carotenoids. Ann. NY Acad. Sci. 691:61–67.

Blazer, V. S., and R. E. Wolke. 1984a. The effects of α-tocopherol on the immune response and non-specific resistance factors of rainbow trout (*Salmo gairdneri* Richardson). Aquaculture 37:1–9.

Blazer, V. S., and R. E. Wolke. 1984b. Effect of diet on the immune response of rainbow trout (*Salmo gairdneri*). Can. J. Fish. Aquat. Sci. 41:1244–1247.

Blom, J. H., K. Dabrowski, and J. Ebeling. 2000. Vitamin C requirements of the angelfish *Pterophylum scalare*. J. World Aquacult. Soc. 31:115–118.

Boonyaratpalin, M. 1998. Nutrition of *Penaeus merguiensis* and *Penaeus indicus*. Rev. Fish. Sci. 6:69–78.

Boonyaratpalin, M., and N. Unprasert. 1989. Effect of pigments from different sources on colour changes and growth of red *Oreochromis niloticus*. Aquaculture 79:375–380.

Boonyaratpalin, M., and W. Phromkunthong. 2001. Bioavailability of ascorbyl phosphate calcium in hybrid catfish, *Clarias macrocephalus* (Gunther) × *Clarias gariepinus* (Burchell) feed. Aquac. Res. 32(Suppl. 1):126–134.

Braekkan, O. R., O. Ingebrigoten, and H. Myklestead. 1969. Interconversion of vitamins A and A2 in fishes. Int. Z. Vitaminforech. 39:123–130.

Brandt, T. M., C. W. Doyoe, and P. A. Seib. 1985. Alternate sources of vitamin C for channel catfish. Prog. Fish Cult. 47:55–59.

Brown, P. B. 1988. Vitamin D Requirement of Juvenile Channel Catfish Reared in Calcium-Free Water. Ph.D. Dissertation, Texas A&M University, College Station, Texas.

Brunson, M. W., H. R. Robinette, P. R. Bowser, and T. L. Wellborn. 1983. Nutritional gill disease associated with starter feeds for channel catfish fry. Prog. Fish Cult. 45:119–120.

Burtle, G. J., and R. T. Lovell. 1989. Lack of response of channel catfish (*Ictalurus punctatus*) to dietary myo-inositol. Can. J. Fish. Aquat. Sci. 46:218–222.

Castledine, A. J., C. Y. Cho, S. J. Slinger, B. Hicks, and H. S. Bayley. 1978. Influence of dietary biotin level on growth, metabolism and pathology of rainbow trout. J. Nutr. 108:698–711.

Catacutan, M. R., and C. R. Lavilla-Pitogo. 1994. L-ascorbyl-2-phosphate Mg as a source of vitamin C for juvenile *Penaeus monodon*. Isr. J. Aquacult. Bamidgeh 46:40–47.

Cavalli, R. O., F. M. M. Batista, P. Lavens, P. Sorgeloos, H. J. Nelis, and A. P. De Leenheer. 2003. Effect of dietary supplementation of vitamin C and E on maternal performance and larval quality of the prawn *Macrobrachium rosenbergii*. Aquaculture 227:131–146.

Cerezuela, R., A. Cuestaa, J. Meseguera, and M. Á. Esteban. 2009. Effects of dietary vitamin D_3 administration on innate immune parameters of seabream (*Sparus aurata* L.). Fish Shellfish Immun. 26:243–248.

Chan, A. C. 1993. Partners in defense, vitamin E and vitamin C. Can. J. Physiol. Pharm. 71:725–731.

Chavan, S. L., H. S. Dhaker, and S. K. Barve. 2003. Influence of dietary biotin level on growth and survival of fry of *Liza parsia*. Appl. Fish. Aquac. 3:42–44

Chavez de Martinez, M. C. 1990. Vitamin C requirement of the Mexican native cichlid (*Cichlasoma urophthalmus*, Gunther). Aquaculture 86:409–416.

Chen, H. Y., and C. F. Chang. 1994. Quantification of vitamin C requirements for juvenile shrimp (*Penaeus monodon*) using polyphosphorylated L-ascorbic acid. J. Nutr. 124:2033–2038.

Chen, H. Y., and G. Hwang. 1992. Estimation of the dietary riboflavin required to maximize tissue riboflavin concentration in juvenile shrimp (*Penaeus monodon*). J. Nutr. 122:2474–2478.

Chen, H. Y., F. C. Wu, and S. Y. Tang. 1991. Thiamin requirement of juvenile shrimp (*Penaeus monodon*). J. Nutr. 121:1984–1989.

Chen, H. Y., F. C. Wu, and S. Y. Tang. 1994. Sensitivity of transketolase to the thiamin status of juvenile marine shrimp (*Penaeus monodon*). Comp. Biochem. Phys. A 109:655–659.

Chen, R., R. Lochmann, A. Goodwin, K. Praveen, K. Dabrowski, and K. J. Lee. 2004. Effects of dietary vitamin C and E on alternative complement activity, hematology, tissue composition, vitamin concentrations and response to heat stress in juvenile golden shiner (*Notemigonus crysoleucas*). Aquaculture 242:553–569.

Chen, S., and A. Li. 1994. Investigation on nutrition of vitamin A for shrimp *Penaeus chinensis*: 1. Effects of vitamin A on shrimp's growth and visual organ. Acta. Zool. Sin. Dongwu. Xuebao. 40:266–273.

Chien, Y. H., and S. C. Jeng. 1992. Pigmentation of kuruma shrimp, *Penaeus japonicus* Bate, by various pigment sources and levels and feeding regimes. Aquaculture 102:333–346.

Chien, Y. H., C. H. Pan, and B. Hunter. 2003. The resistance to physical stresses by *Penaeus monodon* juveniles fed diets supplemented with astaxanthin. Aquaculture 216:177–191.

Cho, C. Y., and B. Woodward. 1990. Dietary pantothenic acid requirements of young rainbow trout (*Oncorhynchus mykiss*) (abstract). FASEB J. 4:3747.

Choubert, G., J. M. Blanc, and H. Poisson. 1998. Effects of dietary keto-carotenoids (canthaxanthin and astaxanthin) on the reproductive performance of female rainbow trout *Oncorhynchus mykiss* (Walbaum). Aquacult. Nutr. 4:249–254.

Christiansen, R., and O. J. Torrissen. 1997. Effects of dietary astaxanthin supplementation on fertilization and egg survival in Atlantic salmon (*Salmo salar* L.). Aquaculture 153:51–62.

Christiansen, R., Ø. Lie, and O. J. Torrissen. 1994. Effect of astaxanthin and vitamin A on growth and survival during first feeding of Atlantic salmon *Salmo salar* L. Aquac. Res. 25:903–914.

Christiansen, R., G. Struksnæs, R. Estermann, and O. J. Torrissen. 1995. Assessment of flesh colour in Atlantic salmon, *Salmo salar* L. Aquac. Res. 26:311–321.

Ciereszko, A., and K. Dabrowski. 1995. Sperm quality and AA concentration in rainbow trout semen are affected by dietary vitamin C: A cross-season study. Biol. Reprod. 52:982–988.

Clerton, P., D. Troutauda, V. Verlhacb, J. Gabaudanb, and P. Deschaux. 2001. Dietary vitamin E and rainbow trout (*Oncorhynchus mykiss*) phagocyte functions: Effect on gut and on head kidney leucocytes. Fish Shellfish Immun. 11:1–13.

Coates, J. A., and J. E. Halver. 1958. Water-soluble vitamin requirements of silver salmon. Special Scientific Report Fisheries 281. Washington, DC: Bureau of Sport Fisheries and Wildlife.

Comacho, A. S. 1978. Characterization of Thiamine Deficiency in Channel Catfish Fed Heated and Nonheated Catfish Processing Waste. Ph.D. Dissertation, Auburn University, Auburn, Alabama.

Cowey, C. B., and B. Woodward. 1993. The dietary requirement of young rainbow trout (*Oncorhynchus mykiss*) for folic acid. J. Nutr. 123:1594–1600.

Cowey, C. B, J. W. Adron, and D. Knox. 1975. Studies on the nutrition of marine flatfish. The thiamin requirement of turbot, *Scophthalmus maximus*. Brit. J. Nutr. 34:383–390.

Cowey, C. B., J. W. Adron, M. J. Walton, J. Murray, A. Youngson, and D. Knox. 1981. Tissue distribution, uptake and requirement for a α-tocopherol of rainbow trout (*Salmo gairdneri*) fed diets with a minimal content of unsaturated fatty acids. J. Nutr. 111:1556–1567.

Cowey, C. B., J. W. Adron, and A. Youngson. 1983. The vitamin E requirement of rainbow trout (*Salmo gairdneri*) given diets containing polyunsaturated fatty acids derived from fish oil. Aquaculture 30:85–93.

Craig, S. R., and D. M. Gatlin, III. 1996. Dietary choline requirement of juvenile red drum (*Sciaenops ocellatus*). J. Nutr. 126:1696–1700.

Cuesta, A., M. A. Esteban, J. Ortuño, and J. Meseguer. 2001. Vitamin E increases natural cytotoxic activity in seabream (*Sparus aurata* L.). Fish Shellfish Immun. 11:293–302.

Cuesta, A., M. A. Esteban, and J. Meseguer. 2002. Natural cytotoxic activity in seabream (*Sparus aurata* L.) and its modulation by vitamin C. Fish Shellfish Immun. 13:97–109.

D'Abramo, L. R., and N. A. Baum. 1981. Choline requirement of the microcrustacean *Moina macrocopa*: A purified diet for continuous culture. Biol. Bull. 161:357–365.

D'Abramo, L. R., C. A. Moncreiff, F. P. Holcomb, J. L. Montanez, and R. K. Buddington. 1994. Vitamin C requirement of the juvenile freshwater prawn, *Macrobrachium rosenbergii*. Aquaculture 128:269–275.

Dabrowski, K. 1990. Ascorbic acid status in the early life of whitefish (*Coregonus lavaretus* L.). Aquaculture 84:61–70.

Dabrowski, K., and A. Ciereszko. 1996. AA protects against male infertility in a teleost fish. Experientia 52:97–100.

Dabrowski, K., and G. Kock. 1989. Absorption and interaction with minerals of ascorbic acid and ascorbic sulfate in digestive tract of rainbow trout. Can. J. Fish. Aquat. Sci. 46:1952–1957.

Dabrowski, K., S. Hinterleitner, C. Sturmbauer, N. El-Fiky, and W. Wieser. 1988. Do carp larvae require vitamin C? Aquaculture 72:295–306.

Dabrowski, K., H. Segner, R. Dallinger, S. Hinterleitner, C. Sturmbauer, and W. Wieser. 1989. Rearing of roach larvae; the vitamin C, minerals interrelationship and nutrition-related histology of the liver and intestine. J. Anim. Physiol. An. N. 62:188–202.

Dabrowski, K., D. Lackner, and C. Doblander. 1990. Utilization of ascorbic sulfate in rainbow trout. Effect of dietary ascorbate on the tissue ascorbic acid, dehydroascorbic acid, ascorbic sulfate and ascorbic sulfate sulfahydrolase. Can. J. Fish. Aquat. Sci. 47:1518–1525.

Deng, D. F., and R. P. Wilson. 2003. Dietary riboflavin requirement of juvenile sunshine bass (*Morone chrysops* ♀ × *Morone saxatilis* ♂). Aquaculture 218:695–701.

Deng, D. F., G. I. Hemre, and R. P. Wilson. 2002. Juvenile sunshine bass (*Morone chrysops* ♀ × *Morone saxatilis* ♂) do not require dietary *myo*-inositol. Aquaculture 213:387–393.

Deshimaru, O., and K. Kuroki. 1976. Studies on a purified diet for prawn. VII. Adequate dietary levels of ascorbic acid and inositol. B. Jpn. Soc. Soc. Sci. Fish. 42:571–576.

Deshimaru, O., and K. Kuroki. 1979. Requirement of prawn for dietary thiamin, pyridoxine, and choline chloride. B. Jpn. Soc. Sci. Fish. 45:363–367.

Desjardins, L. M., J. D. Castell, and J. C. Kean. 1985. Synthesis of dehydroascorbic acid by subadult lobsters (*Homarus americanus*). Can. J. Fish. Aquat. Sci. 42:370–373.

Doolan, B. J., M. A. Booth, G. L. Allan, and P. L. Jones. 2009. Effects of dietary astaxanthin concentration and feeding period on the skin pigmentation of Australian snapper *Pagrus auratus* (Bloch & Schneider, 1801). Aquac. Res. 40:60–68.

Draper, H. H., and A. S. Csallany. 1969. A simplified hemolysis test for vitamin E deficiency. J. Nutr. 98:390–394.

Duncan, P. L., and R. T. Lovell. 1991. Effect of folic acid on growth, survival and hematology in channel catfish (*Ictalurus punctatus*). Twenty-second Annual Conference of the World Aquaculture Society, June 16–20, 1991, San Juan, Puerto Rico.

Duncan, P. L., R. T. Lovell, C. E. Butterworth, L. F. Freeberg, and T. Tamura. 1993. Dietary folate requirement determined for channel catfish, *Ictalurus punctatus*. J. Nutr. 123:1888–1897.

Dupree, H. K. 1966. Vitamins essential for growth of channel catfish, *Ictalurus punctatus*. Technical Paper No. 7. Washington, DC: U.S. Bureau of Sport Fisheries and Wildlife.

Dupree, H. K. 1968. Vitamin E requirement of channel catfish. Pp. 220–222 in Progress in Sport Fishery Research, 1967. Resource Publication No. 77. Washington, DC: U.S. Bureau of Sport Fisheries and Wildlife.

Dupree, H. K. 1970. Dietary requirement of vitamin A acetate and beta carotene. Pp. 148–150 in Progress in Sport Fishery Research, 1969. Resource Publication No. 88. Washington, DC: U.S. Bureau of Sport Fisheries and Wildlife.

Durve, V. S., and R. T. Lovell. 1982. Vitamin C and disease resistance in channel catfish. Can. J. Fish. Aquat. Sci. 39:948–951.

Eckhert, C. D., and A. R. Kemmerer. 1974. Effect of vitamin A and methionine on growth and amino acid composition of body protein in rainbow trout fingerlings. J. Nutr. 104:865–870.

El Naggar, G. O., and R. T. Lovell. 1991. L-Ascorbyl-2-monophosphate has equal antiscorbutic activity as L-ascorbic acid but L-ascorbyl-2-sulfate is inferior to L-ascorbic acid for channel catfish. J. Nutr. 121:1622–1626.

Eo, J., and K. J. Lee. 2008. Effect of dietary ascorbic acid on growth and non-specific immune responses of tiger puffer, *Takifugu rubripes*. Fish Shellfish Immun. 25:611–616.

Eya, J. C. 1996. "Broken-skull disease" in African catfish *Clarias gariepinus* is related to a dietary deficiency of ascorbic acid. J. World Aquacult. Soc. 27:493–498.

Feng, L., W. He, J. Jiang, Y. Liu, and X. Q. Zhou. 2009. Effects of dietary pyridoxine on disease resistance, immune responses and intestinal microflora in juvenile Jian carp (*Cyprinus carpio* var. Jian). Aquacult. Nutr. doi:10.1111/j.1365-2095.2009.00660.x.

Fernández-Palacios, H., M. S. Izquierdo, and L. Robaina. 2005. Efecto de distintas dietas para reproductores de dorada (*Sparus aurata*) sobre la calidad de sus puestas. Informes Técnicos del Instituto Canario de Ciencias marinas no. 12.

Fournier, V., M. F. Gouillou-Coustans, and S. J. Kaushik. 2000. Hepatic ascorbic acid saturation is the most stringent response criterion for determining the vitamin C requirement of juvenile European sea bass (*Dicentrarchus labrax*). J. Nutr. 130:617–620.

Furuita, H., H. Tanaka, T. Yamamoto, N. Suzuki, and T. Takeuchi. 2003. Supplemental effect of vitamin A in diet on the reproductive performance and egg quality of the Japanese flounder *Paralichthys olivaceus* (T & S). Aquac. Res. 34:461–468.

Furuita, H., T. Ishida, T. Suzuki, T. Unuma, T. Kurokawa, T. Sugita, and T. Yamamoto. 2008. Vitamin content and quality of eggs produced by broodstock injected with vitamins C and E during artificial maturation in Japanese eel *Anguilla japonica*. Aquaculture 289:334–339.

Gabaudan, J., and R. W. Hardy. 2000. Vitamin sources for fish feeds. Pp. 961–965 in Encyclopedia of Aquaculture, R. R. Stickney, ed. New York: John Wiley & Sons, Inc.

Gabaudan, J., and V. Verlhac. 2001. Critical review of the requirements of ascorbic acid in cold and cool water fishes (salmonids, percids, plecoglossids, and flatfishes). Pp. 33–48 in Ascorbic Acid in Aquatic Organisms, K. Dabrowski, ed. Boca Raton, FL: CRC Press.

Gatlin, D. M., III, W. E. Poe, and R. P. Wilson. 1986a. Effects of singular and combined dietary deficiencies of selenium and vitamin E on fingerling channel catfish (*Ictalurus punctatus*). J. Nutr. 116:1061–1067.

Gatlin, D. M., W. E. Poe, R. P. Wilson, A. J. Ainsworth, and P. R. Bowser. 1986b. Effects of stocking density and vitamin C status on vitamin E-adequate and vitamin E-deficient fingerling channel catfish. Aquaculture 56:187–195.

George, J. C., B. J. Barnett, C. Y. Cho, and S. J. Slinger. 1981. Vitamin D_3 and muscle function in the rainbow trout. Cytobios 31:7–18.

Gong, H., A. L. Lawrence, D. H. Jiang and D. M. Gatlin, III. 2003. Effect of dietary phospholipids on the choline requirement of *Litopenaeus vannamei* juveniles. J. World Aquacult. Soc. 34:289–299.

Goswami, U. C. 1984. Metabolism of cryptoxanthin in freshwater fish. Brit. J. Nutr. 25:575–581.

Gouillou-Coustans, M. F., P. Bergot, and S. J. Kaushik. 1998. Dietary ascorbic acid needs of common carp *Cyprinus carpio* larvae. Aquaculture 161:453–461.

Graff, I. E., S. Høie, G. K. Totland and Ø. Lie. 2002. Three different levels of dietary vitamin D_3 fed to first-feeding fry of Atlantic salmon (*Salmo salar* L.): Effect on growth, mortality, calcium content and bone formation. Aquacult. Nutr. 8:103–111.

Grahl-Madsen, E., and Ø. Lie. 1997. Effects of different levels of vitamin K in diets for cod (*Gadus morhua*). Aquaculture 151:269–274.

Grant, B. F., P. A. Seib, M. L. Liao, and J. A. Corpron. 1989. Polyphosphorylated L-ascorbic acid: A stable form of vitamin C for aquaculture feeds. J. World Aquacult. Soc. 20:143–157.

Griffin, M. E., K. A. Wilson, M. R. White, and P. B. Brown. 1994. Dietary choline requirement of juvenile hybrid striped bass. J. Nutr. 124:1685–1689.

Guary, M., A. Kanazawa, N. Tanaka, and H. J. Ceccaldi. 1976. Nutritional requirements of prawn: VI. Requirement for ascorbic acid. Mem. Fac. Fish. Kagoshima Univ. 25:53–57.

Halver, J. E. 1957. Nutrition of salmonids fishes. 4. Water-soluble vitamin requirements of Chinook salmon. J. Nutr. 62:225–243.

Halver, J. E. 1972. The vitamins. Pp. 29–103 in Fish Nutrition, J. E. Halver, ed. New York: Academic Press.

Halver, J. E., L. M. Ashley, and R. R. Smith. 1969. Ascorbic acid requirements of coho salmon and rainbow trout. T. Am. Fish. Soc. 90:762–771.

Halver, J. E., R. R. Smith, B. M. Tolbert, and E. M. Baker. 1975. Utilization of ascorbic acid in fish. Ann. NY Acad. Sci. 258:81–102.

Hamre, K., and Ø. Lie. 1995. α-Tocopherol levels in different organs of Atlantic salmon (*Salmo salar* L.): Effects of smoltification, dietary levels of n-3 polyunsaturated fatty acid and vitamin E. Comp. Biochem. Phys. A 111A:547–554.

Hardie, L. J., T. C. Fletcher, and C. J. Secombes. 1990. The effect of vitamin E on the immune response of the Atlantic salmon (*Salmo salar* L.). Aquaculture 87:1–13.

Hardy, R. W., J. E. Halver, and E. L. Brannon. 1979. Effect of dietary protein level on the pyridoxine requirement and disease resistance of Chinook salmon. Pp. 253–260 in Finfish Nutrition and Fish Feed Technology, Volume 1, J. E. Halver and K. Tiews, eds. Berlin, Germany: Heeneman GmbH.

Hari, B., and B. M. Kurup. 2002. Vitamin C (Ascorbyl 2 polyphosphate) requirement of freshwater prawn *Macrobrachium rosenbergii* (de Man). Asian Fish. Sci. 15:145–154.

Harlıo˘glu, M. M., and O. Barım. 2004. The effect of dietary vitamin E on the pleopodal egg and stage-1 juvenile numbers of freshwater crayfish *Astacus leptodactylus* (Eschscholtz, 1823). Aquaculture 236:267–276.

Hashimoto, Y., S. Arai, and T. Nose. 1970. Thiamin deficiency symptoms experimentally induced in the eel. B. Jpn. Soc. Sci. Fish. 36:791–797.

He, H., and A. L. Lawrence. 1991. Estimation of dietary pyridoxine requirement for the shrimp, *Penaeus vannamei* (Abstr.). Paper presented at the 22nd Annual conference, World Aquaculture Society, June 16–20, 1991, San Juan, Puerto Rico.

He, H., and A. L. Lawrence. 1993a. Vitamin E requirements of *Penaeus vannamei*. Aquaculture 118:245–255.

He, H., and A. L. Lawrence. 1993b. Vitamin C requirements of the shrimp *Penaeus vannamei*. Aquaculture 114:305–316.

He, H., A. L. Lawrence, and R. Liu. 1992. Evaluation of dietary essentiality of fat-soluble vitamins, A, D, E and K, for penaeid shrimp (*Penaeus vannamei*). Aquaculture 103:177–185.

He, W., X. Q. Zhou, L. Feng, J. Jiang, and Y. Liu. 2009. Dietary pyridoxine requirement of juvenile Jian carp (*Cyprinus carpio* var. Jian). Aquacult. Nutr. 15:402–408.

Heikkila, R. E., and L. Manzino. 1987. Ascorbic acid, redox cycling, lipid peroxidation, and the binding of dopamine receptor antagonists. Ann. NY Acad. Sci. 498:63–76.

Hemre, G. I., A. Mangor-Jensen, and Ø. Lie. 1994. Broodstock nutrition in turbot (*Scophthalmus maximus*) effect of dietary vitamin E. Fiskeridir. Skr., Ser. Ernær., 8:21–29.

Hemre, G. I., D. F. Deng, R. P. Wilson, and M. H. G. Berntssen. 2004. Vitamin A metabolism and early biology responses in juvenile sunshine bass (*Morone chrysops* × *M. saxatilis*) fed graded levels of vitamin A. Aquaculture 235:645–658.

Henrique, M. M. F., E. F. Gomes, M. F. Gouillou-Coustans, A. Oliva-Teles, and S. J. Davies. 1998. Influence of supplementation of practical diets with vitamin C on growth and response to hypoxic stress of seabream, *Sparus aurata*. Aquaculture 161:415–426.

Hepher, B. 1988. Vitamins. Pp. 224–234 in Nutrition of Pond Fishes. New York: Cambridge University Press.

Hernandez, L. H. H., S. I. Teshima, M. Ishikawa, S. Alam, S. Koshio, and Y. Tanaka. 2005. Dietary vitamin A requirements of juvenile Japanese flounder *Paralichthys olivaceus*. Aquacult. Nutr. 11:3–9.

Hernandez, L. H. H., S. I. Teshima, S. Koshio, M. Ishikawa, Y. Tanaka, and S. Alam. 2007. Effects of vitamin A on growth, serum anti-bacterial activity and tansaminase activities in the juvenile Japanese flounder, *Paralichthys olivaceus*. Aquaculture 262:444–450.

Hilton, J. W., and H. W. Ferguson. 1982. Effect of excess vitamin D_3 on calcium metabolism in rainbow trout *Salmo gairdneri* Richardson. J. Fish. Biol. 21:373–379.

Hilton, J. W., C. Y. Cho, and S. J. Slinger. 1977. Factors affecting the stability of supplemental ascorbic acid in practical trout diets. J. Fish. Res. Board Can. 34:683–687.

Hilton, J. W., C. Y. Cho, and S. J. Slinger. 1978. Effect of graded levels of supplemental ascorbic acid in practical diets fed to rainbow trout (*Salmo gairdneri*). J. Fish. Res. Board Can. 35:431–436.

Holub, B. J., B. Bregeron, and T. Woodward. 1982. The effect of inositol deficiency on the hepatic neutral lipid and phospholipid composition of rainbow trout. J. Nutr. 112(6):xxi.

Hosokawa, H. 1989. The Vitamin Requirements of Fingerling Yellowtail, *Seriola quinqueradiata*. Ph.D. Dissertation, Kochi University, Japan.

Hsiao, S. M., and W. C. Mak. 1978. Artificial fertilization and incubation of fertilized eggs pond reared ayu. China Fish. Mon. 305:2–11.

Hsu, T. S., and S. Y. Shiau. 1997. Comparison of L-ascorbyl-2-polyphosphate with L-ascorbyl-2-sulfate in meeting vitamin C requirements of juvenile grass shrimp, *Penaeus monodon*. Fisheries Sci. 63:958–962.

Hsu, T. S., and S. Y. Shiau. 1998. Comparison of vitamin C requirement for maximum growth of grass shrimp, *Penaeus monodon*, using L-ascorbyl-2-monophosphate-Na and L-ascorbyl-2-monophosphate-Mg. Aquaculture 163:203–213.

Hsu, T. S., and S. Y. Shiau. 1999a. DL-α-tocopheryl acetate is not the only tissue storage form of vitamin E in tilapia fed dietary DL-α-tocopheryl acetate. Fisheries Sci. 65:959–960.

Hsu, T. S., and S. Y. Shiau. 1999b. Tissue storage of vitamin E in grass shrimp *Penaeus monodon* fed dietary DL-α-tocopheryl acetate. Fisheries Sci. 65:169–170.

Hsu, T. S., and S. Y. Shiau. 1999c. Influence of dietary ascorbate derivatives on tissue copper, iron and zinc concentrations in grass shrimp, *Penaeus monodon*. Aquaculture 179:457–464.

Hu, C. J., S. M. Chen, C. H. Pan, and C. H. Huang. 2006. Effects of dietary vitamin A or β-carotene concentrations on growth of juvenile hybrid tilapia, *Oreochromis niloticus* × *O. aureus*. Aquaculture 253:602–607.

Huang, J., L. Tian, X. Wu, H. Yang, and Y. Liu. 2010. Effects of dietary riboflavin levels on antioxidant defense of the juvenile grouper *Epinephelus coioides*. Fish Physiol. Biochem. 36:55–62.

Huang, J. H., S. G. Jiang, H. Z. Lin, F. L. Zhou, and L. Ye. 2008. Effects of dietary highly unsaturated fatty acids and astaxanthin on the fecundity and lipid content of pond-reared *Penaeus monodon* (Fabricius) broodstock. Aquac. Res. 39:240–251.

Huang, J. W., L. X. Tian, Z. Y. Du, H. J. Yang, and Y. J. Liu. 2007. Effects of dietary thiamin on the physiological status of the grouper *Epinephelus coioides*. Fish Physiol. Biochem. 33:167–172.

Hughes, S. G. 1984. Effect of excess dietary riboflavin on growth of rainbow trout. J. Nutr. 114:1660–1663.

Hughes, S. G., G. L. Rumsey, and J. G. Nichum. 1981a. Riboflavin requirement of fingerling rainbow trout. Prog. Fish Cult. 43:167–172.

Hughes, S. G., R. C. Riis, J. G. Nichum, and G. L. Rumsey. 1981b. Biomicroscopic and histologic pathology of the eye in riboflavin deficient rainbow trout (*Salmo gairdneri*). Cornell Vet. 71:269–279.

Hung, S. S. O. 1989. Choline requirement of hatchery-produced juvenile white sturgeon (*Acipenser transmontanus*). Aquaculture 78:183–194.

Hung, S. S. O., C. Y. Cho, and S. J. Slinger. 1980. Measurement of oxidation in fish oil and its effect on vitamin E nutrition of rainbow trout (*Salmo gairdneri*). Can. J. Fish. Aquat. Sci. 37:1248–1253.

Hung, S. S. O., C. Y. Cho, and S. J. Slinger. 1981. Effect of oxidized fish oil, DL-α-tocopheryl acetate and ethoxyquin supplementation on the vitamin E nutrition of rainbow trout (*Salmo gairdneri*) fed practical diets. J. Nutr. 111:648–657.

Ibiyo, L. M. O., J. O. Atteh, J. S. Omotosho, and C. T. Madu. 2007. Vitamin C (ascorbic acid) requirements of *Heterobranchus longifilis* fingerlings. Afr. J. Biotechnol. 6:1559–1567.

Ikeda, S., Y. Ishibashi, O. Murata, T. Nasu, and T. Harada. 1988. Qualitative requirements of the Japanese parrotfish for water-soluble vitamins. B. Jpn. Soc. Sci. Fish. 54:2029–2035.

Ishibashi, Y., S. Ikeda, O. Murata, T. Nasu, and T. Harada. 1992. Optimal supplementary ascorbic acid level in the Japanese parrot fish diet. Nippon Suisan Gakk. 58:267–270.

Jiang, M., W. Wang, H. Wen, F. Wu, Z. Zhao, A. Liu, and W. Liu. 2007. Effects of dietary vitamin K3 on growth, carcass composition and blood coagulation time for grass carp fingerling (*Ctenopharyngodon idellus*) [in Chinese]. Freshwater Fish. 37:61–64.

John, M. J., and C. L. Mahajan. 1979. The physiological response of fishes to deficiency of cyanocobalamin and folic acid. J. Fish Biol. 14:127–133.

John, T. M., J. C. Gorge, J. W. Hilton, and S. J. Slinger. 1979. Influence of dietary ascorbic acid on plasma lipid levels in the rainbow trout. Int. J. Vitam. Nutr. Res. 49:400–405.

Jurss, K. 1978. Effect of pyridoxine deficiency and starvation on activities of aminotransferase in liver and muscle of rainbow trout (*Salmo gairdneri*). Zool. Jahrb. Allg. Zool. 82:141–149.

Jurss, K., and L. Jonas. 1981. Electron microscopic and biochemical investigations on the pyridoxine deficiency of the rainbow trout (*Salmo gairdneri* Richardson). Zool. Jahrb. Allg. Zool. 85:181–196.

Kaewsrithong, J., T. Ohshima, H. Ushio, R. Nagasaka, M. Maita, and M. Sawada. 2001. Effects of an excess dose of dietary α-tocopherol on hydroperoxide accumulation and erythrocyte osmotic fragility of sweet smelt *Plecoglossus altivelis* (Temminck et Schlegel). Aquac. Res. 32 (Suppl. 1):191–198.

Kanazawa, A. 1985. Prawn Feed. Taipei, Taiwan: American Soybean Association.

Kanazawa, A., S. Teshima, and N. Tanada. 1976. Nutritional requirements for choline and inositol. Mem. Fac. Fish. Kagoshima Univ. 28:27–33.

Kanazawa, A., S. I. Teshima, S. Koshio, M. Higashi, and S. Itoh. 1992. Effect of L-ascorbyl-2-phosphate-Mg on the yellowtail, *Seriola quinqueradiata* as a vitamin C source. Nippon Suisan Gakk. 58:337–341.

Karges, R. G., and B. Woodward. 1984. Development of lamellar epithelial hyperplasia in gills of pantothenic acid-deficient rainbow trout, *Salmo gairdneri* Richardson. J. Fish. Biol. 25:57–62.

Kashiwada, K., S. Teshima, and A. Kanazawa. 1970. Studies on the production of B vitamins by intestinal bacteria of fish. 5. Evidence of the production of vitamin B_{12} by microorganisms in the intestinal canal of carp, *Cyprinus carpio*. B. Jpn. Soc. Sci. Fish. 36:421–424.

Kashiwada, K., A. Kanazawa, and S. Teshima. 1971. Studies on the production of B vitamins by intestinal bacteria. 6. Production of folic acid by intestinal bacteria of carp. Mem. Fac. Fish. Kagoshima Univ. 20:185–189.

Katsuyama, M., and T. Matsuno. 1988. Carotenoid and vitamin A and metabolism of carotenoids, beta-carotene, canthaxanthin astaxanthin zeaxanthin lutein and tunaxanthin in tilapia *Tilapia nilotica*. Comp. Biochem. Phys. B 90B:134–139.

Ketola, H. G. 1976. Choline metabolism and nutritional requirement of lake trout (*Salvelinus namaycush*). J. Anim. Sci. 43:474–477.

Khajarern, J., and S. Khajarern. 1997. Stability and bioavailability of vitamin C-glucose in *Clarias* hybrid catfish (*Clarias gariepinus* × *Clarias macrocephalus*). Aquaculture 151:219–224.

Kiron, V., J. Puangkaew, K. Ishizaka, S. Satoh, and T. Watanabe. 2004. Antioxidant status and non-specific immune responses in rainbow trout (*Oncorhynchus mykiss*) fed two levels of vitamin E along with three lipid sources. Aquaculture 234:361–379.

Kissil, G. W., C. B. Cowey, J. W. Adron, and R. H. Richards. 1981. Pyridoxine requirements of the gilthead bream (*Sparus aurata*). Aquaculture 23:243–255.

Kitamura, S., S. Ohara, T. Suwa, and K. Nakagawa. 1965. Studies on vitamin requirements of rainbow trout, *Salmo gairdneri*. 1. On the ascorbic acid. B. Jpn. Soc. Sci. Fish. 31:818–826.

Kitamura, S., T. Suwa, S. Ohara, and K. Nakagawa. 1967a. Studies on vitamin requirements of rainbow trout. 3. Requirement for vitamin A and deficiency symptoms. B. Jpn. Soc. Sci. Fish. 33:1126–1131.

Kitamura, S., T. Suwa, S. Ohara, and K. Nakagawa. 1967b. Studies on vitamin requirements of rainbow trout. 2. The deficiency symptoms of fourteen kinds of vitamin. B. Jpn. Soc. Sci. Fish. 33:1120–1125.

Kocabas, A. M., and D. M. Gatlin, III. 1999. Dietary vitamin E requirement of hybrid striped bass (*Morone chrysops* female × *M. saxatilis* male). Aquacult. Nutr. 5:3–7.

Krossøy, C., R. Waggbø, P. G. Fjelldal, A. Wargelius, E. J. Lock, I. E. Graff, and R. Ørnsrud. 2009. Dietary menadione nicotinamide bisulphite (vitamin K_3) does not affect growth or bone health in first-feeding fry of Atlantic salmon (*Salmo salar* L.). Aquacult. Nutr. 15:638–649.

Kumar, V., B. R. Pillai, P. K. Sahoo, J. Mohanty, and S. Mohanty. 2009. Effect of dietary astaxanthin on growth and immune response of giant freshwater prawn *Macrobrachium rosenbergii* (De Man). Asian Fish. Sci. 22:61–69.

Lall, S. P., and D. E. M. Weerakoon. 1990. Vitamin B_6 requirement of Atlantic salmon (*Salmo salar*) (abstract). FASEB J. 4:3749.

Lall, S. P., G. Olivier, J. A. Hines, and H. W. Ferguson. 1988. The role of vitamin E in nutrition and immune response of Atlantic salmon (*Salmo salar*). Bull. Aqua. Assoc. Can. 88–2:76.

Lall, S. P., G. Olivier, D. E. M. Weerakoon, and J. A. Hines. 1990. The effect of vitamin C deficiency and excess on immune response in Atlantic salmon (*Salmo salar* L.). Pp. 427–441 in Proceedings of the Fish Nutrition Meeting, Toba, Japan, M. Takeda and T. Watanabe, eds. Tokyo, Japan: Japan Translation Center.

Lavens, P., G. Merchie, X. Ramos, AL-H. Kujan, A. Van Hauwaert, A. Pedrazzoli, H. Nelis, and A. De Leenheer. 1999. Supplementation of ascorbic acid 2-monophosphate during the early postlarval stages of the shrimp *Penaeus vannamei*. Aquacult. Nutr. 5:205–209.

Lee, B. J., K. J. Lee, S. J. Lim, and S. M. Lee. 2009. Dietary *myo*-inositol requirement for olive flounder, *Paralichthys olivaceus* (Temminch et Schlegel). Aquac. Res. 40:83–90.

Lee, K. J., and K. Dabrowski. 2004. Long-term effects and interactions of dietary vitamin C and E on growth and reproduction of yellow perch, *Perca flavescens*. Aquaculture 230:377–389.

Lee, K. J., K. W. Kim, and S. C. Bai. 1998. Effects of different dietary levels of L-ascorbic acid on growth and tissue vitamin C concentration in juvenile Korean rockfish, *Sebastes schlegeli* (Hilgendorf). Aquac. Res. 29:237–244.

Lee, M. H., and S. Y. Shiau. 2002. Dietary vitamin C and its derivatives affect immune responses in grass shrimp, *Penaeus monodon*. Fish Shellfish Immun. 12:119–129.

Lee, M. H., and S. Y. Shiau. 2003. Increase of dietary vitamin C improves haemocyte respiratory burst response and growth of juvenile grass shrimp, *Penaeus monodon*, fed with high dietary copper. Fish Shellfish Immun. 14:305–315.

Lee, M. H., and S. Y. Shiau. 2004. Vitamin E requirements of juvenile grass shrimp, *Penaeus monodon*, and effects on immune responses. Fish Shellfish Immun. 16:475–485.

Lee, P. H. 1987. Carotenoids in Cultured Channel Catfish. Ph.D. Dissertation, Auburn University, Auburn, Alabama.

Lehmitz, R., and L. Spannhof. 1977. Transketolase activity and thiamin deficiency in the kidney of rainbow trout, *Salmo gairdneri*, fed raw herring. Arch. Tierernahr. 27:287–295.

Leith, D., J. Holmes, and S. Kaattari. 1990. Effects of vitamin nutrition on the immune response of hatchery-reared salmonids. Final Report, Project 84–45A and 84–45B. Portland, OR: Bonneville Power Administration.

Li, J., L. Zhang, K. S. Mai, Q. H. Ai, J. Wan, C. Zhang, and J. Zhang. 2010. Estimation of dietary biotin requirement of Japanese seabass, *Lateolabrax japonicus* C. Aquacult. Nutr. 16:231–236.

Li, M. H., M. R. Johnson, and E. H. Robinson. 1993. Elevated dietary vitamin C concentrations did not improve resistance of channel catfish, *Ictalurus punctatus*, against *Edwardsiella ictaluri* infection. Aquaculture 117:303–312.

Li, M. H., D. J. Wise, and E. H. Robinson. 1998. Effect of dietary vitamin C on weight gain, tissue ascorbate concentration, stress response, disease resistance of channel catfish *Ictalurus punctatus*. J. World Aquacult. Soc. 29:1–8.

Li, P., X. Wang, and D. M. Gatlin, III. 2008. RRR-α-tocopheryl succinate is a less bioavailable source of vitamin E than all-rac-α-tocopheryl acetate for red drum, *Sciaenops ocellatus*. Aquaculture 280:165–169.

Li, W., X. Q. Zhou, L. Feng, Y. Liu, and J. Jiang. 2010. Effect of dietary riboflavin on growth, feed utilization, body composition and intestinal enzyme activities of juvenile Jian carp (*Cyprinus carpio* var. Jian). Aquacult. Nutr. 16:137–143.

Li, Y., and R. T. Lovell. 1985. Elevated levels of dietary ascorbic acid increase immune responses in channel catfish. J. Nutr. 115:123–131.

Lie, Ø., A. Sandvin, and R. Waagbø. 1994. Transport of alpha-tocopherol in Atlantic salmon (*Salmo salar*) during vitellogenesis. Fish Physiol. Biochem. 13:241–247.

Lightner, D. V., L. B. Colvin, C. Brand and D. A. Donald. 1977. Black death, a disease syndrome of penaeid shrimp related to a dietary deficiency of ascorbic acid. Proc. World Maricult. Soc. 8:611–623.

Lightner, D. V., B. Hunter, P. C. Magarelli, Jr., and L. B. Colvin. 1979. Ascorbic acid:nutritional requirement and role in wound repair in penaeid shrimp. Proc. World Maricult. Soc. 10:513–519.

Lim, C., and R. T. Lovell. 1978. Pathology of the vitamin C deficiency syndrome in channel catfish (*Ictalurus punctatus*). J. Nutr. 108:1137–1146.

Lim, C., B. R. LeaMaster, and J. A. Brock. 1993. Riboflavin requirement of fingerlings red hybrid tilapia grown in seawater. J. World Aquacult. Soc. 24:451–458.

Lim, C., B. R. LeaMaster, and J. A. Brock. 1995. Pyridoxine requirement of fingerling red hybrid tilapia growth in seawater. J. Appl. Aquacult. 5:49–60.

Limsuwan, T., and R. T. Lovell. 1981. Intestinal synthesis and absorption of vitamin B_{12} in channel catfish. J. Nutr. 111:2125–2132.

Lin, M. F., and S. Y. Shiau. 2004. Requirements of vitamin C (L-ascorbyl-2-monophosphate-Mg and L-ascorbyl-2-monophosphate-Na) and its effects on immune responses of grouper, *Epinephelus malabaricus*. Aquacult. Nutr. 10:327–333.

Lin, M. F., and S. Y. Shiau. 2005a. Dietary L-ascorbic acid affects growth, non-specific immune responses and disease resistance in juvenile grouper, *Epinephelus malabaricus*. Aquaculture 244:215–221.

Lin, M. F., and S. Y. Shiau. 2005b. Requirements of vitamin C (L-ascorbyl-2-sulfate and L-ascorbyl-2-polyphosphate) and its effects on immune responses of grouper, *Epinephelus malabaricus*. Aquacult. Nutr. 11:183–189.

Lin, Y. H., and S. Y. Shiau. 2005. Dietary vitamin E requirements of grouper, *Epinephelus malabaricus*, under two lipid levels, and their effects on immune responses. Aquaculture 248:235–244.

Lin, Y. H., and S. Y. Shiau. 2009. Mutual sparing of dietary requirements for alpha-tocopherol and selenium in grouper, *Epinephelus malabaricus*. Aquaculture 294:242–245.

Lin, Y. H., J. Y. Wu, and S. Y. Shiau. 2010. Dietary cobalt can promote gastrointestinal bacterial production of vitamin B_{12} in sufficient amounts to supply growth requirements of grouper, *Epinephelus malabaricus*. Aquaculture 302:89–93.

Liu, A. L., H. Wen, M. Jiang, Z. Y. Zhao, F. Wu, and W. Liu. 2007. Dietary pantothenic acid requirement of grass carp (*Ctenopharyngodon idella*) fingerlings. Fisheries Sci. 26:263–266.

Liu, T. B., A. J. Li, and J. M. Zhang. 1993. Studies on vitamin nutrition for the shrimp Penaeus chinensis: 10. Studies on the choline chloride and inositol requirements in the shrimp *Penaeus chinensis*. J. Ocean Univ. Qingdao 23:67–74.

Liu, T., Z. Jiameng, and A. J. Li. 1995. Studies on the optimal requirements of pantothenic acid, biotin, folic acid and vitamin B12 in the shrimp *Penaeus chinensis*. J. Fish. Sci. China 2:48–55.

Liu, Y., W. N. Wang, A. L. Wang, J. M. Wang, and R. Y. Sun. 2007. Effects of dietary vitamin E supplementation on antioxidant enzyme activities in *Litopenaeus vannamei* (Boone, 1931) exposed to acute salinity changes. Aquaculture 265:351–358.

Lovell, R. T., and J. C. Buston. 1984. Biotin supplementation of practical diets for channel catfish. J. Nutr. 114:1092–1096.

Lovell, R. T., and G. O. El Naggar. 1990. Vitamin C activity for L-ascorbic acid, L-ascorbyl-2-sulfate, and L-ascorbyl-2-phosphate Mg for channel catfish. Proceedings of the Fish Nutrition Meeting, Toba, Japan, M. Takeda and T. Watanabe, eds. Tokyo, Japan: Japan Translation Center.

Lovell, R. T., and Y. P. Li. 1978. Essentiality of vitamin D in diets of channel catfish (*Ictalurus punctatus*). T. Am. Fish. Soc. 107:809–811.

Lovell, R. T., and C. Lim. 1978. Vitamin C in pond diets for channel catfish. T. Am. Fish. Soc. 107:321–325.

Lovell, R. T., and T. Limsuwan. 1982. Intestinal synthesis and dietary nonessentiality of vitamin B_{12} for *Tilapia nilotica*. T. Am. Fish. Soc. 111:485–490.

Lovell, R. R., T. Miyazaki, and S. Rabegnator. 1984. Requirement for α-tocopherol by channel catfish fed diets low in polyunsaturated triglycerides. J. Nutr. 114:894–901.

Mæland, A., R. Waagbø, K. Sandnes, and B. Hjeltnes. 1998. Biotin in practical fish-meal based diet for Atlantic salmon *Salmo salar* L. fry. Aquacult. Nutr. 4:231–247.

Magarelli, P. C., Jr., B. Hunter, D. V. Lightner, and L. B. Colvin. 1979. Black death: An ascorbic acid deficiency disease in penaeid shrimp. Comp. Biochem. Phys. A 63A:103–108.

Mahajan, C. L., and N. K. Agrawal. 1979. Vitamin C deficiency in *Channa punctatus* Bloch. J. Fish Biol. 15:613–622.

Mahajan, C. L., and N. K. Agrawal. 1980. Nutritional requirements of ascorbic acid by Indian major carp *Cirrhina mrigala* during early growth. Aquaculture 19:37–48.

Mai, K., L. Xiao, Q. Ai, X. Wang, W. Xu, and W. Zhang. 2009. Dietary choline requirement for juvenile cobia, *Rachycentron canadum*. Aquaculture 289:124–128.

Maranesi, M., M. Marchetti, D. Bochicchio, and L. Cabrini. 2005. Vitamin B_6 supplementation increases the docosahexaenoic acid concentration of muscle lipids of rainbow trout (*Oncorhynchus mykiss*). Aquac. Res. 36:431–438.

Margolis, L. 1953. The effect of fasting on the bacterial flora of the intestine of fish. J. Fish. Res. Board Can. 10:62–63.

Martins, M. L. 1995. Effect of ascorbic acid deficiency on the growth, gill filament lesions and behavior of pacu fry (*Piaractus mesopotamicus* Holmberg, 1987). Braz. J. Med. Biol. Res. 28:563–568.

Masumoto, T., R. W. Hardy, and E. Casillas. 1987. Comparison of transketolase activity and thiamin pyrophosphate levels in erythrocytes and liver of rainbow trout (*Salmo gairdneri*) as indicators of thiamin status. J. Nutr. 117:1422–1426.

Mathews, C. K., and K. E. van Holde. 1990. Biochemistry. Redwood City, CA: Benjamin/Cummings.

Mayer, F. L., P. M. Mehrle, and P. L. Crutcher. 1978. Interactions of toxaphene and vitamin C in channel catfish. T. Am. Fish. Soc. 107:326–333.

McLaren, B. A., E. Keller, D. J. O'Donnell, and C. A. Elvehjem. 1947. The nutrition of rainbow trout. 1. Studies of vitamin requirements. Arch. Biochem. Biophys. 15:169–178.

Mengqing, L., J. Wenjuan, C., Qing, and W. Jialin. 2004. The effect of vitamin A supplementation in broodstock feed on reproduction performance and larval quality in *Penaeus chinensis*. Aquacult. Nutr. 10:295–300.

Merchie, G., P. Lavens, V. Storch, U. Ubel, H. Nelis, A. Deleenheer, and P. Sorgeloos. 1996. Influence of dietary vitamin C storage on turbot (*Scophthalmus maximus*) and European seabass (*Dicentrarchus labrax*) nursery stages. Comp. Biochem. Phys. A 114A:117–121.

Messager, J. L. 1986. Influence de l'acide ascorbique sur l'hypertyrosinemie granulomateuse du turbot d'elevage, *Scophthalmus maximus* L. Pp. 381–390 in Pathology in Marine Aquaculture, C. P. Vivares, J. R. Bonami, and E. Jaspers, eds. European Aquaculture Society, Special Publication 9. Breden, Belgium: European Aquaculture Society.

Messager, J. L., D. Ansquer, R. Metailler, and J. Person-Le Ruyet. 1986. Induction experimentale de l'hypertyrosinemie granulomateuse chez le trubot d'elevage (*Scophthalmus maximus*) par une alimentation carencee en acide ascorbique. Ichtyophysiol. Acta 10:201–214.

Misra, C. K., B. K. Das, S. C. Mukherjee, and J. Pradhan. 2007. Effects of dietary vitamin C on immunity, growth and survival of Indian major carp *Labeo rohita*, fingerlings. Aquacult. Nutr. 13:35–44.

Mishra, S., and P. K. Mukhopadhyay. 1996. Ascorbic acid requirement of catfish fry *Clasrias batrachus* (Linn). Indian J. Fish. 43:157–162.

Moccia, R. D., S. S. O. Hung, S. J. Slinger, and H. W. Ferguson. 1984. Effect of oxidized oil, vitamin E and ethoxyquin on the histopathology and haematology of rainbow trout, *Salmo gairdneri*. J. Fish. Dis. 7:269–282.

Moe, Y. Y., S. Koshio, S. I. Teshima, M. Ishikawa, Y. Matsunaga, and A. Panganiban. 2004. Effect of vitamin C derivatives on the performance of larval kuruma shrimp, *Marsupenaeus japonicus*. Aquaculture 242:501–512.

Moe, Y. Y., S. Koshio, M. Ishikawa, S. I. Teshima, A. Panganiban, M. Thu, F. R. Michael and T. Ren. 2005. Vitamin C requirement of kuruma shrimp postlarvae, *Marsupenaeus japonicus* (Bate), using l-ascorbyl-2-monophosphate-Na-Ca. Aquac. Res. 36:739–745.

Montero, D., L. Tortb, L. Robainaa, J. M. Vergarac, and M. S. Izquierdoc. 2001. Low vitamin E in diet reduces stress resistance of gilthead seabream (*Sparus aurata*) juveniles. Fish Shellfish Immun. 11:473–490.

Montoya, N., and C. Molina. 1995. Optimum supplemental level of L-ascorbyl-2-phosphate-Mg to diet for white shrimp *Penaeus vannamei*. Fisheries Sci. 61:1045–1046.

Moren, M., I. Opstad, M. H. G. Berntssen, J.-L. Z. Infante, and K. Hamre. 2004. An optimum level of vitamin A supplements for Atlantic halibut (*Hippoglossus hippoglossus* L.) juveniles. Aquaculture 235:587–599.

Morito, C. L. H., D. H. Conrad, and J. W. Hilton. 1986. The thiamin deficiency signs and requirement of rainbow trout (*Salmo gairdneri*, Richardson). Fish Physiol. Biochem. 1:93–104.

Morris, P. C., R. T. M. Baker, and S. J. Davies. 1998. Nicotinic acid supplementation of diets for the African catfish, *Clarias gariepinus* (Burchell). Aquac. Res. 29:791–799.

Mukhopadhyay, P. K., D. N. Chattopadhyay, and G. Mitra. 2003. Broodstock nutrition, the key to quality seed production. Infofish International 3:25–31.

Murai, T., and J. W. Andrews. 1974. Interactions of dietary α-tocopherol, oxidized menhaden oil and ethoxyquin on channel catfish (*Ictalurus punctatus*). J. Nutr. 104:1416–1431.

Murai, T., and J. W. Andrews. 1977. Vitamin K and anticoagulant relationships in catfish diets. B. Jpn. Soc. Sci. Fish. 43:785–794.

Murai, T., and J. W. Andrews. 1978a. Riboflavin requirement of channel catfish fingerlings. J. Nutr. 108:1512–1517.

Murai, T., and J. W. Andrews. 1978b. Thiamin requirement of channel catfish fingerlings. J. Nutr. 108:176–180.

Murai, T., and J. W. Andrews. 1979. Pantothenic acid requirement of channel catfish fingerlings. J. Nutr. 109:1140–1142.

Murai, T., J. W. Andrews, and J. C. Bauerfeind. 1978. Use of L-ascorbic acid, ethocel coated ascorbic acid and ascorbic 2-sulfate in diets of channel catfish (*Ictalurus punctatus*). J. Nutr. 108:1761–1766.

Mustin, W. G., and R. T. Lovell. 1992. Na-L-ascorbyl-2-monophosphate as a source of vitamin C for channel catfish. Aquaculture 105:95–100.

Navarre, O., and J. E. Halver. 1989. Disease resistance and humoral antibody production in rainbow trout fed high levels of vitamin C. Aquaculture 79:207–221.

Nayak, S. K., P. Swaina, and S. C. Mukherjeeb. 2007. Effect of dietary supplementation of probiotic and vitamin C on the immune response of Indian major carp, *Labeo rohita* (Ham.). Fish Shellfish Immun. 23:892–896.

Ng, W. K., G. Serrini, Z. Zhang, and R. P. Wilson. 1997. Niacin requirement and inability of tryptophan to act as a precursor of NAD^+ in channel catfish, *Ictalurus punctatus*. Aquaculture 152:273–285.

Ng, W. K., C. N. Keembiyehetty, and R. P. Wilson. 1998. Bioavailability of niacin from feed ingredients commonly used in feeds for channel catfish, *Ictalurus punctatus*. Aquaculture 161:393–404.

Ng, W. K., Y. Wang, P. Ketchimenina, and K. H. Yuen. 2004. Replacement of dietary fish oil with palm fatty acid distillate elevates tocopherol and tocotrienol concentrations and increases oxidative stability in the muscle of African catfish, *Clarias gariepinus*. Aquaculture 233:423–437.

Niu, J., L. X. Tian, Y. J. Liu, K. S. Mai, H. J. Yang, C. X. Ye, and W. Gao. 2009. Nutrient values of dietary ascorbic acid (L-ascorbyl-2-polyphosphate) on growth, survival and stress tolerance of larval shrimp, *Litopenaeus vannamei*. Aquacult. Nutr. 15:194–201.

NRC (National Research Council). 1983. Nutrient Requirements of Warmwater Fishes and Shellfishes, Revised Edition. Washington, DC: National Academy Press.

Ogino, C. 1965. B vitamin requirements of carp, *Cyprinus carpio*. 1. Deficiency symptoms and requirements of vitamin B5. B. Jpn. Soc. Sci. Fish. 31:546–551.

Ogino, C. 1967. B vitamin requirements of carp. 2. Requirements for riboflavin and pantothenic acid. B. Jpn. Soc. Sci. Fish. 33:351–354.

Ogino, C., N. Uki, T. Watanabe, Z. Iida, and K. Ando. 1970a. B vitamin requirements of carp. 4. Requirement of choline. B. Jpn. Soc. Sci. Fish. 36:1140–1146.

Ogino, C., T. Watanabe, J. Kakino, N. Iwanaga, and M. Mizuno. 1970b. B vitamin requirements of carp. 3. Requirement for biotin. B. Jpn. Soc. Sci. Fish. 36:734–740.

Okada, S., S. A. Nur-E-Borhan, and Y. Yamaguchi. 1994. Carotenoid composition in the exoskeleton of commercial black tiger prawn. Fisheries Sci. 60:213–215.

Olsen, R. E., and A. Mortensen. 1997. The influence of dietary astaxanthin and temperature on flesh colour in Arctic charr *Salvelinus alpinus* L. Aquac. Res. 28:51–58.

Olson, J. A. 1989. Biological action of carotenoids. J. Nutr. 119:94–95.

Ørnsrud, R., I. E. Graff, S. Høie, G. K. Totland and G. I. Hemre. 2002. Hypervitaminosis A in first-feeding fry of the Atlantic salmon (*Salmo salar* L.). Aquacult. Nutr. 8:7–13.

Ortuño, J., M. A. Esteban, and J. Meseguer. 1999. Effect of high dietary intake of vitamin C on non-specific immune response of gilthead seabream (*Sparus aurata* L.). Fish Shellfish Immun. 9:429–443.

Ortuño, J., M. A. Esteban, and J. Meseguer. 2000. High dietary intake of α-tocopherol acetate enhances the non-specific immune response of gilthead seabream (*Sparus aurata* L.). Fish Shellfish Immun. 10:293–307.

Pangantihon-Kühlmann, M. P., O. Millamena, and Y. Chern. 1998. Effect of dietary astaxanthin and vitamin A on the reproduction performance of *Penaeus monodon* broodstock. Aquat. Living Resour. 11:403–409.

Paul, B. N., S. Sarkar, and S. N. Mohanty. 2004. Dietary vitamin E requirement of mrigal, *Cirrhinus mrigala* fry. Aquaculture 242:529–536.

Peng, L. I., and D. M. Gatlin, III. 2009. Dietary vitamin E requirement of the red drum *Sciaenops ocellatus*. Aquacult. Nutr. 15:329–337.

Peres, H., C. Lim, and P. H. Klesius. 2004. Growth, chemical composition and resistance to *Streptococcus iniae* challenge of juvenile Nile tilapia (*Oreochromis niloticus*) fed graded levels of dietary inositol. Aquaculture 235:423–432.

Phillips, A. M., and D. R. Brockway. 1947. The niacin and biotin requirement of trout. T. Am. Fish. Soc. 76:152–159.

Phillips, A. M., A. V. Tunison, H. B. Shaffer, G. K. White, M. W. Sullivan, C. Vincent, D. R. Brockway, and C. M. McCay. 1945. The nutrition of trout. The vitamin requirement of trout. Fish. Res. Bull. 7:1–31.

Phillips, A. M., Jr., H. A. Podoliak, H. A. Poston, D. L. Livingston, H. E. Booke, E. A. Pyle, and G. L. Hammer. 1964. The production of anemia in brown trout. Pp. 66–70 in Fisheries Research Bulletin No. 27. Albany: State of New York Conservation Department.

Phromkunthong, W., M. Boonyaratpalin, and V. Storch. 1997. Different concentrations of ascoryl-2-monophosphate-magnesium as dietary sources of vitamin C for seabass, *Lates calcarifer*. Aquaculture 151:225–243.

Poston, H. A. 1964. Effect of dietary vitamin K and sulfaguanidine on blood coagulation time, microhematocrit, and growth of immature brook trout. Prog. Fish Cult. 26:59–64.

Poston, H. A. 1967. Effect of dietary L-ascorbic acid on immature brook trout. Pp. 45–51 in Fisheries Research Bulletin No 31. Albany: State of New York Conservation Department.

Poston, H. A. 1969a. Effects of massive doses of vitamin D_3 fingerling brook trout. Pp. 45–50 in Fisheries Research Bulletin No. 32. Albany: State of New York Conservation Department.

Poston, H. A. 1969b. The effect of excess levels of niacin on the lipid metabolism of fingerling brook trout. Pp. 9–12 in Fisheries Research Bulletin No. 32. Albany: State of New York Conservation Department.

Poston, H. A. 1971a. Effect of feeding excess vitamin A on the carbohydrate and lipid metabolism and growth of brook trout. Pp. 22–26 in Fisheries Research Bulletin No. 34. Albany: State of New York Conservation Department.

Poston, H. A. 1971b. Effect of excess vitamin K on the growth, coagulation time, and hematocrit values of brook trout fingerlings. Pp. 41–42 in Fisheries Research Bulletin No. 34. Albany: State of New York Conservation Department.

Poston, H. A. 1976a. Relative effect of two dietary water-soluble analogues of menaquinone on coagulation and packed cell volume of blood of lake trout, *Salvelinus namaycush*. J. Fish. Res. Board Can. 33:1791–1793.

Poston, H. A. 1976b. Optimum level of dietary biotin for growth, feed utilization, and swimming stamina of fingerling lake trout (*Salvelinus namaycush*). J. Fish. Res. Board Can. 33:1803–1806.

Poston, H. A., and G. F. Combs, Jr. 1980. Nutritional implications of tryptophan catabolizing enzymes in several species of trout and salmon. P. Soc. Exp. Biol. Med. 163:452–454.

Poston, H. A., and R. N. DiLorenzo. 1973. Tryptophan conversion to niacin in the brook trout (*Salvelinus fontinalis*). P. Soc. Exp. Biol. Med. 144:110–112.

Poston, H. A., and D. L. Livingston. 1969. Effects of massive doses of dietary vitamin E on fingerling brook trout. Pp. 6–12 in Fisheries Research Bulletin No. 33. Albany: State of New York Conservation Department.

Poston, H. A., and D. L. Livingston. 1971. The influence of dietary levels of protein and vitamin A on the liver vitamin A level, lipid metabolism, and growth of brook trout. Pp. 27–34 in Fisheries Research Bulletin No. 34. Albany: State of New York Conservation Department.

Poston, H. A., and J. W. Page. 1982. Gross and histological signs of dietary deficiency of biotin and pantothenic acid in lake trout, *Salvelinus namaycush*. Cornell Vet. 72:242–261.

Poston, H. A., and M. J. Wolfe. 1985. Niacin requirement for optimum growth, feed conversion and protection of rainbow trout, *Salmo gairdneri* Richardson, from ultraviolet-B-irradiation. J. Fish. Dis. 8:451–460.

Poston, H. A., D. L. Livingston, E. A. Pyle, and A. M. Phillips, Jr. 1966. The toxicity of high levels of vitamin A in the diet of brook trout. Pp. 20–24 in Fisheries Research Bulletin No. 29. Albany: State of New York Conservation Department.

Poston, H. A., G. F. Combs, Jr., and L. Leibovitz. 1976. Vitamin E and selenium interrelations in the diets of Atlantic salmon (*Salmo salar*): Gross, histological and biochemical deficiency signs. J. Nutr. 106:892.

Poston, H. A., R. C. Riis, G. L. Rumsey, and H. G. Ketola. 1977. The effect of supplemental dietary amino acids, minerals and vitamins on salmonids fed cataractogenic diets. Cornell Vet. 67:472–509.

Puangkaew, J., V. Kiron, T. Somamoto, N. Okamoto, S. Satoh, T. Takeuchi, and T. Watanabe. 2004. Nonspecific immune response of rainbow trout (*Oncorhynchus mykiss* Walbaum) in relation to different status of vitamin E and highly unsaturated fatty acids. Fish Shellfish Immun. 16:25–39.

Qin, Z. H., J. Li, F. Wang, Q. Liu, and Q. Wang. 2007. Effect of L-ascorbyl-2-polyphosphate on growth and immunity of shrimp (*Penaeus chinensis*) larvae. Fisheries Sci. 26:21.

Ren, T., S. Koshio, M. Ishikawa, S. Yokoyama, F. R. Micheal, O. Uyan, and H. T. Tung. 2007. Influence of dietary vitamin C and bovine lactoferrin on blood chemistry and non-specific immune responses of Japanese eel, *Anguilla japonica*. Aquaculture 267:31–37.

Ren, T., S. Koshio, S. Teshima, M. Ishikawa, A. Panganiban, Jr., O. Uyan, and M. S. Alam. 2008. Effectiveness of L-ascorbyl-2-monophosphate Na/Ca as a vitamin C source for yellowtail *Seriola quinqueradiata* juveniles. Aquacult. Nutr. 14:416–422.

Roberts, M. L., S. J. Davies, and A. L. Pulsford. 1995. The influence of ascorbic acid (vitamin C) on non-specific immunity in the turbot (*Scophthalmus maximus* L.). Fish Shellfish Immun. 5:27–38.

Robinson, E. 1990. Reevaluation of the ascorbic acid (vitamin C) requirement of channel catfish (*Ictalurus punctatus*) (abstract). FASEB J. 4:3745.

Robinson, E. H., and R. T. Lovell. 1978. Essentiality of biotin for channel catfish *Ictalurus punctatus* fed lipid and lipid-free diets. J. Nutr. 108:1600–1605.

Roem, A. J., C. C. Kohler, and R. R. Stickney. 1990. Vitamin E requirements of the blue tilapia, *Oreochromis aureus* (Steindachner), in relation to dietary lipid levels. Aquaculture 87:155–164.

Rosenlund, G., L. Jorgensen, R. Waagbø and K. Sandnes. 1990. Effect of different dietary levels of ascorbic acid in plaice *Pleuronectes platessa* L. Comp. Biochem. Phys. A 96A:395–398.

Ruff, N., R. D. Fitzgerald, T. F. Cross, K. Hamre, and J. P. Kerry. 2003. The effect of dietary vitamin E and C level on market-size turbot (*Scophthalmus maximus*) fillet quality. Aquacult. Nutr. 9:91–103.

Rumsey, G. L. 1991. Choline-betaine requirements of rainbow trout (*Oncorhynchus mykiss*). Aquaculture 95:107–116.

Sakaguchi, H., F. Takeda, and K. Tange. 1969. Studies on vitamin requirements by yellowtail. 1. Vitamin B_6 and vitamin C deficiency symptoms. B. Jpn. Soc. Sci. Fish. 35:1201–1206.

Salte, R., T. Asgard, and K. Liestol. 1988. Vitamin E and selenium prophylaxis against "Hitra Disease" in farmed Atlantic salmon—A survival study. Aquaculture 75:45–55.

Sandel, L. J., and J. C. Daniel. 1988. Effect of ascorbic acid on collagen in RNA levels in short term chondrocyte cultures. Connect. Tissue Res. 17:11–22.

Sandnes, K., Y. Ulgens, O. R. Braekkan, and F. Utne. 1984. The effect of ascorbic acid supplementation in broodstock feed on reproduction of rainbow trout (*Salmo gairdneri*). Aquaculture 43:167–177.

Sandnes, K., O. Torrissen, and R. Waagbø. 1992. The minimum dietary requirement of vitamin C in Atlantic salmon *Salmo salar* fry using Ca ascorbat-2-monophosphate as dietary source. Fish Physiol. Biochem. 10:315–319.

Saroglia, M., and G. Scarano. 1992. Experimental induction of ascorbic acid deficiency in seabass in intensive aquaculture. Bull. Eur. Assn. Fish P. 12:96.

Sato, M., R. Yoshinaka, and S. Ikeda. 1978. Dietary ascorbic acid requirement of rainbow trout for growth and collagen formation. B. Jpn. Soc. Sci. Fish. 44:1029–1035.

Sato, M., T. Kondo, R. Yashinaka, and S. Ikeda. 1982. Effect of dietary ascorbic acid levels on collagen formation in rainbow trout. B. Jpn. Soc. Sci. Fish. 48:553–556.

Sato, M., T. Kondo, R. Yoshinaka, and S. Ikeda. 1983. Effect of water temperature on the skeletal deformity in ascorbic acid deficient rainbow trout. B. Jpn. Soc. Sci. Fish. 49:443–446.

Sato, M., R. Yoshinaka, R. Kuroshima, H. Morimoto, and S. Ikeda. 1987. Changes in water-soluble vitamin contents and transaminase activity of rainbow trout egg during development. B. Jpn. Soc. Sci. Fish. 53:795–799.

Satoh, S., T. Takeuchi, and T. Watanabe. 1987. Requirement of tilapia for α-tocopherol. Nippon Suisan Gakk. 53:119–124.

Sau, S. K., B. N. Paul, K. N. Mohanta, and S. N. Mohanty. 2004. Dietary vitamin E requirement, fish performance and carcass composition of rohu (*Labeo rohita*) fry. Aquaculture 240:359–368.

Sawanboochun, J., W. J. Roy, D. A. Robertson, and J. G. Bell. 2008. The impact of dietary supplementation with astaxanthin on egg quality in Atlantic cod broodstock (*Gadus morhua*, L.). Aquaculture 283:97–101.

Sealey, W. M., and D. M. Gatlin, III. 1999. Dietary vitamin C requirement of hybrid striped bass (*Morone chrysops* ♀ × *M. saxatilis* ♂). J. World Aquacult. Soc. 30:297–301.

Sealey, W. M., and D. M. Gatlin, III. 2002a. Dietary vitamin C and vitamin E interact to influence growth and tissue composition of juvenile hybrid striped bass (*Morone chrysops* ♀ × *M. saxatilis* ♂) but have limited effects on immune responses. J. Nutr. 132:748–755.

Sealey, W. M., and D. M. Gatlin, III. 2002b. In vitro manipulations of vitamin C and vitamin E concentrations alter intracellular O2-production of hybrid striped bass (*Morone chrysops* ♀ × *M. saxatilis* ♂) head-kidney cells. Fish Shellfish Immun. 12:131–140.

Secombes, C. J., S. Chung, and A. H. Jeffries. 1988. Superoxide anion production by rainbow trout macrophages detected by the reduction of ferricytochrome C. Dev. Comp. Immunol. 12:201–206.

Serrini, G., Z. Zhang, and R. P. Wilson. 1996. Dietary riboflavin requirement of fingerling channel catfish (*Ictalurus punctatus*). Aquaculture 139:285–290.

Shaik Mohamed, J. 2001a. Dietary pyridoxine requirement of the Indian catfish, *Heteropneustes fossilis*. Aquaculture 194:327–335.

Shaik Mohamed, J. 2001b. Dietary biotin requirement determined for Indian catfish, *Heteropneustes fossilis* (Bloch), fingerlings. Aquac. Res. 32:709–716

Shaik Mohamed, J., and A. Ibrahim. 2001. Quantifying the dietary niacin requirement of the Indian catfish, *Heteropneustes fossilis* (Bloch), fingerlings. Aquac. Res. 32:157–162.

Shaik Mohamed, J., B. Ravisankar, and A. Ibrahim. 2000. Quantifying the dietary biotin requirement of the catfish, *Clarias batrachus*. Aquacult. Int. 8:9–18.

Shaik Mohamed, J., V. Sivaram, T. S. Christopher Roy, M. Peter Marian, S. Murugadass, and M. Saffiq Hussain. 2003. Dietary vitamin A requirement of juvenile greasy grouper (*Epinephelus tauvina*). Aquaculture 219:693–701.

Shiau, S. Y. 2001. Vitamin C requirement in crustaceans. Pp. 87–95 in Ascorbic Acid in Aquatic Organisms, K. Dabrowski, ed. Boca Raton, FL: CRC Press.

Shiau, S.Y., and Y. Chen. 2000. Estimation of the dietary vitamin A requirement of juvenile grass shrimp, *Penaeus monodon*. J. Nutr. 130:90–94.

Shiau, S. Y., and Y. H. Chin. 1998. Dietary biotin requirement for maximum growth of juvenile grass shrimp, *Penaeus monodon*. J. Nutr. 128:2494–2497.

Shiau, S. Y., and Y. H. Chin. 1999. Estimation of the dietary biotin requirement of juvenile hybrid tilapia, *Oreochromis niloticus* × *O. aureus*. Aquaculture 170:71–78.

Shiau, S. Y., and W. H. Cho. 2002. Choline requirements of grass shrimp, *Penaeus monodon*, are affected by dietary lipid level. Anim. Sci. 75:97–102.

Shiau, S. Y., and H. L. Hsieh. 1997. Vitamin B_6 requirements of tilapia *Oreochromis niloticus* × *O. aureus* fed two dietary protein concentrations. Fisheries Sci. 63:1002–1007.

Shiau, S. Y., and C. W. Hsu. 1999. Dietary pantothenic acid requirement of juvenile grass shrimp, *Penaeus monodon*. J. Nutr. 129:718–721.

Shiau, S. Y., and C. Y. Hsu. 2002. Vitamin E sparing effect by dietary vitamin C in juvenile hybrid tilapia, *Oreochromis niloticus* × *O. aureus*. Aquaculture 210:335–342.

Shiau, S. Y., and T. S. Hsu. 1993. Stability of ascorbic acid in shrimp feed during analysis. Nippon Suisan Gakk. 59:1535–1537.

Shiau, S. Y., and T. S. Hsu. 1994. Vitamin C requirement of grass shrimp, *Penaeus mondon*, as determined with L-ascorbyl-2-monophosphate. Aquaculture 122:347–357.

Shiau, S. Y., and T. S. Hsu. 1995. L-Ascorbyl-2-sulfate has equal antiscorbutic activity as L-ascorbyl-2-monophosphate for tilapia, *Oreochromis niloticus* × *O. aureus*. Aquaculture 133:147–157.

Shiau, S. Y., and T. S. Hsu. 1999. Quantification of vitamin C requirement for juvenile hybrid tilapia, *Oreochromis niloticus* × *O. aureus*, with L-ascorbyl-2-monophosphate-Na and L-ascorbyl-2-monophosphate-Mg. Aquaculture 175:317–326.

Shiau, S. Y., and S. Y. Huang. 2001a. Dietary folic acid requirement for maximal growth of juvenile tilapia, *Oreochromis niloticus* × *O. aureus*. Fisheries Sci. 67:655–659.

Shiau, S. Y., and S. Y. Huang. 2001b. Dietary folic acid requirement determined for grass shrimp, *Penaeus monodon*. Aquaculture 200:339–347.

Shiau, S. Y., and J. Y. Hwang. 1993. Vitamin D requirement of juvenile hybrid tilapia *Oreochromis niloticus* × *O. aureus*. Nippon Suisan Gakk. 59:553–558.

Shiau, S. Y., and J. Y. Hwang. 1994. The dietary requirement of juvenile grass shrimp, *Penaeus monodon*, for vitamin D. J. Nutr. 124:2445–2450.

Shiau, S. Y., and F. L. Jan. 1992a. Dietary ascorbic acid requirement of juvenile tilapia *Oreochromis niloticus* × *O. aureus*. Nippon Suisan Gakk. 58:671–675.

Shiau, S. Y., and F. L. Jan. 1992b. Ascorbic acid requirement of grass shrimp, *Penaeus monodon*. Nippon Suisan Gakk. 58:363.

Shiau, S. Y., and J. S. Liu. 1994a. Quantifying the vitamin K requirement of juvenile marine shrimp, *Penaeus monodon*, with menadione. J. Nutr. 124:277–282.

Shiau, S. Y., and J. S. Liu. 1994b. Estimation of the dietary vitamin K requirement of juvenile *Penaeus chinensis* using menadione. Aquaculture 126:129–135.

Shiau, S. Y., and P. S. Lo. 2000. Dietary choline requirements of juvenile hybrid tilapia, *Oreochromis niloticus* × *O. aureus*. J. Nutr. 130:100–103.

Shiau, S. Y., and P. S. Lo. 2001. Dietary choline requirement of juvenile grass shrimp, *Penaeus monodon*. Anim. Sci. 72:477–482.

Shiau, S. Y., and C. Q. Lung. 1993a. No dietary vitamin B_{12} required for juvenile tilapia *Oreochromis niloticus* × *O. aureus*. Comp. Biochem. Phys. A 105A:147–150.

Shiau, S. Y., and C. Q. Lung. 1993b. Estimation of the vitamin B_{12} requirements of grass shrimp, *Penaeus monodon*. Aquaculture 117:157–163.

Shiau, S. Y., and L. F. Shiau. 2001. Re-evaluation of the vitamin E requirements of juvenile tilapia (*Oreochromis niloticus* × *O. aureus*). Anim. Sci. 72:529–534.

Shiau, S. Y., and S. L. Su. 2004. Dietary inositol requirement determined for juvenile grass shrimp, *Penaeus monodon*. Aquaculture 241:1–8.

Shiau, S. Y., and S. L. Su. 2005. Juvenile tilapia (*Oreochromis niloticus* × *O. aureus*) requires dietary myo-inositol for maximal growth. Aquaculture 243:273–277.

Shiau, S. Y., and G. S. Suen. 1992. Estimation of the niacin requirements for tilapia fed diets containing glucose or dextrin. J. Nutr. 122:2030–2036.

Shiau, S. Y., and G. S. Suen. 1994. The dietary requirement of juvenile grass shrimp (*Penaeus monodon*) for niacin. Aquaculture 125:139–145.

Shiau, S. Y., and M. H. Wu. 2003. Dietary vitamin B_6 requirement determined for grass shrimp, *Penaeus monodon*. Aquaculture 225:397–404.

Shigueno, K. 1988. Use of Mg-L-ascorbyl-2-phosphate as a vitamin C source in shrimp diets. J. World Aquacult. Soc. 19:168–174.

Shimeno, S. 1991. Yellowtail, *Seriola quinqueradiata*. Pp. 181–191 in Handbook of Nutrient Requirements of Finfish, R. P. Wilson, ed. Boca Raton, FL: CRC Press.

Smith, C. E. 1968. Hematological changes in coho salmon fed a folic acid deficient diet. J. Fish. Res. Board Can. 25:151–156.

Smith, C. E., and J. E. Halver. 1969. Folic acid anemia in coho salmon. J. Fish. Res. Board Can. 26:111–114.

Smith, C. E., M. Brin, and J. E. Halver. 1974. Biochemical, physiological, and pathological changes in pyridoxine-deficient rainbow trout (*Salmo gairdneri*). J. Fish. Res. Board Can. 31:1893–1898.

Sobhana, K. S., C. V. Mohan, and K. M. Shankar. 2002. Effect of dietary vitamin C on the disease susceptibility and inflammatory response of mrigal, *Cirrhinus mrigala* (Hamilton) to experimental infection of *Aeromonas hydrophila*. Aquaculture 207:225–238.

Soliman, A. K., and R. P. Wilson. 1992a. Water-soluble vitamin requirements of tilapia. II. Riboflavin requirement of blue tilapia, *Oreochromis aureus*. Aquaculture 104:309–314.

Soliman, A. K., and R. P. Wilson. 1992b. Water-soluble vitamin requirements of tilapia. I. Pantothenic acid requirement of blue tilapia, *Oreochromis aureus*. Aquaculture 104:121–126.

Soliman, A. K., K. Jauncey, and R. J. Roberts. 1994. Water-soluble vitamin requirements of tilapia: ascorbic acid (vitamin C) requirement of Nile tilapia, *Oreochromis niloticus* (L.). Aquacult. Fish. Manage. 25:269–278.

Soliman, A. K., K. Jauncey, and R. J. Roberts. 1986a. The effect of varying forms of dietary ascorbic acid on the nutrition of juvenile tilapias (*Oreochromis niloticus*). Aquaculture 59:1–6.

Soliman, A. K., K. Jauncey, and R. J. Roberts. 1986b. The effect of dietary ascorbic acid supplementation on hatchability, survival rate and fry performance in *Oreochromis mossambicus* (Peters). Aquaculture 59:197–208.

Steffens, W. 1970. The vitamin requirements of rainbow trout (*Salmo gairdneri*). Int. Rev. Ges. Hydrobio. 59:255–282.

Stéphan, G., J. Guillaume, and F. Lamour. 1995. Lipid peroxidation in turbot (*Scophthalmus maximus*) tissue: Effect of dietary vitamin E and dietary *n*-6 or *n*-3 polyunsaturated fatty acids. Aquaculture 130:251–268.

Stickney, R. R., R. B. McGeachin D. H. Lewis, J. Marks, A. Riggs, R. F. Sis, E. H. Robinson, and W. Wurts. 1984. Response of *Tilapia aurea* to dietary vitamin C. J. World Maricult. Soc. 15:179–185.

Storebakken, T., and H. K. No. 1992. Pigmentation of rainbow trout. Aquaculture 100:209–229.

Su, S. L., and S. Y. Shiau. 2004. Requirements of dietary myo-inositol of juvenile grouper, *Epinephelus malabaricus*. J. Fish. Soc. Taiwan 31:313–317.

Sugita, H., C. Miyajima, and Y. Deguchi. 1990. The vitamin B_{12}-producing ability of intestinal bacteria isolated from tilapia and channel catfish. Nippon Suisan Gakk. 56:701.

Sugita, H., C. Miyajima, and Y. Deguchi. 1991a. The vitamin B_{12}-producing ability of the intestinal microflora of freshwater fish. Aquaculture 92:267–276.

Sugita, H., J. Takahashi, C. Miyajima, and Y. Deguchi. 1991b. Vitamin B_{12}-producing ability of the intestinal microflora of rainbow trout (*Oncorhynchus mykiss*). Agr. Biol. Chem. Tokyo 55:893–894.

Suttie, J. W. 1985. Vitamin K. Pp. 225–311 in The Fat-Soluble Vitamins, A. T. Diplock, ed. London, UK: William Heinemann.

Takeuchi, L., T. Takeuchi, and C. Ogino. 1980. Riboflavin requirements in carp and rainbow trout. B. Jpn. Soc. Sci. Fish. 46:733–737.

Takeuchi, T., K. Watanabe, S. Satoh, and T. Watanabe. 1992. Requirement of grass carp fingerlings for α-tocopherol. Nippon Suisan Gakk. 58:1743–1749.

Takeuchi, T., J. Dedi, Y. Haga, T. Seikai, and T. Watanabe. 1998. Effect of vitamin A compounds on bone deformity in larval Japanese flounder (*Paralichthys olivaceus*). Aquaculture 169:155–165.

Teshima, S. I., A. Kanazawa, S. Koshio, and S. Itoh. 1991. L-ascorbyl-2-phosphate-Mg as vitamin C source for the Japanese flounder (*Paralichthys olivaceus*). Pp. 157–166 in Fish Nutrition in Practice, Coll. les Colloq., No. 61S., J. Kaushik and P. Lequet, eds. Paris, France: INRA.

Thompson, I., G. Ghoubert, D. F. Houlihan, and C. J. Secombes. 1995. The effect of dietary vitamin A and astaxanthin on the immunocompetence of rainbow trout. Aquaculture 133:91–102.

Tocher, D. R., G. Mourente, A. Van Der Eecken, J. O. Evjemo, E. Diaz, J. G. Bell, I. Geurden, P. Lavens, and Y. Olsen. 2002. Effects of dietary vitamin E on antioxidant defence mechanisms of juvenile turbot (*Scophthalmus maximus* L.), halibut (*Hippoglossus hippoglossus* L.) and sea bream (*Sparus aurata* L.). Aquacult. Nutr. 8:195–207.

Torrissen, O. J. 1984. Pigmentation of salmonids-effect of carotenoids in eggs and startfeeding diet on survival and growth rate. Aquaculture 43:185–193.

Torrissen, O. J., and R. Christiansen. 1995. Requirements for carotenoids in fish diets. J. Appl. Ichthyol. 11:225–230.

Tsujimura, M., H. Yoshikawa, T. Hasegawa, T. Suzuki, T. Kaisai, T. Suwa, and S. Kitamura. 1978. Studies on the vitamin C activity of ascorbic acid 2-sulfate on the feeding test of newborn rainbow trout. Vitamins (Japan) 52:35–44.

Twibell, R. G., and P. B. Brown. 2000. Dietary choline requirement of juvenile yellow perch (*Perca flavescens*). J. Nutr. 130:95–99.

Verlhac, V., J. Gabaudan, A. Obach, W. Schüep, and R. Hole. 1996. Influence of dietary glucan and vitamin C on non-specific and specific immune responses of rainbow trout (*Oncorhynchus mykiss*). Aquaculture 143:123–133.

Villeneuve, L., E. Gisbert, H. Le Delliou, C. L. Cahu, and J. L. Zambonino-Infante. 2005a. Dietary levels of all-trans retinol affect retinoid nuclear receptor expression and skeletal development in European sea bass larvae. Brit. J. Nutr. 93:791–801.

Villeneuve, L., E. Gisbert, J. L. Zambonino-Infante, P. Quazuguel, and C. L. Cahu. 2005b. Effect of nature of dietary lipids on European sea bass morphogenesis: Implication of retinoid receptors. Brit. J. Nutr. 94:877–884.

Waagbø, R., K. Sandnes, and Ø. Lie. 1998. Effects of inositol supplementation on growth, chemical composition and blood chemistry in Atlantic salmon, *Salmo salar* L., fry. Aquacult. Nutr. 4:53–59.

Wahli, T., W. Meier, and K. Pfister. 1986. Ascorbic acid induced immune-mediated decrease in mortality in *Ichw. thyophthirius multifiliis* infected rainbow trout (*Salmo gairdneri*). Acta Trop. 43:287–289.

Wald, G. 1945–1946. The chemical evolution of vision. Harvey Lect. 41:117–160.

Walton, M. J., C. B. Cowey, and J. W. Andron. 1984. Effects of biotin deficiency in rainbow trout (*Salmo gairdneri*) fed diets of different lipid and carbohydrate content. Aquaculture 37:21–38.

Wang, D., L. Zhao, and Y. Tan. 1995. Requirement of the fingerling grass carp (*Ctenopharyngodon idella*) for choline [in Chinese]. J. Fish. China 19:132–139.

Wang, W. N., L. Wang, and A. L. Wang. 2006. Effect of supplemental l-ascorbyl-2-polyphosphate (APP) in enriched live food on the immune response of *Penaeus vannamei* exposed to ammonia-N. Aquaculture 256:552–557.

Wang, X., K. W. Kim, and S. C. Bai. 2002. Effects of different dietary levels of L-ascorbyl-2-polyphosphate on growth and tissue vitamin C concentrations in juvenile olive flounder, *Paralichthys olivaceus* (Temminck et Schlegel). Aquac. Res. 33:261–267.

Wang, X., K. W. Kim, S. C. Bai, M. D. Huh, and B. Y. Cho. 2003a. Effects of the different levels of dietary vitamin C on growth and tissue ascorbic acid changes in parrot fish (*Oplegnathus fasciatus*). Aquaculture 215:203–211.

Wang, X., K. W. Kim, and S. C. Bai. 2003b. Comparison of L-ascorbyl-2-monophosphate-Ca with L-ascorbyl-2-monophosphate-Na/Ca on growth and tissue ascorbic acid concentrations in Korean rockfish (*Sebastes schlegeli*). Aquaculture 225:387–395.

Wang, X., K. W. Kim, G. J. Park, S. M. Choi, H. K. Jun, and S. C. Bai. 2003c. Evaluation of L-ascorbyl-2-glucose as the source of vitamin C for juvenile Korean rockfish *Sebastes schlegeli* (Hilgendorf). Aquac. Res. 34:1337–1341.

Wang, Z., K. Mai, Z. Liufu, H. Ma, W. Xu, Q. Ai, W. Zhang, B. Tan, and X. Wang. 2006. Effect of high dietary intakes of vitamin E and n-3 HUFA on immune responses and resistance to *Edwardsiella tarda* challenge in Japanese flounder (*Paralichthys olivaceus*, Temminck and Schlegel). Aquac. Res. 37:681–692.

Watanabe, T., F. Takashima, C. Ogino, and T. Hibiya. 1970a. Requirements of young carp for α-tocopherol. B. Jpn. Soc. Sci. Fish. 36:972–976.

Watanabe, T., F. Takashima, C. Ogino, and T. Hibiya. 1970b. Effects of α-tocopherol deficiency on carp. B. Jpn. Soc. Sci. Fish. 36:623–630.

Watanabe, T., T. Takeuchi, M. Wada, and R. Vehara. 1981a. The relationship between dietary lipid levels and α-tocopherol requirement of rainbow trout. B. Jpn. Soc. Sci. Fish. 47:1463–1471.

Watanabe, T., M. Wada, T. Takeuchi, and S. Aria. 1981b. Absence of interconversion of tocopherols in rainbow trout. B. Jpn. Soc. Sci. Fish. 47:1455–1462.

Watanabe, T., T. Takeuchi, and M. Wada. 1981c. Dietary lipid levels and α-tocopherol requirement of carp. B. Jpn. Soc. Sci. Fish. 47:1585–1590.

Wen, H., Z. Zhao, M. Jiang, A. Liu, F. Wu, and W. Liu. 2007. Dietary myo-inositol requirement for grass carp, *Ctenopharyngodon idella* fingerling [in Chinese]. J. Fish. China 14:794–800.

Wen, Z. P., X. Q. Zhou, L. Feng, J. Jiang and Y. Liu. 2009. Effect of dietary pantothenic acid supplement on growth, body composition and intestinal enzyme activities of juvenile Jian carp (*Cyprinus carpio* var. Jian). Aquacult. Nutr. 15:470–476.

Wilson, R. P. 1973. Absence of ascorbic acid synthesis in channel catfish, *Ictalurus punctatus* and blue catfish, *Ictalurus frucatus*. Comp. Biochem. Phys. B 46B:635–638.

Wilson, R. P., and W. E. Poe. 1973. Impaired collagen formation in the scorbutic channel catfish. J. Nutr. 103:1359–1364.

Wilson, R. P., and W. E. Poe. 1988. Choline nutrition of fingerling channel catfish. Aquaculture 68:65–71.

Wilson, R. P., P. R. Bowser, and W. E. Poe. 1983. Dietary pantothenic acid requirement of fingerling channel catfish. J. Nutr. 113:2124–2134.

Wilson, R. P., P. R. Bowser, and W. E. Poe. 1984. Dietary vitamin E requirement of fingerling channel catfish. J. Nutr. 114:2053–2058.

Wilson, R. P., W. E. Poe, and E. H. Robinson. 1989. Evaluation of L-ascorbyl-2-polyphosphate (ASPP) as a dietary ascorbic acid source for channel catfish. Aquaculture 81:129–136.

Woodall, A. N., L. M. Ashley, J. E. Halver, H. S. Olcott, and J. van der Veen. 1964. Nutrition of salmonid fishes. XIII. The α-tocopherol requirement of Chinook salmon. J. Nutr. 84:125–135.

Woodward, B. 1982. Riboflavin supplementation of diets for rainbow trout. J. Nutr. 112:908–913.

Woodward, B. 1983. Sensitivity of hepatic D-amino acid oxidase and glutathione reductase to the riboflavin status of the rainbow trout (*Salmo gairdneri*). Aquaculture 34:193–201.

Woodward, B. 1984. Symptoms of severe riboflavin deficiency without ocular opacity in rainbow trout (*Salmo gairdneri*). Aquaculture 39:275–281.

Woodward B. 1985. Riboflavin requirement for growth, tissue saturation and maximal flavin-dependent enzyme activity in young trout (*Salmo gairdneri*) at two temperatures. J. Nutr. 115:78–84.

Woodward, B. 1990. Dietary vitamin B_6 requirements of young rainbow trout (*Oncorhynchus mykiss*). FASEB J. 4:3748 (abstr.).

Woodward, B., and M. Frigg. 1989. Dietary biotin requirements of young trout (*Salmo gairdneri*) determined by weight gain hepatic biotin concentration and maximal biotin-dependent enzyme activities in liver and white muscle. J. Nutr. 119:54–60.

Wu, D. H., C. X. Liao, and Z. Z. Huang. 1990. Studies on the requirement of grass carp for α-tocopherol in early growth [in Chinese]. J. Chinese Acad. Fish. Sci. 3:42–47.

Wu, F., M. Jiang, Z. Zhao, A. Liu, W. Liu, and H. Wen. 2007a. The dietary niacin requirement of juvenile *Ctenopharyngodon idellus* [in Chinese]. J. Fish. China 29:695–669.

Wu, F., H. Wen, M. Jiang, Z. Zhao, A. Liu, and W. Liu. 2007b. Effects of dietary vitamin B12 on growth, body composition and hemopoiesis of juvenile grass carp (*Ctenopharyngodon idellus*) [in Chinese]. J. Jilin Agric. Univ. 29:695–669.

Xiao, L. D., K. S. Mai, Q. H. Ai, W. Xu, X. J. Wang, W. B. Zhang, and Z. G. Liufu. 2009. Dietary ascorbic acid requirement of cobia, *Rachycentron canadum* Linneaus. Aquacult. Nutr., doi:10.1111/j.1365–2095.2009.00695.x.

Xie, Z., and C. Niu. 2006. Dietary ascorbic acid requirement of juvenile ayu (*Plecoglossus altivelis*). Aquacult. Nutr. 12:151–156.

Yamamoto, Y., and M. Inoue. 1985. Effects of dietary ascorbic acid and dehydroascorbic acid on the acute cadmium toxicity in rainbow trout. B. Jpn. Soc. Sci. Fish. 51:1299–1303.

Yang, Q., X. Zhou, J. Jiang and Y. Liu. 2008. Effect of dietary vitamin A deficiency on growth performance, feed utilization and immune responses of juvenile Jian carp (*Cyprinus carpio* var. Jian). Aquac. Res. 39:902–906.

Yildirim-Aksoy, M., C. Lim, M. H. Li, and P. H. Klesius. 2008. Interaction between dietary levels of vitamin C and E on growth and immune responses in channel catfish, *Ictalurus punctatus* (Rafinesque). Aquac. Res. 39:1198–1209.

Yone, Y., and M. Fujii. 1974. Studies on nutrition of red sea bream. 10. Qualitative requirements for water-soluble vitamins. Rep. Fish. Res. Lab. Kyushu Univ. (Jpn) 2:25–32.

Yone, Y., M. Furuichi, and K. Shitanda. 1971. Vitamin requirements of the red sea bream. 1. Relationship between inositol requirements and glucose levels in diet. B. Jpn. Soc. Sci. Fish. 37:149–155.

Zhao, Z., H. Wen, F. Wu, A. Liu, M. Jiang, and W. Liu. 2008. Dietary folic acid requirement for grass carp fingerling, *Ctenopharyngodon idella* [in Chinese]. J. Shanghai Fish. Univ. 17:187–192.

Zhou, Q., Y. Ding, S. Zheng, S. Su and L. Zhang. 2004. The effect of dietary vitamin C supplementation on growth and anti-disease ability of shrimp, *Penaeus vannamei*. Acta Hydrobiol. Sin. 28:592–598.

10

Feed Additives

Feed or diet additives have been defined as nonnutritive ingredients or nonnutritive components of ingredients that are included in formulations to either influence physical or chemical properties of the diet or affect fish performance or quality of resulting products (Barrows, 2000). The chemical nature of these feed additives is quite diverse, and their use in commercial diet formulations for aquatic species varies considerably. Additives that influence feed quality include pellet binders, preservatives (such as antimicrobial compounds and antioxidants), and feeding stimulants. Such additives are commonly included in commercial formulations to achieve and maintain optimal physical and chemical characteristics. Other additives that may directly affect fish performance or product quality include acidifiers, chemotherapeutants, pigments, exogenous enzymes, immunostimulants, probiotics, and prebiotics. Most of these additives are supplemented to diet formulations to achieve specific purposes. This chapter will review the diverse groups of feed additives and their specific functions in aquatic animal nutrition.

ANTIMICROBIAL AGENTS

A wide array of antibiotics are used as therapeutics in livestock production; however, only three commercial products—oxytetracycline, sulfadimethoxine/ormetoprim, and florfenicol—have been approved by the U.S. Food and Drug Administration (FDA) for use in the farming of fish destined for human consumption after prescribed withdrawal periods (FDA, 2009). Generally, these compounds are most efficiently administered via diet, especially in large culture systems. Typically these compounds are stable during the compression pelleting process and storage. Extrusion processing, however, does inactivate almost all oxytetracycline and has a limited adverse effect on levels of dietary sulfadimethoxine/ormetoprim.

The quantity of antibiotic fed must be controlled. Proper feeding rates and withdrawal times must be followed to reduce the deposition of antibiotics into fish tissues or release into the rearing water that may be discharged into the aquatic environment. Specific administration and withdrawal procedures for the various antibiotics and targeted fish species were established during the registration process. Antibiotics may only be added to feeds in the United States by a licensed manufacturer.

Subtherapeutic concentrations of antibiotics in the diets of fish may influence the composition of bacterial microbiota of the gut and sometimes may stimulate growth and feed efficiency (Viola and Arieli, 1987; Ahmed and Matty, 1989; Rawles et al., 1997). However, concerns about the development of antibiotic resistance and about residue accumulation in fish and the environment have discouraged illegal subtherapeutic use of antibiotics, thereby stimulating an increase in research devoted to alternative methods of preventing or treating bacterial infections in various aquatic species (Shao, 2001; Cabello, 2006).

A variety of antimicrobial compounds may be routinely added to manufactured aquafeeds to prevent the growth of molds and other microbial contaminants. These antimicrobial agents have a particularly important role in semimoist or wet diets because yeast, bacterial, and fungal organisms can grow rapidly at room temperatures (22–25°C) when the dietary moisture content is 12% or more. Antimicrobial agents typically used in fish feeds include benzoic acid, propionic acid, sorbic acid, and calcium, potassium, and sodium salts of these acids (Hardy and Barrows, 2002).

ANTIOXIDANTS

Marine and some vegetable lipids contain high levels of polyunsaturated fatty acids that are quite susceptible to autooxidation. The oxidation products of these lipids include aldehydes and ketones as well as free radicals that then result in an increase in the demand for antioxidant nutrients, such as carotenoids and vitamins C and E in the diet. The presence of oxidized lipid products in the diet may directly affect fish and/or exacerbate deficiencies of antioxidant vitamins, resulting in pathological conditions that include liver degeneration, spleen abnormalities, and anemia (NRC, 1993; Gatlin, 2002). Environmental factors such as temperature as well as dietary factors such as moisture and lipid levels as well as degree of fatty acid unsaturation can affect rates of oxidation. A variety of synthetic antioxidants, such as ethoxyquin, butylated hydroxylanisol (BHA; 2(3)-tertbutyl-4-hydroxyanisole), butylated hydroxytoluene (BHT; 3,5-di-tert-butyl-4-hydroxytoluene), and propyl gallate, are typically added to oils or complete diets to ensure proper protection from oxidation. More information about antioxidant types and mechanisms is available (e.g., Porter, 1980). The maximum concentration of BHA and BHT permitted by the FDA is 0.02% of lipid content; for ethoxyquin, it is 150 mg/kg diet.[1]

BINDING AGENTS

Binders are incorporated into pelleted aquafeeds to improve stability in water, increase pellet firmness, and reduce the amount of fines produced during processing and handling. These binding agents are particularly important in shrimp feeds due to the feeding habits of those organisms that typically result in the feed remaining in water for a relatively long period of time before consumption. Among the most widely used binders are sodium and calcium bentonites, lignosulfonates, hemicellulose, carboxymethylcellulose, alginate, and guar gum (Hardy and Barrows, 2002). More recently, some inert polymeric binders have been introduced, but available information about their composition or toxicity to commonly cultured fish and shrimp is limited. Cereal grains contain starch that, when gelatinized, gives a durable, water-stable pellet. Hence, formulations pelleted by extrusion processing do not require pellet binders; gelatinized starch provides sufficient binding capacity. Certain feed ingredients, such as whey, wheat gluten, pregelatinized starches, and molasses are considered nutritional binders in that they improve pellet stability and also contribute nutritional value. Other binders, including those composed of clays or cellulosic materials, are considered to be inert and have limited or no nutritional value. Such binding agents tend to increase the density of agglomerated particles produced by compression pelleting.

COLOR/PIGMENTATION AGENTS

Fish and shrimp use oxygenated carotenoids (xanthophylls) to produce pigmentation of their flesh, skin, and eggs (Shahidi et al., 1998; Higuera-Ciapara et al., 2006). These pigments cannot be synthesized by fish and shrimp, and therefore must be included in the diet. A variety of other functions have been assigned to carotenoids including: serving as precursors to vitamin A; possessing antioxidant and singlet-oxygen-quenching ability; assisting in behavioral communication among fish species (e.g., male competition for mates and female mate selection); and generally enhancing growth of various fish and crustacean species (Chien and Jeng, 1992; Torrissen and Christiansen, 1995). Although the mobilization of carotenoids from the flesh to skin and ovaries of salmonids during maturation is well documented, possible roles specific to reproduction are still not fully understood. They are hypothesized to be potent antioxidants in eggs at low oxygen tension, and thus important to eggs that incubate a long time, such as salmon and trout eggs.

In salmonids, two oxycarotenoids, astaxanthin (3,3'-dihydroxy-4,4'-diketo-β-carotene) and canthaxanthin (4-4'-diketo-β-carotene), are responsible for the red to orange coloring of flesh, skin, and fins. Astaxanthin is the main carotenoid pigment in tissue of wild salmonids and all crustacean species. This compound is derived mainly from phytoplankton, such as the algae *Spirulina*, and progresses up the food chain. However, under aquacultural conditions, synthetic astaxanthin is primarily used to impart the flesh coloration desired by human consumers. Synthetic canthaxanthin also has been used to pigment salmonids (Hardy and Barrows, 2002; Breithaupt, 2007). The red yeast *Phaffia rhodozyma* is a natural source of astaxanthin (Higuera-Ciapara et al., 2006) as is *Haematococcus* algae (Lorenz and Cysewski, 2000), such that these ingredients may be included in aquatic diets for pigmentation purposes. Feedstuffs of plant origin generally do not contain pigments that impart the desired color to salmon flesh. The major plant carotenoids are lutein (3R,3'R,6'R-β,ε-carotene-3,3'-diole) and zeaxanthin (3R,3'R-β,β-carotene-3,3'-diol), as found in alfalfa, yellow corn, and algae. Dietary lutein produces a yellow color, whereas dietary zeaxanthin imparts a yellow-orange color to body tissues. Table 10-1 summarizes xanthophyll and astaxanthin contents of ingredients that have been used in aquatic feeds.

Some pigmentation of cultured fish may be undesirable. For example, yellow pigment deposition in the flesh of channel catfish is generally considered undesirable because of consumer preference for a white fillet. When dietary levels of xanthophyll reach 11 mg/kg diet, yellow pigmentation is observed in channel catfish flesh (Lovell, 1998). Li et al. (2007) evaluated the effects of different carotenoid pigments including β-carotene, lutein, zeaxanthin, canthaxanthin, and astaxanthin at 100 mg/kg diet on fillet coloration of channel catfish. Visual yellow color intensity score was highest in filets of catfish fed lutein, followed by zeaxanthin, astaxanthin, and canthaxanthin, and lowest for fish fed the basal and β-carotene-supplemented diets.

Skin pigmentation is important for consumer acceptance in other fish species such as yellowtail, red sea bream, and red porgy (Fujita et al., 1983; Lorenz and Cysewski, 2000; Tejera et al., 2007; Doolen et al., 2009). These fish consume crustaceans and other organisms in nature that provide astaxanthin, but lose their natural coloration when fed formulated diets. Red sea bream deposit the ester form of astaxanthin

[1]21 C.F.R. § 573.380, 582.3169, 582.3173 (1987).

TABLE 10-1 Xanthophyll Content of Plant Materials and Astaxanthin Content of Animal Products Used in Aquatic Feeds

Materials	Scientific Name	mg/kg
Plant Materials		**Xanthophylls**
Alfalfa meal		260–330
Alfalfa juice protein		800
Algae meals		2,000–4,000
Capsicum, Spanish		275
Corn (Maize), yellow		17
Corn (Maize) gluten meal		175–290
Corn gluten feed		17
Corn, distillers dried grains + solubles		4–34
Marigold petal meal		7,000
Seaweeds		340–920
Yeast	*Phaffia rhodozyma*	30–800
Animal Products		**Astaxanthin**
Capelin oil	*Mallotus villosus*	64–94
Copepod	*Calanus finmarchicus*	39–84
Copepod oil	*C. finmarchicus*	520
Crab, freeze dried	*Geryon quinquedens*	76
Crab, red	*Pleuroncodes planipes*	100–160
Crab, red, oil extract	*P. planipes*	1,500
Crab, vacuum dried	*Chinochetes opilio*	5
Crayfish, oil extract	*Procambrus clarkii*	750
Crayfish, meal	*P. clarkii*	137
Krill	*Euphausia pacifica*	100–130
Krill, codried with oil	*E. pacifica*	200
Krill, oil	*E. pacifica*	727
Krill	*Megannyctiphanes norvegica*	46–93
Mackerel	*Scomber scomburs*	6–11
Shrimp	*Pandalus borealis*	20–128
Shrimp, vacuum dried	*P. borealis*	100
Shrimp, steam dried	*P. borealis*	192
Shrimp oil	*P. borealis*	1,095

more efficiently than free astaxanthin (Lorenz and Cysewski, 2000). Gilthead sea bream effectively use various natural and synthetic carotenoids for deposition into the skin but not the flesh (Gomes et al., 2002); however, none of the pigments produced exceptional differences in skin coloration. Goldfish and other ornamental fish also metabolize astaxanthin and other carotenoids for deposition into the skin to impart color (Lorenz and Cysewski, 2000). Goldfish and common carp are able to convert dietary zeaxanthin to astaxanthin (Hata and Hata, 1972), and similar conversions have been reported for *Marsupenaeus japonicus* (Tanaka et al., 1976).

The retention of carotenoids in fish tissues depends on an array of factors that include absorption, transport, metabolism, and excretion of these compounds (as reviewed by Torrissen et al., 1989). The digestibility of astaxanthin found in yeast and shrimp waste meal is typically low. However, ensiling of shrimp byproducts improves the digestibility of astaxanthin by degrading the chitin-calcium-protein-carotenoid complex (carotenoprotein compound) in shrimp shells (Torrissen et al., 1981). Free astaxanthin is absorbed more efficiently than the astaxanthin ester, and the rate of hydrolysis of the astaxanthin ester to produce free astaxanthin in the digestive tract of salmonids is limited. Approximately 90% of astaxanthin in fish flesh is located in free form, while the ester form predominates in skin. Salmonids are unable to oxygenate carotenoids, but deposit ingested oxygenated carotenoids without modification. Wide differences in the accumulation of carotenoids may be caused by differences in absorption of these compounds. Apparently, absorption is enhanced by the incorporation of hydroxyl groups into the carotene skeleton because astaxanthin is deposited at significantly higher concentrations than canthaxanthin in both Atlantic salmon and rainbow trout (Torrissen, 1986; Choubert and Storebakken, 1989; Torrissen, 1989). In salmonids, the absorption of astaxanthin and canthaxanthin is 10 to 20 times more efficient than lutein and zeaxanthin; in contrast, chickens absorb zeaxanthin at three times the rate of astaxanthin (Schiedt et al., 1985).

Astaxanthin is commonly found as the primary pigment in crustaceans with the greatest proportion found in the exoskeleton (Katayama et al., 1972; Meyers and Latscha, 1997). The concentration (intensity) of coloration is directly related to the proximity of the dietary carotenoid compound in the biosynthetic pathways to the astaxanthin end product as demonstrated by D'Abramo et al. (1983) for juvenile lobsters fed formulated diets containing different carotenoid compounds or sources of these compounds. Cultured shrimp that have a blue hue rather than a dark green color associated with wild populations are considered to be consuming diets with insufficient levels of carotenoids. Recommended levels are between 50 and 100 ppm of astaxanthin to maintain threshold tissue levels that impact the color for consumer acceptance (Meyers and Latscha, 1997), but levels are also dependent upon the availability of the carotenoid compound(s) itself (themselves). A dietary level of 230 mg of canthaxanthin per 100 g of formulated diet (2300 ppm) for crustacean larvae, *Macrobrachium rosenbergii* and *Litopenaeus vannamei* (Kovalenko et al., 2002; D'Abramo et al., 2006), resulted in a level of pigmentation equivalent to that obtained with the feeding of live *Artemia* nauplii. Sources of carotenoids that have been used in crustacean diets include Antarctica krill (Maoka et al., 1985), *Spirulina*, at 3.0% of the diet (Liao et al. 1993); the yeast *P. rhodozyma* (Meyers and Sanderson, 1992); and the microalga *Heamatococcus pluvialis* (Choubert and Heinrich, 1993).

Torrissen (1990) concluded that dietary carotenoids are required in crustacean as well as fish diets, suggesting that these compounds have a metabolic role similar to vitamins A and D. Miki (1991) found that carotenoids have an antioxidant activity that is 100 times that of α-tocopherol and that carotenoids protect membrane lipids from peroxidation.

ENZYMES

Numerous feedstuffs of plant origin contain various forms of nutrients that are not readily digested by nonruminant animals because specific digestive enzymes are lacking. Supplementation of exogenous enzymes to diets containing such feedstuffs has been implemented in recent years, especially with terrestrial livestock such as swine and poultry, to increase the utilization of these nutrients (Bedford, 2000). For aquatic species, phosphorus in the form of phytate phosphorus, the primary storage form of phosphorus in plant feedstuffs, is a primary concern because of its poor utilization. This results in excretion by fish, potentially leading to enrichment of receiving waters. Up to 70% of the phosphorus in plant feedstuffs may be in the form of phytate, the primary storage form in seeds (grains and oilseeds, for example).

Due to concerns over limiting phosphorus waste in aquaculture, an increase in the utilization of dietary phytate through supplementation of the microbial enzyme phytase has been most extensively pursued. Numerous studies have demonstrated that dietary supplementation of phytase from *Aspergillus niger* can significantly improve the availability of phosphorus in selected fish species (Gatlin and Li, 2008). Generally, an addition of 500–1,000 units of phytase per kilogram diet has sufficiently increased phytate utilization as measured by enhanced bone ash, bone phosphorus, phosphorus retention, and decreased fecal phosphorus. Apparent availability of other minerals such as calcium, magnesium, iron, and zinc is typically increased with the addition of phytase. Improved protein digestibility and/or nitrogen utilization realized through supplementation of dietary phytase has not been consistently observed in aquatic species, although dietary phytic acid has the potential to interfere with protein digestion (e.g., Spinelli et al., 1983; Storebakken et al., 1998).

The improvement of the utilization of dietary protein by supplementation of other exogenous enzymes is highly desirable; however, attempts to date have been rather limited. Kolkovski (2001) reviewed the potential use of enzymes in diet formulations for larval and juvenile fish and concluded that addition of proteolytic enzymes has not consistently improved diet digestibility or certain measures of fish performance. More research in this area is warranted because the contribution of exogenous proteinase and carbohydrase enzymes from many prey organisms can substantially contribute to digestive processes in fish (Kuz'mina and Golovanova, 2004).

Evaluation of the effect of supplementation of other exogenous enzymes such as carbohydrase enzymes, including α-amylase, xylanase, and β-glucanase, has been primarily conducted with swine and poultry (Bedford, 2000). Enzymes such as α-amylase have been supplemented to the diet of terrestrial livestock to increase the digestibility of starch in nonviscous grains such as corn and sorghum, whereas the addition of other enzymes may reduce the adverse effects of nonstarch polysaccharides (NSP) in viscous grains such as barley, oats, rye, and wheat. These enzymes include β-mannanase that hydrolyzes β-mannans in many plant feedstuffs, xylanase that hydrolyzes xylans found principally in wheat, and β-glucanase that hydrolyzes β-glucans that are prominent in barley (Lobo, 1998). The NSP in these viscous grains typically increase intestinal viscosity, reduce rates of digestion, and contribute to the production of wet, viscous manure. Therefore, addition of appropriate carbohydrases to diets containing viscous grains can facilitate more rapid and complete digestion of NSP. Other enzymes, such as α-galactosidase, can hydrolyze oligosaccharides in soybean meal and thus increase the digestibility of that feedstuff (Lobo, 1998).

Incorporation of these enzymes into fish diets may enhance digestion of organic constituents, thus reducing the biochemical oxygen demand in aquaculture systems as well as settleable solids in effluent waters. However, evaluation of these enzymes with fish has been extremely limited to date. In one of the few studies with fish, Glencross et al. (2003) evaluated the addition of α-galactosidase to diets containing lupin meal because of its relatively high level of α-galactosyl homologues of sucrose such as raffinose, stachyose, and verbascose. Supplementation of α-galactosidase significantly improved protein digestion and tended to increase the digestibility of energy and nitrogen-free extract by rainbow trout. Similarly, the addition of a carbohydrase mixture to a diet containing canola meal increased diet digestibility and growth of black tiger prawn, *Penaeus monodon*, apparently caused by the removal of the negative effects of glucosides and other antinutritional factors in canola meal (Buchanan et al., 1997). Such application of carbohydrases to other plant feedstuffs in formulated diets containing undesirable NSP may increase their nutritive value to fish and shellfish. Therefore, in the future, more research in this area should be conducted with the goal of improving the nutritive value of various plant feedstuffs for aquacultured organisms.

One principal constraint to dietary inclusion of exogenous enzymes is that most of these enzymes are not stable at temperatures above approximately 95°C; therefore, they cannot withstand the harsh conditions of preconditioning prior to compression pelleting or extrusion processing. Encapsulation is a technique currently being explored to improve the heat resistance of supplemental enzymes. Currently, post-extrusion application of liquid enzymes is required until more thermally tolerant forms become available. Vacuum infusion is one technique that may allow efficient application of enzymes after pellet formation.

ORGANIC ACIDS

Organic acids including acetic, butyric, citric, formic, lactic, malic, proprionic, and sorbic acids and their salts have been used as acidifiers in animal feeds. These compounds have been shown to exert various beneficial effects such as increased animal performance by limiting the growth of microorganisms in the feed, increasing the availability of nutrients and altering the animal's gastrointestinal tract function and energy metabolism (Luckstadt, 2008). Limited studies thus far with fish have indicated that, similar to terrestrial livestock, dietary organic acids can exert strong antimicrobial effects and potentially increase growth, nutrient utilization, and disease resistance (e.g., Ng et al., 2009).

To date most of the studies concerning organic acids and fish have focused on utilization of phosphorus and other minerals. For example, dietary supplementation of citric acid and sodium citrate was able to improve phosphorus digestibility of fish meal to rainbow trout (Sugiura et al., 1998). Formic acid at supplemention levels of 4 and 10 ml/kg diet significantly improved phosphorus digestibility and retention in rainbow trout (Vielma and Lall, 1997). Supplementing citric acid at levels ranging from 4 to 30 g/kg to diets containing fish or fish bone meals improved phosphorus utilization by rainbow trout and red sea bream (Vielma et al., 1999; Sarker et al., 2005). Pandey and Satoh (2008) observed that 1% citric acid supplementation significantly improved phosphorus retention in rainbow trout fed a low-fish meal-based diet. Supplementing organic acids have also been observed to improve phosphorus digestibility of diets by agastric fish species (Leng et al., 2006). The positive effect of organic acids is probably due to the solubilization of bone minerals in fish meal, as well as a chelating effect that reduces the antagonistic interaction between calcium and phosphorus that could precipitate calcium and phosphorus at the intestinal brush border (Sugiura et al., 1998; Sarker et al., 2005). Baruah et al. (2005) observed that supplementation (30 g/kg diet) of citric acid significantly increased bone phosphorus content in Indian carp (*Labeo rohita*) fed a soybean meal-based diet. These authors also observed significant interaction between citric acid and phytase enzyme on bone phosphorus content, indicating a synergetic effect of these two feed additives.

High levels of free form organic acids in feeds for swine can result in reduced feed intake, mucosal damage to the gastrointestinal tract, bone demineralization, and metabolic acidosis (Partanen and Mroz, 1999; Gauthier, 2002), which could ultimately have a negative effect on the growth performance of the animal. Sugiura et al. (1998) observed that citric acid levels up to 5% (50 g/kg diet) did not affect feed intake of rainbow trout but resulted in marked reduction of feed intake in goldfish (*Carassius auratus*). Organic acids also can possibly disturb acid:base balance and mineral homeostasis but more research on aquatic organisms is warranted in this regard.

FEEDING STIMULANTS/PALATABILITY ENHANCERS

Feedstuffs of marine origin, such as fish meal, krill meal, shrimp meal, fish solubles, fish oil, and various protein hydrolysates, are noted for the high palatability that they render as ingredients to feed for aquatic organisms (Barrows, 2000). When plant feedstuffs are substituted for these highly palatable marine feedstuffs, diet acceptance diminishes and performance responses of several cultured species, especially those with carnivorous feeding habits, decrease accordingly (Hardy and Barrows, 2002). Therefore, palatability enhancement of diets containing plant feedstuffs is an area of investigation that has received substantial attention during recent years as reduced inclusion of marine feedstuffs is pursued to decrease diet costs as well as the significant dependence on such ingredients.

A variety of natural and synthetic ingredients have been effective in overcoming the adverse effects of reduced diet palatability caused by the lack of marine feedstuffs. Such ingredients may be termed palatability enhancers, gustatory stimulants, or attractants, although, in the strictest sense, attractants are defined as substances that orient animals located at a distance toward a potential prey item (Guillaume and Metailler, 2001). An ingredient that leads to ingestion is also commonly referred to as a stimulant. Compounds that are known to stimulate diet intake by fish have been characterized to be of low molecular weight, nonvolatile, water-soluble, and nitrogenous (Carr, 1982). Guillaume and Metailler (2001) further classified fish attractants into three primary groups consisting of (1) L-amino acids, (2) betaine or other molecules with a pentavalent nitrogen atom, and

(3) nucleosides and nucleotides, with the phosphorylated nucleotides being more effective. The feedstuffs of marine origin previously stated as palatability enhancers neatly fit into these categories because various fish meals and fish solubles are rich in nucleotides, fish protein hydrolysates contain free amino acids, and invertebrate meals are rich in nitrogenous bases. Many of these feeding stimulants are typically found in high concentrations in tissue extracts of natural prey items of various fish and shrimp species. Feeding stimulants are frequently species specific, and mixtures are generally more effective than single compounds in increasing feed intake. The amino acid glycine has been shown to be a feeding stimulant in several carnivorous fish species at levels up to 2% of diet, as has betaine and various combinations of betaine, glycine, and other amino acids (Gatlin and Li, 2008). A mixture of L-alanine, L-serine, inosine 5′-monophosphate, and betaine included at 2 and 4% of diet, as well as other amino acid attractant mixtures included in diets based primarily on plant feedstuffs, have resulted in improved weight gain and feed intake of carnivorous fish.

Dietary nucleotides have most recently received considerable attention as immunomodulating compounds for various fish species (addressed in the next section); however, these compounds also may actually influence diet intake of fish. Inosine and inosine 5′-monophosphate have been effective in improving diet consumption by different fish species (Gatlin and Li, 2007).

IMMUNOSTIMULANTS

Heightened attention has recently been directed to the development of nutritional strategies and dietary supplements that positively influence health and immunity of aquacultured organisms to reduce dependence on chemotherapeutics to treat diseases. In particular, supplementation of nonnutritional immunostimulants into diets has become recognized as an effective practice for enhancing immunocompetence and disease resistance of various aquatic species. Compounds identified as having immunostimulatory properties are very diverse in their chemical composition and mode(s) of immunostimulation. Those compounds that have been shown to have beneficial effects on some aquatic species are presented as broadly defined groups in the following subsections.

Microbial, Plant, and Animal Fractions

Inactivated natural microbes or microbial products such as β-glucans, lipopolysaccharides, and peptidoglycans can stimulate the cell-mediated immune system. Some of these products can be delivered orally without complete degradation in the digestive system and taken up by the intestinal mucosa and thus may be used as potential immunostimulants for aquaculture.

β-glucans have been extensivly recognized as immunostimulants in various aquatic animals (reviewed by Sealey and Gatlin, 2001; Gatlin, 2002). A variety of β-glucans has been extracted from many microbial fractions and products such as brewers yeast and barley. These compounds generally have been shown to influence nonspecific immune responses such as blood neutrophil oxidative radical production and superoxide anion production in activated macrophages. These changes may lead to significant reductions in mortality of fish exposed to different pathogenic organisms. Discoveries of β-glucan receptors such as Dectin-1 and Toll-like receptors on leukocyte surfaces of vertebrates have provided some understanding about mechanisms of immune responses induced by β-glucans (reviewed by Brown and Gordon, 2003). Currently consistent conclusions about the efficacy of long-term administration of yeast glucans are lacking; however, several research groups have generated both in vitro and in vivo evidence that inappropriate use of β-glucans may suppress immunity at the cellular and organismal levels (reviewed by Gatlin, 2002).

Lactoferrin is an iron-containing glycoprotein released from neutrophil granules during inflammation. It is present in milk or other exocrine secretions of mammals and appears to have antibacterial, antiviral, antifungal, antiinflammatory, antioxidant, and immunomodulatory activities in vertebrates. This protein can modulate several physiological processes such as expression of heat shock proteins; however, the digestion, absorption, and transport of this protein when supplemented to a diet remain undefined. Yet, several studies have confirmed the beneficial effects of dietary lactoferrin supplementation on immune functions such as cellular innate responses, including respiratory burst and natural cytotoxic activity of fish (Esteban et al., 2005) and phenoloxidase activity and bacterial agglutination titers in crustaceans (Chand et al., 2006). Enhanced resistance of fish and shrimp to environmental stressors such as high temperature (Yokoyama et al., 2005) and low salinity (Koshio et al., 2000), as well as infectious bacterial (e.g., Sakai et al., 1993; Chand et al., 2006) and viral diseases (e.g., Zhong et al., 2002), also have been effected through lactoferrin supplementation. Effective dietary levels reported in several studies have ranged from 100 to 1,000 mg/kg diet.

Some substances obtained from seaweeds, mainly polysaccharides, have modified the immune response and increase protection against infectious diseases in some fish (reviewed by Chiu et al., 2008). Alginates containing high percentages of mannuronic acid polymers (86–99.9% M-blocks, termed high-M alginates) produce an immunomodulatory effect on Atlantic halibut (Skjermoa and Berghb, 2004) and Atlantic cod (Skjermoa et al., 2006). Sodium alginate also has been shown to exert immunostimulating effects and confer disease resistance to grouper (Chiu et al., 2008; Yeh et al., 2008) and Pacific white shrimp (Cheng et al., 2005).

Chitin is a polymer of glucosamine and is found in shells or cell walls of invertebrates, fungi, and yeasts. This compound is the main component of crustacean exoskeletons, making up 50–80% of organic compounds and consisting of calcium oxide and protein units. Chitosan, an aminopolysaccharide, is prepared from shellfish chitin by treatment with alkali. Both chitin and chitosan exerted immunostimulatory effects on common carp (Gopalakannan and Arul, 2006).

The influences of numerous other dietary immunostimulants on fish and shrimp health, including peptidoglycan, lipopolysacchardes, sulfated polysaccharide, 3, 3′, 5-triiodo-L-thyronine, β-hydroxy-β-methylbutyrate, and certain herbal products such as essential oils have been studied to a limited extent, but additional supporting research is warranted. For example, the phenolic compounds carvacrol and thymol extracted from the oregano essential oils, have been shown to have considerable antimicrobial properties, and their supplementation in the diet increased resistance of channel catfish to *Aeromonas hydrophila* (Zheng et al., 2009)

Nucleosides and Nucleotides

Nucleosides and nucleotides traditionally have not been considered to be essential nutrients because they are metabolically synthesized, and no signs of deficiency are observed when they are excluded from the diet. Therefore, the nutritional value of nucleosides, nucleotides, and oligonucleotides continues to be debated. Recently, this opinion has been modified by the published results of successive investigations that suggest nucleotide requirements increase during injury and/or wound repair, and deficiency may impair liver, heart, intestine, and immune functions (Gatlin and Li, 2007). Initial efforts in exploration of nucleotide nutrition in fish can be traced to the early 1990s. Worldwide, heightened attention was aroused by the reports of Burrells et al. (2001a,b) about the role of nucleotides in formulated aquatic feeds by influencing immune responses and enhancing resistance of Atlantic salmon to various pathogens. However, a complete understanding of the molecular mechanism(s) associated with physiological and immunological responses induced by dietary nucleotides is still lacking. It is currently known that tissues such as the intestinal mucosa, bone marrow, hematopoietic cells, lymphocytes, and the brain have limited capacity for de novo synthesis of nucleotides and derive their supply from the salvage pathway. Therefore, the endogenous supply of nucleotides may be inadequate, and a dietary supply may be needed for optimal functioning, especially of the immune system, under stressful conditions such as sepsis and trauma (Gatlin and Li, 2007). Improved resistance to various bacterial, viral, rickettsial, and ectoparasitic organisms has been reported for several different aquatic species with dietary supplementation of various nucleotide preparations ranging from 0.62 to 5 g/kg diet. Different components of the specific and nonspecific immune systems including antibody response, neutrophil oxidative radical production, and serum lysozyme, as well as intracellular and extracellular superoxide anion production of head kidney macrophages, have been affected by the provision of dietary nucleotides.

Synthetic Chemicals

Levamisole is an antihelminthic compound used for the treatment of nematodes in humans and animals. FK-565 (heptanoyl-y-D-glutamyl-(L)-meso-diaminopimelyl-(D)-alanine) is a peptide related to lactoyl tetrapeptide isolated from cultures of *Streptomyces olivaceogriseus* and has been shown to be active against microbial infection in mice (Mine et al., 1983). Both of these synthetic chemicals, levamisole and FK-565, have been reported to provide immunomodulatory effects in fish (reviewed by Sakai, 1999; Li et al., 2006).

PROBIOTICS AND PREBIOTICS

Probiotics, originally defined as live, microbial dietary supplements that improve health of humans and terrestrial livestock, have recently been a topic of interest in aquaculture (Gatesoupe, 1999; Gatlin, 2002; Irianto and Austin, 2002; Burr et al., 2005). These dietary supplements alter the microbiota in the gastrointestinal (GI) tract, which is recognized as playing important roles in growth, digestion, immunity, and disease resistance of the host organism. Although some probiotics are designed to treat the aquatic environment for competitively excluding potential pathogens, most probiotics are delivered to hosts by dietary supplementation.

In recent years, many studies have shown the potential value of the use of probiotics in aquaculture through diet manipulation to enhance growth, immunity, and resistance of aquatic animals to various diseases. However, not all investigations of probiotics have yielded positive results. Besides different strains of bacteria, other organisms such as bacteriphages, fungi, and microalgae also have been added to diets as probiotics. It is anticipated that dietary supplementation of probiotics will become increasingly important as an effective strategy to improve the health of aquacultured organisms. One possible constraint to the use of probiotics is maintenance of viability during diet manufacture. Prebiotics and inactivated probiotics are two other related dietary supplements that have emerged during recent years. Prebiotics are defined as nondigestible dietary ingredients that beneficially affect the host by selectively stimulating the growth of and/or activating the metabolism of health-promoting bacteria in the GI tract (Gibson and Roberfroid, 1995). A variety of compounds have already been established as having prebiotic effects in terrestrial food animals and some limited number of fish and shrimp species. These compounds include oligosaccharides such as mannanoligosaccharides, fructooligosaccharides, and galactooligosaccharies, as well as some commercial products containing mixtures of autolyzed brewers yeast, dairy sugars, and fermentation products (Gatlin and Li, 2008). Dietary fatty acids and carbohydrates have been previously shown to alter the microbiota of the GI tract of fish and thus may have potential application as prebiotics (Ringø et al., 1998; Ringø and Olsen, 1999). Use of these supplements remains in its infancy in aquaculture. However, additional literature on the effects of prebiotics on aquatic species, including specific responses of microbes in the GI tract, continues to grow. Some prebiotics have been shown to enhance weight gain, feed efficiency, and some immune responses and resistance to some bacterial and parasitic infections. Although not all prebiotic investigations with fish and shrimp have yielded positive responses, negative effects of prebiotic supplementation have not been reported.

Inactivated probiotics also have shown protection against infectious diseases in some fish including rainbow trout (Irianto and Austin, 2002) and goldfish (Irianto et al., 2003). In these studies, formalin-inactivated cells that are added to the diet have assisted in controlling bacterial infection. Based on these studies, the classification of inactivated cells as probiotics, immunostimulants, or oral vaccines is uncertain; however, the benefit of using dead cells as diet additives for aquaculture should receive further consideration. Development of prebiotics and combined prebiotic/probiotic supplements may have many applications to aquaculture and need to be further evaluated.

HORMONES

Various natural and synthetic hormones have been used in aquaculture for a variety of purposes including inducement of spawning, sex reversal or production of monosex populations, as well as growth enhancement. A wide variety of hormones have been evaluated in fish growth experiments including growth hormone, thyroid hormones, gonadotropin, prolactin, insulin, and various steroids, however, none of these hormones has been approved by the FDA for use in fish destined for human consumption and thus are not supplemented to feeds in purified form.

Experimental feeding of synthetic androgens has been shown to enhance growth and improve feed conversion of some species, especially salmonids (Donaldson et al., 1979; Higgs et al., 1982; Matty, 1986). Approximately 20 fish species have shown anabolic responses to steroids (Donaldson et al., 1979; Matty, 1986). Some warmwater species, however, such as channel catfish (Gannam and Lovell, 1991a,b), have responded negatively to the feeding of androgens. Prolonged steroid treatment for growth acceleration may cause detrimental side effects including early gonadal development, skeletal deformity, increased susceptibility to infections, and pathological changes in the liver, kidney, and digestive tract (Zohar, 1989; Gannam and Lovell, 1991a,b).

Sex steroids also have been administered in the diet to reverse the sex of some species of salmonids, carps, and tilapias (Pandian and Sheela, 1995). These hormones are typically dissolved in ethanol and sprayed onto the diet, which is administered to first-feeding fish for approximately 1 month to produce same-sex populations. Administration of 17-α-methyltestosterone to tilapia by such methods produces predominantly male fish that grow more rapidly than females and also limits unwanted reproduction. This practice must be conducted under an approved New Animal Drug Application from the FDA if the fish are destined for human consumption. Studies have shown that the 17-α-methyltestosterone is stable in feed for several months if stored at 4°C or lower but declines linearly with time at higher temperatures (Barry et al., 2007); it is rapidly depleted from viscera and carcass of fish (Johnstone et al., 1983) such that it poses no threat to human food safety, and its sensitivity to photo-oxidation and bacterial degradation will not result in adverse environmental effects (Green and Teichert-Coddington, 2000).

REFERENCES

Ahmed, T. S., and A. J. Matty. 1989. The effects of antibiotics on growth and body composition of carp (*Cyprinus carpio*). Aquaculture 77:211–220.

Barrows, F. T. 2000. Feed additives. Pp. 335–340 in Encyclopedia of Aquaculture, R. R. Stickney, ed. Hoboken, NJ: John Wiley and Sons.

Barry, T. P., A. Marwah and P. Marwah. 2007. Stability of 17α-methyltestosterone in fish feed. Aquaculture 271:523–529.

Baruah, K., A. K. Pal, N. P. Sahu, K. K. Jain, S. C. Mukherjee, and D. Debnath. 2005. Dietary protein level, microbial phytase, citric acid and their interactions on bone mineralization of *Labeo rohita* (Hamilton) juveniles. Aquac. Res. 36:803–812.

Bedford, M. 2000. Exogenous enzymes in monogastric nutrition—their current value and future benefits. Anim. Feed Sci. Tech. 86:1–13.

Breithaupt, D. E. 2007. Modern application of xanthophylls in animal feeding—a review. Trends Food Sci. Tech. 18:501–506.

Brown G. D., and S. Gordon. 2003. Fungal β-glucans and mammalian immunity. Immunity 19:11–315.

Buchanan, J., H. Z. Sarac, D. Poppi, and R. T. Cowan. 1997. Effects of enzyme addition to canola meal in prawn diets. Aquaculture 151:29–35.

Burr, G., D. M. Gatlin, III, and S. Ricke. 2005. Microbial ecology of the gastrointestinal tract of fish and the potential application of prebiotics and probiotics in finfish aquaculture. J. World Aquacult. Soc. 36:425–436.

Burrells, C., P. D. William, and P. F. Forno. 2001a. Dietary nucleotides: A novel supplement in fish feeds. 1. Effects on resistance to diseases in salmonids. Aquaculture 199:159–169.

Burrells, C., P. D. William, P. J. Southage, and S. L. Wadsworth. 2001b. Dietary nucleotides: A novel supplement in fish feeds. 2. Effects on vaccination, salt water transfer, growth rate and physiology of Atlantic salmon. Aquaculture 199:171–184.

Cabello, F. C. 2006. Heavy use of prophylactic antibiotics in aquaculture: A growing problem for human and animal health and for the environment. Environ. Microbiol. 8:1137–1144.

Carr, W. E. S. 1982. Chemical stimulation of feeding behavior. Pp. 259–273 in Chemoreception in Fishes, T. J. Hara, ed. Amsterdam, the Netherlands: Elsevier.

Chand, R. K., P. K. Sahoo, J. Kumari, B. R. Pillai, and B. K. Mishra. 2006. Dietary administration of bovine lactoferrin influences the immune ability of the giant freshwater prawn *Macrobrachium rosenbergii* (de Man) and its resistance against *Aeromonas hydrophila* infection and nitrite stress. Fish Shellfish Immun. 21:119–129.

Cheng, W., C. H. Liu, C. M. Kuo, and J. C. Chen. 2005. Dietary administration of sodium alginate enhances the immune ability of white shrimp *Litopenaeus vannamei* and its resistance against *Vibrio alginolyticus*. Fish Shellfish Immun. 18:1–12.

Chien, Y.-H., and S.-C. Jeng. 1992. Pigmentation of kuruma prawn, *Penaeus japonicas* Bate, by various pigment sources and levels and feeding regimes. Aquaculture 102:333–346.

Chiu, S. T., R. T. Tsai, J. P. Hsu, C. H. Liu, and W. Cheng. 2008. Dietary sodium alginate administration to enhance the non-specific immune responses, and disease resistance of the juvenile grouper *Epinephelus fuscoguttatus*. Aquaculture 277:66–72.

Choubert, G., and O. Heinrich. 1993. Carotenoid pigments of the green alga *Haematococcus pluvialis*: assay on rainbow trout, *Oncorhynchus mykiss*, pigmentation in comparison with synthetic astaxanthin and canthaxanthin. Aquaculture 112:217–226.

Choubert, G., and T. Storebakken. 1989. Dose response to astaxanthin and canthaxathin pigmentation of rainbow trout fed various carotenoid concentrations. Aquaculture 81:69–77.

D'Abramo, L. R., N. A. Baum, C. E. Bordner, and D. E. Conklin. 1983. Carotenoids as a source of pigmentation in juvenile lobsters fed a purified diet. Can. J. Fish. Aquat. Sci. 40:699–704.

D'Abramo, L. R., E. I. Perez, R. Sangha, and A. Puello-Cruz. 2006. Successful culture of the larvae of *Litopenaeus vannamei* fed a microbound formulated diet exclusively from either stage PZ2 or M1 to PL1. Aquaculture 261:1356–1362.

Donaldson, E. M., U. H. M. Fagerlund, D. A. Higgs, and J. R. McBride. 1979. Hormonal enhancement of growth in fish. Pp. 455–597 in Fish Physiology, Vol. 8, W. S. Hoar, D. J. Randall, and J. R. Brett, eds. New York: Academic Press.

Doolen, B. J., M. A. Booth, G. L. Allan, and P. L. Jones. 2009. Changes in skin colour and cortisol response of Australian snapper *Pagrus auratus* (Bloch & Schneider, 1801) to different background colours. Aquac. Res. 40:542–550.

Esteban, M. A., A. Rodriguez, A. Cuesta, and J. Meseguer. 2005. Effects of lactoferrin on non-specific immune responses of gilthead seabream (*Sparus auratus* L.). Fish Shellfish Immun. 18:109–124.

FDA (U.S. Food and Drug Administration). 2009. Aquaculture: Approved Drugs, 06/18/2009. Available online at: *http://www.fda.gov/Animal Veterinary/DevelopmentApprovalProcess/Aquaculture/ucm132954.htm*. Accessed on May 11, 2010.

Fujita, T., M. Satake, S. Hikichi, M. Takeda, S. Shimeno, H. Kuwabara, W. Miki, K. Yamaguchi, and S. Konosu. 1983. Pigmentation of cultured yellowtail with krill oil. B. Jpn. Soc. Sci. Fish. 49:1595–1600.

Gannam, A. L., and R. T. Lovell. 1991a. Effects of feeding 17 alpha-methyltestosterone, 11-ketotesterone, 17 beta-estradiol and 3,5,3'-triiodothyronine to channel catfish, *Ictalurus punctatus*. Aquaculture 92:377–388.

Gannam, A. L., and R. T. Lovell. 1991b. Growth and bone development in channel catfish fed 17-α-methyltestosterone in production ponds. J. World Aquacult. Soc.22:95–100.

Gatesoupe, F. J. 1999. The use of probiotics in aquaculture. Aquaculture 180:147–165.

Gatlin, D. M., III. 2002. Nutrition and fish health. Pp. 671–702 in Fish Nutrition, 3rd edition, J. E. Halver and R.W. Hardy, eds. San Diego, CA: Academic Press.

Gatlin, D. M., III, and P. Li. 2007. Nucleotides. Pp. 193–209 in Dietary Supplements for the Health and Quality of Cultured Fishes, H. Nakagawa, M. Sato, and D. M. Gatlin III, eds. Oxon, UK: CABI.

Gatlin, D. M., III, and P. Li. 2008. Use of Diet Additives to Improve Nutritional Value of Alternative Protein Sources. Pp. 501–522 in Alternative Protein Sources in Aquaculture Diets, C. Lim, C. D. Webster, and C.-S. Lee, eds. New York: Haworth Press.

Gauthier, R. 2002. The mode of action of acidifiers and the interest they generate in the growing-finishing phase. P. 16 in Current Developments in Pig Production. Maisons-Alfort, France: French Association of Swine Practitioners.

Gibson, G. R., and M. B. Roberfroid. 1995. Dietary modulation of the human colonic microbiota: Introducing the concept of prebiotics. J. Nutr. 125:1401–1412.

Glencross, B. D., T. Boujard, and S. J. Kaushik. 2003. Influence of oligosaccharides on the digestibility of lupin meals when fed to rainbow trout, *Oncorhynchus mykiss*. Aquaculture 219:703–713.

Gomes, E., J. Dias, P. Silva, L. Valente, J. Empis, L. Gouveia, J. Bowen, and A. Young. 2002. Utilization of natural and synthetic sources of carotenoids in the skin pigmentation of gilthead seabream (*Sparus aurata*). Eur. Food Res. Technol. 214:287–293.

Gopalakannan, A., and V. Arul. 2006. Immunomodulatory effects of dietary intake of chitin, chitosan and levamisole on the immune system of *Cyprinus carpio* and control of *Aeromonas hydrophila* infection in ponds. Aquaculture 255: 179–187.

Green, B. W., and D. R. Teichert-Coddington. 2000. Human food safety and environmental assessment of the use of 17α-methyltestosterone to produce male tilapia in the United States. J. World Aquacult. Soc. 31:337–357.

Guillaume, J., and R. Metailler. 2001. Raw materials and additives used in fish foods. Pp. 281–295 in Nutrition and Feeding of Fish and Crustaceans, J. Guillaume, S. Kaushik, P. Bergot, and R. Metailler, eds. Chichester, UK: Praxix Publishing.

Hardy, R. W., and F. T. Barrows. 2002. Diet formulation and manufacture. Pp. 505–600 in Fish Nutrition, J. E. Halver and R. W. Hardy, eds. San Diego, CA: Academic Press.

Hata, M., and M. Hata. 1972. Carotenoid pigments in goldfish. 4. Carotenoid metabolism. B. Jpn. Soc. Sci. Fish. 38:331–338.

Higgs, D. A., U. H. M. Fagerlund, J. G. Eales, and J. R. McBride. 1982. Application of thyroid and steroid hormones as anabolic agents in fish culture. Comp. Biochem. Phys. B 73B:143–146.

Higuera-Ciapara, I., L. Felix-Valenzuela, and F. M. Goycoolea. 2006. Astaxanthin: A review of its chemistry and applications. Crit. Rev. Food Sci. 46:185–196.

Irianto, A., and B. Austin. 2002. Use of probiotics to control furunculosis in rainbow trout, *Oncorhynchus mykiss* (Walbaum). J. Fish Dis. 25:333–342.

Irianto, A. P. A., W. Robertson, and B. Austin. 2003. Oral administration of formalin-inactivated cells of *Aeromonas hydrophila* A3-51 controls infection by atypical *A. salmonicida* in goldfish, *Carassius auratus* (L.). J. Fish Dis. 26:117–120.

Johnstone, R., D. J. Macintosh and R. S. Wright. 1983. Elimination of orally administered 17α-methyltestosterone by *Oreochromis mossambicus* (tilapia) and *Salmo gairdneri* (rainbow trout) juveniles. Aquaculture 35:249–257.

Katayama, T., T. Katama, and C. O. Chichester. 1972. The biosynthesis of astaxanthin in the prawn Penaeus japonicus Bate (Part II). Int. J. Biochem. 3:363–368.

Kolkovski, S. 2001. Digestive enzymes in fish larvae and juveniles—Implications and applications to formulated diets. Aquaculture 200:181–201.

Koshio, S., S. Yokoyama, M. Ishikawa, H. Hayasawa, and K. Oshida. 2000. Effect of bovine lactoferrin on growth survival and tolerance to stress of kuruma prawn *Penaeus japonicus*. Pp. 443–449 in Lactoferrrin: Structure, Function and Applications, K. Shimazaki, ed. Amsterdam, the Netherlands: Elsevier.

Kovalenko, E. E., L. R. D'Abramo, C. L. Ohs, and R. K. Buddington. 2002. A successful microbound diet for the larval culture of freshwater prawn (*Macrobrachium rosenbergii*). Aquaculture 220:769–780.

Kuz'mina, V. V., and I. L. Golovanova. 2004. Contribution of prey proteinases and carbohydrases in fish digestion. Aquaculture 234:347–360.

Leng, X. J., F. Lun, X.Q. Li, Z.Q. Wang, and K. J. Xu. 2006. Effects of citric acid on growing performance and nutrients digestibility of allogynogenetic crucian carp. J. Shanghai Fish. Univ. 15:178–181.

Li, M. H., E. H. Robinson, D. F. Oberle, and P. V. Zimba. 2007. Effects of various dietary carotenoid pigments on fillet appearance and pigment absorption in channel catfish *Ictalarus punctatus*. J. World Aquacult. Soc. 38:557–563.

Li, P., X. Wang, and D. M. Gatlin III. 2006. Evaluation of levamisole as a feed additive for growth and disease management of hybrid striped bass *Morone chrysops* × *M. saxatilis*. Aquaculture 251:201–209.

Liao, W. L., S. A. Nur-E-Borhan, S. Okada, T. Matsui, and K. Yamaguchi. 1993. Pigmentation of a cultured black tiger prawn by feeding with a *Spirulina*-supplemented diet. Nippon Suisan Gakk. 59:165–169.

Lobo, J. P. 1998. How can enzymes benefit me? Feed Manag. 49:12–16.

Lorenz, R. T., and G. R. Cysewski. 2000. Commercial potential for *Haematococcus* microalgae as a natural source of astaxanthin. Trends Biotechnol. 18:160–167.

Lovell, T. 1998. Nutrition and Feeding of Fish, 2nd edition. Boston, MA: Kluwer Academic Publishers.

Luckstadt, C. 2008. The use of acidifiers in fish nutrition. CAB Reviews: Perspectives in Agriculture, Veterinary Science, Nutrition and Natural Resources. Vol. 3, no. 44. doi: 10.1079/PAVSNNR20083044.

Maoka, T., M. Katsuyama, N. Kaneko, and T. Matsuno. 1985. Stereochemical investigation of carotenoids in the Antarctic krill *Euphuasia superba*. B. Jpn. Soc. Sci. Fish. 51:1671–1673.

Matty, A. J. 1986. Nutrition, hormones and growth. Fish Physiol. Biochem. 2:141–150.

Meyers, S. P., and T. Latscha. 1997. Carotenoids. Pp. 164–193 in Crustacean Nutrition, Advances in World Aquaculture, Vol. 6, L. R. D'Abramo, D. E. Conklin, and D. M. Akiyama, eds. Baton Rouge, LA: World Aquaculture Society.

Meyers, S. P., and G. W. Sanderson. 1992. Natural pigments for salmon feed. Feed Manag. 43:12–16, 20.

Miki, W. 1991. Biological functions and activities of animal carotneoids. Pure Appl. Chem. 63:141–146.

Mine, Y., Y. Yokota, Y. Wakai, S. Fukuda, M. Nishida, S. Gota, and S. Kuwahara. 1983. Immunoactive peptide, FK-156 and FK-565: 1. Enhancement of host resistance to microbial infection in mice. J. Antibiot. 36:1045–1050.

Ng, W.-K., C.-B. Koh, K. Sudesh, and A. Siti-Zahrah. 2009. Effects of dietary organic acids on growth, nutrient digestibility and gut microflora of red hybrid tilapia, *Oreochromis* sp., and subsequent survival during a challenge test with *Streptococcus agalactiae*. Aquac. Res. 40:1490–1500.

NRC (National Research Council). 1993. Nutrient Requirements of Fish. Washington, DC: National Academy Press.

Pandey, A., and S. Satoh. 2008. Effects of organic acid on growth and phosphorus utilization in rainbow trout *Oncorhynchus mykiss*. Fish. Sci. 74(4):867–874.

Pandian, T. J., and S. G. Sheela. 1995. Hormonal induction of sex reversal in fish. Aquaculture 138:1–22.

Partanen, K., and Z. Mroz. 1999. Organic acids as an alternative for prophylactic medication of pig diets. Nutr. Res. Rev. 12:117–145.

Porter, W. L.1980. Recent trends in food applications of antioxidants. Pp. 295–365 in Autoxidation in Food and Biological Systems, M. G. Simic and M. Karel, eds. New York: Plenum.

Rawles, S. D., A. Kocabas, and D. M. Gatlin, III. 1997. Dietary supplementation of Terramycin and Romet-30 does not enhance growth of channel catfish but does influence tissue residues. J. World Aquacult. Soc. 28:392–401.

Ringø, E., and R. E. Olsen. 1999. The effect of diet on aerobic bacterial flora associated with intestine of Arctic charr, *Salvelinus alpinus* L. J. Appl. Microbiol. 86:22–28.

Ringø, E., H. R. Bendiksen, S. J. Gausen, A. Sundsfjord, and R. E. Olsen. 1998. The effect of dietary fatty acids on lactic acid bacteria associated with the epithelial mucosa and from faecalia of Arctic charr, *Salvelinus alpinus* L. J. Appl. Microbiol. 85:855–864.

Sakai, M. 1999. Current research status of fish immunostimulants. Aquaculture 172:63–92.

Sakai, M., T. Obubo, S. Atsuta, and M. Kobayashi. 1993. Enhancement of the resistance to bacterial infection in rainbow trout, *Oncorhynchus mykiss* (Walbaum), by oral administration of bovine lactoferrin. J. Fish Dis. 16:239–247.

Sarker, S. A., S. Satoh, and V. Kiron. 2005. Supplementation of citric acid and amino acid-chelated trace element to develop environment-friendly feed for red sea bream, *Pagrus major*. Aquaculture 248:3–11.

Schiedt, K., F. J. Leuenberger, M. Vecchi, and E. Glinz. 1985. Absorption, retention, and metabolic transformations of carotenoids in rainbow trout, salmon and chicken. Pure Appl. Chem. 57:685–692.

Sealey, W. M., and D. M. Gatlin, III. 2001. Overview of nutritional strategies affecting the health of marine fish. Pp. 103–118 in Nutrition and Fish Health, C. Lim and C. D. Webster, eds. Binghamton, NY: The Haworth Press Inc.

Shahidi, F., Metusalach, and J. A. Brown. 1998. Carotenoid pigments in seafoods and aquaculture. Critical Rev. Food Sci. 38:1–67.

Shao, Z. J. 2001. Aquaculture pharmaceuticals and biologicals: current perspectives and future possibilities. Adv. Drug Deliver. Rev. 50:229–243.

Skjermoa, J., and Ø. Berghb. 2004. High-M alginate immunostimulation of Atlantic halibut (*Hippoglossus hippoglossus* L.) larvae using *Artemia* for delivery, increases resistance against vibriosis. Aquaculture 238:107–113.

Skjermoa, J., T. R. Størseth, K. Hansen, A. Handå, and G. Øie. 2006. Evaluation of β-(1→3, 1→6)-glucans and high-M alginate used as immunostimulatory dietary supplement during first feeding and weaning of Atlantic cod (*Gadus morhua* L.). Aquaculture 261:1088–1101.

Spinelli, J., C. R. Houle, and J. C. Wekell. 1983. The effect of phytates on the growth of rainbow trout (*Salmo gairneri*) fed purified diets containing varying quantities of calcium and magnesium. Aquaculture 30:71–83.

Storebakken, T., K. D. Shearer, and A. J. Roem. 1998. Availability of protein, phosphorus and other elements in fish meal, soy-protein concentrate and phytase-treated soy-protein-concentrate-based diets to Atlantic salmon, *Salmo salar*. Aquaculture 161:365–379.

Sugiura, S. H., F. M. Dong, and R. W. Hardy. 1998. Effects of dietary supplements on the availability of minerals in fish meal; preliminary observations. Aquaculture 160:283–303.

Tanaka, Y., H. Matsuguchi, and T. Katayama. 1976. The biosynthesis of astaxanthin. XVI. The carotenoids in crustacea. Comp. Biochem. Phys. B 54B:391–393.

Tejera, N., J. Rosa Cejas, C. Rodríguez, B. Bjerkeng, S. Jerez, A. Bolaños, and A. Lorenzo. 2007. Pigmentation, carotenoids, lipid peroxides and lipid composition of skin of red porgy (*Pagrus pagrus*) fed diets supplemented with different astaxanthin sources. Aquaculture 270:218–230.

Torrissen, O. J. 1986. Pigmentation of salmonids—A comparison of astaxanthin and canthaxanthin as pigments sources for rainbow trout. Aquaculture 53:271–278.

Torrissen, O. J. 1989. Pigmentation of salmonids—Interactions of astaxanthin and canthaxanthin on pigment deposition in rainbow trout. Aquaculture 81:129–136.

Torrissen, O. J. 1990. Biological activities of carotenoids in fishes. The current status of fish nutrients in aquaculture. Pp. 387–399 in Proceedings of the Third International Symposium on Feeding and Nutrition of Fish, M. Takeda and T. Watanabe, eds. Tokyo, Japan: Tokyo University of Fisheries.

Torrissen, O. J., and R. Christiansen. 1995. Requirements for carotenoids in fish diets. J. Appl. Ichthyol. 11:225–230.

Torrissen, O. J., E. Tidemann, F. Hansen, and J. Raa. 1981. Ensiling in acid: A method to stabilize astaxanthin in shrimp processing byproducts and improve uptake of this pigment by rainbow trout. Aquaculture 26:77–83.

Torrissen, O. J., R. W. Hardy, and K. D. Shearer. 1989. Pigmentation of salmonids—Carotenoid deposition and metabolism. Rev. Aquat. Sci. 1:209–225.

Vielma, J., and S. P. Lall. 1997. Dietary formic acid enhances apparent digestibility of minerals in rainbow trout, *Oncorhynchus mykiss* (Walbaum). Aquacult. Nutr. 3:265–268.

Vielma, J., K. Ruohonen, and S. P. Lall. 1999. Supplemental citric acid and particle size of fish bone-meal influence the availability of minerals in rainbow trout *Oncorhynchus mykiss* (Walbaum). Aquacult. Nutr. 5:65–71.

Viola, S., and Y. Arieli. 1987. Non-hormonal growth promoters for tilapia and carp I. Screening tests in cages. Isr. J. Aquacult. Bamidgeh 42:91–94.

Yeh, S. P., C. A. Chang, C. Y. Chang, C. H. Liu, and W. Cheng. 2008. Dietary sodium alginate administration affects fingerling growth and resistance to *Streptococcus* sp. and iridovirus, and juvenile non-specific immune responses of the orange-spotted grouper, *Epinephelus coioides*. Fish Shellfish Immun. 25:19–27.

Yokoyama, S., S. Koshio, N. Takakura, K. Oshida, M. Ishikawa, F. J. Gallardo-Cigarroa, and S. Teshima. 2005. Dietary bovine lactoferrin enhances tolerance to high temperature stress in Japanese flounder *Paralichthys olivaceus*. Aquaculture 249:367–373.

Zheng, Z. L., J. Y. W. Tan, H. Y. Liu, X. H. Zhou, X. Xiang, K. Y. Wang. 2009. Evaluation of oregano essential oil (*Origanum heracleoticum* L.) on growth, antioxidant effect and resistance against *Aeromonas hydrophila* in channel catfish (*Ictalurus punctatus*). Aquaculture 292:214–219.

Zhong, J., Y. Wang, and Z. Zhu. 2002. Introduction of the human lactoferrin gene into grass carp (*Ctenopharyngodon idellus*) to increase resistance against GCH virus. Aquaculture 214:93–101.

Zohar, Y. 1989. Endocrinology and fish farming: Aspects in reproduction, growth, and smoltification. Fish Physiol. Biochem. 7:395–405.

11

Antinutritional Factors and Adventitious Toxins in Feeds

The terms "antinutritional factor" (ANF) and "antinutrients" refer to endogenous substances in foods and feedstuffs that produce negative effects on health and nutrient balance when ingested by animals or humans. Some plants or plant-derived feedstuffs containing antinutrients may cause acute signs of poisoning, but subtle effects produced only by prolonged ingestion of a given plant are more common under normal feeding conditions. Such effects might include disturbance of digestive processes and growth, decreased feed efficiency, pancreatic hypertrophy, hypoglycaemia, liver dysfunction, goiterogenesis, and immune suppression. Various animal byproducts also contain antinutrients that, if not properly preserved or processed, may adversely affect animals. The effects vary with species of animal, age or stage of development, size, sex, state of health and plane of nutrition, and any stress factors that might be superimposed on these variables.

Plant and animal feed ingredients currently used in aquaculture or with the potential for use as a nutrient source for fish are listed in Table 11-1, along with the known antinutrients found in each. Unknown factors are not listed for obvious reasons. However, it is likely that components of plants not harmful to traditional farm animals or model animals such as rats and mice are harmful to various fish species in cultivation. Moreover, it is possible that hitherto unidentified compounds in plants are important in fish nutrition and health.

A review article by Krogdahl et al. (2009), a risk evaluation regarding use of plant ingredients in diets for aquacultured fish by the Norwegian Scientific Committee for Food Safety (Hemre et al., 2009), and the earlier comprehensive review by Francis et al. (2001a) have been among the main sources for the information presented below. Most of the compounds listed in Table 11-1 are addressed to varying degrees in the present review focusing on knowledge gained from experiments with purified antinutrients in fish. Hardly any information has been found on effects of antinutrients in shrimp.

The term "adventitious toxin" refers in this context to compounds that are natural or accidental contaminants derived from human sources, with the exception of pathogenic agents, that are present in and/or on the product intended

TABLE 11-1 Important Antinutrients Present in Some Commonly Used Potential Fish Feed Ingredients

Plant Feedstuff	Antinutrient Present
Soybean meal	Proteinase inhibitors, lectins, phytic acid, saponins, phytoestrogens, antivitamins, phytosterols, allergens
Kidney beans	Proteinase inhibitors, amylase and lipase inhibitors, lectins, phytic acid, saponins, phytoestrogens, antivitamins, phytosterols, allergens
Rapeseed meal	Proteinase inhibitors, glucosinolates, phytic acid, tannins, erucic acid
Lupin seed meal	Proteinase inhibitors, saponins, phytoestrogens, alkaloids
Pea seed meal	Proteinase inhibitors, lectins, tannins, cyanogens, phytic acid, saponins, antivitamins
Sunflower oil cake	Proteinase inhibitors, saponins, arginase inhibitor
Cottonseed meal	Phytic acid, phytoestrogens, gossypol, antivitamins, cyclopropenoid acid
Alfalfa leaf meal	Proteinase inhibitors, saponins, phytoestrogens, antivitamins
Mustard oil cake	Glucosinolates, tannins
Sesame meal	Phytic acid, proteinase inhibitors
Raw fish	Thiaminase
Crustaceans	Chitin, fluorine

SOURCE: Francis et al. (2001b).

for animal feed and that present potential danger to animal or human health or to the environment or could adversely affect fish and shrimp production. Examples are pesticides, mycotoxins, environmental pollutants or substances such as oxidized fatty acids, and polyamines that are produced during storage and processing of the feed ingredients, such as oxidized fatty acids and polyamines. Further, feed ingredients may contain elevated levels of inherent substances, e.g., some feed materials of plant origin contain elevated levels of heavy metals due to geological characteristics of the soil. Table 11-2 lists various compounds that potentially need attention in fish feeds (Hemre et al., 2009). The following sections provide detailed information about the most prominent ANFs and adventitious toxins encountered in feedstuffs used for aquatic feeds.

ANTINUTRIENTS IN PLANT FEEDSTUFFS

Enzyme Inhibitors

Most plant seeds contain inhibitors protecting their major components from unintended degradation, i.e., inhibitors of protein, starch, and lipid. They are simple or more complex proteins that are denatured by treatments such as heating,

TABLE 11-2 Adventitious Toxins and Other Undesirable Substances That May Contaminate Fish Feed (Undesirable Substances by Category)[a]

Item	Substance
Ions or elements	Arsenic, Lead, Fluorine, Mercury, Nitrites, Cadmium
Products produced under storage or processing	Aflatoxin B_1 and other mycotoxins
	Hydrocyanic acid
	Theobromine
	Volatile mustard oil
	Vinyl thiooxazolidone
	Rye ergot (*Claviceps purpurea*)
	Weed seeds and unground and uncrushed fruits containing alkaloids, glucosides, or other toxic substances separately or in combination (including *Lolium temulentum* L., *Lolium remotum* Schrank, and *Dantura stramonium* L.)
	Castor oil plant (*Ricinus communis* L.)
	Crotalaria (*Crotalaria* spp.)
	Oxidized fatty acids, aldehydes, and ketones
	Biogenic amines
Organohalogenated compounds	Aldrin
	Dieldrin
	Camphechlor (toxaphene)
	Chlordane
	Dichlorodiphenyltrichloroethane (DDT)
	Endosulfan
	Endrin
	Heptachlor
	Hexacholorobenzene (HCB)
	Hexacholorocyclo-hexane (HCH)
	Dioxins and dioxin-like PCBs
Botanical impurities	Apricots (*Prunus armeniaca* L.)
	Bitter almonds (*Prunus dulcis* [Mill] D.A. Webb var. *amara* [DC.] Focke) = (*Prunus amygdalus Batsch* var. *amara* [DC.] Focke)
	Unhusked beech mast (*Fagus silvatica* L.)
	Camelina (*Camelina sativa* L. Crantz)
	Mowrah, Bassia, Madhuca (*Madhuca longifolia* L. Macbr.) = (*Bassia longifolia* L.) = (*Illiped malabrorum* Engl.) = Gmelin (*Madhuca indica*) = (*Bassia latifolia* Roxb.) = (*Illipe latifolia* [Roscb.] F. Mueller)
	Purghera (*Jatropha curcas* L.)
	Croton (*Croton tiglium* L.)
	Indian mustard (*Brassica juncea* L. Czern. and Coss. spp.) = (*intergrifolia* [West.] Thell.)
	Sareptian mustard (*Brassica juncea* L. Czern. and Coss. spp. *Juncea*)
	Chinese mustard (*Brassica juncea* L. Czern. and Coss. spp.), (*juncea* var. *lutea* Batalin)
	Black mustard (*Brassica nigra* L. Koch)
	Ethiopian mustard (*Brassica carinata* A. Braun)

[a]Adapted from Directive 2002/32/EC, European Commission; Commission of the European Communities (2003).

alcohol extraction, and fermentation. However, complete inactivation is usually not intended as it induces side effects that make nutrients unavailable to the animals.

Proteinase inhibitors, i.e., inhibitors of trypsin, chymotrypsin, elastase, enterokinase, and carboxypeptidases, are proteins that form stoichiometric, reversible complexes with the enzymes and inhibit their activity in the gastrointestinal (GI) tract (Liener, 1980; Bhat et al., 1981). The specificity of proteinase inhibitors varies. Some inhibit only one type of enzyme, e.g., trypsin or carboxypeptidase; others inhibit two or three. Proteinase inhibitors, even at low dietary levels, have been found to reduce apparent digestibility in rainbow trout (*Oncorhynchus mykiss*) not only of protein but also of lipids (Berg-Lea et al., 1989; Krogdahl et al., 1994; Olli et al., 1994a). The proteinase inhibitors seem to stimulate pancreatic enzyme secretion. However, after longer-term feeding, the pancreas may no longer manage to compensate for the presence of active inhibitors by increasing secretion.

Amylase inhibitors, proteins that inhibit amylase in monogastric animals, are found in starch-rich plant parts such as beans, grains, and tubers (Whitaker, 1997; Gibbs and Alli, 1998; Sawada et al., 2002). Studies of effects of amylase inhibitors in fish are limited. Fernandez et al. (2001) and Natarajan et al. (1992) have shown that an amylase inhibitor from wheat is active against amylase from sea bream, carp, and tilapia. Salmonid and cod amylase are affected by an inhibitor from beans (Froystad et al., 2006). Effects of amylase inhibitors in diets for Atlantic salmon seem to be limited, however, because the amylase of this species seems to have a defective substrate anchor, reducing its catalytic ability. Its secretion from the pancreas also seems limited. In a recent feeding trial with Atlantic salmon in which an amylase inhibitor concentrate from common beans was included in a pelleted diet at 2 g/kg, no effect was observed on starch digestibility or digestibility of protein, total lipid, and fatty acids (Chikwati, 2007). This level should be in the upper range of levels present in extruded fish diets with significant inclusion of beans and other plant feedstuffs.

Lipase inhibitors are present in many oil seeds in which they appear to be important in the prevention of autodigestion of the lipid reserves by the seed lipases prior to seed germination (Gargouri et al., 1984; Wang and Huang, 1984; Bau et al., 1997; Gupta et al., 2003). The lipase inhibitors are protein in nature and have been described in seeds of soybean (Gargouri et al., 1984; Wang and Huang, 1984; Satouchi et al., 1998, 2002), sunflower (Wang and Huang, 1984; Chapman, 1987) and peanut (Wang and Huang, 1984). Lipase inhibitor proteins have also been demonstrated in a number of cereals including wheat (Borel et al., 1989; Cara et al., 1992; Tani et al., 1994, 1995; O'Connor et al., 2003), oat (O'Connor et al., 2003), millet, barley, and sorghum (Cara et al., 1992).

The lipase inhibitor from soybean is considered to prevent lipase function through interference with its binding to the surface of the emulsified lipid droplets (Gargouri et al., 1984; Wang and Huang, 1984). For lipase inhibitors from sunflower seed, however, a direct interaction between the enzyme and the inhibitor by formation of a tightly bound inhibitor protein-lipase complex was postulated as the likely mode of interaction between the inhibitor and the lipase (Chapman, 1987). The lipase inhibitors in wheat appear to behave in a comparable manner to the soybean lipase inhibitors (Borel et al., 1989).

Presence of inhibitory activity in heat-treated soybean flour suggests a degree of thermostability of the lipase inhibitors and their possible presence in solvent-extracted soybean meal. Lipase inhibitor activity may partly account for the impaired lipid digestibility in Atlantic salmon fed diets with an inclusion of solvent-extracted soybean meal (Olli and Krogdahl, 1994; Olli et al., 1994b; Storebakken et al., 1998a; Refstie et al., 2000, 2001).

No feeding trials have been reported that investigated dietary effects of plant lipase inhibitors in fish, despite the economic implications of poor utilization of the expensive lipid. Therefore, in vivo biological activity in various fish species is not known, and no maximum inclusion level can be estimated. Effects of lipase inhibitors from cereals on fish are possible confounders to the effects on fish macronutrient digestibilities that are attributed to cereal fiber.

Lectins

Lectins (previously known as agglutinins or hemagglutinins) are a group of soluble, heterogeneous (glyco) proteins that possess at least one noncatalytic domain that binds reversibly to a specific mono- or oligosaccharide (Peumans and Van Damme, 1995). They were discovered first in plants (Stillmark, 1888), but later also in microorganisms and animals.

The specific recognition of carbohydrates is one of lectins' characteristics used in their classification. Thus, plant lectins have been classified according to the mono- or oligosaccharide that prevents the lectin from agglutinating erythrocytes or precipitating glycoconjugates in vitro (from Van Damme et al., 1998).

The legume (bean) lectins are a large family of homologous proteins (Sharon and Lis, 1990). Some seeds contain as much as 20 g per kg on a dry matter basis, although most legume seeds contain 1–5 g per kg. On a protein basis, seed lectins account for 0.1–5% of total protein, but up to 50% in some *Phaseolus* species. It appears that variations in lectin levels occur between different cultivars within a species, as well as with stress factors and ecological and climatological conditions that the plants are exposed to during growth.

Lectins are widespread in plants and crops and are partially resistant to heat denaturation, as well as to breakdown during passage through the digestive tract (Hara et al., 1984; Pusztai et al., 1990; Bardocz et al., 1995). They interact with endogenous surface receptors of the intestinal epithelial cells of all higher and lower animals that ingest them. During migration along the crypt-villus axis, the maturing intestinal epithelial cells continuously change in terms of glycation profile of components in the cellular membrane. This variability in glycation patterns helps explain the variability in biological effects of different lectins in different species of animals (Roth, 1987; Pusztai et al., 1995). Because of their ability to bind to glycoconjugates on animal cell receptors, lectins can have striking biological effects on intestinal function, as well as systemic effects depending on the function of the reseptors. The potency of the lectin's effect on cell metabolism appears to be correlated with the strength of its binding to a ligand, which in turn is dependent not only on its defined sugar/carbohydrate specificity, but also on other structural intricacies of the ligand (Grant et al., 1983).

Binding of soybean lectin (also called soybean agglutinin; SBA) to carbohydrate moieties of glycoconjugates, specifically to N-acetyl-d-galactosamine, on the intestinal brush border membrane of Atlantic salmon and rainbow trout has been demonstrated (Hendricks et al., 1990; Buttle et al., 2001), in which higher maximum binding and lower dissociation constants were observed in the distal intestine relative to the more proximal areas (Hendriks et al., 1990). Inclusion of soybean lectin at a level of 35 g per kg feed, an inclusion level equivalent to a 600 g solvent-extracted soybean meal per kg feed, reportedly caused histological changes in the distal intestine of Atlantic salmon (Buttle et al., 2001). The work of Bakke-McKellep and coworkers (2008) indicated that the lectin from kidney beans (*Phaseolus vulgaris*) inhibits glucose transport into the intestinal epithelium of Atlantic salmon when tissue is exposed in vitro. The soybean lectin did not show similar inhibition of glucose absorption.

An involvement of lectin in soybean-meal-induced enteritis, first described in Atlantic salmon in 1990 (van den Ingh and Krogdahl, 1990; van den Ingh et al., 1991) and later in rainbow trout (Rumsey et al., 1993, 1994) and common carp (Uran et al., 2008), is questioned because lectins have low solubility in the alcohol mix that removes the potent factor(s). Moreover, the characteristics of the histological alterations reported by Buttle et al. (2001) resulting from inclusion of soybean lectins were different from those observed in the soybean-meal-induced enteritis. However, Francis et al. (2001b) suggested that lectins' deleterious effects are more potent when in the presence of other ANFs.

Maximum inclusion level of lectins from various plant protein sources have not been determined and will most likely vary considerably among specific lectins and fish species, as indicated by the discussion above.

Saponins

Saponins are a highly diverse group of glycosides produced primarily by plants, particularly legumes, but also by a few lower marine animals and some bacteria. More than 100 families of plants are known to contain saponins and considerable heterogeneity of chemical structures exists. Saponins are heat-stabile, alcohol-soluble compounds. They are amphipathic molecules, containing a hydrophobic steroidal or triterpenoid aglycone to which one or more (hydrophilic) sugar chains are attached. The sugar moiety (oligosaccharide) usually contains glucose, galactose, glucuronic acid, xylose, rhamnose, or methyl pentose glycosidically linked to the hydrophobic aglycone (sapogenin).

Saponins are reported to have diverse biological effects. They show antifungal, anticancer, and antiviral activity, immune stimulating and antioxidant properties, inhibitory effects on protein digestion and vitamin absorption, and glucocorticoid-like effects (reviewed by Francis et al., 2002). Saponins form micelles and can intercalate into cholesterol-containing membranes. Saponins also affect functions of mammalian intestinal epithelia by increasing the permeability, facilitating uptake of substances that are normally not absorbed, such as allergens (Johnson et al.1986; Gee et al., 1996).

Orally administered saponins that are incorporated into cell membranes will eventually be lost in the normal process of intestinal epithelial replacement (Sjolander and Cox, 1998). Saponins are also lost due to binding with cholesterol, forming an insoluble complex that cannot be absorbed (Malinow et al., 1977). In the only published work on the fate of orally administered saponins in fish, soyasaponins appeared not to be degraded during passage through the GI tract of Atlantic salmon (Knudsen et al., 2006).

The work of Knudsen et al. (2007, 2008) strongly indicates that saponins play a role in soybean-meal-induced enteritis, but they require the presence of other unknown components. Supplementation of a semipurified soy saponin product to a fish meal-based diet (1.7 and 2.6 g/kg diet) did not alter intestinal morphology. However, when the diet contained lupin kernel meal, the saponin supplementation caused symptoms similar to those developed in salmonids fed soybean meal. In vitro studies of the gut wall in the lupin + saponin fed fish showed increased permeability. The authors concluded that soybean saponins increased the intestinal epithelial permeability but did not, per se, induce enteritis.

At this time, data are insufficient to reliably predict safe levels of saponins in diets for fish. In a study by Twibell and Wilson (2004) supplementation of a semipurified diet with 0.26% purified saponins did not cause any alteration in growth or feed intake. Caution should be exercised when deciding acceptable levels. Reduced feed intake has been observed in fish fed diets containing high levels of saponins due to effects on palatability. Specific responses or sensitivities of different fish species to the various saponins have not been adequately addressed, and herbivorous or omnivorous fish may exhibit different tolerances to saponins than carnivorous fish. It should be noted that substantial chemical heterogeneity of saponins exists, and different saponins exhibit varying levels of biological activity (Yoshiki et al., 1998; Oda et al., 2000).

Phytoestrogens

Phytoestrogens are reported to be present in several plant-based fish feed ingredients including soy, lupin, cottonseed, linseed, safflower, and alfalfa (reviewed by Francis et al., 2001b). They include isoflavones, coumestans, and lignans. Plant estrogens are mostly isoflavones that occur as glycosides with a sugar moiety attached to one or more hydroxyl groups located at various positions (Pellisero et al., 1991a). The total isoflavone content of soybeans has been reported to vary and can reach levels as high as 4,200 mg/kg (Wang and Murphy, 1994a), but considerable variation exists influenced by variety, location, and growing conditions (Wang and Murphy, 1994b).

One of the concerns regarding estrogenic effects of phytoestrogens is their effect on reproductive performance, but only a few studies have addressed this issue. Pelissero et al. (1991a) demonstrated increased vitellogenin in blood plasma of juvenile male and nonvitellogenic females fed diets containing soybean meal compared to individuals fed a casein-based control diet in Siberian sturgeon (*Acipenser baeri*). Pelissero et al. (1991b) subsequently reported direct evidence for estrogenic activity of daidzein, biochanin A, genistein, equol, and coumestrol after intraperitoneal injection in this species. However, the activity was considerably less that the effect of estradiol. The affinity of genistein for blood plasma binding proteins in fish has been observed to be much lower than endogenous estrogens as shown by Tollefsen et al. (2004) and Milligan et al. (1998). Milligan et al. (1998) concluded that the phytoestrogens coumestrol, genistein, and daidzein are unlikely to produce effects by displacing endogenous steroids from plasma binding proteins in rainbow trout. No differences in blood plasma estradiol concentrations were found in female rainbow trout fed diets containing 500 or 1,000 mg aglycone genistein per kilogram for 1 year (Bennetau-Pelissero et al., 2001). Reproductive performance was affected, but only in females fed the diet containing genistein at 500 mg/kg. However, male rainbow trout had decreased levels of 11-ketotestosterone and testosterone and reduced sperm motility but increased sperm volume. Spermatocrit was reduced in males at the higher dietary level of genistein. Inudo et al. (2004) found no effect on fecundity or fertility of Japanese medaka (*Oryzias latipes*) fed diets containing up to 58.5 mg/kg genistein and 37.3 mg/kg daidzein over a 28-day period, although vitellogenesis was induced in males.

Isoflavonoids in plants occur mainly as glycosides, but the vast majority of research has focused on aglycone forms. It is not known whether, and to what extent, glycoside forms are converted in the fish GI tract, and if they elicit the same changes when fish are fed aglycone isoflavonoids. The information necessary to understand the significance of dietary isoflavonoids on other physiological processes in fish is incomplete. D'Souza et al. (2005) fed rainbow trout genistein aglycone and found accumulation in muscle although no dose–response or time relationship was found. Evidence in mammals suggests that enterohepatic recycling and reabsorption after glycoside hydrolysis and deglucuronidation by intestinal microbes is necessary to achieve a significant physiological role of dietary isoflavonoids (Gatlin et al., 2007). Additional research is needed to clarify the metabolism of native forms of estrogenic compounds found in plant feedstuffs by fish.

There is insufficient information regarding the effects of dietary phytoestrogens in fish to reliably predict safe inclusion levels.

Glucosinolates/Goitrogens

Cruciferous foods and feeds such as rapeseed, as well as soybeans, have been associated with disrupted thyroid hormone production, or goiter, in animals and humans. The causative agents (goitrogens) of the thyroid disruption are glucosinolates. Content and composition of glucosinolates vary due to plant species, agronomic practices, and climatic conditions. Major metabolites of glucosinolates are thiocyanates, isothiocyanates, nitriles, 5-vinyl-2-oxazolidinethione, and 5-vinyl-1,3-oxyzolodine-2-thione (VOT). Different glycosinolate profiles among the rapeseed meals induce varying levels of glucosinolate metabolites in animal tissues (Burel et al., 2001; Francis et al., 2001b; Tripathi and Mishra, 2007). Hydrolysis of the glucosinolates can be catalyzed by thioglucosidases present in the feedstuffs themselves, separated from the substrate by physical barriers. Damage to plant tissue will release the enzyme and initiate the hydrolysis that eventually will result in toxic products. Also, some microbes of the GI tract possess thioglucosidases, which will add to the hydrolysis (Combourieu et al., 2001; Shapiro et al., 2001; Fuller et al., 2007).

The major deleterious effects of glucosinolate ingestion in animals in general are reduced palatability, and decreased growth and production (reviewed by Tripathi and Mishra, 2007). The thiocyanates interfere with iodine availability, whereas VOT is responsible for the morphological and physiological changes of the thyroid. The nitriles are known to affect liver and kidney functions.

Tripathi and Mishra (2007) reported that the general response of fish to dietary inclusion of canola and rapeseed meals is favorable, but all reports recommend a limited inclusion level depending on the glucosinolate content of the meals. Diets containing 0.22–2.18 mmol/kg of glucosinolates did not affect intake, growth, or thyroid hormone levels in red sea bream (*Pagrus auratus*) (Glencross et al., 2004a,b); at 2.18 mmol/kg glucosinolates reduced growth approximately by 15%. The rainbow trout (*Oncorhynchus mykiss*), when fed diets containing 1.4–19.3 mmol/kg of glucosinolate, showed a significant reduction in growth and changes in thyroid histology (Burel et al., 2000). A strong growth depression was observed at glucosinolate intake of 30–47 μmol/kg fish body weight per day. Diets containing 1.4 μmol/g did not cause growth-depressing effects but decreased plasma T3 levels. The authors suggested that a total glucosinolate content of 1.4 µmol/g of diet is considered a safe upper limit, although this may induce low-level goitrogenecity.

Tannins

Tannins are water-soluble phenolic compounds with molecular weights between 500 and 3,000. They are grouped into two classes, the hydrolyzable and the non-hydrolyzable or condensed (Marquardt, 1989). Hydrolyzable tannins are polyesters of phenolic acids such as gallic acid, hexahydroxydiphenic acid, and/or their derivates and D-glucose or quinic acid. The condensed tannins are polymers of flavan-3-ols, flavan-3,4-diols, or related flavanol residues linked via carbon-carbon bonds, which do not have carbohydrate cores like the hydrolyzable tannins (Mueller-Harvey and McAllan, 1992). The seed coat of brown legumes and rapeseeds are rich in condensed tannins, which are brown pigments coloring the seeds (Rao and Prabhavathi, 1982; Francis et al., 2001b). Such condensed tannins may disturb digestion by binding digestive enzymes or complexing to feed components such as proteins or minerals, and also have an astringent, bitter flavor (Liener, 1989; Marquardt, 1989; Mueller-Harvey and McAllan, 1992; Sandoval and Carmona, 1998).

Effects of tannins in fish have not been investigated extensively. Common carp have been shown to tolerate 20 g condensed (quebracho) tannin powder per kilogram diet without any effect on feed intake and growth (Becker and Makkar, 1999). It is, however, not known how well commercially available purified tannins serve as models for condensed tannins naturally occurring in plant seeds. In light of this, broad bean (*Vicia faba*) meal with a high content of condensed tannin had lower in vitro protein digestibility than soybean, and the differences were more pronounced when simulating carp digestion compared to rainbow trout (Grabner and Hofer, 1985). This indicates species differences in tannin interactions with other dietary components in the gut.

As the information about tannins in fish is so limited, caution should be exercised in incorporating feedstuffs containing high amounts of tannins in fish feeds. If tannin-rich seeds are to be used in aquafeeds, the dark seed coats should be removed, and only dehulled seeds (kernels) used. Most plant seeds used in salmonid feeds are, however, light colored. When feeding such ingredients, tannins should not be problematic.

Phytic Acid

Phytic acid (myoinositol, 1, 2, 3, 4, 5, 6 hexakis-dihydrogen phosphate) and phytate (a mixed cation salt of phytic acid) are found in significant amounts in plant material and are generally regarded as the primary storage form of both phosphate and inositol in seeds. Common plant feedstuffs such as soybean and rapeseed meal contain 10–15 and 50–75 g phytate per kilogram, respectively (Francis et al., 2001b). Phytic acid has 12 hydrogens (on the 6 phosphate groups) that are titratable in water. These negatively charged sites bind mainly K^+ and Mg^{2+} and also form salts and precipitate with other cations including Ca^{2+}, Mn^{2+}, Zn^{2+}, Ba^{2+}, or Fe^{3+} at alkaline pH (Lott et al., 2000; Hídvégi and Lásztity, 2002). Phytic acid also complexes with protein at low pH, below the isoelectric point of the protein, and when phytic acid is soluble (Hídvégi and Lásztity, 2002). Thus, phytic acid-protein complexes are only partially hydrolysed by pepsin (Spinelli et al., 1983), and phytic acid furthermore reduces the stability of trypsin (Caldwell, 1992).

In fish, a supplement of 5 g phytic acid per kilogram diet slowed growth by 10% in rainbow trout (Spinelli et al., 1983), while a dietary supplement of 26 g phytic acid per kilogram dramatically depressed growth in Chinook salmon (Richardson et al., 1985). The latter study also showed an increased incidence of cataracts when feeding diets high in phytic acid but low in Zn. Supplementation of 5–10 g phytic acid per kilogram diet furthermore slowed growth in common carp (Hossain and Jauncey, 1993). In line with this, improved growth and higher digestibility and retention of protein and mineral elements was shown in rainbow trout when comparing dephytinized with standard soy protein meal (Vielma et al., 2004), soy protein concentrate (Vielma et al., 2002, 2004), or rapeseed protein concentrate (Teskeredžić et al., 1995).

However, when comparing standard to dephytinized soy protein concentrate at 50% dietary inclusion in Atlantic salmon, Storebakken et al. (1998b) found no difference in growth. Still, in line with the expected effects of dietary phytic acid, the apparent digestibility and retention of both protein and phosphorus as well as whole-body concentrations of ash, Ca, Mg, and Zn were higher in fish fed the dephytinized soy product. Supporting information was reported by Denstadli et al. (2006) who found a tolerance for sodium phytate with regard to feed intake and growth of Atlantic salmon. The apparent digestibilities of Mg and Zn were reduced by dietary phytic acid in a dose-dependent manner, which resulted in lower whole-body concentrations of Ca and Mg and a lower concentration of Zn in the vertebral column, although apparent digestibility and retention of protein was little affected (Helland et al., 2006). When feeding as much as 21 g phytic acid per kilogram diet, the intestinal trypsin activity was also significantly depressed in Atlantic salmon (Denstadli et al., 2006).

Richardson et al. (1985) found abnormalities in thyroid, kidney, and alimentary tract morphology of rainbow trout fed 26 g phytic acid per kilogram diet. Furthermore, the pyloric caeca were found to be abnormally hypertrophied and showing cytoplasmic vacuolation. Supporting this, Hossain and Jauncey (1993) found hypertrophy and vacuolization of the cytoplasm of the intestinal epithelium in common carp fed 10 g phytic acid per kilogram diet. Thus, dietary phytic acid can affect tissue mass and potentially functionality of digestive organs in fish, and it may potentially be toxic at very high dietary levels. Feeding up to 21 g phytic acid per kilogram diet did not, however, induce any histomorphological changes in the distal intestine of Atlantic salmon (Denstadli et al., 2006). Although this dietary level of phytic acid lowered the mineral content of the fish, it did not induce any skeletal malformations within 80 days of feeding and a tripling of the body weight, starting at 36 g (Helland et al., 2006).

Gossypol

Gossypol ($C_{30}H_{30}O_8$) is a yellow, lipid soluble polyphenolic aldehyde derived from the cotton plant (genus *Gossypium*, family Malvaceae). It is the major toxic constituent in cottonseeds and is found in discrete pigment glands located in various parts of the plant. Gossypol content of cottonseeds varies from a trace to about 6%. It is affected by plant species and variety, as well as by environmental factors such as climate, soil type, and fertilization. Gossypol is a natural component of all except the "glandless" variety of cotton. Gossypol occurs in bound and free forms. Free gossypol, including gossypol derivatives and transformation or breakdown products, can be toxic to single-stomached animals. The free form is found as two enantiomers (Huang et al., 1987), and the ratios vary among varieties and processing methods (Gamboa et al., 2001). Gossypol and its metabolites exert pro- and antioxidant potential. Considerable evidence points to oxidative stress, formation of reactive oxygen species, and DNA scission, characteristics of redox-cycling by electron transfer in biosystems (Kovacic, 2003) as mechanisms of action. Furthermore, gossypol binds to microsomal membranes, inhibits DNA synthesis, and causes depletion of iron and glutathione in mammalian cells (Gawai et al., 1995).

Gossypol present in feed ingredients is bound to amino acids in peptides, especially lysine, and has high affinity for dietary iron. Bound gossypol is not readily absorbed and may prevent lysine absorption and induce lysine deficiency. Free gossypol is readily absorbed from the GI tract. Conjugation, metabolism, and urinary excretion of gossypol are limited; most is eliminated in the feces.

The toxic effects of gossypol have been documented in several species of fish (reviewed by Li and Robinson, 2006). Different levels of toxicity have been reported for different species, and even within a species various toxicity levels have been reported. The reasons for these differences may be the various conditions, fish ages and sizes, diet formulations, and analytical methods used. The toxic effects that have been reported include decreased feed consumption, growth, hematocrit, hemoglobin, and reproductive capacity. Histological lesions have been found in liver, kidney, spleen, and gonads. A study of effects of gossypol on reproduction in rainbow trout revealed negative effects and transfer of the compound to the offsprings (Blom et al., 2001). Despite several reports of toxic effects, death from gossypol toxicosis has rarely been reported in fish.

Yildirim et al. (2003) reported immunomodulation (improved macrophage chemotaxis ratio, serum lysozyme activity) and increased resistance of juvenile channel catfish (*Ictalurus punctatus*) to *Edwardsiella ictaluri* challenges at levels of 900 mg gossypol per kilogram or higher of a casein-gelatin based diet. However, in a later study, no differences to *E. ictaluri* challenge were observed in juvenile catfish fed up to 800 mg gossypol per kilogram of a soybean-meal-based diet (Yildirim-Aksoy et al., 2004).

Gossypol is not expected to cause toxicity if cottonseed meal from "glandless" varieties of cotton plants is utilized in feeds. It has also been suggested that cottonseed meals containing gossypol should not be problematic given the expected low inclusion levels of cottonseed meal in fish feeds (Li and Robinson, 2006). The Commission of the European Communities limits the amount of free gossypol in feed materials. The maximum allowable level in complete feeds for fish is 20 mg/kg (Commission of the European Communities, 2003).

Cyclopropene Fatty Acids—Sterculic and Malvalic Acids

Two fatty acids containing a cyclopropene ring characterized as 8-(2-octyl-cyclopropen-1-yl)-octanoic acid or sterculic acid and 7-(2-octyl-cyclopropen-1-yl)-heptanoic acid or malvalic acid are found in a large number of seed oils from plant families of the order Malvales. The most important member of this order in the context of aquaculture is cotton. The cyclopropene fatty acids are found in commercial cottonseed oil and cottonseed meal. These fatty acids inhibit desaturases in many animal species, including fish, with alterations in lipid metabolism and increased accumulation of saturated fatty acids as a result (Raju and Reiser, 1967). The cyclopropene ring is highly reactive, and the mechanism of action seems to be an irreversible reaction between the ring structures and thiol moieties on the enzymes (Raju and Reiser, 1967).

Similar effects on lipid metabolism have been observed in rainbow trout (Roehm et al., 1970; Malevski et al., 1974; Struthers et al., 1975; reviewed by Hendricks, 2002). Rainbow trout also show alterations in the mixed function oxidase system (Eisele et al., 1978) and other liver enzymes (Selivonchick et al., 1981). The effects also comprise structural alterations in the liver with altered histological appearance, abnormal accumulation of glycogen, necrosis, and fibrous accumulation in hepatocytes. When rainbow trout were fed diets with 7.5% cottonseed oil and a level of 90 mg/kg sterculic acid, one-third had developed hepatic neoplasm at 12 months, whereas two-thirds showed similar symptoms when fed diets with 25% glandless cotton seeds (Hendricks et al., 1980a). Interaction effects between the cyclopropene fatty acids and aflatoxin, another liver carcinogen, have been studied. Synergistic effects were observed that were dependent on the aflatoxin species used. The AFB_1 was much more potent than the AFQ_1 (Sinnhuber et al., 1974; Hendricks et al., 1980b). The main body of information regarding effects of the cyclopropene fatty acids in fish originates from experiments with rainbow trout. However, carcinogenesis has also been observed in sockeye salmon, but only when supplied together with aflatoxin AFB_1. In channel catfish, sterculic acid has been observed to cause growth retardation (Hendricks, 2002).

Because of chances of transferring cyclopropene fatty acids from the broodstock to the offspring, caution should be exercised if cotton seed meal is included in broodstock diets. Moreover, as young fish are more vulnerable than older, care should also be taken before cottonseed is included in starter feeds (Hendricks, 2002).

Glycoalkaloids

Potatoes and other tubers contain secondary metabolites such as the steroid glycoalkaloids (SGA) α-solanine and α-chaconine. The two compounds differ only in glycosylation. The latter contains L-mannose instead of D-glucose. Excessive SGA contents (i.e., 20 mg/100 g fresh weight or more) cause an unacceptable bitter taste, and tubers become unfit for human consumption. Steroid glycoalcaloid content varies quantitatively and qualitatively in response to photoperiod, wavelength, and intensity of light, soil moisture, stage of growth, storage conditions, growing season, air and soil temperatures, and wounding. Potatoes grown in a hot dry climate were reported to contain more glycoalkaloids than those grown in a high-altitude, cooler climate. Thus the exposure of potato plants to physiological stress and especially to extreme climatic conditions causes accumulation of SGA in the tubers. The role of SGA in the defense mechanism of potatoes is not clear. Potatoes also contain other proteins and metabolites that protect the tuber against attack from insects and microorganisms (reviewed by Friedman, 2006).

The main effects of SGAs on mammals are cell membrane permeabilization and cholinesterase activity inhibition. Due to the latter, SGAs can cause prolongation of myorelaxant, anesthetic and analgesic actions, and affect ester prodrug activation in blood serum. The main potato GAs, α-solanine and α-chaconine, in certain concentrations can provoke serious food poisoning, while the principal tomato alkaloid, tomatine, seems to be less toxic, at least for humans (Morris and Lee, 1984).

Potato protein concentrate (PPC) is a candidate plant protein concentrate for fish. Standard PPC is produced as a co-product in the processing of potatoes for potato starch production. Potatoes contain proteins that are dissolved in the potato juice. To retrieve this protein, the potato juice is heated, resulting in precipitation of the protein. After centrifuging and drying, the precipitate results in PPC. High-quality PPC contains ≥ 85% crude protein (CP; N × 6.25) on a dry matter (DM) basis, and the amino acid composition of the protein is well balanced for fish. However, experimental testing of standard PPC in diets for rainbow trout resulted in severe appetite loss, even at dietary inclusion levels as low as 5%. This was attributed to the solanidine glycoalkaloids in the PPC, among which α-solanine and α-chaconine are the best known (Refstie and Tiekstra, 2003). Previous work has shown that rainbow trout embryos exhibit a toxic response to chaconine, solasidine, repin, and solanine, while Japanese medaka (*Oryzias latipes*) embryos were only affected by the compounds chaconine and solanine (Crawford and Kocan, 1993).

No estimate can be suggested for the maximum level of alkaloids in fish diets due to lack of relevant scientific information. It can, however, be stated that the potato alkaloids are very toxic and should not be present in feed.

Arginase Inhibitors

An inhibitor of the enzyme arginase occurs in sunflower seeds and soybeans (Reifer and Morawska, 1963). The inhibitor is an unstable nitrogen-derivative of the phenolic compound chlorogenic acid (CGA; Morawska-Muszynska and Reifer, 1965; Reifer and Augustyniak, 1968). The compound rapidly decomposes in both acidic and alkaline conditions.

Sunflower and soybean are the two plant protein sources with significant levels of CGAs. Chlorogenic acid levels in sunflower seed kernels are in the range of 11–45 g/kg (mean, 28 g/kg) (Dorrell, 1976; Dreher and Holm, 1983), and caffeoylquinic acid (5-CQA) is the dominating compound. The CGA content in sunflower seeds varies with crop variety and is highly correlated to seed oil content. Levels do not appear to be significantly affected by differences in climate (Dorrell, 1976). Levels of 10 g/kg (Pratt and Birac, 1979) are present in soybeans. Caffeoylquinic acid not only inhibits arginase but also other enzymes, such as glycerol-3-phosphate dehydrogenase and the translocase T1, of the hepatic and renal glucose-6-phosphatase systems (Arion et al., 1997, 1998). With regard to arginase, 5-CQA was shown to possess inhibitor activity in a linear, dose-dependent manner, while the N-5-CQA exhibited inhibitory activity following a logarithmic function of the inhibitor concentration (Reifer and Augustyniak, 1968). The nature of the inhibitor-enzyme interaction is not fully established. The arginase inhibition by N-5-CQA can be reversed and blocked by organic reducing agents such as L-cysteine, reduced glutathione, 2-mercaptoethanol, and ascorbate (Morawska-Muszynska and Reifer, 1965; Reifer and Augustyniak, 1968). The arginase inhibitor N-5-CQA is unstable and inactivated by oxidation. Presence of arginase stabilizes the inhibitor and protects it from oxidation by chlorogenic acid oxidase (Muszynska and Reifer, 1970) and possibly by air as well.

Arginases in fish are reported to participate in L-arginine metabolism and biosynthesis (Wright, 1995; Jenkinson et al., 1996; Berge et al., 1997), alternative activation of macrophages (Joerink et al., 2006), and biochemical survival strategies in harsh environments (Randall et al., 1989; Wright and Land, 1998; Steele et al., 2001). Arginase inhibition reduces fish cellular proliferation and differentiation by disrupting polyamine biosynthesis in many tissues. The intestine, where proliferation, differentiation, and migration form the basis of homeostasis and adaptive responses of the enterocytes, is particularly vulnerable. Furthermore, inhibition of arginase activity may result in overproduction of nitric oxide with consequent nitric oxide-mediated cytotoxicity, and related complications.

There is a deficiency of data on the bioavailability of the arginase inhibitor in any animal species, on the possible in vivo modification of CGA into the N-derivative with arginase inhibitory activity, and on the effects of CGA and related phenolic compounds as antinutritional factors in fish or other animal species. A number of studies have been conducted on the nutritional value of sunflower meal (SFM) in diets for fish species. Dietary SFM inclusion levels of up to 40% for rainbow trout (*Oncorhynchus mykiss*; Sanz et al., 1994; Gill et al., 2006), 64% for European eel (*Anguilla anguilla*; Garcia-Gallego et al., 1998), 20% for tilapia (*Tilapia rendalii*), at least 22.7% for Atlantic salmon (*Salmo salar* L.; Gill et al., 2006), and 12% for gilthead sea bream (*Sparus aurata*; Gill et al., 2006; Lozano et al., 2007) have been reported to cause no adverse effects on growth and feed utilization performance. However, dietary inclusion of more than 20% SFM has been associated with a reduction in crude protein digestibility of Atlantic salmon (Gill et al., 2006; Aslaksen et al., 2007), reflecting reduced digestibilities of most amino acids. This finding is consistent with reported effects of chlorogenic, caffeic, and quinic acids reducing, in vitro, the bioavailability of amino acids including lysine, cysteine, methionine (Pierpoin, 1970), alanine, phenyalanine, and glutamic acid (Sripad et al., 1982).

Quinolizidine Alkaloids

Alkaloids are secondary metabolites produced by plants often serving a role in plant defense. They are derived from amino acids, containing nitrogen in a heterocyclic ring. Quinolizidine alkaloids are found in a wide variety of plants and are the predominant antinutrient in lupins (genus *Lupinus*). The levels and specific alkaloid profiles vary among the more than 300 species of lupins (Wink et al., 1995). While lupins also contain other alkaloids, the quinolizidine alkaloids are of most concern for animal health.

Different varieties of lupins contain different alkaloid profiles influenced by variety, location, and growing conditions (Sujak et al., 2006). The predominant alkaloids in *L. albus* are lupanine (70%), albine (15%), 13α-hydroxylupanine (8%), and multiflorine (3%). In *L. angustifolius*, lupanine (70%), 13α-hydroxylupanine (12%), and angustifoline (10%) predominate, and in *L. luteus*, the predominant alkaloids are lupinine (60%), sparteine (30%), and p-coumaryl-lupinine (5%) (Wink et al., 1995). Selective breeding has produced varieties of lupins with low alkaloid content (commonly called "sweet lupins"). Currently low alkaloid varieties contain less than 600 mg/kg DM with some cultivars consistently containing less than 100 mg/kg (Glencross, 2001). Wild type, high-alkaloid varieties can contain up to 40,000 mg alkaloids per kilogram of seed meal.

There are few reports regarding direct biological effects of quinolizidine alkaloids in fish. In mammals, quinolizidine alkaloids appear to cause toxicity through neurological effects leading to loss of motor coordination and muscular control. All of the major quinolizidine alkaloids found in lupin exhibit antimuscarinic acetylcholine receptor activity, while several also exhibit antinicotinic activity (Wink et al., 1998). Lupinine and sparteine also inhibit Alpha$_2$ adrenegic receptors, while sparteine also inhibits butylcholine esterase (Wink et al., 1998). Direct evidence for similar effects in fish is lacking.

The bitter taste of the lupin alkaloids is believed to be responsible for decreased feed intake in fish fed diets containing high levels of alkaloids. Low-alkaloid varieties of lupins do not appear to cause palatability problems, but reduce growth performance at higher inclusion levels (Glencross et al., 2004a,b). Glencross et al. (2006) added gramine, an indole alkaloid found in lupin, to rainbow trout diets at levels up to 10,000 mg/kg and reported decreased feed intake above 500 mg alkaloid per kilogram diet and above. Fish fed the highest levels consumed too little food to meet maintenance protein and energy requirements. The authors noted increasing densities of melano-macrophage centers in the head kidney at increasing levels of dietary gramine but attributed the differences to starvation due to a lack of evidence for a direct toxic effect of alkaloids. No significant lesions were observed in the liver, kidney, spleen, pyloric caeca, or intestine.

Chien and Chiu (2003) reported that alkaloid removal from *L. angustifolius* seed meal in diets (670 g meal per kilogram diet) for juvenile tilapia (*Oreochromis niloticus* × *O. aureus*) increased feed efficiency but did not affect growth. The authors offered no explanation for this effect. However, the alkaloid content of seed before (40 mg/kg) and after (20 mg/kg) extraction were both low. Additionally, the authors noted that all of the experimental diets were deficient in methionine, lysine, and threonine, but no supplementation was made. Therefore, whether the observed effects are attributable to the alkaloid content is questionable.

At this time, there are insufficient data to reliably predict safe levels of lupin alkaloids in diets for fish. Caution should be exercised when determining inclusion levels. Reduced feed intake has been observed in fish fed diets containing high levels of alkaloids due to effects on palatability. Based on the results of Glencross et al. (2006), the highest demonstrated level of dietary gramine that did not reduce feed intake in rainbow trout was 100 mg/kg (reduced intake was observed at the next higher inclusion, 500 mg/kg diet). Also, it is not known whether different fish species exhibit varying sensitivities to lupin alkaloids or whether safety levels of dietary lupin of one species or cultivar differ.

Cyanide-Releasing Compounds

Cyanogenic phytochemicals are molecules present in plants that release hydrogen cyanide (HCN). More than 3,000 species of higher plants with diverse taxonomic designations are known to be cyanogenic, including a significant number of food plants of global economic importance such as cassava, linseed, sorghum, peas, maize, barley, wheat, peanut, and rapeseed (Poulton, 1990; Jones, 1998; Lechtenberg and Nahrstedt, 1999; IPCS, 2004). The source of HCN has been identified only in approximately 10% of the known cyanogenic plants (Poulton, 1990). At least 60 cyanogenic glycosides are known (Lechtenberg and Nahrstedt, 1999), and all are *O*-β-glycosidic derivatives of α-hydroxynitriles (cyanohydrins) that, in turn, are mostly derived from any of five hydrophobic protein amino acids—tyrosine, phenyalanine, valine, leucine, and isoleucine—or the nonprotein amino acid cyclopentenylglycine (Poulton, 1990). In plant tissues, the cyanogenic glycosides are stored intact separated from hydrolytic enzymes that degrade them to liberate hydrogen cyanide.

Hydrogen cyanide is a colorless, highly volatile, flammable liquid that is completely miscible in water, partly dissociating to liberate the cyanide ion (CN^-) and forming a weak acidic solution. It is toxic to plants, animals, and humans. Its toxicity stems from its ability to link with the cofactors iron, manganese, or copper of various metalloproteins, especially cytochrome C oxidase of the mitochondrial respiratory chain (EFSA, 2007). Uptake of HCN in fish can be through ingestion, the skin, or the gill respiratory epithelium, following which it rapidly distributes throughout the body through binding to metalloproteins. The enzyme cytochrome c oxidase is the main site of rapid lethal cyanide action, and the brain is the major target organ of the HCN-induced cytotoxic hypoxia (Eisler, 1991). The result is cessation of cellular respiration causing cytotoxic hypoxia and tissue anoxia and a reactive shift from aerobic to anaerobic metabolism. Respiratory arrest due to the depression of the central nervous system by a combination of cytotoxic hypoxia and lactic acidosis is the cause of death (Eisler, 1991; Beasley and Glass, 1998).

There are presently no publications on dietary effects of HCN or on experiments involving other forms of oral dosing of HCN to fish. The reported toxicity of HCN to fish is in relation to cyanide in water, mainly in ecotoxicological experiments studying contamination of natural fish habitats by cyanide in industrial effluents. These exposures result in chronic cyanide effects in the fish that affect fecundity and other aspects of fish reproduction. Eisler (1991) and EFSA (2007) presented detailed reviews. The only report on exposure of fish to cyanide through the diet is a growth study conducted with Nile tilapia fed soaked and unsoaked sun-dried cassava meal with total cyanide levels of 9.9 and 71.1 mg/kg, respectively (Ng and Wee, 1989). The cyanide did not depress growth in the fish as the two diets showed similar growth at similar levels of inclusion. No relevant information is available on maximum dietary levels in fish.

Erucic Acid

Erucic acid, cis-13-docosenoic acid, is a monounsaturated fatty acid produced by members of the brassica family of plants. Especially high levels, up to 60%, are found in some rapeseed and mustard oils. This fatty acid has gained attention both as an antinutrient and as a possible positive bioactive compound. Epidemiological studies indicate that a high proportion of erucic acid in dietary fat prevents coronary heart disease (Rastogi et al., 2004). On the other hand, erucic acid may be involved in detrimental effects on the heart tissue. Rats have shown accumulation of lipid and tissue lesions in the heart when fed oils with high concentration of erucic acid (McCutcheon et al., 1976). Whether erucic acid is the main causative agent for the possible positive and negative effects has not been established (Food Standards Australia New Zealand, 2003). Some rapeseed oils contain several biologically active compounds of importance for the effects either alone or in interaction with the erucic acid. Negative effects of high erucic acid rapeseed oil have been observed in coho salmon fed diets with 6% and 12% rapeseed oil (Hendricks, 2002). The fish showed growth depression, mortalities, and histopathology in the skin, gills, kidney, and heart. The heart epicardium accumulated lipid in the connective tissue, but no necrosis was observed in the cardiac muscle. Rapeseed oil has been an ingredient in fish diets in several other studies (reviewed by Hendricks, 2002). In none of these, negative indications are given. As a consequence of the potential for positive as well as negative characteristics of rapeseed oil, both high and low erucic acid rape plants have been developed through plant breeding. Double low rapeseed oil, with low levels of both erucic acid and glucosinolates, are now the dominating varieties used in fish feed. The oil has become highly valuable as a replacement for marine oils in fish diets. It is rich in monounsaturated fatty acids, C18:1, and low in n-6 fatty acids. The fatty acid ratio n-3:n-6 is therefore favorable compared to other plant oils used as replacers for marine oils. Replacement with rapeseed oil secures a fatty acid profile of the product more in line with recommendations for humans than replacement with, for example, soybean or other plant oils.

Allergens

Feed or food allergens are defined as substances that react with IgE antibodies and induce allergic sensitization/reactions, usually via mast cell degranulation and histamine release. No common structure can predict whether an antigen is an allergen, but generally allergens resist enzymatic hydrolysis in the GI tract (see review by Aalberse, 1997).

Whether fish react allergically, i.e., with a type I hypersensitivity reaction, has not been demonstrated. With the exception of Perciformes (tilapia, sea bass, and sea bream; Mulero et al., 2007), mast cells of salmonids and most other orders of finfish investigated do not contain histamine (Reite, 1965, 1972; Dezfuli et al., 2000; Mulero et al., 2007), and the fish do not react to intravascular injection of histamine (Reite, 1972). Also, teleosts do not appear to have an analogous structure to monomeric IgE, only tetrameric IgM and possibly monomeric IgD (Wilson et al., 1997; Choi et al., 2007). On the other hand, the distal intestine of at least some fish appears to be sensitive to antigen stimulation, and strong immune responses are achieved with antigens delivered to this region (Ellis, 1995).

A few protein components of some legume seeds and cereals elicit antigenic effects in animals. These compounds are capable of inducing intestinal mucosal lesions, abnormalities in the villi, specific and nonspecific immune responses, and abnormal movement of digesta through the gut (D'Mello, 1991; Lalles and Peltre, 1996). Soybean protein contains compounds such as glycinin and beta conglycinin, which act as allergens to several animals and humans. Rumsey et al. (1993, 1994) reported that high levels of immunologically active glycinin and betaconglycinin in different soy preparations seemed to negatively affect growth performance in rainbow trout. They assumed that the comparatively under-investigated effects of allergens can provide answers to why conventionally processed soybean, in which the proteinase inhibitors and lectins have been largely inactivated, results in poor growth of salmonid fish. Haemagglutination inhibition assays by the same authors showed that normal processing measures like toasting and defatting did not significantly reduce antigenicity levels in soybean meal. However, the evidence was not directly conclusive that it indeed was glycinin, beta-conglycinin, or any other allergen/antigen in the soybeans caused the inflammation. Kaushik et al. (1995) found no antigenic proteins in soy protein concentrate but glycinin and beta-conglycinin were detected in defatted, toasted soy flour. Following a 12-week growth trial feeding rainbow trout the two soy products at various levels, neither soy product elicited detectable levels of antibodies against soybean protein in the sera of the fish. Growth or apparent digestibility coefficients for dry matter, protein, or growth were not significantly different for the different soy products. Thus, rainbow trout do not appear to react allergically to soybean meal. Van den Ingh et al. (1991, 1996), Bæverfjord and Krogdahl (1996), and Krogdahl et al. (2000) observed enteritis-like changes in the distal intestine of Atlantic salmon fed diets containing solvent-extracted soybean meal or an alcohol extract of soybean meal. The antigenic compounds present in feed may trigger a variety of nonspecific and specific immune responses in the fish intestine (Bæverfjord and Krogdahl, 1996; Bakke-McKellep et al., 2000, 2007), which might lead to a reduction in growth.

The presence of allergens or antigens that fish react to in common plant-derived feed ingredients remains a matter of controversy. Recently, however, the involvement of cells positive for an antibody against human CD3-epsilon was observed in the distal intestine of Atlantic salmon exhibiting the inflammatory response against soybean meal (Bakke-McKellep et al., 2007), suggesting that putative T cells are involved in the enteropathy. Furthermore, significant up-regulation of CD4 and CD8-beta mRNA in the tissue indicated a mixed population of T cells present. This suggests that the inflammation has similarities to a type IV hypersensitivity reaction, involving cytotoxic (CD8+) as well as T helper cells (CD4+). The specific pathogenesis and cause of the soybean meal-induced enteropathy in salmonids remains to be discovered.

COMBINED EFFECTS OF PLANT ANTINUTRIENTS

More detailed knowledge of interactions between ANFs would be particularly useful, as many of the feed materials that have the potential to be used as fish feed ingredients contain more than one antinutrient. Furthermore, mixes of different ingredients, with their differing ANF profiles, are often used in practical feed formulations. Studies are needed to evaluate the effects of mixtures of antinutrients, preferably in proportions similar to those in various plant-derived feedstuffs.

Combined effects can be beneficial or detrimental. Tannins are known to interact with other antinutrients. For example, interaction between tannins and lectins removed the inhibitory action of tannins on amylase (Fish and Thompson, 1991), while interactions between tannins and cyanogenic glycosides (Goldstein and Spencer, 1985) reduced the deleterious effects of the latter. Moreover, simultaneous consumption of saponin and tannin resulted in the loss of their individual toxicity to rats (Freeland et al., 1985). This is considered to be due to chemical reactions between them, leading to the formation of tannin-saponin complexes, inactivating the biological activity of both. On the other hand, Alvarez and Torres-Pinedo (1982) suggested that soybean lectin bound to rabbit jejunal enterocyte apical membranes had a potentiating effect on saponin's detrimental influence on epithelial barrier function. Knudsen et al. (2008) demonstrated that purified saponins from soybeans supplemented to a fish meal-based diet did not elicit inflammatory responses in the distal intestine of Atlanic salmon or changes in fecal dry matter, nor did lupin kernel meal from sweet lupins (*L. angustifolius*). However, the combination of soy saponins and lupin kernel meal did cause an inflammatory response in the distal intestine similar to the one observed in the defatted (extracted) soybean meal control group. This combination of saponins and lupin meal also increased intestinal permeability compared to the fish meal control group, but not to the same extent as that measured in the soybean meal control group. The authors concluded that the combination of saponins and an unidentified factor or factors in the lupin meal caused the inflammation and increased tissue permeability. However, changes in the intestinal microbiota due to the different feed ingredients and an interaction between the saponins and microbiota could not be ruled out.

Bakke-McKellep et al. (2008) found that the combination of soybean lectin (SBA) and soybean trypsin inhibitor severely reduced intestinal epithelial function of Atlantic salmon, as assessed by glucose absorption in an in vitro system. The two antinutrients when given alone did not affect glucose absorption in intestinal tissue when compared to that of tissue not exposed to ANFs. Interestingly, phytohemagglutinin (PHA), a lectin from kidney beans, reduced glucose absorption in a similar manner as the SBA/trypsin inhibitor combination in the same experiment. Thus, on an individual basis, PHA appears to be a more toxic lectin to Atlantic salmon enterocytes than SBA.

Studies by Iwashita et al. (2008, 2009) underline the importance of interactions between antinutrients. Saponins and lectins fed to rainbow trout in semipurified diets did not cause morphological alterations in the intestine when given alone. However, together they induced development of hyperplastic connective tissue in the mucosal folds of the distal intestine. The alterations differed from and were not as severe as the typical soybean-induced enteritis observed in salmonids and some other fish species (Bæverfjord and Krogdahl, 1996; Uran et al., 2008). Interestingly, supplementation with taurocholate prevented the alterations.

ANTINUTRIENTS IN ANIMAL FEEDSTUFFS

Antivitamins—Thiaminase

The term antivitamins relates to effects observed in feeding trials that might be explained by reduced utilization of vitamins. The components are in general not well defined, and information on effects in fish is lacking (Francis et al., 2001b). According to Francis et al. (2001b) all are heat labile. Antivitamins, except thiaminase, are not dicussed in more detail herein.

Several fish species, such as herring, mackerel, and carp, as well as mussels, clams, and shrimp, contain components with thiaminase-like activity (NRC, 1982). Thiaminase in carp liver was the first to be described from fish (Bos and Kozik, 2000). It showed a molecular weight of 55 kD and seemed to contain a single peptide chain. A thiaminase from the sea water fish Eisturalia petimb has also been described (Nishimune et al., 2008) and observed to be a protein with one active and one inactive subfragment with a total molecular weight of 100 kD. Thiaminases appear to be dependent on a cosubstrate, an amine or sulfahydryl-containing compound such as pyridine, proline, or cysteine (Bos and Kozic, 2000) and are inactivated by heat treatment (Zajicek et al., 2009). The enzyme is active in live fish and in raw fish products. It inactivates thiamin by cleavage of the rather unstable bridge between the two ring structures of the molecule. Several cases of thiamin deficiency were reported in the early days of salmon aquaculture. Thiamine deficiency also occurs in fish in the wild feeding on prey fish with high levels of thiaminase (Fisher et al., 1998). The enzyme is concentrated in the digestive tract and has been isolated from the liver (Nishimune et al., 2008), but may also originate from the intestinal microbiota (Honeyfield et al., 2002). The adverse effects of thiaminase are well known in fur animal production as well. Fish feed made with fresh products from species with high thiaminase activity must not be stored for any length of time before being fed. Extensive use of fisheries bicatch in aquaculture feed may require supplemental thiamin to prevent thiamin deficiency (Ishihara et al., 1974; Anglesea and Jackson, 1985).

Chitin

Chitin, a linear polymer of N-acetyl glucosamine units, is a major constituent of the exoskeleton of crustaceans. The chitin-rich exoskeleton constitutes up to 20% of their gross energy. If undigested, the chitin in feces represents a great loss of energy from the diet, potentially increasing feed intake and lowering dietary energy utilization. Chitin digestibility may vary among fish species as chitinase activity seems to vary (Krogdahl et al., 2005). However, present information on chitin digestibility is limited. Salmonids show very low chitin digestibilities (Lindsay et al., 1984; Olsen et al., 2006). Undigested chitin in the gut can interfere with nutrient digestibility. Chitin has high water-binding capacities and the ability to bind fatty acids as well as bile acids. These characteristics may explain the negative effect on nutrient digestibility and growth observed in rainbow trout (Lindsay et al., 1984) and hybrid tilapia, *Oreochromis niloticus* × *Oreochromis aureus* (Shiau and Yu, 1999), fed diets containing large amounts of chitin. Similar observations have been made in Atlantic salmon (Olsen et al., 2006). Chitin also has been shown to have immunostimulatory properties in some aquatic species (see Chapter 10).

Fluorine

Krill and some other potential marine protein sources for aquaculture have high concentrations of fluorine, which has been of some concern for the industry. However, results from a study with Atlantic cod, Atlantic salmon, rainbow trout, and Atlantic halibut (*Hippoglossus hippoglossus*) indicated that fluorine from krill (*Thysanoessa inermis* and *Euphausia superba*) and the amphipod *Themisto libelulla* is not harmful for these species (Moren et al., 2007). No organs showed any increase in fluorine levels, and neither growth nor health parameters were negatively altered. The amphipod meal had the highest fluorine level (4,000 ± 800 mg/kg), while meal from *Thysanoessa inermis* and *Euphausia superba* showed about one-fourth of that level (780 ± 160 mg/kg and 1,160 ± 230 mg/kg, respectively).

ADVENTITIOUS TOXINS

Oxidized Lipid

Marine lipids contain large quantities of polyunsaturated fatty acids that are quite susceptible to autooxidation. The oxidation products of these lipids include aldehydes and ketones as well as free radicals that increase the demand for antioxidants. Thus, the presence of oxidized lipid products in the diet may directly affect fish and exacerbate deficiencies of antioxidant vitamins, resulting in various pathological conditions, including liver degeneration, spleen abnormalities, and anemia (NRC, 1993; Hardy, 2001). Synthetic antioxidants such as ethoxyquin are typically added to marine lipids to ensure they are properly protected from oxidation.

Biogenic Amines

Level of biogenic amines is used as a quality criterion for freshness of raw materials processed to fish meal and meals from the land animal slaughter industry. Accumulation of biogenic amines in general parallels bacterial growth (Chin and Koehler, 1986). Biogenic amines are produced by decarboxylation of free amino acids through the action of anaerobic bacteria. The amino acids phenylalanine, lysine, histidine, tyrosine, and methionine are the precursors to the following biogenic amines, respectively: phenethylamine, cadaverine, histamine, tyramine, and spermidine and spermine (reviewed by Sander et al., 1996). Fish meals produced from fish containing high levels of histamin can be aqutely toxic to chickens, and chickens have been used to evaluate suitability of raw materials for fish feed. Atlantic salmon (Anderson et al., 1997), Atlantic halibut (Aksnes and Mundheim, 1997), and different species of tropical shrimp (Ricque et al., 1998) have been found to reduce feed intake and growth when fed diets with fish meal made from stale fish. Opstvedt et al. (2000) fed Atlantic salmon diets based either on high-quality fish meal supplemented with cadaverine, histamine, putrescine, and tyramine or a diet with fish meal from stale fish. The Atlantic salmon reduced feed intake, growth and feed utilization when fed the diet with the low-quality fish meal, but the biogenic amines seemed not to be the cause of these effects. A similar experiment has been carried out with blue shrimp (*Litopenaeus stylirostris*) with similar results as in Atlantic salmon (Tapia-Salazar et al., 2004). Survival, feed consumption, and final weight were reduced in shrimp fed the diet with fish meal made from stale fish, but not because of the content of biogenic amines. An experiment conducted to evaluate the toxicity of biogenic amines in rainbow trout may indicate that rainbow trout are less sensitive to fish meal made from spoiled fish than Atlantic salmon (Fairgrieve et al., 1994). The study showed that fish meal acutely toxic to chicken did not cause any acute toxicity, mortality, or effects on feed intake in rainbow trout juveniles after 16 weeks of feeding, neither did diets based on casein supplemented with histamine, putresine, and cadaverine. A stomach distention was, however, observed in fish fed high levels of histamin. At the present state of knowledge, biogenic amines do not seem to cause toxicity symptoms in fish or shrimp, but other components of spoiled fish can be detrimental.

Mycotoxins

Mycotoxins are a diverse group of potentially toxic metabolites produced by a variety of fungal species that often contaminate feedstuffs and prepared fish diets (NRC, 1993; Manning, 2001). It is estimated that 25% or more of the world's crops are contaminated with mycotoxins and contamination of feedstuffs, with multiple mycotoxins frequently observed (CAST, 2003). Fungal growth can occur in the field, at harvest, or during storage, transport, or processing of feedstuffs. It is primarily related to environmental conditions, namely temperature and moisture. Generally, the growth of mycotoxin-producing fungi that affect grains and oilseeds can be inhibited by maintaining moisture levels below 12%. Inappropriate processing and storage conditions increase the risk of mycotoxin production and may restrict the use of potential nutrient sources for fish. Increased utilization of byproducts from the biofuel industry as feed ingredients in recent years has also contributed to an increased risk of exposing farmed fish to mycotoxins (Hooft and Bureau, 2010). Present knowledge regarding the effects of mycotoxins on fish is limited compared to that available for domesticated terrestrial animals. In animals, symptoms of mycotoxicoses include a reduction in feed intake; digestive disturbances; endocrine, immunological, and neurological effects; cancer in various organs; and death (Roberts, 2002; Santacroce et al., 2008). Sensitivity to mycotoxins varies greatly both among and within species, and depends on life stage, nutritional, and health status and environmental conditions.

The most common toxin-producing fungal organisms contaminating feedstuffs used in the preparation of aquaculture feeds include those of the genera *Aspergillus*, *Penicillium*, and *Fusarium*. The toxins of greatest concern in fish nutrition are aflatoxin B_1, ochratoxin A, fumonisin B_1, the trichothecenes, in particular deoxynivalenol (DON) and T-2 toxin, and zearalenone. Some mycotoxins are products of fungi from more than one genus, such as ochratoxin and cyclopiazonic acid, which are produced by both *Aspergillus* and *Penicillium* fungal species (Varga et al., 2001).

Four different aflatoxins are produced by *Aspergillus* molds. Aflatoxin B_1 was reported to be extremely toxic to rainbow trout more than 30 years ago. Liver carcinomas and death were observed when rainbow trout were fed aflatoxin B_1 at extremely low concentrations (0.4 to 20 μg/kg) (Halver and Mitchell, 1969). Levels above 80 μg/kg caused acute toxic effects, including hepatic necrosis, brachial edema, and hemorrhage (Roberts, 2002). Rainbow trout are considered to be more susceptible to aflatoxin B_1 compared to other salmonids and much more sensitive than warmwater species such as channel catfish and tilapia (Manning, 2001). Groundnut, copra, palm kernel, cottonseed, maize, and products derived from these crops are considered to be the main sources of aflatoxin contamination in animal feeds (Santacroce et al., 2008).

Ochratoxin A (OTA), produced by both *Aspergillus* and *Penicillium* species, is considered to be the second largest group of mycotoxins world-wide. Its main mechanism of action seems to be interference with cellular function by inhibition of synthesis of the phenylalanine-tRNA and mitochondrial ATP production and by induction of oxidative stress. Animal studies indicate that OTA is a nephro- and hepatotoxin, an immune suppressant, a potent teratogen, and a carcinogen (reviewed by El-Sayed et al., 2009). Investigations of toxicological effects of OTA in fish and shrimp are few. Rainbow trout showed increased mortality and degeneration and necrosis of kidneys and liver after intraperitoneal injection of acute doses (Doster et al., 1973). In agreement with these observations, channel catfish fed OTA-contaminated diets, 2 and 4 mg/kg, showed reduced weight gain and feed efficiency, pathological lesions in liver and kidney, and increased mortality (Manning et al., 2003, 2005). Sea bass have also shown sensitivity to OTA with an estimated acute lethal dose of 285 μg/kg body weight (El-Sayed et al., 2009).

The fumonisins comprise at least 28 different compounds categorized into three groups, A, B, and C. Those compounds belonging to group B are considered to be the most toxic. Fumonisin B_1 is widespread and the most thoroughly studied (Sulyok et al., 2007). Its primary mechanism of action is interference with sphingolipid metabolism. Sphingolipids are components of cell walls essential for cell survival, differentiation, stress resistance, and apoptosis (Hannun and Obeid, 2002; Spiegel and Milstien, 2002). Reduction in feed intake and feed utilization, impaired antibody production, hepatic lesions, and alterations in hematological parameters have been observed in young channel catfish fed diets containing fumonisin B_1 at levels between 20 and 720 mg/kg (Lumlertdacha et al., 1995; Yildirim et al., 2000). At the lowest level, only effects on feed intake were observed, while a high mortality rate was also observed at the highest levels. Nile tilapia appear equally sensitive to fumonisin B_1 as do channel catfish (Tuan at al., 2003), whereas carp respond more severely, also showing neurotoxic and degenerative alteration in the brain (Petrinec et al., 2004; Kovacic et al., 2009).

Trichothecenes are a large group of chemically related mycotoxins produced primarily by *Fusarium* species. They are commonly divided into four groups, A, B, C, and D, according to their chemical properties. Among the most commonly studied are T-2 toxin, deoxynivalenol (DON) and diacetoxyscirpenol (DAS). Trichothecenes are powerful inhibitors of protein synthesis, which act on the 60S ribosomal subunit (reviewed by Rocha et al., 2005). They affect cell division, inhibit mitochondrial function, and induce apoptosis. They easily pass body barriers such as the skin and mucosa of the lungs and intestines, and do not need activation in the body to be harmful. Tissues with high cell turnover are the most sensitive to trichothecenes. The gross toxic effects of trichothecenes on animals include growth retardation, reduced ovarian function and reproductive disorders, alteration of the immune response, feed refusal, and vomiting (reviewed by Rocha et al., 2005). Manning et al. (2003) observed pathological alterations in the stomach tissue, head kidney, and trunk kidney of channel catfish receiving T-2 toxin at levels between 1.3 and 5.0 mg/kg in the diet. Reduced resistance to *Edwardsiella ictaluri* was also observed at inclusion levels of 1 and 2 mg/kg (Manning et al., 2005). Additionally, reduced growth, edema, and intestinal and hematological alterations were observed in rainbow trout exposed to levels of T-2 toxin up to 15 mg/kg diet (Poston et al., 1982). Woodward et al. (1983) reported that diets containing graded levels of DON ranging from 1.0 to 12.9 mg/kg from artificially infected corn caused progressively greater reductions in live weight gain, feed intake, and feed efficiency of juvenile rainbow trout. Recently, Hooft (2010) reported significant decreases in important growth and nutrient utilization in young rainbow trout (24 g) fed practical diets containing from 0.3 to 2.6 mg/kg DON from naturally contaminated corn. The effects were not solely a result of a reduction in feed intake and no differences were observed in the apparent digestibility coefficients of crude protein and gross energy of fish fed diets containing 0.3 (control) to 2.0 mg/kg DON. However, histopathological changes of the livers of fish fed diets containing 1.4 and 2.6 mg/kg were noted (Hooft, 2010). The variability in sensitivities among several species of fish to DON has been attributed to differences in the abilities of the intestinal microbes of different species to transform DON to its less toxic metabolites (Guan et al., 2009).

Zearalenone, produced by several *Fusarium* species, is a potent estrogenic compound that disturbs reproduction in many animal species. It is widely found in grains. Intraperitoneal injection of zearalenone induced vitellogenin as well as zona radiata proteins production in rainbow trout (Arukwe et al., 1999; Celius et al., 2000) and affected sperm number and quality in carp (Sandor and Ványi, 1990).

Even though knowledge on the effects of mycotoxins in fish is limited, it is clear that using mycotoxin-contaminated feed ingredients in fish feeds can cause severe health problems in cultivated fish, and subsequently, potentially substantial economical losses. Of concern are also the possibilities of transfer of mycotoxins from feed to the fish flesh. It is well known from terrestrial animal production that some mycotoxins are transferred to meat, milk, and eggs if present in the feed (Galtier, 1998; Goyarts, 2007). Proper storage of feedstuffs and finished diets is critical to prevent contamination with mycotoxins. These conditions include low moisture (< 12%), low relative humidity, and limited oxygen. Additional precautions may also be taken including supplementation of mold inhibitors such as propionic acid to prepared diets.

Pesticides

A wide range of pesticides are in use in today's agriculture. Hence, feed ingredients potentially contain residues of such compounds. There are, however, limited data available on the occurrence of pesticides in feed materials and feeds for fish. Highest contents of pesticides are found in ingredients of animal origin, especially in fish meal and fish oil. The number of studies of the dietary toxicity of pesticides in fish is limited. Fish seem highly sensitive to waterborne exposure to organochlorine pesticides, and some studies indicate that fish are also sensitive to oral exposure but possibly with less severe effects (Petri et al., 2006; Glover et al., 2007). There is a need to thoroughly survey their presence in feed ingredients of plant origin, to conduct studies of their potential toxicity to fish, and to limit potentially harmful levels in formulated feeds.

Other Contaminants

A variety of compounds, both of natural and industrial origin, may contaminate feedstuffs and/or prepared diets for fish. Examples of such compounds include polycyclic aromatic hydrocarbons (PAHs), dioxins, dioxin-like polychlorinated biphenyls (PCBs) and nondioxin-like PCBs, polybrominated diphenyl ethers (PBDEs), chlorinated paraffins, fluoride organic tin compounds, perfluorooctane sulfonate (PFOS), and metals and other mineral elements such as arsenic, cadmium, lead, and mercury (Hendricks, 2002; Alexander et al., 2007; Ikonomou et al., 2007; Friesen et al., 2008). The adverse effects of these compounds may include reduced growth, tissue lesions, and even mortality, but the severity of these conditions vary considerably depending on dietary concentration of the contaminant as well as the fish species. Care in monitoring the quality of feedstuffs to be used in aquafeeds will generally limit the potential adverse effects of these contaminants. Additional information on minerals is found in Chapter 7.

UNKNOWN COMPOUNDS

The existence of as yet undiscovered antinutrients, antigens, or toxins that can compromise the function of fish cannot be ruled out and should be kept in mind. The continuous search for new ingredients for fish feeds brings new nutrient sources into focus, such as single-cell proteins, krill, and marine algae. The complexity of feed ingredients and the difficulties often encountered when fractionating as well as separating and analyzing the constituents often result in chemical changes, new compounds, or even destruction of certain compounds, precluding their identification and characterization. Limitations in the sensitivity of many methods of analysis may prevent the identification of potentially harmful compounds present at very low levels.

CONCLUSIONS

Knowledge regarding responses to various feed ingredients as well as purified ANFs and adventitious components give some insights into their effects and their possible roles in production and health of the cultivated organisms. Recent findings suggest combinations of various ANFs, including the specific ANFs mentioned above, have particular significance in causing detrimental effects to intestinal structure, function, and defense mechanisms at lower levels than what the individual ANFs would elicit. Moreover, it cannot be ruled out that as yet unidentified components play important roles regarding the reduction in performance and nutrient utilization often observed with feeds high in plant or other unconventional feed ingredients. Health problems are also reported related to such components, in particular those causing intestinal symptoms. Also the gut microbiota may be an important mediator of effects of the antinutrients and adventitious compounds. More research is needed to reveal consequences of various ANFs, individually and in combinations, as well as the role of the microbiota in the GI tract. Long-term effects of antinutrients merit investigation to aid in predicting how a feedstuff or combinations of feedstuffs may affect target aquatic species and identify steps needed to reduce or prevent detrimental health effects.

REFERENCES

Aalberse, R. C. 1997. Food allergens. Environ. Toxicol. Phar. 4:55–60.

Aksnes, A., and H. Mundheim. 1997. The impact of raw material freshness and processing temperature for fish meal on growth, feed efficiency and chemical composition of Atlantic halibut (*Hippoglossus hippoglossus*). Aquaculture 149:87–106.

Alexander J., L. Frøyland, G. I., Hemre, B. K. Jacobsen, E. Lund, H. M. Meltzer, and J. Utne Skåre. 2007. A comprehensive assessment of fish and other seafood in the Norwegian diet, Opinion of The Norwegian Scientific Committee for Food safety. Oslo, Norway: VKM (Norwegian Scientific Committee for Food Safety).

Alvarez, J. R., and R. Torres-Pinedo. 1982. Interactions of soybean lectins, soyasaponins, and glycinin with rabbit jejunal mucosa in vitro. Pediatr. Res. 16:728–731.

Anderson, J. S., D. A. Higgs, R. M. Beames and M. Rowshandeli. 1997. Fish meal quality assessment for Atlantic salmon (*Salmo salar*, L.) reared in sea water. Aquacult. Nutr. 3:25–38.

Anglesea, J. D., and A. J. Jackson.1985. Thiaminase activity in fish silage and moist fish feed. Anim. Feed Sci. Tech.13:39–46.

Arion, W. J., W. K. Canfield, F. C. Ramos, P. W. Schindler, H. J. Burger, H. Hemmerle, G. Schubert, P. Below, and A. W. Herling. 1997. Chlorogenic acid and hydroxynitrobenzaldehyde: New inhibitors of hepatic glucose 6-phosphatase. Arch. Biochem. Biophys. 339:315–322.

Arion, W. J., W. K. Canfield, F. C. Ramos, M. L. Su, H. J. Burger, H. Hemmerle, G. Schubert, P. Below, and A. W. Herling. 1998. Chlorogenic acid analogue S 3483: A potent competitive inhibitor of the hepatic and renal glucose-6-phosphatase systems. Arch. Biochem. Biophys. 351:279–285.

Arukwe, A., T. Grotmol, T. B. Haugen, F. R. Knudsen, and A. Goksoyr. 1999. Fish model for assessing the in vivo estrogenic potency of the mycotoxin zearalenone and its metabolites. Sci. Total Environ. 236:153–161.

Aslaksen, M. A., O. F. Kraugerud, M. Penn, B. Svihus, V. Denstadli, H. Y. Jørgensen, M. Hillestad, Å. Krogdahl, and T. Storebakken. 2007. Screening of nutrient digestibilities and intestinal pathologies in Atlantic salmon, *Salmo salar*, fed diets with legumes, oilseeds, or cereals. Aquaculture 272:541–555.

Bæverfjord, G., and Å. Krogdahl. 1996. Development and regression of soybean meal induced enteritis in Atlantic salmon, *Salmo salar* L., distal intestine: A comparison with the intestines of fasted fish. J. Fish Dis. 19:375–387.

Bakke-McKellep, A. M., C. McL Press, G. Bæverfjord, Å Krogdahl, and T. Landsverk. 2000. Changes in immune and enzyme histochemical phenotypes of cells in the intestinal mucosa of Atlantic salmon, *Salmo salar* L., with soybean meal-induced enteritis. J. Fish Dis. 23:115–127.

Bakke-McKellep, A. M., M. H. Penn, P. M. Salas, S. Refstie, S. Sperstad, T. Landsverk, E. Ringo, and Å. Krogdahl. 2007. Effects of dietary soyabean meal, inulin and oxytetracycline on intestinal microbiota and epithelial cell stress, apoptosis and proliferation in the teleost Atlantic salmon (*Salmo salar* L.). Brit. J. Nutr. 97:699–713.

Bakke-McKellep, A. M., M. Penn, E. Chikwati, E. Hage, C. Chunfang, and Å. Krogdahl. 2008. In vitro effects of various antinutritional factors on intestinal glucose absorption and histology in Atlantic salmon, Abstract. XIII International Symposium on Fish Nutrition and Feeding, June 1–5, Florianópolis, Brasil.

Bardocz, S., G. Grant, S. W. B. Ewen, T. J. Duguid, D. S. Brown, K. Englyst, and A. Pusztai. 1995. Reversible effect of phytohaemagglutinin on the growth and metabolism of rat gastrointestinal tract. Gut 37:353–360.

Bau, H. M., C. Villaume, J. P. Nicolas, and L. Mejean. 1997. Effect of germination on chemical composition, biochemical constituents and antinutritional factors of soya bean (*Glycine max*) seeds. J. Sci. Food Agr. 73:1–9.

Beasley, D. M. G., and W. I. Glass. 1998. Cyanide poisoning: Pathophysiology and treatment recommendations. Occup. Med-C. 48:427–431.

Becker, K., and H. P. S. Makkar. 1999. Effects of dietary tannic acid and quebracho tannin on growth performance and metabolic rates of common carp (*Cyprinus carpio* L.). Aquaculture 175:327–335.

Bennetau-Pelissero, C., B. Breton, B. Bennetau, G. Corraze, F. Le Menn, B. Davail-Cuisset, C. Helou, and S. J. Kaushik. 2001. Effect of genistein-enriched diets on the endocrine process of gametogenesis and on reproduction efficiency of the rainbow trout *Oncorhynchus mykiss*. Gen. Comp. Endocr. 121:173–187.

Berge, G. E., E. Lied, and H. Sveier. 1997. Nutrition of Atlantic salmon (*Salmo salar*): The requirement and metabolism of arginine. Comp. Biochem. Phys. A 117:501–509.

Berg-Lea, T., L. E. Brattås, and Å. Krogdahl. 1989. Soybean proteinase inhibitors affect nutrient digestion in rainbow trout. Pp. 99–102 in Recent Advances of Research in Antinutritional Factors in Legume Seeds, J. Huisman, T. F. B. van der Pool, and I. Liener, eds. Wageningen, the Netherlands: Pudoc.

Bhat, P. G., R. T. Jacob, and T. N. Pattabiraman. 1981. Enzyme inhibitors from plants: Enterokinase inhibitors in tubers and seeds. J. Biosciences 4:371–378.

Blom, J. H., K. J. Lee, J. Rinchard, K. Dabrowski, and J. Ottobre. 2001. Reproductive efficiency and maternal-offspring transfer of gossypol in rainbow trout (*Oncorhynchus mykiss*) fed diets containing cottonseed meal. J. Anim. Sci. 79:1533–1539.

Borel, P., D. Lairon, E. Termine, R. Grataroli, and H. Lafont. 1989. Isolation and properties of lipolysis inhibitory proteins from wheat germ and wheat bran. Plant Food. Hum. Nutr. 39:339–348.

Bos, M., and A. Kozik. 2000. Some molecular and enzymatic properties of a homogeneous preparation of thiaminase I purified from carp liver. J. Protein Chem. 19:75–84.

Burel, C., T. Boujard, A. M. Escaffre, S. J. Kaushik, G. Boeuf, K. A. Mol, S. Van der Geyten, and E. R. Kuhn. 2000. Dietary low glucosinolate rapeseed meal affect thyroid status and nutrient utilization in rainbow trout (*Oncorhynchus mykiss*). Brit. J. Nutr. 83:653–664.

Burel, C., T. Boujard, S. J. Kaushik, G. Boeuf, K. A. Mol, S. Van der Geyten, V. M. Darras, E. R. Kuhn, B. Pradet-Balade, B. Querat, A. Quinsac, M. Krouti, and D. Ribaillier. 2001. Effects of rapeseed meal-glucosinolates on thyroid metabolism and feed utilization in rainbow trout. Gen. Comp. Endocr. 124:343–358.

Buttle, L. G., A. C. Burrells, J. E. Good, P. D. Williams, P. J. Southgate, and C. Burrells. 2001. The binding of soybean agglutinin (SBA) to the intestinal epithelium of Atlantic salmon, *Salmo salar* and rainbow trout, *Oncorhynchus mykiss*, fed high levels of soybean meal. Vet. Immunol. Immunop. 80:237–244.

Caldwell, R. A. 1992. Effect of calcium and phytic acid on the activation of trypsinogen and the stability of trypsin. J. Agr. Food Chem. 40:43–47.

Cara, L., P. Borel, M. Armand, H. Lafont, G. Lesgards, and D. Lairon. 1992. Milling and processing of wheat and other cereals affect their capacity to inhibit pancreatic lipase in vitro. J. Food Sci. 57:466–469.

CAST (The Council for Agricultural Science and Technology). 2003. Mycotoxins: Risks in Plant, Animal, and Human Systems. Task Force Report from The Council for Agricultural Science and Technology. Ames, IA: CAST.

Celius, T., J. B. Matthews, J. P. Giesy and T. R. Zacharewski. 2000. Quantification of rainbow trout (*Oncorhynchus mykiss*) zona radiata and vitellogenin mRNA levels using real-time PCR after in vivo treatment with estradiol-17[beta] or [alpha]-zearalenol. J. Steroid Biochem. 75:109–119.

Chapman, G. W., Jr. 1987. A proteinaceous competitive inhibitor of lipase isolated from Helianthus annuus seeds. Phytochemistry 26:3127–3131.

Chien, Y. H., and Y. H. Chiu. 2003. Replacement of soybean (Glycine max (L.) Merrill) meal by lupin (Lupinus angustifolius) seed meal in diet for juvenile tilapia (*Oreochromis niloticus* x *O. aureus*) reared indoors. Aquac. Res. 34:1261–1268.

Chikwati, E. 2007. Effects of Soy Saponins and Phytosterols on Structure, Function and Dietary Lipid Digestion and Uptake by Atlantic Salmon Intestine. Masters thesis, Norwegian School of Veterinary Science, Oslo, Norway.

Chin, E. D., and P. E. Koehler. 1986. Effect of salt concentration and incubation temperature on formation of histamine, phenethylamine, t tamine, and tyramine during miso fermentation. J. Food Protect. 49:423–427.

Choi, D. H., H. N. Jang, D. M. Ha, J. W. Kim, C. H. Oh, and S. H. Choi. 2007. Cloning and expression of partial Japanese flounder (*Paralichthys olivaceus*) IgD. J. Biochem. Mol. Biol. 40:459–466.

Combourieu, B., L. Elfoul, A. M. Delort, and S. Rabot. 2001. Identification of new derivatives of sinigrin and glucotropaeolin produced by the human digestive microflora using H-1 NMR spectroscopy analysis of in vitro incubations. Drug Metab. Dispos. 29:1440–1445.

Commission of the European Communities. 2003. Commission directive 2003/100/EC of 31 October 2003 amending Annex I to Directive 2002/32/EC of the European Parliament and of the Council on Undesirable Substances in Animal Feed. Official J. EU L 285:33–37.

Crawford, L., and R. M. Kocan. 1993. Steroidal alkaloid toxicity to fish embryos. Toxicol. Lett. 66:175–181.

D'Mello, F. J. P. 1991. Antigenic proteins. Pp. 108–125 in Toxic Substances in Crop Plants, F. J. P. D'Mello, C. M. Duffus, and J. H. Duffus, eds. Science Park, Cambridge, UK: The Royal Society of Chemistry, Thomas Graham House.

D'Souza, N., D. I. Skonberg, M. E. Camire, K. E. Guthrie, J. Malison, and L. Lima. 2005. Influence of dietary genistein levels on tissue genistein deposition and on the physical, chemical, and sensory quality of rainbow trout, *Oncorhynchus mykiss*. J. Agr. Food Chem. 53:3631–3636.

Denstadli, V., A. Skrede, Å. Krogdahl, S. Sahlstrøm, and T. Storebakken. 2006. Feed intake, growth, feed conversion, digestibility, enzyme activities and intestinal structure in Atlantic salmon (*Salmo salar* L.) fed graded levels of phytic acid. Aquaculture 256:365–376.

Dezfuli, B. S., S. Arrighi, C. Domeneghini, and G. Bosi. 2000. Immunohistochemical detection of neuromodulators in the intestine of *Salmo trutta* L. naturally infected with *Cyathocephalus truncatus* Pallas (Cestoda). J. Fish Dis. 23:265–273.

Dorrell, D. G., 1976. Chlorogenic acid content of meal from cultivated and wild sunflowers. Crop Sci. 16:422.

Doster, R. C., G. H. Arscott, and R. O. Sinnhube. 1973. Comparative toxicity of ochratoxin A and crude aspergillus-ochraceus culture extract in Japanese-Quail (Coturnix-Coturnix-Japonica). Poultry Sci. 52:2351–2353.

Dreher, M. L., and E. T. Holm. 1983. A high performance liquid chromatographic method for chlorogenic acid determination in sunflower seeds. Statistics 10:100.

EFSA (European Food Safety Authority). 2007. Opinion of the Scientific Panel on Contaminants in the Food Chain on a request from the Commission related to cyanogenic compounds as undesirable substances in animal feed. The EFSA Journal 434:1–67.

Eisele, T. A., J. E., Nixon, N. E. Pawlowski, and R. O. Sinnhuber. 1978. Effects of dietary cyclopropene fatty acids on the mixed function oxidase system of the rainbow trout (*Salmo gairdneri*). J. Environ. Pathol. Tox. 1:773–778.

Eisler, R. 1991. Cyanide hazards to fish, wildlife and invertebrates: A synoptic review, Biological Report 85 (1.23). Contaminant Hazard Reviews 1–35, 1–58. Laurel, MD: Patuxent Wildl. Res. Center, U.S. Geological Survey.

Ellis, A. E. 1995. Recent development in oral vaccine delivery systems. Fish Pathol. 30:293–300.

El-Sayed, Y. S., R. H. Khalil, and T. T. Saad. 2009. Acute toxicity of ochratoxin-A in marine water-reared sea bass (*Dicentrarchus labrax* L.). Chemosphere 75:878–882.

Fairgrieve, W. T., M. S. Myers, R. W. Hardy, and F. M. Dong. 1994. Gastric abnormalities in rainbow trout (*Oncorhynchus mykiss*) fed amine-supplemented diets or chicken gizzard-erosion-positive fish meal. Aquaculture 127(2-3):219–232.

Fernandez, I., M. Diaz, T. Martinez, and F. J. Moyano. 2001. A quick method for the assessment of activity and inhibition of fish amylases. Aquacult. Nutr. 7:19–23.

Fish, B. C., and L. U. Thompson. 1991. Lectin tannin interactions and their influence on pancreatic amylase activity and starch digestibility. J. Agr. Food Chem. 39:727–731.

Fisher, J. P., S. B. Brown, G. W. Wooster, and P. R. Bowser. 1998. Maternal blood, egg and larval thiamin levels correlate with larval survival in landlocked Atlantic salmon (*Salmo salar*). J. Nutr. 128:2456–2466.

Food Standards Australia New Zealand. 2003. Erucic acid in food: A Toxicological Review and Risk Assessment, June 2003. Technical report series No. 21. Canberra, Australia: Food Standards Australia New Zealand.

Francis, G., H. P. S. Makkar, and K. Becker. 2001a. Antinutritional factors in plant-derived alternate fish ingredients and their effects in fish. Aquaculture 199:197–227.

Francis, G., H. P. S. Makkar, and K. Becker. 2001b. Effects of cyclic and regular feeding of a *Quillaja* saponin supplemented diet on growth and metabolism of common carp (*Cyprinus carpio* L.). Fish Physiol. Biochem. 24:343–350.

Francis, G., H. P. S. Makkar, and K. Becker. 2002. Dietary supplementation with a *Quillaja* saponin mixture improves growth performance and metabolic efficiency in common carp (*Cyprinus carpio* L). Aquaculture 203:311–320.

Freeland, W. J., P. H. Calcott, and L. R. Anderson. 1985. Tannins and saponin: Interaction in herbivore diet. Biochem. Syst. Ecol. 13(Ž2):189–193.

Friedman, M. 2006. Potato glycoalkaloids and metabolites: Roles in the plant and in the diet. J. Agr. Food Chem. 54:8655–8681.

Friesen, E. N., M. G. Ikonomou, D. A. Higgs, K. P. Ang, and C. Dubetz. 2008. Use of terrestrial based lipids in aquaculture feeds and the effects on flesh organohalogen and fatty acid concentrations in farmed Atlantic salmon. Environ. Sci. Technol. 42:3519–3523.

Froystad, M. K., E. Lilleeng, A. Sundby, and Å. Krogdahl. 2006. Cloning and characterization of alpha-amylase from Atlantic salmon (*Salmo salar* L.). Comp. Biochem. Phys. A 145:479–492.

Fuller, Z., P. Louis, A. Mihajlovski, V. Rungaparnestry, B. Ratcliffe, and A. J. Duncan. 2007. Influence of cabbage processing methods and prebiotic manipulation of colonic microflora on glucosinolate breakdown in man. Brit. J. Nutr. 98:364–372.

Galtier, P. 1998. Biological fate of mycotoxins in animals. Rev. Med. Vet-Toulouse 149:549–554.

Gamboa, D. A., M. C. Calhoun, S. W. Kuhlmann, A. U. Haq, and C. A. Bailey. 2001. Tissue distribution of gossypol enantiomers in broilers fed various cottonseed meals. Poultry Sci. 80:920–925.

Garcia-Gallego, M., H. Akharbach, and M. De la Higuera. 1998. Use of protein sources alternative to fish meal in diets with amino acids supplementation for the European eel (*Anguilla anguilla*). Anim. Sci. 66:285–292.

Gargouri, Y., R. Julien, G. Pieroni, R. Verger, and L. Sarda. 1984. Studies on the inhibition of pancreatic and microbial lipases by soybean proteins. J. Lipid Res. 25:1214–1221.

Gatlin, D. M., III, F. T. Barrows, P. Brown, K. Dabrowski, T. G. Gaylord, R. W. Hardy, E. Herman, G. Hu, A. Krogdahl, R. Nelson, K. Overturf, M. Rust, W. Sealey, D. Skonberg, E. J. Souza, D. Stone, R. Wilson, and E. Wurtele. 2007. Expanding the utilization of sustainable plant products in aquafeeds: A review. Aquac. Res. 38:551–579.

Gawai, K. R., C. Cox, J. Jackson, and R. R. Dalvi. 1995. Changes in the activity of metabolic and non-metabolic liver enzymes in rats following co-administration of gossypol with phenobarbital. Pharmacol. Toxicol. 76:289–291.

Gee, J. M., G. M. Wortley, I. T. Johnson, K. R. Price, A. A. J. J. Rutten, G. F. Houben, and A. H. Penninks. 1996. Effects of saponins and glycoalkaloids on the permeability and viability of mammalian intestinal cells and on the integrity of tissue preparations in vitro. Toxicol. in Vitro 10:117–128.

Gibbs, B. F., and I. Alli. 1998. Characterization of a purified [alpha]-amylase inhibitor from white kidney beans (*Phaseolus vulgaris*). Food Res. Int. 31:217–225.

Gill, N., D. A. Higgs, B. J. Skura, M. Rowshandeli, B. S. Dosanjh, J. Mann, and A. L. Gannam. 2006. Nutritive value of partially dehulled and extruded sunflower meal for post-smolt Atlantic salmon (*Salmo salar* L.) in sea water. Aquac. Res. 37:1348–1359.

Glencross, B. 2001. Feeding lupins to fish: A review of the nutritional and biological values of lupins in aquaculture feeds. North Beach, Australia: Department of Fisheries, Government of Western Australia.

Glencross, B., W. Hawkins, and J. Curnow. 2004a. Nutritional assessment of Australian canola meal. Part I. Evaluation of canola oil extracted methods and meal processing conditions on the digestible value of canola meal fed to the red seabream (*Pagrus auratus*, Paulin). Aquac. Res. 35:15–24.

Glencross, B., W. Hawkins, and J. Curnow. 2004b. Nutritional assessment of Australian canola meal. Part II. Evaluation of canola oil extracted methods on the protein value of canola meal fed to the red seabream (*Pagrus auratus*, Paulin). Aquac. Res. 35:25–34.

Glencross, B., D. Evans, N. Rutherford, W. Hawkins, P. McCafferty, K. Dods, B. Jones, D. Harris, L. Morton, M. Sweetingharn, and S. Sipsas. 2006. The influence of the dietary inclusion of the alkaloid gramine, on rainbow trout (*Oncorhynchus mykiss*) growth, feed utilisation and gastrointestinal histology. Aquaculture 253:512–522.

Glover, C. N., D. Petri, K.-E. Tollefsen, N. Jørum, R. D. Handy, and M. H. G. Berntssen. 2007. Assessing the sensitivity of Atlantic salmon (*Salmo salar*) to dietary endosulfan exposure using tissue biochemistry and histology. Aquat. Toxicol. 84:346–355.

Goldstein, W. S., and K. C. Spencer. 1985. Inhibition of cyanogenesis by tannins. J. Chem. Ecol. 11:847–857.

Goyarts, T., S. Dänickse, H. Valenta and K. H. Ueberschär. 2007. Carry-over of *Fusarium* toxins (deoxynivalenol and zearalenone) from naturally contaminated wheat to pigs. Food Addit. Contam. 24:364–380.

Grabner, M., and R. Hofer. 1985. The digestibility of the proteins of broad bean (*Vicia faba*) and soybean (*Glycine max*) under in vitro conditions stimulating the alimentary tracts of rainbow trout (*Salmo gairdneri*) and carp (*Cyprinus carpio*). Aquaculture 48:111–122.

Grant, G., L. J. More, N. H. McKenzie, J. C. Stewart, and A. Pusztai. 1983. A survey of the nutritional and haemagglutination properties of legume seeds generally available in the UK. Brit. J. Nutr. 50:207–214.

Guan, S., J. W. He, C. Young, H. Zhu, X. Z. Li, C. Ji, and T. Zhou. 2009. Transformation of trichothecene mycotoxins by microorganisms from fish digesta. Aquaculture 290:290–295.

Gupta, A., H. R. Sadeghipour, and S. C. Bhatla. 2003. Subcellular detection of lipase activity in plant protoplasts using fluorescence microscopy. Plant Growth Regul. 41:259–264.

Halver, J. E., and I. A. Mitchell. 1969. Trout hepatoma research conference papers. Bur. Sport Fish. Wildlife Rep. 70. Washington, DC: U.S. Department of the Interior.

Hannun Y. A., and L. M. Obeid. 2002. The Ceramide-centric universe of lipid-mediated cell regulation: Stress encounters of the lipid kind. J. Biol. Chem. 277:25847–25850.

Hara, T., Y. Mukunoki, I. Tsukamoto, M. Miyoshi, and K. Hasegawa. 1984. Susceptibility of Kintoki bean lectin to digestive enzymes in vitro and its behavior in the digestive organs of mouse in vivo. J. Nutr. Sci. Vitaminol. 30:381–394.

Hardy, R. W. 2001. Nutritional deficiencies in commercial aquaculture: Likelihood, onset, and identification. Pp. 131–147 in Nutrition and Fish Health, C. Lim and C. D. Webster, eds. London, UK: The Haworth Press, Inc.

Helland, S., V. Denstadli, P. E. Witten, K. Hjelde, T. Storebakken, A. Skrede, T. Åsgård, and G. Bæverfjord. 2006. Hyper dense vertebrae and mineral content in Atlantic salmon (*Salmo salar*) fed diets with graded levels of phytic acid. Aquaculture 261:603–614.

Hemre, G. I., H. Amlund, M. Aursand, A. M. Bakke, R. E. Olsen, E. Ringø, and B. Svihus. 2009. Criteria for safe use of plant ingredients in diets for aquacultured fish, Opinion of the Panel on Animal Feed of the Norwegian Scientific Committee for Food Safety. Oslo, Norway: VKM.

Hendricks, J. D. 2002. Adventitious toxins. Pp. 601–649 in Fish Nutrition, 3rd Edition, J. E. Halver and R. W. Hardy, eds. Amsterdam, the Netherlands: Elsevier.

Hendricks, J. D., R. O. Sinnhuber, P. M. Loveland, N. E. Pawlowski, and J. N. Nixon. 1980a. Hepatocarcinogenicity of glandless cottonseeds and cottonseed oil to rainbow trout (*Salmo gairdneri*). Science 208:309–311.

Hendricks, J. D., R. O. Sinnhuber, J. E. Nixon, J. H. Wales, M. S. Masri, and D. P. Hsieh. 1980b. Carcinogenic response of rainbow trout (*Salmo gairdneri*) to aflatoxin Q1 and synergistic effect of cyclopropenoid fatty acids. J. Natl. Cancer I. 64:523–528.

Hendriks, H. G. C. J. M., T. S. G. A. M. van den Ingh, Å. Krogdahl, J. Olli, and J. F. J. G. Koninkx. 1990. Binding of soybean agglutinin to small intestinal brush border membranes and brush border membrane enzyme activities in Atlantic salmon (*Salmo salar*). Aquaculture 91:163–170.

Hídvégi, M., and R. Lásztity. 2002. Phytic acid content of cereals and legumes and interaction with proteins. Period. Polytech. Chem. 46:59–64.

Honeyfield, D. C., J. P. Hinterkopf, and S. B. Brown. 2002. Isolation of thiaminase-positive bacteria from alewife. T. Am. Fish. Soc. 131:171–175.

Hooft, J. 2010. The Effects of Feed-borne Fusarium Mycotoxins on the Performance and Health of Rainbow Trout (*Oncorhynchus mykiss*). Masters Thesis, University of Guelph, Guelph, Ontario, Canada.

Hooft, J., and D. P. Bureau. 2010. Mycotoxins in Feed Affect Fish Health, Performance. The Global Aquacult. Advocate Magazine March/April: 31–32.

Hossain, M. A., and K. Jauncey. 1993. The effect of varying dietary phytic acid, calcium and magnesium levels on the nutrition of common carp, *Cyprinus carpio*. Pp. 705–715 in Fish Nutrition in Practice, S. J. Kaushik and P. Luquent, eds. Proc. IV International Symposium on Fish Nutrition and Feeding, June 24–27, 1991, Biarritz, France.

Huang, L., D. K. Zheng, and Y. K. Si. 1987. Resolution of acemic gossypol. J. Ethnopharmacol. 20:13–20.

Ikonomou, M. G., D. A. Higgs, M. Gibbs, J. Oakes, B. Skura, S. McKinley, S. K. Balfry, S. Jones, R. Withler, and C. Dubetz. 2007. Flesh quality of market-size farmed and wild British Columbia salmon. Environ. Sci. Technol. 41:437–443.

Inudo, M., H. Ishibashi, N. Matsumura, M. Matsuoka, T. Mori, S. Taniyama, K. Kadokami, M. Koga, R. Shinohara, T. H. Hutchinson, T. Iguchi, and K. Arizono. 2004. Effect of estrogenic activity, and phytoestrogen and organochlorine pesticide contents in an experimental fish diet on reproduction and hepatic vitellogenin production in medaka (*Oryzias latipes*). Comparative Med. 54:673–680.

IPCS (International Panel on Chemical Safety). 2004. Hydrogen cyanide and cyanides: Human health aspects. Cicads 61. Geneva, Switzerland: World Health Organization.

Ishihara, T., M. Yasuda, S. Kashiwag, and M. Yagi. 1974. Studies on thiaminase-I in marine fish. 5. Preventive effect of thiamine on nutritional disease of yellowtail fed anchovy. B. Jpn. Soc. Sci. Fish. 40:675–682.

Iwashita, Y., T. Yamamoto, H. Furuita, T. Sugita, and N. Suzuki. 2008. Influence of certain soybean antinutritional factors supplemented to a casein-based semipurified diet on intestinal and liver morphology in fingerling rainbow trout *Oncorhynchus mykiss*. Fish. Sci. 74:1075–1082.

Iwashita, Y., N. Suzuki, H. Matsunari, T. Sugita, and T. Yamamoto. 2009. Influence of soya saponin, soya lectin, and cholyltaurine supplemented to a casein-based semipurified diet on intestinal morphology and biliary bile status in fingerling rainbow trout *Oncorhynchus mykiss*. Fish. Sci. 75:1307–1315.

Jenkinson, C. P., W. W. Grody, and S. D. Cederbaum. 1996. Comparative properties of arginases. Comp. Biochem. Phys. B 114:107–132.

Joerink, M., H. F. J. Savelkoul, and G. F. Wiegertjes. 2006. Evolutionary conservation of alternative activation of macrophages: Structural and functional characterization of arginase 1 and 2 in carp (*Cyprinus carpio* L.). Mol. Immunol. 43:1116–1128.

Johnson, I. T., J. M. Gee, K. Price, C. Curl, and G. R. Fenwick. 1986. Influence of saponins on gut permeability and active nutrient transport invitro. J. Nutr. 116:2270–2277.

Jones, D. A. 1998. Why are so many food plants cyanogenic? Phytochemistry 47:155–162.

Kaushik, S. J., J. P. Cravedi, J. P. Lalles, J. Sumpter, B. Fauconneau, and M. LaRoche. 1995. Partial or total replacement of fish-meal by soybean protein on growth, protein-utilization, potential estrogenic or antigenic effects, cholesterolemia and flesh quality in rainbow-trout, *Oncorhynchus-mykiss*. Aquaculture 133:257–274.

Knudsen, D., O. Ron, G. Baardsen, J. Smedsgaard, W. Koppe, and H. Froklaer. 2006. Soyasaponins resist extrusion cooking and are not degraded during gut passage in Atlantic salmon (*Salmo salar* L.). J. Agr. Food Chem. 54:6428–6435.

Knudsen, D., P. Uran, A. Arnous, W. Koppe, and H. Frokiaer. 2007. Saponin-containing subfractions of soybean molasses induce enteritis in the distal intestine of Atlantic salmon. J. Agr. Food Chem. 55:2261–2267.

Knudsen, D., F. Jutfelt, H. Sundh, K. Sundell, W. Koppe, and H. Frokiaer. 2008. Dietary soya saponins increase gut permeability and play a key role in the onset of soyabean-induced enteritis in Atlantic salmon (*Salmo salar* L.). Brit. J. Nutr. 100:120–129

Kovacic, P. 2003. Mechanism of drug and toxic actions of gossypol: focus on reactive oxygen species and electron transfer. Curr. Med. Chem. 10:2711–2718.

Kovacic, S., S. Pepeljnjak, Z. Petrinec, and M. S. Klaric. 2009. Fumonisin B-1 neurotoxicity in young carp (*Cyprinus Carpio* L.). Arh. Hig. Rada Toksiko. 60:419–426.

Krogdahl, Å., T. Berg-Lea, and J. J. Olli. 1994. Soybean proteinase inhibitors affect intestinal trypsin activities and amino acid digestibilities in rainbow trout (*Oncorhyncus mykiss*). Comp. Biochem. Phys. A 107:215–219.

Krogdahl, Å., A. M. Bakke-McKellep, K. H. Røed, and G. Bæverfjord. 2000. Feeding Atlantic salmon (*Salmo salar* L.) soybean products: Effects on disease resistance (furunculosis), and lysozyme and IgM levels in the intestinal mucosa. Aquacult. Nutr. 6:77–84.

Krogdahl, Å., G. I. Hemre, and T. P. Mommsen. 2005. Carbohydrates in fish nutrition: Digestion and absorption in postlarval stages. Aquacult. Nutr.11:103–122.

Krogdahl, Å., M. H. Penn, J. Thorsen, S. Refstie, and A. M. Bakke. 2009. Important antinutrients in plant feedstuffs for aquaculture: An update on recent findings regarding responses in salmonid. Aquac. Res. 41:333–344.

Lalles, J. P., and G. Peltre. 1996. Biochemical features of grain legume allergens in humans and animals. Nutr. Rev. 54:101–107.

Lechtenberg, M., and A. Nahrstedt. 1999. Cyanogenic glycosides. Pp. 147–191 in Naturally Occurring Glycosides, R. Ikan, ed. Chichester, UK: John Wiley & Sons.

Li, M. H. H., and E. H. Robinson. 2006. Use of cottonseed meal in aquatic animal diets: A review. N. Am. J. Aquacult. 68:14–22.

Liener, I. 1980. Toxic Constituents of Plant Foodstuffs. New York: Academic Press.

Liener, I. E., 1989. Antinutritional factors in legume seeds: State of the art. Pp. 6–14 in Recent Advances of Research in Antinutritional Factors in Legume Seeds, J. Huisman, T. F. B. van der Poel, and I. E. Liener, eds. Wageningen, the Netherlands: Pudoc Wageningen.

Lindsay, G. J. H., M. J. Walton, J. W. Adron, T. C. Fletcher, C. Y. Cho, and C. B. Cowey. 1984. The growth of rainbow trout (*Salmo gairdneri*) given diets containing chitin and its relationship to chitinolytic enzymes and chitin digestibility. Aquaculture 37:315–334.

Lott, J. N. A., I. Ockenden, V. Raboy, and G. D. Batten. 2000. Phytic acid and phosphorus in crop seeds and fruits: A global estimate. Seed Sci. Res. 10:11–33.

Lozano, N. B. S., A. T. Vidal, S. Martínez-Llorens, S. N. Mérida, J. E. Blanco, A. M. López, M. P. Torres, and M. J. Cerdá. 2007. Growth and economic profit of gilthead sea bream (*Sparus aurata*, L.) fed sunflower meal. Aquaculture 272:528–534.

Lumlertdacha, S., R. T. Lovell, R. A. Shelby, S. D. Lenz, and B. W. Kemppainen. 1995. Growth, hematology, and histopathology of channel catfish, ictalurus-punctatus, fed toxins from fusarium-moniliforme. Aquaculture 130:201–218.

Malevski, Y., J. H. Wales, and M. W. Montgomery. 1974. Liver damage in rainbow trout, *Salmo gairdneri*, fed cyclopropenoid fatty acids. J. Fish. Res. Board Can. 31:1397–1400.

Malinow, M. R., P. Mclaughlin, L. Papworth, C. Stafford, G. O. Kohler, A. L. Livingston, and P. R. Cheeke. 1977. Effect of alfalfa saponins on intestinal cholesterol absorption in rats. Am. J. Clin. Nutr. 30:2061–2067.

Manning, B. B. 2001. Mycotoxins in fish feeds. Pp. 267–287 in Nutrition and Fish Health, C. Lim and D. Webster, eds. London, UK: The Haworth Press.

Manning, B. B., R. M. Ulloa, M. H. H. Li, E. H. Robinson, and G. E. Rottinghaus. 2003. Ochratoxin A fed to channel catfish (*Ictalurus punctatus*) causes reduced growth and lesions of hepatopancreatic tissue. Aquaculture 219:739–750.

Manning, B. B., J. S. Terhune, M. H. Li, E. H. Robinson, D. J. Wise, and G. E. Rottinghaus. 2005. Exposure to feedborne mycotoxins T-2 toxin or ochratoxin a causes increased mortality of channel catfish challenged with Edwardsiella ictaluri. J. Aquat. Anim. Health 17:147–152.

Marquardt, R. R. 1989. Dietary effects of tannins, vicine and convicine. Pp. 141–155 in Recent Advances of Research in Antinutritional Factors in Legume Seeds, J. Huisman, T. F. B. van der Poel, and I. E. Liener, eds. Wageningen, the Netherlands: Pudoc Wageningen.

McCutcheon, J. S., T. Umermura, M. K. Bhatnager, and B. L. Walker. 1976. Cardiopathogenicity of rapeseed oils and oil blends differing in erucic, linoleic, and linolenic acid content. Lipids 11:545–552.

Milligan, S. R., O. Khan, and M. Nash. 1998. Competitive binding of xenobiotic oestrogens to rat alpha-fetoprotein and to sex steroid binding proteins in human and rainbow trout (*Oncorhynchus mykiss*) plasma. Gen. Comp. Endocr. 112:89–95.

Morawska-Muszynska, G., and I. Reifer. 1965. Preparation and properties of the arginase inhibitor from sunflower seeds. Acta Biochim. Pol. 12:187–194.

Moren, M., M. K. Malde, R. E. Olsen, G. I. Hemre, L. Dahl, O. Karlsen, and K. Julshamn. 2007. Fluorine accumulation in Atlantic salmon (*Salmo salar*), Atlantic cod (*Gadus morhua*), rainbow trout (*Oncorhyncus mykiss*) and Atlantic halibut (*Hippoglossus hippoglossus*) fed diets with krill or amphipod meals and fish meal based diets with sodium fluoride (NaF) inclusion. Aquaculture 269:525–531.

Morris, S. C., and T. H. Lee. 1984. The toxicity and teratogenicity of solanaceae glycoalkaloids, particularly those of the potato (*Solanum tuberosum*)—A review. Food Technol. Aust. 36:118–124.

Mueller-Harvey, I., and A. B. McAllan. 1992. Tannins: Their biochemistry and nutritional properties. Pp. 151–217 in Advances in Plant Cell Biochemistry and Biotechnology, I. M. Morrison, ed. London, UK: JAI Press Ltd.

Mulero, I., M. P. Sepulcre, J. Meseguer, A. García-Ayala, and V. Mulero. 2007. Histamine is stored in mast cells of most evolutionarily advanced fish and regulates the fish inflammatory response. P. Nat. Acad. Sci. USA 104:19434–19439.

Muszynska, G., and I. Reifer. 1970. The arginase inhibitor from sunflower seeds: Purification and inhibitory properties. Acta. Biochim. Pol. 17:247–252.

Natarajan, M., B. Ross, and L. G. Ross. 1992. Susceptibility of carp and tilapia [alpha]-amylase to purified wheat amylase inhibitor. Aquaculture 102:265–274.

Ng, W. K., and K. O. K. L. Wee. 1989. The nutritive value of cassava leaf meal in pelleted feed for Nile tilapia. Aquaculture (Amsterdam) 83:45–58.

Nishimune, T., Y. Watanabe, and H. Okazaki. 2008. Studies on the polymorphism of thiaminase I in seawater fish. J. Nutr. Sci. Vitaminol. 54:339–346.

NRC (National Research Council). 1982. Nutrient Requirements of Mink and Foxes. Washington, DC: National Academy Press.

NRC. 1993. Nutrient Requirements of Fish. Washington, DC: National Academy Press.

O'Connor, C. J., D. Sun, B. G. Smith, and L. D. Melton. 2003. The inhibitory effects of brans and their aqueous extracts on the lipolysis of tributyrin catalyzed by calf pregastric lipase. J. Food Sci. 68:1818–1825.

Oda, K., H. Matsuda, T. Murakami, S. Katayama, T. Ohgitani, and M. Yoshikawa. 2000. Adjuvant and haemolytic activities of 47 saponins derived from medicinal and food plants. Biol. Chem. 381:67–74.

Olli, J. J., and Å Krogdahl. 1994. Nutritive value of four soybean products as protein sources in diets for Rainbow trout (*Onorhyncus mykiss*, Walbaum) reared in fresh water. Acta Agr. Scand. 44:185–192.

Olli, J. J., K. Hjelmeland, and Å. Krogdahl. 1994a. Soybean trypsin inhibitors in diets for Atlantic salmon (*Salmo salar*, L): Effects on nutrient digestibilities and trypsin in pyloric caeca homogenate and intestinal content. Comp. Biochem. Phys. A 109:923–928.

Olli, J. J., Å. Krogdahl, T. S. G. A. M van den Ingh, and L. E. Brattås. 1994b. Nutritive value of four soybean products in diets for Atlantic salmon (*Salmo salar*, L). Acta Agr. Scand. A-An 44:50–60.

Olsen, R. E., J. Suontama, E. Langmyhr, H. Mundheim, E. Ringo, W. Melle, M. K. Malde, and G. I. Hemre. 2006. The replacement of fish meal with Antarctic krill, Euphausia superba in diets for Atlantic salmon, *Salmo salar*. Aquacult. Nutr. 12:280–290.

Opstvedt, J., H. Mundheim, E. Nygård, H. Aase, and I. H. Pike. 2000. Reduced growth and feed consumption of Atlantic salmon (*Salmo salar* L.) fed fish meal made from stale fish is not due to increased content of biogenic amines. Aquaculture 188:323–337.

Pelissero, C., F. Lemenn, and S. Kaushick. 1991a. Estrogenic effect of dietary soya bean meal on vitellogenesis in cultured Siberian sturgeon *Acipenser baeri*. Gen. Comp. Endocr. 83:447–457.

Pelissero, C., B. Bennetau, P. Babin, F. Lemenn, and J. Dunogues. 1991b. The estrogenic activity of certain phytoestrogens in the Siberian sturgeon *Acipenser baeri*. J. Steroid Biochem. 38:293–299.

Petri, D., C. N. Glover, S. Ylving, K. Kolas, G. Fremmersvik, R. Waagbo, and M. H. G. Berntssen. 2006. Sensitivity of Atlantic salmon (*Salmo salar*) to dietary endosulfan as assessed by haematology, blood biochemistry, and growth parameters. Aquatic Toxicology 80:207–216.

Petrinec Z., S. Pepeljnjak, S. Kovacic, and A. Krznaric. 2004. Fumonisin B-1 causes multiple lesions in common carp (*Cyprinus carpio*). Deut. Tierarztl. Wochenschrift 111:358–363.

Peumans, W. J., and E. J. M. Van Damme. 1995. Lectins as plant defense proteins. Plant Physiol. 109:347–352.

Pierpoin, W. S. 1970. Formation and behaviour of o-quinones in some processes of agricultural importance. Rep. Rothamsted Exp. Stn., Harpenden, Herts, UK 2:199–218.

Poston, H. A., J. L. Coffin, and G. F. Combs. 1982. Biological effects of dietary T-2 toxin on rainbow-trout, *Salmo-gairdneri*. Aquat. Toxicol. 2:79–88.

Poulton, J. E. 1990. Cyanogenesis in plants. Plant Physiol. 94:401.

Pratt, D., and P. M. Birac. 1979. Source of antioxidant activity of soybeans and soy products. J. Food Sci. 44:541–544.

Pusztai, A., S. W. B. Ewen, G. Grant, W. J. Peumans, E. J. M. Van Damme, L. Rubio, and S. Bardocz. 1990. The relationship between survival and binding of plant lectins during small intestinal passage and their effectiveness as growth factors. Digestion 46:308–316.

Pusztai, A., S. W. B. Ewen, G. Grant, W. J. Peumans, E. J. M. Van Damme, M. E. Coates, and S. Bardocz. 1995. Lectins and also bacteria modify the glycosylation of gut surface receptors in the rat. Glycoconjugate J. 12(1):22–35.

Raju P. K., and R. Reiser. 1967. Inhibition of fatty acyl desaturase by cyclopropene fatty acids. J Biol Chem 242:379–384.

Randall, D. J., C. M. Wood, S. F. Perry, H. Bergman, G. M. O. Maloiy, T. P. Mommsen, and P. A. Wright. 1989. Urea excretion as a strategy for survival in a fish living in a very alkaline environment. Nature 337:165–166.

Rao, B. S. N., and T. Prabhavathi. 1982. Tannin content of foods commonly consumed in India and its influence on ionisable iron. J. Sci. Food Agr. 33:89–96.

Rastogi, T., K. S. Reddy, M. Vazm, D. Spiegelman, D. Prabhakaran, W. C. Willett, M. Stampfer, and A. Ascherio. 2004. Diet and risk of ischemic heart disease in India. Am. J. Clin. Nutr. 79:582–592.

Refstie, S., and H. A. J. Tiekstra. 2003. Potato protein concentrate with low content of solanidine glycoalkaloids in diets for Atlantic salmon (*Salmo salar*). Aquaculture 216:283–298.

Refstie, S., Ø. J. Korsøen, T. Storebakken, G. Bæverfjord, I. Lein, and A. J. Roem. 2000. Differing nutritional responses to dietary defatted soybean meal in rainbow trout (*Oncorhynchus mykiss*) and Atlantic salmon (*Salmo salar*). Aquaculture 190:49–63.

Refstie, S., T. Storebakken, G. Bæverfjord, and A. J. Roem. 2001. Long-term protein and lipid growth of Atlantic salmon (*Salmo salar*) fed diets with partial replacement of fish meal by soy protein products at medium or high lipid level. Aquaculture 193:91–106.

Reifer, I., and H. Augustyniak. 1968. Preliminary identification of the arginase inhibitor from sunflower seeds. B. Acad. Pol. Sci. Biol. 16:139–144.

Reifer, I., and G. Morawska. 1963. An arginase inhibitor from sunflower seeds (*Helianthus annuus*). Acta Biochim. Pol. 10:413–417.

Reite, O. B. 1965. A phylogenetical approach to the functional significance of tissue mast cell histamine. Nature 206:1334–1336.

Reite, O. B. 1972. Comparative physiology of histamine. Physiol. Rev. 52:778–819.

Richardson, N. L., D. A. Higgs, R. M. Beames, and J. R. McBride. 1985. Influence of dietary calcium, phosphorus, zinc and sodium phytate level on cataract incidence, growth and histopathology in juvenile Chinook salmon (*Oncorhynchus tsawytscha*). J. Nutr. 115:553–567.

Ricque-Marie, D., M. I. Abdo-De La Parra, L. E. Cruz-Suarez, G. Cuzon, M. Cousin, and I. H. Pike. 1998. Raw material freshness, a quality criterion for fish meal fed to shrimp. Aquaculture 165:95–109.

Roberts, R. J. 2002. Nutritional pathology. Pp. 453–504 in Fish Nutrition, 3rd Edition, J. E. Halver and R. W. Hardy, eds. Amsterdam, the Netherlands: Elsevier.

Rocha, O., K. Ansari, and F. M. Doohan. 2005. Effects of trichothecene mycotoxins on eukaryotic cells: A review. Food Addit. Contam. 22:369–378.

Roehm J. N., D. J. Lee, J. H. Wales, S. D. Polityka, and R. O. Sinnhuber. 1970. The effect of dietary sterculic acid on the hepatic lipids of rainbow trout. Lipids 5:80–84.

Roth, J. 1987. Subcellular organization of glycosylation in mammalian cells. Biochim. Biophys. Acta 906:405–436.

Rumsey, G. L., S. G. Hughes, and R. A. Winfree. 1993. Chemical and nutritional evaluation of soya protein preparations as primary nitrogen sources for rainbow trout (*Oncorhynchus mykiss*). Anim. Feed Sci. Tech. 40:135–151.

Rumsey, G. L., A. K. Siwicki, D. P. Anderson, and P. R. Bowser. 1994. Effect of soybean protein on serological response, non-specific defense mechanisms, growth, and protein utilization in rainbow trout. Vet. Immunol. Immunop. 41:323–339.

Sander, J. E., T. Cai, N. Dale, and L. W. Bennett. 1996. Development of biogenic amines during fermentation of poultry carcasses. J. Appl. Poultry Sci. 5:161–166

Sándor, G., and A. Ványi. 1990. Mycotoxin research in the Hungarian Central Veterinary Institute. Acta Vet. Hung. 38:61–68

Sandoval, W., and A. Carmona. 1998. Comparative study of condensed and hydrolysable tannin interaction with isolated proteins and enzymes. Pp. 39–42 in Recent Advances of Research in Antinutritional Factors in Legume Seeds and Rapeseed, A. J. M. Jansman, G. D. Hill, J. Huisman, and A. F. B. van der Poel, eds. Wageningen, the Netherlands: Wageningen Pers.

Santacroce, M. P., M. C. Conversano, E. Casalino, O. Lai, C. Zizzadoro, G. Centoducati, and G. Crescenzo. 2008. Aflatoxins in aquatic species: Metabolism, toxicity and perspectives. Rev. Fish Biol. Fisher. 18:99–130.

Sanz, A., A. E. Morales, M. De La Higuera, and G. Cardenete. 1994. Sunflower meal compared with soybean meal as partial substitutes for fish meal in rainbow trout (*Oncorhynchus mykiss*) diets: Protein and energy utilization. Aquaculture (Amsterdam) 128:287–300.

Satouchi, K., K. Hirano, O. Fujino, M. Ikoma, T. Tanaka, and K. Kitamura. 1998. Lipoxygenase-1 from soybean seed inhibiting the activity of pancreatic lipase. Biosci. Biotech. Bioch. 62:1498–1503.

Satouchi, K., Y. Kodama, K. Murakami, T. Tanaka, H. Iwamoto, and M. Ishimoto. 2002. A lipase-inhibiting protein from lipoxygenase-deficient soybean seeds. Biosci. Biotech. Bioch. 66:2154–2160.

Sawada, S., Y. Takeda, and M. Tashiro. 2002. Primary structures of alpha- and beta-subunits of alpha-amylase inhibitors from seeds of three cultivars of Phaseolus beans. J. Protein Chem. 21:9–17.

Selivonchick D. P., J. L. Williams, and H. W. Schaup. 1981. Alteration of liver microsomal proteins from rainbow trout (*Salmo gairdneri*) fed cyclopropenoid fatty acids. Lipids 16:211–214.

Shapiro, T. A., J. W. Fahey, K. L. Wade, K. K. Stephenson, and P. Talalay. 2001. Chemoprotective glucosinolates and isothiocyanates of broccoli sprouts: Metabolism and excretion in humans. Cancer Epidem. Biomar. 10:501–508.

Sharon, N., and H. Lis. 1990. Legume lectins—a large family of homologous proteins. FASEB J. 4:3198–3208.

Shiau, S. Y., and Y. P. Yu. 1999. Dietary supplementation of chitin and chitosan depresses growth in tilapia, *Oreochromis niloticus* × *O. aureua*. Aquaculture 179:439–446.

Sinnhuber, R. O., D. J. Lee, J. H Wales, M. K. Landers, and A. C. Keyl. 1974. Hepatic carcinogenesis of aflatoxin M1 in rainbow trout (*Salmo gairdneri*) and its enhancement by cyclopropene fatty acids. J. Natl. Cancer I. 53:1285–1288.

Sjolander, A., and J. C. Cox. 1998. Uptake and adjuvant activity of orally delivered saponin and ISCOM (TM) vaccines. Adv. Drug Deliver. Rev. 34:321–338.

Spiegel, S., and S. Milstien. 2002. Sphingosine 1-phosphate, a key cell signaling molecule. J. Biol. Chem. 277:25851–25854.

Spinelli, J., C. R. Houle, and J. C. Wekell. 1983. The effect of phytates on the growth of rainbow trout (*Salmo gairdneri*) fed purified diets containing varying quantities of calcium and magnesium. Aquaculture 30:71–83.

Sripad, G., V. Prakash, and M. S. N. Rao. 1982. Extractability of polyphenols of sunflower seed in various solvents. J. Biosciences 4:145–152.

Steele, S. L., T. D. Chadwick, and P. A. Wright. 2001. Ammonia detoxification and localization of urea cycle enzyme activity in embryos of the rainbow trout (*Oncorhynchus mykiss*) in relation to early tolerance to high environmental ammonia levels. J. Exp. Biol. 204:2145–2154.

Stillmark, H. 1888. Über Ricin ein giftiges Ferment aus den Samen von *Ricinus communis* L. und einige anderen Euphorbiaceen. Tartu: Inaugural Dissertation Dorpat.

Storebakken, T., I. S. Kvien, K. D. Shearer, B. Grisdale-Helland, S. J. Helland, and G. M. Berge. 1998a. The apparent digestibility of diets containing fish meal, soybean meal or bacterial meal fed to Atlantic salmon (*Salmo salar*): Evaluation of different faecal collection methods. Aquaculture 169:195–210.

Storebakken, T., K. D. Shearer, and A. J. Roem. 1998b. Availability of protein, phosphorus and other elements in fish meal, soy-protein concentrate and phytase-treated soy-protein-concentrate-based diets to Atlantic salmon, *Salmo salar*. Aquaculture 161:365–379.

Struthers, B. J., D. J. Lee, and R. O. Sinnhuber. 1975. Altered lipid metabolism in liver of rainbow trout fed cyclopropenoid fatty acids. Exp. Mol. Pathol. 23(2):181–187.

Sujak, A., A. Kotlarz, and W. Strobel. 2006. Compositional and nutritional evaluation of several lupin seeds. Food Chem. 98:711–719.

Sulyok, M., R. Krska, and R. Schuhmacher. 2007. A liquid chromatography/tandem mass spectrometric multi-mycotoxin method for the quantification of 87 analytes and its application to semi-quantitative screening of moldy food samples. Anal. Bioanal. Chem. 389:1505–1523.

Tani, H., H. Ohishi, and K. Watanabe. 1994. Purification and characterization of proteinaceous inhibitor of lipase from wheat flour. J. Agr. Food Chem. 42:2382–2385.

Tani, H., H. Ohishi, and K. Watanabe. 1995. Kinetics studies of a wheat-derived lipase inhibitor. J. Agr. Food Chem. 43:2796–2797.

Tapia-Salazar, M., L. E. Cruz-Suarez, D. Ricque-Marie, I. H., Pike, T. K. Smith, A. Harris, E. Nygard, and J. Opstvedt. 2004. Effect of fishmeal made from stale versus fresh herring and of added crystalline biogenic amines on growth and survival of blue shrimp Litopenaeus stylirostris fed practical diets. Aquaculture 242:437–453.

Teskeredžić, Z., D. A. Higgs, B. S. Dosanjh, J. R. McBride, R. W. Hardy, R. M. Beames, J. D. Jones, M. Simell, T. Vaara, and R. B. Bridges. 1995. Assessment of undephytinized and dephytinized rapeseed protein concentrate as sources of dietary protein for juvenile rainbow trout (*Oncorhynchus mykiss*). Aquaculture 131:261–277.

Tollefsen, K. E., J. Ovrevik, and J. Stenersen. 2004. Binding of xenoestrogens to the sex steroid-binding protein in plasma from Arctic charr (*Salvelinus alpinus* L.). Comp. Biochem. Phys. C 139:127–133.

Tripathi, M. K., and A. S. Mishra. 2007. Glucosinolates in animal nutrition: A review. Anim. Feed Sci. Tech. 132:1–27.

Tuan, N. A., B. B. Manning, R. T. Lovell, and G. E. Rottinghaus. 2003. Responses of Nile tilapia (*Oreochromis niloticus*) fed diets containing different concentrations of moniliformin or fumonisin B-1. Aquaculture 217:515–528.

Twibell, R. G., and R. P. Wilson. 2004. Preliminary evidence that cholesterol improves growth and feed intake of soybean meal-based diets in aquaria studies with juvenile channel catfish, *Ictalurus punctatus*. Aquaculture 236:539–546.

Uran, P. A., A. A. Goncalves, J. J. Taverne-Thiele, J. W. Schrama, J. A. J. Verreth, and J. H. W. M. Rombout. 2008. Soybean meal induces intestinal inflammation in common carp (*Cyprinus carpio* L.). Fish Shellfish Immun. 25:751–760.

van Damme, E. J. M., W. J. Peumans, A. Pusztai, and S. Bardocz. 1998. Handbook of Plant Lectins: Properties and Biomedical Applications. West Sussex, UK: John Wiley & Sons.

van den Ingh, T. S., and Å. Krogdahl. 1990. Negative effects of antinutritional factors from soybeans in Salmonidae [in Dutch]. Tijdschr. Diergeneesk. 15:935–938.

van den Ingh, T. S. G. A. M., Å. Krogdahl, J. Olli, H. G. C. J. M. Hendricks, and J. F. J. G. Koninkx. 1991. Effects of soybean-containing diets on the proximal and distal intestine in Atlantic salmon (*Salmo salar*): A morphological study. Aquaculture 9:297–305.

van den Ingh, T. S. G. A. M., J. J. Olli, and Å. Krogdahl. 1996. Alcohol-soluble components in soybeans cause morphological changes in the distal intestine of Atlantic salmon, *Salmo salar* L. J. Fish Dis. 19:47–53.

Varga, J., K. Rigo, J. Teren, and A. Mesterhazy. 2001. Recent advances in ochratoxin research–I. Production, detection and occurrence of ochratoxins. Cereal Res. Commun. 29(1–2):85–92.

Vielma, J., K. Ruohonen, and M. Peisker. 2002. Dephytinization of two soy proteins increases phosphorus and protein utilisation by rainbow trout, *Oncorhynchus mykiss*. Aquaculture 204:145–156.

Vielma, J., K. Ruohonen, J. Gabaudan, and K. Vogel. 2004. Top-spraying soybean meal-based diets with phytase improves protein and mineral digestibilities but not lysine utilisation in rainbow trout, *Oncorhynchus mykiss* (Walbaum). Aquac. Res. 35:955–964.

Wang, H. J., and P. A. Murphy. 1994a. Isoflavone content in commercial soybean foods. J. Agr. Food Chem. 42:1666–1673.

Wang, H. J., and P. A. Murphy. 1994b. Isoflavone composition of American and Japanese soybeans in Iowa—Effects of variety, crop year, and location. J. Agr. Food Chem. 42:1674–1677.

Wang, S. M., and A. H. C. Huang. 1984. Inhibitors of lipase activities in soybean and other oil seeds. Plant Physiol. 76:929.

Whitaker, J. R. 1997. Protease and alpha-amylase inhibitors of higher plants. Antinutrients and phytochemicals in food. ACS Symposium Series, 210th National Meeting of the American Chemical Society, August 20–25, 1995, Chicago, IL.

Wilson, M., E. Bengten, N. W. Miller, L. W. Clem, L. DuPasquier, and G. W. Warr. 1997. A novel chimeric Ig heavy chain from a teleost fish shares similarities to IgD. P. Natl. Acad. Sci. USA. 94:4593–4597.

Wink, M., C. Meissner, and L. Witte. 1995. Patterns of quinolizidine alkaloids in 56 species of the genus *Lupinus*. Phytochemistry 38:139–153.

Wink, M., T. Schmeller, and B. Latz-Bruning. 1998. Modes of action of allelochemical alkaloids: Interaction with neuroreceptors, DNA, and other molecular targets. J. Chem. Ecol. 24:1881–1937.

Woodward, B., L. G. Young, and A. K. Lun. 1983. Vomitoxin in diets for rainbow-trout (*Salmo gairdneri*). Aquaculture 35:93–101.

Wright, P. 1995. Induction of ornithine-urea cycle enzymes and nitrogen metabolism and excretion in rainbow trout (*Oncorhynchus mykiss*) during early life stages. J. Exp. Biol. 198:127–135.

Wright, P. A., and M. D. Land. 1998. Urea production and transport in teleost fishes. Comp. Biochem. Phys. A 119:47–54.

Yildirim, M., B. B. Manning, R. T. Lovell, J. M. Grizzle, and G. E. Rottinghaus. 2000. Toxicity of moniliformin and fumonisin B-1 fed singly and in combination in diets for young channel catfish *Ictalurus punctatus*. J. World Aquacult. Soc. 31:599–608.

Yildirim, M., C. Lim, P. J. Wan, and P. H. Klesius. 2003. Growth performance and immune response of channel catfish (*Ictalurus puctatus*) fed diets containing graded levels of gossypol-acetic acid. Aquaculture 219:751–768.

Yildirim-Aksoy, M., C. Lim, P. Wan, and P. H. Klesius. 2004. Effect of natural free gossypol and gossypol-acetic acid on growth performance and resistance of channel catfish (*Ictalurus punctatus*) to *Edwardsiella ictaluri* challenge. Aquacult. Nutr. 10:153–165.

Yoshiki, Y., S. Kudou, and K. Okubo. 1998. Relationship between chemical structures and biological activities of triterpenoid saponins from soybean. Biosci. Biotech. Bioch. 62:2291–2299.

Zajicek, J. L., L. Brown, S. B. Brown, D. C., Honeyfield, J. D. Fitzsimons, and D. E. Tillitt. 2009. Variations of thiaminase I activity pH dependencies among typical Great Lakes forage fish and *Paenibacillus thiaminolyticus*. J. Aquat. Anim. Health 21:207–216.

12

Digestibility and Availability

The terms "digestibility" and "availability" refer to the amount or proportion of nutrients or categories of nutrients, such as crude protein, that disappears from a meal as it passes through the digestive system and is excreted (egested) in feces. Consequently, digestibility is primarily a measure of disappearance of nutrients. Digested nutrients are presumed available to the organism for growth and metabolism, although this is not always the case. Digestibility is the term used for nutrients that are subject to the digestive process, such as proteins that are hydrolyzed to amino acids prior to absorption. The bioavailability of nutrients (e.g., amino acids) has been defined as the proportion in which ingested nutrients from a particular source is absorbed in a form that can be utilized in metabolism by animals (Batterham, 1992; Ammerman et al., 1995; Lewis and Bayley, 1995). Bioavailability of nutrients is assessed using a wide variety of methodological approaches. Bioavailability implies metabolic utilization of the nutrients, a topic beyond the scope of this chapter.

After food or feed is ingested by an organism, it passes through the digestive tract where complex compounds are broken down to simpler compounds and nutrients are absorbed through a variety of mechanisms (simple diffusion, facilitate diffusion, active transport) into the body. Materials that are not absorbed are excreted, or egested, as feces. Thus, measuring nutrient or energy in foods or feeds and in feces provides the information needed to calculate digestibility. Digestible energy (DE) intake, for example, refers to the difference between the total gross energy consumed and the total gross energy egested in feces of fish or shrimp after feeding. A complete description of energy partitioning is provided in Chapter 4.

After nutrients are absorbed they are metabolized, with some of the energy in nutrients excreted as nitrogenous compounds such as NH_3^+ and urea in the urine or via the gills. When the energy content of these materials are measured and subtracted from digestible values, the resulting value is called metabolizable energy (ME). In past NRC fish publications, both DE and ME values were presented, but due to concerns about the accuracy of ME values associated with the methodology used to measure them, plus the difficulty of obtaining them, only DE values are presented here. Moreover, DE values are now used exclusively in research and commercial fish feed formulation.

METHODS USED IN DIGESTIBILITY DETERMINATION

Determining digestibility of food and feeds in animals requires collection of fecal material. Direct and indirect methods are used to collect feces. Both involve feeding test feed ingredients singly or, more commonly, as a component of a diet. The direct method used for fish involves feeding a measured quantity of feed and collecting all feces resulting from that feeding. Total collection of feces in aquatic animals is challenging since it is difficult to ensure separate collection of fecal material egested by the animals from uneaten feed particles and urinary and gill excretion. Smith (1971, 1976) developed a confinement system to achieve this with fish. The primary advantage of the direct method is that there are no losses of nutrients or feed components in feces. Everything is collected, even water-soluble compounds, which eliminates a potential source of error associated with sampling fish/shrimp feces. The disadvantage of the direct method is that the fish need to be confined in a static water system and force-fed. This results in stress on the fish that may affect digestive and metabolic processes and may yield digestibility values that are not credible.

The indirect method of digestibility determination is widely used with most species of farmed fish and shrimp. This method relies on the collection of a representative sample of feces that is free of uneaten feed particles and the use of a nontoxic, inert, indigestible digestion indicator, such as chromic oxide or yttrium oxide, added to the feed. The indicator passes through the digestive tract at the same rate as food and is unaffected by the digestive process, e.g., not absorbed. Digestibility of nutrients is estimated based on relative enrichment of the feces with the digestion indicator compared to the level present in the feed. Rather than total feces collection, the indirect method relies on collecting a sample of feces from continuous feeding, with the assumption that the collected sample is representative of feces excreted over a period of time. Fish or shrimp can be held in normal fish rearing tanks or specialized tanks purpose-built for passive feces collection. The indirect method has several advantages over the direct method. They include minimum stress on fish or shrimp associated with a rearing/holding tank environment and the fact that many fish or shrimp can be used in a single replicate tank rather than a single fish or shrimp, an advantage for schooling species or those more secure in a group. Other advantages include flexibility in size (age) of fish or shrimp used and the fact that fish and shrimp are not force-fed but rather consume feed voluntarily.

Feces are collected by active or passive methods (see below). The apparent digestibility coefficient (ADC) of nutrients in feeds is calculated using the concentration of the indirect marker (chromic oxide in this example) in feed and feces using the following equation:

$$\text{Apparent Digestibility Coefficient} = 1 - \frac{Cr_2O_3 \text{ in feed}}{Cr_2O_3 \text{ in feces}}$$

The apparent digestibility coefficient of a specific nutrient, proximate category, or energy in feeds can be calculated as follows:

$$\text{Apparent Digestibility Coefficient} =$$

$$1 - \frac{Cr_2O_3 \text{ in feed}}{Cr_2O_3 \text{ in feces}} \times \frac{\text{Nutrient content of feces}}{\text{Nutrient content of feed}}$$

Feces are composed of the undigested food components and unreabsorbed residues of body origin. These residues are the remains of mucosal cells, digestive enzymes, mucoproteins, and other secretions released into the digestive tract by the animal, together with the residues of the microflora that inhabit the digestive tract (Nyachoti et al., 1997). Fecal losses not directly originating from ingested food but from the animal itself are referred to as endogenous gut losses. The term "apparent digestibility coefficient" is used to acknowledge the fact that values obtained using either the direct or indirect method are not corrected for endogenous gut losses.

There is interest in quantifying the endogenous gut losses to calculate the "true" digestibility of nutrients in feeds and ingredients. In terrestrial animals, a protein-free diet is sometimes fed in digestibility trials to correct for endogenous nitrogen losses (ENL). However, fish and shrimp will not accept (or will eat very little of) a protein-free diet, making it very difficult to calculate meaningful estimates of ENL. Moreover, there is evidence that the amount of ENL produced by animals receiving a protein-free diet differs significantly from that of animals fed diets containing protein. Several dietary constituents (fiber, antinutritional factors) can enhance ENL (Nyachoti et al., 1997). For these reasons, it is reasonable to doubt the accuracy of "true" protein digestibility coefficients calculated using estimates of ENL obtained from animals fed protein-free diets. Accurate estimation of ENL may require the use of more sophisticated techniques (see Nyachoti et al., 1997, for review). This type of work remains to be carried out with fish.

In fish maintaining a high feed intake, the contribution of ENL to total fecal nitrogen is probably small. Under these conditions, the difference between the "true" and apparent digestibility of protein is negligible (Hardy, 1997). The use of nutritionally balanced reference and test diets as proposed by Cho et al. (1982) generally allows the fish to maintain high feed intake and good growth rate, which in turn allows the measurement of apparent digestibility values that are reliable and repeatable.

Fecal collection with the indirect method can be done in an active or passive manner. Active feces collection methods include manual stripping, dissection, or vacuum-removal. Stripping involves manually expelling small volumes of feces from the distal portion of the intestine of lightly anesthetized fish by applying gentle pressure on the abdominal cavity. Typically, 20–30 fish are used per replicate tank, with fish large (> 100 g) enough to produce a significant quantity of feces. Care should be taken to avoid squeezing fish too strongly to prevent injury to the fish or expelling partially digested food or body fluids into the collection pan. Dissection is accomplished by killing fish and removing fecal material from the distal segment of the intestine. This is best done after freezing the intestine (Hemre et al., 2003). Vacuum-removal involves using a small tube inserted into the anus of the fish and applying vacuum pressure to remove fecal material. Each of these methods can result in contamination of feces samples with partially digested food or endogenous material.

Passive feces collection methods rely on the collection of fecal material naturally egested by the animal. Fecal material is collected from tank effluent water by screening or filtering tank effluent water, or by settling suspended fecal material in low-flow apparatus. Feces can also be collected by siphoning or netting. Special tanks, referred to as the Guelph or modified Guelph system, have been designed to facilitate feces collection by settling (Cho and Slinger, 1979; Hajen et al., 1993). In these systems, water flows are adjusted so that voided feces are quickly swept out of the fish tank into a vertical settling column where water flow is slow, allowing fecal particles to settle. Effluent water exits gently at the top of the column. Typically, fish are fed at an established time, such as late in the day, tanks and collection columns are cleaned to ensure that no uneaten feed is present, and feces collect overnight and are removed in the morning. Feces can also be collected by directing tank effluent water containing fecal material over a movable screen system that catches and deposits feces onto a tray (Choubert et al., 1979). Ogino et al. (1973) developed a system called "TUF" to trap fecal particles in a column though which effluent water containing fecal material passed.

Active methods are not an option to collect feces from shrimp. Therefore, fecal material is collected from tanks by siphoning, netting, or settling of tank effluent water in which fecal material is suspended using modifications of the Guelph system. Some researchers use relatively few shrimp per tank; many researches replicate tanks per diet and conduct studies for long periods to collect enough fecal material to analyze, at least 2 g of dry material, or about 30 g of wet fecal material (Smith et al., 2007). Others use many shrimp per tank and fewer replicate tanks per diet (Davis et al., 2002). Most studies collect fecal material at set times after feeding (Cruz-Suarez et al., 2007; Smith et al., 2007), but some collect it as soon as it appears (Davis et al., 2002). The longer the fecal material is exposed to water, the greater the chance for soluble nutrients to be lost and the more likely calculated values will overestimate actual digestibility. Despite the potential problems associated with leaching of nutrients from fecal material, coprophagia, cannibalism, exuviae, and small amounts of fecal material, published values for in vivo digestibility coefficients of common feed ingredients determined for shrimp do not differ greatly from those measured using fish.

Potential Sources of Error

Active methods for sampling feces can result in expulsion of partially digested material. This technique generally leads to an underestimation of the digestibility of nutrients (protein in particular) because of contamination of feces with endogenous material that would otherwise be reabsorbed before the feces are excreted (Cho et al., 1982; Hajen et al., 1993; Vandenberg and de la Noüe, 2001).

Passive methods for sampling feces can result in losses of soluble nutrients, such as free amino acids, leading to an overestimation of digestibility because the nutrient content of sampled material is lower than in naturally expelled feces. Since soluble nutrients are generally readily absorbable, this is usually not a large source of error. Differences of approximately 5–9% between active and passive collection methods have been reported in the literature.

Differences in accuracy of ADC values associated with feces collection methods used with fish have been the subject of much discussion (Windell et al., 1978a). It is difficult to compare ADC of feeds and practical ingredients such as fish meal across studies because differences in ADC values are often found even in the same laboratory. This is largely the result of actual differences that exist among some feed ingredients from different sources or batches. This difference is especially true for rendered animal protein ingredients and fish meal, although differences exist among sources of many other practical ingredients. In contrast, differences in ADC values for purified ingredients, such as casein, are relatively small across laboratories regardless of feces collection method used. Small analytical errors in laboratory analysis of feeds and ingredients can also contribute to variation in ADC values. Some anecdotal information suggests that the stability of feces in seawater is less than in freshwater, thus leading to an overestimation of ADC values due to loss of soluble nutrients from settled feces prior to collection. Supplementing basal feeds with an ingredient, such as guar gum, that improves the stability of fecal solids may improve the accuracy of ADC values determined using fecal settling methods (Brinker, 2007).

Analysis of feeds and feces follows normal analytical practices. Analyses are usually done on dry samples. Further analysis to determine proximate composition, energy content, or contents of specific nutrients, e.g., amino acids or minerals, is required to calculate their ADC values. The concentration of the digestion indicator is also measured. The amount of fecal material needed for analysis depends upon how many chemical analyses are being conducted. Fish feces are approximately 90% moisture, so collecting 50 g of wet feces will yield 5 g of dry material, enough for complete proximate, energy, and nutrient analysis. Smaller amounts are suitable if analytical methods requiring small sample amounts are used.

DIGESTIBILITY OF FEED INGREDIENTS

Measuring digestibility of feed ingredients follows the same feeding and feces collection protocols as measuring digestibility of feeds, but, because most feed ingredients cannot be fed alone, ingredients are combined with basal diets, usually in a 70:30 ratio (70% basal diet combined with 30% ingredient). Basal diets used in digestibility studies vary with species of fish, but were originally composed mainly of casein, gelatin, and dextrin. Feeds composed of practical ingredients are now more widely used to measure digestibility of feed ingredients because high feed consumption is important in digestibility assessment, both to ensure normal digestion and also to produce sufficient fecal material for analysis for collection and analysis. Also, many fish species do not readily consume casein-gelatin diets. However, in certain situations, using highly refined ingredients in basal diets is desirable because it reduces complications associated with measuring apparent digestibility of nutrients present at low or unequal levels in basal diets and test ingredients. For example, if the level of phosphorus is high in a basal diet and low in a test ingredient, the ADC of phosphorus in the test ingredient is more difficult to measure accurately. Ingredients that enhance palatability of casein-gelatin diets, such as krill or betaine, can sometimes increase feed intake. Another option is to replace casein and gelatin with a refined protein such as squid protein, fish fillet protein, or a high-protein, functional ingredient such as wheat gluten meal.

ADC values for nutrients and proximate categories can

easily be determined in complete feeds by adding an inert indicator to the feeds. This approach is useful to measure the effects of pelleting conditions on digestibility or to test for additivity of ADC values of individual ingredients (discussed below).

Interpretation and Limitations of Digestibility Measurements

To calculate digestibility values of proximate categories or individual nutrients in feed ingredients, one must know their concentrations as well as the concentrations of the inert indicator in feed and fecal samples. One must also know the ADC values for the basal diet. The ADC values are calculated on a dry-weight basis, making it crucial that both feed and feces samples are completely dry. Since ADC values are determined mathematically, small differences in moisture content of basal diets, feed, and ingredient samples and small errors in analytical determinations can result in large errors in ADC values. Accuracy is also affected by small deviations from the 70:30 ratio of basal diet and ingredient caused by errors made during mixing in weighing the two or by differences in moisture content between the basal diet and test ingredient. The effects of these inevitable errors are amplified by the mathematical equations used to calculate ADC values, making it possible to calculate values more than 100% or less than 0%, both impossible. The mathematical equation used to calculate ADC values has been reviewed, critiqued, and modified to account for some of these errors (Sugiura et al., 1998; Forster, 1999; Bureau and Hua, 2006). The preferred equation to calculate ADC values is:

$$ADC_{\text{test ingredient}} = ADC_{\text{test diet}} + ((ADC_{\text{test diet}} - ADC_{\text{ref. diet}}) \times (0.7 \times D_{\text{ref}} / 0.3 \times D_{\text{ingredient}}))$$

where D_{ref} is the percentage of nutrient or kcal/g gross energy of the reference diet, and $D_{ingredient}$ is the percentage of nutrient or kcal/g gross energy of the ingredient.

In the formula above, values of 70% and 30% were used for levels of the reference diet and ingredient in the combined test diet. In practice, any proportion of the reference diet and ingredient can be used providing that the test diet can be delivered to the fish in some way (it can be pelleted or made into a physical particle), fish consume the diet, and the proportions are such that normal digestion occurs.

Researchers should use the same basal diet formulation and obtain ingredients from the same source to allow digestibility results from studies done at different times in their laboratory to be compared and also to check consistency in protocols and ADC values. Researchers should report formulations and ADC values for their basal diet to allow other researchers to interpret ADC results. Theoretically, using different basal diets should not affect ADC values of feed ingredients, but it may in practice. It is also wise to avoid over interpreting ADC values. A difference in ADC protein values of 1–2% for the same ingredient between laboratories, for example, has little practical significance. Similarly, such a difference between two ingredients evaluated in the same laboratory study may be statistically significant, but have no biological significance. On the other hand, differences of 1–2% among feeds in which a progressive substitution of a single ingredient is being made are likely significant.

Factors Affecting Digestibility

Feed intake, fish size, and water temperature are experimental variables that may affect digestibility. Feed processing conditions also affect ADC values for some feed constituents. Although a number of studies suggested that digestive and absorptive efficiencies decrease as meal size increases (Pandian, 1967; Solomon and Brafield, 1972; Windell et al., 1978b), most studies show that feeding levels have no impact on digestibility of nutrients, assuming that feed intake is not below levels needed to support reasonable growth rates. This conclusion is supported by studies showing that intestinal absorptive capacity greatly exceeds nutrient loads in the gut and is not likely to be a limiting factor for nutrient digestibility.

Early studies reported that water temperature had no effect on ADC of protein and energy for rainbow trout (Windell et al., 1978b; Cho and Slinger, 1979; Cho and Kaushik, 1990). However, more recent studies provide convincing evidence that water temperature affects ADC of protein, lipid, starch, and energy of feed fed to rainbow trout (Choubert et al., 1982; Brauge et al., 1995; Médale et al., 1991, 1999; Azevedo et al., 1998; Bureau et al., 2008; Moreira et al., 2008; Hua and Bureau, 2009a,b). Studies by Choubert et al. (1982) and Azevedo et al. (1998) suggest a linear increase of about 1% in ADC of protein and energy with increasing temperature from 6 to 15°C in rainbow trout. Differences in experimental protocols, coupled with the fact that as water temperature decreases, feeding level also decreases, thereby increasing proportion of nitrogen and energy in feces from endogenous losses, may explain the differences in apparent digestibility reported by different researchers. The effects of water temperature on lipid and starch digestibility are mainly associated with the physicochemical properties of the dietary components.

Digestibility of Proteins

Proteins are complex compounds varying in size, structure, solubility, and, as a result, digestibility (Table 12-1). Proteins are not absorbed as such, but rather the amino acids that make up proteins are absorbed. Thus, the digestibility of proteins depends on the extent to which they can be hydrolyzed to free amino acids. This, in turn, largely depends on the type of protein and the degree to which linkages between amino acid chains (disulfide bonds between cysteine on amino acid chains) or between amino acids and other compounds, such as sugars (glycoproteins, for example), can be hydrolyzed. Proteins such as connective tissue (collagen), feathers, hair, hooves, and scales contain more disulfide bonds than do soluble proteins, such as albumin, and as a result are more difficult to digest. Complex protein ingredients, such as fish meal or poultry byproduct meal, contain varying proportions of proteins that are difficult to digest (hydrolyze) and less difficult to digest. The digestibility of complex protein ingredients is the sum of the digestibility of the various proteins comprising the ingredient. Hence, processing of feed ingredients to partially break down or remove proteins that are difficult to digest improves protein digestibility. An extreme example is poultry feathers, which are virtually indigestible without being first pressure cooked under strong alkaline conditions to hydrolyze disulfide bonds in feather proteins. Protein digestibility of hydrolyzed feather meal is sufficiently high to make it a useful feed ingredient. Proteins with low ADC values, e.g., cartilaginous proteins, can be removed from animal or fish meal by screening (Rathbone et al., 2001) or air-classification.

Plant proteins present a different challenge to digest in that they are typically encased in starch, preventing the action of digestive enzymes. Increasing digestibility of plant proteins involves grinding seeds (grains and oilseeds) to release protein-surrounded starch. Heat treatment enhances digestibility of plant proteins, such as soybean meal, by reducing the activity of trypsin inhibitors. However, excessive heat treatment lowers protein digestibility by creating new linkages in proteins that are resistant to digestion in plant proteins (Arndt et al., 1999) as well as in fish meal (Opstvedt et al., 1984; Arndt et al., 1999). The Maillard (Browning) reaction is the most common chemical change that occurs when plant proteins are heated in the presence of reducing sugars, such as glucose, that lowers digestibility (Scott et al., 1982). Lysine, arginine, histidine, and tryptophan contain reactive epsilon amino groups that form bonds that are not hydrolyzed by digestive enzymes. Chemical analysis shows that the amino acids are present although they are not digestible. Using the available lysine method of Carpenter (1960) solves this problem, at least in salmonid feeds (Opstvedt et al., 1984).

Digestibility of Amino Acids

The ADC values for proteins are the fractional sums of ADC values for amino acids and other nitrogenous compounds in feed ingredients (Table 12-2). However, it is important to note that ADC values for each amino acid are not necessarily the same as that for protein in a given ingredient; some are higher and some are lower (Storebakken et al., 2000). Further, differences in amino acid digestibility are found in some ingredients processed in different ways. For example, the digestibility of lysine in blood meals differs depending on the process used to dry the meal (Ehab and Bureau, 2007).

Competitive interactions between lysine and arginine are well known in terrestrial animals, especially birds, and result from increases in the arginine-degrading enzyme arginase activity in the kidney (Scott et al., 1982). Increasing dietary arginine level prevents growth depression in birds. Because lysine and arginine share an active transport system in the intestine, excessive levels of one may affect absorption of the other. Some evidence of lysine-arginine interactions has been reported at the absorptive and metabolic levels in fish (Kaushik and Fauconneau, 1984; Chiu et al., 1987; Davies et al., 1997; Berge et al., 2002), but nearly all research shows an absence of lysine-arginine interaction sufficient to affect fish growth (Robinson et al., 1981; Tibaldi et al., 1994; Berge et al., 2002).

Digestibility of Energy

Protein (amino acids), lipids, and carbohydrates all contribute energy to feeds. Hence the digestibility of energy is the fractional sum of ADC values for protein, lipid, and carbohydrates in a feed ingredient or feed (Table 12-3).

Digestibility of Carbohydrates

Carbohydrates are mixtures of highly digestible sugars, moderately digestible gelatinized starch, poorly digestible compounds such as raw starch and chitin, and indigestible compounds such as insoluble carbohydrates (Stone, 2003). Feed ingredients and complete feeds vary widely in the proportion of these materials and thus digestibility (Table 12-4). Analysis of different carbohydrates is challenging, making it difficult to assign to estimate digestibility.

Significant differences in the digestibility of raw and gelatinized starch have been reported in a large number of studies (Krogdahl et al., 2005). A metaanalysis of data from published studies indicated that digestibility of raw and gelatinized starches differed significantly, and their ADC values are negatively affected by dietary inclusion levels (Hua and Bureau, 2009a). This metaanalysis also revealed a significant effect of water temperature on starch digestibility and interspecies differences in the ability to digest starch. Finally, feces collection method was found to significantly contribute to the variability of starch digestibility reported across studies. Davis and Arnold (1995) reported higher degrees of starch gelatinization for extruded whole wheat, corn flour, rice flour, and milo, but they did not always find higher digestibility of the ingredients for *Litopenaeus vannamei*. Venou et al. (2003) found that extrusion increased ADC values of feeds in which corn was the carbohydrate source but did not find differences in ADC values when wheat was the carbohydrate source. However, for most plant protein sources, extrusion increases starch digestibility. The ADC value for starch was 95.7 for sea bass fed extruded

TABLE 12-1 Apparent Digestibility of Protein in Selected Feed Ingredients for Several Fish Species and Shrimp

Feedstuff	International Feed No.	Salmon	Rainbow Trout	Atlantic Cod	European Sea Bass	Hybrid Striped Bass	Silver Perch
Alfalfa meal	1-00-023	—	61[a]	—	—	—	—
Blood meal	5-00-381	30[c]	84–99[d,e,f]	—	91[g]	86[h]	90[i]
Casein	5-01-162	100[n]	92–95[a,o]	—	—	—	—
Canola meal	5-06-145	79[c]	—	79[r]	—	43[s]	83[i]
Corn, grain	4-02-935	—	95[v]	—	—	—	—
Corn gluten meal	5-28-242	92[aa]	92–97[f,aa]	86[r]	91[bb]	—	95[i]
Cottonseed meal	5-01-621	—	82–88[gg]	—	—	84[h]	83[i]
Feather meal, hydrolyzed	5-03-795	71–80[c,aa]	77–87[d,e,aa]	62[r]	—	—	93[i]
Fish meal, not specified		—	—	—	—	—	89–94[i]
Fish, anchovy meal	5-01-985	91[aa]	94–97[f,aa]	92[r]	—	—	—
Fish, herring meal	5-02-000	91–95[c,aa]	95[aa]	93[r]	—	—	—
Fish, menhaden meal	5-02-009	83–88[c,aa]	86–90[f,aa]	—	—	88[h]	—
Krill meal	5-16-423	—	—	96[r]	—	—	—
Lupin meal		100[n]	85–97[n,o,oo]	90[r]	94[bb]	—	97[i]
Meat and bone meal	5-00-388	85[v]	83–88[d]	—	—	73[h]	73[i]
Meat meal		—	—	—	—	—	—
Peanut meal	5-03-650	—	—	—	—	80[s]	98[i]
Poultry byproduct meal	5-03-798	74–94[c,aa]	83–96[d,e,aa]	80[r]	97[g]	55[s]	—
Poultry meal		—	—	—	—	—	85[i]
Rapeseed meal	5-03-871	—	91[oo]	—	—	—	—
Rice bran	4-03-928	—	64[f]	—	—	71[h]	—
Shrimp meal		—	—	67[r]	—	—	—
Sorghum, grain	4-04-383	—	—	—	—	—	78[i]
Soybean meal	5-04-612	77–94[c,n]	90–99[n,o]	92[r]	90[g]	80[h]	95[i]
Soy protein concentrate		90[n]	98–100[n,o]	99[r]	97[bb]	—	—
Soy protein isolate		97[n]	98[n,o]	97[r]	—	—	—
Squid meal		—	—	—	—	—	—
Starch, corn		—	—	—	—	—	—
Wheat flour	4-05-199	98[aa]	82–100[f,aa]	—	—	—	—
Wheat gluten	5 05 221	99[aa]	100[f,aa]	100[r]	98[g]		100[i]
Wheat middling	4-05-205	86[aa]	68–91[f,aa]	—	—	92[h]	—
Yeast, brewer's	7-05-527	—	57[e]	—	—	54[s]	—

NOTES: Dashes indicate that data were not available. IGIR, digestibility was determined by indicator, Guelph or its modified systems, and ingredient fed in reference diet; ISIR, determined by indicator, stripping method, and ingredient fed in reference diet; ISPIR, determined by indicator method, feces collected by siphoning, ingredient fed in reference diet; IEIM, determined by indicator method, feces collected by surgical excision, ingredient fed in mixed diet; IEIS, determined by indicator method, feces collected by surgical excision, single ingredient fed; IEIM, determined by indicator method, feces collected by surgical excision, ingredient fed in mixed diet.

Salmon includes Atlantic, Chinook, and Coho salmon.

[a]Smith (1977) and Smith et al. (1980) by metabolism chamber, single ingredients fed; [b]Popma (1982) by indicator method, feces collected by frequent removal from water, ingredient fed in mixed diet; [c]Hajen et al. (1992) by indicator method, feces collected from water, ingredient fed in mixed diet; [d]Bureau et al. (1999) by IGIR; [e]Cheng et al. (2004) by IGIR; [f]Gaylord et al. (2009); [g]da Silva and Oliva-Teles (1998) by IGIR; [h]Sullivan and Reigh (1995) by ISIR; [i]Allan et al. (2000) by IGIR; [j]McGoogan and Reigh (1996) by ISIR; [k]Lupatsch et al. (1997) by indicator method, feces collected by stripping, single ingredi-

faba bean, indicating a positive effect of extrusion for this product (Adamidou et al., 2009). Similarly, Theissen et al. (2003) reported an increase in ADC_{starch} values from 24.7 to 100 in rainbow trout following extrusion of dehulled peas.

Starch is added to feeds from a wide variety of plant sources. In addition, several nonstarch components, such as lipids and protein, are associated with the starch granules. These components may influence digestion of starch through reducing contact between starch and digestive enzymes or affecting properties of the starch granule during feed processing (Svihus et al., 2005). The classification of starch into raw versus gelatinized starch is based on the degree of gelatinization, a rather simplified representation of a group of heterogeneous compounds. Starch of different origins is highly variable in terms of the primary structure of starch, starch granule size, and amylase:amylopectin ratio, all of which can contribute to differences in digestibility (Svihus et al., 2005). Gaylord et al. (2009) reported ADC_{starch} values

Protein (%)											
Red Drum	Channel Catfish	Nile/Blue Tilapia	Siberian Sturgeon	Haddock	Gilthead Sea Bream	Cobia	Yellowtail	Largemouth Bass	Pacu	Rockfish	Penaid Shrimp
—	—	66[b]	—	—	—	—	—	—	—	—	—
100[j]	—	—	—	—	90[k]	—	—	—	—	87[l]	66–71[m]
—	97[p]	—	—	—	—	—	95[q]	—	—	—	96.4[m]
—	—	—	—	83[t]	—	—	—	—	—	—	80[u]
—	60[w]	75–84[b,x]	—	—	—	—	—	—	86[y]	—	66–83[z]
—	—	89–97[x,cc]	—	92[t]	90[dd]	94[ee]	50[q]	94[ff]	96[y]	92[l]	59[m]
76–85[j,hh]	83[ii]	82[jj]	88[kk]	—	75[dd]	—	—	—	—	81[l]	83[ll]
—	74[w]	79[jj]	91[kk]	—	58[dd]	—	—	—	—	79[l]	64[m]
—	—	86–90[jj,mm]	95[kk]	—	83[k]	96[ee]	89[q]	88[ff]	85[y]	95[l]	88[m]
—	—	91[cc]	—	—	—	—	—	—	—	95[l]	83–89[m,nn]
—	—	—	—	96[t]	96[dd]	—	—	—	—	—	90[m]
77–96[j,hh]	88[ii]	85[b]	—	—	—	—	—	—	—	—	84–89[m]
—	—	—	—	—	—	—	—	—	—	—	81[m]
—	—	—	—	—	—	—	—	—	—	—	93–97[pp]
74–79[j,hh]	78[b]	78[jj]	85[kk]	—	72[dd]	87[ee]	—	—	—	—	60–88[ll,qq]
—	—	—	—	—	79–90[k,dd]	—	80[q]	—	—	91[l]	75[rr]
—	—	—	—	—	—	90[ee]	—	—	—	—	93[rr]
—	—	74–90[jj,mm]	90[kk]	—	82[dd]	—	—	82[ff]	—	—	79[m]
—	—	87[x]	—	—	80[k]	91[ee]	—	—	—	—	—
—	—	85[x]		—	—	89[ee]	—	—	—	—	—
77[j]	—	—	—	—	—	—	—	—	—	—	58[ll]
—	—	—	—	74[t]	—	—	—	—	—	—	58[ll]
77[j]	—	86[x]	—	—	—	—	—	—	—	—	66–83[z]
80–86[j,hh]	93[ii]	87–94[b,cc]	92[kk]	92[t]	87–91[k,dd]	91[ee]	—	94[ff]	91[y]	84[l]	89–97[ll,rr,ss,tt]
—	—	—	—	—	—	—	87[q]	—	—	—	93[tt]
—	—	—	—	—	—	—	—	—	—	—	94[m]
—	—	—	—	—	85[k]	—	—	—	—	—	82–87[m,ll]
—	—	—	—	—	—	—	—	—	—	—	81–96[ss,uu]
—	—	80[x]	—	—	82[k]	—	—	—	—	95[l]	67–87[z]
—	—	—	—	—	—	—	—	—	—	—	96[m]
87[j]	72[w]	75[mm]	—	—	—	—	—	—	—	—	81[ll]
—	—	—	—	—	—	—	—	—	—	78[l]	—

ent fed; [l]Lee (2002) by IGIR; [m]Lemos et al. (2009); [n]Glencross et al. (2004) by IGIR; [o]Glencross et al. (2005) by ISIR; [p]Saad (1989) by IEIM; [q]Masumoto et al. (1996) by IGIR; [r]Tibbetts et al. (2006) by IGIR; [s]Gaylord et al. (2004) by ISIR; [t]Tibbetts et al. (2004) by IGIR; [u]Cruz-Suarez et al. (2001) by ISPIR; [v]Cho et al. (1982) by IGIR; [w]Cruz (1975) by IEIS; [x]Sklan et al. (2004) by ISPIR; [y]Abimorad et al. (2008) by IGIR; [z]Davis and Arnold (1995); [aa]Sugiura et al. (1998) by IGIR; [bb]Kaushik (2002) by IGIR; [cc]Koprucu and Ozdemir (2005) by IGIR; [dd]Nengas et al. (1995) by IGIR; [ee]Zhou et al. (2004) by IGIR; [ff]Portz and Cyrino (2004) by IGIR; [gg]Cheng and Hardy (2002b) by IGIR; [hh]Gaylord and Gatlin III (1996) by ISIR; [ii]Wilson and Poe (1985) by IEIM; [jj]Guimaraes et al. (2008) by IGIR; [kk]Liu et al. (2009) by indicator method, feces collected by ladling out water, ingredient fed in mixed diet; [ll]Brunson et al. (1997) by ISPIR; [mm]Hanley (1987) by IEIS; [nn]Ezquerra et al. (1997) by in vitro hydrolysis; [oo]Burel et al. (2000) by IGIR; [pp]Smith et al. (2007) by ISPIR; [qq]Forster et al. (2003) by IGIR; [rr]Yang et al. (2009) by IGIR; [ss]Akiyama et al. (1989) by ISPIR; [tt]Cruz-Suarez et al. (2009) by ISPIR; [uu]Cousin et al. (1996) by ISPIR.

in various barley cultivars in extruded diets fed to rainbow trout ranging from 41 to 77. ADC_{starch} values among the cultivars increased with decreasing amylose and increasing amylopectin level. In the same study, Gaylord et al. (2009) reported ADC_{starch} values of 66 for waxy wheat and 51 for soft, white wheat, which had a higher amylose and lower amylopectin level than did the waxy wheat. Arnesen and Krogdahl (1995) found that oat starch even in raw form was highly digestible to rainbow trout, in contrast to what has been observed for wheat starch (Bergot and Breque, 1983). Arnesen and Krogdahl (1995) found that fish seemed to utilize starch of mixed sources more efficiently than a single source, suggesting a positive interaction between starches from different sources. Aslaksen et al. (2007) observed that starch in legumes is less digestible than starch in cereals to Atlantic salmon. The digestibility of starch appears to be lower in saltwater than freshwater for Atlantic salmon, whereas no significant effect of water salinity was observed in rainbow trout (Hua and Bureau, 2009a).

TABLE 12-2 Amino Acid Availability and Protein Digestibility Values of Selected Feed Ingredients for Atlantic Salmon,[a] Rainbow Trout,[b,c,d] Striped Bass,[e,f] Channel Catfish,[g] Nile Tilapia,[h,i] Gilthead Sea Bream,[j] Siberian Sturgeon,[k] Largemouth Bass,[l] Pacu,[m] Rockfish,[n] Yellowtail,[o] Silver Perch,[p] and Pacific White Shrimp[q,r,s,t,u]

Feedstuff Fish Species Common Name	International Feed No.	Crude Protein (%)	ARG (%)	CYS (%)	HIS (%)	ILE (%)	LEU (%)	LYS (%)	MET (%)	PHE (%)	THR (%)	TRP (%)	TYR (%)	VAL (%)
Barley, protein concentrate														
Rainbow trout		92	93	—	89	80	91	85	83	91	87	—	89	89
Barley, waxbar														
Rainbow trout		57	91	—	67	61	79	68	64	82	62	—	76	72
Blood meal	5-00-381													
Gilthead sea bream		90	97	—	95	80	96	94	89	95	96	—	—	96
Rainbow trout		88	97	—	87	86	87	90	91	89	90	—	87	81
Rockfish		87	90–91	48–81	90–99	65–77	96–99	86–91	95–98	96–98	89–93	—	87–92	98–99
Silver perch		90	93	87	94	80	93	93	92	93	94	—	93	92
Striped bass		63	70	—	66	38	71	73	64	69	59	—	69	64
Canola meal	5-06-145													
Atlantic salmon		—	97	97	95	87	85	92	100	89	93	—	93	84
Rainbow trout		79	92	—	96	85	92	88	87	90	90	—	90	87
Silver perch		83	92	80	92	85	88	86	89	88	88	—	90	85
Striped bass		43	61	—	42	45	53	54	62	54	54	—	54	37
Casein	5-01-162													
Yellowtail		95	96	96	98	97	98	98	98	97	95	—	98	95
Pacific white shrimp		99	99	—	99	99	100	100	—	99	99	—	99	99
Corn, grain	4-02-935													
Channel catfish		—	—	82	90	68	88	97	71	82	70	—	78	74
Pacu		86	93	—	92	81	91	82	80	90	81	—	—	81
Rainbow trout		65	99	—	89	86	94	73	64	71	73	—	80	90
Corn, gluten meal	5-28-241													
Atlantic salmon		95	100	91	95	90	88	100	94	91	92	—	92	91
Cobia		94	96	—	93	95	95	97	96	96	96	—	96	96
Largemouth bass		94	98	—	92	95	89	96	83	93	96	80	—	97
Nile tilapia		90	89	87	90	90	88	88	94	90	89	—	86	89
Pacu		96	96	—	94	89	94	89	90	94	89	—	—	89
Rainbow trout		92	99	—	96	91	97	91	92	95	93	—	95	94
Rockfish		92	89–92	80	83–90	93–95	99	84–87	98–99	96–97	88–93	—	80–97	91–92
Striped bass		91	95	87	83	92	96	87	91	—	93	—	—	93
Yellowtail		50	48	47	51	45	47	48	50	47	43	—	51	40
Pacific white shrimp		88	88	34	82	73	65	89	71	76	70	—	55	71
Cottonseed meal	5-01-621													
Channel catfish		—	91	—	82	72	76	71	76	84	77	—	73	76
Nile tilapia		82	90	88	82	86	81	69	82	70	74	88	84	81
Rainbow trout		75–88	87–94	73	87–93	83–85	85–89	84	82–89	86–90	82–90	96	89–90	85–87
Rockfish		81	86–87	63	68–81	78–80	83–85	77–79	70–82	86	80–81	—	87	82
Siberian sturgeon		88	95	—	83	81	84	81	81	89	82	—	—	83
Silver perch		83	91	79	87	74	75	60	74	83	78	—	83	77
Feather meal	5-03-795													
Nile tilapia		79	85	82	78	82	80	86	95	78	75	78	68	74
Rainbow trout		85	92	—	88	87	88	91	85	91	88	—	86	85
Siberian sturgeon		91	96	—	74	96	95	86	89	94	91	—	—	95
Silver perch		93	96	92	93	94	94	90	96	94	93	—	91	93

TABLE 12-2 Continued

Feedstuff Fish Species Common Name	International Feed No.	Crude Protein (%)	ARG (%)	CYS (%)	HIS (%)	ILE (%)	LEU (%)	LYS (%)	MET (%)	PHE (%)	THR (%)	TRP (%)	TYR (%)	VAL (%)
Fish meal, not specified														
Cobia		96	98	—	97	97	97	98	96	95	97	—	97	95
Gilthead sea bream		83	93	—	85	88	92	91	91	88	93	—	—	90
Largemouth bass		88	93	—	86	89	86	96	83	91	88	82	—	92
Nile tilapia		89	91	96	88	92	87	95	93	91	86	93	86	81
Pacu		85	95	—	94	88	93	89	90	93	88	—	—	88
Silver perch		89–94	91–98	89–100	96–98	93–97	94–97	95–97	91–98	92–96	93–97	—	93–96	93–97
Yellowtail		89	93	90	93	90	91	93	92	89	89	—	90	86
Pacific white shrimp		85–92	91–93	72–79	89	87	81–88	91–93	90–95	82–87	79–84	—	75	85–88
Fish meal, anchovy	5-01-985													
Atlantic salmon		77	82	—	76	82	82	81	—	80	81	81	87	82
Nile tilapia		91	92	90	92	91	91	91	94	90	90	—	91	90
Rainbow trout		97	99	—	96	100	94	97	99	94	97	—	95	100
Rockfish		95	93–98	93–94	88–97	89–96	98–99	90–95	98–99	95–97	96–98	—	93–99	96–97
Fish meal, herring	5-02-000													
Atlantic salmon		83–94	94–96	86–96	88–94	89–95	89–94	90–96	88–92	89–93	93–100	57–93	90–96	88–94
Striped bass		88	94	64	89	90	93	92	89	—	90	—	—	90
Fish meal, menhaden	5-02-009													
Atlantic salmon		86	87–91	92	85–91	89–91	89–90	84–88	84	87–89	85–88	73–89	86–92	86–90
Channel catfish		78–89	91	—	85	87	89	86	83	87	87	—	89	87
Rainbow trout		86	92	—	92	90	94	95	94	92	91	—	92	90
Fish meal, sardine														
Rainbow trout		90	98	—	98	96	99	99	97	98	98	—	97	97
Lupin meal														
Pacific white shrimp		94	97	94	96	97	96	93	100	95	92	—	98	94
Meat and bone meal	5-00-388													
Cobia		87	93	—	88	91	92	85	93	92	92	—	92	91
Channel catfish		—	88	—	82	81	82	87	80	85	76	—	83	81
Nile tilapia		78	87	93	85	77	82	83	95	84	79	81	76	82
Siberian sturgeon		85	86	—	62	84	87	82	85	87	83	—	—	85
Silver perch		72	74	68	76	75	78	76	82	75	76	—	81	74
Pacific white shrimp		74	88	75	90	90	89	93	91	88	86	91	88	87
Meat meal														
Gilthead sea bream		79	89	—	75	82	83	86	86	80	88	—	—	81
Rockfish		91	92–95	64–75	79–90	81–90	85–91	89–90	94–99	83–91	83–89	—	88–92	82–89
Yellowtail		80	82	44	86	76	78	85	84	78	74	—	76	72
Pacific white shrimp		75	77	63	76	70	70	76	74	73	67	—	58	67
Peanut meal	5-03-650													
Cobia		90	97	—	92	95	96	96	94	95	95	—	95	95
Channel catfish		—	98	—	89	93	95	94	91	96	93	—	95	93
Silver perch		98	96	98	95	92	92	90	98	94	90	—	97	93
Striped bass		80–85	92–93	71	65–68	83–88	88	85–94	82–94	88	81–85	—	85	80–87
Pacific white shrimp		93	94	85	92	93	94	92	97	94	80	—	79	92
Poultry byproduct meal	5-03-798													
Largemouth bass		82	91	—	93	86	89	91	71	88	86	52	—	83
Nile tilapia		90	91	97	97	91	88	96	97	95	86	93	91	81
Siberian sturgeon		90	94	—	76	94	95	91	95	95	91	—	—	94
Striped bass		55	74	—	59	60	67	61	67	62	58	—	67	61
Poultry meal														
Cobia		91	94	—	91	92	93	92	93	91	93	—	93	92
Gilthead sea bream		80	91	—	80	82	84	86	91	82	89	—	—	83
Silver perch		85	89	91	91	85	87	89	92	86	88	—	86	85

continued

TABLE 12-2 Continued

Feedstuff Fish Species Common Name	International Feed No.	Crude Protein (%)	ARG (%)	CYS (%)	HIS (%)	ILE (%)	LEU (%)	LYS (%)	MET (%)	PHE (%)	THR (%)	TRP (%)	TYR (%)	VAL (%)
Rapeseed meal	5-03-871													
Cobia		89	94	—	86	91	92	93	91	90	91	—	91	90
Rice bran	4-03-928													
Channel catfish		—	94	—	83	88	91	95	88	90	88	—	94	89
Rainbow trout		64	86	—	60	57	78	68	61	72	74	—	74	71
Pacific white shrimp		76	85	—	83	73	75	81	—	75	73	—	76	76
Shrimp meal														
Pacific white shrimp		75	82	—	75	82	82	86	—	76	84	—	77	79
Soybean meal	5-04-604													
Atlantic salmon		88	87	—	86	79	76	84	94	79	85	50	83	77
Channel catfish		—	97	—	88	80	84	94	85	84	88	—	83	79
Cobia		91	93	—	89	91	91	93	92	90	90	—	90	89
Gilthead sea bream		87	96	—	89	89	90	92	89	89	91	—	—	89
Largemouth bass		94	98	—	91	97	98	96	80	95	96	87	—	99
Nile tilapia		87	89	86	89	86	87	83	84	86	87	—	86	84
Pacu		91	96	—	94	89	93	90	89	94	88	—	—	88
Rainbow trout		89	98	—	100	98	98	93	90	95	95	—	95	99
Rockfish		84	86–88	71–76	74–81	76–83	80–86	82–88	71–90	78–84	81–86	—	84–85	83–87
Siberian sturgeon		92	98	—	82	94	95	93	89	95	94	—	—	94
Silver perch		95	98	94	97	95	95	97	96	96	96	—	96	95
Striped bass		93	97	70	92	94	95	95	92	—	91	—	—	93
Pacific white shrimp		89	93–99	51	89–95	87–96	86–94	90–97	85–97	87–95	74–94	—	78	82–95
Soy protein concentrate	5-08-038													
Rainbow trout		99	98	—	97	92	97	96	96	96	93	—	96	93
Yellowtail		87	90	87	93	88	87	91	87	89	83	—	89	80
Pacific white shrimp		—	95	—	91	90	88	93	93	90	87	—	—	89
Soy protein isolate														
Pacific white shrimp			98		96	95	94	97	95	95	94		—	94
Squid meal														
Gilthead sea bream		87	93	—	88	91	92	93	93	90	94	—	—	91
Wheat, flour	4-05-199													
Rainbow trout		82	42	—	71	87	91	71	85	92	96	—	73	95
Wheat, middlings	4-05-205													
Channel catfish		—	95	—	95	88	90	96	83	93	89	—	89	90
Rainbow trout		68	79	—	96	90	96	90	72	90	90	—	90	93
Wheat, gluten	5-05-221													
Nile tilapia		100	97	—	95	92	97	95	97	97	94	—	96	94
Pacific white shrimp		89	88	47	88	93	91	83	93	93	83	—	76	91
Yeast, brewer's	7-05-527													
Rainbow trout			69	43	65	47	56	64	76	57	42	56	—	52
Rockfish		78	81–87	76–86	63–76	72–76	74–81	70–80	61–83	74–81	71–74	—	73–87	70–73
Striped bass		54	75	—	54	49	69	100	94	58	67	—	66	50

NOTES: Dash indicates data were not available. ARG, arginine; CYS, cystine; HIS, histidine; ILE, isoleucine; LEU, leucine; LYS, lysine; MET, methionine; PHE, phenylalanine; THR, threonine; TRP, tryptophan; TYR, tyrosine; VAL, valine. IGIR, determined by indicator, Guelph or its modified systems, and ingredient fed in reference diet; ISIR, by indicator, stripping method, and ingredient fed in reference diet; ISPIR, by indicator method, feces collected by siphoning, and ingredient fed in reference diet.

[a]Anderson et al. (1992, 1995) by ISIR; [b]Gaylord et al. (2009) by ISIR; [c]Cheng et al. (2002b) by IGIR; [d]Cheng et al. (2004) by IGIR; [e]Gaylord et al. (2004) by ISIR; [f]Small et al. (1999) by indicator method, feces collected by stripping, ingredient fed in mixed diet comparing with a protein-free diet; [g]Wilson et al. (1981) by indicator method, surgical excision, ingredient fed in mixed diet comparing with a protein-free diet; [h]Koprucu and Ozdemir (2005) by IGIR; [i]Guimaraes et al. (2008) by IGIR; [j]Lupatsch et al. (1997) by indicator method, feces collected by stripping, single ingredient fed; [k]Liu et al. (2009) by indicator method, feces collected by ladling out water, ingredient fed in mixed diet; [l]Portz and Cyrino (2004) by IGIR; [m]Abimorad et al. (2007) by IGIR; [n]Lee (2002) by IGIR; [o]Masumoto et al. (1996) by IGIR; [p]Allan et al. (2000) by IGIR; [q]Akiyama et al. (1989) by ISPIR; [r]Forster et al. (2003) by IGIR; [s]Smith et al. (2007) by ISPIR; [t]Cruz-Suarez et al. (2009) by ISPIR; [u]Yang et al. (2009) by IGIR.

Digestibility of Fiber

Fiber is a proximate analysis category that includes nonstarch polysaccharides such as cellulose, hemicellulose, β-glucans, pectins, and gums (Krogdahl et al., 2005). Glucosamine from the exoskeleton of crustaceans such as shrimp is also considered fiber in proximate analysis. Fiber is essentially indigestible to nearly all fish species, although there are exceptions such as grass carp, *Ctenopharyngodon idella*. Cellulase activity was detected in the hepatopancrease and intestine of grass carp and activity levels increased with increasing amounts of cellulose in the diet (Das and Tripathi, 1991). However, activity was reduced when antibiotics were added to the diet, suggesting that intestinal cellulase activity was of microbial origin. Unlike terrestrial vertebrates, the gastrointestinal tract of fish is in constant contact with the environment, making it difficult to determine whether cellulase activity in the gut of fish is produced by colonizing bacteria or by bacteria from rearing water or present on ingested plant material (Saha et al., 2006).

Cellulose is sometimes used in research diets to adjust for differences in dietary energy content among dietary treatments. Hansen and Storebakken (2007) showed that in extruded diets, up to 15% cellulose had no effect on nutrient digestibility for Atlantic salmon.

Digestibility of Lipids

Lipids are a complex class of materials with many different compounds differing in structure and composition (Sargent et al., 2002). Lipid sources contain five lipid classes: triacyglyercols, wax esters, phospholipids, sphingolipids, and sterols. All except sterols are composed of two or three fatty acids connected to a short carbon backbone. Fatty acids must be hydrolyzed from their carbon backbone to be absorbed. The different classes of lipids are associated with different roles and with different sources. For example, triacglycerols are storage lipids in animals and plants (seeds). Phospholipids are found in cell membranes. Wax esters are abundant in marine zooplankton. Lipid class likely plays a minor role in determining the digestibility of lipid sources (Table 12-4). For example, wax esters from marine zooplankton are readily digested by salmon (Olsen et al., 2004; Bogevik et al., 2008).

Fatty acids are categorized on the basis of degree of saturation, e.g., saturated fatty acids (SFA) having no double bonds along the carbon chain, monosaturated fatty acids (MFA) having one double bond, and polyunsaturated fatty acids having two or more double bonds (PUFA). Evidence suggests that ADC values decrease with increasing proportion of SFA in a lipid source in both warmwater and coldwater fish species (Austreng et al., 1979; Cho and Kaushik, 1990; Olsen and Ringø, 1998; Menoyo et al., 2003; Ng et al., 2003, 2004; Bahurmiz and Ng, 2007). However, combining lipid sources high in SFA level with lipids high in MFA or PUFA, such as soy, rapeseed, or fish oils, increased the digestibility of SFA, similar to observations in poultry (Sibbald, 1978). Hua and Bureau (2009b) analyzed published studies on lipid digestibility in rainbow trout and Atlantic salmon to develop a model that estimates digestibility of dietary lipids. In essence, the model adds the fractional digestibility values of different categories of fatty acids (SFA, MFA, and PUFA) to arrive at a total ADC value, then adds factors to account for the positive effects of MFA, PUFA, and water temperature on SFA digestibility. This model adds a level of refinement and accuracy to estimations of lipid digestibility.

Digestibility of Minerals

Digestibility of minerals is mainly influenced by chemical form, e.g., chlorides, sulfates, or oxides. In general, the digestibility of minerals is higher for forms that dissociate readily in the acidic environment of the stomach, such as chlorides, than those that do not, such as oxides. For this reason, mineral supplements should be chemical forms that dissociate, especially for stomachless fish species.

Phosphorus presents unique challenges to fish feed formulators, especially when feeds contain plant-derived ingredients. Phosphorus in fish feeds can be present in organic compounds, such as phospholipids or phytate phosphorus, bone (calcium phosphate), or as feed supplements, e.g., monobasic or dibasic calcium phosphate, and sodium phosphate. The ADC values for each form are different (Table 12-5). Further, phosphorus digestibility in feeds can be influenced by other dietary ingredients. For example, phosphorus digestibility is reduced from 70% to 25% with increasing calcium (bone) content (Sugiura et al., 2000). Ingredients that add bone to feeds, such as fish meal, especially those ingredients produced from seafood processing wastes or bony species of fish such as blue whiting (Albrektsen et al., 2009), reduce phosphorus digestibility in feeds. Phytate phosphorus, the storage form of phosphorus in seeds (grains, oilseeds, lupins), is essentially indigestible to fish and all monogastric animals. Adding microbial phytase to feeds hydrolyzes phytate phosphorus, increasing its availability (Rodehutscord and Pfeffer, 1995; Jackson et al., 1996; Hughes and Soares, 1998; Forster et al., 1999; Vielma et al., 2000; Suguira et al., 2001; Cheng and Hardy, 2002a). Interactions that lower phosphorus digestibility are responsible for nonadditivity of phosphorus ADC values for ingredients when combined in a feed. Hua and Bureau (2006) and Hua et al. (2008) described a model to estimate phosphorus digestibility in fish feeds. Similar to the lipid model, the phosphorus model adds the digestibility of phosphorus fractions in a feed or feed ingredient, and adds factors to account for interactions among ingredients that lower phosphorus digestibility.

The ADC values of zinc are also affected by other dietary ingredients, as ADC values for zinc are reduced in feeds that contain high-ash fish meal and/or plant protein concentrates having high phytate levels (Richardson et al., 1985; Shearer

TABLE 12-3 Apparent Digestibility of Energy in Selected Diet Ingredients of Several Fish Species and Pacific White Shrimp (*Litopenaeus vannamei*)

Feedstuff	International Feed No.	Atlantic Salmon	Rainbow Trout	Atlantic Cod	European Sea Bass	Hybrid Striped Bass	Silver Perch
Blood meal	5-00-381	—	80–99[a,b]	—	92[c]	—	100[d]
Casein	5-01-162	100[i]	92[j]	—	—	—	—
Canola meal	5-06-145	—	56–75[k]	61[l]	—	—	58[d]
Corn, grain	4-02-935	—	—	—	—	41[n]	—
Corn gluten meal	5-28-242	—	—	83[l]	87[s]	—	100[d]
Cottonseed meal	5-01-621	—	—	—	—	65–73[n,x]	53[d]
Feather meal, hydrolyzed	5-03-795	—	—	—	—	—	—
Fish meal, not specified		—	99[j]	—	—	—	89–98[d]
Fish, anchovy meal	5-01-985	—	—	86[l]	—	—	—
Fish, herring meal	5-02-000	—	—	93[l]	—	—	—
Fish, menhaden meal	5-02-009	—	—	—	—	96[n]	—
Krill meal	5-16-423	—	—	96[l]	—	—	—
Lupin meal		70[i]	64[j]	75[l]	87[s]	—	51–70[d]
Meat and bone meal	5-00-388	—	68–83[a]	—	—	80[n]	75–81[d]
Meat meal		—	—	—	—	—	—
Peanut meal	5-03-650	—	—	—	—	—	77[d]
Poultry byproduct meal	5-03-798	—	75–87[a,b]	71[l]	97[c]	—	—
Poultry meal		—	—	—	—	—	94[d]
Feather meal, hydrolyzed	5-03-795	—	76–80[a,b]	59[l]	—	—	100[d]
Rapeseed meal	5-03-871	—	76[cc]	—	—	—	—
Rice bran	4-03-928	—	—	—	—	47–93[n,x]	—
Shrimp meal		—	—	41[l]	—	—	—
Sorghum, grain	4-04-383					44[x]	38[d]
Soybean meal	5-04-612	89[i]	72[j]	88[l]	69[c]	55[n]	78[d]
Soy protein concentrate		100[i]	87[j]	95[l]	88[s]	—	—
Soy protein isolate		100[i]	96[j]	92[l]	—	—	—
Squid meal		—	—	—	—	—	—
Wheat flour	4-05-199	—	—	—	—	—	31–53[d]
Wheat gluten		—	—	95[l]	93[c]	—	94[d]
Wheat middling	4-05-205	—	—	—	—	61[n]	—
Yeast, brewer's		—	—	—	—	—	—

NOTE: Dashes indicate that data were not available. IGIR, digestibility was determined by indicator, Guelph or its modified systems and ingredient fed in reference diet; ISIR, determined by indicator, stripping method, and ingredient fed in reference diet; ISPIR, determined by indicator method, feces collected by siphoning, ingredient fed in reference diet; IEIM, determined by indicator method, feces collected by surgical excision, ingredient fed in mixed diet; IEIS, determined by indicator method, feces collected by surgical excision, single ingredient fed; IEIM, determined by indicator method, feces collected by surgical excision, ingredient fed in mixed diet.

[a]Bureau et al. (1999) by IGIR; [b]Cheng et al. (2004) by IGIR; [c]da Silva and Oliva-Teles (1998) by IGIR; [d]Allan et al. (2000) by IGIR; [e]McGoogan and Reigh

and Hardy, 1987; Gatlin and Phillips, 1989; Brown et al., 1992). Feed producers should consider these interactions when formulating fish feeds.

Measuring mineral digestibility is difficult for marine fish because of the intake of nondietary sources from drinking and as a result of active excretion of minerals into the intestine associated with maintenance of osmotic balance.

Effect of Feed Processing

Wilson and Poe (1985) found that extrusion processing increased the digestibility of energy, but had no effect on the digestibility of protein as compared with pellet processing for catfish diets. Similar results were found by Booth et al. (2000) for silver perch. Most of the improvement in ADC values for energy can be attributed to the positive effects of extrusion on starch gelatinization. Varying extrusion conditions (temperature, pressure) had no effect on ADC of protein, amino acids, or energy of a fish meal/wheat diet for rainbow trout (Sørensen et al., 2002). In contrast, Booth et al. (2000) and Cheng and Hardy (2003) demonstrated improved ADC protein values for silver perch fed a practical diet and for full-fat soybean meal in rainbow trout. Barrows et al. (2007) reported no effect of extrusion on ADC values for

Energy (%)										
Red Drum	Channel Catfish	Nile Tilapia	Siberian Sturgeon	Haddock	Gilthead Sea Bream	Cobia	Largemouth Bass	Pacu	Rockfish	Pacific White Shrimp
58[e]	—	—	—	—	58–90[f,g]	—	—	—	84–86[h]	—
—	—	—	—	—	—	—	—	—	—	—
—	—	—	—	60[m]	—	—	—	—	—	—
56[e]	57[o]	61[p]	—	—	—	—	—	76[q]	—	60[r]
—	—	83–89[p,t]	—	81[m]	80[g]	94[u]	77[v]	86[q]	89[h]	87[w]
70[y]	80[o]	—	72[k]	—	39[g]	—	—	—	41–55[h]	61[r]
—	—	—	80[k]	—	64[g]	96[u]	—	—	73–85[h]	—
—	—	89[p]	93[k]	—	80[f]	—	78[v]	75[q]	90–99[h]	87[w]
—	—	92[t]	—	—	—	—	—	—	93–96[h]	—
—	—	—	—	92[m]	94[g]	—	—	—	—	—
60–92[e,y]	92[o]	—	—	—	—	—	—	—	—	75[r]
—	—	—	—	—	—	—	—	—	—	—
—	—	—	—	—	—	—	—	—	—	77[z]
86[y]	76[o]	—	75[k]	—	69[g]	90[u]	—	—	—	62–85[r,w]
—	—	—	—	—	78[f]	—	—	—	90–93[h]	76[w]
—	—	—	—	—	—	84[u]	—	—	—	82[w]
72[y]	—	59[aa]	86[k]	—	80[g]	—	85[v]	—	—	—
—	—	79[p]	—	—	78–80[f]	91[u]	—	—	—	—
—	67[bb]	—	—	—	—	—	—	—	—	—
—	—	57[p]	—	—	—	83[u]	—	—	—	—
12[e]	—	—	—	—	—	—	—	—	—	—
—	—	—	—	70[m]	—	—	—	—	—	44[r]
53[e]		69[p]								
63[y]	72[bb]	84[t]	80[k]	92[m]	72[f]	87[u]	75[v]	78[q]	61–64[h]	76–81[r,w]
—	—	—	—	—	—	—	—	—	—	—
—	—	—	—	—	—	—	—	—	—	—
—	—	—	—	—	87[f]	—	—	—	—	—
62[y]	—	72[p]	—	—	88[f]	—	—	—	39–46[h]	66[r]
—	—	—	—	—	—	—	—	—	—	86–100[r,w]
34[e]	—	58[aa]	—	—	—	—	—	—	—	52[r]
—	—	—	—	—	—	—	—	—	60–66[h]	—

(1996) by ISIR; [f]Lupatsch et al. (1997) by indicator method, feces collected by stripping, single ingredient fed; [g]Nengas et al. (1995) by IGIR; [h]Lee (2002) by IGIR; [i]Glencross et al. (2004) by IGIR; [j]Glencross et al. (2005) by ISIR; [k]Liu et al. (2009) by indicator method, feces collected by ladling out water, ingredient fed in mixed diet; [l]Tibbetts et al. (2006) by IGIR; [m]Tibbetts et al. (2004) by IGIR; [n]Sullivan and Reigh (1995) by ISIR; [o]Wilson and Poe (1985) by IEIM; [p]Sklan et al. (2004) by ISPIR; [q]Abimorad et al. (2007) by IGIR; [r]Brunson et al. (1997) by ISPIR; [s]Kaushik (2002) by IGIR; [t]Koprucu and Ozdemir (2005) by IGIR; [u]Zhou et al. (2004) by IGIR; [v]Portz and Cyrino (2004) by IGIR; [w]Yang et al. (2009) by IGIR; [x]Rawles and Gatlin, III (2000) by ISIR; [y]Gaylord and Gatlin, III (1996) by ISIR; [z]Smith et al. (2007) by ISPIR; [aa]Hanley (1987) by IEIS; [bb]Cruz (1975); [cc]Burel et al. (2000) by IGIR.

corn gluten meal. These studies clearly show that the positive effects of extrusion on ADC values for protein or energy depend upon the ingredient or diet formulation. Particle size of feed ingredients does not affect ADC values in silver perch or rainbow trout (Booth et al., 2000; Zhu et al., 2008).

Conflicting findings are reported on the effects of extrusion on ADC values for minerals in feed ingredients. Cheng and Hardy (2003) reported that extrusion processing lowered ADC phosphorus values in full-fat soybean meal, presumably by inactivating endogenous phytase that would have liberated phosphorus in the gut. The opposite was found for extruded rapeseed meal (Burel et al., 2000).

To summarize the effects of extrusion processing on ADC values of fish feed ingredients or fish feeds, the most important effect is on starch gelatinization. This effect depends on the temperature, time, and pressure used during conditioning and extrusion, but the effects vary with starch source.

Interactions Affecting Additivity of ADC Values

The ADC values of protein, amino acids, energy, and lipid for individual ingredients can be used to predict the digestibility of a formulated feed by adding fractional ADC values for each ingredient (Lupatsch et al., 1997). This

TABLE 12-4 Apparent Digestibility of Lipid and Carbohydrate in Selected Feed Ingredients for Fish Species and Pacific White Shrimp (*Litopenaeus vannamei*)

		Lipid (%)										Carbohydrate (%)				
Feedstuff	International Feed No.	Rainbow Trout	Red Drum	Channel Catfish	Nile/ Blue Tilapia	Hybrid Striped Bass	Haddock	Gilthead Sea Bream	Cobia	Largemouth Bass	Pacific White Shrimp	Rainbow Trout	Channel Catfish	Nile/ Blue Tilapia	Gilthead Sea Bream	Pacific White Shrimp
Alfalfa meal	1-00-023	71[a]	—	51[b]	—	—	—	—	—	—	—	—	12[c]	27[d]	—	—
Barley		—	—	—	82[e]	—	—	—	—	—	—	—	—	61[e]	—	—
Blood meal	5-00-381	88–99[f]	—	—	—	—	—	—	—	—	—	—	—	—	—	—
Canola meal	5-06-145	—	—	—	—	—	87[g]	—	—	—	—	—	—	—	—	100[h]
Corn, grain	4-02-935	—	—	76[b]	76–90[d,e]	82[i]	—	—	—	—	—	—	66[c]	45–58[d,e]	—	—
Corn gluten meal	5-28-242	—	—	—	90–94[e,j]	—	57[g]	83[k]	96[l]	83[m]	92[n]	—	—	80[e]	—	—
Cottonseed meal	5-01-621	60–79[o]	75[p]	88[b]	—	92[i]	—	50[k]	—	—	—	84[q]	17[c]	—	—	—
Feather meal, hydrolyzed	5-03-795	40–83[f]	—	83[b]	—	—	—	—	—	—	—	—	—	—	—	—
Fish meal, not specified		—	—	—	90[e]	—	—	95[r]	96[l]	98[m]	93[n]	—	—	—	—	—
Fish, anchovy meal	5-01-985	—	—	—	98[j]	—	—	—	—	—	—	—	—	—	—	—
Fish, herring meal	5-02-000	97[a]	—	—	—	—	98[g]	—	—	—	—	—	—	—	—	—
Fish, menhaden meal	5-02-009	—	68[p]	97[b]	98[d]	95[i]	—	—	—	—	—	—	—	—	—	—
Meat and bone meal	5-00-388	77–93[a,f]	67[p]	—	—	88[i]	—	90[k]	92[l]	—	84[n]	—	—	—	—	—
Meat meal		—	—	—	—	—	—	88[r]	—	—	—	—	—	—	—	—
Peanut meal	5-03-650	—	—	—	—	—	—	—	94[l]	—	95[n]	—	—	—	—	—
Poultry byproduct meal	5-03-798	92[f]	59[p]	—	—	—	—	84[k]	—	98[m]	—	—	—	—	—	—
Poultry meal		—	—	—	78[e]	—	—	95[r]	92[l]	—	90[n]	—	—	—	—	—
Rapeseed meal	5-03-871	—	—	—	92[e]	—	—	—	94[l]	—	—	—	—	50[e]	—	—
Sorghum	4-04-383	—	—	—	83[e]	—	—	—	—	—	—	—	—	70[e]	—	—
Soybean meal	5-04-612	—	63[p]	81[b]	90–92[e,j]	54[i]	83[g]	63[k]	92[l]	93[m]	92[n]	—	—	54–65[d,e]	49[r]	—
Squid meal		—	—	—	—	—	—	83[r]	—	—	—	—	—	—	—	—
Starch, corn (cooked)		—	—	—	—	—	—	—	—	—	—	52[q]	—	66[s]	—	—
Starch, corn		—	—	—	—	—	—	—	—	—	82[t]	24[q]	—	55[s]	—	—
Starch, potato		—	—	—	—	—	—	—	—		80[t]	—	—	—	—	—
Wheat flour	4-05-199	—	88[p]	—	80[e]	—	—	—	—	—	—	—	—	—	77[r]	—
Wheat middling	4-05-205	—	—	—	—	84[i]	—	—	—	—	—	—	—	—	—	—
Wheat grain	4-05-268	—	—	96[b]	—	66[i]	—	—	—	—	—	—	59[c]	61–72[d,e]	—	—
Wheat gluten	5-05-221	—	—	—	—	—	—	—	—	—	93[n]	—	—	—	—	—

NOTE: Dashes indicate that data were not available. IGIR, digestibility was determined by indicator, Guelph or its modified systems and ingredient fed in reference diet; ISIR, determined by indicator, stripping method, and ingredient fed in reference diet; ISPIR, determined by indicator method, feces collected by siphoning, ingredient fed in reference die; IEIM, determined by indicator method, feces collected by surgical excision, ingredient fed in mixed diet; IEIS, determined by indicator method, feces collected by surgical excision, single ingredient fed.

[a]Cho et al. (1982) by IGIR; [b]Cruz (1975) by IEIS; [c]Cruz (1975) by IEIS; [d]Popma (1982) by ISPIR; [e]Sklan et al. (2004) by ISPIR; [f]Bureau et al. (1999) by IGIR; [g]Tibbetts et al. (2004) by IGIR; [h]Cruz-Suarez et al. (2001) by ISPIR; [i]Rawles and Gatlin III (2000) by ISIR; [j]Koprucu and Ozdemir (2005) by ISIR; [k]Nengas et al. (1995) by IGIR; [l]Zhou et al. (2004) by IGIR; [m]Portz and Cyrino (2004) by IGIR; [n]Yang et al. (2009) by IGIR; [o]Cheng and Hardy (2002b) by IGIR; [p]Gaylord and Gatlin III (1996) by ISIR; [q]Smith (1977) and Smith et al. (1980) determined by metabolism chamber, single ingredient fed; [r]Lupatsch et al. (1997) determined by indicator method, feces collected by stripping, single ingredient fed; [s]Saad (1989) by IEIM; [t]Cousin et al. (1996) by ISPIR.

TABLE 12-5 Phosphorus Availability of Selected Feed Ingredients for Several Fish Species and Pacific White Shrimp (*Litopenaeus vannamei*)

		Phosphorus (%)										
Source	International Feed No.	Coho Salmon	Rainbow Trout	Red Drum	European Sea Bass	Nile Tilapia	Channel Catfish	Common Carp	Cobia	Yellowtail	Largemouth Bass	Pacific White Shrimp
Animal products												
Blood meal	5-00-381	—	88–99[a]	—	—	—	—	—	—	—	—	—
Casein	5-01-162	—	90[b]	—	—	—	90[c]	97[b]	—	—	—	—
Egg albumin		—	—	—	—	—	—	71[b]	—	—	—	—
Feather meal, hydrolyzed	5-03-795	75[d]	62[d]	—	—	—	—	—	—	—	—	—
Fish meal, not specified		—	—	—	49–63[e]	—	—	—	—	—	72[f]	—
Fish meal, anchovy	5-01-985	47[d]	50[d]	—	—	28[g]	—	—	—	—	—	—
Fish meal, brown		—	74[b]	—	—	—	—	24[b]	—	—	—	—
Fish meal, herring	5-02-009	57[d]	44[d]	—	—	—	—	—	—	—	—	—
Fish meal, menhaden	5-02-009	40[d]	37[d]	48[h]	—	—	60[i]	—	—	—	—	—
Fish meal, white		—	66[b]	—	—	—	—	18[b]	—	—	—	—
Meat and bone meal	5-00-388	—	58–93[a]	66[h]	—	—	—	—	62[j]	—	—	76[k]
Poultry byproduct meal	5-03-798	68[d]	64[d]	27[h]	—	—	—	—	—	—	94[f]	—
Poultry meal		—	—	—	—	—	—	—	62[j]	—	—	76[k]
Inorganic phosphates												
Calcium, monobasic	6-01-082	—	94[b]	—	65[e]	—	94[i]	94[b]	—	92[l]	—	46[m]
Calcium, dibasic	6-01-080	—	71[b]	—	66[e]	—	65[i]	46[b]	—	59[l]	—	19[m]
Calcium, tribasic	6-01-084	—	64[b]	—	42[e]	—	—	13[b]	—	49[l]	—	10[m]
Potassium, monobasic		—	—	—	—	—	—	—	—	96[l]	—	68[m]
Sodium, monobasic	6-04-288	—	—	—	—	—	90[i]	—	—	95[l]	—	68[m]
Plant products												
Corn gluten meal	5-28-242	16[d]	9[d]	—	—	28[g]	25[i]	—	70[j]	—	83[f]	71[k]
Cottonseed meal	5-01-621	—	54[n]	40[h]	—	—	—	—	—	—	—	—
Peanut meal	5-03-650	—	—	—	—	—	—	—	58[j]	—	—	59[k]
Rapeseed meal	5-03-871	—	26–42[o]	—	—	—	—	—	56[j]	—	—	—
Rice bran	4-03-928	—	19[b]	—	—	—	—	25[b]	—	—	—	—
Soybean meal	5-04-612	28[d]	22[d]–57[p]	47[h]	—	30[g]	29[c]	—	59[j]	—	88[f]	64[k]
Soy protein concentrate	5-08-038	30	59[p]	—	—	—	—	—	—	—	—	—
Soy protein isolate		—	42[p]	—	—	—	—	—	—	—	—	—
Wheat flour	4-05-199	50[d]	47[d]	—	—	—	—	—	—	—	—	—
Wheat germ	5-05-218	—	58[b]	—	—	—	—	57[b]	—	—	—	—
Wheat gluten	5-05-221	57[d]	75[d]	—	—	—	—	—	—	—	—	72[k]
Wheat middlings	4-05-205	41[d]	55[d]	—	—	—	28[i]	—	—	—	—	—
Yeast, brewer's	7-05-527	—	69[q]	—	—	—	—	93[b]	—	—	—	—

NOTES: Dash indicates data were not available. IGIR, digestibility was determined by indicator, Guelph or its modified systems and ingredient fed in reference diet; ISIR, determined by indicator, stripping method and ingredient fed in reference diet.

[a]Bureau et al. (1999) by IGIR; [b]Ogino et al. (1979); [c]Wilson et al. (1982); [d]Sugiura et al. (1998) by IGIR; [e]Pimentel-Rodrigues and Oliva-Teles (2007); [f]Portz and Cyrino (2004) by IGIR; [g]Koprucu and Ozdemir (2005); [h]Gaylord and Gatlin, III (1996) by ISIR; [i]Lovell (1978); [j]Zhou et al. (2004) by IGIR; [k]Yang et al. (2009); [l]Sarker et al. (2009); [m]Davis and Arnold (1994); [n]Cheng and Hardy (2002b) by IGIR; [o]Burel et al. (2000); [p]Glencross et al. (2004) by IGIR; [q]Cheng et al. (2004) by IGIR.

practice assumes that ADC values measured in ingredients are unaffected by feed processing. However, ADC values for carbohydrates (NFE in proximate analysis) measured in ingredients are not necessarily additive when ingredients are combined in a feed. To complicate matters further, ADC values of certain specific subcategories can be influenced by dietary levels of other subcategories, specific nutrients, feed processing conditions, or, in the case of lipids, by water temperature. Thus, digestibility of lipids, carbohydrates, and phosphorus in feeds can be influenced in this manner, resulting in nonadditivity of fractional ADC values. Newly developed models take into account the fractional ADC values of subcategories of lipid and phosphorus sources (Hua and Bureau, 2006, 2009b; Hua et al., 2008). The models yield more accurate estimates of ADC values of feeds using ADC values of ingredients and their constituents.

REFERENCES

Abimorad, E. G., D. J. Carneiro, and E. C. Urbinati. 2007. Growth and metabolism of pacu (piaractus mesopotamicus holmberg 1887) juveniles fed diets containing different protein, lipid and carbohydrate levels. Aquac. Res. 38(1):36–44.

Abimorad, E. G., G. H. Squassoni, and D. J. Carneiro. 2008. Apparent digestibility of protein, energy, and amino acids in some selected feed ingredients for pacu *Piaractus mesopotamicus*. Aquacult. Nutr. 14:374–380.

Adamidou, S., I. Nengas, M. Alexis, E. Foundoulaki, D. Nikolopoulou, P. Campbell, I. Karacostas, G. Rigos, G. J. Bell, and K. Jauncey. 2009. Apparent nutrient digestibility and gastrointestinal evacuation time in European seabass (*Dicentrarchus labrax*) fed diets containing different levels of legumes. Aquaculture 289:106–112.

Akiyama, D. M, S. R. Coelho, A. L. Lawrence, and E. H. Robinson. 1989. Apparent digestibility of feedstuffs by the marine shrimp *Penaeus vannamei* BOONE. Nippon Suisan Gakk. 55:91–98.

Albrektsen, S., B. Hope, and A. Aksnes. 2009. Phosphorous (P) deficiency due to low P availability in fishmeal produced from blue whiting (*Micromesistius poutassou*) in feed for under-yearling Atlantic salmon (*Salmo salar*) smolt. Aquaculture 296:318–328.

Allan, G. L., S. Parkinson, M. A. Booth, D. A. Stone, S. J. Rowland, J. Frances, and R. Warner-Smith. 2000. Replacement of fish meal in diets for Australian silver perch, *Bidyanus bidyanus*: I. Digestibility of alternative ingredients. Aquaculture 186:293–310.

Ammerman, C. B., D. H. Baker, and A. J. Lewis. 1995. Bioavailability of nutrients for animals: Amino acids, minerals and vitamins. New York: Academic Press.

Anderson, J. S., S. P. Lall, D. M. Anderson, and J. Chandrasoma. 1992. Apparent and true availability of amino acids from common feed ingredients for Atlantic salmon (*Salmo salar*) reared in seawater. Aquaculture 108:111–124.

Anderson, J. S., S. P. Lall, D. M. Anderson, and M. A. McNiven. 1995. Availability of amino acids from various fish meals fed to Atlantic salmon (*Salmo salar*). Aquaculture 138:291–301.

Arndt, R. E., R. W. Hardy, S. H. Sugiura, and F. M. Dong. 1999. Effects of heat treatment and substitution level on palatability and nutritional value of soy defatted flour in feeds and coho salmon, *Oncorhynchus kisutch*. Aquaculture 180:129–145.

Arnesen, P., and Å. Krogdahl. 1995. Nutrient digestibilities, weight gain and plasma and liver levels of carbohydrate in rainbow trout, *Oncorhynchus mykiss* (Walbaum), fed diets with oats. Aquacult. Nutr. 1:201–211.

Aslaksen, M. A., O. F. Kraugerud, M. Penn, B. Svihus, V. Denstadli, H. Y. Jorgensen, M. Hillestad, Å. Krogdahl, and T. Storebakken. 2007. Screening of nutrient digestibilities and intestinal pathologies in Atlantic salmon, *Salmo salar*, fed diets with legumes, oilseeds, or cereals. Aquaculture 272:541–555.

Austreng, E., A. Skrede, and A. Eldegard. 1979. Effect of dietary fat source on the digestibility of fat and fatty acids in rainbow trout and mink. Acta Agr. Scand. 29:119–126.

Azevedo, P. A., C. Y. Cho, S. Leeson, and D. P. Bureau. 1998. Effects of feeding level and water temperature on growth, nutrient and energy utilization and waste outputs of rainbow trout (*Oncorhynchus mykiss*). Aquat. Living Resour. 11(4):227–238.

Bahurmiz, O. M., and W. K. Ng. 2007. Effects of dietary palm oil source on growth, tissue fatty acid composition and nutrient digestibility of red hybrid tilapia, *Oreochromis* sp., raised from stocking to marketable size. Aquaculture 262:382–392.

Barrows, F. T., D. A. J. Stone, and R. W. Hardy. 2007. The effects of extrusion conditions on the nutritional value of soybean meal for rainbow trout (*Oncorhynchus mykiss*). Aquaculture 265:244–252.

Batterham, E. S. 1992. Availability and utilization of amino acids for growing pigs. Nutr. Res. Rev. 5:1–18.

Berge, G. E., H. Sveier and E. Lied. 2002. Effects of feeding Atlantic salmon (*Salmo salar* L.) imbalanced levels of lysine and arginine. Aquacult. Nutr. 8:239–248.

Bergot, F., and J. Breque. 1983. Digestibility of starch by rainbow trout: Effects of the physical state of starch and of the intake level. Aquaculture 34:203–212.

Bogevik, A. S., D. R. Tocher, H. Langmyhr, R. Waagbo, and R. E. Olsen. 2008. Atlantic salmon (*Salmo salar*) postsmolts adapt lipid digestion according to elevated dietary wax esters from *Calanus finmarchicus*. Aquacult. Nutr. 15:94–103.

Booth, M. A., G. L. Allen, and R. Warner-Smith. 2000. Effects of grinding, steam conditioning and extrusion of a practical diet on digestibility and weight gain of silver perch, *Bidyanus bidyanus*. Aquaculture 182:287–299.

Brauge, C., G. Corraze, and F. Médale. 1995. Effect of dietary levels of lipid and carbohydrate on growth performance, body composition, nitrogen excretion and plasma glucose levels in rainbow trout reared at 8 or 18°C. Reprod. Nutr. Dev. 35:277–290.

Brinker, A. 2007. Guar gum in rainbow trout (*Oncorhynchus mykiss*) feed: The influence of quality and dose on stabilization of fecal solids. Aquaculture 267:315–327.

Brown, M. L., G. R. Nematipour, and D. M. Gatlin, III. 1992. Dietary protein requirement of juvenile sunshine bass at different salinities. Prog. Fish. Cult. 54:148–156.

Brunson, J. F., R. P. Romaire, and R. C. Reigh. 1997. Apparent digestibility of selected ingredients in diets for white shrimp *Penaeus setiferus* L. Aquacult. Nutr. 3:9–16.

Bureau, D. P., and K. Hua. 2006. Letter to the editor of Aquaculture. Aquaculture 252:103–105.

Bureau, D. P., A. M. Harris, and C. Y. Cho. 1999. Apparent digestibility of rendered animal protein ingredients for rainbow trout (*Oncorhynchus mykiss*). Aquaculture 180:293–310.

Bureau, D. P., K. Hua, and A. M. Harris. 2008. The effect of dietary lipid and long-chain n-3 PUFA levels on growth, energy utilization, carcass quality and immune function of rainbow trout, *Oncorhynchus mykiss*. J. World Aquacult. Soc. 39:1–21.

Burel, C., T. Boujard, F. Tulli, and S. J. Kaushik. 2000. Digestibility of extruded peas, extruded lupin and rapeseed meal in rainbow trout (*Oncorhynchus mykiss*) and turbot (*Psetta maxima*). Aquaculture 188:285–298.

Carpenter, K. J. 1960. The estimation of the available lysine in animal-protein foods. Biochem J. 77:604–610.

Cheng, Z. J., and R. W. Hardy. 2002a. Effect of microbial phytase on apparent nutrient digestibility of barley, canola meal, wheat and wheat middlings, measured *in vivo* using rainbow trout (*Oncorhynchus mykiss*). Aquacult. Nutr. 8:271–277.

Cheng, Z. J., and R. W. Hardy. 2002b. Apparent digestibility coefficients and nutritional value of cottonseed meal for rainbow trout (*Oncorhynchus mykiss*). Aquaculture 212:361–372.

Cheng, Z. J., and R. W. Hardy. 2003. Effects of extrusion and expelling processing, and microbial phytase supplementation on apparent digestibility coefficients of nutrients in full-fat soybeans for rainbow trout (*Oncorhynchus mykiss*). Aquaculture 218:501–514.

Cheng, Z. J., R. W. Hardy, and N. J. Huige. 2004. Apparent digestibility coefficients of nutrients in brewer's and rendered animal by-products for rainbow trout (*Oncorhynchus mykiss* (Walbaum)). Aquac. Res. 35:1–9.

Chiu, Y. N., R. E. Austic, and G. L. Rumsey. 1987. Interactions among dietary minerals, arginine and lysine in rainbow trout (*Salmo gairdneri*). Fish Physiol. Biochem. 4:45–55.

Cho, C. Y., and S. J. Kaushik. 1990. Nutritional energetics in fish: Energy and protein utilization in rainbow trout (*Salmo gairdneri*). Pp.132–172 in Aspects of Food Production, Consumption and Energy Values, World Review of Nutrition and Dietetics, G. H. Bourne, ed. Basel, Switzerland: Karger.

Cho, C. Y., and S. J. Slinger. 1979. Apparent digestibility measurement in feedstuffs for rainbow trout. Pp. 239–247 in Proceedings of the World Symposium on Finfish Nutrition and Fish Feed Technology, Vol. II, June 20–23, 1978, Hamburg, Germany.

Cho, C. Y., S. J. Slinger, and H. S. Bayley. 1982. Bioenergetics of salmonid fishes: Energy intake, expenditure and productivity. Comp. Biochem. Phys. B 73B:25–41.

Choubert, G., Jr., J. de la Noüe, and P. Luquet. 1979. Continuous quantitative automatic collector for fish feces. Prog. Fish Cult. 41:64–67.

Choubert, G., Jr., J. de la Noüe, and P. Luquet. 1982. Digestibility in fish: Improved device for the automatic collection of feces. Aquaculture 29:185–189.

Cousin, M., G. Cuzon, and J. Guillaume. 1996. AQUACOP. Digestibility of starch in *Penaeus vannamei*: In vivo and in vitro study on eight samples of various origin. Aquaculture 140:361–372.

Cruz, E. M. 1975. Determination of Nutrient Digestibility in Various Feedstuffs for Channel Catfish. Ph.D. dissertation, Auburn University, Auburn, AL.

Cruz-Suarez, L. E., D. Ricque-Marie, M. Tapia-Salazar, I. M. McCallum, and D. Hickling. 2001. Assessment of differently processed feed pea (*Pisum sativum*) meals and canola meal (*Brassica* sp.) in diets for blue shrimp (*Litppenaeus stylirostris*). Aquaculture 196:87–104.

Cruz-Suarez, L. E., M. Nieto-Lopez, C. Guajardo-Barbosa, M. Tapia-Salazar, U. Scholz, and D. Ricque-Marie. 2007. Replacement of fish meal with poultry by-product meal in practical diets for *Litopenaeus vannamei*, and digestibility of the tested ingredients and diets. Aquaculture 272:466–476.

Cruz-Suarez, L. E., M. Tapia-Salazar, D. Villarreal-Cavazos, J. Beltran-Rocha, M. G. Nieto-Lopez, A. Lemme, and D. Ricque-Marie. 2009. Apparent dry matter, energy, protein and amino acid digestibility of four soybean ingredients in white shrimp *Litopenaeus vannamei* juveniles. Aquaculture 292:87–94.

Da Silva, J. G., and A. Oliva-Teles. 1998. Apparent digestibility coefficients of feedstuffs in seabass (*Dicentrarchus labrax*) juveniles. Aquat. Living Resour. 11:187–192.

Das, K. M., and S. D. Tripathi. 1991. Studies on the digestive enzymes of grass carp, *Ctenopharyngodon idella* (Val.). Aquaculture 92:21–32.

Davies, S. J., P. C. Morris and R. T. M. Baker. 1997. Partial substitution of fish meal and full-fat-soya bean meal with wheat gluten and influence of lysine supplementation in diets for rainbow trout *Oncorhynchus mykiss* (Walbaum). Aquac. Res. 28:317–328.

Davis, D. A., and C. R. Arnold. 1994. Estimation of apparent phosphorus availability from inorganic phosphorus sources for *Penaeus vannamei*. Aquaculture 127:245–254.

Davis, D. A., and C. R. Arnold. 1995. Effects of two extrusion processing conditions on the digestibility of four cereal grains for *Penaeus vannamei*. Aquaculture 133:287–294.

Davis, D. A., C. R. Arnold, and I. McCallum. 2002. Nutritional value of feed peas (*Pisum sativum*) in practical diet formulations for *Litopenaeus vannamei*. Aquacult. Nutr. 8:87–94.

Ehab, E.-H. R., and D. P. Bureau. 2007. Comparison of the bioavailability of lysine in blood meals of various origins to that of L-lysine HCl for rainbow trout (*Oncorhynchus mykiss*). Aquaculture 262:402–409.

Ezquerra, J. M., F. L. Garcia-Carreno, and O. Carrillo. 1997. In vitro digestibility of dietary protein sources for white shrimp (*Penaeus vannamei*). Aquaculture 163:123–136.

Forster, I. P. 1999. A note on the method of calculating coefficients of digestibility for nutrients provided by single ingredients to feeds of aquatic animals. Aquacult. Nutr. 5:143–145.

Forster, I., D. A. Higgs, B. S. Dosanjh, M. Rowshadeli, and J. Parr. 1999. Potential for dietary phytase to improve the nutritive value of canola protein concentrate and decrease phosphorus output in rainbow trout (*Oncorhynchus mykiss*) held in 11°C fresh water. Aquaculture 179:109–125.

Forster, I. P., W. Dominy, L. Obaldo, and A. G. J. Tacon. 2003. Rendered meal and bone meals as ingredients of diets for shrimp *Litopenaeus vannamei* (Boone, 1931). Aquaculture 219:655–670.

Gatlin, D. M., III, and H. F. Phillips. 1989. Dietary calcium, phytate and zinc interactions in channel catfish. Aquaculture 79:259–266.

Gaylord, T. G., and D. M. Gatlin, III. 1996. Determination of digestibility coefficients of various feedstuffs for red drum (*Sciaenops ocellatus*). Aquaculture 139:303–314.

Gaylord, T. G., S. D. Rawles, and D. M. Gatlin, III. 2004. Amino acid availability from animal, blended, and plant feedstuffs for hybrid striped bass (*Morone chrysops* × *M. saxatilis*). Aquacult. Nutr. 10:345–352.

Gaylord, T. G., F. T. Barrows, S. D. Rawles, K. Liu, P. Bregitzer, A. Hang, D. E. Obert, and C. Morris. 2009. Apparent digestibility of nutrients and energy in extruded diets from cultivars of barley and wheat selected for nutritional quality in rainbow trout, *Oncorhynchus mykiss*. Aquacult. Nutr. 15:306–312.

Glencross, B. D., C. G. Carter, N. Duijster, D. R. Evans, K. Dods, P. McCafferty, W. E. Hawkins, R. Maas, and S. Sipsas. 2004. A comparison of the digestibility of a range of lupin and soybean protein products when fed to either Atlantic salmon (*Salmo salar*) or rainbow trout (*Oncorhynchus mykiss*). Aquaculture 237:333–346

Glencross, B., D. Evans, K. Dods, P. McCafferty, W. Hawkins, R. Maas, and S. Sipsas. 2005. Evaluation of the digestible value of lupin and soybean protein concentrates and isolates when fed to rainbow trout, *Oncorhynchus mykiss*, using either stripping or settlement faecal collection methods. Aquaculture 245:211–220.

Guimaraes, I. G., L. E. Pezzato, and M. M. Barros. 2008. Amino acid availability and protein digestibility of several protein sources for Nile tilapia, *Oreochromis niliticus*. Aquacult. Nutr. 14:396–404.

Hajen, W. E., D. A. Higgs, R. M. Beames, and B. S. Dosanjh. 1992. Digestibility of various feedstuffs by post-juvenile Chinook salmon (*Oncorhynchus tshawylsche*) in sea water. 2. Measurement of digestibility. Aquaculture 112:333–348.

Hajen, W. E., R. M. Beames, D. A. Higgs, and B. S. Dosanjh. 1993. Digestibility of various feedstuffs by post-juvenile Chinook salmon in sea water. 1. Validation of technique. Aquaculture 112:321–332.

Hanley, F. 1987. The digestibility of foodstuffs and the effects of feeding selectivity on digestibility determinations in tilapia, *Oreochromis niloticus* (L). Aquaculture 66:163–179.

Hansen, J. Ø., and T. Storebakken. 2007. Effects of dietary cellulose level on pellet quality and nutrient digestibilities in rainbow trout (*Oncorhynchus mykiss*). Aquaculture 272:458–465.

Hardy, R. W. 1997. Understanding and using apparent digestibility coefficients in fish nutrition. Aquacult. Mag. 23:84–89.

Hemre, G.-I., Ø. Karlsen, A. Mangor-Jensen, and G. Rosenlund. 2003. Digestibility of dry matter, protein, starch and lipid by cod, *Gadus morhua*: Comparison of sampling methods. Aquaculture 225:225–232.

Hua, K., and D. P. Bureau. 2006. Modeling digestible phosphorus fractions of salmonid fish feeds. Aquaculture 254:455–465.

Hua, K., and D. P. Bureau. 2009a. A mathematical model to explain variations in estimates of starch digestibility and predict digestible starch content of salmonid fish feeds. Aquaculture 294:282–287.

Hua, K., and D. P. Bureau. 2009b. Development of a model to estimate digestible lipid content of salmonid fish feeds. Aquaculture 286:271–276.

Hua, K., C. F. M. de Lange, A. J. Niimi, G. Cole, R. D. Moccia, M. Z. Fan, and D. P. Bureau. 2008. A factorial model to predict phosphorus waste output of rainbow trout (*Oncorhynchus mykiss*). Aquac. Res. 39:1059–1068.

Hughes, K. P., and J. H. Soares. 1998. Efficacy of phytase on phosphorus utilization in practical diets fed to striped bass *Morone saxatillis*. Aquacult. Nutr. 4:122–142.

Jackson, L. S., M. H. Li, and E. H. Robinson. 1996. Use of microbial phytase in channel catfish *Ictalurus punctatus* diets to improve utilization of phytate phosphorus. J. World Aquacult. Soc. 27:297–302.

Kaushik, S. J. 2002. European Sea Bass, *Dicentrarchus labrax*. Pp. 28–39 in Nutrient Requirements and Feeding of Finfish for Aquaculture, C. D. Webster and C. E. Lim, eds. Wallingford, UK: CABI Publishing.

Kaushik, S. J., and B. Fauconneau. 1984. Effects of lysine administration on plasma arginine and on some nitrogenous catabolites in rainbow trout. Comp. Biochem. Physiol. 79A:459–462.

Koprucu, K., and Y. Ozdemir. 2005. Apparent digestibility of selected feed ingredients for Nile tilapia (*Oreochromis niloticus*). Aquaculture 250:308–316.

Krogdahl, A., G.-I. Hemre, and T. P. Mommsen. 2005. Carbohydrates in fish nutrition: Digestion and absorption in postlarval stages. Aquacult. Nutr. 11:103–122.

Lee, S. M. 2002. Apparent digestibility coefficients of various feed ingredients for juvenile and grower rockfish (*Sebastes schlegeli*). Aquaculture 207:79–95.

Lemos, D., A. L. Lawrence, and A. J. Siccardi, III. 2009. Prediction of apparent protein digestibility of ingredients and diets by *in vitro* pH-stat degree of protein hydrolysis with species-specific enzymes for juvenile Pacific white shrimp *Litopenaeus vannamei*. Aquaculture 295:89–98.

Lewis, A. J., and H. S. Bayley. 1995. Bioavailability of Nutrients for Animals: Amino Acids. New York: Academic Press.

Liu, H., X. Wu, M. Zhao, M. Xue, L. Guo, Y. Zheng, and Y. Yu. 2009. Nutrients apparent digestibility coefficients of selected protein sources for juvenile Siberian sturgeon (*Acipenser baerii* Brandt), compared by two chromic oxide analyses methods. Aquacult. Nutr. 15:650–656.

Lovell, R. T. 1978. Dietary phosphorus requirement of channel catfish (*Ictalurus punctatus*). T. Am. Fish. Soc. 107:617–621.

Lupatsch, I., G. W. M. Kissil, D. Sklan, and E. Pfeffer. 1997. Apparent digestibility coefficients of feed ingredients and their predictability in compound diets for gilthead seabream, *Sparus aurata* L. Aquacult. Nutr. 3:81–89.

Masumoto, T., T. Ruchimat, Y. Ito, H. Hosokawa, and S. Shimeno. 1996. Amino acid availability values for several protein sources for yellowtail (*Seriola quinqueradiata*). Aquaculture 146:109–119.

McGoogan, B. B., and R. C. Reigh. 1996. Apparent digestibility of selected ingredients in red drum (*Sciaenops ocellatus*) diets. Aquaculture 141:233–244.

Médale, F., P. Aguirre, and S. J. Kaushik. 1991. Utilization of dietary carbohydrates by rainbow trout at two water temperatures. Pp. 392–395 in Energy Metabolism of Farm Animals, European Association for Animal Production Pub. 58, C. Wenk and M. Boessinger, eds. Wageningen, the Netherlands: Pudoc.

Médale, F., J. M. Poli, F. Vallée, and D. Blanc. 1999. Comparaison de l'utilisation digestive et métabolique d'un régime riche en glucides par la carpe à 18°C et 25°C. Cybium 23:139–152.

Menoyo, D., C. J. Lopez-Boge, J. M. Bautista, and A. Obach. 2003. Growth, digestibility and fatty acid utilization in large Atlantic salmon (*Salmo salar*) fed varying levels on n-3 and saturated fatty acids. Aquaculture 225:295–307.

Moreira, I. S., H. Peres, A. Couto, P. Enes, and A. Oliva-Teles. 2008. Temperature and dietary carbohydrate level effects on performance and metabolic utilisation of diets in European sea bass (*Dicentrarchus labrax*) juveniles. Aquaculture 274:153–160.

Nengas, I., M. N. Alexis, S. J. Davies, and G. Petichakis. 1995. Investigation to determine digestibility coefficients of various raw materials in diets for gilthead sea bream, *Sparus auratus* L. Aquac. Res. 26:185–194.

Ng, W. K., P. J. Campbell, J. R. Dick, and J. G. Bell. 2003. Interactive effects of dietary palm oil concentration and water temperature on lipid digestibility in rainbow trout (*Oncorhynchus mykiss*). Lipids 38:1031–1038.

Ng, W. K., T. Sigholt, and J. G. Bell. 2004. The influence of environmental temperature on the apparent nutrient and fatty acid digestibility in Atlantic salmon (*Salmo salar* L.) fed finishing diets containing different blends of fish oil, rapeseed oil and palm oil. Aquac. Res. 35:1228–1237.

Nyachoti, C. M., C. F. M. de Lange, B. W. McBride, and H. Schulze. 1997. Significance of endogenous gut nitrogen losses in the nutrition of growing pigs: A review. Can. J. Anim. Sci. 149–163.

Ogino, C., J. Kakino, and M. S. Chen. 1973. Determination of metabolic fecal nitrogen and endogenous nitrogen excretion of carp. B. Jpn. Soc. Sci. Fish. 39:519–523.

Ogino, C., T. Takeuchi, H. Takeda, and T. Watanabe. 1979. Availability of dietary phosphorus in carp and rainbow trout. B. Jpn. Soc. Sci. Fish. 45:1527–1532.

Olsen, R. E., and E. Ringø. 1998. The influence of temperature on the apparent nutrient and fatty acid digestibility of Arctic charr, *Salvelinus alpines* L. Aquac. Res. 29:695–701.

Olsen, R. E., R. J. Henderson, J. Sountama, G. I. Hemre, E. Ringø, W. Melle, and D. R. Tocher. 2004. Atlantic salmon, *Salmo salar*, utilizes wax ester-rich oil from *Calanus finmarchicus* effectively. Aquaculture 240:433–339.

Opstvedt, J., R. Miller, R. W. Hardy, and J. Spinelli. 1984. Heat-induced changes in sulfhydryl groups and disulfide bonds in fish protein and their effect on protein and amino acid digestibility in rainbow trout (*Salmo gairdneri*). J. Agr. Food Chem. 32:929–935.

Pandian, T. J. 1967. Intake, digestion, absorption and conversion of food in the fishes *Megalops cyprinoides* and *Ophiocephalus striatus*. Mar. Biol. 1:16–32.

Pimentel-Rodrigues, A., and A. Oliva-Teles. 2007. Phosphorus availability of inorganic phosphates and fish meals in European sea bass (*Dicentrarchus labrax* L.) juveniles. Aquaculture 267:300–307.

Popma, T. J. 1982. Digestibility of Selected Feedstuffs and Naturally Occurring Algae by Tilapia. Ph.D. Dissertation, Auburn University, Auburn, AL.

Portz, L., and J. Cyrino. 2004. Digestibility of nutrients and amino acids of different protein sources in practical diets by largemouth bass *Micropterus salmoides* (Lacepede, 1802). Aquac. Res. 35:312–320.

Rathbone, C. K., J. K. Babbitt, F. M. Dong, and R. W. Hardy. 2001. Performance of juvenile coho salmon *Oncorhynchus kisutch* fed diets containing meals from fish wastes, deboned fish wastes, or skin-and-bone by-product as the protein ingredient. J. World Aquacult. Soc. 32:21–29.

Rawles, S. D., and D. M. Gatlin, III. 2000. Nutrient digestibility of common feedstuffs in extruded diets for Sunshine Bass *Morone chrysops* ♀ × *M. saxatilis* ♂. J. World. Aquacult. Soc. 31:570–579.

Richardson, N. L., D. A. Higgs, R. M. Beames, and J. M. McBride. 1985. Influence of dietary calcium, phosphorus, zinc and sodium phytate level on cataract incidence, growth and histopathology in juvenile Chinook salmon (*Oncorhynchus tshawytscha*). J. Nutr. 115:553–567.

Robinson, E. H., R. P. Wilson, and W. E. Poe. 1981. Arginine requirement and apparent absence of a lysine-arginine antagonism in fingerling channel catfish. J. Nutr. 111:46–52.

Rodehutscord, M., and E. Pfeffer. 1995. Effects of supplemental microbial phyatse on phosphorus digestibility and utilization in rainbow trout (*Oncorhynchus mykiss*). Water Sci. Technol. 31:143–147.

Saad, C. R. 1989. Carbohydrate Metabolism in Channel Catfish. Ph.D. Dissertation, Auburn University, Auburn, AL.

Saha, S., R. N. Roy, S. K Sen, and A. K. Ray. 2006. Characterization of cellulase-producing bacteria from the digestive tract of tilapia, *Oreochromis mossambica* (Peters) and grass carp, *Ctenopharyngodon idella* (Valenciennes). Aquac. Res. 37:380–388.

Sargent, J. R., D. R. Tocher, and J. G. Bell. 2002. The lipids. Pp. 181–257 in Fish Nutrition, 3rd Edition, J. E. Halver and R. W. Hardy, eds. New York: Academic Press, Inc.

Sarker, P. K., H. Fukada, and T. Masumoto. 2009. Phosphorus availability from inorganic phosphorus sources in yellowtail (*Seriola quinqueradiata* Temminck and Schlegel). Aquaculture 289:113–117.

Scott, M. L., M. C. Nesheim, and R. J. Young. 1982. Nutrition of the Chicken, 3rd edition. Ithaca, NY: M. L. Scott and Associates.

Shearer, K. D., and R. W. Hardy. 1987. Phosphorus deficiency in rainbow trout fed a diet containing deboned fillet scrap. Prog. Fish Cult. 49:192–197.

Sibbald, I. R. 1978. The true metabolizable energy values of mixtures of tallow with either soybean oil or lard. Poultry Sci. 57:473–477.

Sklan, D., T. Prag, and I. Lupatsch. 2004. Apparent digestibility coefficients of feed ingredients and their prediction on diets for tilapia *Oreochromis niloticus* × *Oreochromis aureus* (Teleostei, Cichlidae). Aquac. Res. 35:358–364.

Small, B. C., R. E. Austic, and J. H. Soares, Jr. 1999. Amino acid availability of four practical feed ingredients fed to striped bass *Morone saxatilis*. J. World. Aquacult. Soc. 30:58–64.

Smith, D. M., S. J. Tabrett, B. D. Glencross, S. J. Irvin, and M. C. Barclay. 2007. Digestibility of lupin kernel meals in feeds for the black tiger shrimp, *Penaeus monodon*. Aquaculture 264:353–362.

Smith, R. R. 1971. A method for measuring digestibility and metabolizable energy of fish feeds. Prog. Fish Cult. 33:132–134.

Smith, R. R. 1976. Metabolizable energy of feedstuffs for trout. Feedstuffs 48:16–21.

Smith, R. R. 1977. Recent research involving full-fat soybean meal in salmon diets. Salmonid 1:8–18.

Smith, R. R., M. C. Peterson, and A. C. Allred. 1980. The effect of leaching on apparent digestion coefficients in determining digestibility and metabolizable energy of feedstuffs for salmonids. Prog. Fish Cult. 42:195–199.

Solomon, D. J., and A. G. Brafield. 1972. The energetic of feeding, metabolism and growth of perch (*Perca fluviatlilis* L.). J. Anim. Ecol. 41:699–718.

Sørensen, M., K. Ljøkjel, T. Storebakken, K. D. Shearer, and A. Skrede. 2002. Apparent digestibility of protein, amino acids and energy in rainbow trout (*Oncorhynchus mykiss*) fed a fish meal based diet extruded at different temperatures. Aquaculture 211:215–225.

Stone, D. A. J. 2003. Dietary carbohydrate utilization by fish. Rev. Fish. Sci. 11(4):337–369.

Storebakken, T., K. D. Shearer, G. Baeverfjord, B. G. Nielsen, T. Asgard, T. Scott, and A. DeLaporte. 2000. Digestibility of macronutrients, energy and amino acids, absorption of elements and absence of intestinal enteritis in Atlantic salmon, *Salmo salar*, fed diets with wheat gluten. Aquaculture 184:115–132.

Sugiura, S. H., F. M. Dong, C. K. Rathbone, and R. W. Hardy. 1998. Apparent protein digestibility and mineral availabilities in various feed ingredients for salmonid feeds. Aquaculture 159:177–202.

Sugiura, S. H., J. K. Babbitt, F. M. Dong, and R. W. Hardy. 2000. Utilization of fish and animal by-product meals in low-pollution feeds for rainbow trout, *Oncorhynchus mykiss* (Walbaum). Aquac. Res. 31(7):585–593.

Sugiura, S. H., J. Gabaudan, F. M. Dong, and R. W. Hardy. 2001. Dietary microbial phytase supplementation and the utilization of phosphorus, trace minerals and protein by rainbow trout [*Oncorhynchus mykiss* (Walbaum)] fed soybean meal-based diets. Aquac. Res. 32:583–592.

Sullivan, J. A., and R. C. Reigh. 1995. Apparent digestibility of selected feedstuffs in diets for hybrid striped bass (*Morone saxatilis* × *Morone chrysops*). Aquaculture 138:313–322.

Svihus, B., A. K. Uhlen, and O. M. Harstad. 2005. Effect of starch granule structure, associated components and processing on nutritive value of cereal starch: A review. Anim. Feed Sci. Tech. 122:303–320.

Thiessen, D. L., G. L. Campbell, and P. D. Adelizi. 2003. Digestibility and growth performance of juvenile rainbow trout (*Oncorhynchus mykiss*) fed with pea and canola products. Aquacult. Nutr. 9:67–75.

Tibaldi, E., F. Tulli and D. Lanari. 1994. Arginine requirement and effect of different arginine and lysine levels for fingerling sea bass (*Dicentrarchus labrax*). Aquaculture 127:207–218.

Tibbetts, S. M., S. P. Lall, and J. E. Milley. 2004. Apparent digestibility of common feed ingredients by juvenile haddock, *Melanogrammus aeglefinus* L. Aquac. Res. 35:643–651.

Tibbetts, S. M., J. E. Milley, and S. P. Lall. 2006. Apparent protein and energy digestibility of common and alternative feed ingredients by Atlantic cod, *Gadus morhua* (Linnaeus, 1759). Aquaculture 261:1314–1327.

Vandenberg, G., and J. de la Noüe. 2001. Apparent digestibility comparison in rainbow trout (*Oncorhynchus mykiss*) assessed using three methods of faeces collection and three digestibility markers. Aquacult. Nutr. 7:237–245.

Venou, M. N., N. Alexis, E. Fountoulaki, I. Enegas, M. Apostolopoulou, and I. Castritsi-Cathariou. 2003. Effect of extrusion of wheat and corn on gilthead bream (*Sparus aratus*) growth, nutrient utilization efficiency, rates of gastric evacuation and digestive enzyme activities. Aquaculture 225:207–223.

Vielma, J., T. Makinen, P. Ekholm, and J. Koskela. 2000. Influence of dietary soy and phytase levels on performance and body composition of large rainbow trout (*Oncorhynchus mykiss*) and algal availability of phosphorus load. Aquaculture 183:349–362.

Wilson, R. P., and W. E. Poe. 1985. Apparent digestibility of protein and energy in feed ingredients for channel catfish. Prog. Fish Cult. 47:154–158.

Wilson, R. P., E. H. Robinson, and W. E. Poe. 1981. Apparent and true availability of amino acids from common feed ingredients for channel catfish. J. Nutr. 111:923–929

Wilson, R. P., E. H. Robinson, D. M. Gatlin, III, and W. E. Poe. 1982. Dietary phosphorus requirement of channel catfish. J. Nutr. 112:1197–1202.

Windell, J. T., J. W. Foltz, and J. A. Sarokon. 1978a. Methods of fecal collection and nutrient leaching in digestibility studies. Prog. Fish Cult. 40:51–55.

Windell, J. T., J. W. Foltz, and J. A. Sarokok. 1978b. Effect of fish size, temperature and amount fed on nutrient digestibility of a pelleted diet by rainbow trout, (*Salmo gairdneri*). T. Am. Fish. Soc. 107:613–616.

Yang, Q., X. Zhou, Q. Zhou, B. Tan, S. Chi, and X. Dong. 2009. Apparent digestibility of selected feed ingredients for white shrimp *Litopenaeus vannamei*, Boone. Aquac. Res. 1–9.

Zhou, Q. C., B. P. Tan, K. S. Mai, and Y. J. Liu. 2004. Apparent digestibility of selected feed ingredients for juvenile cobia *Rachycentron canadum*. Aquaculture 241:441–451.

Zhu, S., S. Chen, R. W. Hardy, and F. T. Barrows. 2008. Digestibility, growth and excretion response of rainbow trout (*Oncorhynchus mykiss* Walbaum) to feeds of different ingredient particle sizes. Aquac. Res. 32:885–893.

13

Nutrient Delivery and Feeding Practices

Nutrient delivery and meeting requirements of fish and shrimp species entail numerous and very specific challenges unknown to other livestock production sectors. Feeds and feeding strategies are critical aspects in fish and shrimp culture and are influenced by a wide variety of factors including species, strains, and life stages, characteristic of the feed used as well as environmental conditions and husbandry practices. Diet characteristics, such as type (live, fresh, frozen, or compounded feeds), physical characteristics (size of feed particles, fineness of grind, density, or specific gravity), and palatability, should be carefully considered for various life stages and species. Feed allowance, ration size, and feeding frequency are also important factors influencing nutrient intake and retention and, consequently, growth, health, and quality of the animal. The choice of feed type (floating or sinking) and method of feeding depends on the species, culture system, equipment used to deliver feed, and manpower. There is no single feeding protocol or method that fits all species and production systems. Consequently, this chapter covers various aspects of feeding, nutrient delivery, as well as approaches and models to improve efficiency of nutrient delivery and utilization, cost effectiveness, and environmental sustainability of production systems.

FEEDING EARLY LIFE STAGES

Newly hatched fish and crustacean species include a wide range of physiological forms, which for simplicity will be generically described as larvae. As described in Chapter 14, the successful use of formulated feeds is related to a variety of factors, which include the species, size of the larvae, and the type or developmental stage of the digestive system. Larvae that have rudimentary digestive systems are generally the most difficult to feed and often require live feeds whereas those with more developed digestive systems can often be reared solely on manufactured feeds.

Live Foods

The preferred live organisms to feed to larval fish and crustaceans are those in their natural diets, including both phytoplankton and zooplankton. Phytoplankton makes up the base of aquatic food webs and is consequently essential as food for larval rearing of many species. Phytoplankton is a critical food source for all stages of bivalve molluscs, larval stages of some crustacean species, early growth stages of some fish species, and the culture of numerous zooplankton species used as food for larval fish and shrimp (Couteau, 1996). Various algae species (live, preserved as paste, or dried) used to culture zooplankton differ in nutrient content, quality, and ease of culture. The rotifer (*Brachionus* spp.) and brine shrimp (*Artemia* spp.) are the primary types of zooplankton produced in mass quantities as live food for larvae. Phytoplankton and zooplankton from different geographic origins and cultured under different conditions may differ in nutritional quality, primarily n-3 and n-6 polyunsaturated fatty acids (PUFA) content (Watanabe et al., 1983; Volkman et al., 1989; Reitan et al., 1997). Many fish larvae are very sensitive to a deficiency of n-3 PUFA (Watanabe et al., 1983; Gatesoupe and Le Millinaire, 1985; Rainuzzo et al., 1997, Sargent et al., 1999); thus, the composition and quantity of fatty acids in zooplanktonic food affect larval growth and survival. Enrichment of zooplankton to ensure adequate levels of PUFA may be accomplished by two procedures (Sorgeloos, 1980): (1) the zooplankton can be fed for a period of 24 hours on live food source such as marine algae (e.g., *Chlorella* spp.) or yeast containing a high concentration of PUFA, or (2) the zooplankton nauplii can be exposed to a prepared nutrient source such as microparticulate feeds, preserved algae pastes, or a suspension of lipid rich in PUFA (e.g., fish oil and an emulsifying compound) for 3 to 12 hours before being offered to the larvae.

Microalgae are commonly utilized in the production of zooplankton as an enrichment medium. In addition, algae may be utilized in the culture system in which case the algae can serve as a nutrient source to the larvae as well as the live food organisms (Reitan et al., 1997). The use of algae in intensive culture systems has a number of advantages, including it most likely modifies the bacterial population of the water, enhances water quality, stabilizes the nutrient content of live foods, modifies light levels, stimulates the digestive process in the cultured organism, and contributes to the establishment of early gut microbiota (Reitan et al., 1997; Liao et al., 2001; Palmer et al., 2007).

Because food of optimal size, quantity, and quality must be available to the larvae at all times, the management of live food production systems as well as levels of live foods in the culture system must be suitable. Larvae can obtain live food in either extensive or intensive culture systems. From a nutritional perspective, intensive culture systems where live food is added to the system provide more control of the quality and quantity of live foods than do extensive systems where live food is produced naturally. Certain prey densities seem to be effective across a number of larval fish species (e.g., 10–20 rotifers per ml); however, optimum prey density may vary with the species, ontogeny, size of prey, and culture system (Lee and Ostrowski, 2001). The best way to determine optimal densities is to monitor both number of larval prey at intake and density of prey in the culture system to avoid under and over feeding (Palmer et al., 2007). Underfeeding retards larval growth and development; whereas, overfeeding can result in reduced capture success and can also lower water quality. In the case of *Artemia*, they may outgrow and hence compete with the culture species (Lee and Ostrowski, 2001). Lastly, because there is no food for zooplankton in intensive larval rearing systems, the nutritional quality of the zooplankton will decrease over time.

Prepared Larval and Fry Feeds

Prepared feeds for larvae are typically called microparticulate diets. Despite improvements in larval diet quality, only a few larval species can be reared exclusively on prepared feeds (Jones et al., 1993; Liao et al., 2001). Rather they are fed together with live foods, a practice called cofeeding. Microparticulate diets for larvae must meet the nutritional requirements of the species, be of a size appropriate for ingestion, and possess physical properties, such as buoyancy, texture, and color, that are appropriate for larvae. Nutritional components of prepared diets for larvae differ significantly from those in feeds for juvenile stages primarily in the need for phospholipids, essential fatty acids, and free amino acids or short peptides (Jones et al., 1997; Jones, 1998; Cahu and Infante, 2001). Limited information exists on the differences in quantitative nutrient requirements between larval fish and juveniles, except for the common understanding that larval fish have a higher metabolic rate and thus benefit from a higher availability of nutrients and energy in their diet (Dabrowski, 1986).

Optimum diet particle size increases in proportion to fish size and should not exceed 20% of the mouth opening (Dabrowski and Bardega, 1984). Frequent feeding is important when rearing larvae as well as fry or postlarval stages; food can be offered 10 to 24 times per day or almost continuously and/or in excess (Charlon and Bergot, 1984; Charlon et al., 1986). When utilizing prepared larval feed, the nutrient content and delivery of the feed is important, but of equal importance is the effect of prepared feeds on water quality. High surface area to volume ratio of microparticulate diets as well as the need to offer feed in excess of intake, increases leaching of nutrients into the water leading to deterioration of water quality and increased bacterial loading of the culture system (Jones et al., 1993) if not managed properly.

Feed formulations for larvae and fry[1] generally contain high-quality marine ingredients. These feeds are often used for fry of other species that are large enough at first feeding to ingest relatively large feed particles. Larval diet containing single-cell protein and freeze-dried animal tissues have proven successful with the stomachless larvae of common carp, grass carp, and silver carp (Dabrowski et al., 1984; Dabrowski and Poczyczynski, 1988). Microparticulate diets for marine fish and shrimp larvae have utilized micropulverized meals from fish, crab, squid, mussel, yeast, chicken egg yolk, short-necked clam, and krill (Kanazawa, 1986; Kanazawa et al., 1989; Kanazawa, 1990; Cahu and Infante, 2001) as well as semipurified ingredients such as casein and gelatin (Paibulkichakul et al., 1998). The actual formulation will also depend on the manufacturing methods used such as the amounts and types of binders included.

PRODUCTION DIETS AND FEED MANAGEMENT

Fish cultivation differs from terrestrial animal production in several key aspects, including the fact that feeding is controlled by the aquaculturist. There are relatively few aquaculture systems in which feed intake is completely controlled by the fish. Thus, management of feeding is a crucial aspect of fish culture success. It is important that fish are fed such that their potential for growth is realized, but not such that feed is wasted. Uneaten feed represents an economic loss, is difficult to recover and account for, and can contribute significantly to degradation of the rearing environment and to waste output of the operation. Rearing animals in aquatic environments greatly complicates monitoring of growth, standing biomass, and management of feed delivery and waste outputs compared to other livestock species (e.g., poultry and swine). Without accurate estimates of fish growth, and standing biomass in a rearing system, it is difficult to calculate feed

[1]A larva is an organism from the beginning of exogenous feeding to metamorphosis into juvenile. At the larval stage the animal differs greatly in appearance and behavior from a juvenile or an adult. A fry is a term used to describe a young fish at the postlarval stage. All stages from hatching to the fingerling stage can potentially be covered by "fry." Source: *www.aqualex.org*.

efficiency values amongst operations or between production cycles. Consequently the proper delivery of feed is critical from both economic and environmental points of view.

FEED UTILIZATION AND FISH GROWTH

Growth or the deposition of new tissue is a reflection of genetic potential, nutrient intake, and environmental conditions. The maximum potential for growth is genetically determined and can only be reached under suitable environmental conditions with appropriate nutrient intake. Growth is the net result of nutrients that are retained by the animal once undigested nutrients, metabolic wastes are processed, and maintenance and voluntary activity have been met. Biomass (live weight) gain in animals is the result of deposition of water, protein, fat, and minerals as well as a small amount of other compounds. The quantity of nutrients deposited per unit of live weight gain is not constant but changes with a variety of factors such as the weight of the animal, nutrient content of the feed and physiological state. Figure 13-1 shows the changes in the relative composition of rainbow trout of various sizes from various feeding trials. Similar results have been reported by Shearer (1994) and Lupatsch et al. (1998). The dry matter (i.e., protein + lipid + ash) content of fish increases quite dramatically as the fish grow. Consequently, one unit of biomass gain (e.g., 1 gram or 1 kilogram of biomass gain) means different amounts of dry matter deposited as the fish grow. Numerous other factors (nutrient composition of the feed, feeding level, physiological and environmental factors) can also affect the composition of fish. A change in biomass is, therefore, not of "absolute" or "constant" composition as gain may represent varying quantities of a nutrient either as the animal grows or as conditions change.

The most commonly used measure of animal biomass gain in response to feed utilization is feed conversion ratio (FCR). This is most precisely defined as dry weight feed intake per wet body weight gain. The reciprocal ratio, live body weight gain to dry feed intake, is termed feed efficiency (FE) and is often expressed as a percentage. At the farm level, a true FCR is typically not calculated. Instead an observed FCR or a pond conversion ratio (S, as described by Swingle [1968]) is often calculated based on the quantity of feed utilized on the farm and the quantity of product sold over a given time period (feed added per unit net fish produced). This is not a true measure of FCR but a measure of feed offered and harvested biomass at the population level. Such values represent the true FCR of surviving animals as well as contributions of natural food sources (resulting in a reduction in apparent FCR) as well as feed that was not consumed, animals lost during the production cycle (mortality, theft, escape) resulting in an apparent increase in FCR. Consequently, on farm observed FCR (feed in:product out) is often imprecisely defined and is in general more a measure of the overall efficiency of a given aquaculture operation than a measure of the nutritional quality of the feed.

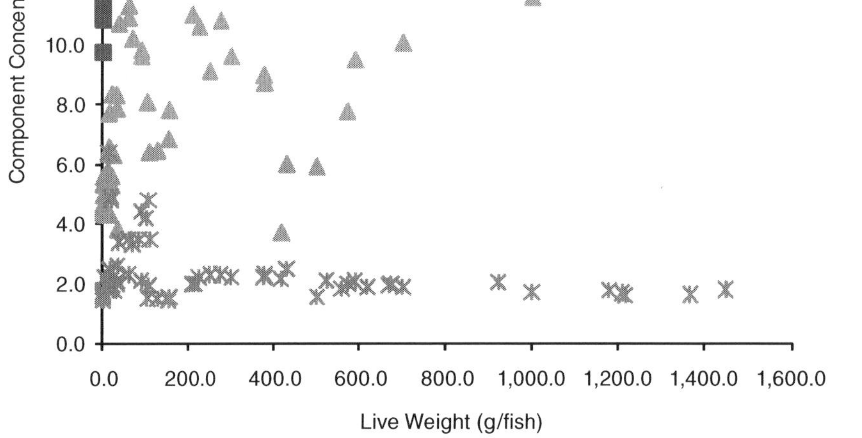

FIGURE 13-1 Proximate components of rainbow trout, *Oncorhynchus mykiss*, of increasing weight (data derived from the model described in Chapter 4).

A number of studies have been conducted to model the impact of feed inputs and various endogenous (genetic, sex) and exogenous (environmental, husbandry) factors on feed efficiency. The results and models derived from these studies have been widely presented and generally accepted. Yet numerous contradictions exist among these studies. For example, some studies suggest that maximum FE is achieved at feeding levels below that required for maximum growth (G_{max}) (Brett and Groves, 1979). Conversely, Talbot (1993) suggested that maximum FE is attained at maximum feed intake (R_{max}) and G_{max}. The differences between these studies are largely related to the fact that parameters used in these analyses and models (mass of feed offered, biomass gain, G_{max}, R_{max}, growth rate, FE, FCR, etc.) are all relative and/or semiquantitative parameters. Discussion of the limitations of a number of these parameters (growth rate, G_{max}, R_{max}) can be found in Chapter 4 and elsewhere (Cho, 1992; Azevedo et al., 1998; Cho and Bureau, 1998; Bureau and Hua, 2008).

The amount of feed required by a fish to achieve a certain amount of weight gain depends primarily on the nutrient density of the feed and feed intake, which in turn determines daily intake. Consequently, a larger quantity of a low nutrient dense feed (or a feed with low digestibility) will be required as compared to a higher nutrient dense feed (or feed with a higher digestibility) in order to achieve the same growth performance, assuming the feeds are nutritionally adequate and balanced. As a simple example, feeding 100% of the ration of diet with 30% digestible protein and 3,000 kcal digestible energy (DE)/kg will result in the same daily intake of digestible protein and energy as feeding 75% of the ration using a diet with 40% digestible protein and 4,000 kcal DE/kg (assuming the feeds are appropriately balanced). Clearly, as the digestibility or nutrient density of a diet is reduced, more feed is required to sustain the same biomass gain, resulting in an increased FCR or a reduced FE.

Part of the variability in feed composition is also related to the fact that aquaculture is a field in rapid evolution. Feeds used in trout and salmon production have evolved very significantly over just three decades. In the early 1980s, for example, commercial rainbow trout feeds used were relatively low in protein (e.g., 35–40%) and fat (e.g., 8–10%) but rich in starch (35–40%). Feed conversion ratios (FCR, feed/gain) of 2 to 2.5 were common for market size rainbow trout (approximately 1 kg live weight). Today, the use of lower carbohydrate, higher digestible energy density feeds (e.g., 34–55% crude protein, 18–32% fat, 10–20% starch) allows FCR of about 0.8 to 1.2 for market size rainbow trout.

Because both "feed" and "fish" are highly variable in chemical composition, FE or FCR can only be considered "semiquantitative" parameters and never "absolute" indicators of efficiency of conversion of dietary nutrients (inputs) into fish biomass. Caution should be exercised when comparing FE or FCR across studies, facilities, seasons, sampling intervals, etc. Describing and predicting FE:FCR is achievable with different feeds and as a function of fish size, environment, etc., but it is clearly not a simple task. Probably the best use of FCR values is for economic analyses at the farm level because it allows an estimation of the overall economic efficiency of the feed, feeding practices, and management practices as combined factors.

Population Estimates

Determining the size of a population and the total biomass in a culture system are critical components of proper feed management. In terrestrial systems, live animals as well as mortalities can be directly counted to determine the population. In aquatic production systems the animal is reared in a wide variety of environments and housing infrastructures such as tanks, raceways, ponds, or net pens. Under the best of conditions, animal inventory is rarely fully visible to the caretakers due to water turbidity, light reflection, and diffraction. Logistical issues such as the capture, handling and counting of free swimming animals limit the frequency and accuracy of monitoring mean animal weights and survival. Consequently, determining the number of animals, size distribution, and total biomass are problematic and one of the primary reasons for inefficient culture practices and feeding.

Population numbers are best managed in highly confined production systems (tanks, raceways, and cages) where the animals can be quantified on a regular basis and mortalities are easier to identify. The determination of the population in confined systems is typically done by physically moving the animals from one culture container to another or through nonintrusive measures such as encouraging them to pass by or through a counting device. Such devices can use fluctuations in water conductivity, optical imagery or hydroacoustic to determine fish numbers (Eatherley et al., 2005). The most common measure of fish biomass is to subsample the population; however, this can cause stress and damage to the fish and is labor intensive. Newer technologies including various forms of digital imagery (e.g., stereo photography) can be utilized to predict fish size (Ruff et al., 1995; Beddow et al., 1996; Van Rooij and Videler, 1996; Petrell et al., 1997; Costa et al., 2006) without removing the fish from the culture environment.

In pond production systems, determining the population becomes quite problematic as crowding or capturing fish by seining is very inefficient, and size grading is often poorly achieved or problematic. The population in a pond can be quantified, but it usually requires drain harvesting, which is seldom cost effective over relatively short time periods. Hence, the only two points in time when the animals are accurately quantified in pond systems is during stocking and when a pond is drain harvested.

In all production systems, accuracy of stocking and survival of the animals is critical to proper feed management. In production systems when the animal is not easily quantified (e.g., pond production systems) proper transportation and acclimation of animals during the stocking process is particularly critical. If the animals are improperly quantified at stocking or if mortality occurs after stocking and is not quantified, population estimates are inaccurate and feed management decisions are likely to overestimate requirements. In production systems that are drain harvested after one season, such as shrimp or single batch fish system, estimates of survival are often based on subsampling (e.g., cast netting for shrimp), direct observation of mortality, and historical results. When dealing with pond systems that are not drain harvested on a regular basis (e.g., multibatch catfish culture), inventory estimates are quite problematic. Inefficiencies in estimating population numbers, as well as the retention of fish larger than the targeted size (fish that escape top harvesting and grow beyond the minimal market size) lead to considerable inefficiencies of production systems. In general, the more frequently these systems are top harvested the more accurate the population estimates, and the shorter the period between drain harvesting the better the overall inventory control.

Feeding Programs

The key to optimum utilization of nutritionally complete diets is the proper delivery of the feed in adequate levels and over a suitable time period to meet the nutritional requirements of the species for a given growth rate. This may mean providing all the feed that an animal will consume (satiation) or providing a restricted ration (less than satiation). Satiation feeding typically refers to providing feed up to the point of fulfilling the desire to eat. Hence, one is providing all the feed the animals will consume, albeit within a distinct number of meals or time periods (e.g., all the feed that can be consumed within 15 minutes two times a day). Maintaining accurate records of feed inputs, growth, number of fish, biomass loading of the culture system, as well as environmental factors that may influence growth and feed consumption are essential to optimize feed utilization. Growth and survival are often difficult to forecast, making the determination of feeding level complex. Feed inputs are also dependent on the nutrient content and type of feed utilized. Characteristics to consider include the nutrient density of the diet, processing of the diet (grinding, forming, and top coating), size and type (floating/sinking) of pellet, number of feedings per day, as well as feed delivery mechanisms and quantity of feed to be delivered at each feeding. For each type of aquaculture system feeding practices differ; consequently, the concepts of feeding are far more important than any single protocol.

Pellet Characteristics

To minimize economic inefficiencies and pollution loading of the culture system the feed must be in a form that promotes quick and complete consumption. The size and form of the feed will depend on the cultured species, its size, culture conditions, and feed delivery. The particle size of the feed should progressively increase with the size of the animal. Typically feeds are selected that are 20–50% of the size of the fish's mouth, or, in the case of shrimp, a size easily handled by the shrimp (Obaldo and Tacon, 2001). Providing a feed that is too large will result in poor feed consumption and increased handling, while offering one that is too small makes it difficult for the fish to eat enough, promotes leaching of nutrients, and complicates capture of the feed by the culture species. Another consideration when selecting the size of the feed is that under culture conditions fish are not all the same size. If you switch to a larger size too soon, the larger fish will easily consume the feed while the smaller fish may struggle, leading to more variability in fish size. Consequently, under some conditions a mixture of feed sizes are used to transition fish to a larger particle size. To avoid stressful competition among fish, enough pellets should be delivered to allow all fish to feed at the same time.

The density of the feed plays a critical role in how it is offered because it determines if the feed will sink or float. Sinking feeds have sufficient density to sink immediately when placed in water and are primarily used in fish, shrimp, and early fry production. These feeds are used when the feed should pass through the water column and should not remain on the surface. Sinking feeds allows fish feeding in the water column or shrimp feeding on the bottom to easily obtain the pellets. Floating feeds are those that are designed to primarily remain on the surface for some period of time. They include both flake feeds and expanded extruded feeds. Extrusion is the process of choice for most production fish feeds as it offers the possibility to better control product density and to produce diets with high lipid content. Hence, most floating and slow sinking feeds are produced using extrusion technology. For crustacean species such as shrimp as well as numerous fish species, water-stable sinking feeds are produced by the pelleting process. Irrespective of the process to produce the feed, it must be in a form that is of suitable size and density.

Pellet durability is another critical issue because feed has to be stored and moved prior to its consumption. Most feeds are moved by bulk handling systems, transported in large containers or bags, and quite often fed using mechanical or pneumatic conveying systems. If the pellet breaks down prior to feeding, the fines are considered an economic loss because they may interfere with conveyor systems, they are not readily consumed by the culture species, and they contribute to the pollution loading of the culture system. A pellet that is too soft will break down during transport, and one that is too hard or contains too high a level of binder may not be consumed or result in other problems. Responses are probably species-specific: however, for rainbow trout, high levels of binders, pellet durability, and water stability can all negatively influence feed intake, evacuation rates, and digestibility (Hilton et al., 1981; Storebakken, 1985; Baeverfjord et al., 2006). Conversely, inadequate pellet stability has led to the release of oils into the stomach of trout resulting in regurgitation (Baeverfjord et al., 2006).

Feed Delivery Systems

Feed may be delivered to the culture species by various means. Such systems may be as simple as storing feed in a bag, weighing a ration, and simply dispersing it by hand or as complex as a computer-automated system using bulk bins, automatic conveying of the feed, and dispersion of the feed into the culture system combined with feed back to measure the fish's response. Such feeding systems are broadly categorized as (a) nondemand systems in which the ration size and time of delivery is predetermined (b) demand or self-feeding systems for which the stock activates or provides feedback for feed delivery. In both cases, similar equipment may be utilized, which includes feeding by hand or through the use of mechanical feeders using rotating belts or disks, vibration, or pneumatic forces to deliver the feed. A description of various feeding systems is presented by Lekang (2007).

Feeding for Semiintensive Systems

Extensive and semiintensive culture systems have been utilized for thousands of years. Such systems are used for subsistence farming as well as commercial production of fry, fingerling, and food fish with limited nutrient inputs. Organisms produced in these systems typically feed low on the food chain (e.g., Pacific white shrimp, carp, and tilapia). They are amenable to culture under a wide variety of conditions and are capable of using suboptimal feeds because nutrient intake is supplemented from natural food organisms that are found in the culture systems. In terms of standing crop, production is increased as one moves from natural foods enhanced with fertilization, to single ingredients like rice bran in combination with inorganic fertilization, to farm-made or incomplete feeds, and finally to commercially produced complete feeds (Tacon and De Silva, 1997; El-Sayed, 2006).

In the case of culture conditions and species that are capable of utilizing natural foods that are present in culture systems (e.g., pond culture of shrimp, tilapia, and carp) it is prudent to encourage the use of natural foods because this will reduce the investment for prepared feeds. Stable carbon isotope analysis has proven to be a useful tool to demonstrate the relative contribution of various food sources for species such as Pacific white shrimp (Anderson et al., 1987), golden shiner (Lochmann and Phillips, 1996), as well as the common carp and tilapia (Schroeder, 1983). In such cases natural productivity, contributed from 40–80% of carbon deposition in the culture species. This clearly indicates the potential contribution of natural food sources even under semiintensive and in some cases intensive conditions. In well-fertilized production systems, natural foods are capable of supporting a given level of biomass often referred to as critical standing crop (El-Sayed, 2006). Until the critical standing crop is reached, the use of prepared feeds does not increase production. Hence, eliminating or reducing feed inputs results in more efficient utilization of resources. For example in the semiintensive culture of tilapia, fingerling fish are often stocked into well-fertilized ponds in which natural foods are plentiful. Research conducted by Diana et al. (1996), demonstrated that at an initial stocking density of 3 fish/m^2, fish obtained sufficient natural food to grow rapidly until they were around 100 g, after which time prepared feeds had to be applied to maximize production. Delaying feeding in well-fertilized ponds does not reduce fish yield but does reduce feed costs provided appropriate stocking densities are used (El-Sayed, 2006). Feeding level is another important factor in semiintensive systems. Because natural foods are present, feeding to satiation is not likely to increase production but increases production costs. Diana et al. (1994) reported that reducing feed inputs in tilapia production ponds (3 fish/m^2) from satiation to 50% of satiation did not reduce production. This feeding level encouraged the fish to forage for natural foods and improved the observed conversion of feed inputs to harvest biomass. The consumption of natural food sources should be encouraged when possible because this will reduce feed related costs. However, the quantity and quality of natural foods are quite variable within a site, between sites and across time; hence, site-specific management practices have to be developed if the contribution of natural foods is to be optimized.

Feeding Intensive Production Systems

Except for nonprofit culture operations (state, federal, and tribal hatcheries) raising fish for stock rehabilitation, intensive production systems are used for commercial production and are thus driven by economic considerations. Intensive systems typically lack natural foods, and therefore nutritionally complete feeds are used. Economic returns from investments in feeds and feed management are primary driving forces for management decisions. However, government regulations, waste production, and consumer demands for specific products (e.g., pigmented fillets) also can greatly influence the selection of feeds, their nutrient and ingredient compositions, and production practices.

Proper delivery of feed is the most complex element of commercial aquaculture. Feeding levels can be calculated fairly precisely if the population size, mean weight, and the projected growth can be determined with sufficient accuracy. Biomass gain (i.e., weight gain) is determined by the growth potential of the animal and environmental factors, such as water temperature, photoperiod, salinity, weather conditions, etc. Consequently, biomass increase and time needed to reach market weight cannot always be accurately predicted based on previous performance and/or published recommendations or predictive models.

The same factors that hinder accurate assessments of

inventory also complicate feeding. Feed is often delivered to a fish population that is only partly visible or, in many cases, delivered based on the fish's response to feed inputs (e.g., by satiation feeding or the use of mechanical demand feeders) without an accurate knowledge of the biomass. Feed delivered but not consumed is wasted and difficult to recover and quantify except through the use of specialized equipment and systems. Overfeeding leads to economic loss and greater waste outputs. Feeding to near satiation or as much as 70% below satiation does not result in reduced efficiencies of nutrient retention (Azevedo et al., 1998). However, underfeeding limits the ability of the fish to express their growth potential, which leads to economic loss (opportunity cost) and may influence profitability.

When culture conditions allow the use of floating feed it provides a management tool for the visual observation of fish feeding behavior and feed consumption. The response of the fish to offered feed is a critical management tool because it gives an indication of water-quality conditions (low dissolved oxygen or high ammonia levels will often result in reduced feed intake), the health or disease status of the fish (sick fish often reduce feed intake or refuse to feed), as well as fish appetite (feeding response slows as satiation is approached). Satiation feeding is the process of offering feed until the fish no longer exhibit a strong feeding response. This is an easy way to provide the maximum ration that fish will consume on a given day and feed offering. This can be accomplished by visually observing feeding response or through the use of self-feeders (manual and automated feed delivery systems that rely on input from the fish). In the case of visual observation, the feed is offered over a fixed time period (e.g., 15–30 minutes), during which time the fish are allowed to consume as much feed as they will eat. Feeding activity should be aggressive during the entire period in which feed is available to the fish. As the fish consume feed, their feeding activity will slow until they become lethargic or disinterested in the feed and ultimately cease feeding.

Slow sinking feeds can also be fed to apparent satiation by direct observation in clear water systems or using a variety of feedback systems typically used in sea cages. Various devices have been developed for automatic observation and control of feed input for fish in sea cages employing video cameras, infrared photoelectric sensors, and sonar. In these systems feeding of fish in the cages starts with an automatic feeder delivering a certain quantity of feed at certain intervals. Feeding continues until feed pellets are observed (via video monitoring or sonar) passing through the schooling fish indicating fish are reaching satiation. In one system, feed reaching the bottom of the cage is collected and lifted to the surface by a pump for collection and recording and is possibly reused. Such feedback allows dynamic feeding according to appetite. Similarly, when self-feeding mechanisms are properly utilized, the fish will trigger the release of feed until they have consumed as much as they would like to eat.

Feeding fish with a floating feed (or using some other sort of feedback system) has both advantages and disadvantages. The observation of feeding response is an invaluable tool that provides insight into how well the fish are doing and allows daily adjustments of feed inputs based on actual demand. The use of floating feeds and feeding to satiation is a very popular method of feeding that typically results in good growth and reasonable feed utilization. This is a common method employed in production systems with inadequate inventory control, such as multibatch catfish production systems or mixed sex pond culture of tilapia. One advantage is that the fish will regulate their feed intake based on the current environmental conditions, size, and, to a certain extent, nutrient requirements. Hence, as long as environmental conditions are adequate, the population will consume an adequate amount of feed for good growth. Knowledge of the population in terms of biomass, fish size, and number is not required. Feeding to satiation typically results in maximum growth rates, but it is difficult to execute and can result in higher FCR values because in some species at certain lifestages and under certain conditions, nutrient (protein) deposition and associated biomass gain tend to level off at high feeding levels (Alanärä, 1992; Musiri and Lovell, 1993; Bureau et al., 2006). Furthermore, under commercial conditions it is easy to overfeed as the feeding response or intensity of feeding diminishes, resulting in uneaten feed being wasted. Feeding to satiation remains one of the most popular feeding strategies in fish culture because in many operations, growth rate and total production have a stronger impact of profitability of the enterprise than does efficiency of conversion of feed inputs.

In cases where inventory control is difficult (e.g., multibatch production systems) and one does not desire to feed to satiation, floating feeds can be utilized to estimate the level of feed inputs required to reach satiation. Feed inputs can be set at a level below satiation (e.g., 90%) for a period of time, after which feed input levels are reevaluated with another satiation feeding followed by another adjustment of feed inputs. Similarly, in situations where feed inputs are predetermined using feed tables or computer programs that use a variety of techniques to predict feed inputs, the use of floating feeds allows the observation of feeding response and the estimation of feed not consumed (visual estimates of feed remaining on the surface). The use of floating feed to visualize feeding is an important management tool in a number of feeding strategies and should be utilized when ever practical.

Quite often feed tables or computer-generated rates are utilized to estimate the required daily ration. As daily intake can vary because of a variety of factors such as the nutrient density and digestibility of the diet, environmental factors, genetic potential, and disease occurrences, calculated rations should only be considered a guide. The advantage of "predetermining" the ration to be fed is that nutrient intake is controlled and is typically set below satiation. This may not produce the fastest growth rate but will often produce the best FCR, albeit not always. As an example, Table 13-1

TABLE 13-1 General Daily Feeding Rates (% of Body Weight) and Frequency Guide for the Production of Channel Catfish, Common Carp, and Nile Tilapia Using a 32% Protein Floating Feed at 28°C in Single Batch Production Systems[a]

	Channel Catfish (*Ictalurus punctatus*)		Common Carp (*Cyprinus carpio*)		Nile Tilapia (*Oreochromis niloticus*)	
Fish Mean Weight (g)	Feed Rate (%)	Frequency (× per day)	Feed Rate (%)	Frequency (× per day)	Feed Rate (%)	Frequency (× per day)
25	4.0	2	4.5	3	4.5	3
50	3.5	2	4.0	3	3.7	3
75	3.2	2	3.6	3	3.4	3
100	3.0	2	3.3	3	6.2	3
150	2.8	1	3.1	2	3.0	2
200	2.5	1	3.0	2	2.8	2
250	2.2	1	2.6	2	2.5	2
300	2.0	1	2.4	2	2.3	2
400	1.6	1	2.1	2	2.0	2
500	1.4	1	1.7	2	1.7	2
600	1.2	1	1.4	2	1.4	2

[a]Schmittou et al. (1998).
Feed allowances should be adjusted for temperature shifts based on the following schedule.
At ≤ 15°C feed at 1% rate 1 × per day only 3 × per week.
At 16–19°C feed at 60% of calculated allowance 1 × per day every day.
At 20–24°C, feed at 80% of the calculated allowance 1 or 2 × per day every day.
At 25–29°C, feed at 100% of the calculated allowance and maximum frequency every day.
At 30–32°C, feed at 80% calculated allowance.
At ≥ 33°C, feed only what fish are observed to consume.

presents site-specific feed inputs for various sizes of catfish, tilapia, and carps offered a 28% protein floating feed and reared under intensive culture conditions with a temperature of around 28°C. The daily ration is expressed as a percentage of the body weight (% body weight × weight = daily ration) with the total ration divided into one or more feedings. Daily feed rate is decreased as the animal size increases and the need for multiple feedings is also reduced. As nutrient intake is directly proportional to growth rate, this is logical as the relative growth rate (percentage increase per day) of an animal is reduced as the animal gets larger. Table 13-2 is another example of a feed table in which daily ration is adjusted based on fish weight and temperature. In this table both daily feed ration and DE requirements are presented. In this case, for any given size of fish, daily feed and energy requirements increase with temperature as long as the animal is within the optimal temperature range for growth. Temperatures exceeding these levels will result in stress and reduction or cessation of growth and consequently feed intake. Table 13-3 is an example of feeding rates for a single batch of channel catfish grown from advanced fingerlings to marketable size in temperate outdoor ponds in the southern United States. The feed rate (percentage body weight) does not shift as much as one would expect based on the fish weight. This is because of the interaction of fish weight and temperature on feed intake and is typical for temperate pond systems. Feed requirement is consequently the results of genetic potential and health and physiological status of the

TABLE 13-2 Example of Daily Digestible Energy (DE) and Feed Requirement of Rainbow Trout (*Oncorhynchus mykiss*) of Different Sizes Calculated Using a Nutritional Energetic Model (see Chapter 4)

	5°C		10°C		15°C	
Live Weight (g/fish)	DE Requirement (kcal/fish per day)	Feed (g/100 g fish)	DE Requirement (kcal/fish per day)	Feed (g/100 g fish)	DE Requirement (kcal/fish per day)	Feed (g/100 g fish)
1	0.09	2.05	0.19	4.34	0.31	6.88
5	0.29	1.18	0.60	2.42	0.85	3.89
10	0.44	0.98	0.97	1.94	1.37	3.05
25	0.84	0.75	1.80	1.48	2.43	2.35
50	1.69	0.59	2.96	1.23	4.46	1.85
75	1.92	0.56	3.85	1.13	5.78	1.70
100	2.45	0.53	4.91	1.06	7.36	1.58
150	3.31	0.49	6.63	0.99	10.3	1.47
200	4.23	0.47	8.20	0.95	12.7	1.41
250	4.38	0.47	10.0	0.92	15.5	1.38
300	6.10	0.45	12.2	0.91	18.9	1.35
350	7.14	0.45	13.9	0.90	20.8	1.35
400	8.09	0.45	15.7	0.90	25.0	1.34
450	9.14	0.45	17.8	0.89	27.4	1.34
500	10.0	0.45	20.0	0.90	30.0	1.34
600	12.1	0.45	23.9	0.90	35.8	1.35
700	14.4	0.46	28.5	0.91	42.7	1.37
800	16.6	0.46	33.8	0.93	50.6	1.40
900	19.1	0.47	37.8	0.94	59.8	1.43
1,000	21.3	0.48	44.5	0.97	65.0	1.44

NOTE: Fish raised at different temperatures at a thermal-unit growth coefficient (growth rate) of 0.220 and fed a nutritionally adequate diet with 4,500 kcal DE/kg. See Chapter 4 for details on calculation of digestible energy requirement.

TABLE 13-3 Example of Feed Consumption Rates for Channel Catfish (*Ictalurus punctatus*) Reared in Outdoor Ponds from Advanced Fingerlings to Marketable Size[a]

Date	AM Water Temperature (°C)	Fish Weight (g)	Feeding Rate (% body weight)
May 1	18.9	50	2.1
May 15	22.2	62	3.4
June 1	21.1	82	2.9
June 15	27.2	111	3.2
July 1	27.2	143	2.7
July 15	27.8	176	2.4
August 1	27.8	233	1.8
August 15	27.2	285	2.0
September 1	25.0	335	1.5
September 15	25.0	382	1.3
October 1	18.9	462	1.1

[a]Adapted from Robinson et al. (2001).

animal, environmental conditions, husbandry practices, and a plethora of other factors.

In Table 13-1, the number of feedings is reduced as the fish gets larger. The need for multiple feedings is often driven by the physical capacity of the stomach to hold feed and nutrient demands of growth. From a practical standpoint, the number of feedings utilized is tempered by nutrient loading (e.g., more small feedings are utilized in closed systems to spread out nutrient loading over time) as well as feed-related labor costs and the expectation of improved growth rates. Feeding frequency varies with size or life stage from up to 24 times per day for newly hatched larvae and fry to 3–4 times per day for fingerlings and 1–3 times per day for growout production.

In terms of growout of juveniles to marketable size, feeding frequency or time of fed application seem to have limited effect on food size of catfish when reared in ponds, cages, and raceways (Li and Lovell, 1992; Webster et al., 1992; Jarboe and Grant, 1996; Wu et al., 2004). Other species such as trout and salmon respond better to an increased number of feedings, for example, 2–3 feedings per day (Thomassen and Fjaera, 1996; Ruohonen et al., 1998). Multiple daily feedings are more likely to improve the growth rates of species with small stomachs and continuous foraging behavior (e.g., common carp, tilapia, and shrimp). For example, during the growout of shrimp, increasing the number of feedings will result in improved production (Robertson et al., 1993; Carvalho and Nunes, 2006). A rather large improvement is made from 1 to 2 feedings per day, but improvements are reduced as the number of feedings increases and are typically minor after 4 feedings per day. Given the labor and fuel costs associated with feeding shrimp reared in outdoor ponds, commercial producers typically feed 2 times a day during the pond growout phase because increased feeding frequencies are often not cost effective.

In outdoor pond production systems, labor and fuel costs often play a role in determining commercial feeding practices, often limiting the number of feedings to 1–2 times per day. Feed can be broadcast by hand, but in most cases feed is distributed by pneumatic mechanical feeders, which blow the feed onto the water surface. Feed should be distributed over a relatively large area to allow equal access of the fish to the feed. In cases where the fish are concentrated in a relatively small area such as a cage or tank, labor and fuel costs are often reduced, and there is an opportunity to utilize self-feeding and automated systems. In such cases, feed inputs can be regulated either through the use of self-feeding systems (demand feeders), automated feedback systems, preprogrammed feeding regimens, or by hand feeding using visual observations of the animals. Irrespective of the method and number of feedings, observation of the response to feed is a critical component of all feed management strategies.

Predicting Growth, Nutrient Input, and Feed Utilization

Bioenergetics approaches have been effectively used to predict growth, feed ration, FCR, and waste outputs of fish fed diets of varying nutritional composition and reared under different environments (Cho et al., 1991, 1994; Cho, 1992; Cho and Bureau, 1998; Kaushik, 1998; Lupatsch and Kissil, 1998, 2005; Lupatsch et al., 1998; 2001; Cui and Xie, 1999; Bureau et al., 2002, 2003; Azevedo et al., 2005; Papatryphon et al., 2005; Zhou et al., 2005; Bureau and Hua, 2008). Fish growing at different rates will also retain nutrients at different rates; thus, they will have different energy and feed requirements. Consequently, the daily energy requirement should be calculated based on a given level of performance (e.g., expected or achievable level of growth), feed composition (nutrient density), and life stage. This can be done using factorial approaches (Cho and Bureau, 1998; Lupatsch et al., 1998; Booth et al., 2010; Pirozzi et al., 2010.) that divides the organism's energy and/or nutrient requirements into its different components or fractions.

This concept have been most recently incorporated into the "Fish-PrFEQ" model (Cho and Bureau, 1998), which predicts energy requirement of fish based on expected level of performance, dietary DE content, and expected body energy deposition. Further development, or various adaptations, of the Fish-PrFEQ model have been proposed by Kaushik (1998), Bureau et al. (2002, 2003), Papatryphon et al. (2005), and Zhou et al. (2005). The main premise is that "animals will seek to eat a sufficient amount of an appropriately balanced diet to achieve their target or preferred performance unless limited by constraints or overridden by an externally managed intervention" (Oldham et al., 1997). The Fish-PrFEQ model estimates a DE requirement to achieve certain "desired" performance (determined by fish genetic potential and prevailing conditions). The calculated DE requirement is translated into a feed requirement by considering the DE content (kcal/g feed) of the feed being fed.

Bioenergetics models are very useful and practical because they allow modeling of feed requirement and prediction of feed efficiency simply on the basis of relatively straightforward estimations of the DE requirement of the fish, rather than through a detailed analysis of nutrient deposition and nutritional composition of the feed. Comparison of feeds can also be done simply on the basis of their DE content. The suitability of comparing feeds on the basis of their DE content has been demonstrated at numerous occasions, although a number of studies have shown that comparisons need to be made carefully, notably when dealing with feeds with high digestible carbohydrate content (Bureau et al., 1998) or comparing different fish species or fish at different life stages (Azevedo et al., 2004a,b).

The user of bioenergetics models should always remember that bioenergetics is a "methodology" or "system." Animals do not metabolize "energy" per se, but metabolize "nutrients." The amount of energy effectively "retained" or "harnessed" during life processes (efficiency of energy utilization) is governed by which and/or how these nutrients are utilized. More importantly, most nutrients have very specific roles, and many nutrients cannot be substituted by one another. Studies indicate that utilization of energy-yielding nutrients is highly dependent on the type of nutrients (not simply their "energy" value), the balance between these nutrients, and the genetic and physiological state of the fish (Azevedo et al., 2004a,b; Encarnação et al., 2006). Considering nutrient inputs simply on the basis of their "free energy" content is, therefore, largely irrational.

Arguably the most significant limitation of bioenergetics models is that they are based on "hierarchy of energy allocation." Growth is the surplus of energy after all other components of the energy budget have been covered or satisfied (Kitchell et al., 1977). Bioenergetics models consequently predict that when body energy gain (recovered energy [RE] according to the NRC [1981] nomenclature) is nil (RE = 0), growth (live weight gain) and feed efficiency (gain/feed) should also be nil. Bioenergetics models consequently assume that the relationship between "energy deposition" and growth (biomass gain) is absolute, an assumption that has been shown to be inaccurate. The RE reflects the "weighed average" of the deposition of energy-yielding nutrients (mostly protein and lipids) in the body of the animal. Volumes of evidence indicate that animals (pig, chicken, fish) fed a ration allowing an RE = 0 can still deposit body protein (positive nitrogen balance and associated energy gain) and gain weight, while mobilizing body lipids. Live weight gain is consequently driven by protein deposition because there are 3 to 6 grams of water associated with each gram of tissue protein deposited (Cho and Kaushik, 1990; Bureau et al., 2002). Lipid gain results in no or insignificant live weight gain because, in fish, lipids are stored in tissues by substituting water (Shearer, 1994).

The current bioenergetic models are very useful and have been utilized with a wide range of fish and crustacean species. However, they are not robust enough to describe adequately and predict efficiency of conversion of dietary inputs into biomass under a wide variety of conditions (Bureau and Hua, 2008). Hence, there is a need for more advanced models that describe metabolic uses of specific nutrients as opposed to simple aggregates of nutrient in the form of energy. To meet this need there is a trend to move toward nutrient-flow models. In these models, nutrient transformations are represented at the whole animal level based on biological principles. Such models will be more robust toward nutrient utilization and allow the evaluation of the relationships between nutrition, genetics, health, animal management, growth performance, and nutrient excretion to be established.

Irrespective of the mechanism of the model, the use of biological models to predict current and future growth, survival, and feed inputs are excellent tools to improve the management efficiency and profitability of production systems. Quite often outputs from biological models lend themselves to merging with financial (cash) and economic (cash and noncash resource) analyses to facilitate future production and or production strategies. For example, Cacho et al. (1990) utilized a combined model to evaluate the effects of dietary protein and ration size on cost structures for channel catfish production. If suitable models can be built that utilize past inputs and predict future outputs for the biological response of the animal (growth, size distribution, feed utilization, waste outputs, and survival), the culture system (e.g., level of supplemental aeration) as well as variable and fixed costs, then a more rational decision can be made with regard to nutrient content of the feed, feed inputs, as well as other management and marketing strategies.

POLLUTION LOADING AND WASTE MANAGEMENT

Feeds constitute the major pollutant source for aquaculture operations, either directly as uneaten feed or indirectly as biogenic waste (Cho, 1991; Cho et al., 1991, 1994; Cho and Bureau, 2001; Gatlin and Hardy, 2002; Bureau and Hua, 2008). These pollutants include uneaten feed, solid waste, and dissolved waste products, which contribute to nutrient loading of the culture system and the environment if water is discharged. Fecal and metabolic wastes produced by fish can, in general, only be partially recovered at the farm. Nutrient loading is particularly problematic in cage culture operations because recovery of fecal and metabolic wastes that disperse rapidly in the environment is extremely difficult. The main concern is the release of nitrogen (N), phosphorus (P), and solid organic matter wastes. Nitrogen and P are limiting nutrients for algae growth in marine and freshwater ecosystems, respectively. The effects of these pollutants are immediate and broad, causing both environmental concern and possible economic inefficiencies in the production systems due to increased costs associated with the biological and chemical processing of the waste. Impacts of nutrient loading can be mediated by minimizing the generation of the waste products, intercepting and removing pollutants from the water or through biological processing. The most direct method of reducing waste products is achieved by manipulation of the nutrient density and digestibility of the diet, resulting in a low pollution or nutrient dense feed in combination with proper feeding strategies (Cho and Bureau, 2001).

Reducing waste outputs of aquaculture operations is considered a key element for the long-term sustainability of aquaculture in many parts of the world. It is difficult for fish culture operations and governmental regulators to predict, or set goals for reducing, environmental impacts without first having access to objective estimates of the amount of waste associated with production (actual or planned). Directly monitoring and estimating waste outputs from effluent of aquaculture facilities is an inaccurate and costly process (Cho et al., 1991, 1994). It is also extremely difficult for certain types of facilities, such as cage culture operations (Reid, 2004). Because aquaculture wastes are ultimately from biological and dietary origins, the use of nutrient mass balance offers a simple and economical alternative to chemical methods of estimating waste outputs. Using a nutritional mass balance approach, Cho et al. (1991, 1994) demonstrated that estimation of waste outputs of fish culture operations could be accurately and economically made, with great flexibility, based on feed inputs and feed components utilization by the fish. Nutrient mass balance models have been developed and proven to be very useful in practical fish culture operations to predict body weight gain, feed requirements, and solid N and P waste outputs (Cho et al., 1991, 1994; Bureau et al., 2003; Papatryphon et al., 2005) as well as waste transport (Dudley et al., 2000).

Solid waste outputs from animals consist primarily of undigested carbohydrates and minerals. Protein and lipids are typically low because the digestibility of these nutrients is generally quite good. Solids carry 7–33% of the total N and 30–84% of the total P in the wastewater and should be removed as soon as possible (Cripps and Bergheim, 2000) if the culture technology permits it. Organic or solid wastes (fecal material and wasted feed) settling to the sediment can have an impact on the benthic ecosystem. Such impacts are of concern to regulators and have resulted in considerable efforts in developing models to predict waste outputs (Cho and Bureau, 2001; Islam, 2005).

Containment and collection of wastes is very difficult and costly; hence, minimization of waste outputs is a critical component of waste management (Cripps and Bergheim, 2000). Key concepts for reducing the impact of waste products from aquaculture feeds include selecting highly digestible ingredients, reducing low and indigestible components of the feed, and minimizing P levels of the diet. Reducing solid waste outputs can be obtained by increasing the digestibility of the ingredients and reducing the carbohydrate content, which in turn means increasing the protein and energy content of the diet, resulting in a more nutrient dense diet. The release of N waste products can also be improved by minimizing the catabolism of proteins and promoting the deposition or retention of amino acids. To do this the diet has to be formulated properly with regard to amino acid balance as well as digestible energy to protein ratio. Imbalanced amino acid profiles, excessive levels of amino acids, or low levels of nonprotein energy will typically increase the catabolism of amino acids and the release of nitrogenous waste products (Cho and Bureau, 2001). Similar concepts can be applied to the release of P into the environment.

Using the previously mentioned concepts, the trout industry has been able to make impressive reduction in waste outputs. Gatlin and Hardy (2002) estimated that today's trout diet formulations result in about 70% less urinary and 50% less fecal loss of P as compared to what occurred in the 1980s. Similar reductions in N excretion and organic waste production have also been achieved (Bureau and Hua, 2010). The improvements in nutrient retention and waste production can only be achieved if nutrient requirements are well established and digestibility coefficients for ingredients determined. Hence, it is critical that we precisely determine nutrient requirement and availability values for primary culture species.

CONCLUSIONS

The selection of the proper feed in combination with appropriate feed management strategies are critical keys to obtaining efficient aquaculture production. Matching the potential growth of the culture species with proper management of the feed ration and nutrient density of the feed to provide the proper daily delivery of nutrients is one of the most difficult tasks to effectively achieve. Yet it is often the one task most likely to be misunderstood and/or overlooked in terms of importance. Small errors in overfeeding, or delivering feed that the animal cannot consume, leads to direct loss of profits and increased nutrient loading of the culture system. Feeding fish that exceed minimal market sizes, because harvest inefficiencies, often leads to poor conversion of feed to final product because of both inefficiencies of growing larger fish as well as increased likelihood that part of the population will be lost due to disease or some other problem. Keeping the best records of feed inputs, population numbers, and growth rates are critical control points in feed management. In addition, feed management must also be tied to nutrient loading of the culture system, and consideration must be given to oxygen consumption after feeding as well as to the level of waste products directly and indirectly released by the animals.

REFERENCES

Alanärä, A. 1992. The effect of time-restricted demand feeding on feeding activity, growth and feed conversion in rainbow trout (*Oncorhynchus mykiss*). Aquaculture 108:357–368.

Anderson, R. K., P. L. Parker, and A. Lawrence. 1987. A $^{13}C/^{12}C$ tracer study of the utilization of presented feed by a commercially important shrimp *Penaeus vannamei* in a pond growout system. J. World Aquacult. Soc. 18:148–155.

Azevedo, P. A., C. Y. Cho, S. Leeson, and D. P. Bureau. 1998. Effects of feeding level and water temperature on growth, nutrient and energy utilization and waste outputs of rainbow trout (*Oncorhynchus mykiss*). Aquat. Living Resour. 11:227–238.

Azevedo, P. A., S. Leeson, C. Y. Cho, and D. B. Bureau. 2004a. Growth, nitrogen, and energy utilization by four juvenile salmonid species: Diet, species, and size effects. Aquaculture 234:393–314.

Azevedo, P. A., S. Leeson, C. Y. Cho, and D. B. Bureau. 2004b. Growth and feed utilization of large size rainbow trout and Atlantic salmon reared in freshwater: diet and species effects, and responses over time. Aquacult. Nutr. 10:401–411.

Azevedo, P. A., J. van Milgen, S. Leeson, and D. P. Bureau. 2005. Comparing efficiency of metabolizable energy utilization by rainbow trout (*Oncorhynchus mykiss*) and Atlantic salmon (*Salmo salar*) using factorial and multivariate approaches. J. Anim. Sci. 83:842–851.

Baeverfjord, G., S. Refstie, P. Krogedal, and T. Åsgård. 2006. Low feed pellet water stability and fluctuating water salinity cause separation and accumulation of dietary oil in the stomach of rainbow trout (*Oncorhynchus mykiss*). Aquaculture 261:1335–1345.

Beddow, T. A., L. G. Ross, and J. A. Marchant. 1996. Predicting salmon biomass remotely using a digital stereo-imaging technique. Aquaculture 146:189–203.

Booth, M. A., G. L. Allan, and I. Pirozzi. 2010. Estimation of digestible protein and energy requirements of yellowtail kingfish *Seriola lalandi* using a factorial approach, *Aquaculture* in press, doi: 10.1016/j.aquaculture.2010.07.019.

Brett, J. R., and T. D. D. Groves. 1979. Physiological energetics. Pp. 279–352 in Fish Physiology, Vol. 8, W. S. Hoar, D. J. Randall, and J. R. Brett, eds. New York: Academic Press.

Bureau, D. P., and K. Hua. 2008. Models of nutrient utilization by fish and potential applications for fish culture operations. Pp. 442–461 in Mathematical Modelling in Animal Nutrition, J. France, and E. Kebreab, eds. Wallingford, UK: CAB International.

Bureau, D. P., and K. Hua. 2010. Towards effective nutritional management of waste outputs in aquaculture, with particular reference to salmonid aquaculture operations. Aquac. Res. 41:777–792.

Bureau, D. P., J. B. Kirkland, and C. Y. Cho. 1998. The partitioning of energy from digestible carbohydrate by rainbow trout (*Oncorhynchus mykiss*). Pp. 163–166 in Energy Metabolism of Farm Animals, K. J. McCracken, E. F. Unsworth, and A. R. G. Wylie, eds. Wallingford, UK: CAB International Press.

Bureau, D. P., S. J. Kaushik, and C. Y. Cho. 2002. Bioenergetics. Pp. 1–53 in Fish Nutrition, J. E. Halver and R. W. Hardy, eds. San Diego, CA: Academic Press.

Bureau, D. P., S. Gunther, and C. Y. Cho. 2003. Chemical composition and preliminary theoretical estimates of waste outputs of rainbow trout reared on commercial cage culture operations in Ontario. N. Am. J. Aquacult. 65:33–38.

Bureau, D. P., K. Hua, and C. Y. Cho. 2006. Effect of feeding level on growth and nutrient deposition in rainbow trout (*Oncorhynchus mykiss* Walbaum) growing from 150 to 600 g. Aquac. Res. 37:1090–1098.

Cacho, O. J., U. Hatch, and H. Kinnucan. 1990. Bioeconomic analysis of fish growth: Effects of dietary protein and ration size. Aquaculture 88:223–238.

Cahu, C. L., and J. L. Z. Infante. 2001. Substitution of live food by formulated diets in marine fish larvae. Aquaculture 200:161–180.

Carvalho, E. A., and A. J. P. Nunes. 2006. Effects of feeding frequency on feed leaching loss and grow-out patterns of the white shrimp *Litopenaeus vannamei* fed under a diurnal feeding regime in pond enclosures. Aquaculture. 252:494–502.

Charlon, N., and P. Bergot. 1984. Rearing system for feeding fish larvae on dry diets. Trial with carp (*Cyprinus carpio* L.) larvae. Aquaculture 41:1–9.

Charlon, N., H. Durante, A. M. Escaffre, and P. Bergot. 1986. Alimentation artificielle des larves de carpe (*Cyprinus carpio*). Aquaculture 54:83–88.

Cho, C. Y. 1991. Prediction model for energy requirement of salmonid fishes. Pp. 446–449 in Proceedings of the 12th Symposium of Energy Metabolism of Farm Animals, EAAP Publ. No. 58, C. Wenk and M. Boessinger, eds. Zurich, Switzerland: Kartause Ittingen.

Cho, C. Y. 1992. Feeding systems for rainbow trout and other salmonids with reference to current estimates of energy and protein requirements. Aquaculture 100:107–123.

Cho, C. Y., and D. P. Bureau. 1998. Development of bioenergetic models and the Fish-PrFEQ software to estimate production, feeding ration and waste output in aquaculture. Aquat. Living Resour. 11:199–210.

Cho, C. Y., and D. P. Bureau. 2001. A review of diet formulation strategies and feeding systems to reduce excretory and feed wastes in aquaculture. Aquac. Res. 32:349–360.

Cho, C. Y., and S. J. Kaushik.1990. Nutritional energetics in fish: Energy and protein utilization in rainbow trout (*Salmo gairdneri*). World Rev. Nutr. Diet. 61:132–172.

Cho, C. Y., J. D. Hynes, K. R. Wood, and H. K. Yoshida. 1991. Quantification of fish culture wastes by biological (nutritional), and chemical (luminological) methods; the development of high nutrient dense (HND) diets. Pp. 37–50 in Nutritional Strategies and Aquaculture Waste, Proceedings of the First International Symposium on Nutritional Strategies in Management of Aquaculture Waste, C. B. Cowey, and C. Y. Cho, eds. Guelph, ON: University of Guelph.

Cho, C. Y., J. D. Hynes, K. R. Wood, and H. K. Yoshida. 1994. Development of high-nutrient-dense, low-pollution diets and prediction of aquaculture wastes using biological approaches. Aquaculture 124:293–305.

Costa, C., A. Loy, S. Cataudella, D. Davis, and M. Scardi. 2006. Extracting fish size using dual underwater cameras. Aquacult. Eng. 35:218–227.

Couteau, P. 1996. Micro-Algae. Pp. 7–47 in Manual on the Production and Use of Live Food for Aquaculture, FAO Fisheries Technical Paper 361, P. Lavens and P. Sorgeloos, eds. Rome, Italy: FAO.

Cripps, S. J., and A. Bergheim. 2000. Solids management and removal for intensive land-based aquaculture production systems. Aquacult. Eng. 22:33–56

Cui, Y., and S. Xie. 1999. Modelling growth in fish. Pp. 413–434 in Feeding Systems and Feed Evaluation Models, M. K. Theodorou, and J. France, eds. Wallingford, UK: CAB International.

Dabrowski, K. 1986. Ontogenetical aspects of nutritional requirements in fish. Comp. Biochem. Phys. A 85A:639–655.

Dabrowski, K., and R. Bardega. 1984. Mouth-size and recommendation of feed size preferences in three cyprinid fish. Aquaculture 40:41–46.

Dabrowski, K., and P. Poczyczynski. 1988. Mass rearing of coregonid larvae and juveniles on dry diet. Aquaculture 69:307–316.

Dabrowski, K., P. Bergot, and S. Kaushik. 1984. Rearing of coregonid larvae using dry and live food. Preliminary data. Aquaculture 41:11–20.

Diana, J. S., C. K. Lin, and K. Jaiyam. 1994. Supplemental feeding of tilapia in fertilized ponds. J. World Aquacult. Soc. 25:497–506.

Diana, J. S., C. K. Lin, and Y. Yi. 1996. Timing of supplemental feeding for tilapia production. J. World Aquacult. Soc. 27:410–419.

Dudley, R. W., V. G. Panchang, and C. R. Newell. 2000. Application of a comprehensive modeling strategy for the management of net-pen aquaculture waste transport. Aquaculture 187:319–349.

Eatherley, D. M. R., J. L. Thorley, A. B. Stephen, I. Simpson, J. C. MacLean, and A. F. Youngson. 2005. Trends in Atlantic salmon: The Role of Automatic Fish Counter Data in Their Recording. Scottish Natural Heritage Commissioned Report No. 100 (ROAME No. F01NB02). Edinburgh, Scotland: Scottish Natural Heritage.

El-Sayed, A-F. M. 2006. Tilapia Culture. Cambridge, MA: GABI Publishing.

Encarnação, P., C. F. M. de Lange, and D. P. Bureau. 2006. Diet energy source affect lysine utilization for protein deposition in rainbow trout (*Oncorhynchus mykiss*). Aquaculture 261:1371–1381.

Gatesoupe, F. J., and C. Le Millinaire. 1985. Adaptation de la qualité alimentaire des filtreurs-proies aux besoins nutritifs des larves de poissons marins. Coll. Franco-Japenais d'Océanogr. Marseille. 8:51–63.

Gatlin, D. M., III, and R. W. Hardy. 2002. Manipulations of diets and feeding to reduce losses of nutrient in intensive aquaculture. Pp. 151–165 in Aquaculture and the Environment in the United States, J. Tomasso, ed. Baton Rouge, LA: U.S. Aquaculture Society.

Hilton, J. W., C. Y. Cho, and S. J. Slinger. 1981. Effect of extrusion processing and steam pelleting diets on pellet durability, pellet water absorption, and physiological response of rainbow trout (*Salmo gairdneri* R.). Aquaculture 25:185–194.

Islam, S. 2005. Nitrogen and phosphorus budget in coastal and marine cage aquaculture and impacts of effluent loading on ecosystem: Review and analysis towards model development. Mar. Pollut. Bull. 50:48–62.

Jarboe, H. H., and W. J. Grant. 1996. Effects of feeding time and frequency on growth of channel catfish *Ictalurus punctatus* in closed recirculating raceway systems. J. World Aquacult. Soc. 27:235–239.

Jones, D. A. 1998. Crustacean larval microparticulate diets. Rev. Fish. Sci. 6:41–54.

Jones, D. A., M. S. Kamarudin, and L. L. Vay. 1993. The potential for replacement of live feeds in larval culture. J. World Aquacult. Soc. 24:199–210.

Jones, D. A., A. B. Yule, and D. L. Holland. 1997. Larval nutrition. Pp. 353–389 in Crustacean Nutrition, D. Akiyama, D. Conklin, and L. R. D'Abramo, eds. Baton Rouge, LA: World Aquaculture Society.

Kanazawa, A.1986. New developments in fish nutrition. Pp. 9–14 in The First Asian Fisheries Forum, J. L. Maclean, L. B. Dizon, and L. V. Hsillos, eds. Manila, Philippines: Asian Fisheries Society.

Kanazawa, A. 1990. Microparticluate feeds for Penaeid larvae. Adv. Trop. Aquacult. 9:395–404.

Kanazawa, A., S. Koshio, and S. Teshima. 1989. Growth and survival of larval red sea bream Pagrus major and Japanese flounder *Paralichthys divaceus* fed microbound diets. J. World Aquacult. Soc. 20:31–37.

Kaushik, S. J. 1998. Nutritional bioenergetics and estimation of waste production in nonsalmonids. Aquat. Living Resour. 11:211–217.

Kitchell, J. F., D. J. Stewart, and D. Weininger. 1977. Applications of a bioenergetics model to yellow perch (*Perca flavescens*) and walleye (*Stizostedion vitreum vitreum*). J. Fish. Res. Board Can. 34:1922–1935.

Lee, C. S., and A. C. Ostrowski. 2001. Current status of marine finfish larviculture in the United States. Aquaculture 200:89–109.

Lekang, O. I. 2007. Aquaculture Engineering. Oxford, UK: Blackwell Publishing Ltd.

Li, M., and R. T. Lovell. 1992. Comparison of satiate feeding and restricted feeding of channel catfish with various concentrations of dietary protein in production ponds. Aquaculture 103:165–175.

Lochmann, R., and H. Phillips. 1996. Stable isotopic evaluation of the relative assimilation of natural and artificial foods by golden shiner *Notemigonus crysoleucas* in ponds. J. World Aquacult. Soc. 27:168–177.

Liao, I. C., H. M. Su, and E. Y. Chang. 2001. Techniques in finfish larviculture in Taiwan. Aquaculture 200:1–31.

Lupatsch, I., and G. W. Kissil. 1998. Predicting aquaculture waste from gilthead seabream (*Sparus aurata*) culture using a nutritional approach. Aquat. Living Resour. 11:265–268.

Lupatsch, I., and G. W. Kissil. 2005. Feed formulations based on energy and protein demands in white grouper (*Epinephelus aeneus*). Aquaculture 248:83–95.

Lupatsch, I., G. W. Kissil, D. Sklan, and E. Pfeffer. 1998. Energy and protein requirements for maintenance and growth in gilthead seabream (*Sparus aurata* L.). Aquacult. Nutr. 4:165–173.

Lupatsch, I., G. W. Kissil and D. Sklan 2001. Optimization of feeding regimes for European seabass *Dicentrarchus labrax*: A factorial approach. Aquaculture 202:289–302.

Musiri, P., and R. T. Lovell. 1993. Comparison of satiate and restricted feeding of channel catfish with diets of varying protein quality in production ponds. J. World Aquacult. Soc. 24:459–465.

NRC (National Research Council). 1981. Nutrient Requirements of Fish. Washington, DC: National Academy Press.

Obaldo, L. G., and A. G. J. Tacon. 2001. Manufacturing different diet sizes and its effect on pellet water stability and growth of three size classes of Pacific white shrimp, *Litopenaeus vannamei*. J. Appl. Aquacult. 11:57–66.

Oldham, J. D., G. Emmans, and I. Kyriazakis. 1997. Limits and limitations to nitrogen use in farm animals. P. Nutr. Soc. 56:535–545.

Palmer, P. J., M. J. Burke, C. J. Palmer, and J. B. Burke. 2007. Developments in controlled green-water larval culture technologies for estuarine fishes in Queensland, Australia and elsewhere. Aquaculture 272:1–21.

Paibulkichakul, C., S. Piyatiratitivorakul, P. Kittakoop, V. Viyakarn, A. W. Fast, and P. Menasveta. 1998. Optimal dietary levels of lecithin and cholesterol for black tiger prawn *Penaeus monodon* larvae and postlarvae. Aquaculture 167:273–281.

Papatryphon, E., J. Petit, H. M. G. Van Der Werf, K. J. Sadasivam, and K. Claver. 2005. Nutrient-balance modeling as a tool for environmental management in aquaculture: The case of trout farming in France. Environ. Manage. 35:161–174.

Petrell, R. J., X. Shi, R. K. Ward, A. Naiberg, and C. R. Savage 1997. Determining fish size and swimming speed in cages and tanks using simple video techniques. Aquacult. Eng. 16:63–84.

Pirozzi, I., M. A. Booth, and G. L. Allan. 2010. A factorial approach to deriving dietary specifications and daily feed intake for mulloway, *Argyrosomus japonicus*, based on the requirements for digestible protein and energy. Aquaculture 302:235–242. doi: 10.1016/j.aquaculture.2010.02.032.

Rainuzzo, J. R., K. I. Reitan, and Y. Olsen. 1997. The significance of lipids at early stages of marine fish: A review. Aquaculture 155:103–115.

Reid, G. K. 2004. Evaluation of Nutrient Loading Dynamics from Tank and Cage Cultured Rainbow Trout. Ph.D. Dissertation, University of Guelph, Guelph, Canada.

Reitan, K. I., J. R. Rainuzzo, G. Oie, and Y. Olsen. 1997. A review of the nutritional effects of algae in marine fish larvae. Aquaculture 1500:207–221.

Robertson, L., A. L. Lawrence, and F. L. Castille. 1993. Effect of feeding frequency and feeding time on growth of *Penaeus vannamei* (Boone). Aquac. Res. 24:1–6.

Robinson, E. H., M. H. Li, and B. B. Manning. 2001. A Practical Guide to Nutrition, Feed, and Feeding, 2nd Revision. Mississippi Agricultural and Forestry Experiment Station Bulletin No. 1113. Mississippi State: Mississippi Agricultural and Forestry Experiment Station.

Ruff, B. P., J. A. Marchant, and A. R. Frost. 1995. Fish sizing and monitoring using a stereo image analysis system applied to fish farming. Aquacult. Eng. 14:155–173.

Ruohonen, K., J. Vielma, and D. J. Grove. 1998. Effects of feeding frequency on growth and food utilisation of rainbow trout *Oncorhynchus mykiss* fed low-fat herring or dry pellets. Aquaculture 165:111–121.

Sargent, J., L. McEvoy, A. Estevez, G. Bell, M. Bell, J. Henderson, and D. Tocher. 1999. Lipid nutrition of marine fish during early development: Current status and future directions. Aquaculture 179:217–229.

Schmittou, R. H., Z. Jian, and M. C. Cremer. 1998. Principles and Practices of 80:20 Pond Fish Farming. St. Louis, MO: American Soybean Association.

Schroeder, G. L. 1983. Sources of fish and prawn growth in polyculture ponds as indicated by δ C analysis. Aquaculture 35:29–42.

Shearer, K. D. 1994. Factors affecting the proximate composition of cultured fishes with emphasis on salmonids. Aquaculture 119:63–88.

Sorgeloos, P. 1980. The use of the brine shrimp Artemia in aquaculture. Pp. 25–46 in The Brine Shrimp Artemia: Ecology, Culture and Use in Aquaculture, Vol. 3, G. Persoon and P. Sorgeloos, eds. Wetteren, Belgium: Universa Press.

Storebakken, T. 1985. Binders in fish feeds: I. Effect of alginate and guar gum on growth, digestibility, feed intake and passage through the gastrointestinal tract of rainbow trout. Aquaculture 47:11–26.

Swingle, H. S. 1968. Estimaton of standing crops and rates of feeding fish in ponds. FAO Fish. 44:416–423.

Tacon, A. G. J., and S. S. DeSilva. 1997. Feed preparation and feed management strategies within semi-intensive fish farming systems in the tropics. Aquaculture 51:379–404.

Talbot, C. 1993. Some aspects of the biology of feeding and growth in fish. P. Nutr. Soc. 52:403–416.

Thomassen J. M., and S. O. Fjaera. 1996. Studies of feeding frequency for Atlantic salmon (*Salmo salar*). Aquaculture Engineering. 15:149–157.

Van Rooij, J. M., and J. J. Videler. 1996. A simple field method for stereophotographic length measurement of free-swimming fish: Merits and constraints. J. Exp. Mar. Biol. Ecol. 195:237–249.

Volkman, J. K., S. W. Jeffrey, P. D. Nichols, G. I. Rogers, and C. D. Garland. 1989. Fatty acid and lipid composition of 10 species of microalgae used in mariculture. J. Exp. Mar. Biol. Ecol. 128:219–240.

Watanabe, T., C. Kitajima, and S. Fujita. 1983. Nutritional value of live organisms used in Japan for mass propagation of fish: A review. Aquaculture 34:115–143.

Webster, C. D., J. H. Tidwell, and D. H. Yancey. 1992. Effect of protein level and feeding frequency on growth and body composition of cage-reared channel catfish. Prog. Fish Cult. 54:92–96.

Wu, G., I. P. Saoud, C. Miller, and D. A. Davis. 2004. The effect of feeding regimen on mixed-size pond-growth channel catfish, *Ictalurus punctatus*. J. Appl. Aquacult. 15:115–125.

Zhou, A., S. Xie, W. Lei, X. Zhu, and Y. Yang. 2005. A bioenergetic model to estimate feed requirement of gibel carp, *Carassius auratus gibelio*. Aquaculture 248:287–297.

14

Larval Nutrition

Information about the nutrient requirements of larval fish and crustaceans remains in an infant stage. The understanding of larval nutrition leading to successful culture is the foundation for the domestication of farmed freshwater and marine species. The newly hatched nauplii of *Artemia* and species of rotifers and copepods have generally served as an excellent source of food for larvae of many species of fish and crustaceans. Live *Artemia* nauplii are obtained through hatching of cysts that are collected from the natural environment and subject to periodic, unpredictable shortages that cannot supply the demand. As a result, prices increase, leading to overall increases in production costs. In addition, variation in the nutritional quality of hatched *Artemia* nauplii caused by temporal or spatial differences in cyst collections remains a chronic problem. That inherent problem has been partially overcome by the use of different methods designed to eliminate presumed nutritional deficiencies associated with both *Artemia* and rotifers (Clawson and Lovell, 1992).

Although noteworthy progress has been achieved in the development of larval microdiets during the past 25 years, live foods continue to be the choice, and considerable information about use, culture, and empirically derived nutritional value is available (Conceicão et al., 2010). However, use of live food dramatically restricts the understanding of quantitative nutrient requirements and interaction of nutrients in larvae. Additionally, the understanding of larval nutrition is also complicated by changing nutrient requirements for the different stages of larval development as the anatomy of the digestive tract and corresponding presence of particular enzymes change, often over several weeks.

The nutrient enhancement of live foods has provided some understanding of essential nutrients. However, knowledge of quantitative nutrient requirements and nutrient interactions can only be accurately determined and substantially increased when formulated diets fed exclusively achieve the growth rates and survival and physiological indicators that are achieved with the feeding of live food. Formulated diets to determine nutritional requirements must consist of chemically defined ingredients whereby control of nutrient composition can be exercised. Information derived from the use of these experimental diets can ultimately be transferred with confidence to the development of formulations that can be practically used for commercial manufacture of feeds. The results of efforts during the present decade have demonstrated that the difficult challenges of producing successful formulated diets (Langdon, 2003) can be overcome, and these diets have proved to serve as exclusive sources of nutrients throughout the larval cycle of some species. Yet, the culture of larvae of many species of fish and crustaceans still remains precariously dependent upon the availability and use of live plant and/or animal food. This chapter focuses on the knowledge of larval physiology as it relates to the development of successful formulated diets whereby nutrient requirements can be determined and this information applied to the formulation and manufacture of diets for commercial use. As appropriate, recommendations concerning requirements derived from experiments that have used nutrient-enhanced live food or formulated diets are presented.

DIGESTIVE ENZYMES

Satisfaction of nutrient requirements is definitely influenced by the type and quantity of digestive enzymes produced by the different stages (age) of larval development. Jones et al. (1997a) observed ontogenetic changes in the type and level of activity of enzymes in larval crustaceans. These changes occurred in response to changes in the type of food consumed, i.e., plant, animal, or a combination. Trypsin enzyme activity dominates in larval stages of decapod larvae followed by amylase activity. Comparatively lower levels of trypsin activity appear to correspond to longer gut retention times combined with efficient assimilation.

At one time, the prevailing explanation for the perennial lack of success in the culture of larvae was insufficient enzyme activity. Researchers speculated that the enzyme manufacturing capacity within the gut was far lower than what was needed and that effective digestion was only accomplished through the assistance of exogenous enzymes that originated from the sources of live food (Lauff and Hoffer, 1984; Munilla-Moran et al., 1990; Kolkovski et al., 1993, 1997a). However, Lovett and Felder (1990) found that contribution of enzyme activity from *Artemia* prey was very low compared to that measured in larvae of *Penaeus setiferus* larvae. Cahu and Zambonino Infante (1997) found that the lack of good growth in European sea bass (*Dicentrarchus labrax*) larvae (15–40 days old) fed a formulated diet was not due to a lack of sufficient endogenous enzyme activity. Garcia-Ortega et al. (1998) observed a minimal contribution of enzymes from consumed prey to the larval gut. Lazo et al. (2000a) found that the activity of digestive enzymes for larval red drum (*Sciaenops ocellatus*) was unaffected by diet, the availability of prey, and the contribution coming from exogenous sources. For trypsin the proportion of enzymatic activity that could be attributed to prey increased with age but only reached 17%. Lazo et al. (2007) also found that alkaline proteases initially appear in red drum larvae and acid proteases appear much later when the stomach has formed. Trypsin-like, lipase and amylase enzymes are present before the onset of exogenous feeding. Lazo et al. (2007) thus speculated that the activity of some enzymes during the course of development is controlled by genetic mechanisms. However, the type of enzymes produced during the gut metamorphosis does present some challenges. The ineffectiveness may reside in a lack of understanding of the quantity and specificity of the enzymes that are present during different stages of larval development. The digestive capacities of larvae relative to the ingredients of a formulated feed may be species- or stage-specific, corresponding to the development/maturation of the different parts of the gut. Puello-Cruz et al. (2002) found that trypsin activity is stage development and species specific, presumably corresponding to the onset of the development of a functional stomach. Although a sufficient quantity of enzymes may be present, the specificity may limit effectiveness to certain dietary sources of nutrients. It appears that although the array of digestive enzymes available are different and qualitatively restricted, particularly during early stages of development, a sufficient number and concentration of enzymes eventually are endogenously produced and present to achieve good growth and survival through the different stages of metamorphosis.

The inability to survive and grow when fed formulated diets can be attributed to either physical or chemical characteristics that are incompatible with the enzymatic capacity of very young larvae. For example, efforts to culture the larvae of *Macrobrachium rosenbergii* using microbound diets that were produced in a variety of ways proved unsuccessful (Ohs et al., 1998). These diets were readily consumed and the guts were full; however, growth and survival were significantly lower than that achieved with live food. This diet was a modification of a similar egg albumin-based diet that was successful in the axenic culture of successive generations of the microcrustacean *Moina macrocopa* (D'Abramo, 1979). When the egg albumin was substituted with egg yolk as the primary source of protein (Kovalenko et al., 2002), growth and survival achieved with the experimental microbound diet fed exclusively from stage 5 through metamorphosis to postlarva stage was essentially equivalent to that achieved with live *Artemia* larvae. Although minor changes in nutrient composition occurred with the use of egg yolk, it appears that the egg albumin, despite possessing an excellent amino acid profile, may have been chemically modified such that protein could not be efficiently digested by *M. rosenbergii* to provide the essential amino acids. Whether this unique response is characteristic of other crustacean species is unknown.

In marine fish, a very high activity of peptidases has been detected in the early stages of development (Cahu and Zambonino Infante, 2001). This condition suggests that during early larval development, the provision of protein in the form of hydrolysates would be an effective vehicle to satisfy amino acid requirements. Thereafter, as the gut develops, a variety of enzymes originating from the enterocytes that line the digestive track begin to produce a large complement of other enzymes.

Sequential changes in qualitative enzyme activity are consistent with the strictly carnivorous feeding activity of early larval stages of the freshwater prawn *Macrobrachium rosenbergii* (Kamarudin et al., 1994). The ontogenetic change in qualitative and quantitative enzyme activity has also been described for the larvae of the Atlantic white shrimp *Penaeus setiferus* (Lovett and Felder, 1990). Digestive capacity corresponds to the anatomical development of the digestive system, which in turn can be related to changes in habitat and diet during metamorphosis (Lovett and Felder, 1989). Digestive capacity may also be based upon the quality of food presented/consumed (Zambonino Infante and Cahu, 1994a,b). Larval fish appear to be more effective in metabolizing dietary phospholipid rather than triglycerides. The differential utilization is founded on different efficiencies of transport of these lipid classes.

Descriptions of the profiles of digestive enzyme activity during development are now being complemented by investigations into the realm of dietary modulation of digestive enzymes (Zambonino Infante and Cahu, 2007). Expression of enzyme activity is influenced by certain nutrients (molecules). Nutrients may be principal factors in the regulation of some developmental genes in larvae. For example, dietary levels of vitamin A have been shown to play a role in the maturation of the intestine and pancreas of the European sea bass (Villeneuve et al., 2005). Gisbert et al. (2005) found that the quantity and quality of dietary lipid affected the patterns of lipid absorption in intestinal mucosa of the European sea bass.

RELATIONSHIP OF LARVAL STAGE, DURATION OF GUT RETENTION, AND LEVEL OF ENZYME ACTIVITY

The rate of passage of food through the gut of a larval fish or crustacean combined with the digestibility of the nutrients will definitely influence the relative effectiveness of a formulated diet to provide an adequate supply of essential nutrients for growth and survival. For live food, rates of food consumption decrease during ontogeny and average time of food retention correspondingly increases as observed in larval lobsters (*Homarus gammarus*) (Kurmaly et al., 1990). A trend of decreasing levels of trypsin exists as the feeding habits of crustacean larvae change from herbivorous to carnivorous for different species and through different stages of larval development within a species (LeVay et al., 2001). Herbivorous decapod larvae consume food with a comparatively low energy content and accordingly exhibit high enzyme activity and rapid turnover that result in a comparatively low assimilation efficiency (Jones et al., 1997a). Lowest levels of enzyme activity correspond to the highest times of retention of food in the gut. Therefore, as the amount of food consumed per unit of time decreases, the retention time in the gut increases but the level of enzymatic activity decreases. Carnivorous larvae consume less food per unit of time and therefore have comparatively longer gut retention times. The food, exposed to a low enzyme activity, is highly digestible with a correspondingly high energy content. Therefore, formulated diets fed to carnivorous stages of larvae must have a high digestibility. For some early stages of larvae that are herbivorous and have higher rates of consumption, retention time through the gut is low but the enzyme activity is comparatively higher. Therefore, to develop effective types of formulated feed for different stages of larval development, these stage-dependent differences in digestive physiology (i.e., enzyme activity and rate of consumption) have to be considered. For example, diets for early, herbivorous stages of larvae may not need to be as highly digestible or could contain less protein as long as it is highly digestible and of the proper quality (amino acid profile).

The digestive capacity of the gut during ontogeny is also a factor. Achieving good digestibility of formulated diets for carnivorous fish and crustaceans is a difficult challenge because, although sufficient enzyme activity is present in the primordial gut, the quality of enzymes is restricted. Ingredient sources of dietary nutrients need to be compatible with the existing digestive capacity, particularly during early development of the gut. Larval diets cannot be simply ground pieces of a commercial diet that has proved successful for growth and survival of large specimens.

NUTRITIONAL ENRICHMENT OF LIVE FOOD

Much of the information about nutritional requirements of larvae has been acquired through the nutritional enhancement of live foods, specifically rotifers and nauplii of *Artemia*. The fatty acid and vitamin content of rotifers has been modified by short term (< 8 hours) exposure to emulsions of marine oils. This approach has less control of the nutritional composition and often results in high levels of lipid. The long-term enrichment technique is based upon a continuous provision of nutrients through a vehicle that is consistent in composition. A wide array of products have been used to enrich vitamin content of rotifers and are based upon the use of oil-soluble vitamins or derivatives of water-soluble vitamins, such as ascorbyl palmitate in conjunction with commercial products, such as emulsions and spray dried whole algal cells, designed for fatty acid enrichment (Merchie et al., 1997). Condensed *Chlorella* spp. paste has been used to supplement vitamins. Culture Selco® (CS) and AlgaMac®, commercially available enrichment products, are used to deliver specific dietary levels of DHA and EFA to rotifers and *Artemia*. For rotifers, the accumulated levels are much higher than those that can be achieved by feeding mixtures of algae and/or baker's yeast (Léger et al., 1989).

FORMULATED DIETS

The value of formulated diets that serve as complete replacements for live food is obvious because of both cost and the lack of consistent nutrient quality of live food. Formulated diets that can achieve consistent and reliable production equivalent to that of live food still do not exist and have particularly been an impediment to the progress of marine fish culture throughout the world.

Research that has addressed the development and evaluation of nutritionally complete larval diets has been conducted for over three decades. Most diets, at best, have served as supplements rather than complete substitutions (Kumlu and Jones, 1995). Advances in the nutritional and physical quality of formulated diets have resulted in some success such that the preweaning period using live prey species has decreased dramatically for certain cultured species. The lack of success might in part be due to imposed biases that are partly associated with trying to realize an ideal that may not be attainable. Successful future development of practical microparticulate diets for larvae may only be achieved through the recognition that a useful diet is simple to produce and that technical success is based upon the acceptance of compromise in meeting all of the desirable characteristics of a diet. A microparticulate diet needs to have an inherent flexibility that affords a broad utility for many species. Moreover, attempts to mimic the ingredient composition of larval diets with that of successful diets for juvenile culture may not be fruitful because of lack of applicability due to significant differences in physiology in developmental forms. In addition, it is commonly understood that larval fish and crustaceans have a metabolic rate that is higher than corresponding juveniles and consequently benefit from the provision of higher concentrations of dietary nutrients and energy (Dabrowski, 1986).

During the past decade, significant progress has been achieved in the culture of larvae through the development of microparticulate diets, characterized as either microbound, microencapsulated, or microcoated (Tucker, 1998). Ingredients of microencapsulated feeds are surrounded by a membrane or capsule wall that consists of a cross-linked protein or lipid. Microencapsulated particles can be produced by spraying droplets that coat dietary ingredients. Another method of preparation sprays the droplets that contain the ingredients (including a binder) into a liquid solution that activates the binder. In some cases, preparation is expensive, and the limited success achieved has been attributed to poor digestibility and the comparatively high levels of non-nutritional ingredients that constitute the capsule (Yúfera et al., 2005). Yúfera et al. (2005) modified the internal gelation microencapsulation procedure by "entrapping" the ingredients of a diet in a matrix of calcium-alginate.

Microbound diets are the most simple and commonly used form of microdiet. Dietary ingredients are mixed with a binder, generally agar, calcium alginate, carageenan, casein, gelatin, zein, or carboxymethylcellulose to achieve a binding that will impart a desired physical stability. Binding is generally activated by either temperature or a chemical reaction. The diets are most often dried, and then ground and sieved to the desired size for feeding. Care must be exercised in the use of binders because although physical integrity is achieved, nutrient utilization may be correspondingly reduced (Gawlicka et al., 1996). Microcoated diets are microbound diets coated with either a lipid or lipoprotein layer to reduce leaching of water-soluble ingredients. Some efforts have been devoted to the encapsulation of water-soluble nutrients through the use of lipid walls (Onal and Langdon, 2004). The leaching of water-soluble nutrients contained within these capsules is substantially reduced. The capsules, in turn, have been embedded into microparticles. Using such a technique to effectively deliver water-soluble nutrients in large quantities remains in the developmental stage, but has great potential for the determination of nutrient requirements and nutrient delivery. Delivery of compounds such as hormones or antibiotics is also an attractive possibility.

Microdiets have been prepared using equipment from the pharmaceutical industry. A marumerizer combined with a radial discharge extruder (microextrusion marumerization, MEM) or without the extruder (particle-assisted rotational agglomeration, PARA) have been used to produce water-stable and palatable particles (≈ 500 µm), achieving some success with feeding larval walleye *Sander vitreus* (Barrows and Lellis, 2006). Particles produced by the PARA process were smaller and less dense than those produced by the MEM process. These microdiets have a promising application for the determination of qualitative and quantitative nutrient requirements of larvae.

Some microparticulate diets have been successfully used in commercial hatcheries but only as partial replacement of live feed (Kumlu, 1999). Commercially available liquid feeds are essentially microparticles suspended in a medium.

Physical Form of the Diet

Effective provision of nutrients to larvae is critically dependent upon the physical form of the microparticulate diet. Successful consumption of the diet may be dependent upon the moisture content of the diet, which affects its physical form. Ideally, a diet that has low moisture content would be most desirable because it offers advantages when stored frozen for an extended period of time. In contrast, high-moisture diets naturally limit the duration of storage and shelf life. In addition, the rate of leaching of water-soluble nutrients would be amplified in formulated larval diets containing high moisture content because these diets are more prone to physical disintegration. A dry diet has many appealing benefits, but this bias, based upon an overriding desire for convenience, may be impractical to realize good consumption. High-moisture diets may also offer a level of palatability and eventual consumption that cannot be achieved with dry diets.

For dry diets, the method of drying may also influence the performance of a formulated diet. Teshima and Kanazawa (1983) found differences in growth and survival of larvae of *Marsupenaeus japonicus* when fed carrageenan-microbound diets dried under different conditions. The ideal of producing a diet that reflects as much as possible the relative amounts of macronutrients found in the natural diet may be unattainable because establishing the physical integrity of a diet often requires the addition of carbohydrates. Fernandez-Díaz and Yúfera (1995) found that physical and chemical breakdown of dietary microcapsules consisting of dietary protein and carbohydrates was variable based on the action of a cross-linking agent.

Shape and size are other important characteristics. The size of formulated diets designed for fish and filter-feeding crustacean larvae is generally confined to particles that are small enough to be entirely consumed without the assistance of mouthparts. Larvae of gilthead sea bream (*Sparus aurata*) have been observed to ingest inert particles that are 60 to 80% of the width of the mouth (Fernandez-Díaz et al., 1994). As larvae of the mud crab *Scylla serrata* metamorphosed through four stages, from zoea I stage to megalopa, particle size preference increased from < 150 µm to 400–600 µm (Genodepa et al., 2004a). However, size of particles may not always be a defining factor, being species-specific because of feed consumption characteristics.

Although diets that contain high levels of moisture may not be amenable to the production of particles less than 150 µm, the restriction of particle size imposed by high-moisture formulated diets may be inconsequential for some species of crustacean larvae that are raptorial feeders. These species possess mouthparts or appendages that permit the grasping and manipulating of particles that are physically modified into sizes that are consumed.

Shape of the particle may also influence sinking rate (buoyancy) and consumption of formulated diets. Some species may prefer irregular size particles, whereas the consumption of food by other species may be limited to smooth particles. These types of species-specific preferences suggest that videography of the feeding of larva forms will have to become an important part of the evaluation of the consumptive "appeal" of formulated diets. The diet may have the necessary size and nutrient profile but still remain unacceptable based on shape. Considerable effort has been devoted to establishing diets that are neutrally buoyant to afford as much access to the diet in the water column. However, achieving this ideal may be unnecessary. Although some diets may have comparatively high rates of sinking, maintenance in the water column can be achieved with appropriate aeration (creating an upwelling water movement) or design of the culture tank. Reduction in the rate of sinking may be possible through a decrease in the density of the particles achieved, for example, by a lipid coating. The lack of buoyancy may pose no threat to successful culture of some species that have been observed to feed readily off the bottom. Early weaning of larvae to microparticulate diets may instill a feeding behavior whereby rapid recognition and ingestion would offset loss due to rapid sinking through the water column. The lack of ingestion of formulated diets may in fact be the lack of visual stimuli. Some of those stimuli may reside in the movement or color of the live prey, characteristics that are absent or not normal in formulated diets.

Feeding Larval Fish

Larval stage is defined relative to the stage of metamorphosis of external and physiological characteristics, from hatch until the juvenile stage. External characteristics and major organ functions of juveniles match those of the adult. For practical nutritional needs, larval fish and crustaceans can be divided into three groups according to morphology of the alimentary tract and the kind and sequence of enzymes secreted into the gut (Dabrowski, 1984). The first group includes such fish as salmonids and channel catfish, as well as crustaceans such as crayfish, which appear to have a functional stomach and differentiated digestive system before changing from endogenous to external nutrients. The second group includes fish such as striped bass and many marine species of fish and shrimp, for which the initial larval stages have a very rudimentary digestive tract with no functional stomach or well-differentiated gastric glands; during ontological development, the animal undergoes complex metamorphosis of the digestive system. The third larval group consists of those that develop a functional digestive tract but remain stomachless throughout life, such as carps. Species that have structurally and functionally differentiated alimentary tracts at the time of first feeding pose less of a problem with initial feeding. These species (e.g., salmonids, catfish, tilapia) are often reared on high-quality fry feeds with a high degree of success. Larvae with less differentiated or developed digestive systems at first feeding are more difficult to feed and usually require live feeds as a part of their diet.

Larval fish and crustaceans undergo different phases of larval metamorphosis and at a certain phase can be weaned to dry, prepared diets (Holt, 1993; Fontagné et al., 2000; Cahu et al., 2003; Curnow et al., 2006). Most larvae can be weaned to artificial feeds late in the larval cycle but few can be weaned from first feeding (Le Ruyet et al., 1993; Palmer et al., 2007). For example, striped bass, which complete metamorphosis in 21 to 42 days, cannot use dry diets at day 5, when initial feeding begins, but they can after day 15 (Baragi and Lovell, 1986; Tuncer et al., 1990). Common carp can be transferred to commercial dry diets at the size of 15–30 mg (Bryant and Matty, 1980), whereas larval whitefish must obtain a size of 50 mg to be weaned to dry diets (Dabrowski and Poczyczynski, 1988). The transition from live to dry diet is a gradual process and is tolerated to varying degrees depending on the species. Successful weaning of the Senegal sole (*Solea senegalensis*) to formulated diets was accomplished through the cofeeding of live food and formulated diets (Cañavate and Fernández-Díaz, 1999). Two critical components of successful weaning through cofeeding are the timing and the proper proportions of the live food and formulated feeds. Red drum larvae can be weaned early, generally in the first week of life. Dry feeds are readily accepted by first feeding larvae and a standard feeding protocol is dry microparticulate diet and live enriched rotifers for 5 days, followed by microparticulate diet (of increasing size) only (Holt, 1993). For red drum, cofeeding of microparticulate diets in the presence of algae has also been demonstrated as a successful technique to allow the rearing of larval red drum from the initiation of feeding with results similar to those of live foods (Lazo et al., 2000b). Such results demonstrate that as we improve our knowledge of feeding stimulants and larval nutrition, replacement of live feeds will eventually be realized.

There are a number of reasons for the reduced ability of some larvae, particularly marine fish larvae, to utilize prepared diets at first feeding, which include lower ingestion rates (Lauff and Hoffer, 1984; Kolkovski et al., 1993, 1997b) and poorly digested (low enzyme activity or insufficient stimulation of enzyme secretion) and inadequate nutrition (Kolkovski et al., 1993; Teshima et al., 2000; Lazo et al., 2000b). The inability of some larval fish to immediately use prepared diets may also be due to the absence of hormones or their regulators, or growth factors, that are provided in live foods (Lauff and Hofer, 1984; Baragi and Lovell, 1986) and aid in digestion (Ronnestad et al., 2007). Other dietary factors in live food might inhibit or stimulate hormone action in larvae. Thyroid hormones appear to play an important role in larval fish metamorphosis and differentiation of the digestive system, and in growth. The results of experiments that have investigated the role of thyroid hormones are both positive and negative and these findings may be because of dietary level and method of delivery (Power et al., 2001). De Jesus et al. (1998) demonstrated that the rate of metamorphosis of grouper (*Epinephelus coioides*) larvae was positively influenced when cultured in thyroid solutions. The hormone appears to be of maternal origin. Thyroid levels may be influenced by diet components (Miwa and Inui, 1987; Specker, 1988; Inui et al., 1989). Although in many cases dry diets may not be able to be used at first feeding, there are numerous benefits to the inclusion of dry diets during the larval rearing process, even if only as a partial replacement for live foods.

Successful Forms of Larval Diets

Fernandez-Díaz et al. (1994) emphasized the importance of using the proper size particles for effective consumption by the different stages of larvae. For the use of such small particles, meals or any other ingredients should be ground as fine as possible so that they are homogenously distributed (i.e., nutrients in these ingredients are available in the intended proportions within each particle). Care should be exercised because heat processing associated with fine grinding may alter certain dietary ingredients and their delivery of nutrients. Ingredients such as micropulverized meals from fish, crab, squid, mussel, yeast, chicken egg yolk, short-necked clam, and krill have been used (Kanazawa, 1990; Cahu and Zambonino Infante, 2001), as well as chemically defined ingredients such as casein and gelatin (Paibulkichakul et al., 1998). The comparatively short gut retention time of carnivorous larvae, particularly during early stages of metamorphosis, precludes the use of poorly digestible diets because rapid digestion is critical. This may be the reason that the greatest success has been with herbivorous forms of crustacean larvae (Bautista et al., 1989; Koshio et al., 1989). Microcoated and microencapsulated diets have proved to be highly successful for species such as penaeid prawns that filter-feed during the zoeal stages and for larval fish that are gulp feeders (Jones et al., 1979a). For herbivorous species of penaeid shrimp, rapid rates of transit time through the gut combined with the highest levels of enzyme activity (Jones et al., 1997b) enable nutrient requirements to be met with diets that are not highly digestible.

Teshima et al. (1993) created successful carrageenan-bound microbound diets with casein as the primary protein source for nutritional studies with larval *M. japonicus*. For a carnivorous shrimp species, Kovalenko et al. (2002) successfully reared larvae of the freshwater prawn *M. rosenbergii* from stage V through the post-larva stage by feeding a microparticulate diet as an exclusive substitute for *Artemia* nauplii. Both survival and growth of the formulated diet fed larvae were equivalent to what was achieved with the live *Artemia* nauplii containing high levels of long-chain polyunsaturated fatty acids (LC-PUFA) of the n-3 series PUFA. Megalopa of the mud crab *Scylla serrata* were successfully cultured to the crab stage using a microbound diet that contained 3% zein as the binder (Genodepa et al., 2004b). Survival for the megalopa fed either the microbound diet or 100% *Artemia* nauplii was not significantly different. D'Abramo et al. (2006) successfully reared the larvae of *Litopenaeus vannamei* from either stage PZ2 or M1 to PL1, again using a microparticulate microbound diet that served as a complete substitute for *Artemia* nauplii, and a mix of rotifers and algae. No diet-dependent differences in growth, survival, and trypsin activity were observed. A microencapsulated diet fed to gilthead sea bream larvae yielded a growth rate and survival that was similar to larvae fed rotifers (Yúfera et al., 1999). However, the applicability of such a diet in this study and many others is questionable because the feeding trial was only conducted from 8 to 15 days posthatching. Therefore, the success of formulated diets as a complete replacement of live food is often described or limited to certain stages of larval development. Holme et al. (2006) with larvae of the mud crab (*S. serrata*) found that a zein-based, microbound diet was much more effective for later megalopa stages than the early zoea III stage. Developmental indices and survival of larval stages PIII through PII of *L. vannamei* were essentially equivalent when fed with a live diet (algae and *Artemia* nauplii) or a microencapsulated diet exclusively (Pedroza-Islas et al., 2004).

NUTRIENT REQUIREMENTS

Attempts to grow marine fish larvae with microparticulate diets exclusively have generally been unsuccessful. Most of the efforts have been devoted to weaning from live food to microparticulate diets. Therefore, the understanding of nutritional requirements remains in its infancy. Caution needs to be exercised relative to the interpretation of published results about nutrient requirements for some fish and shrimp larvae or the effectiveness of particular ingredients. Conclusions are often misleading because they are based on survival responses that are too low to permit a satisfactory level of confidence. For those experiments that have exclusively used formulated microparticulate diets, information that has been derived is primarily qualitative in nature, and many investigations consist of only a small proportion of the duration of the larval cycle, thereby providing limited information. Some quantitative requirements are recommended, but many of the diets retain some physical shortcomings that contribute to leaching or poor consumption, thereby contributing to a reduction in effective transfer of nutrients to the larvae. The lack of a confident means to quantify ingestion also introduces questions about the accuracy of the determinations of nutrient requirements. In addition, the quality of the larvae must be good to conduct these experiments so the quality of broodstock nutrition can be a confounding variable. Most of the research devoted to an understanding of nutrient requirements has been conducted through the nutrient modification of live food and has focused on lipid requirements. Larval fish and shrimp develop rapidly, using dietary LC-PUFA as well as adequate phospholipids for the synthesis of cellular biomembranes. Therefore, the nutrition of broodstock becomes critical such that ova must be adequately supplied with these nutrients to promote the normal growth and health of larvae. Sargent et al. (1999) suggested a total quantitative dietary lipid requirement for marine fish larvae of approximately 10%, because that was the level found in the egg yolk sac of marine species of fish larvae.

The comparatively high growth rates of fish and shrimp larvae result in critical qualitative and quantitative nutrient requirements that must be satisfied through the provision of the proper source of these nutrients. For example, an excellent dietary source of protein for the juvenile form may not be correspondingly effective for larval forms because of lack of a similar nature of digestive activity.

Protein and Amino Acids

It is assumed that larvae require the same 10 essential amino acids (EAA) that have been described for juveniles/adults of species of fish and shrimp due to the inability to synthesize them. Information about quantitative requirements for EAA is lacking. Protein sources with amino acid profiles that suggest deficient quantities can be nutritionally enhanced by the addition of pure EAA (Cahu et al., 1999). Yet, the benefits of adding free amino acids to formulated larval diets to provide a balanced array of EAA to enhance growth remains inconclusive. For example, microcapsules containing an array of EAA were effective as live *Artemia* when fed to larvae of gilthead sea bream for 10 days (Aragão et al., 2007). For some amino acids, 40 to 60% was lost after the first 5 minutes of immersion in water.

Kanazawa (1990) recommended that the quantity of protein in larval feed should range between 23 and 57%. The results of the study of Durruty et al. (2002) for *L. vannamei* and *L. setiferus* suggest that protein requirements may increase by 2-fold, 30% to 60%, from the zoea to the mysis stage. When growth of larval Japanese flounder fed a commercial diet that contained 55.8 to 61.2% crude protein was compared to growth attained with experimental diets containing 40, 50, and 60%, results suggested that the dietary protein level that would achieve optimum growth needed to contain at least 60% crude protein (Bai et al., 2001). In this experiment live food was fed with each of the formulated diets for the first 45 days post hatch followed by the formulated diets fed exclusively for the next 38 days.

The actual requirements of EAA for different species of larval fish and shrimp will require an understanding of the factors that impact their bioavailability (Conceicão et al., 2003). Understanding bioavailability of EAA will lead to knowledge of what would be effective protein sources for larvae and if a supplementation of specific EAA will be needed. The addition of dietary free amino acids may help to enhance attraction and consumption by serving as chemoattractants (Kolkovski et al., 1997c).

Protein in the form of a hydrolysate seems to enhance growth rates of fish larvae as reported for goldfish (Szlaminska et al., 1993), sea bass (Cahu and Zambonino Infante, 1995), and summer flounder *Paralichthys dentatus* (Lian et al., 2008). These hydrolysates most probably serve as chemoattractants that enhance ingestion rates. Kolkovski (2008) provided a summary of hydrolysates used as feed attractants and protein sources in microdiets.

Carvalho et al. (1999) observed an improvement in the growth of carp larvae when fish protein hydrolysate accounted for 50% of the total supply of dietary nitrogen. A combination of growth and survival of first feeding larvae of common carp was best when a 1:1 mixture of intact protein (casein) and a fish protein hydrolysate was used in experimental diets (Carvalho et al., 1997). The benefit of protein in the form of dietary peptides may lie in the high level of peptidase activity in early larval development, possibly combined with the existence of membrane transporters that are specific for peptides to override high transit time through the gut. The investigation of Kotzamanis et al. (2007) suggested that growth and gut microbiota can be influenced by different molecular fractions and levels of fish protein hydrolysates, supporting the earlier published work of Savoie et al. (2006) with wolffish (*Anarhichas minor*). However, larvae of the Atlantic halibut (*Hippoglossus hippoglossus*) did not benefit from the inclusion of prehydrolyzed proteins in microparticulate feed (Kvale et al., 2007). The lack of a positive response was speculated to be due to the species-specific slow feeding behavior of the larvae, thereby allowing the loss of important dietary amino acids to leaching.

Lipid and Fatty Acids

Fish and crustacean larvae appear to require levels of dietary lipid that generally exceed those reported for juvenile forms. This requirement is presumably based on the high growth rates and the correspondingly high levels of energy that need to be metabolized under these conditions. The amounts of dietary lipid that have been recommended for larvae of different species of marine fish range from 10 to 37%.

The relative nutritive value of (LC-PUFA) for larval fish species can be determined through starvation studies. Koven et al. (1989) starved larvae of gilthead sea bream and found a hierarchical pattern of loss, n-6 > n-9 > n-3. For the n-3 LC-PUFA, docosahexaenoic (DHA) was preferentially conserved over eicosapentaenoic (EPA). For turbot (*Scophthalmus maximus*) larvae, Rainuzzo et al. (1994) found that EPA was used while DHA and 20:4 n-6 arachidonic acid (ARA) were conserved. Using microencapulsated diets containing labeled palmitic acid, Jones et al. (1979b) found evidence that larvae of *M. japonicus* have limited ability to elongate n-3 and n-6 PUFA to n-3 and n-6 LC-PUFA.

Juvenile forms of fish species generally require 0.5 to 1.0% of dietary n-3 LC-PUFA, and the requirement of larval fish for these fatty acids is probably much higher, commonly greater than 1%. For fatty acids, Sargent et al. (1997) provided a comprehensive review of polyunsaturated fatty acid nutrition of larval marine fish, concentrating on LC-PUFA. Limited acceptance of formulated diets in the past has essentially confined knowledge of fatty acid requirements derived from live food enriched with LC-PUFA. For live food, efforts have commonly focused on enhancement of the content of two LC-PUFA, docosahexaenoic acid (DHA, 22:6n-3) and eicosapentaenoic acid (EPA, 20:5n-3), and the corresponding responses of larvae. For example, the algal species *Isochrysis galbana* and *Pavlova lutheri* are commonly cultured to feed rotifers because their high levels of 22:6n-3 are transferred to the rotifers. The alga *Nannochloris occulata* is commonly fed as a good source of 20:5n-3. The nutritionally enhanced rotifers are in turn fed to the fish larvae. The algae can also serve as a nutrient source (Reitan et al., 1997).

In the natural diets of marine fish larvae, sources of LC-PUFA are consumed as part of phospholipid compounds and the 22:6n-3 to 20:5n-3 ratio is commonly 2:1. For live foods, *Artemia* strains are often deficient in both 20:5n-3 and 22:6 n-3 (Léger et al., 1986; Navarro et al., 1992) and are therefore enriched through the feeding of emulsions of marine fish oils. Enrichment has been effectively accomplished through the use of lecithin and 22:6n-3 derived from fish oils (McEvoy et al., 1995, 1996). Dietary lecithins derived from a marine (bonito eggs), rather than a terrestrial source (soy lecithin), are more effective, as first detailed by the results reported for larval sea bream (Kanazawa, 1985). A level of 1.0% of pure phosphatidylcholine (PC) or phosphatidylinositol (PI), extracted from bonito eggs and soybean, respectively, was found to be most effective in growth and survival of larval *M. japonicus* (Kanazawa et al., 1985a). Pure PC derived from chicken egg and pure phosphatidylethanolamine (PE) derived from bonito eggs and ovine brain did not yield an equivalent effect.

Live copepods represent a natural phospholipid-derived source of LC-PUFA, but the composition of copepods is subject to algal diet (Graeve et al., 1994) and can also change depending on the developmental stage of the copepod. Shields et al. (1999) found that feeding natural copepods was superior to *Artemia* nauplii relative to survival, pigmentation, and retinal morphology of halibut larvae. These enhanced responses were suggested to be related to the provision of higher levels of dietary DHA. However, the cost effectiveness and practicality of copepod culture remain subjects of concern. The required higher levels of DHA are related to its role in the development of neural tissues, including the brain, and the retina (Mourente et al., 1991; M. V. Bell et al., 1995). Rodriguez et al. (1994) found that higher levels of DHA in rotifers fed to larvae of gilthead sea bream improved growth and survival. In a later study, Rodriquez et al. (1997) observed higher rates of growth of larval sea bream when fed rotifers having a DHA:EPA ratio of 1.5 relative to those fed rotifers containing ratios of less than 0.6.

Investigations have revealed that 22:6n-3 is a very important fatty acid in larval nutrition relative to growth and survival. Dietary 22:6n-3 has also been found to be effective in enhancing the health of larvae, improving tolerance to stress in larval red sea bream *Pagrus major* (Kanazawa, 1997). Gapasin et al. (1998) demonstrated that LC-PUFA enriched rotifers *Branchionus plicatilis* fed to milkfish (*Chanos chanos*) larvae enhanced growth and resistance to salinity stress.

Quantitative requirements for LC-PUFA (DHA + EPA) range between 0.5 and 1.0%, and a dietary level of 1.0% of LC-PUFA appears to be a good foundation from which to work. There is some evidence that 22:6n-3 can be biosynthesized from 20:5n-3 through elongation and desaturation, but the rate of synthesis appears to be insufficient to compensate for dietary deficiencies of 22:6n-3. Experimental results have shown that the n-3 LC PUFA are preferentially conserved presumably due to their role in the structure of cellular membranes. Therefore, to particularly allow for the use of these fatty acids for growth, saturated and monounsaturated fatty acids need to be provided to effectively serve as readily catabolized sources of energy.

The importance of dietary ARA for growth and survival of juvenile turbot was first identified by Castell et al. (1994). J. G. Bell et al. (1995) demonstrated its dietary value in relationship to the fatty acid composition of phospholipids and the respective production of prostaglandins. Enhancement of dietary levels of ARA to 1.0% and 1.8% improved growth and survival, respectively, for larvae of the gilthead sea bream (Bessonart et al., 1999). Koven at al. (2001) found that the stress created by handling was reduced in the presence of consumed rotifers enriched with ARA.

Sources of neutral lipids in microparticulate diets have included fish meal, cod liver oil, menhaden oil, and roe oil, whereas sources of phospholipids have been soybean or marine derived. Formulated diets generally use neutral lipid in the form of triacylgylcerols and generally the ratio of 22:6n-3 to 20:5n-3 is approximately 1:1, not the previously stated 2:1 that is characteristic of natural live diets of larvae. For example, a positive effect on growth (measured as standard length) and survival associated with the feeding of rotifers having higher ratios of DHA/EPA, as effected by different enrichments, was suggested in studies conducted with larval Atlantic cod *Gadus morhua* (Park et al., 2006). To achieve what would be considered a ratio generally found in live food consumed in the natural marine environment, additions of 22:6n-3 would be necessary.

Phospholipids

The beneficial effect of dietary phospholipids for larvae of *Plecoglossus altivelis* was first reported by Kanazawa et al. (1981). Culture of most larvae still requires the use of live foods that are commonly rich in phospholipids. Accordingly, controlled experiments to determine nutritional requirements are difficult because dietary phospholipids have already been consumed for a period of time before weaning to formulated diets. Studies using highly purified phospholipids are rare, and generally phospholipids of plant origin are used rather than marine sources derived from the prey of marine larval fish.

In the absence of phospholipids in the form of soybean lecithin, zoea larvae of *M. japonicus* had 100% mortality (Kanazawa et al., 1985a). The study of Kanazawa et al. (1985a) with larvae of *M. japonicus* suggests a comparatively noteworthy growth and survival response to the addition of 3.5–6% soybean lecithin, containing 23.6% PC. Survival and final weight of first feeding larvae of carp improved with the addition of 2% dietary phospholipids from different sources (Guerden et al., 1995a,b). A dietary requirement for phospholipid for larval goldfish (Szlaminska et al., 1993) and red sea bream (Kanazawa et al., 1983) using 5% dietary soybean lecithin has also been reported. Controlled studies and other studies with less control of the quality of dietary phospholipid suggest that the two most active (efficient) phospholipid molecules in the diets of larval fish and shrimp are PC and PI. Yet, the species-dependent differential effects of PC vs. PI are not fully understood. Kanazawa et al. (1985b) observed that PC (42% purity) or PC + PI (92% of all phospholipids) was more effective than PE (63% of all phospholipids) for growth and survival of larval ayu. The effectiveness of different phospholipids was also demonstrated by Kanazawa et al. (1985a) for larval *M. japonicus*. Teshima et al. (1986) found that soybean PE (95% PE) was inferior to soybean PI (60% of all PL), which in turn was inferior to PC (68% of total phospholipids) for growth and survival of larval *M. japonicus*. Kanazawa (1993) also found that the addition of PC rather than PI improved the growth of the Japanese flounder (*Paralichthys olivaceus*). Required dietary levels of phospholipid for larval forms would be at least equivalent and are commonly acknowledged as most probably exceeding the levels reported for juvenile forms of fish and crustacean species. Kanazawa (1993) found that required levels of dietary soybean lecithin in rock bream (*Oplegnathus fasciatus*) decreased from 5% for larvae to 3% for juveniles (PC + PI = 1.6 to 1.0%, respectively). Teshima et al. (1986) observed the best metamorphosis in larvae of the Kuruma shrimp *M. japonicus* fed diets containing 3.0% soybean PC. This result falls within the reported range of the phospholipid requirement (1 to 3%, dry weight), suggested by Coutteau et al. (1997) for marine larval species. The lower levels for marine larval species are probably attributed to higher levels of PC within the source. Reported optimal dietary levels for fish and shrimp larvae range from 3.5 to 7.0% when soybean lecithin is used as a phospholipid source. Generally, the PC level of soybean lecithin is approximately 25%, therefore suggesting actual dietary PL requirements in the form of PC to be 0.83 to 1.75%. Camara et al. (1997) used purified soybean PC (95% pure) and found that dietary levels of 1.5 to 3.0% were optimal for the metamorphosis of larval *M. japonicus*.

Phospholipids are known to be components of lipoproteins that are used for the transport of lipids in both crustaceans and fish. For juvenile fish and shrimp, dietary phospholipid has been shown to be important in contributing to higher turnover rates of cholesterol from the gut to the midgut gland (shrimp) and ultimately to the circulatory system.

Both fish and crustaceans possess the ability to biosynthesize phospholipids, but the dietary requirement is suggested to be the result of an insufficient rate of synthesis to meet the demands of high growth rates and metabolism. Dietary phospholipids may also play a role in the effective absorption of dietary cholesterol and triglycerides (Jones et al., 1997b). Not all phospholipids or combinations of phospholipids can satisfy the dietary requirement. The location of the LC-PUFA molecule, connected by an ester bond at the sn-2 position of the phospholipid molecule, appears to be critical for the expression of the activity of the phospholipid molecule. However, evidence is lacking as to whether after consumption of the food containing the LC-PUFA-laden phospholipid, the LC-PUFA molecule is assimilated as either part of or apart from the phospholipid molecule.

The phospholipid benefit is related to the molecule per se and not the result of an effective provision of other required nutrients such as choline, inositol, or specific fatty acids that are provided by the specific molecule. The benefit of dietary phospholipid in larval diets has also been attributed to the molecules being more digestible and contributing to the stability (physical integrity) of formulated diets in which it is an ingredient. Dietary phospholipids have been shown to have a growth-inhibiting effect when provided at higher levels.

Caution needs to be exercised when applying the information concerning the phospholipid requirement because there are different sources of phospholipids containing different proportions of those compounds that are considered effective. The lowest requirements would most probably be associated with marine-derived sources rather than plant-derived sources because the amount of PC would be higher.

Cholesterol and Other Sterols

Cholesterol, a nutrient repeatedly demonstrated to be required by juveniles of all crustacean species studied (see Chapter 6), has been shown to be required by larval forms of crustaceans. The study of sterol requirements for crustacean larvae was initiated by Teshima et al. (1982), who demonstrated that dietary cholesterol was required for the normal metamorphosis and survival of *M. japonicus*.

The requirement for cholesterol is specific and cannot be satisfied through the provision of other dietary sterols. No sparing of dietary cholesterol for larval (1.0% level) or juvenile (0.05% level) *M. japonicus* with dietary sitosterol at ratios of 1:1 to 1:100 (cholesterol:sitosterol) was found by Teshima et al. (1989). None of an array of sterol compounds (cholesterol precursors) served as total replacements for dietary cholesterol for larval *M. japonicus* (Teshima et al., 1983).

Vitamins

The nutrient requirements *M. japonicus* for several water-soluble vitamins and one fat-soluble vitamin (tocopherol) were reported from results of early studies conducted by Kanazawa (1990) using microbound, microparticulate diets. He acknowledged that the requirements were probably overestimates because of leaching of the dietary nutrients. Precise requirements of vitamins and minerals are difficult to determine and are considered to be overestimates because of the leaching of the microparticulate diet and rates of leaching that are influenced by solubility in water. Merchie et al. (1997) used stable phosphate esters of ascorbic acid (vitamin C) to determine that a dietary level of 20 mg/kg of ascorbic acid was sufficient for normal growth and survival of the larvae of European sea bass and turbot. Accuracy in estimating vitamin and mineral requirements will require some type of encapsulation technique to eliminate leaching or continuous feeding of small amounts of diet rather than sporadic feeding of large amounts of diet throughout the duration of the investigation. For practical diets, formulation of vitamin mixes that provide levels of water-soluble vitamins that are in excess of what is required may be the most effective strategy to deliver required amounts. The alternatives, encapsulation or continuous feeding, are labor intensive and questionably cost effective.

CONCLUSIONS

When nutritionally complete, formulated diets that completely substitute for live food are successfully developed for the entire larval stage of a fish or crustacean species, then the ability to determine the nutritional requirements accurately will increase dramatically. Most of the information about quantitative nutrient requirements is not very definitive and subject to wide ranges, a testimony to the need to develop better diets. Therefore, the nutrient requirements stated in this review should be accepted with caution and serve as a foundation for future investigations that are designed to exercise better control. Poor survival caused by inadequacies in the physical form of the diet or poor acceptance of the diet will need to be overcome before studies investigating nutrient requirements can be confidently accepted. The determination of nutrient requirements of larvae appear to be best achieved with a water-stable diet that is fed either continuously or at specific intervals/pulses (Rabe and Brown, 2000) to minimize leaching of water-soluble nutrients, thereby assuring accuracy. The ideal binder would be nutritionally inert and also not affect the delivery and utilization of nutrients. Effective formulation will also have to incorporate results from investigations that show the influence of dietary nutrients themselves on larval development including organogenesis and associated enzyme activity. This knowledge will undoubtedly help to increase survival, which has chronically been a bottleneck to successful larval culture of marine organisms and the expansion of aquaculture enterprises. Progress in understanding nutrient requirements may require species-specific standardized practices. Although the goal of the determination of nutrient requirements will be the same, the road to progress and attaining knowledge may require different types of diets. Previously, it seemed that culture success could only be achieved through the use of a combination of live food and formulated feeds. However, recent advances, as exemplified by the work of Lazo et al. (2000b), Kovalenko et al. (2002), and D'Abramo et al. (2006), suggest that exclusive feeding of formulated diets can be achieved very early in development. Even weaning to the exclusive feeding of formulated diets to later stages of larval development represents a significant costs savings in food and labor on a commercial level. The increase in the successful use of formulated microparticulate diets lays the foundation for the ability to accurately determine nutrient requirements through manipulation of the nutrient composition of the diets. The evaluation of nutrient requirements will definitely necessitate an accurate determination and understanding of microdiet intake derived from the use of effective inert markers (Teshima et al., 2000).

Fish and crustacean larvae definitely require high-energy diets to meet their special developmental needs, and the source of this energy is both protein and lipid. When compared to juveniles of the same species, larvae require more dietary protein and lipid because of higher metabolic and growth rates. Therefore, a formulated diet should be very high in energy. Required levels of protein seem to fall in the range of 50 to 60%. There is sufficient evidence to indicate the importance of both dietary phospholipids and LC-PUFA, whether provided through enriched live foods, formulated diets, or a combination. Collectively, the results of an array of investigations suggest the inclusion of phospholipids, principally PC, at a dietary level of approximately 3%. At least 1% LC-PUFA in a triglyceride form is recommended, and a ratio of 2:1 for 22:6n-3 to 20:5n-3 is an ideal goal for provision of these LC-PUFA to marine larvae. Microbound microparticulate diets seem the easiest to produce and manipulate to a form that will meet both the physical and nutritional requirements of a variety of larval species and presumably the most practical to transition to commercial production. Future progress in the determination of nutrient requirements of larval fish and crustaceans is critically based on effective formulated diets that yield growth and survival equivalent to what is achieved with live food. These diets need to be amendable for production of different sizes and different nutrient composition so that they can meet specific physiological needs during ontogeny and be used throughout most or the entire larval phase of culture. In addition, accurate physiological indices will need to be used to comparatively assess performance relative to dietary quality and quantity of nutrients.

REFERENCES

Aragão, C., L. E. C. Conceicão, M. Lucuisse, M. Yúfera, and M. T. Dinis. 2007. Do dietary amino acid profiles affect performance of larval gilthead seabream? Aquat. Living Resour. 20:155–161.

Bai, S., Y.-T. Cha, and X. Wang. 2001. Preliminary study on the protein requirement of larval Japanese flounder *Paralichthys olivaceus*. N. Am. J. Aquacult. 63:92–98.

Baragi, V., and R. T. Lovell. 1986. Digestive enzyme activities in striped bass from first feeding through larva development. T. Am. Fish. Soc. 115:478–484.

Barrows, F. T., and W. A. Lellis. 2006. Effect of diet processing method and ingredient substitution on feed characteristics and survival of larval walleye, Sander vitreus. J. World Aquacult. Soc. 37(2):154–160.

Bautista, M. N., O. M. Millamena, and A. Kanazawa. 1989. Use of kappa-carrageenan microbound diet (C-MBD) as feed for *Penaeus japonicus* larvae. Mar. Biol. 103:169–173.

Bell, J. G., J. D. Castell, D. R. Tocher, F. M. MacDonald, and J. R. Sargent. 1995. Effect of different arachidonic acid:docosahexenoic acid ratios on phosopholipid fatty acid compositions and prostaglandin production in juvenile turbot (*Scophthalmus maximus*). Fish Physiol. Biochem. 14:139–151.

Bell, M. V., R. S. Batty, J. R. Dick, K. Fretwell, J. C. Navarro, and J. R. Sargent. 1995. Dietary deficiency of docosahexaenoic acid impairs vision at low light intensities in juvenile herring (*Clupea harengus* L.). Lipids 30:373–476.

Bessonart, M., M. S. Izquierdo, M. Salhi, C. M. Hernandez-Cruz, M. M. Gonzalez, and H. Fernandez-Palacios. 1999. Effect of dietary arachidonic acid on growth and fatty acid composition of gilthead sea bream (*Sparus aurata* L.) larvae. Aquaculture 179:265–275.

Bryant, P. L., and A. J. Matty. 1980. Optimisation of Artemia feeding rate for carp (*Cyprinus carpio* L.) larvae. Aquaculture 21:203–212.

Cahu, C. L., and J. L. Zambonino Infante. 1995. Maturation of the pancreatic and intestinal digestive functions in sea bass (*Dicentrarchus labrax*): Effect of weaning with different protein sources. Fish Physiol. Biochem. 14:431–437.

Cahu, C. L., and J. L. Zambonino Infante. 1997. Is the digestive capacity of marine fish larvae sufficient for compound diet feeding? Aquacult. Int. 5:151–160.

Cahu, C. L., and J. L. Zambonino Infante. 2001. Substitution of live food by formulated diets in marine fish larvae. Aquaculture 200:161–180.

Cahu, C. L., J. Guillaume, S. Kaushik, P. Bergot, and R. Metailler. 1999. Nutrition and feeding of penacid shrimp larvae. Pp. 253–263 in Nutrition and Feeding of Fish and Crustaceans, J. Guillaume, S. Kaushik, P. Bergot, and R. Metailler, eds. London, UK: Springer.

Cahu, C. L., J. L. Z. Infante, and V. Barbosa. 2003. Effect of dietary phospholipid level and phospholipid:neutral lipid value on the development of sea bass (*Dicentrarchus labrax*) larvae fed a compound diet. Brit. J. Nutr. 90:21–28.

Camara, M. R., P. Coutteau, and P. Sorgeloos. 1997. Dietary phosphatidylcholine requirements in larval and post larval *Penaeus japonicus*. Batc. Aquacult. Nutr. 3:39–47.

Cañavate, J. P., and C. Fernández-Díaz. 1999. Influence of co-feeding larvae with live and inert diets on weaning the sole *Solea senegalensis* onto commercial dry feeds. Aquaculture 174:255–263.

Carvalho, A. P. C., A.-M. Escaffre, A. Oliva Teles, and P. Bergot. 1997. First feeding of common carp larvae on diets with high levels of protein hydrolysates. Aquacult. Int. 5:361–367.

Carvalho, A. P. C., A. M. Escaffre, A. Oliva Teles, and P. Bergot. 1999. Protein hydrolysate vs. fish meal in compound diets for 10-day old sea bass *Dicentrarchus labrax* larvae. Aquaculture 171:109–119.

Castell, J. D., J. G. Bell, D. R. Toucher, and J. R. Sargent. 1994. Effects of purified diets containing different combinations of arachidonic and docosahexenoic acid on survival, growth, and fatty acid composition of juvenile turbot (*Scophthalmus maximus*). Aquaculture 128:315–333.

Clawson, J. A., and R. T. Lovell. 1992. Improvement of nutritional value of *Artemia* for hybrid striped bass/white bass (*Morone saxatilis* × *M. chrysops*) larvae by n-3 HUFA enrichment of nauplii with menhaden oil. Aquaculture 108:125–134.

Conceicão, L. E. C., H. Grasdalen, and I. Ronnestad. 2003. Amino acid requirements of fish larvae and post-larvae: new tools and recent findings. Aquaculture 227:221–232.

Conceicão, L. E. C., M. Yúfera, P. Makridis, S. Morais, and M. T. Dinis. 2010. Live feeds for early stages of fish rearing. Aquac. Res. 41(5):613–640.

Coutteau, P., I. Guerden, M. R. Camara, P. Bergot, and P. Zorruelos. 1997. Review of the dietary effects of phospholipids in fish and crustacean larviculture. Aquaculture 155:149–164.

Curnow, J., J. King, J. Bosmans, and S. Kolkovski. 2006. The effect of reduced Artemia and rotifer use facilitated by a new microdiet in the rearing of barramundi *Lates calcarifer* (BLOCH) larvae. Aquaculture 257:204–213.

D'Abramo, L. R. 1979. Dietary fatty acid and temperature effects on the productivity of the cladoceran *Moina macrocopa*. Biol. Bull. 157:234–248.

D'Abramo, L. R., E. I. Perez, R. Sangha, and A. Puello-Cruz. 2006. Successful culture of larvae of *Litopenaeus vannamei* fed a microbound formulated diet exclusively from either stage PZ2 or M1 to PL1. Aquaculture 261:1356–1362.

Dabrowski, K. 1984. The feeding of fish larvae. Present state of the art and perspectives. Reprod. Nutr. Dev. 24:807–833.

Dabrowski, K. 1986. Ontogenetical aspects of nutritional requirements in fish. Comp. Biochem. Phys. A 85A:639–655.

Dabrowski, K., and P. Poczyczynski. 1988. Mass rearing of coregonid larvae and juveniles on dry diet. Aquaculture. 69:307–316.

de Jesus, E. G. T., J. D. Toledo, and M. S. Simpas. 1998. Thyroid hormones promote early metamorphosis in grouper (*Epinephelus coioides*) larvae. Gen. Comp. Endocr. 112:10–16.

Durruty, C., L. Maldonado, G. Gaxiola, T. Garcia, and R. Pedroza. 2002. Requerimientos de proteina dietética en larvas de *Litopenaeus setiferus* y *Litopenaeus vannamei*. P. 104 in Book of Abstracts, 6th International Symposium on Aquatic Nutrition, Cancún, Mexico.

Fernandez-Díaz, C., and M. Yúfera. 1995. Capacity of gilthead seabream *Sparus aurata* L. larvae to break down dietary microcapsules. Aquaculture 134:269–278.

Fernandez-Díaz, C., E. Pascual, and M. Yúfera. 1994. Feeding behavior and prey selection size of gilthead sea bream, *Sparus aurata*, larvae fed on inert and live food. Mar. Biol. 118:323–328.

Fontagné, S., J. Robin, G. Corrazea, and P. Bergota, 2000. Growth and survival of European sea bass (*Dicentrarchus labrax*) larvae fed from first feeding on compound diets containing medium-chain triacylglycerols. Aquaculture 190:261–271.

Gapasin, R. S. J. 1998. Enrichment of live food with essential fatty acids and vitamin C: Effects on milkfish (*Chanos chanos*) larval performance. Aquaculture 162:269–286.

Garcia-Ortega, A., J. Verreth, P. Coutteau, H. Segner, E. A. Huisman, and P. Sorgeloos. 1998. Biochemical and enzymatic characterization of decapsulated cysts and nauplii of the brine shrimp *Artemia* at different developmental stages. Aquaculture 161:501–514.

Gawlicka, A., L. McLaughlin, S. S. O. Hung, and J. de la Novie. 1996. Limitations of carrageenan microbound diets for feeding white sturgeon, *Acipenser transmontanus* larvae. Aquaculture 141:245–265.

Genodepa, J., P. C. Southgate, and C. Zeng. 2004a. Diet particle size preference and optimal ration for mud crab, *Scylla serrata*, larvae fed microbound diets. Aquaculture 230:493–505.

Genodepa, J., C. Zeng, and P. C. Southgate. 2004b. Preliminary assessment of a microbound diet as an Artemia replacement for mud crab, *Scylla serrata*, megalopa. Aquaculture 236:497–509.

Gisbert, E., L. Villenueve, J. L. Zambonino Infante, P. Quazuguel and C. L. Cahu. 2005. Dietary phospholipids are more efficient than neutral lipids for long chain polyunsaturated supply in European sea bass *Dicentrarchus labrax* larval development. Lipids 40:609–618.

Graeve, M., G. Kattner, and W. Hagen. 1994. Diet induced changes in the fatty acid composition of Arctic copepods: Experimental evidence of trophic markers. J. Exp. Mar. Biol. Ecol. 182:97–100.

Guerden, I., N. Charlon, D. Marion, and P. Bergot. 1995a. Dietary phospholipids and body deformities in carp *Cyprinus carpio* larvae. Pp. 162–165 in Larvi '95, Fish and Shellfish Symposium, European Aquaculture Society Special Publication 24, P. Lavens, E. Jaspers, and I. Roelants, eds. Ghent, Belgium: European Aquaculture Society.

Guerden, I., J. Radunz-Neto, and P. Bergot. 1995b. Essentiality of dietary phospholipids for carp *Cyprinus carpio* larvae. Aquaculture 131:303–314.

Holme, M.-H., C. Zeng, and P. C. Southgate. 2006. Use of microbound diets for larval culture of the mud crab, Scylla serrata. Aquaculture 257:482–490.

Holt, G. J. 1993. Feeding larval red drum on microparticulate diets in a closed recirculating water system. J. World Aquacult. Soc. 24(2):225–230.

Inui, Y., M. Tagawa, S. Miwa, and T. Hirano. 1989. Effect of bovine TSH on the tissue thyroxine level and metamorphosis in premetamorphic flounder larvae. Gen. Comp. Endocr.74:406–410.

Jones, D. A., A. Kanazawa, and S. A. Rahman. 1979a. Studies on the presentation of artificial diets for rearing the larvae of *Penaeus japonicus* Bate. Aquaculture 17:33–43.

Jones, D. A., A. Kanazawa, and K. Ono. 1979b. Studies on the nutritional requirements of the larval stages of *Penaeus japonicus* using microencapsulated diets. Mar. Biol. 54:261–267.

Jones, D. A., M. Kumlu, L. LeVay, and D. J. Fletcher. 1997a. The digestive physiology of herbivorous, omnivorous and carnivorous crustacean larvae: A review. Aquaculture 155:285–295.

Jones, D. A., A. B. Yule, and D. L. Holland. 1997b. Larval nutrition. Pp. 353–389 in Crustacean Nutrition, L. R. D'Abramo, D. E. Conklin, and D. M. Akiyama, eds. Baton Rouge, LA: World Aquaculture Society.

Kamarudin, M. S., D. A. Jones, L. Le Vay, and A. Z. Abidin. 1994. Ontogenetic changes in digestive enzyme activity during larval development of *Macrobrachium rosenbergii*. Aquaculture 123:323–333.

Kanazawa, A. 1985. Essential fatty acid and lipid requirement in fish. Pp. 281–298 in Nutrition and Feeding of Fish, C. B. Cowey, A. M. Mackie, and J. G. Bell, eds. London: Academic Press.

Kanazawa, A. 1990. Microparticulate feeds for penaeid larvae. Adv. Trop. Aquacult. Actes de Colloque 9:395–404.

Kanazawa, A. 1993. Essential phospholipids of fish and crustaceans. Pp. 519–530 in Fish Nutrition in Practice, S. J. Kaushik, and P. Luquet, eds. Proceedings of the Fourth International Symposium on Fish Nutrition and Feeding, June 24–27, 1991, Biarritz, France.

Kanazawa, A. 1997. Effects of docosahexenoic acid and phospholipids on stress tolerance of fish. Aquaculture 155:129–134.

Kanazawa, A., S. Teshima, T. Inamori, T. Iwashita, and A. Nagao. 1981. Effect of phospholipids on growth, survival rate and incidence of malformation in larval ayu. Mem. Fac. Fish. Kagoshima University 30:301–309.

Kanazawa, A., S. Teshima, S. Inamori, and S. Matsubara. 1983. Effects of dietary phospholipids on growth of larval red sea bream and knife jaw. Mem. Fac. Fish. Kagoshima University 30:301–309.

Kanazawa, A., S. Teshima, and M. Sakamoto 1985a. Effects of dietary lipids, fatty acids and phospholipids on growth and survival of the prawn (*Penaeus japonicus*) larvae. Aquaculture 50:39–49.

Kanazawa, A., S. Teshima, and M. Sakamoto 1985b. Effects of dietary bonito-egg phospholipids on growth and survival of the larval ayu, *Plecoglossus altivelis*. ZX. Angew Ichthyol. 4:165–170.

Kolkovski, S. 2008. Advances in marine fish larvae diets. Pp. 20–45 in Avances en Nutricion Acuicola IX., eds. L. E. Cruz Suárez, D. Ricque Marie, M. T. Salazar, M. G. Nieto López, D. A. Villarreal Cavazos, J. P. Lazo and M. T. Viana. IX Simposio Internacional de Nutrición Acuícola, 24–27 Noviembre, Universidad Autonóma de Nuevo León, Monterey, Nuevo León, México.

Kolkovski, S., A. Tandler, G. W. Kissil, and A. Gertler. 1993. The effect of dietary exogenous digestive enzymes on ingestion, assimilation, growth, and survival of gilthead seabream (*Sparus aurata*, Sparidae, Linneaus) larvae. Fish Physiol. Biochem. 12:203–209.

Kolkovski, S., A. Tandler, and M. S. Izquierdo. 1997a. Effects of live food and dietary digestive enzymes on the efficiency of microdiets for seabass (*Dicentrarchus labrax*) larva. Aquaculture 148:313–332.

Kolkovski, S., A. Arieli, and A. Tandler. 1997b. Visual and chemical cues stimulate microdiet ingestion in seabream larvae. Aquacult. Int. 5:527–536.

Kolkovski, S., W. M. Koven and A. Tandler. 1997c. The mode of action of Artemia in enhancing utilization of microdiet by gilthead seabream *Sparus aurata* larvae. Aquaculture 155:193–205.

Koshio, S., A. Kanazawa, S. Teshima, and J. D. Castell. 1989. Nutritive evaluation of crab protein for larval *Penaeus japonicus* fed microparticulate diets. Aquaculture 81:145–154.

Kotzamanis, Y. P., E. Gisbert, F. J. Gatesoupe, J. Zambonino, and C. Cahu. 2007. Effects of different dietary levels of fish protein hydrolysates on growth, digestive enzymes, gut microbiota, and resistance to Vibrio anguillarum in European sea bass (*Dicentrarchus labrax*) larvae. Comp. Biochem. Phys. A 147:205–214.

Kovalenko, E. E., L. R. D'Abramo, C. L. Ohs, and R. K. Buddington. 2002. A successful microbound diet for the larval culture of freshwater prawn (*Macrobrachium rosenbergii*). Aquaculture 210:385–395.

Koven, W. M., G. Kissil, and A. Tandler. 1989. Lipid and (n-3) requirement of *Sparus aurata* larvae during starvation and feeding. Aquaculture 79:185–189.

Koven, W. M., Y. Bar, S. Lutzky, I. Ben-Atia, R. Weiss, M. Harel, P. Behrens, and P. Tandler. 2001. The effect of dietary arachidonic acid (20:4n-6) on growth, survival, and resistance to handling stress in gilthead seabream (*Sparus sparus*) larvae. Aquaculture 193:107–122.

Kumlu, M. 1999. Feeding and digestion in larval decapod crustaceans. Turk. J. Biol. 23:215–229.

Kumlu, M., and D. A. Jones. 1995. The effect of live and artificial diets on growth, survival, and trypsin activity in larvae of *Penaeus indicus*. J. World Aquacult. Soc. 26(4):406–415.

Kurmaly, K., D. A. Jones, and A. B. Yule. 1990. Acceptability and digestion of diets fed to larval stages of *Homarus gammarus* and their role of dietary conditioning behaviour. Mar. Biol. 106:181–190.

Kvale, A., A. Nordgreen, S. K. Tonheim, and K. Hamre. 2007. The problem of meeting dietary protein requirement in intensive aquaculture of marine fish larvae with emphasis on Atlantic halibut (*Hippoglossus hippoglossus* L.). Aquacult. Nutr. 13:170–185.

Langdon, C. 2003. Microparticle types for delivering nutrients to marine fish larvae. Aquaculture 227:259–275.

Lauff, M., and R. Hoffer. 1984. Proteolytic enzymes in fish development and the importance of dietary enzymes. Aquaculture 37:335–346.

Lazo, J. P., G. J. Holt, and C. R. Arnold. 2000a. Ontogeny of pancreatic enzymes in larval red drum *Sciaenops ocellatus*. Aquacult. Nutr. 6(3):183–192.

Lazo, J. P., M. T. Dinis, G. J. Holt, C. Faulk, and C. R. Arnold. 2000b. Co-feeding microparticulate diets with algae: toward eliminating the need of zooplankton at first feeding in larval red drum (*Sciaenops ocellatus*). Aquaculture 188:339–351.

Lazo, J. P., R. Mendoza, G. J. Holt, C. Aguilera, and C. R. Arnold. 2007. Characterization of digestive enzymes during larval development of red drum (*Sciaenops ocellatus*). Aquaculture 265:194–205.

Léger, P., D. A. Bengston, K. L. Simpson, and P. Sorgeloos. 1986. The use and nutritional value of Artemia as a food source. Oceanogr. Mar. Biol. 24:521–623.

Léger, P., D. Grymonpre, E. Van Ballaer, and P. Sorgeloos. 1989. Advances in the enrichment of rotifers and *Artemia* as food sources in marine larviculture. Aquaculture Europe'89, Short Communications and Abstracts, Special Publication 10:141–142.

Le Ruyet, J. P., J. C. Alexandre, L. Thébaud, and C. Mugnier. 1993. Marine fish larvae feeding: Formulated diets or live prey? J. World Aquacult. Soc. 24:211–224.

Le Vay, L., D. A. Jones, A. C. Puello-Cruz, R. S. Sangha, and C. Ngamphongsai. 2001. Digestion in relation to feeding strategies exhibited by crustacean larvae. Comp. Biochem. Phys. A 128:623–630.

Lian, P., C. E. Lee, and D. A. Bengston. 2008. Development of a squid-hydrolysate-based larval diet and its feeding performance on summer flounder, *Paralichthys dentatus*, larvae. J. World Aquacult. Soc. 39:196–204.

Lovett, D. L., and D. L. Felder. 1989. Ontogeny of gut morphology in the white shrimp *Penaeus setiferus* (Decapoda, Penaeidae). J. Morphol. 201:253–272.

Lovett, D. L., and D. L. Felder. 1990. Ontogenetic change in digestive enzyme activity of larval and postlarval white shrimp *Penaeus setiferus* (Crustacea, Decapoda, Penaeidae). Biol. Bull. 178:144–159.

McEvoy, L. A., J. C. Navarro, F. Amat, and J. R. Sargent. 1995. Optimisation of a phospholipid-enhanced enrichment emulsion for *Artemia*. Pp. 141–144 in Larvi '95. Fish and Shellfish Symposium, European Aquaculture Society Special Publication 24, P. Lavens, E. Jaspers, and I. Roelants, eds. Gent, Belgium: European Aquaculture Society.

McEvoy, L. A., J. C. Navarro, F. Hontoria, F. Amat, and J. R. Sargent. 1996. Two novel *Artemia* enrichment diets containing polar lipid. Aquaculture 144:339–352.

Merchie, G., P. Lavens, and P. Sorgeloos. 1997. Optimization of dietary vitamin C in fish and crustacean larvae: A review. Aquaculture 155:165–181.

Miwa, S., and Y. Inui. 1987. Effect of various doses of thyroxine and triiodothyronine on the metamorphosis of flounder (*Paralichthys divaceus*). Gen. Comp. Endocr. 67:356–363.

Mourente, G., D. Toucher, and J. Sargent. 1991. Specific accumulation of docosahexaenoic acid (22:6n-3) in brain lipids during development of juvenile turbot *Scophthalmus maximus* L. Lipids 26:871–877.

Munilla-Moran, R., J. R. Stark, and A. Barbour. 1990. The role of exogenous enzymes on the digestion of the cultured turbot larvae, *Scophthalmus maximus* L. Aquaculture 88:337–350.

Navarro, J. C., F. Amat, and J. R. Sargent. 1992. Fatty acid composition of coastal and inland *Artemia* spp. populations from Spain. Aquaculture 102:219–230.

Ohs, C. L., L. R. D'Abramo, R. K. Buddington, H. R. Robinette, and J. M. Roethke. 1998. Evaluation of a spray-dried artificial diet for larval culture of freshwater prawn *Macrobrachium rosenbergii*, and striped bass, *Morone saxatilis*. Aquacult. Nutr. 4:73–82.

Onal, U., and C. Langdon. 2004. Lipid spray beads for delivery of riboflavin to first-feeding fish larvae. Aquaculture 233:477–493.

Paibulkichakul, C., S. Piyatiratitivorakul, P. Kittakoop, V. Viyakarn, A. W. Fast, and P. Menasveta. 1998. Optimal dietary levels of lecithin and cholesterol for black tiger prawn *Penaeus monodon* larvae and postlarvae. Aquaculture 167:273–281.

Palmer, P. J., M. J. Burke, C. J. Palmer, and J. B. Burke. 2007. Developments in controlled green-water larval culture technologies for estuarine fishes in Queensland, Australia and elsewhere. Aquaculture 272:1–21.

Park, H. G., V. Puvanendran, A. Kellett, C. C. Parrish, and J. A. Brown. 2006. Effect of enriched rotifers on growth, survival, and composition of larval Atlantic cod (*Gadus morhua*). ICES J. Mar. Sci. 63:285–295.

Pedroza-Islas, R., P. Gallardo, E. J. Vernon-Carter, T. García-Galano, C. Rosas, C. Pascual, and G. Gaxiola. 2004. Growth, survival, quality and digestive enzyme activities of larval shrimp fed microencapsulated, mixed and live diets. Aquacult. Nutr. 10:167–173.

Power, D. M., L. Llewellyn, M. Faustino, M. A. Nowell, B. Th. Björnsson, I. E. Einarsdottir, A. V. M. Canario, and G. E. Sweeny. 2001. Thyroid hormones in growth and development of fish. Comp. Biochem. Phys. C 130:447–459.

Puello-Cruz, A. C., R. S. Sangha, D. A. Jones, and L. LeVay. 2002. Trypsin enzyme activity during larval development of *Litopenaeus vannamei* (Boone) fed on live feeds. Aquac. Res. 33:333–338.

Rabe, J., and J. A. Brown. 2000. A pulse feeding strategy for rearing larval fish: an experiment with yellowtail flounder. Aquaculture 191:289–302.

Rainuzzo, J. R., K. I. Reitan, L. Jorgensen, and Y. Olsen. 1994. Lipid composition in turbot larvae fed live feed cultured by emulsion of different lipid classes. Comp. Biochem. Phys. A 107:699–710.

Reitan, K. I., J. R. Rainuzzo, G. Oie, and Y. Olsen. 1997. A review of nutritional effects of algae in marine fish larvae. Aquaculture 155:207–221.

Rodriguez, C., J. A. Perez, M. S. Izquierdo, J. Mora, A. Lorenzo, and H. Fernandez-Palacios. 1994. Essential fatty acid requirements of larval gilthead sea bream (*Sparus aurata* L.) Aquacult. Fish. Manage. 25:295–304.

Rodriquez, C., J. A. Perez, M. Diaz, and M. S. Izquierdo. 1997. Influence of EPA/DHA ratio in rotifers on gilthead seabream (*Sparus aurata*) larval development. Aquaculture 150:77–89.

Rønnestad, I., Y. Kamisaka, L. E. C. Conceicão, S. Morais and S. K. Tonheim. 2007. Digestive physiology of marine fish larvae: Hormonal control and processing capacity for proteins, peptides and amino acids. Aquaculture 268:82–97.

Sargent, J. R., L. A. McEvoy, and J. G. Bell. 1997. Requirements, presentation, and sources of polyunsaturated fatty acids in marine fish larval feeds. Aquaculture 155:117–127.

Sargent, J. R., L. McEvoy, A. Estevez, G. Bell, M. Bell, J. Henderson, and D. Tocher. 1999. Lipid nutrition of marine fish during early development: current status and future directions. Aquaculture 179:217–229.

Savoie, A., N. R. Le Francois, C. Cahu, P. U. Blier, and I. Andreassen. 2006. Do protein hydrolysates improve survival and growth of newly-hatched spotted wolffish (Anarichas minor), a non-metamorphic aquaculture fish species? Aquaculture 261:782–788.

Shields, R. J., J. G. Bell, F. S. Luizi, B. Gara, N. R. Bromage, and J. Sargent. 1999. Natural copepods are superior to enriched *Artemia* nauplii as feed for halibut larvae (*Hippoglossus hippoglossus*) in terms of survival, pigmentation and retinal morphology: Relation to dietary essential fatty acids. J. Nutr. 129:1186–1194.

Specker, J. L. 1988. Preadaptative role of thyroid hormones in larval and juvenile salmon: Growth, the gut and evolutionary considerations. Am. Zool. 28:337–349.

Szlaminska, M., A. M. Escaffre, N. Charlon, and P. Bergot. 1993. Preliminary data on semi-synthetic diets for goldfish (*Carassius auratus* L.) larvae. Pp. 607–612 in Fish Nutrition in Practice, S. J. Kaushik and P. Luquet, eds. Proceedings of the Fourth International Symposium on Fish Nutrition and Feeding, June 24–27, 1991, Biarritz, France.

Teshima, S., and A. Kanazawa. 1983. Effects of several factors on growth and survival of the prawn larvae reared with microparticulate diets. B. Jpn. Soc. Sci. Fish. 49:1893–1896.

Teshima, S., A. Kanazawa, H. Sasada, and M. Kawasaki. 1982. Requirements of the larval prawn for cholesterol and soybean phospholipids. Mem. Fac. Fish. Kagoshima University 31:193–199.

Teshima, S., A. Kanazawa, and H. Sasada. 1983. Nutritive value of dietary cholesterol and other sterols to larval prawn, *Penaeus japonicus* Bate. Aquaculture 31:159–167.

Teshima, S., A. Kanazawa, and Y. Kakuta. 1986. Growth, survival, and body composition of the prawn larvae receiving several dietary phospholipids. Mem. Fac. Fish. Kagoshima University 35:17–27.

Teshima, S., A. Kanazawa, S. Koshio, and N. Kondo. 1989. Nutritive value of sitosterol for prawn, *Penaeus japonicus*. Nippon Suisan Gakk. 55:153–157.

Teshima, S., M. Ishikawa, and S. Koshio. 1993. Recent developments in nutrition and microparticulate diets of larval prawns. Isr. J. Aquacult. Bamidgeh 45(4):175–184.

Teshima, S., M. Ishikawa, and S. Koshio. 2000. Nutritional assessment and feed intake of microparticulate diets in crustacean and fish. Aquac. Res. 31:691–702.

Tucker, J. W., Jr. 1998. Marine Fish Culture. Boston, MA: Kluwer Academic Publishers.

Tuncer, H., R. M. Harrell, and E. D. Houde. 1990. Acceptance and consumption of food by striped bass and hybrid larvae. J. World Aquacult. Soc. 21(3):225–235.

Villeneuve, L., E. Gisbert, H. Le Delliou, C. L. Cahu, and J. L. Zambonino Infante. 2005. Dietary levels of all-trans retinal affect retinoid nuclear receptor expression and skeletal development in European sea bass larvae. Brit. J. Nutr. 93:1–12.

Yúfera, M., E. Pascual, and C. Fernandez-Díaz. 1999. A highly efficient microencapsulated food for rearing larvae of marine fish. Aquaculture 177:249–256.

Yúfera, M., C. Fernández-Díaz, and E. Pascual. 2005. Food microparticles for larval fish prepared by internal gelation. Aquaculture 248:253–262.

Zambonino Infante, J. L., and C. L. Cahu. 1994a. Development and response to a diet change of some digestive enzymes in sea bass (*Dicentrarchus labrax*) larvae. Fish Physiol. Biochem. 12:399–408.

Zambonino Infante, J. L., and C. L. Cahu. 1994b. Influence of diet on pepsin and some pancreatic enzymes in sea bass (*Dicentrarchus labrax*) larvae. Comp. Biochem. Phys. A 109A, 209–212.

Zambonino Infante, J. L., and C. L. Cahu. 2007. Dietary modulation of some digestive enzymes and metabolic processes in developing marine fish: Application to diet formulation. Aquaculture 268:98–105.

15

Ingredients, Formulation, and Processing

The practical application of fish nutrition is to produce feeds that support growth, health, and welfare of farmed aquatic animals. This objective is achieved by selecting appropriate feed ingredients, deciding how they should be combined to meet the nutritional requirements of farmed aquatic animals, and processing the combination or mixture of ingredients into a physical form suitable for practical use. Each step in the process of making fish and shrimp feeds requires specific information, judgment, and compromise. Complete information must be available for feed ingredients being considered as components of feeds, including proximate composition, nutrient content, quality and potential variability among sources or producers, antinutrient content, contaminant level, and the digestibility of nutrients to farmed fish or shrimp. The nutritional requirements of the fish or shrimp at the life stage for which the feed is being prepared must be known. Potential interactions among feed ingredients that might influence the bioavailability of essential nutrients must be considered, as well as the physical characteristics of a feed ingredient mixture that might affect how it can be processed into pellets. Processing feed ingredient mixtures into pellets also requires careful consideration. Physical characteristics of pellets, such as hardness and durability, water stability, and porosity, are determined by the blend of ingredients used in a feed mixture and by the processing techniques used during conditioning of the ingredient mixture and pelleting. Further, processing techniques increase the availability of some nutrients in feeds but decrease the availability of others. The specific application in which a feed pellet will be used also affects the choice of processing method, the ingredients used in the feed mixture, and the way in which pelleting equipment is operated. For example, the buoyancy of feed pellets can be varied to produce pellets that float or sink. Overall, therefore, feed manufacture is a complicated undertaking that cannot be reduced to a series of steps that can then be applied to multiple situations. In nearly all cases of feed manufacturing, judgment and compromise are made. However, several straightforward considerations can serve as a guide to feed manufacturing that illustrate both the complexity of manufacturing feed and the relatively simple principles upon which feed production is based. Readers seeking more information on fish feed manufacturing are encouraged to consult Hardy and Barrows (2002).

FEED INGREDIENTS

Feed ingredients are selected and combined to supply energy and essential nutrients such as amino acids, vitamins, minerals, and essential fatty acids to support fish growth, health, and reproduction, as well as product quality. Ingredients are also selected and combined based on how they affect the physical characteristics of pellets. Ingredient cost, palatability, physical characteristics, and availability also factor into ingredient selection for feeds. Many feed ingredients are byproducts of human food production, such as soybean meal and corn gluten meal, both byproducts of cooking oil production. Fish meal, in contrast, is now produced exclusively for use in animal feeds. Most ingredients used in fish feeds are also used in livestock or poultry feeds, but some are used exclusively in fish feeds, such as squid liver meal.

Ideally, each ingredient selected as a component of feeds has a specific role to play in the feed. Protein supplements, defined as protein-rich (> 35% crude protein) ingredients, are selected to supply protein and/or specific amino acids to a blended protein mixture. Oil sources are selected to supply dietary energy, essential fatty acids, and, in feeds for crustaceans, sterols. Starch sources, such as ground wheat or corn (maize) starch, are added to provide dietary energy and to act as nutritional binders, plus to allow extruded pellets to expand during starch gelatinization. However, all ingredients are complex mixtures of nutrients, nonnutrients, bioactive compounds, and, in some cases, compounds that interfere with digestion or metabolism, e.g., antinutrients. Antinutrients are mainly found in plant products. Given their complexity, feed ingredients possess both benefits and risks to fish and animals. Risks can be minimized by processing or supplementation; however, to produce high-quality, nutritious feeds, producers must understand both the benefits and risks of ingredients. For more information on antinutrients in feed ingredients, see Chapter 11, as well as Francis et al. (2001), Gatlin et al. (2007), and Krogdahl (2010).

Feed ingredients are sourced from marine resources (fish meal, fish oil, hydrolysates), plant seeds (grains, oilseeds, pulses, and others), rendered animal proteins (poultry byproduct meals, meat and bone meal, blood meal, and others), seafood processing byproducts, and single-cell organisms. Each feed ingredient has a specific definition that includes a description of its source, how it is processed, stage of maturity in the case of forages, and nutritional composition. Information on specific feed ingredients and how they are regulated can be obtained from the Association of American Feed Control Officials (AAFCO, *www.aafco.org*) or in AAFCO's official publication (AAFCO, 2010). Other sources of information on feed ingredient definitions and composition include Hertrampf and Piedad-Pascual (2000) and the Atlas of Nutritional Data on United States and Canadian Feeds (NRC, 1972). All sources have advantages and disadvantages as ingredients in feeds. Advantages are generally associated with nutrient content, while disadvantages include antinutrient contents, presence of contaminants, propensity for becoming moldy and developing aflatoxins, poor or variable quality, potential for oxidation, sustainability or cost.

Premixes are used to supply vitamins and minerals to feeds; these concentrated mixes are added in small amounts to provide appropriate levels of essential vitamins and minerals. Generally, vitamin premixes are added to meet all essential vitamin requirements in feeds used in intensive aquaculture systems where natural food is not available. The potential contribution of vitamins present in feed ingredients is normally not considered. In semiintensive aquaculture systems, vitamin premixes are not expected to supply all essential vitamin requirements.

Additives are supplied to feeds in small amounts to increase digestion of specific feed components, impart color, alter physiology of fish, supply essential nutrients, increase feed intake, and prevent microbial spoilage during storage of feed. Example of additives used to increase feed component digestibility are enzymes, e.g., phytase and amylase. Carotenoid pigments are supplied to feeds to impart color to skin, muscle, or eggs. Examples include astaxanthin, *Haematococcus pluvialis* (an algal species), krill, and crustacean processing byproducts. Supplements added to alter physiological functions include immunostimulants, prebiotics, probiotics, and products designed to enhance smolting and enhance the success of seawater transfer in salmon. Feed-grade amino acids are added to feeds to balance levels of lysine, methionine, or threonine. Feed intake can be improved in some feeds by adding palatability-enhancing materials, e.g, fish hydrolysate fractions, betaine, or mixtures of phosphorylated nucleotides. Feed additives are covered in more detail in Chapter 10.

FEED FORMULATION

Feed formulation is the process of combining feed ingredients to meet the nutritional needs of farmed animals, birds, or fish to produce a mixture that can be pelleted, shipped, and stored; is relatively water-stable; supports growth, health, and wellness; and is economical to use. Formulation begins with establishing target levels for protein, energy, and essential nutrients in the feed. These levels are based upon the established or estimated requirements of the target species. The second step is to select ingredients that are appropriate sources of nutrients and appropriate choices based on the criteria discussed above. The nutrient content in each ingredient being considered must be known, and, if available, the apparent digestibility of nutrients in each ingredient. Most feed formulation is done using least-cost computer programs that calculate the best and most economical combination of ingredients to meet the dietary levels of nutrients specified for a feed. The cost of each ingredient must be known to produce a least-cost feed. Least-cost formulations are somewhat a misnomer in that they are the lowest cost formulation within the constraints imposed on the formulation. For example, a relatively expensive feed ingredient may be required at a low percentage to ensure optimum fish growth. In this case, a minimum level of this ingredient is established, even though removing this minimum constraint would lower the cost of the feed. Feed formulation using least-cost formulation programs is increasingly referred to as "precision formulation" to address this point.

Feed formulation programs are useful, time-saving tools, but they are also notorious for computing extremely efficient blends of ingredients that would fail if made because they cannot be pelleted or sustain rapid fish growth and health. To avoid problems generated by least-cost formulation programs, upper and lower limits are placed on feed ingredients to ensure that the resulting mixture is practical. For example, fish and/or plant oils are excellent sources of dietary energy. Although there are limits to the amount of oils that can be added to feeds, least-cost formulations do not automatically take this limitation into consideration. Likewise, starch is needed to make durable, water-stable pellets, so there has to be a minimum amount of starch in a feed mixture. As mentioned above, for some farmed fish species, a minimum percentage of fish meal has to be included in the feed; without it, growth rates are reduced. In the case of antinutrients in some feed ingredients, allowances must be made to ensure that levels are below the threshold known to reduce fish growth or affect health. Experienced feed formulators are able to set upper and lower limits for ingredients and for dietary nutrient levels in the final feed, and also to make allowances for partial losses of essential nutrients, such as vitamins, that occur during feed pelleting and storage. Another potential pitfall associated with the use of feed formulation programs is failure to take into account changes in dietary energy, protein, or digestibility values when using requirement levels expressed as mg (or g) per kg diet. Experienced operators make adjustments in nutrient requirement levels, especially essential amino acids and phosphorus, to account for differences in dietary energy levels in feed formulations. Establishing minimum levels in feed formulation programs of essential nutrients on a mg or g/kg diet basis can lead to deficiencies in the final formulation unless such adjustments are made.

Least-cost formulation is based on two assumptions that those operating the computer programs often do not realize. First, there is no "ideal" formulation. Many possible formulations can meet the specifications established for nutrient content of the ingredient blend. The key to making a useful ingredient blend are the upper and lower limits placed on ingredients and on nutrient levels. Second, feed formulation programs "assume" that nutrients in various feed ingredients are equal in value, unless otherwise specified. In other words, total lysine or phosphorus in fish meal is equal to lysine or phosphorus in soybean meal unless the apparent digestibility or bioavailability of the nutrient is accounted for. It is wise to remember these assumptions when using least-cost feed formulation programs.

FEED MANUFACTURING

Feed manufacturing is the physical process of forming feed ingredient mixtures into particles used to feed fish or shrimp. Another term for feed manufacturing is pelleting, but this term excludes other types of feeds produced for larval fish, such as flaked feeds, microbound feeds, microextruded marumerized (MEM) pellets, feeds produced by particle-assisted rotational agglomerization (PARA), spray beadlets, microencapsulated feeds, and complex feeds, which are feed particles containing smaller particles in which various nutrients are enclosed (Hardy and Barrows, 2002). Regardless of the technology used or the kind of feed produced, the aim of the manufacturing process is to use physical and mechanical forces to make particles that are practical to ship, store, and use, plus are acceptable to fish. For the most part, commercial (other than larval) feeds are manufactured as pellets by cooking extrusion, compression pelleting, or cold extrusion.

Ingredient mixtures, regardless of the type of pelleting, undergo a series of steps in the process of feed manufacturing that include grinding, mixing, conditioning, pelleting, cooling and drying, top-dressing, packaging (sacking), storing, and shipping. Briefly, grinding is done to reduce particle sizes of ingredients or mixtures prior to pelleting to increase surface area of ingredient particles, and to reduce differences in the average size of particles from different feed ingredients. If the average particle size of feed ingredients is dissimilar, ingredients can segregate during mixing and affect the homogeneity of a mixture. Grinding is also important to reduce particle size to facilitate conditioning and pelleting, and to ensure that particles are small enough to ensure that individual feed pellets contain all nutrients. Mixtures for fish feeds are ground to smaller particle sizes than is necessary for livestock or poultry feeds. For most fish feeds, the mixture is ground to pass through a 200 μm screen. Because some ingredients used in fish feeds containing relatively high lipid contents, grinding to this small screen size can cause the screens to become blocked. To minimize this, grinders that use high air flow to force particles through screens, called windswept pulverizers, are often used in commercial feed production. Ball mills are another approach to reduce particle size of ingredients containing > 5% lipid, such as fish meal. Feed producers can grind individual ingredients and/or feed mixtures. Concerning mixing, the goal is to produce a homogenous blend of dry ingredients. Mixing is an art in itself in that overmixing (mixing for too long a period) can be just as deleterious as undermixing because particles begin to segregate based on material density. Appropriate mixing times can be determined by adding iron filings to the mix, taking samples at intervals and determining the weight of iron filings in samples removed from the mixer using a magnet to remove the filings. Mixers can be batch mixers or continuous mixers; most fish feed producers prefer batch systems.

After mixing, feed mixtures are conditioned in a chamber to which steam and physical agitation are introduced. Conditioning prepares a feed mixture for pelleting, increasing the moisture content, heating the mixture, and adding energy to activate gluten proteins. This process occurs at ambient pressure. After 30 seconds to 5 minutes of conditioning, depending on the feed formulation, type of pelleting and rate of steam addition, the mixture enters either a compression pelleting chamber or an extrusion barrel to form the mixture into pellets. Compression pelleting involves the use of a static roller to force a feed mixture that has been exposed to dry steam for 10–30 seconds through tapered holes in a rotating metal die that resembles a doughnut or a tire. Heat, moisture, and pressure compress the mixture into dense threads that are cut off by a stationary knife as they emerge from the outside surface of the rotating die. Pellet diameter is determined by the diameter of holes in the die, and pellet length is determined by the adjustment of the knife. Compression pellets are often referred to as sinking pellets because of their high density compared to extruded pellets. A variation of the usual compression pelleting system is to install an expander between the conditioner and compression pelleter. The expander is simply a cone through which the feed mixture is forced, creating pressure that gelatinizes starch before pelleting to increase starch digestibility. After pellets exit the pelleting die, they move along a belt through a drying chamber where forced air removes moisture and cools the hot pellets. No added heat is required to reduce the moisture level to < 12%, the target moisture level to prevent mold growth.

Cooking-extrusion pelleting differs from compression pelleting in that more conditioning (added steam and agitation), higher moisture in the feed mixture (~ 20–23%), and much more force and pressure along the course of the extruder barrel are involved. The extruder barrel is pressurized and additional steam is added to the mixture as it moves through the barrel. Inside the barrel is a screw on which the flights become closer, resulting in increased pressure and force. This keeps the added moisture in a liquid state. As a result, starch gelatinization is more complete. When the feed mixture passes through a die shaped like a plate with holes at the end of the barrel, pressure drops to ambient external air pressure and entrapped moisture instantly converts to a gas (steam), creating micropockets within pellet strands that are then cut to length with a rotating knife. Pellet expansion from micropockets reduces bulk density of the pellets. Pellet density can be varied by adjusting conditions (pressure, moisture, heat) in the barrel of the extruder to produce floating, neutrally buoyant, or slowly sinking pellets. Pellets are then conveyed through a forced air, heated dryer to reduce moisture to < 12%.

Both compressed pellets and extrusion pellets can be coated with liquid additives, a process called top-dressing. Typically, fish or plant oil blends are added to pellets after pelleting; adding too much to the mixture before pelleting interferes with pellet compression. Other liquid additives, such as probiotics or enzymes such as phytase, can be added by top-dressing. Extruded pellets can soak up higher amounts of oil than compressed pellets due to differences in density. Vacuum chambers are used to produce high-fat extruded salmon feeds. These chambers are operated on batches by lowering the pressure, then adding the fat, which replaces air in the pellets, soaking fat throughout the pellet rather than adding to the surface, as is the case with compressed pellets.

Cold extrusion simply refers to extrusion without addition of heat or steam. For cold extrusion to produce stable pellets, binders must be included in the feed mixture, and the mixture must contain 28–32% moisture. Cold-extruded pellets should be used within a short period unless they are dried or frozen. The first commercial salmon feeds, such as the Oregon moist pellet, were produced by cold extrusion. Marumerized pellets are also produced by cold extrusion (see below).

Larval feed manufacturing presents several challenges. First, feed mixtures must be very finely ground because the size of feed particles is very small. Second, particles must be water-stable and neutrally buoyant. Third, the techniques used to make water-stable particles must not reduce nutrient digestibility. Fourth, particles must not foul rearing water. Finally, particles must be recognized, palatable, and accepted by fish as food.

Larval feeds can be categorized as microbound, on-sized, microencapsulated, and complex feeds. Microbound feeds are those that use special combinations of feed ingredients to produce extruded or compressed pellets that are then crumbled and sized by screening to yield particles within specific size ranges. Flaked feeds can also be broken up and sized to various small particle sizes. Such particles are commonly used as starter feeds for salmonids and other farmed species not having a larval stage. For larval fish, microbound, crumbled feeds are sometimes effective, but there are limitations to their effectiveness. On-sized feeds, in contrast, are micropellets produced by techniques used to make small pharmaceutical products. They are produced using MEM or PARA processes. Microencapsulated feeds are produced by coating a feed mixture containing nutrients with an impermeable coating or a coating designed to dissolve (controlled release). Examples include cross-linked proteins and lipid-walled microcapsules, the latter often referred to as a "complex feed." These are micropellets produced by MEM or PARA processes that contain even smaller microcapsules produced by other methods distributed within the micropelleted material. The advantage of complex-feed particles is that the interior particle components can be made to have different properties than the main body of the particle. For example, small vehicles can be made to deliver water-soluble nutrients that would otherwise leach from particles, or designed to release nutrients in the intestine after passing through the stomach.

FEED QUALITY ASSESSMENT

Feed production is an accurate but not always exact process. Ingredients vary in composition and quality from batch to batch. Levels of some essential nutrients are reduced during pelleting, drying, and storage. To account for this, feed manufacturers always formulate feeds to supply a slight excess of protein, lipid and other essential nutrients to ensure that sufficient levels of essential nutrients are present when the feed is used. Feed manufacturers also retain samples of feeds for quality testing and in case any dispute about feed quality arises after the feed is used by fish farmers. Feeds are routinely tested for proximate composition and the status of lipid oxidation. Shrimp feeds are also routinely tested for water stability. Abusive storage conditions, such as exposure to moisture or excessive heat, or prolonged storage beyond the manufacturers' recommended shelf life of feeds are the primary causes of feed quality problems at the farm. Feeds showing any signs of mold upon visual inspection or oxidation detected by heat production of feed in bags or by smell should be discarded to prevent fish health problems.

Feed ingredient adulteration, substitution, or mislabeling, either by accident or deliberately, is a concern to feed manufacturers. Adulteration refers to the addition of material to an ingredient to increase its economic value without increasing its nutritional value. An example is addition of melamine to increase the apparent protein content of an ingredient. Melamine contains 66.64% nitrogen on a molecular weight basis. Hence, adding melamine to a feed ingredient increases its nitrogen content, leading to an inflated protein content when Kjeldahl nitrogen is used to analyze protein content ($N \times 6.25$). Suspected adulteration with melamine is easily detected if an amino acid analysis is conducted because the sum of amino acids will not match total protein in the ingredient when analyzed by Kjeldahl nitrogen. Other kinds of adulteration, such as adding soybean hulls to soybean meal, can easily be detected by ingredient protein content. If analysis results of a common ingredient are widely different from tabled values, adulteration, substitution, or mislabeling is a possibility. Another useful technique to check ingredient quality is feed microscopy. Although training and experience is necessary to use feed microscopy effectively, it remains a powerful tool. Further details are available in the Manual of Feed Microscopy and Quality Control, Third edition (Khajarern and Khajarern, 1999).

ENVIRONMENTAL AND SUSTAINABILITY CONCERNS

Environmental and sustainability concerns have had a large impact on fish feeds over the past decade. Excess phosphorus in feeds, for example, leads to excessive excretion of phosphorus, thereby contributing to eutrophication of rivers and lakes receiving fish farm effluent water. Regulations limiting phosphorus levels in fish farm effluent water have led to more sophisticated feed formulation to match available phosphorus levels in feeds with dietary requirements of fish and to limit levels of unavailable phosphorus in feeds. In contrast, regulating phosphorus levels in feeds to achieve reductions in phosphorus levels in farm effluent water has limited innovative solutions associated with changes in feed formulation and phase-feeding strategies. Protein is the most expensive component of feeds, and protein metabolized for energy results in low retention of dietary protein as tissue protein and excessive nitrogen excretion into the aquatic environment. Feeds are increasingly being formulated to supply dietary energy needs of fish with nonprotein sources, resulting in higher protein retention and lower nitrogen losses. Fish farmers have an economic interest in minimizing nutrient losses to the environment and increasing their use to support fish growth. Sophisticated models have been developed to allow fish farms to predict phosphorus and nitrogen losses when different feeds are used (Hua and Bureau, 2006; Dumas et al., 2007; Bureau and Hua, 2010).

Marine resources used to produce fish meal and fish oil are finite resources that have been fully utilized for decades. Aside from higher recovery and utilization of seafood processing byproducts, there is no prospect of increasing fish meal and fish oil production from wild stocks of marine fish (Naylor et al., 2009). Therefore, continued growth of feed production depends on the development and use of alternative sources of protein and oil from sustainable sources. Fortunately, a growing body of knowledge exists to support rationale replacement of fish meal and fish oil with sustainable alternatives in major farmed species, such as salmonids, shrimp, and many marine species (see Chapter 16). However, lack of information on dietary nutrient requirements of many important farmed species and on the effects of ingredient substitution in production feeds limits the extent to which alternative sources of protein and oil can be used. As mentioned earlier, fish meal is a complex material containing a wide array of essential nutrients and biologically active compounds, many of which are absent in plant proteins. Moreover, plant proteins possess negative properties, such as antinutrients and nonsoluble carbohydrates, which have to be overcome to avoid adverse effects on fish growth, health, or reproduction. Replacing fish oil with alternative lipid sources in fish feeds generally does not affect fish performance as long as essential fatty acid requirements are met (see Table 19-3). However, the fatty acid profile of fish tissues reflects the fatty acid profile of the diet, so care must be taken to avoid lowering the content of long-chain polyunsaturated fatty acids by excessive use of alternative lipids in fish feeds. At present, fish oil is the only practical source of long-chain polyunsaturated fatty acids (PUFA) such as eicosapentaenoic acid (EPA) and docosahexaenoic acid (DHA) for farmed fish and crustacean feeds (see Chapters 6 and 16). Further investment in research and development is needed to allow higher levels of replacement of marine resources with plant-derived feed ingredients and possible single-cell proteins to improve the sustainability of aquaculture production.

REFERENCES

AAFCO (Association of American Feed Control Officials). 2010. Official Publication. Oxford, IN: Association of American Feed Control Officials.

Bureau, D. P., and Hua, K. 2010. Towards effective nutritional management of waste outputs in aquaculture. Aquac. Res. 41:777–792.

Dumas, A., C. F. M. de Lange, J. France, and D. P. Bureau. 2007. Quantitative description of body composition and rates of nutrient deposition in rainbow trout (*Oncorhynchus mykiss*). Aquaculture 273:165–181.

Francis, G., H. P. S. Makkar, and K. Becker. 2001. Antinutritional factors present in plant-derived alternate fish feed ingredients and their effects in fish. Aquaculture 199:197–227.

Gatlin, D. M., III, F. T. Barrows, D. Bellis, P. Brown, J. Campen, K. Dabrowski, T. G. Gaylord, R. W. Hardy, E. Herman, G. Hu, Å. Krogdahl, R. Nelson, K. Overturf, M. Rust, W. Sealey, D. Skonberg, E. Souza, D. Stone, R. Wilson, and E. Wurtele. 2007. Expanding the utilization of sustainable plant products in feeds—a review. Aquac. Res. 38:551–579.

Hardy, R. W., and F. T. Barrows. 2002. Diet formulation and manufacturing. Pp. 505–600 in Fish Nutrition, 3rd Edition, J. E. Halver and R. W.Hardy, eds. New York: Academic Press, Inc.

Hertrampf, J. W., and F. Piedad-Pascual. 2000. Handbook on Ingredients for Aquaculture Feeds. Dordrecht, the Netherlands: Kluwer Academic Publishers.

Hua, K., and D. P. Bureau. 2006. Modeling digestible phosphorus content of salmonid fish feeds. Aquaculture 254:455–465.

Khajarern, J., and S. Khajarern. 1999. Manual of Feed Microscopy and Quality Control, 3rd Edition. American Soybean Association and U.S. Grains Council. Available online at: *http://www.asasea.com/index.php?language=en&screenname=__docs_Feed%20Microsopy%20 Manuals*. Accessed on November 18, 2010.

Krogdahl, Å. 2010. Important antinutrients in plant feedstuffs for aquaculture. An update on recent findings regarding responses in salmonids. Aquac. Res. 41:333–344.

Naylor, R., R. Hardy, D. Bureau, A. Chiu, M. Elliott, A. Farrell, I. Forster, D. Gatlin, R. Goldberg, K. Hua, and P. Nichols. 2009. Feeding aquaculture in an era of finite resources. P. Natl. Acad. Sci. USA 106(36):15103–15110.

NRC (National Research Council). 1972. Atlas of Nutritional Data on United States and Canadian Feeds. Washington, DC: National Academy Press.

16

Replacement of Marine Resources: Effects of Alternative Ingredients and Strategies on Nutritional Quality

The last National Research Council publication focused on nutritional requirements for optimal growth and prevention of pathology in fish (NRC, 1993). Since then, a number of major changes dictate that in addition to quantity of fish and seafood produced, the nutrient content and thus the nutritional quality of the product to the human consumer is a major consideration in aquaculture. Some changes have been driven largely by consumers who have become more aware of the link between what they eat and subsequent health. Of great importance for consumers in the developed world, fish are a unique and rich source of n-3 or "omega-3" long-chain polyunsaturated fatty acids (LC-PUFA), eicosapentaenoic acid (EPA), and docosahexaenoic acid (DHA), which have well-known and almost universally accepted beneficial effects on neural development in young children and on a range of human pathologies, including cardiovascular and inflammatory diseases, and neurological disorders (Brouwer et al., 2006; Eilander et al., 2007; Ruxton et al., 2007). These benefits have been embraced by the industry, which has been quick to market fish and seafood as a "healthful" option and component of the diet providing important and, in some cases, unique sources of particular health-promoting nutrients, with omega-3 fatty acids the most prominent. The health benefits of "omega-3" are almost entirely due to EPA and DHA and, therefore, it is important that the levels of those fatty acids are maintained in farmed fish and seafood. Highly pertinent to delivering this aim is the second major change since 1993, which is that the supplies of two of the most important feed ingredients supplying EPA and DHA in fish feeds—marine fish meal (FM) and fish oil (FO)—are rapidly becoming limiting to the growth of aquaculture production (Naylor et al., 2009).

LIMITATIONS TO SUPPLY AND USE OF MARINE RESOURCES, FISH MEAL, AND FISH OIL

With declining capture fisheries worldwide, aquaculture now supplies an increasing proportion of food fish, 47% of the total in 2006, and is the fastest growing food sector with production increasing by an average of almost 9% per year (FAO, 2009). Paradoxically, however, diets for farmed carnivorous and marine finfish and crustaceans have primarily been based on marine resources, specifically low-value, pelagic fish species that have traditionally been major feed ingredients. In some areas, largely Asia, these fish are fed directly to carnivorous marine fish such as groupers, Asian sea bass, cobia, snappers, tuna, etc. as so-called "trash fish" feeds, although efforts continue to promote the use of manufactured extruded feeds. However, pelleted feeds have been traditionally based on FM and FO, derived from industrial, feed-grade fisheries (also called reduction fisheries) of small pelagic species such as anchovies, sardines, herring, and mackerel. The reduction fisheries have, at best, reached their sustainable limits and, over the past 30 years, around 20–25 million tonnes (Mt) of feed fish have been caught annually, which reduce to about 6–7 Mt of FM and 1.0–1.4 Mt of FO. In 1992, aquaculture consumed around 15% and 20% of global FM and FO supplies, respectively (Tacon, 2005). In 2006, aquaculture consumed an estimated 68% of the available FM and almost 89% of the available FO (Tacon and Metian, 2008a). There is no realistic prospect of FM and FO production being increased in the future, and, indeed, there is increasing competition for these small pelagic species for direct human consumption (Tacon and Metian, 2009a,b). In addition, environmental effects such as El Niño can have a severe impact, with the last major event in 1998 reducing FO production to under 0.8 Mt.

In addition to limiting supplies, there is now increasing awareness among consumers and governments of the ecological consequences and environmental impact of human activities. As a result, there are concerns over the use of marine products for animal feeds, with aquaculture especially perceived to be driving exploitation of marine fisheries (Naylor et al., 2000). The efficiency of aquaculture activities in converting dietary nutrients into human food is also receiving considerable attention and is often calculated as a fish-in fish-out (FIFO) ratio, the precise calculation of which is the subject of some debate (see Kaushik and Troell, 2010). However, this debate has also focused attention on the great differences within global aquaculture, because the use of FM and especially FO is greatest for the high-value, high-trophic level carnivorous species generally farmed in the economically developed countries (Tacon et al., 2010). Reflecting this, the FIFO ratio, especially of salmon farming, has received considerable attention. Although the reported values are often controversial, a recent report estimated that they range from about 1.4 to 4.9 for the major carnivorous species including salmonids, marine fish, and shrimp (Tacon and Metian, 2008a). These values are showing a decreasing trend, but the limits imposed by the finite FM and FO supplies described above mean the continued high use of marine products in aquaculture is increasingly perceived as unsustainable and overexploitative of natural resources, adding further pressure to reduce dependency on marine meals and oils (Naylor et al., 2009). However, it should be noted that substitutes for FM and FO also carry an environmental impact. Life cycle assessment/analysis (LCA), the investigation and evaluation of the environmental impacts of a given product or service caused or necessitated by its existence, is now increasingly being applied to fish feeds (Papatryphon et al., 2004). The assessment covers extraction of raw materials, production and transformation of the primary ingredients, manufacturing of the feed, use of feed at the farm, transportation, and the production and use of energy resources. An analysis of salmonid feeds revealed that the use of marine fishery resources and nutrient emissions at the farm contribute most to the potential environmental impacts (Papatryphon et al., 2004).

Consumers also have a greater awareness of risks associated with foods and dietary habits. Various potential food safety risks associated with the use of compound aquaculture feeds, because of endogenous contaminants within the feed ingredients or because of contamination of the feed during storage, such as heavy metals, persistent organic pollutants (POP), *Salmonellae*, mycotoxins, veterinary drug residues, agricultural and other chemicals, and excess minerals (Tacon and Metian, 2008b). Several of these, especially POP, are associated with the use of FM and FO. Particular issues have arisen with heavy metals in wild fish and the lipid-soluble POP in so-called "oily" fish, such as salmon, that deposit substantial amounts of lipid in the flesh. As a result, there have been major perceived health scares around consumption of fish, including mercury levels in shark and swordfish and POP levels in farmed salmon (Hites et al., 2004). Although the content of POP in farmed salmon was shown to be well below all national and international limits (as set by the U.S. Food and Drug Administration, UK and European Food Safety Authority, World Health Organization, and European Union) (Bell and Waagbø, 2008), the negative reporting impacted significantly on sales and, possibly, consumers' perception of farmed fish. Irrespective of the scientific debate over the potential risks, the levels of contaminants allowed in animal feeds and human food are strictly controlled and likely to show a decreasing trend, and continued reduction of contaminant levels in farmed fish is a high priority for everyone involved in fish production.

Therefore, the strictly limited supply and ever rising demand resulting in increasing prices, environmental pressures to use more sustainable resources, and restrictions on contaminants in animal feeds demand that, for aquaculture to continue to expand, sustainable alternatives to FM and FO are required.

SUBSTITUTION OF FISH MEAL

Alternatives to Fish Meal

The protein component of aquaculture feeds is the single most important and expensive dietary component, especially for carnivorous/piscivorous and marine fish that tend to have higher dietary protein requirements than freshwater fish (Wilson, 2002). Formerly, FM was the major dietary protein source, commonly making up between 20–60% of fish diets in general (Watanabe, 2002). Good quality FM is an ingredient that has excellent nutritional properties. It is highly digestible and contains high levels of most essential amino acids. Fish meal also contains numerous essential or conditionally essential nutrients, such as n-3 LC-PUFA and minerals, and supplies other nutrients that contribute to its overall value as an ingredient in fish feeds. However, well-conducted, systematic research has shown that economical feeds can be formulated with low levels of FM and FO, provided feeds contain an array of supplements. Therefore, the dietary requirement of farmed fish for high-quality protein can probably be met by sources other than FM, including terrestrial animal products, single-cell proteins, and plant meals, assuming that energy and essential nutrient levels are sufficient and antinutritional factors controlled (Glencross et al., 2007). However, the production of successful FM-free feeds requires the use of highly processed and expensive ingredients (soy protein concentrate, canola protein concentrate, land animal proteins, krill meal, amino acids, feed attractants). Until recently, these diets were more expensive to produce than diets containing FM, but an increase in FM prices coupled with production of feed-grade protein concentrates has changed the price constraints formerly associated with using food-grade protein concentrates. Nevertheless, the use of a small amount of FM (e.g., 5–20%) in grower feeds for salmonids and marine fish appears to make economical and practical sense at this point in time. Indeed, a variety of protein sources have been part of fish feed formulations since the inception of manufactured feeds, so the current situation is a case of increasing use of alternate protein sources rather than fundamental change (Hardy, 1989; Rumsey, 1993).

Processed animal protein ingredients (often referred to as

land animal products, or LAP) such as blood meal, feather meal, meat and bone meal, and poultry byproducts meal, compare favorably on a cost-per-unit protein with many other protein sources used in fish feeds. The LAP are produced using a variety of raw materials and cooking and drying conditions. They are the most directly comparable substitute for FM because their amino acid compositions are more similar to FM than most plant products, plus they contain essential minerals, phospholipids, and cholesterol depending upon the product. However, they are highly variable in terms of nutritional content. The outbreak of bovine spongiform encephalopathy (BSE) in the United Kingdom in the 1980s led to a ban on the use of animal products in aquaculture feeds within the European Union for many years. Although this ban has been relaxed on some products such as blood meals, there is still only limited use of LAP in aquaculture feeds in Europe, mainly due to consumer and retailer resistance based on perceived risks of including animal byproducts in animal feeds.

Various plant protein sources are commonly used in aquaculture feeds, including meals from oilseeds (soybean, rapeseed/canola, sunflower, cottonseed), grains (wheat and corn glutens), and legumes (peas, beans, peanut and lupins). There are two main factors to consider in formulating feeds with high levels of plant meals: energy density, as some meals have high carbohydrate content (nonstarch polysaccharides being the main issue) that is of little nutritional value for piscivorous/carnivorous species, and amino acid content, because some essential amino acids, such as lysine and methionine, are generally deficient in plant protein sources (Gatlin et al., 2007). Further, a wide range of antinutritional factors, including protease inhibitors, lectins, saponins, and phytic acid, are present in plant feedstuffs and must be controlled (Hardy and Barrows, 2002; Krogdahl et al., 2010). Defatted soybean meal is consistently available and cost effective and, importantly, has a favorable amino acid profile compared with many other plant protein sources. It is also generally accepted, both qualitatively and quantitatively, by most fish species (Watanabe, 2002). However, a further concern in some regions of the world, especially Europe, regarding plant proteins is the presence of genetically modified (GM) products, especially those derived from soybean, canola, and maize (Pusztai and Bardocz, 2006). No differences in fish performance have been found between fish fed diets containing GM products and their non-GM equivalents (Hemre et al., 2004, 2007; Li et al., 2008; Sissener et al., 2009a,b). However, a study in Atlantic salmon showed that, although short transgenic sequences were detected in gut tissues, no transgenic fragments were found in liver, muscle, or brain (Sanden et al., 2004).

Considering the wide range of plant feedstuffs available, blending various plant protein sources is likely to be a viable strategy for replacing dietary FM. Energy density can be maintained by avoiding high inclusion levels of meals with lower protein:carbohydrate (NSP) ratios; the use of protein concentrates, amino acid composition, and content can be balanced with crystalline amino acid supplements; and antinutritional factors can be controlled using heat treatments, phytase, and genetic improvement of plants in the longer term. These issues are covered in considerably more depth in Chapters 4, 6, and 11.

Fish meals also supply significant amounts of oil including cholesterol and phospholipids, the latter at higher levels than FO. Thus, lipid contents in FM can vary from 5–13% of weight, with a triacylglycerol (TAG)/phospholipid ratio of around 2:1 (de Koning, 2005), and phospholipid can account for 5–25% of the total lipid of FM-based feeds depending on lipid content and formulation (Johnson and Barnett, 2003). Plant meals generally have much lower levels of residual phospholipids, mainly due to the original seeds/beans having much lower phospholipid contents. Soybean has the highest levels with full-fat soybean meal (FFSM or soy flour) having about 20–25% lipid with around 0.3–0.6% phospholipid, primarily phosphatidylcholine (PC), phosphatidylethanolamine (PE), phosphatidylinositol (PI), and phosphatidic acid (PA). Individual phospholipid class compositions for FM reflect the phospholipid composition of fish (PC, PE, PI, PS, PA, and lyso-phospholipids) (Tocher, 1995). Few data are available on the lipid and phospholipid contents and compositions of other plant meals. Cholesterol is the predominant sterol in animals, whereas plant products, both oils and meals, contain virtually no cholesterol, usually < 5% of total sterols (Tocher et al., 2008). Thus, feeds formulated with FM and FO will provide at least 1 g of cholesterol per kg feed, and substitution with plant products will greatly reduce the level of cholesterol supplied by the feed. In contrast, other sterols (phytosterols) are present in plant tissues, including β-sitosterol and campesterol with smaller amounts of Δ^5-avenasterol, stigmasterol, and brassicasterol (Padley et al., 1994).

Effects of Fish Meal Substitution on Growth of Fish

Fish meal levels in fish and shrimp feeds for most species have decreased by approximately half in the past decade. Although substitution of FM is a greater challenge with salmonids and marine fish than with many freshwater species, such as cyprinids and tilapia, replacing a large proportion of FM with alternative sources can be accomplished without deleterious effects on growth performance, nutrient utilization, and health of salmonid fish (Kaushik et al., 1995; Rodehutscord et al., 1995; Espe et al., 2006, 2007). Indeed, FM-free feeds have been shown to support trout performance equivalent to FM-based feeds when using a combination of other protein sources of plant or animal origin (Lee et al., 2002; Barrows et al., 2007). However, other studies have resulted in significantly lower growth performance of the fish, despite the fact that the experimental diets appeared to be nutritionally adequate. Health issues such as intestinal inflammation or enteritis have been observed in salmon and, to a lesser extent, in trout when fed soybean meal (Baeverfjord and Krogdahl, 1996; Bureau et al., 1998). Differences in feed formulation and choice of alternative protein sources are the likely cause of different results (Kaushik et al., 1995;Teskeredzic et al., 1995; Medale et al., 1998; Watanabe et al., 1998; Glencross et al., 2004; Morris et al., 2005). The duration of experimental feeding trials may also be a factor that explains differing results with FM replacement. For instance, in a 6-month trial, replacement of FM above 75% resulted in growth reduction in rainbow trout, despite the addition of crystalline amino acids, although growth rates only became depressed after 3 months, which is similar to the length of many nutritional trials (de Francesco et al., 2004). This may be due to depletion of stores of some unknown nutrient that is deficient in the FM substitutes. Reduced growth performance in salmonids may also be a consequence of reduced feed intake and impact on gut integrity rather than effects on apparent protein and amino acid digestibilities (Kaushik et al., 2004; Espe et al., 2006). Some studies appear to suggest that changes in palatability and physical properties of feeds may cause the reduced feed intake, particularly during initial experimental periods (Gomes et al., 1995; de Francesco et al., 2004; Kaushik et al., 2004; Espe et al., 2006). Atlantic salmon appear less able than rainbow trout to tolerate high levels of plant proteins, possibly related to differences in digestive capacity and sensitivity to antinutritional factors, although the evidence is not clear (Refstie et al., 2000; Glencross et al., 2004; Mundheim et al., 2004; Espe et al., 2006).

Complete replacement of FM with maize gluten and soy protein concentrate significantly reduced feed intake and growth in European sea bass (Dias, 1999). In a 12-week trial in sea bream, slight growth reductions when fed diets containing 50–75% plant protein ingredients and a 30% growth reduction with complete replacement of FM were associated with a marked reduction in feed intake (Gomez-Requeni et al., 2004). Reduced feed intake has also been observed with turbot fed either maize gluten or lupin (Burel et al., 2000a,b). However, in turbot fed for 12 weeks on a mixture of lupin, wheat, and maize gluten with supplementary crystalline amino acids, growth rate was only compromised with 90–100% replacement (Fournier et al., 2004). Furthermore, sea bass fed up to 98% of dietary protein as plant meals showed no reduction in voluntary feed intake or growth over a 12-week period (Kaushik et al., 2004) and, in a longer-term trial in sea bream, 75% plant protein inclusion did not reduce growth (Sitja-Bobadilla et al., 2005).

It is probable that unknown nutrient deficiencies (e.g., arachidonic acid [ARA], taurine, cholesterol, phospholipids, or sphingolipids) and differences in digestible nutrient contents of diets are important factors in several of these studies, particularly those performed in Europe where the use of certain feed ingredients, especially LAP, has been restricted. Most economical alternative feed ingredients (e.g., soybean meal, corn gluten meal, canola meal, meat and bone meal, feather meal, and animal fats) have significant limitations and cannot be used individually at very high levels in the diet of most fish species. Therefore, formulating successful cost-effective feeds, relying less on FM and FO, requires not only the use of combinations of economical ingredients and a complete understanding of the nutrient requirement of the animal, but also several other, less well-defined factors, including tolerance to antinutritional factors, interactions between nutrients, and palatability of the finished feed. Thus, better knowledge of the composition, availability of nutrients, variability, and limitations of various feed ingredients is equally important. It is likely that a combination of plant-derived and other alternative feed ingredients will be required to replace FM and that feed supplements, such as amino acids, flavorings, and phytase, will be necessary to produce aquafeeds lacking FM that support growth rates necessary for economic production of farmed fish (Gatlin et al., 2007; Webster et al., 2007). Meeting all the indispensible amino acid requirements of the animal is the essential factor. Indeed recently, it was shown that the marine fish Senegal sole (*Solea senagalensis*) could be grown effectively on diets devoid of FM provided the dietary amino acid composition was appropriately formulated (Silva et al., 2009).

In summary, studies have shown that economical feeds can be formulated with very low to no FM provided great attention is paid to the formulation process. It is likely that nutrient deficiencies, differences in digestible nutrient contents of diets and possibly reduced feed intake associated with altered palatability can explain the poor performance of fish fed low FM diets observed in many published studies.

Effects of Fish Meal Substitution on Nutrient Content of Fish

A large body of literature describes the effects of substitution of FM with alternative protein sources in feeds for fish with studies encompassing many different alternatives, fish species, and levels of substitution, but relatively few have monitored the influence on nutrient content or flesh quality. However, Gatlin et al. (2007) reviewed the studies that had investigated the effects of plant protein sources on fish fillet quality and found that significant effects were reported in around 40% of the studies. Although color was the attribute most affected by protein source, texture and flavor could also be influenced, particularly when the level of FM substitution was high. In contrast, several studies reported no differences in consumer acceptability of fish fillets when the level of plant protein inclusion was in the range of 30–60% of the feed. Therefore, the data conflict. It is clear that the effects of different protein sources, including plant products, as substitutes for FM in fish feeds, and their effects on nutritional and product quality, requires some further investigation.

SUBSTITUTION OF FISH OIL

Alternatives to Fish Oil

Lipid (oil or fat) in dietary formulations serves two important roles, as a high-density energy supply and as a source of essential fatty acids (EFA). The available alternative oils can generally very effectively supply energy, but few have favorable EFA compositions. Thus, the almost unique feature of marine FOs in fish feeds is as a source of LC-PUFA, EPA, and DHA, which together can satisfy the EFA requirements of all fish species (Tocher, 2003). The limited global supply of FO, rather than FM, is currently the more serious problem for two reasons. First, it is far more imminent, and shortages are already causing concern in the aquaculture feed industry, whereas FM will not be limiting for a few more years (FAO, 2009). Second, as described above, a number of alternative protein sources and strategies can replace FM, whereas there are no ideal substitutes for marine FO as suppliers of LC-PUFA.

Alternative Marine Oils

In marine ecosystems, LC-PUFA are synthesized by the primary producers, predominantly phytoplankton, and are incorporated into the lipid of zooplankton and subsequently fish. There are a few alternative marine oils from species at a similar trophic level to fish, including squid and seal oils, but they are produced in only small quantities. They cannot be regarded as any more sustainable than FO and have equally challenging ecological issues. Other alternative sources of marine LC-PUFA include harvesting lower trophic levels, specifically zooplankton (Olsen et al., 2010). Potential zooplankton species include copepods, assumed to be the most abundant; euphausiids, including *Euphasia superba* in the Antarctic, *E. pacifica* in the Pacific, and *Meganyctiphanes norvegica* in the Arctic; and amphipods. Several problems are associated with harvesting zooplankton, not least the technical difficulties in trawling small animals and the consequent environmental impact of the high energy demand, and the need to process the catch immediately because zooplankton have very low stability and deteriorate quickly after catch (Sæther et al., 1986). However, there are also biological problems because of great variability in content and composition of zooplankton lipids, and their fatty acid composition, based on species, season, and life cycle. Calanoid copepod oil is rich in wax esters rather than TAG, and so there are also possible chemical problems associated with the high content of fatty alcohols. The greatest drawback is ecological. At present, the harvest is very low and generally not permitted by local governments, except for exploratory fishing at very low biomass. Current knowledge of biomass, annual production, and the level of fish and marine mammal consumption is limited, and so large-scale commercial harvest cannot be demonstrated as sustainable. However, with the existing biomass and potential in mind, research programs are aimed at obtaining good estimates of biomass and production in many countries. Therefore, it is possible that some harvest may be allowed in the future.

A further potential source of marine oil is mesopelagic fish that live in the intermediate pelagic water masses between the euphotic zone at 100 m depth and the deep bathypelagic zone where no light is visible at 1,000 m. Most species make extensive vertical migrations into the epipelagic zone at night where they prey on plankton and each other. Despite their huge biomass (Gjøsæter and Kawaguchi, 1980), fishing of many mesopelagic fish species failed in the 1970s. One reason may be that, like calanoid copepods, these species are often rich in wax esters that are not suitable for human nutrition as they are mildly laxative and can cause loss of hair and skin damage (seborrhea) (Place, 1992). However, wax esters represent a major storage of lipid in the marine environment (Lee et al., 2006), and many fish, including Atlantic salmon, appear to effectively utilize copepod oil. Thus, wax ester-rich oils may have the potential to be useful source of marine lipid (Olsen et al., 2004).

Other potential sources of marine oils include fisheries bycatches and byproducts. Bycatches, which are nontarget animals, are considered waste and generally discarded at sea. They are estimated at around 7 Mt or around 8% of global catch (Kelleher, 2005). However, in Asia, low-value bycatches are now being landed and used for animal feeds. More than 6 Mt of "trash fish" have been estimated to be used globally in aquaculture (Tacon et al., 2006). The main fisheries producing discards are shrimp (in particular tropical) and demersal finfish trawl, accounting for more than 50% of all discards. The northern Pacific is the area with the highest discard rate, contributing to 40% of all discards. Processing byproducts from pelagic fisheries, including all fatty fish, like salmon, herring, and mackerel, and farmed fish are other potential sources of FO. The amount of waste produced is largely dependent on the type of processing, with the production of fillets resulting in around 50% waste. In 2006, fisheries of Atlantic herring, mackerel, and various salmonids were 2.25, 0.56, and 0.89 Mt (FAO, 2009), respectively, potentially producing 1.9 Mt of waste that, with an average lipid content of 20%, could generate 0.38 Mt of LC-PUFA-rich FO. It has been estimated that seafood processing waste may amount to 1 Mt in Alaska (Crapo and Bechtel, 2003) and around 0.65 Mt in Norway (Rubin, 2007), and globally the volumes are probably comparable to bycatches. The increased value of byproducts, combined with an aim to exploit these resources, suggests utilization is probably increasing. In 2007 in Norway, almost 0.5 Mt was processed, which included all farmed fish trash and rejects and a significant portion of the commercial fishing waste. There are limitations to the use of byproducts because refeeding (e.g., salmonid-based products to salmonids) is not permitted and processing waste from the major farmed freshwater species like carp are probably of more limited use because the levels of n-3 LC-PUFA may be lower.

Single-Cell and Algal Oils

The vast majority of marine LC-PUFA originate in single-celled organisms, predominantly microalgae such as diatoms and dinoflagellates, some closely related algae-like microbes, and bacteria. Single cell oils (SCO) are essentially microbial lipids produced by a range of bacteria, fungi (including yeasts), and heterotrophic and photosynthetic microalgae. For instance, various oleaginous zygomycetes have potential medical or nutritional interest due to their specific fatty acid compositions including LC-PUFA such as ARA, produced by the fungus *Mortierella alpina* (Kendrick and Ratledge, 1992; Certik and Shimizu, 2000). Thraustochytrids such as schizochytrium, marine microheterotrophs aligned taxonomically with heterokont algae, can be grown using fermentor-like technology producing relatively high biomass and contents of oils rich in DHA (Bajpai et al., 1991; Kendrick and Ratledge, 1992). Replacement of dietary FO with thraustochytrid oil had no effect on growth over a 9-week trial in Atlantic salmon parr, and tissue DHA levels were significantly increased (Miller et al., 2007). However, production is not currently on a scale or at a cost required to replace FO, and so thraustochytrid oil and/or biomass can only be used in specialized situations such as in marine fish larval feeds and enrichers. The only other significant algal oil/meal presently available is from the marine heterotrophic dinoflagellate alga *Crypthecodinium cohnii*, which also produces a DHA-rich lipid and is currently used spray-dried in products for larval nutrition. Production of photoautotrophic (photosynthetic) microalgae, including *Isochrysis galbana*, *Nannochloropsis oculata*, and *Pavlova lutheri*, in plastic bioreactors using light as a source of energy, is currently restricted to providing live feeds for rotifers (*Brachionus* spp.) and brine shrimp (*Artemia* spp.) or so-called "green water" for marine larval fish and shellfish culture. As such it is at a local, hatchery-based level, and there is no major commercial production. Therefore, production of marine phytoplankton is currently limited by bioreactor system technology that is not yet capable of producing algal biomass on the scale required to replace FO. As a result, there are currently no good sources of EPA-rich SCO. Further research and development is required to improve production systems and the genetic modification of strains.

Finally, there is currently great interest in algae and other microorganisms for the production of biofuels. The rather different market for fuels may drive technological advances at a quicker pace and, if successful, could develop into a large industry. Depending upon organism/species and culture conditions, it is possible that post-oil extraction byproducts from this potential industry, such as cellular material, may have nutrient and/or fatty acid compositions that could have useful applications in aquaculture feeds.

Transgenics

Considerable advances have been made in the past 10 years in elucidating the molecular mechanisms of LC-PUFA synthesis in organisms ranging from microorganisms to vertebrates, including the major primary producers, microalgae and higher plants. The isolation and cloning of the molecular machinery, the genes and cDNAs, have opened up the possibility of genetically engineering the trait for n-3 LC-PUFA (EPA and DHA) synthesis in oil-producing organisms including oleaginous yeasts and oilseed plants. For instance, the oleaginous yeast, *Yarrowia lipolytica*, can accumulate oil up to 40% of the dry weight, although normally the only PUFA is 18:2n-6. Metabolic engineering has produced a strain that can produce TAG with an EPA content of > 50% (Zhu et al., 2009). However, at the moment, large-scale production of low-cost n-3 LC-PUFA from microorganisms requires some technical advances in the production systems, whether fermentor-based or not, and significant infrastructure development. In contrast, the technology and infrastructure for oilseed production and processing is already globally available, and transgenic oilseeds may ultimately offer the most effective and cost effective solution.

Higher plants do not normally synthesize LC-PUFA (Table 16-1), but over the past 15 years considerable research effort has gone into bioengineering this trait into oilseed crops. Early work investigated the expression of various Δ6 desaturases in soybean and safflower, resulting in γ-linolenic acid (GLA, 18:3n-6) contents in the seed oils as high as 70% of total fatty acids (see Damude and Kinney, 2008). However, expression of an ω-3 (Δ15) desaturase gene along with the Δ6 gene is required to produce stearidonic acid (SDA, 18:4n-3). Recently, coexpression of borage Δ6 and *Arabidopsis* ω-3 desaturase genes in soybean seed resulted in SDA contents as high as 30% (Eckert et al., 2006). An SDA-rich GM-soybean oil has been developed and is likely to be the first transgenic seed oil containing n-3 PUFA to be commercially available (Monsanto, 2008) (Table 16-1).

For production of LC-PUFA, EPA, and DHA in oilseeds, there are several possible strategies, including recreating in plants the microbial or animal pathways, or the polyketide (PKS) pathway as found in some thraustochytrids. Recreating the microbial pathway was initially demonstrated by Qi et al. (2004), who expressed three genes from microalgae and fungus in *Arabidopsis* resulting in around 3% EPA in leaves. Subsequently, production of EPA in soybean seeds was achieved by coexpression of Δ6 desaturase from *Mortierella* with n-3 desaturases from *Arabidopsis* and the freshwater mold *Saprolegnia* with contents reaching 10% of seed fatty acids (Kinney et al., 2004). Replacing *Mortierella* Δ6 desaturase with *Saprolegnia* Δ6 increased seed EPA to 20%. When a Δ4 desaturase gene from *Thraustochytrium* was added, more than 3% DHA was observed in soybean embryo oil (Kinney et al., 2004).

Production of EPA (~ 2%) was also demonstrated in lin-

TABLE 16-1 Fatty Acid Compositions (Percentage of Total Fatty Acids) of Major Vegetable Oils and Animal Fats[a]

Oil/Fat	16:0	18:0	18:1n-9	18:2n-6	18:3n-3	Other Significant Fatty Acids
Borage	10	4	18	36	tr[b]	26 (18:3n-6), 6 (C20/22)
Buglossoides	4	2	7	11	39	5 (18:3n-6), 18 (18:4n-3)
Camelina	5	2	12	15	47	19 (C20 monos)
Coconut	8	3	6	2	0	76 (sats < 16:0)
Corn	11	2	24	58	1	
Cottonseed	23	2	17	52	0.2	
Echium	7	4	16	15	32	11 (18:3n-6), 14 (18:4n-3)
Evening primrose	8	2	11	68	tr	10 (18:3n-6)
Linseed/Flaxseed	5	4	20	13	53	
Olive	11	2	73	8	0.6	
Palm	44	4	37	9	0	
Palm kernel	8	2	16	2	0	72 (sats < 16:0)
Peanut	10	2	45	32	0	7 (LC-sat/mono)
Perilla	6	2	12	16	64	
Rapeseed/Canola	3	2	60	20	12	
SDA-soybean	12	4	15	21	11	7 (18:3n-6), 26 (18:4n-3)
Sesame	9	6	38	45	0.6	
Soybean	10	4	23	51	7	
Sunflower	6	5	20	66	0	
Lard	24	14	41	10	1	3 (16:1)
Poultry	22	6	37	20	1	6 (16:1)
Tallow	25	19	36	3	1	4 (16:1)

[a]Padley et al. (1986, 1994); Gunstone and Harwood (2007).
[b]tr = trace.

seed (flax) transformed with genes from the diatom *Phaeodactylum tricornutum* and the fungus *Physcomitrella patens*, although a build-up of GLA and SDA was observed (Abbadi et al., 2004). This was suggested to be due to a "substrate-dichotomy bottleneck" that required fatty acids to be shuttled between phospholipid substrates, required for desaturations, and acyl-CoA pools that are required for elongations (Napier, 2007). Thus, addition of a lysophosphatidyl acyltransferase from *Thraustochytrium* and an n-3 desaturase increased EPA production in *Brassica juncea* seeds up to 15% of total fatty acids (Wu et al., 2005). Further, addition of an elongase from rainbow trout and the Δ4 desaturase from *Thraustochytrium* resulted in production of small amounts of DHA (~1.5%), and also proved that the use of animal genes may be possible. Indeed, using animal desaturases that use acyl-CoA substrates was suggested as a mechanism for avoiding the substrate dichotomy bottleneck (Domergue et al., 2005; Robert et al., 2005). Thus, expression of a bifunctional Δ6/Δ5 acyl-CoA desaturase from zebrafish along with an acyl-CoA elongase from the nematode *Caenorhabditis elegans* in *Arabidopsis* resulted in up to 2.5% EPA in the seed oil (Robert et al., 2005). Retransformation of these plants with Δ5 elongase and Δ4 desaturase genes from *Pavlova salina* resulted in trace amounts of DHA (0.5%) in the seed oil.

Assembling LC-PUFA pathways with multiple enzymes is complicated (Napier, 2007). An alternative is to express polyketide (PKS) complexes, which have most of the activities necessary for EPA or DHA synthesis in a single polypeptide. Expressing three genes from *Schizochytrium* encoding subunits of PKS that catalyze the synthesis of DHA from malonyl-CoA in *Arabidopsis* resulted in a DHA content of around 1% of seed fatty acids (Metz et al., 2006), but this has subsequently been increased to approximately 2.5% (Metz et al., 2007).

Just as transgenics can be used to modify feedstuffs (GM oilseeds) to suit the metabolism of the fish, it can equally be used to modify the metabolism of the fish to suit the available feeds. Stable transformed zebrafish were produced by microinjection of 1-cell embryos with either Δ6-desaturase-like, Δ5-desaturase-like, or elongase-like genes of masu salmon (*Oncorhynchus masou*) (Alimuddin et al., 2005, 2007, 2008). The levels of EPA and DHA in the transgenic zebrafish were slightly, but significantly, increased over wildtype controls. Therefore, molecular tools and technologies that could transform the aquaculture feed industry are already available and developing rapidly. However, application of these tools will be dependent upon public acceptance of the technology, which currently varies considerably throughout the world, with European consumers being particularly opposed to GM products in the food chain.

Animal Fats

The major animal fats in terms of tonnages are beef/sheep tallow and pork lard, with more than 16 Mt being produced in 2008 (Table 16-2). There was also around 7 Mt of butter

TABLE 16-2 World Oil and Fat Production 2008[a]

Oil/Fat	Million Tonnes (Mt)
Palm	42.4
Soybean	37.7
Rapeseed/Canola	19.4
Sunflower	10.1
Cottonseed	5.1
Palm kernel	4.9
Peanut	4.5
Olive	2.9
Corn	2.4
Coconut	3.1
Sesame	0.8
Linseed/Flaxseed	0.6
Castor	0.6
Total vegetable oil	**134.5**
Tallow	8.6
Lard	7.7
Butter	7
Fish	1.1
Total animal fat/oil	**24.4**

[a]Gunstone (2010).

produced. Despite the globally high level of poultry production and consumption, there are few data on poultry fat/oil production, but it is mainly produced as a byproduct of the animal rendering industry and extracted from discarded poultry parts. The fatty acid compositions of animal fats are generally dominated by high-saturated (up to 50%) and monounsaturated fatty acids (Table 16-1), but they vary quite widely largely dependent upon the dietary history of the animals. For instance, poultry fat can contain up to 20% PUFA and low levels of EPA and DHA, and Australian poultry fat can contain > 50% monounsaturated fatty acids, predominantly 18:1n-9, because of the practice of feeding canola (rapeseed) grains (J. G. Bell, personal communication, May 2009). In general, animal fats can be regarded as lacking n-3 LC-PUFA. Therefore, the high saturated fat, low PUFA, and lack of LC-PUFA, and presence of trans fatty acids (Padley et al., 1986, 1994; Gunstone and Harwood, 2007) dictate that animal fats generally cannot be considered as good substitutes for dietary FO in terms of their fatty acid profile for most species, although they could be useful energy sources if incorporated into dietary formulations at appropriate levels in some species. Tallow may also be a reasonable source of conjugated linoleic acid (CLA).

Vegetable Oils

None of the above potential oil sources is currently an ideal replacement for FO. All have one or more ecological, environmental, technological, nutritional, sustainability, or other issue that limits their supply and use. In contrast, oils from terrestrial plants, specifically the oilseed crops producing vegetable oil (VO), have almost no restraints to supply, although their production and use also raise environmental and nutritional issues, respectively. In 2008, global FO production was around 1.1 Mt. In contrast, production of VO exceeded 135 Mt with the four major oils: palm (42.4 Mt), soybean (37.7), rapeseed/canola (19.4), and sunflower (10.1), accounting for more than 80% of the total (Table 16-2) (Gunstone, 2010).

The major difference between FO and VO is fatty acid composition, as higher plants generally do not produce fatty acids with chain length greater than C_{18}, and so the C_{20} and C_{22} LC-PUFA are not components of any VO. The fatty acid compositions of the majority of VO are dominated by four fatty acids, 16:0, 18:0, 18:1n-9, and 18:2n-6, although the major fatty acid varies. Therefore, saturated fatty acid-rich oils include palm oil (16:0) and the short-chain or lauric (12:0)-rich oils, coconut and palm kernel. Monoenoic acid (18:1n-9)-rich oils include rapeseed and olive oils, and linoleic acid (18:2n-6)-rich oils include soybean, sunflower, cottonseed, corn, and sesame (Table 16-1). Linolenic acid (18:3n-3) is much less common. The only major VO rich in this fatty acid is linseed (flax), although it still only represents 0.5% of world VO production (Table 16-2). Camelina and perilla, other 18:3n-3-rich oils, are being produced in much lower volumes. Soybean and rapeseed oils contain reasonable amounts of 18:3n-3, but with 18:2n-6:18:3n-3 ratios of almost 7 and more than 2, respectively. The data in Tables 16-1 and 16-2 can be used to calculate the overall abundance of individual fatty acids by multiplying their percentage content in an oil by the tonnage of that oil. This shows that the 18:2n-6:18:3n-3 ratio in world oil supply post hydrogenation is around 24. Some oilseeds express a Δ6 desaturase, and so a few oils are available that contain Δ6-desaturated fatty acids, such as 18:3n-6 in borage and evening primrose oils, and SDA (18:4n-3) in echium and bugglossoides oils.

Despite the limitations in fatty acid compositions, VO are currently the only readily available, cost effective, and sustainable alternative lipid source for aquaculture diets. In the past dozen years or so, considerable research has been performed to investigate the effects of complete and partial replacement of dietary FO with VO with freshwater species such as catfish, carp, and tilapia; salmonids including Atlantic salmon, rainbow trout, and Arctic charr; and marine species including gilthead sea bream, European sea bass, turbot, and cod being most prominent (Turchini et al., 2009). The VO most commonly used have included rapeseed/canola, soybean, palm, sunflower, corn, olive, linseed, and echium (Turchini et al., 2009). The number of publications in the scientific literature is now in the hundreds and, until recently, literature reviews were limited (G. Bell et al., 2005; Bell and Waagbø, 2008; Tocher, 2009; Turchini et al., 2009). The following sections summarize the results from the many trials drawing from a recent and extensive review volume that provides a comprehensive reference base for this subject (Turchini et al., 2010).

Impact of Replacement of Dietary Fish Oil with Vegetable Oils on Growth of Fish

Perhaps surprisingly, despite the number of studies, the effects of substitution of FO with VO on growth performance are not entirely clear. In a recent review, Turchini et al. (2009) concluded that ". . . if the EFA requirements are met, a significant proportion (60–75%) of dietary FO can be substituted in almost all finfish species studied." Perhaps the biggest question with FO substitution is, therefore, if EFA requirements are otherwise met, why cannot 100% of FO be substituted in all species? Many factors can affect the results in substitution trials in addition to species and the fatty acid compositions of the replacing oils and the overall diet, including the other dietary ingredients (such as the level of FM and other protein sources), duration of the trials, size of the fish, and possibly environmental conditions. Species is a major factor. Although fish can be placed into several groups based on ecology, particularly water temperature (cold, temperate, and warm) and feeding habit (piscivorous vs. omnivorous/herbivorous) (Turchini et al., 2009), there are only two groups based on the biochemistry of EFA requirements: those that require LC-PUFA and those whose requirements can be satisfied by C_{18} PUFA (Tocher, 2010).

Based on the majority of studies to date, the overall conclusion is that 100% of dietary FO can be substituted with VO in fish whose EFA requirements can be satisfied by C_{18} PUFA (freshwater and salmonid species) without affecting growth performance or feed efficiency (Turchini et al., 2009). However, the majority of data were obtained in trials where a large proportion of dietary protein was supplied by FM. A dietary level of 40% FM would generally supply lipid to at least 4% total diet and n-3 LC-PUFA to around 1–1.5% of diet, sufficient to satisfy EFA requirements. Therefore, replacement of FO must be considered in combination with the level of dietary FM. Dual-replacement of FO and FM is required to fully assess the effects of reducing dietary LC-PUFA levels (Torstensen et al., 2008). This has been described in trout where there was no effect on growth with 100% substitution of FO with VO when the dietary level of FM was 40%, but growth retardation was observed when FM was reduced to 20% (Drew et al., 2007). With species that require LC-PUFA (e.g., most marine fish), the situation is slightly more complicated, with the maximum level of FO replacement possible without affecting growth performance in the range 60–75%. This difference in maximum level of substitution in comparison to freshwater/salmonid fish has generally been explained by the qualitative differences in EFA requirement. Thus, VO can supply EFA (18:3n-3 and/or 18:2n-6) to freshwater/salmonid fish, but not to marine fish that require EPA and/or DHA (Tocher, 2010). However, as the majority of studies, such as those summarized above, have used FM-based diets that generally would not be considered limiting in terms of n-3 LC-PUFA level, growth retardation at high levels of FO substitution requires more explanation. Possibly these results reflect the difficulty of estimating EFA requirements in fish. Most EFA requirement studies used relatively low lipid levels compared to those now used commercially, with the consequence that the levels of PUFA or LC-PUFA tested in terms of percentage of total diet weight may not be as high as those used commercially during the high-FM/high-FO era. This suggests a minimum EFA requirement to prevent deficiency pathologies. This level is essentially what the EFA requirement studies have measured, but there could be another level for optimal growth (Tocher, 2010). This means there is a range between these two levels where increasing EFA can give increasing growth. Little direct evidence supports this hypothesis because the required studies have not been done, but the fact that effects on growth have been observed when dietary LC-PUFA levels are reduced, but still exceed EFA requirement levels, suggests it is the case. The effect has not been clearly observed in salmonids because they can produce EPA and DHA from 18:3n-3; any difference between minimal and optimal EFA levels would be masked somewhat by endogenous synthesis of LC-PUFA.

Impact of Replacement of Dietary Fish Oil with Vegetable Oils on Nutrient Content of Fish: n-3 LC-PUFA Levels

The final fatty acid composition of fish is the consequence of dietary input and endogenous metabolism that encompasses the processes of digestion and absorption, which combined are measured as digestibility, conversion to other fatty acids or fatty acid derivatives, oxidation, and deposition. Irrespective of species and endogenous metabolism, substitution of FO with VO almost invariably results in decreased proportions of LC-PUFA, ARA, EPA, and DHA, and increased percentages of C_{18} fatty acids, 18:1n-9, 18:2n-6, and 18:3n-3 (Table 16-3). The precise changes depend on the particular oils in question. If northern hemisphere FO are being replaced, the levels of long-chain monoenes, 20:1 and 22:1, will also be reduced. The relative changes in C_{18} fatty acids will depend upon substituting VO, with 18:1n-9 particularly increased with rapeseed/canola or olive oils, 18:2n-6 increased greatly when soybean, sunflower, or corn oils are used, and 18:3n-3 increased when linseed oil is used. The quantitative extent of the changes also depends upon the level and duration of substitution (Tocher et al., 2010). In addition, as dietary fatty acid composition influences depot lipid, TAG, more than membrane phospholipids, the effects of diet are reflected in different species and tissues to different extents depending upon fat content and the relative proportions of TAG and phospholipid present (Tocher, 2003). For example, the effects of FO substitution are generally quantitatively less in lean species with low flesh lipid contents, such as cod, compared to an oily fish such as salmon (Table 16-3). Furthermore, the correlation between dietary and tissue fatty acid compositions in salmon is greater in flesh, where more lipid is deposited, than in liver.

TABLE 16-3 Effect of Complete or Partial Replacement of Dietary Fish Oils by Vegetable Oils on Fatty Acid Compositions (Percentage of Weight) of Total Lipid of Flesh of Salmonids and Marine Fish[a]

	Atlantic Salmon[b] (*Salmo salar*)		Arctic Charr[c] (*Salvelinus alpinus*)		Atlantic Salmon[d] (*Salmo salar*)		Gilthead Sea Bream[e] (*Sparus auratus*)		Atlantic Cod[f] (*Gadus morhua*)		European Sea Bass[g] (*Dicentrarchus labrax*)	
Fatty Acid	FO	LO	FO	EO	FO	VO	FO	LO	FO	EO	FO	VO
14:0	4.7	0.7	3.3	2.6	4.9	0.8	4.6	2.6	1.2	0.7	2.6	1.3
16:0	12.9	8.3	15.5	14.0	13.0	12.9	19.0	15.1	15.3	13.1	16.8	15.2
18:0	1.9	3.6	2.5	2.9	2.6	3.1	4.0	4.0	3.3	4.2	3.8	4.1
Total saturated	19.9	12.7	22.1	20.1	21.1	17.0	28.0	21.6	20.0	18.2	24.5	21.7
16:1n-7	7.5	0.8	4.9	3.9	4.4	0.9	7.4	4.2	1.8	0.7	4.6	2.8
18:1n-9	16.0	18.6	15.9	16.9	2.7	2.5	15.6	19.9	10.9	11.4	14.7	21.1
18:1n-7	3.8	1.3	2.4	2.0	14.6	41.3	2.8	2.1	2.3	1.6	2.6	2.1
20:1n-9	16.6	1.6	5.0	3.8	11.4	2.7	2.2	1.3	2.7	1.9	2.1	2.0
22:1n-11	9.9	1.1	5.6	3.9	12.3	1.3	2.3	1.4	1.6	0.9	1.0	0.9
Total monoenes	57.1	24.2	36.7	32.8	48.2	49.0	31.8	29.8	20.5	17.6	26.0	29.7
18:2n-6	3.9	13.1	4.4	8.4	3.3	14.4	6.2	12.0	2.0	9.9	3.8	8.5
18:3n-6	0.1	0.0	0.2	3.0	tr	tr	0.1	nd	0.5	3.8	0.2	0.2
20:3n-6	0.2	0.1	0.3	0.9	0.1	0.7	nd	nd	0.1	0.6	0.3	0.4
20:4n-6	0.3	0.1	0.7	0.5	0.7	0.9	0.8	0.5	1.6	0.7	0.9	0.6
Total n-6 PUFA	4.9	14.0	6.3	13.3	4.5	17.0	7.1	12.7	5.3	15.6	6.8	10.9
18:3n-3	0.8	38.7	3.3	8.9	1.1	8.2	1.7	16.6	2.0	11.8	1.1	11.0
18:4n-3	1.5	1.2	2.7	5.2	1.5	1.2	1.6	1.1	1.4	4.0	1.2	0.7
20:4n-3	1.2	1.8	1.2	1.3	1.4	0.9	0.4	0.7	0.8	0.6	0.4	0.3
20:5n-3	4.3	1.3	6.7	5.2	4.6	1.6	11.0	6.0	12.8	7.9	9.3	5.7
22:5n-3	1.5	0.4	1.5	1.2	2.1	0.6	2.8	1.8	2.0	1.0	1.5	0.9
22:6n-3	8.1	3.1	18.5	12.9	12.7	3.4	13.0	8.2	33.3	22.1	17.3	11.7
Total n-3 PUFA	17.5	49.1	34.1	35.0	23.4	16.0	31.6	35.3	52.4	47.5	31.8	31.1

NOTE: EO, echium oil; FO, fish oil; LO, linseed oil; PO, palm oil; RO, rapeseed (canola) oil; tr, (trace); nd, (not determined).
[a]Adapted from Tocher et al. (2010).
[b]Bell et al. (2004); initial weight 127 g; 100% replacement for 40 weeks.
[c]Tocher et al. (2006); initial weight 5 g; 80% replacement for 16 weeks.
[d]Torstensen et al. (2005); initial weight 0.16 g; 100% replacement, RO/PO/LO (3.7:2:1) for 22 months.
[e]Izquierdo et al. (2003); initial weight 10 g; 60% replacement for 101 days.
[f]Bell et al. (2006); initial weight 4 g; 100% replacement for 18 weeks.
[g]Mourente and Bell (2006); initial weight 5 g; 60% replacement, LO/PO/RO (7:3:2) for 64 weeks.

When dietary formulations contain high 18:3n-3, as in linseed oil, evidence of desaturation of 18:3n-3 through increased levels of SDA and 20:4n-3 have been reported in salmon and freshwater fish. In contrast, in sea bass and sea bream, there was no increase in the percentages of any desaturated intermediates such as SDA or 20:4n-3 when fed linseed oil. Irrespective of species, provision of dietary 18:3n-3 does not compensate for the lack of dietary LC-PUFA and, indeed, the greater the concentration of 18:3n-3 in the diet, the greater the decrease in tissue EPA and DHA levels. Echium oil has been investigated as an alternative to dietary FO in both salmonid and marine fish, although the rationale for its use and potential benefit is somewhat different in the two groups (Bell et al., 2006; Tocher et al., 2006). However, feeding echium oil was not able to maintain the levels of EPA and DHA in the salmonid Arctic charr, although the reduction in their levels was less than observed in other trials on salmonids fed linseed oil at similar levels. Similarly, the levels of 20:4n-3 in cod tissues were not increased by feeding echium oil, so little evidence suggests echium offered any advantage over linseed oil and other 18:3n-3-rich oils. In conclusion, dietary echium oil has not delivered benefits that its fatty acid composition had suggested.

Substituting FO with VO has, therefore, two separate effects on fatty acid composition of fish: reducing the levels of health-promoting n-3 LC-PUFA and increasing the levels of C_{18} PUFA. It is important to note that C_{18} PUFA should not be regarded equally. A major issue in human lipid nutrition, implicated in many inflammatory-based pathologies, is the very high n-6 PUFA:n-3 PUFA ratio in the human diet, particularly in the developed world, with ratios of 10–20:1 in comparison with 1–2:1 during the majority of human evolution (Simopoulos, 2006). This change has largely been driven by the agricultural/industrial revolutions of the past 250 years, which greatly increased production of 18:2n-6-rich grains and, more recently, oilseeds. There is currently great debate about 18:2n-6 and its effect in the human diet, but the majority view is that the great excess is detrimental (Whelan, 2008; Ramsden et al., 2009). This considered, it is desirable to avoid excessive levels of 18:2n-6 in fish. Therefore, more than simply providing a potential substrate for LC-PUFA synthesis in species that have the pathway, 18:3n-3 is required in the dietary VO formulations for all fish to limit and balance 18:2n-6. Indeed, there is evidence that 18:3n-3 itself can have benefits in human nutrition (Brenna et al., 2009). Similarly, even if SDA is not effective in maintaining EPA in fish fed VO, increased levels of SDA itself in fish may be useful because it is recognized as having potential benefits in human nutrition (Whelan, 2009). In order to achieve a more balanced composition, blending of VO is required. A blend of rapeseed, palm, and linseed in a ratio of around 4:2:1 provided levels of saturated (including 16:0 from palm oil), monounsaturated (albeit 18:1n-9 rather than $C_{20/22}$ monoenes), and polyunsaturated (C_{18} rather than LC-PUFA) fatty acids as found in northern hemisphere FO (G. Bell et al., 2005; Torstensen et al., 2008). The use of rapeseed and linseed ensured 18:2n-6 was as low as possible and balanced by a similar level of 18:3n-3. However, few feed production plants have facilities to utilize three or more oils, so blends of two or at maximum three oils (including FO) are the current industry standard. In summary, irrespective of how effective VO blends are in balancing dietary fatty acid compositions, they cannot prevent reduction in n-3 LC-PUFA, which can potentially impact on nutritional quality of the fish product for the human consumer.

Other Impacts of Replacement of Dietary Fish Oil with Vegetable Oils

Cholesterol and Phospholipid

As described above, plants in general do not produce much cholesterol. Therefore, VO contain very little cholesterol (40–200 mg per kg), not only posing potential problems for shrimp, but also possibly having an impact on fish (Tocher et al., 2008). Studies investigating gene expression in salmon using microarray analyses of liver transcriptomes have shown that several genes of the cholesterol synthesis pathway are upregulated in fish fed VO compared to fish fed FO (Leaver et al., 2008; Taggart et al., 2008). Because tissue levels of cholesterol were not reduced, it is speculated that the upregulation of cholesterol biosynthesis effectively compensates for the lower intake (Leaver et al., 2008). The trigger affecting gene transcription could be reduced dietary cholesterol but could also be the presence of VO-derived phytosterols that have been reported in fish fed VO as replacements for FO (Miller et al., 2008). This area may require further consideration in the future as dietary cholesterol concentrations decline, and phytosterols increase, as the proportion of plant meals and oils in dietary formulations increases.

Unless crude oils are used, the majority of phospholipids in fish feeds will be derived from meals and other products that are primarily protein sources. The phospholipid contents of refined oils, both FO and VO, are very low because they are removed during the degumming process. Crude soybean, sunflower, and rapeseed/canola oils contain around 1.5–3.1%, 0.5–1.0%, and less than 0.5% total phospholipid, respectively, which is removed to form the commercial by-products, lecithins. They can have variable phospholipid contents and class compositions but commonly contain 50–60% total phospholipid, and PC, PE, PI, and PA in approximately equal proportions (Tocher et al., 2008). Diets formulated with defatted plant meals and degummed VO will likely contain less phospholipid than traditional FM/FO diets. This is clearly a potential problem with feeds for larval and juvenile shrimp and larval marine fish (Teshima, 1997; Tocher et al., 2008). However, there is no evidence that intact phospholipid is required by adult fish or shrimp. Reduced levels are not anticipated to be a problem, but should be monitored and, if necessary, phospholipids (lecithins) could be supplemented (Tocher et al., 2008).

Contaminants

In addition to nutrients, fish also receive undesirable compounds through the diet, including toxic elements, such as mercury and arsenic, and organic contaminants (POP), including polychlorinated dibenzo-p-dioxins and furans (dioxin/furans), polychlorinated biphenyls (PCB), and polybrominated diphenyl ether (PBDE) flame retardants (Bell and Waagbø, 2008). The POP are highly lipophilic with biological half-lives of several decades, which means they can particularly accumulate in predators at the top of the food chain. Their persistent nature means they will remain in the biota for a considerable period. The ocean is essentially a sink for these pollutants and thus all marine fish and seafood can contain these contaminants at varying levels depending upon the physiology of the species and geographical location. For instance, although high levels of POP were reported in farmed Atlantic salmon in Europe (Hites et al., 2004), a thorough study on contaminant levels in salmon in British Columbia, Canada, showed that POP were relatively low in both market-size farmed and wild fish (Ikonomou et al., 2007).

Fish is the main source of arsenic in the diet of humans, but approximately 99% is in the form of arsenobetaine and arsenocholine, and only a small amount (< 1%) is toxic inorganic arsenic. Similarly, although mercury (Hg) is the most toxic nonradioactive element, methyl-Hg is the most toxic form, and Hg in fish is largely present as the much less toxic Hg-cysteine. Farmed fish do not contain high levels of Hg but some wild fish, including species such as shark, swordfish, king mackerel, and tilefish (Gulf of Mexico), can accumulate significant levels of Hg (FDA, 2006). In 2004, the U.S. Food and Drug Administration and Environmental Protection Agency advised women who may become pregnant, pregnant women, nursing mothers, and young children to avoid these species. The Hg levels in oily fish including salmon, trout, herring, and mackerel contain about one-sixteenth of the Hg in swordfish, and canned tuna contains about half the Hg as fresh tuna.

Farmed fish fed marine products can accumulate contaminants, which is particularly an issue with the lipid-soluble POP in species such as salmon that deposit substantial amounts of lipid in the flesh. Dioxins/furans occur naturally as products of incomplete organic combustion or via processes including steel, paint, and paper production, but their levels in the environment have fallen since the 1950s. Similarly, since PCBs, found in paints, plasticizers, and dielectric fluids were banned in 1976, their levels in the environment are also declining. However, the PBDE have been used as flame retardants in textiles, furniture, computer equipment, and cables since the late 1980s. Their levels in the environment are increasing despite production of the more toxic tetra-, penta-, hexa-, and nona-BDE being banned in the European Union in 2004 in favor of less persistent deca-PBDE. A total of 210 "dioxin" compounds have been identified, but only 17 dioxin congeners are considered toxic. Thus, dioxin concentrations are normally expressed in terms of toxic equivalents (TEQ) that are generated by multiplying the concentration of individual congeners by World Health Organization toxic equivalency factors (TEF) related to their toxic potential and human health risk. In addition to national bodies, the European Union in 2006 set combined limits for dioxins and dioxin-like PCB in animal feeds and food,[1] although the FDA has currently only set limits for PCB and not dioxins.[2]

Fish receive most of the contaminants from their feed, which can be controlled in farmed fish a variety of ways. One option is simply to source raw materials from less polluted environments. For instance, FO from the southern hemisphere are lower in dioxins and PCB than those from the northern Atlantic. Some environments like the Baltic contain levels of pollutants that often render FO unusable as contaminant levels exceed maximum limits for animal feeds (Bell and Waagbø, 2008). Therefore, a second option is to decontaminate the oils and meals. The very low level of POP in fish (parts per trillion for dioxins) is highlighted by the fact that decontamination removes 2 g of dioxins and dioxin-like PCB (DL-PCB) from 1 Mt of fish. Several processing plants operate in Europe and, although capacity is currently limited and adds cost, the ability to render contaminated FO useable will be invaluable in a market of limited supply. A third option is to replace FO with non- or less contaminated oils such as VO. Substitution of 100% dietary FO with a VO blend through the 2-year growth cycle reduced dioxin and DL-PCB levels in salmon flesh from 2.01 to 0.68 TEQ/kg, a 66% reduction (G. Bell et al., 2005). In similar trials, salmon fed 100% FO for 2 years had flesh dioxin/DL-PCB levels of 1.76 TEQ/kg that reduced to 0.63 and 0.30 TEQ/kg flesh in fish fed 75 or 100% VO (J. G. Bell et al., 2005; Berntssen et al., 2005). Levels of PBDE were also reduced by 72% in flesh of salmon fed 75% VO compared to fish fed 100% FO (Bell and Waagbø, 2008). A study in Canada has reported similar reductions in flesh POP levels with replacement of FO by flaxseed oil in feed for Atlantic salmon (Friesen et al., 2008). In all these trials, the diets were formulated with FM, so contaminant levels could be reduced further by partial FM replacement. Following a 24-week finishing diet period, where fish were fed 100% FO to restore EPA and DHA levels, the flesh dioxin/DL-PCB concentration in the fish previously cultured on 100% VO was still around 50% lower than the value seen in fish fed FO throughout (J. G. Bell et al., 2005). Formulating finishing diets with decontaminated FO would be a refinement to the strategy to preserve FO stocks and reduce contaminant levels while maintaining n-3 LC-PUFA levels in farmed fish (Bell et al., 2008; Tocher, 2009). However, it should be noted that although VO are very low in contaminants of marine origin, they could contain other contaminants such as pesticides and herbicides that equally require monitoring.

Organoleptic and Other Product Quality Issues

The effects of substituting dietary FO with VO on an extensive range of sensory and quality attributes in Atlantic salmon, rainbow trout, gilthead sea bream, and European sea bass have been determined (RAFOA, 2005). Sensory attributes included organoleptic properties as measured by taste panel analyses of raw (sashimi), cooked, and smoked products, and electronic nose analyses of volatiles. The product quality and shelf-life properties determined included freshness as measured using the Quality Index Method (QIM) for each species, color, fillet texture, liquid-holding capacity, and lipid oxidation. High levels of VO in the diet had only minor effects on organoleptic properties of raw, cooked, or smoked salmon. In preference tests, fish fed VO tended to be more favored than those fed FO. Likewise, dietary VO had only minor effects on organoleptic properties in cooked and smoked rainbow trout as well as cooked sea bream. However, in very fresh sea bass, fish fed FO tended to be scored superior to fish fed VO. Feeding a finishing diet containing only FO eliminated the differences in organoleptic properties. Electronic nose analyses discriminated between fish fed FO and fish fed VO with best discrimination achieved in rainbow trout. Feeding a finishing diet resulted in loss of much of the discrimination. No correlations were made between

[1]Commission Regulation (EC) No 1881/2006, [Official Journal L 364 of 20.12.2006, p.5].

[2]FDA Code of Federal Regulations, CFR Title 21, 109.30 and 509.30.

electronic nose results and organoleptic/sensory or quality analyses. Replacing FO with VO generally had only minor effects on product quality and storage stability and, in all species studied, storage time on ice was more important that dietary treatment (RAFOA, 2005). However, in a recent study fillets from tilapia fed crude palm oil-based diets exhibited significantly higher oxidative stability during frozen storage, compared to fish fed diets containing FO (Ng and Bahurmiz, 2009). Compared to fish fed FO and perhaps as expected (Rørå et al., 2005), lipid oxidation in salmon decreased with inclusion of VO in the diet, including fish fed linseed oil rich in 18:3n-3, which is expected to be more prone to oxidation. Lipid oxidation products were much lower in trout and the marine fish with lower flesh lipid contents. There were few effects of diet, although there were trends for lower values in fish fed VO compared to fish fed FO.

Effects on Fish Health

Several studies demonstrate that dietary VO can affect fish health and welfare. However, although a range of effects have been reported, they are often not consistent between species or indeed between different trials in the same species. Clearly many interacting and confounding factors are involved that precludes too many generalizations. The following is a summary of the range of effects reported in different studies in various species with representative, but not exhaustive, references. It is important to note that often there are also studies where similar effects were not observed. The data were primarily compiled from a recent comprehensive review with an extensive reference list (Montero and Izquierdo, 2010).

Many of the reported effects could be a direct consequence of altered cellular fatty acid compositions. Specifically, reduced dietary EPA and increased dietary 18:2n-6 affect both n-3 PUFA:n-6 PUFA and, more importantly, EPA:ARA ratios, with the latter having potential consequences for eicosanoid metabolism (Tocher, 2003). Indeed, increased production of n-6-derived eicosanoids and decreased production of n-3-derived eicosanoids have been reported in fish fed VO (Gjøen et al., 2004; Ganga et al., 2005; Mourente et al., 2007). The effect of VO substitution on eicosanoid production will depend on both the fatty acid composition of the diet, that is the oils used, and the ability of the species to produce LC-PUFA. The rapid turnover of enterocytes in the gastrointestinal tract makes them particularly susceptible to dietary fatty acids, resulting in alteration of their membrane fatty acid composition, morphology, paracellular permeability, epithelial transport functions, and gut microbiota that can all affect their function in digestion, absorption, and utilization of nutrients; immune defense; and smoltification (Olsen et al., 1999, 2000; Cahu et al., 2000; Caballero et al., 2002; Jutfelt et al., 2007). Effects of VO feeding on liver histology, including development of steatosis, are among the most consistent (Caballero et al., 2002; Mourente et al., 2007).

Immune cell function has been shown to be affected by dietary VO, with phagocytic and bacteriocidal activities of macrophages, and respiratory burst and superoxide production in macrophages and circulating leukocytes, all reduced in fish fed VO compared to fish fed FO (Montero et al., 2003; Bell et al., 2006; Lin and Shiau, 2007; Mourente et al., 2007). In contrast, other studies have reported increased functionality in immune cells from fish fed VO (Gjøen et al., 2004; Balfry et al., 2006). In either case, the effects may be due to changes in cellular fatty acid compositions because VO have been shown to reduce LC-PUFA levels, and increase 18:3n-3 and 18:2n-6 levels, of immune cells (Montero et al., 2003; Mourente et al., 2007). Proliferation, oxidative burst, and production of cytokines are influenced by immune cell fatty acid composition in mammals, with phagocytosis strongly correlated with n-3/n-6 PUFA ratio in human peripheral blood leukocytes (Kew et al., 2003).

Other effects of VO feeding on immune functions include decreased numbers of circulating leukocytes (Mourente et al., 2005, 2007; Petropoulos et al., 2009), although other studies have not reported this result (Balfry et al., 2006; Bell et al., 2006). With respect to effects of VO feeding on the humoral system, lysozyme activity appears to be unaffected (Montero et al., 2003; Bell et al., 2006; Mourente et al., 2007; Petropoulos et al., 2009), but alternative complement activity (Montero et al., 2003; Lin and Shiau, 2007), antibody production (Kiron et al., 1995), and leptin (Ganga et al., 2005) were all reduced. Little is known about effects on viral infections, but dietary VO increased hepatic expression of Mx (protein of resistance to mixovirus) (Montero et al., 2008), as previously observed in mammals. Generally, substitutions by single VO appear to affect immune parameters more than blends of VO (Montero et al., 2003; Balfry et al., 2006; Lin and Shiau, 2007; Montero et al., 2008). The n-3 PUFA-rich VO have stronger negative effects on salmonid immune systems, whereas n-6 PUFA-rich VO have greater negative effects in marine fish (Montero et al., 2008). However, the effects of VO replacement on immune parameters do not necessarily reduce disease resistance, with both increased (Fracalossi and Lovell, 1994; Lodemel et al., 2001; Gjøen et al., 2004) and decreased (Thompson et al., 1996; Bransden et al., 2003) resistance reported, although no major effects on immunization/vaccination efficiency have been observed (Kiron et al., 1995; Bransden et al., 2003; Balfry et al., 2006).

Fish welfare may also be affected by VO substitution as post-stress plasma cortisol levels were increased by 18:3n-3-rich linseed oil, whereas VO rich in 18:2n-6 had no significant effect (Montero et al., 2003). Consistent with this, ACTH-induced cortisol release was higher in head kidney cells from fish fed linseed oil (Ganga et al., 2010). Finally, dietary VO inclusion could be relevant in the development of atherosclerosis, cardiac lesions, and cataract development (Bell et al., 1993; RAFOA, 2005; Seierstad et al., 2005).

In conclusion, it is clear that substitution of dietary FO with VO has a range of effects that could influence fish health and welfare. However, they are rather variable depending upon many factors including fish species, VO used, level and duration of substitution, other dietary components, and environmental conditions. So far, no major pathologies have been identified, but further studies for each species are required to fully understand the impact that major changes in dietary formulation may have.

Strategies for Minimizing the Impact of Fish Oil Substitution on Nutrient Content and Nutritional Quality of Fish

The potential detrimental impact of substitution of dietary FO with oils such as VO that lack n-3 LC-PUFA can be minimized by a number of strategies, including the choice of both substituting oil and FO, level and duration of substitution, and feeding regimes including the use of "finishing" feeds. The inclusion of fish in the human diet is beneficial from two directions; it not only increases dietary n-3 LC-PUFA intake, but also its inclusion replaces something else in the diet, such as red meat. Exactly the same principle is involved in choosing oils to replace FO in aquaculture feeds. It is not only the inclusion of certain fatty acids that is important, but also the fact that including them reduces the use of others.

In this context, the replacing oil or oil blend should ideally satisfy some general criteria, including a high monounsaturated fatty acid content. This is not simply because monounsaturates will satisfy one of the important roles of dietary lipid, as good energy sources, but also to reduce the level of C_{18} PUFA. A relatively low C_{18} PUFA content is also desirable. Oils with high C_{18} PUFA (> 50%) will contain a significant amount of these fatty acids in the sn-2 position that is retained throughout metabolism, so these oils are more effective in competing with, and replacing, LC-PUFA in membrane phospholipids. For this reason, oils with high C_{18} PUFA should be used sparingly. Furthermore, a high n-3/n-6 ratio with 18:2n-6 content limited as much as possible is recommended. In terms of nutritional quality, it should be an aim to maintain the high n-3/n-6 ratio of fish as the human diet is imbalanced with very high levels of n-6 PUFA, especially 18:2n-6 (Simopoulos, 2006). Therefore, the replacement oil/blend should contain 18:3n-3, not simply for species that can potentially convert it to EPA and DHA, but also for fish that cannot, because including 18:3n-3 will help to retain n-3/n-6 ratio and limit 18:2n-6 inclusion. Simply substituting fish oil with high levels of oils rich in 18:3n-3, such as flax/linseed, is counterproductive due to the reasons given above.

In terms of single oils, rapeseed (canola) satisfies these criteria as well as any VO. However, some evidence may suggest blends are better in terms of health and welfare of the fish (Montero and Izquierdo, 2010). This advantage is possibly related to TAG stereochemistry (molecular species) such that, although overall fatty acid compositions are similar, the individual TAG molecules are different and can have differential effects on phospholipid composition.

Irrespective of how effective alternative oils are in balancing fatty acid compositions, they cannot prevent reduction in n-3 LC-PUFA, but this can be minimized by appropriate feeding strategies. Several studies have investigated the use of "finishing diets" containing FO to restore levels of n-3 LC-PUFA in the flesh (Bell and Waagbø, 2008). This strategy has some limitations as the process follows simple dilution kinetics, so the law of diminishing returns applies (Jobling, 2003; Robin et al., 2003). Therefore, it can be reasonably successful depending upon the length of time the fish are fed the finishing diet. For salmon, this could be from 4 to 6 months due to the high lipid and TAG content of the flesh, but can be shorter in marine fish with low flesh lipid contents (Bell and Waagbø, 2008). Although this process is quite successful in restoring EPA and DHA levels, up to 80% of values in fish fed FO, "wash-out" of 18:3n-3 and especially 18:2n-6 is only partial (Bell and Waagbø, 2008). In terms of reducing FO use, this strategy is also only partly successful because the finishing period with near market-size fish requires high feed input. However, over a 2-year production cycle for salmon, feeding 100% VO for most of the time followed by a 5-month finishing period reduced FO consumption by 50% (Bell, J. G., unpublished data). At lower inclusion levels, dietary VOs have correspondingly less impact on flesh fatty acid compositions (Robin et al., 2003). Therefore, an alternative to finishing diets is the use of lower levels of FO substitution throughout the entire growth phase, with VO and FO blends chosen to limit LC-PUFA reduction in the flesh (Pratoomyot et al., 2008). Thus, South American FO contain higher levels of n-3 LC-PUFA and so can deliver similar levels of n-3 LC-PUFA at lower inclusion levels than North Atlantic FO, with VO chosen to compensate for lower monounsaturated and higher saturated fatty acid levels in southern FO. Choice of FO is also important in different feeding strategies. With a high replacement/finishing feed strategy, high-energy long-chain monounsaturated fatty acid-rich FO (e.g., capelin, herring, and mackerel) can be used in the energy-demanding ongrowing phase, with high n-3 LC-PUFA FO (e.g., anchovy, sardine, and menhaden) used in the finishing feed for restoring EPA and DHA. As these southern hemisphere FO usually have lower organic contaminant levels than most northern hemisphere FO, their use closer to market may be advantageous, with decontaminated northern FO an alternative. For the alternative strategy of lower level replacement throughout with no finishing diet, the high n-3 LC-PUFA FO are the clear choice.

The above section has been dictated by the current lack of suitable, sustainable sources of n-3 LC-PUFA as alternatives to the traditional FO sources of n-3 LC-PUFA. It is highly probable that this situation will change in the coming years with increasing availability of sustainable and cost-effective sources of n-3 LC-PUFA, albeit most likely via transgenic approaches delivering novel products from GM organisms. Effective application of these products as ingredients for aquaculture feeds will provide a new and unique set of challenges.

CONCLUSIONS

Commercial aquaculture feeds generally contain appreciable amounts of FM and FO. However, during recent years it has become evident that the increasing demands on these resources will outstrip supply unless levels in feeds are reduced. As a consequence, a considerable amount of research is being devoted to identifying, developing, and evaluating alternative ingredients. Although progress has been made, there are currently insufficient substitutes to meet the demands of an expanding aquaculture industry, and this remains a very active area of research.

Compounding the issue is that when fish are fed diets containing FO they are a unique and rich source of n-3 LC-PUFA that are important components of the human diet. When FO are substituted by VO the content of these unique fatty acids is reduced.

REFERENCES

Abbadi, A., F. Domergue, J. Bauer, J. A. Napier, R. Welti, U. Zahringer, P. Cirpus, and E. Heinz. 2004. Biosynthesis of very-long-chain polyunsaturated fatty acids in transgenic oilseeds: Constraints on their accumulation. Plant Cell 16:2734–2748.

Alimuddin, G. Yoshizaki, V. Kiron, S. Satoh, and T. Takeuchi. 2005. Enhancement of EPA and DHA biosynthesis by over-expression of masu salmon Δ6-desaturase-like gene in zebrafish. Transgenic Res. 14:159–165.

Alimuddin, G. Yoshizaki, V. Kiron, S. Satoh, and T. Takeuchi. 2007. Expression of masu salmon Δ5-desaturase-like gene elevated EPA and DHA biosynthesis in zebrafish. Mar. Biotechnol. 9:92–100.

Alimuddin, V. Kiron, S. Satoh, T. Takeuchi, and G. Yoshizaki. 2008. Cloning and over-expression of masu salmon (*Oncorhynchus masou*) fatty acid elongase-like gene in zebrafish. Aquaculture 282:13–18.

Baeverfjord, G., and Å. Krogdahl. 1996. Development and regression of soybean meal induced enteritis in Atlantic salmon, *Salmo salar* L., distal intestine: A comparison with the intestines of fasted fish. J. Fish. Dis. 19:375–387.

Bajpai, P. K., P. Bajpai, and O. P. Ward. 1991. Optimization of production of docosahexaenoic acid (DHA) by *thraustochytrium aureum* ATCC 34304. J. Am Oil Chem. Soc. 68:509–514.

Balfry, S. K., J. Oakes, M. Rowshandeli, G. Deacon, B. J. Skura, and D. A. Higgs. 2006. Efficacy of an equal blend of canola oil and poultry fat as an alternate dietary lipid source for Atlantic salmon (*Salmo salar* L.) in seawater. II: Effects on haematology and immunocompetence. Aquac. Res. 37:192–199.

Barrows, F., T. G. Gaylord, D. J. Stone, and C. E. Smith. 2007. Effect of protein source and nutrient density on growth efficiency of rainbow trout (*Oncorhynchus mykiss*). Aquac. Res. 38:1747–1758.

Bell, G., B. Torstensen, and J. Sargent. 2005. Replacement of marine fish oils with vegetable oils in feeds for farmed salmon. Lipid Technol. 17:7–11.

Bell, J. G., and R. Waagbø. 2008. Safe and nutritious aquaculture produce: benefits and risks of alternative sustainable aquafeeds. Pp. 185–225 in Aquaculture in the Ecosystem, M. Holmer, K. D. Black, C. M. Duarte, N. Marba, and I. Karakassis, eds. London, UK: Springer Verlag BV.

Bell, J. G., A. H. McVicar, J. R. Sargent, and K. D. Thompson. 1993. Dietary sunflower, linseed and fish oils affect phospholipid fatty acid compostition, development of cardiac lesions, phospholipase activity and eicosanoid production in Atlantic salmon (*Salmo salar*). Prostag. Leukotr. Ess. 49:665–673.

Bell, J. G., R. J. Henderson, D. R. Tocher, and J. R. Sargent. 2004. Replacement of dietary fish oil with increasing levels of linseed oil; modification of flesh fatty acid compositions in Atlantic salmon (*Salmo salar*) using a fish oil finishing diet. Lipids 39:223–232.

Bell, J. G., F. McGhee, J. R. Dick, and D. R Tocher. 2005. Dioxin and dioxin-like polychlorinated biphenyls (PCBs) in Scottish farmed salmon: Effects of replacement of dietary marine fish oil with vegetable oils. Aquaculture 243:305–314.

Bell, J. G., F. Strachan, J. Good, and D. R. Tocher. 2006. Effect of dietary echium oil on growth, fatty acid composition and metabolism, gill prostaglandin production and macrophage activity in Atlantic cod (*Gadus morhua* L.). Aquac. Res. 37:606–617.

Bell, J. G., M. Sprague, E. A. Bendiksen, J. R. Dick, F. Strachan, J. Pratoomyot, M. H. G. Berntssen, and D. R. Tocher. 2008. Using decontaminated fish oil or a vegetable/fish oil blend to reduce organic contaminant concentrations in diets and flesh of farmed Atlantic salmon (*Salmo salar*). Organohalogen Compd. 70:894–897.

Berntssen, M. H. G., A.–K. Lundebye, and B. E. Torstensen. 2005. Reducing the levels of dioxins and dioxin-like PCBs in farmed Atlantic salmon by substitution of fish oil with vegetable oil in the feed. Aquacult. Nutr. 11:219–231.

Bransden, M. P., C. G. Carter, and P. D. Nichols. 2003. Replacement of fish oil with sunflower oil in feeds for Atlantic salmon (*Salmo salar* L.): Effect on growth performance, tissue fatty acid composition and disease resistance. Comp. Biochem. Phys. B 135:611–625.

Brenna, J. T., N. Salem Jr., A. J. Sinclair, and S. C. Cunnane. 2009. α-Linolenic acid supplementation and conversion to n-3 long-chain polyunsaturated fatty acids in humans. Prostag. Leukotr. Ess. 80:85–91.

Brouwer, I. A., A. Geelen, and M. B. Katan. 2006. *n*-3 fatty acids, cardiac arrhythmia and fatal coronary heart disease. Prog. Lipid Res. 45:357–367.

Bureau, D. P., A. M. Harris, and C. Y. Cho. 1998. The effect of purified alcohol extracts from soy products on feed intake and growth of Chinook salmon *Oncorhynchus tsawytsscha* and rainbow trout *Oncorhynchus mykiss*. Aquaculture 161:27–43.

Burel, C., T. Boujard, S. J. Kaushik, G. Boeuf, S. Van Der Geyten, K. A. Mol, E. R. Kuhn, A. Quinsac, M. Krouti, and D. Ribaliller. 2000a. Potential of plant protein sources as fish meal substitutes in diets for turbot (*Psetta maxima*): Growth, nutrient utilisation and thyroid status. Aquaculture 188:363–382.

Burel, C., T. Boujard, F. Tulli, and S. J. Kaushik. 2000b. Digestibility of extruded peas, extruded lupin, and rapeseed meal in rainbow trout (*Oncorhynchus mykiss*) and turbot (*Psetta maxima*). Aquaculture 188:285–298.

Caballero, M. J., A. Obach, G. Rosenlund, D. Montero, M. Gisvold, and M. S. Izquierdo. 2002. Impact of different dietary lipid sources on growth, lipid digestibility, tissue fatty acid composition and histology of rainbow trout, *Oncorhynchus mykiss*. Aquaculture 214:253–271.

Cahu, C. L., J. L. Zambonino Infante, G. Corraze, and D. Coves. 2000. Dietary lipid level affects fatty acid composition and hydrolase activities of intestinal brush border membrane in seabass. Fish Physiol. Biochem. 23:165–172.

Certik, M., and S. Shimizu. 2000. Kinetic analysis of oil biosynthesis by an arachidonic acid-producing fungus, mortierella alpina 1S-4. Appl. Microbiol. Biot. 54:224–230.

Crapo, C., and P. J. Bechtel. 2003. Utilization of Alaska's seafood processing by-products. Pp. 105–119 in Advances in Seafood By-Products 2002 Conference Proceedings, P. J. Bechtel, ed. Fairbanks: Alaska Sea Grant College Program, University of Alaska.

Damude, H. G., and A. J. Kinney. 2008. Enhancing plant seed oils for human nutrition. Plant Physiol. 147:962–968.

De Francesco, M., G. Parisi, F. Medale, S. J. Kaushik, and B. M. Poli. 2004. Effect of long term feeding with a plant protein mixture based diet on growth and body/fillet quality traits of large rainbow trout (*Oncorhynchus mykiss*). Aquaculture 263:413–429.

De Koning, A. J. 2005. Properties of South African fish meal: A review. S. African J. Sci. 101:21–25.

Dias, J. 1999. Lipid Deposition in Rainbow Trout (*Oncorhynchus mykiss*) and European Seabass (*Dicentrarchus labrax* L.): Nutritional Regulation of Hepatic Lipogenesis. Ph.D. Dissertation, Univ. Porto (Portugal) and Univ. Bordeaux I (France).

Domergue, F., A. Abbadi, U. Zahringer, H. Moreau, and E. Heinz. 2005. In vivo characterization of the first acyl-CoA delta6-desaturase from a member of the plant kingdom, the microalga *Ostreococcus tauri*. Biochem. J. 389:483–490.

Drew, M. D., A. E. Ogunkoya, D. M. Janz, and A. G. Van Kessel. 2007. Dietary influence of replacing fish meal and oil with canola protein concentrate and vegetable oils on growth performance, fatty acid composition and organochlorine residues in rainbow trout (*Oncorhynchus mykiss*). Aquaculture 167:260–268.

Eckert, H., B. LaVallee, B. J. Schweiger, A. J. Kinney, E. B. Cahoon, and T. Clemente. 2006. Co-expression of the borage Δ6 desaturase and the Arabidopsis Δ15 desaturase results in high accumulation of stearidonic acid in the seeds of transgenic soybean. Planta 224:1050–1057.

Eilander, A., D. C. Hundscheid, S. J. Osendarp, C. Trander, and P. L. Zock. 2007. Effects of n-3 long chain polyunsaturated fatty acid supplementation on visual and cognitive development throughout childhood: A review of human studies. Prostag. Leukotr. Ess. 76:189–203.

Espe, M., A. Lemme, A. Petri, and A. El-Mowafi. 2006. Can Atlantic salmon grow on diets devoid of fish meal? Aquaculture 255:255–262.

Espe, M., A. Lemme, A. Petri, and A. El-Mowafi. 2007. Assessment of lysine requirement for maximal protein accretion in Atlantic salmon using plant protein diets. Aquaculture 263:168–178.

FAO (Food and Agricultural Organization of the United Nations). 2009. The State of World Fisheries and Aquaculture 2008 (SOFIA). Rome, Italy: FAO Fisheries and Aquaculture Department.

FDA (U.S. Food and Drug Administration). 2006. Mercury levels in commercial fish and shellfish. Available online at: *http://www.cfsan.fda.gov/~frf/sea-mehg.html*. Accessed on May 19, 2010.

Fournier, V., C. Huelvan, and E. Debruyeres. 2004. Incorporation of a mixture of plant feedstuffs as a substitute for fish meal in diets of juvenile turbot (*Psetta maxima*). Aquaculture 236:451–465.

Fracalossi, D. M., and R. T. Lovell. 1994. Dietary lipid sources influence responses of channel catfish (*Ictalurus punctatus*) to challenge test with the pathogen *Edwardsiella ictaluri*. Aquaculture 119:287–298.

Friesen, E. N., M. G. Ikonomou, D. A. Higgs, K. P. Ang, and C. Dubetz. 2008. Use of terrestrial based lipids in aquaculture feeds and the effects on flesh organohalogen and fatty acid concentrations in farmed Atlantic salmon. Environ. Sci. Technol. 42:3519–3523.

Ganga, R., J. G. Bell, D. Montero, L. Robaina, M. J. Caballero, and M. S. Izquierdo. 2005. Effect of dietary lipids on plasma fatty acid profiles and prostaglandin and leptin production in gilthead seabream (*Sparus aurata*). Comp. Biochem. Phys. B 142:410–418.

Ganga, R., J. G. Bell, D. Montero, E. Atalah, Y. Vraskou, L. Tort, L. Fernandez, and M. S. Izquierdo. 2010. ACTH-stimulated cortisol-release by the head-kidney-interrenal tissue from sea bream (*Sparus aurata*) fed with linseed and soybean oil. Br. J. Nutr. Endocrinol. doi:10.1017/S0007114510003430.

Gatlin, D. M., III, F. T. Barrows, P. Brown, K. Dabrowski, G. T. Gibson, R. W. Hardy, H. Elliot, G. Hu, Å. Krogdahl, R. Nelson, K. Overturf, M. Rust, W. Sealey, D. Skonberg, E. J. Souza, D. Stone, R. Wilson, and E. Wurtele. 2007. Expanding the utilization of sustainable plant products in aquafeeds: A review. Aquac. Res. 38:551–579.

Gjøen, T., A. Obach, C. Røsjø, B. G. Helland, G. Rosenlund, E. Hvattum, and B. Ruyter. 2004. Effect of dietary lipids on macrophage function, stress susceptibility and disease resistance in Atlantic salmon (*Salmo salar*). Fish Physiol. Biochem. 30:149–161.

Gjøsæter, J., and K. Kawaguchi. 1980. A Review of the World Resources of Mesopelagic Fish, FAO Technical Report No.193. Rome, Italy: FAO Fisheries and Aquaculture Department.

Glencross, B. D., C. G. Carter, N. Duijster, D. R. Evans, K. Dods, P. McCafferty, W. E. Hawkins, R. Maas, and S. Sipsas. 2004. A comparison of the digestibility of a range of lupin and soybean protein products when fed to either Atlantic salmon (*Salmo salar*) or rainbow trout (*Oncorhynchus mykiss*). Aquaculture 237:333–346.

Glencross, B. D., M. Booth, and G. L. Allan. 2007. A feed is only as good as its ingredients—a review of ingredient evaluation strategies for aquaculture feeds. Aquacult. Nutr. 13:17–34.

Gomes, E. F., P. Rema, and S. J. Kaushik. 1995. Replacement of fish meal by plant proteins in the diet of rainbow trout (*Oncorhynchus mykiss*): Digestibility and growth performance. Aquaculture 130:177–186.

Gomez-Requeni, P., M. Mingarro, S. Kirchner, J. A. Calduch-Giner, F. Medale, S. A. M. Martin, D. F. Houlihan, S. J. Kaushik, and J. Perez-Sanchez. 2004. Protein growth performance, amino acid utilisation and somatotrophic axis responsiveness to fish meal replacement by plant protein sources in gilthead sea bream (*Sparus aurata*). Aquaculture 232:493–510.

Gunstone, F. D. 2010. The world's oils and fats. Pp. 61–98 in Fish Oil Replacement and Alternative Lipid Sources in Aquaculture Feeds, G. M. Turchini, W. K. Ng, and D. R. Tocher, eds. Boca Raton, FL: Taylor & Francis, CRC Press.

Gunstone, F. D., and J. L. Harwood. 2007. Occurrence and characteristics of oils and fats. Pp. 37–142 in The Lipid Handbook, 3rd edition, F. D. Gunstone, J. L. Harwood, A. J. Dijkstra, eds. Boca Raton, FL: Taylor & Francis, CRC Press.

Hardy, R. W. 1989. Diet preparation. Pp. 473–546 in Fish Nutrition, 2nd Edition, J. E. Halver, ed. New York: Academic Press.

Hardy, R. W., and F. T. Barrows. 2002. Diet formulation and manufacture. Pp. 505–600 in Fish Nutrition, 3rd Edition, J. E. Halver and R. W. Hardy, eds. San Diego, CA: Academic Press.

Hemre, G.-I., M. Sanden, A. M. Bakke-McKellep, A. Sagstad, and Å. Krogdahl. 2004. Growth, feed utilization and health of Atlantic salmon *Salmo salar* L. fed genetically modified compared to non-modified commercial hybrid soybeans. Aquacult. Nutr. 11:157–167.

Hemre, G.-I., A. Sagstad, A. M. Bakke-McKellep, A. Danieli, R. Acierno, M. Maffia, M. Frøystad, Å. Krogdahl, and M. Sanden. 2007. Nutritional, physiological and histological responses in Atlantic salmon *Salmo salar* L. fed diets with genetically modified maize. Aquacult. Nutr. 13:186–199.

Hites, R. A., J. A. Foran, D. O. Carpenter, M. C. Hamilton, B. A. Knuth, and S. J. Schwager. 2004. Global assessment of organic contaminants in farmed salmon. Science 303:226–229.

Ikonomou, M. G., D. A. Higgs, M. Gibbs, J. Oakes, B. Skura, S. McKinley, S. K. Balfry, S. Jones, R. Withler, and C. Dubetz. 2007. Flesh quality of market-size farmed and wild British Columbia salmon. Environ. Sci. Technol. 41:437–443.

Izquierdo, M. S., A. Obach, L. Arantzamendi, D. Montero, L. Robaina, and G. Rosenlund. 2003. Dietary lipid sources for seabream and seabass: Growth performance, tissue composition and flesh quality. Aquacult. Nutr. 9:397–407.

Jobling, M. 2003. Do changes in Atlantic salmon, *Salmo salar* L. fillet fatty acids following a dietary switch represent wash-out or dilution? Test of a dilution model and its application. Aquac. Res. 34:1215–1221.

Johnson, R. B., and H. J. Barnett. 2003. Determination of fat content in fish feed by supercritical fluid extraction and subsequent lipid classification of extract by thin layer chromatography-flame ionization detection. Aquaculture 216:263–282.

Jutfelt, F., R. E. Olsen, B. T. Bjornsson, and K. Sundell. 2007. Parr–smolt transformation and dietary vegetable lipids affect intestinal nutrient uptake, barrier function and plasma cortisol levels in Atlantic salmon. Aquaculture 273:298–311.

Kaushik, S., and M. Troell. 2010. Taking the fish-in fish-out ratio a step further. Aquacult. Eur. 35:15–17.

Kaushik, S. J., J. P. Cravedi, J. P. Lalles, J. Sumpter, B. Fauconneau, and M. Laroche. 1995. Partial or total replacement of fishmeal by soybean protein on growth, protein utilisation, potential estrogenic or antigentic effects, cholesterolemia and flesh quality in rainbow trout, *Oncorhynchus mykiss*. Aquaculture 133:257–274.

Kaushik, S. J., D. Covès, G. Dutto, and D. Blanc. 2004. Almost total replacement of fish meal by plant protein sources in the diet of a marine teleost, the European sea bass (*Dicentrarchus labrax*). Aquaculture 230:391–404.

Kelleher, K. 2005. Discards in the world's marine fisheries: An update. FAO Fisheries Technical Paper. No. 470. Rome, Italy: FAO Fisheries and Aquaculture Department.

Kendrick, A., and C. Ratledge. 1992. Lipids of selected molds grown for production of n-3 and n-6 polyunsaturated fatty acids. Lipids 27:15–20.

Kew, S., T. Banerjee, A. M. Minihane, Y. E. Finnegan, C. M. Williams, and P. C. Calder. 2003. Relation between the fatty acid composition of peripheral blood mononuclear cells and measures of immune cell function in healthy, free-living subjects aged 25–72 y. Am. J. Clin. Nutr. 77:1278–1286.

Kinney, A. J., E. B. Cahoon, H. G. Damude, W. D. Hitz, C. W. Kolar, and Z. B. Liu, inventors. 2004. Enhancing plant seed oils for human nutrition: Production of very long chain polyunsaturated fatty acids in oilseed plants. International Patent Publication WO2004071467.

Kiron, V., H. Fukuda, T. Takeuchi, and T. Watanabe. 1995. Essential fatty acid nutrition and defence mechanisms in rainbow trout *Oncorhynchus mykiss*. Comp. Biochem. Phys. A 111:361–367.

Krogdahl, Å., M. H. Penn, J. Thorsen, S. Refstie, and A. M. Bakke. 2010. Important antinutrients in plant feedstuffs for aquaculture: An update on recent findings regarding responses in salmonids. Aquac. Res. 41:333–344.

Leaver, M. J., L. A. N. Villeneuve, A. Obach, L. Jensen, J. E. Bron, D. R. Tocher, and J. B. Taggart. 2008. Functional genomics reveals increased cholesterol and highly unsaturated fatty acid biosynthesis after dietary substitution of fish oil with vegetable oils in Atlantic salmon (*Salmo salar*). BMC Genomics 9:299. doi:10.1186/1471-2164-9-299.

Lee, K.-J., K. Dabrowski, J. H. Blom, S. C. Bai, and P. C. Stromberg. 2002. Mixture of animal and plant protein sources can completely replace fish meal in a diet for juvenile rainbow trout (*Oncorhynchus mykiss*). J. Anim. Physiol. An. N. 86:201–213.

Lee, R. F., W. Hagen, and G. Kattner. 2006. Lipid storage in marine zooplankton. Mar. Ecol-Prog. Ser. 307:273–306.

Li, M. H., G. F. Hartnell, and E. H. Robinson. 2008. Evaluation of cottonseed meal diets from genetically modified cotton as feed ingredients for channel catfish, *Icatalus puntatus*. Aquacult. Nutr. 14:490–498.

Lin, Y. H., and S. Y. Shiau. 2007. Effects of dietary blend of fish oil with corn oil on growth and non-specific immune responses of grouper *Epinephelus malabaricus*. Aquacult. Nutr. 13:137–144.

Lodemel, J. B., T. M. Mayhew, R. Myklebust, R. E. Olsen, S. Espelid, and E. Ringø. 2001. Effect of three dietary oils on disease susceptibility in Arctic charr (*Salvelinus alpinus* L.) during cohabitant challenge with *Aeromonas salmonicida subsp. salmonicida*. Aquac. Res. 32:935–945.

Medale, F., T. Bouchard, F. Vallee, D. Blanc, M. Mambrini, A. Roem, and S. J. Kaushik. 1998. Voluntary feed intake, nitrogen and phosphorous losses in rainbow trout, *Oncorhynchus mykiss*. Aquaculture 124:117–126.

Metz, J. G., J. H. Flatt, and J. M. Kuner, inventors. 2006. The genes for the enzymes of the polyunsaturated fatty acid polyketide synthase of Schizochytrium and their use in the manufacture of polyunsaturated fatty acids. International Patent Publication WO 2006135866.

Metz, J. G., J. M. Kuner, and J. C. Lippmeier, inventors. 2007. Polyunsaturated fatty acid production in genetically modified organisms using PUFA polyketide synthase systems. International Patent Publication WO2007106903.

Miller, M. R., P. D. Nichols, and C. G. Carter. 2007. Replacement of fish oil with thraustochytrid *schizochytrium* sp. L oil in Atlantic salmon parr (*Salmo salar* L) diets. Comp. Biochem. Phys. A 148:382–392.

Miller, M. R., P. D. Nichols, and C. G. Carter. 2008. The digestibility and accumulation of dietary phytosterols in Atlantic salmon (*Salmo salar* L.) smolt fed diets with replacement plant oils. Lipids 43:549–557.

Monsanto. 2008. Monsanto and the Solae Company to collaborate on soy based omega-3 portfolio. Monsanto Company. Available online at: *www.monsanto.com/features/omega.asp*. Accessed on April 16, 2008.

Montero, D., and M. Izquierdo. 2010. Health and welfare of fish fed vegetable oils as alternative lipid sources to fish oil. Pp. 439–485 in Fish Oil Replacement and Alternative Lipid Sources in Aquaculture Feeds, G. M. Turchini, W. K. Ng, and D. R. Tocher, eds. Boca Raton, FL: Taylor & Francis, CRC Press.

Montero, D., T. Kalinowski, A. Obach, L. Robaina, L. Tort, M. J. Caballero, and M. S. Izquierdo. 2003. Vegetable lipid sources for gilthead seabream (*Sparus aurata*): effects on fish health. Aquaculture 225:353–370.

Montero, D., V. Grasso, M. S. Izquierdo, R. Ganga, F. Real, L. Tort, M. J. Caballero, and F. Acosta. 2008. Total substitution of fish oil by vegetable oils in gilthead seabream (*Sparus aurata*) diets: Effects on hepatic Mx expression and some immune parameters. Fish Shellfish Immun. 24:147–155.

Morris, P. C., P. Gallimore, J. Handley, G. Hide, P. Haughton, and A. Black. 2005. Full-fat soya for rainbow trout (*Oncorhynchus mykiss*) in freshwater: Effects on performance, composition and flesh fatty acid profile in absence of hind gut enteritis. Aquaculture 248:147–161.

Mourente, G., and J. G. Bell. 2006. Partial replacement of dietary fish oil with blends of vegetable oils (rapeseed, linseed and palm oils) in diets for European sea bass (*Dicentrarchus labrax* L.) over a long term growth study: Effects on muscle and liver fatty acid composition and effectiveness of a fish oil finishing diet. Comp. Biochem. Phys. B 145:389–399.

Mourente G., J. E. Good, and J. G. Bell. 2005. Partial substitution of fish oil with rapeseed, linseed and olive oils in diets for European sea bass (*Dicentrarchus labrax* L.): effects on flesh fatty acid composition, plasma prostaglandins E_2 and $F_{2\alpha}$, immune function and effectiveness of a fish oil finishing diet. Aquacult. Nutr. 11:25–40.

Mourente, G., J. E. Good, K. D. Thompson, and J. G. Bell. 2007. Effects of partial substitution of dietary fish oil with blends of vegetable oils, on blood leucocyte fatty acid compositions, immune function and histology in European sea bass (*Dicentrarchus labrax* L.). Brit. J. Nutr. 98:770–779.

Mundheim, H., A. Aksnes, and B. Hope. 2004. Growth, feed efficiency and digestibility in salmon (*Salmo salar* L.) fed different dietary proportions of vegetable protein sources in combination with two fish meal qualities. Aquaculture 237:315–331.

Napier, J. A. 2007. The production of unusual fatty acids in transgenic plants. Annu. Rev. Plant Biol. 58:295–319.

Naylor, R. L., R. J. Goldberg, J. H. Primavera, N. Kautsky, M. C. M. Beveridge, J. Clay, C. Folke, J. Lubchenco, H. Mooney, and M. Troell. 2000. Effect of aquaculture on world fish supplies. Nature 405:1017–1024.

Naylor, R. L., R. W. Hardy, D. P. Bureau, A. Chiu, M. Elliot, A. P. Farrell, I. Forster, D. M. Gatlin, R. J. Goldburg, K. Hua, and P. D. Nichols. 2009. Feeding aquaculture in an era of finite resources. P. Natl. Acad. Sci. USA 106:15103–15110.

Ng, W.-K., and O. M. Bahurmiz. 2009. The impact of dietary oil source and frozen storage on the physical, chemical and sensorial quality of fillets from market-size red hybrid tilapia, Oreochromis sp. Aquaculture 113:1041–1048.

NRC (National Research Council). 1993. Nutrient Requirements of Fish. Washington, DC: National Academy Press.

Olsen, R. E., R. Myklebust, T. Kaino, and E. Ringø. 1999. Lipid digestibility and ultrastructural changes in the enterocytes of Arctic charr (*Salvelinus alpinus* L.) fed linseed oil and soybean lecithin. Fish Physiol. Biochem. 21:35–44.

Olsen, R. E., R. Myklebust, E. Ringø, and T. M. Mayhew. 2000. The influences of dietary linseed oil and saturated fatty acids on caecal enterocytes in Arctic charr (*Salvelinus alpinus* L.): A quantitative ultrastructural study. Fish Physiol. Biochem. 22:207–216.

Olsen, R. E., R. J. Henderson, J. Sountama, G.-I. Hemre, E. Ringø, W. Melle, and D. R. Tocher. 2004. Atlantic salmon, *Salmo salar*, utilizes wax ester-rich oil from *Calanus finmarchicus* effectively. Aquaculture 240:433–449.

Olsen, R. E., R. Waagbø, and S. P. Lall. 2010. Alternative marine resources. Pp. 267–324 in Fish Oil Replacement and Alternative Lipid Sources in Aquaculture Feeds, G. M. Turchini, W. K. Ng, and D. R. Tocher, eds. Boca Raton, FL: Taylor & Francis, CRC Press.

Padley, F. B., F. D. Gunstone, and J. L. Harwood. 1986. Occurrence and characteristics of oils and fats. Pp. 49–170 in The Lipid Handbook, F. D. Gunstone, J. L. Harwood, and F. B. Padley, eds. London, UK: Chapman and Hall.

Padley, F. B., F. D. Gunstone, and J. L. Harwood. 1994. Occurrence and characteristics of oils and fats. Pp. 47–223 in The Lipid Handbook, 2nd edition, F. D. Gunstone, J. L. Harwood, and F. B. Padley, eds. Cambridge, UK: Chapman and Hall.

Papatryphon, E., J. Petit, S. J. Kaushik, and H. M. G. van der Werf. 2004. Environmental impact assessment of salmonid feeds using Life Cycle Assessment (LCA). Ambio 33:316–323.

Petropoulos, I. K., K. D. Thompson, A. Morgan, J. R. Dick, D. R. Tocher, and J. G. Bell. 2009. Effects of substitution of dietary fish oil with a blend of vegetable oils on liver and peripheral blood leucocyte fatty acid composition, plasma prostaglandin E_2 and immune parameters in three strains of Atlantic salmon (*Salmo salar*). Aquacult. Nutr. 15:596–607.

Place, A. R. 1992. Comparative aspects of lipid digestion and absorption—physiological correlates of wax ester digestion. Am. J. Physiol. 263:R464–R471.

Pratoomyot, J., E. Å Bendiksen, J. G. Bell, and D. R. Tocher. 2008. Comparison of effects of vegetable oils blended with southern hemisphere fish oil and decontaminated northern hemisphere fish oil on growth performance, composition and gene expression in Atlantic salmon (*Salmo salar* L.). Aquaculture 280:170–178.

Pusztai, A., and S. Bardocz. 2006. GMO in animal nutrition: Potential benefits and risks. Pp. 513–540 in Biology of Nutrition in Growing Animals, R. Mosenthin, J. Zentek, and T. Zebrowska, eds. Edinburgh, Scotland: Elsevier.

Qi, B., T. Fraser, S. Mugford, G. Dobson, O. Sayanova, J. Butler, J. A. Napier, A. K. Stobart, and C. M. Lazarus. 2004. Production of very long chain polyunsaturated omega-3 and omega-6 fatty acids in plants. Nat. Biotechnol. 22:739–745.

RAFOA (Researching Alternatives to Fish Oils in Aquaculture). 2005. Researching alternatives to fish oil in aquaculture. European Union Framework programme 5 RT & D project, Q5RS-2000-30058. Available online at: *http://www.rafoa.stir.ac.uk/project_results.html*. Accessed on May 19, 2010.

Ramsden, C. E., J. R. Hibbeln, and W. E. Lands. 2009. Letter to the Editor re: Linoleic acid and coronary heart disease, Prostag. Leukotr. Ess. (2008), by W.S. Harris. Prostag. Leukotr. Ess. 80:77–79.

Refstie, S., O. J. Korsoen, T. Storebakken, G. Baeverfjord, I. Lein, and A. J. Roem. 2000. Differing nutritional responses to dietary soybean meal in rainbow trout (*Oncorhynchus mykiss*) and Atlantic salmon (*Salmo salar*). Aquaculture 190:49–63.

Robert, S., S. P. Singh, X. R. Zhou, J. R. Petrie, S. I. Blackburn, P. M. Mansour, P. D. Nichols, Q. Liu, and A. Green. 2005. Metabolic engineering of Arabidopsis to produce nutritionally important DHA in seed oil. Funct. Plant Biol. 32:473–479.

Robin, J., C. Regost, J. Arzel, and S. Kaushik. 2003. Fatty acid profile of fish following a change in dietary fatty acid source: Model of fatty acid composition with a dilution hypothesis. Aquaculture 225:283–293.

Rodehutscord, M. S. Mandel, M. Pack, S. Jacobs, and E. Pfeffer. 1995. Free amino acids can replace protein-bound amino acids in test diets for studies in rainbow trout. J. Nutr. 125:956–963.

Rørå, A. M. B., B. Ruyter, J. Skorve, R. K. Berge, and K.-E. Slinning. 2005. Influence of high content of dietary soybean oil on quality of large fresh, smoked and frozen Atlantic salmon (*Salmo salar*). Aquacult. Int. 13:217–231.

Rubin. 2007. Marine by-products—a huge unexploited potential for value adding. Available online at: *http://www.rubin.no/eng*. Accessed on May 19, 2010.

Rumsey, G. L. 1993. Fish meal and alternative sources of protein. Fisheries 18:14–19.

Ruxton, C. H. S., S. C. Reed, M. J. A. Simpson, and K. J. Millington. 2007. The health benefits of omega-3 polyunsaturated fatty acids: A review of the evidence. J. Hum. Nutr. Diet. 20:275–285.

Sæther, O., T. E. Ellingsen, and V. Mohr. 1986. Lipolysis post mortem in North Atlantic krill. Comp. Biochem. Phys. B 83:51–55.

Sanden, M., I. J. Bruce, M. A. Rahman, and G.-I. Hemre. 2004. The fate of transgenic sequences present in genetically modified plant products in fish feed, investigating the survival of GM soybean DNA fragments during feeding trials in Atlantic salmon, *Salmo salar* L. Aquaculture 237:391–405.

Seierstad, S. L., T. T. Poppe, E. O. Koppang, A. Svindland, G. Rosenlund, L. Frøyland, and S. Larsen. 2005. Influence of dietary lipid composition on cardiac pathology in farmed Atlantic salmon, *Salmo salar* L. J. Fish Dis. 28:677–690.

Silva, J. M. G., M. Espe, L. E. C. Conceicao, J. Dias, and L. M. P. Valente. 2009. Senegalese sole juveniles (*Solea senegalensis* Kaup, 1858) grow equally well on diets devoid of fish meal provided the dietary amino acids are balanced. Aquaculture 296:309–317.

Simopoulos, A. P. 2006. Evolutionary aspects of diet, the omega-6/omega-3 ratio and genetic variation: Nutritional implications for chronic diseases. Biomed. Pharmacother. 60:502–507.

Sissener, N. H., M. Danden, A. M. Bakke, Å. Krogdahl, and G. I. Hemre. 2009a. A long term trial with Atlantic salmon (*Salmo salar* L.) fed genetically modified soy; focusing general health and performance before, during and after the parr-smolt transformation. Aquaculture 298:101–1107.

Sissener, N. H., M. Danden, A. M. Bakke, Å. Krogdahl, and G. I. Hemre. 2009b. An assessment of organ and intestinal histomorphology and cellular stress response in Atlantic salmon (*Salmo salar* L.) fed genetically modified Roundup Ready® soy. Aquaculture 294:108–117.

Sitja-Bobadilla, A., S. Pena-Llopis, P. Gomez-Requeni, F. Medale, S. J. Kaushik, and J. Perez-Sanchez. 2005. Effect of fish meal replacement by plant protein sources on non-specific defence mechanisms and oxidative stress in gilthead sea bream (*Sparus aurata*). Aquaculture 249:387–400.

Tacon, A. G. J. 2005. Salmon aquaculture dialogue: Status of information on salmon aquaculture feed and the environment. International Aquafeed 8:22–37.

Tacon, A. G. J., and M. Metian. 2008a. Global overview on the use of fish meal and fish oil in industrially compounded aquafeeds: Trends and future prospects. Aquaculture 285:146–158.

Tacon, A. G. J., and M. Metian. 2008b. Aquaculture feed and food safety: The role of the Food and Agriculture Organization and the Codex Alimentarius. Ann. NY Acad. Sci. 1140:50–59.

Tacon, A. G. J., and M. Metian. 2009a. Fishing for aquaculture: Non-food use of small pelagic forage fish—A global perspective. Rev. Fish. Sci. 17:305–317.

Tacon, A. G. J., and M. Metian. 2009b. Fishing for feed or fishing for food: Increasing global competition for small pelagic forage fish. Ambio 38:294–302.

Tacon, A. G. J., M. R. Hasan, and R. P. Subasinghe. 2006. Use of fishery resources as feed inputs to aquaculture development: Trends and policy implications. FAO Fisheries Circular No.1018. Rome, Italy: FAO Fisheries and Aquaculture Department.

Tacon, A. G. J., M. Metian, G. M. Turchini, and S. S. DeSilva. 2010. Responsible aquaculture and trophic level implications to global fish supply. Rev. Fish. Sci. 18:94–105.

Taggart, J. B., J. E. Bron, S. A. M. Martin, P. J. Seear, B. Hoyheim, R. Talbot, L. Villeneuve, G. E. Sweeney, D. F. Houlihan, C. J. Secombes, D. R. Tocher, and A. J. Teale. 2008. A description of the origins, design and performance of the TRAITS/SGP Atlantic salmon (*Salmo salar* L.) cDNA microarray. J. Fish Biol. 72:2071–2094.

Teshima, S. I., 1997. Phospholipids and sterols. Pp. 85–107 in Crustacean Nutrition, L. R. D'Abramo, D. E. Conklin, and D. M. Akiyama, eds. Baton Rouge, LA: World Aquaculture Society.

Teskeredzic, Z., D. A. Higgs, B. S. Dosanjh, J. R. McBride, R. W. Hardy, R. M. Beames, J. D. Jones, M. Simell, T. Vaara, and R. B. Bridges. 1995. Assessment of undephytinised and dephytinised rapeseed protein concentrate as sources of dietary protein for juvenile rainbow trout, *Oncorhynchus mykiss*. Aquaculture 131:261–277.

Thompson, K. D., M. F. Tatner, and R. J. Henderson. 1996. Effects of dietary (n-3) and (n-6) polyunsaturated fatty acid ratio on the immune response of Atlantic salmon (*Salmo salar* L.). Aquacult. Nutr. 2:21–31.

Tocher, D. R., 1995. Glycerophospholipid metabolism. Pp. 119–157 in Biochemistry and Molecular Biology of Fishes, Vol. 4, Metabolic and Adaptational Biochemistry, P. W. Hochachka, and T. P. Mommsen, eds. Amsterdam, the Netherlands: Elsevier Press.

Tocher, D. R. 2003. Metabolism and functions of lipids and fatty acids in teleost fish. Rev. Fish. Sci. 11:107–184.

Tocher, D. R. 2009. Issues surrounding fish as a source of ω3 long-chain polyunsaturated fatty acids. Lipid Technol. 21:13–16.

Tocher, D. R. 2010. Fatty acid requirements in ontogeny of marine and freshwater fish. Aquac. Res. 41:717–732.

Tocher, D. R., J. R. Dick, P. MacGlaughlin, and J. G. Bell. 2006. Effect of diets enriched in Δ6 desaturated fatty acids (18:3n-6 and 18:4n-3), on growth, fatty acid composition and highly unsaturated fatty acid synthesis in two populations of Arctic charr (*Salvelinus alpinus* L.). Comp. Biochem. Phys. B 144:245–253.

Tocher, D. R., E. Å. Bendiksen, P. J. Campbell, and J. G. Bell. 2008. The role of phospholipids in nutrition and metabolism of teleost fish. Aquaculture 280:21–34.

Tocher, D. R., D. S. Francis, and K. Coupland. 2010. n-3 polyunsaturated fatty acid-rich vegetable oils and blends. Pp. 209–244 in Fish Oil Replacement and Alternative Lipid Sources in Aquaculture Feeds, G. M. Turchini, W.–K. Ng, and D. R. Tocher, eds. Boca Raton, FL: Taylor & Francis, CRC Press.

Torstensen, B. E., J. G. Bell, G. Rosenlund, R. J. Henderson, I. E. Graff, D. R. Tocher, Ø. Lie, and J. R. Sargent. 2005. Tailoring of Atlantic salmon (*Salmo salar* L.) flesh lipid composition and sensory quality by replacing fish oil with a vegetable oil blend. J. Agr. Food Chem. 53:10166–10178.

Torstensen, B. E., M. Espe, M. Sanden, I. Stubhaug, R. Waagbo, G.–I. Hemre, R. Fontanillas, U. Nordgarden, E. M. Hevroy, P. Olsvik, and M. Berntssen. 2008. Novel production of Atlantic salmon (*Salmo salar*) protein based on combined replacement of fish meal and fish oil with plant meal and vegetable oil blends. Aquaculture 285:193–200.

Turchini, G. M., B. E. Torstensen, and W.–K. Ng. 2009. Fish oil replacement in finfish nutrition. Rev. Aquacult. 1:10–57.

Turchini, G. M., W.–K. Ng, and D. R. Tocher, eds. 2010. Fish Oil Replacement and Alternative Lipid Sources in Aquaculture Feeds. Boca Raton, FL: Taylor & Francis, CRC Press.

Watanabe, T. 2002. Strategies for further development of aquatic feeds. Fish. Sci. 68:242–252.

Watanabe, T., V. Verakunpiriya, K. Watanabe, K. Viswanath, and S. Satoh. 1998. Feeding rainbow trout with non-fishmeal diets. Fisheries Sci. 63:258–266.

Webster, C., C. Lim, and S.–C. Lee. 2007. Use of Alternative Protein Sources in Aquaculture Diets. New York: The Haworth Press.

Whelan, J. 2008. The health implications of changing linoleic acid intakes. Prostag. Leukotr. Ess. 79:165–167.

Whelan, J. 2009. Dietary stearidonic acid is a long chain (n-3) polyunsaturated fatty acid with potential health benefits. J. Nutr. 139:1–6.

Wilson, R. P. 2002. Amino acids and proteins. Pp. 143–179 in Fish Nutrition, 3rd Edition, J. E. Halver and R. W. Hardy, eds. San Diego, CA: Academic Press.

Wu, G., M. Truksa, N. Datla, P. Vrinten, J. Bauer, T. Zank, P. Cirpus, E. Heinz, and X. Qiu. 2005. Stepwise engineering to produce high yields of very long-chain polyunsaturated fatty acids in plants. Nat. Biotechnol. 23:1013–1017.

Zhu, Q., Z. Xue, N. Yadav, P. Sharpe, H. Damude, D. Xie, D. Short, R Hong, A. Kinney, and B. Tyreus. 2009. Commercial production of eicosapentaenoic acid by fermentation of engineered yeast. Pp. 65 in Lipids, Fats and Oils: From Knowledge to Application, book of abstracts, 7th Euro Fed Lipid Congress, October 18–21, 2009, Graz, Austria. Frankfurt, Germany: European Federation for the Science and Technology of Lipids.

17

Critical Research Needs

The committee had the opportunity to review progress in fish and shrimp nutrition since publication of the previous edition in 1993, and to examine the current state of knowledge in focal areas of fish and shrimp nutrition. This effort led to the development of a unique perspective and the realization of significant gaps remaining in knowledge of fish and shrimp nutrition. Many specific needs for research are identified in the text of each chapter. With those gaps in mind, this chapter, Critical Research Needs, was generated. The committee found that relatively few additional requirement values were available to add to Table 18-1 from the 1993 publication Nutrient Requirements of Fish. Limited effort has been directed toward obtaining refined estimates of the previously published nutrient requirements of the most well-established species of farmed fish and shrimp. Further, little research has been conducted to estimate the dietary nutrient requirements of the numerous species of farmed fish and shrimp now under some stage of aquaculture development or production.

Since publication of the 1993 edition of the report, many new analytical techniques have been developed to improve understanding of the roles that specific nutrients play in fish and shrimp physiology. The new area of how and why nutrients can affect gene activity and growth is progressing rapidly, and the techniques for genetic manipulations of desired traits in farmed species are being improved annually. Therefore, these are relatively new areas for productive research in studies investigating fish and shrimp nutrient requirements.

REQUIREMENTS, DELIVERY, AND INTERACTION OF NUTRIENTS

There are several possible explanations for slow progress in defining dietary nutrient requirements of farmed fish and shrimp. First, widespread use of fish meal in fish feeds has allowed manufacturers to produce feeds without having accurate estimates of the essential amino acid requirements of most species. The amino acid profile of high-quality fish meal commonly satisfies these requirements, provided sufficient levels of fish meal are present. Another explanation is that, for the most part, vitamin supplementation levels in fish feeds do not take into account any contribution of vitamins from feed ingredients. Rather, all vitamins are supplemented at required levels or in excess. Fish can obtain many essential minerals from water in which they live. Those that must be supplied to meet dietary requirements are either added as supplements or obtained from dietary ingredients, with fish meal a major contributor. Finally, using fish oil as an ingredient in fish feeds is usually sufficient to avoid essential fatty acid deficiencies. Shifts in fish feed formulation toward lower use levels of fish meal and fish oil, with concomitant shifts to higher use of plant protein and oil sources, increase the importance of having accurate estimates of nutrient requirements for feed formulations to ensure aquaculture production efficiency, improve protein and phosphorus retention, and reduce fecal waste.

There is a critical need to establish refined estimates of essential amino acid requirements for cyprinids, tilapia, marine fish species, and penaeid shrimp. Lysine, methionine, arginine, and leucine requirements have been studied, but estimates for other essential amino acids like threonine, isoleucine, phenylalanine, valine, and histidine are available from the results of only a few studies. More rigorous studies are needed to establish essential amino acid requirements with confidence. There is a critical need to develop a better system for expressing amino acid requirements of fish and shrimp, replacing current modes of expression (% diet, grams/16 g N, grams/kcal, digestible energy [DE], Ideal Protein). The issue of absolute and relative amino acid requirements needs to be investigated in reference to the effect on life stages, growth rate, and feed efficiency of fish and shrimp.

An understanding of whether the main difference in qualitative essential fatty acid requirements (C18 polyunsaturated fatty acids [PUFA] vs. long-chain-PUFA [LC-PUFA] fatty acids) among species is dependent on environmental habitat (marine or brackish or freshwater) or whether feeding habit (carnivorous/piscivorous or herbivorous/omnivorous) is equally or more important needs to be achieved. The influence of dietary lipid content and/or dietary energy level on quantitative essential fatty acid (EFA) requirements needs to be elucidated. Accurate quantitative requirements for EFAs should be determined for the different species used in aquaculture, particularly during critical life stages, such as during larval development and for maturing broodstock.

A more definitive understanding of the role of carbohydrate sources on metabolism and energy partitioning is needed. The relationships of carbohydrate content to lipid and fatty acid content in feeds should also be a topic of investigation for farmed fish and shrimp species. The role of carbohydrate sparing on amino acid and protein requirements needs further investigation and should incorporate modern techniques of energy calculations.

Micronutrient requirements of farmed fish and shrimp are another area where knowledge is lacking. Determining efficient and effective levels for water-soluble and fat-soluble vitamins in the diet to maintain health and physiological condition in addition to normal growth is a promising area of research. Whether dietary micronutrient requirements should be based on the minimum level needed to prevent clinical deficiency signs or impaired growth (equivalent to the minimum daily allowance) or based on other criteria is an open question. Selenium is a perfect example. Selenium requirements of different aquatic species based on different responses including growth, stress tolerance, and disease resistance need to be reevaluated. For all micronutrients, requirements for all life history stages need to be determined. The specific nutrient requirements for brood stock of species reared, together with the specific diet formulations needed, are critical research needs.

Finally, research should focus upon the delivery of nutrients in different types of manufactured feeds to include various forms and isomers of the nutrients that can interact and impact specific nutrient availability.

FISH MEAL AND FISH OIL ALTERNATIVES

Given the predicted feed requirements associated with increasing aquaculture production, global fish meal and fish oil supplies are clearly inadequate to support the demand, and use levels in feeds will have to be reduced. Reducing levels of marine resources in fish feeds will require much more care in formulation and require accurate estimates of dietary requirements for essential nutrients by different species of fish and shrimp. Reduction or elimination of levels of fish meal in feeds requires the use of alternative protein sources derived from grains, oilseeds, rendered animal and poultry byproducts, seafood processing waste, single-cell proteins, algae or plant protein concentrates, and other potential sources. With the exception of proteins from rendered products, most of these sources do not have amino acid profiles that satisfy all essential amino acid requirements of fish, based upon current knowledge and extrapolation to species for which requirement values are not known. Reducing fish meal levels in fish feeds presents another problem because fish meal is a source of essential minerals, especially phosphorus and critical trace minerals, as well as vitamins and semiessential nutrients such as taurine. Adding protein concentrates from grains and oilseeds can add phytate, further lowering levels of available phosphorus in fish feeds. Reducing fish oil levels alters the content of essential fatty acids in feeds as well as levels of fat-soluble vitamins.

DIET FORMULATIONS AND PROCESSING

More rigorous methodology to assess bioavailability of essential amino acids in feed ingredients is necessary. Accurate values for digestibility of amino acids in feed ingredients from alternative protein sources are necessary to formulate efficient diets. A better understanding of requirements for different amino acids for maintenance and catabolism in various species is necessary. Understanding the effect of dietary levels of energy-yielding nutrients (starch, lipids, types of amino acids) on efficiency of utilization of various amino acids is critical.

An extensive search for alternative lipid sources for dietary formulations should be undertaken. Entirely new n-3 LC-PUFA sources to supply essential fatty acids that do not originate from conventional marine resources, such as novel algal preparations, need to be developed and evaluated. The effectiveness of genetically modified products, such as transgenic microorganisms and altered oil-seed plants, for use as ingredients in feeds also should be explored.

Nutrient interactions among amino acids, fatty acids, soluble carbohydrates, and lipids should be investigated as new ingredients, formulations, and processing methods are used to develop diets for fish and shrimp. Comprehensive studies that compare inorganic vs. organic trace minerals on bioavailability and corresponding excretion into environmental systems will be critical as farming of different species in alternative systems, such as recirculating aquaculture systems or in low or zero-exchange shrimp farming, is expanded.

The delivery of water-soluble nutrients in sinking and larval feeds remains a major factor that influences nutrient intake and the corresponding measurement of requirements, especially in species that do not consume feed quickly. Quantification of nutrient retention and waste production for many species is lacking. The limited data restrict efforts to model culture systems to include nutrient utilization and economic return, as well as environmental interactions. Standards for feed quality and farm evaluation criteria are lacking. Important information concerning optimal utilization of nutrients in aquaculture/polyculture situations continues to be lacking. Lastly, research to eliminate the need for natural foods for fish and shrimp larvae and broodstock shrimp is needed to stabilize supplies of seed stock and to improve biosecurity by eliminating the potential for disease associated with live feed.

NUTRIGENOMIC EFFECTS AND METABOLISM

The application of modern molecular techniques is essential to determine the effects of nutrients at the cell and organelle level where basic metabolic reactions occur. Nutrigenomics is the term used to define the use of transcriptomics and proteomics for the study of interactions between gene function and nutrition, and as such is required to determine the biochemical and molecular physiological response mechanisms involved with changes in dietary formulation. Enhanced molecular studies become particularly relevant in trying to delineate the precise biochemical and metabolic effects on animal physiology when measuring the effects of different protein or carbohydrate feed ingredients. When investigating the effects of different fatty acids for energy utilization and incorporation into lipoproteins in intracellular membranes, molecular analysis is undoubtedly the best tool for dissection of these complex metabolic events.

With the evolution of high-throughput sequencing and development of full sequence maps for many fish species, the identification of gene-specific regulation in regards to physiological processes has additional relevance. This is because it not only increases understanding of gene function and nutrient/gene interactions, but also has potential for use in candidate gene selection and selective breeding enhancement. Other specific research areas to be addressed include the regulation of carbohydrates as an energy source and in glucose transport and processing, immunological effects of different dietary ingredients, and the identification of pathway interactions during metabolic processing of different ingredients. Many additional areas of research are needed to improve current knowledge of specific nutrient requirements and of formulation techniques for specific developments in fish husbandry techniques.

CONCLUSIONS

The critical research needs covered above are key areas to be addressed to support continued growth in the aquaculture sector. Continuing research is needed to ensure economic production of fish and shrimp for consumers while allowing producers to cope with changes in feeds associated with increasing use of alternative ingredients. Research on the use of antibiotics, probiotics, and other immunostimulants, is critical, because huge economic, environmental, and food safety issues from diseases are globally encountered. Consumers expect and deserve safe and nutritious farmed fish and shrimp products. Feed composition has a profound effect on the nutritional value of farmed product to consumers, and the critical research needs described above recognize this. Society also has an important stake in continued improvement of both the health and welfare of farmed fish and shrimp, and in reducing the impacts of aquaculture on the aquatic environment. Advancing research efforts are keys to ensuring progress in both areas.

As these critical needs are individually addressed, the results will have a major impact in improving economic aspects of aquaculture production and concurrently improve the health and welfare of fish. The nutritional quality and safety of product will be improved. Finally, environmental stewardship will be reflected by more efficient and effective use of marine resources with minimum impact on the aquatic environment.

18

Nutrient Requirements Tables

The values in the nutrient requirements tables represent minimum requirements for maximum performance of fish and shrimp under experimental conditions. With few exceptions, the data were obtained with juvenile and larval fish and shrimp and under conditions considered to be optimal. However, requirements for some nutrients (e.g., essential fatty acids, vitamins) vary with the developmental stage and possibly with physiological stage. Where several values appear in the literature, the value presented in the table represents a consensus of the committee for the most reasonable estimate of the requirement. Some of the listed dietary requirements for shrimp are much higher than those for fish species. The differences are probably an artifact resulting from the feeding habits of shrimp that can result in a substantial loss of certain nutrients from feeds before ingestion. These values, although probably not needed to meet actual metabolic requirements, are presented because they are the only values reported in the scientific literature.

The stated requirements do not include any surpluses. In practice, however, a margin of safety is commonly added, whereby nutrient levels are increased to compensate for processing and storage losses, variation in composition and bioavailability of nutrients in feed ingredients, possible interactive effects, and variation in requirements caused by environmental effects. If diet formulations contain different digestible nutrient densities (e.g., formulated to contain higher or lower concentrations of digestible energy or protein than the values listed in the table), the other nutrient concentrations should be modified appropriately. The challenges associated with defining and meeting essential nutrient requirements are discussed in the preceding chapters.

The requirements for amino acids, fatty acids, vitamins, and minerals were determined with diets containing purified and chemically defined ingredients that are highly digestible to the organism; therefore, the values in the table represent almost 100% bioavailability to the targeted species. Consequently, nutrient bioavailability must be considered when formulating diets from practical feedstuffs because the bioavailability of nutrients is usually substantially less than the bioavailability of nutrients in purified diets.

For many farmed fish species, no amino acid requirement data are available. In practice, feeds for these species are sometimes formulated to meet the essential amino acid requirements of salmon or trout, based on the assumption that essential amino acid requirements are similar among carnivorous fish species. A similar approach is commonly used in the formulation of diets for shrimp. Refer to Chapter 5 for a detailed discussion of essential amino acid requirements of fish and shrimp. For other essential nutrients, feeds are either over-fortified (vitamins) or formulated to meet requirements that have been determined for similar species.

TABLE 18-1 Nutrient Requirements of Freshwater Fish (dry-matter basis)[a,b]

Item	Atlantic Salmon *Salmo salar*	Common Carp *Cyprinus carpio*	Rohu *Labeo rohita*	Tilapia *Oreochromis* spp.	Channel Catfish *Ictalurus punctatus*	Hybrid Striped Bass *Morone saxatilis* × *Morone chrysops*	Rainbow Trout *Oncorhynchus mykiss*	Pacific Salmon *Oncorhynchus* spp.
Typical Energy and Protein Concentrations[c]								
Digestible energy (kcal/kg diet)	4,400	3,200	3,200	3,400	3,000	4,000	4,200	4,200
Digestible protein (%)	36	32	32	29	29	36	38	40
Nutrient Requirements								
Amino acids (%)								
Arginine	1.8	1.7	1.7	1.2	1.2	1.0	1.5	2.2
Histidine	0.8[d]	0.5	0.9	1.0	0.6	NT	0.8	0.7
Isoleucine	1.1	1.0	1.0	1.0	0.8	NT	1.1	1.0
Leucine	1.5	1.4	1.5	1.9	1.3	NT	1.5	1.6
Lysine	2.4	2.2	2.3	1.6	1.6	1.6	2.4	2.2
Methionine	0.7	0.7	0.7	0.7	0.6	0.7	0.7	0.7
Methionine + cystine	1.1	1.0	1.0	1.0	0.9	1.1	1.1	1.1
Phenylalanine	0.9	1.3	0.9	1.1	0.7	0.9	0.9	0.9
Phenylalanine + tyrosine	1.8	2.0	1.6	1.6	1.6	NT	1.8	1.8
Threonine	1.1	1.5	1.7	1.1	0.7	0.9	1.1	1.1
Tryptophan	0.3	0.3	0.4	0.3	0.2	0.3	0.3	0.3
Valine	1.2	1.4	1.5	1.5	0.8	NT	1.2	1.2
Taurine	NR	NR	NT	NT	NR	NR	NR[e]	NT
Fatty acids (%)								
18:3n-3	1.0	0.5–1.0	NT	NT	1.0–2.0	NR	0.7–1.0	1.0
n-3 LC-PUFA[f]	0.5–1.0	R	NT	R	0.5–0.75	0.5–1.0	0.4–0.5	1.0
18:2n-6	NT	1.0	NT	0.5–1.0	NT	NT	1.0	1.0
Cholesterol (%)	NT	NT	NT	NT	NT	NT	NT	NT
Phospholipids (%)	NT (4.0–6.0)[g]	NT (2.0)[g]	NT	NT	NT	NT	NT (4.0–14.0)[g]	NT
Macrominerals (%)								
Calcium	NR	0.34	NT	R/0.7[h]	R/0.45[h]	NR	NR	NR
Chlorine	NT	NT	NT	0.15	0.17	NT	NT	NT
Magnesium	0.04	0.05	NT	0.06	0.04	NT	0.05	NT
Phosphorus	0.80	0.70	NT	0.40	0.33	0.50	0.70	0.60
Potassium	NT	NT	NT	0.20–0.30	0.26	NT	NT	0.80
Sodium	NR	NT	NT	0.15	0.06	NT	NR	NT
Microminerals (mg/kg)								
Copper	5	3	NT	5	5	NT	3	NT
Iodine	R	NT	NT	NT	1.1	NT	1.1	1
Iron	30–60	150	NT	85	30	NT	NT	NT
Manganese	10	12	NT	7	2.4	NT	12	NT
Selenium	NT	NT	NT	NT	0.25	0.25	0.15	R
Zinc	37	15	NT	20	20	37	15	NT
Fat-soluble vitamins[i]								
A (mg/kg)	NT	1.2	NT	1.8	0.6	0.5	0.75	R
D (μg/kg)	NT	NT	NT	9	12.5	NT	40	NR
E (mg/kg)	60	100	132	60	50	28	50	50
K (mg/kg)	< 10	NT	NT	NT	R	NT	R	R

continued

TABLE 18-1 Continued

Item	Atlantic Salmon *Salmo salar*	Common Carp *Cyprinus carpio*	Rohu *Labeo rohita*	Tilapia *Oreochromis* spp.	Channel Catfish *Ictalurus punctatus*	Hybrid Striped Bass *Morone saxatilis* × *Morone chrysops*	Rainbow Trout *Oncorhynchus mykiss*	Pacific Salmon *Oncorhynchus* spp.
Water-soluble vitamins (mg/kg)								
Thiamin	NT	0.5	NT	NT	1	NT	1	10
Riboflavin	NT	7	NT	6	9	5	4	7
Vitamin B_6	5	6	NT	15	3	NT	3	6
Pantothenic Acid	NT	30	NT	10	15	25	20	20
Niacin	NT	28	NT	26	14	NT	10	150
Biotin	NT	1	NT	0.06	R	NT	0.15	1
Vitamin B_{12}	NT	NR	NT	NR	R	NT	R	0.02
Folacin	NT	NR	NT	1	1.5	NT	1	2
Choline[j]	NT	1,500	NT	1,000	400	500	800	800
Myoinositol[j]	NT	440	NT	400	NR	NR	300	300
Vitamin C[k]	20	45	NT	20	15	22	20	NT

[a]These requirements have been determined with highly purified ingredients in which the nutrients are highly digestible; therefore the values presented represent near 100% bioavailability.

[b]R, Required in diet but quantity not determined; NR, not required under practical conditions (e.g., diets containing ingredients from marine and land animal proteins and fish oil and water of at least medium hardness); and NT, not tested.

[c]Typical digestible energy and digestible crude protein concentrations (digestible N × 6.25) in commercial diets.

[d]Young Atlantic salmon undergoing rapid growth after transfer to seawater appear to require up to 1.4% dietary histidine to prevent ocular pathology (bilateral cataracts).

[e]Although taurine is not a normal amino acid (it is an amino-sulfonic acid that is a derivative of cysteine), it is included here. Taurine is reported in one study to be required by rainbow trout fry fed all-plant protein diets.

[f]20:5n-3 and/or 22:6n-3.

[g]Values in parentheses represent requirements reported for larval/early juvenile stages.

[h]Dietary requirement in the absence of waterborne calcium.

[i]Conversion factors for fat-soluble vitamins are as follows: 10,000 IU ≈ 3,000 μg vitamin A (retinol), 1 IU = 0.025 μg vitamin D (cholecalciferol).

[j]Diet without phospholipids. Please refer to Chapter 9, Vitamins, for a full discussion of choline, inositol, and phospholipids.

[k]As L-ascorbyl-2-monophosphate or L-ascorbyl-2-polyphosphate. Please refer to Chapter 9, Vitamins, for a full discussion of different vitamin C stable forms.

TABLE 18-2 Nutrient Requirements of Marine Fish (dry-matter basis)[a,b]

Item	Asian Sea Bass (Barramundi) *Lates calcarifer*	Cobia *Rachycentron canadum*	European Sea Bass *Dicentrarchus labrax*	Japanese Flounder *Paralichthys olivaceus*	Grouper *Epinephelus* spp.	Red Drum *Sciaenops ocellatus*	Yellowtail *Seriola* spp.
Typical Energy and Protein Concentrations[c]							
Digestible energy (kcal/kg diet)	4,200	4,200	4,000	4,000	4,000	4,000	4,200
Digestible protein (%)	38	38	40	40	42	36	38
Nutrient Requirements							
Amino acids (%)							
Arginine	1.8	NT	1.8	2.0	NT	1.8	1.6
Histidine	NT	NT	NT	NT	NT	NT	NT
Isoleucine	NT	NT	NT	NT	NT	NT	NT
Leucine	NT	NT	NT	NT	NT	NT	NT
Lysine	2.1	2.3	2.2	2.6	2.8	1.7	1.9
Methionine	0.8	0.8	NT	0.9	NT	0.8	0.8
Methionine + cystine	1.2	1.1	1.1	NT	NT	1.2	1.2
Phenylalanine	NT	NT	NT	NT	NT	NT	NT
Phenylalanine + tyrosine	NT	NT	NT	NT	NT	NT	NT
Threonine	NT	NT	1.2	NT	NT	0.8	NT
Tryptophan	NT	NT	0.3	NT	NT	NT	NT
Valine	NT	NT	NT	NT	NT	NT	NT
Taurine[d]	R	R	0.2	R	R	R	R
Fatty acids (%)							
18:3n-3	NT	NT	NR	NT	NT	NR	NT
n-3 LC-PUFA[e]	NT	NT	1.0	1.4	1.0	0.5–1.0	2.0–3.9
18:2n-6	NT	NT	NT	NT	NT	NT	NT
Cholesterol (%)	NT	NT	NT	NT	NT	NT	NT
Phospholipids (%)	NT	NT	2–3 (12)[f]	7 (7)[f]	NT	NT	NT
Macrominerals (%)							
Calcium	NT	NT	NT	NT	NT	NT	NT
Chlorine	NT	NT	NT	NT	NT	R	NT
Magnesium	NT	NT	NT	NT	NT	NT	NT
Phosphorus	NT	NT	0.65	0.60	NT	0.80	NT
Potassium	NT	NT	NT	NT	NT	NT	NT
Sodium	NT	NT	NT	NT	NT	R	NT
Microminerals (mg/kg)							
Copper	NT	NT	NT	NT	5	NT	NT
Iodine	NT	NT	NT	NT	NT	NT	NT
Iron	NT	NT	NT	NT	NT	NT	NT
Manganese	NT	NT	NT	NT	NT	NT	NT
Selenium	NT	NT	NT	NT	0.7	NT	NT
Zinc	NT	NT	NT	NT	NT	20	NT
Fat-soluble vitamins[g]							
A (mg/kg)	NT	NT	31	2.7	0.9	NT	5.6
D (μg/kg)	NT	NT	NT	NT	NT	NT	NR
E (mg/kg)	NT	NT	NT	NT	115	31	119
K (mg/kg)	NT	NT	NT	NT	NT	NT	NR

continued

TABLE 18-2 Continued

Item	Asian Sea Bass (Barramundi) *Lates calcarifer*	Cobia *Rachycentron canadum*	European Sea Bass *Dicentrarchus labrax*	Japanese Flounder *Paralichthys olivaceus*	Grouper *Epinephelus* spp.	Red Drum *Sciaenops ocellatus*	Yellowtail *Seriola* spp.
Water-soluble vitamins (mg/kg)							
Thiamin	NT	NT	NT	NT	NT	NT	11
Riboflavin	NT	NT	NT	NT	NT	NT	11
Vitamin B_6	NT	NT	NT	NT	NT	NT	12
Pantothenic Acid	NT	NT	NT	NT	NT	NT	36
Niacin	NT	NT	NT	NT	NT	NT	12
Biotin	NT	NT	NT	NT	NT	NT	0.67
Vitamin B_{12}	NT	NT	NT	NT	NR	NT	0.05
Folacin	NT	NT	NT	NT	NT	NT	1.2
Choline[h]	NT	700	NT	NT	NT	600	1,000
Myoinositol[h]	NT	NT	NT	NT	350	NT	420
Vitamin C[i]	30	45–54	20	NT	18	15	43–52

[a]These requirements have been determined with highly purified ingredients in which the nutrients are highly digestible; therefore the values presented represent near 100% bioavailability.

[b]R, Required in diet but quantity not determined; NR, not required under practical conditions (e.g., diets containing ingredients from marine and land animal proteins and fish oil and water of at least medium hardness); and NT, not tested.

[c]Typical digestible energy and digestible crude protein concentrations (digestible N × 6.25) in commercial diets.

[d]Although taurine is not a normal amino acid (it is an amino-sulfonic acid that is a derivative of cysteine), it is included here. Taurine is likely required at approximately 0.5%, especially when fish are fed diets contain low levels of marine or animal proteins.

[e]20:5n-3 and/or 22:6n-3.

[f]Values in parentheses represent requirements reported for larval/early juvenile stages.

[g]Conversion factors for fat-soluble vitamins are as follows: 10,000 IU ≈ 3,000 μg vitamin A (retinol), 1 IU = 0.025 μg vitamin D (cholecalciferol).

[h]Diet without phospholipids. Please refer to Chapter 9, Vitamins, for a full discussion of choline, inositol, and phospholipids.

[i]As L-ascorbyl-2-monophosphate or L-ascorbyl-2-polyphosphate. Please refer to Chapter 9, Vitamins, for a full discussion of different vitamin C stable forms.

TABLE 18-3 Nutrient Requirements of Shrimp (dry-matter basis)[a,b]

Item	Kuruma Prawn *Marsupenaeus japonicus*	Fleshy Prawn *Fenneropenaeus chinensis*	Pacific White Shrimp *Litopenaeus vannamei*	Tiger Shrimp *Penaeus monodon*
Typical Energy and Protein Concentrations[c]				
Digestible energy (kcal/kg diet)	4,400	3,200	3,000	3,000
Digestible protein (%)	38	32	30	34
Nutrient Requirements				
Amino acids (%)				
Arginine	1.6	NT	NT	1.9
Histidine	0.6	NT	NT	0.8
Isoleucine	1.3	NT	NT	1.0
Leucine	1.9	NT	NT	1.7
Lysine	1.9	NT	1.6	2.1
Methionine	0.7	NT	NT	0.7
Methionine + cystine	1.0	NT	NT	1.0
Phenylalanine	1.5	NT	NT	1.4
Phenylalanine + tyrosine		NT	NT	NT
Threonine	1.3	NT	NT	1.4
Tryptophan	0.4	NT	NT	0.2
Valine	1.4	NT	NT	NT
Fatty acids (%)				
18:3n-3	NT	0.7–1.0	NR	1.2
n-3 LC-PUFA[d]	1.1	1.0	0.25–0.50	0.9
18:2n-6	NT	NT	NR	1.2
Cholesterol (%)[e]	0.2	NT	0.13	NT
Phospholipids (%)[f]	1.0–1.5	NT	R	1.0–1.5
Macrominerals (%)				
Calcium	NR	NT	R	NR
Chlorine	NT	NT	NT	NT
Magnesium	0.3	NT	0.26–0.35	NT
Phosphorus	1.0	NT	0.3–0.7	0.7
Potassium	1.0	NT	R	1.2
Sodium	NT	NT	NT	NT
Microminerals (mg/kg)				
Copper	R	25	16–32	10–30
Iodine	NT	NT	NT	NT
Iron	R	R	R	R
Manganese	R	NT	R	NT
Selenium	NT	NT	0.2–0.4	NT
Zinc	NT	15	15	NT
Fat-soluble vitamins[g]				
A (mg/kg)	NT	36–54	1.4	2.5
D (μg/kg)	NT	NT	NT	100
E (mg/kg)	NT	NT	100	90
K (mg/kg)	NT	185	NT	35

continued

TABLE 18-3 Continued

Item	Kuruma Prawn *Marsupenaeus japonicus*	Fleshy Prawn *Fenneropenaeus chinensis*	Pacific White Shrimp *Litopenaeus vannamei*	Tiger Shrimp *Penaeus monodon*
Water-soluble vitamins (mg/kg)[h]				
Thiamin	60–120	NT	NT	14
Riboflavin	80	NT	NT	23
Vitamin B_6	120	NT	80–100	72–89
Pantothenic Acid	NT	100	NT	100
Niacin	400	NT	NT	7.2
Biotin	NT	0.4	NT	2
Vitamin B_{12}	NT	0.01	NT	0.20
Folacin	NT	5	NT	2
Choline[i]	600	NT	NT	6,200
Myoinositol[i]	2,000	4,000	NT	3,000
Vitamin C[j]	50–100	600	50–100	50–100

[a]These requirements have been determined with highly purified ingredients in which the nutrients are highly digestible; therefore the values presented represent near 100% bioavailability. Vitamin values for shrimp are likely not true requirements, per se, but are dietary levels required in experimental studies to prevent deficiency signs and promote growth and health. As such, they are allowances to account for leaching and losses during mastication and ingestion.

[b]R, Required in diet but quantity not determined; NR, not required under practical conditions (e.g., diets containing ingredients from marine and land animal proteins and fish oil and water of at least medium hardness); and NT, not tested.

[c]Typical digestible energy and digestible crude protein concentrations (digestible N × 6.25) in commercial diets.

[d]20:5n-3 and/or 22:6n-3.

[e]In the presence of dietary phospholipid.

[f]Expressed as phosphatidylcholine.

[g]Conversion factors for fat-soluble vitamins are as follows: 10,000 IU ≈ 3,000 μg vitamin A (retinol), 1 IU = 0.025 μg vitamin D (cholecalciferol).

[h]Values are likely not true metabolic requirements because of substantial loss of nutrients in pellets associated with the feeding habits of shrimp.

[i]Diet without phospholipids. Please refer to Chapter 9, Vitamins, for a full discussion of choline, inositol, and phospholipids.

[j]As L-ascorbyl-2-monophosphate or L-ascorbyl-2-polyphosphate or other stabilized forms. Please refer to Chapter 9, Vitamins, for a full discussion of different vitamin C stable forms.

TABLE 18-4 Partial Summary of Deficiency Signs and Pathologies Associated with Deficiencies of Essential Nutrients (For a More Complete Description, Consult the Chapters on Specific Nutrients—Chapters 5 to 9)[a]

Nutrient	Time to Onset[b]	Signs of Deficiency[c]	Comments
Amino acids			
Histidine	Several weeks	Lens cataracts	Observed in Atlantic salmon smolts at seawater transfer and warm ocean temperatures
Methionine		Lens cataracts	
Tryptophan	12 weeks	Spinal deformities (scoliosis), cataract	Reported in juvenile Pacific salmon
Essential Fatty Acids		Myocarditis, pale/swollen (fatty) liver, intestinal steatosis, fin erosion, bleeding from gills, lordosis, reduced reproductive potential, and shock syndrome, anemia	
Cholesterol (%)			No requirement in finfish
Phospholipids (%)		Skeletal malformations	
Macrominerals (%)			
Magnesium		Low Mg levels in tissues (bone), spinal deformity, renal calcinosis	
Phosphorus	7–10 weeks	Cessation of feeding, low P levels in skin and bones, operculum and jaw deformities	In rapidly growing rainbow trout
Microminerals			
Copper		Poor growth, low tissue Cu levels	Cataracts reported in common carp
Iodine		Thyroid hyperplasia (goiter)	
Iron		Hypochromic normocytic anemia	
Manganese		Short-body dwarfism	
Selenium		High fry mortality, reduced glutathione peroxidase activity	Symptoms exacerbated with tocopherol deficiency
Zinc		Bilateral lens cataracts, low tissue Zn levels	
Fat-soluble vitamins			
A	> 15 weeks	Exopthalmia, edema, ascites, hemorrhages at base of fins	
D	12 weeks	Lethargy, tetany	
E	12–20 weeks	Anemia, acites, fragile erythrocytes, dermal depigmentation, steatitis	
K	10–14 weeks	Prolonged blood clotting time, hemorrhages in muscle and viscera	
Water-soluble vitamins			
Thiamin	8–12 weeks	Hyperexcitability to sudden stimulus with paralysis or aberrant swimming	
Riboflavin	8–12 weeks	Cataracts (bilateral), photophobia	
Vitamin B_6	3–4 weeks	Convulsions, rapid and gasping breathing, paralysis (tail down, head up position)	
Pantothenic acid	8–12 weeks	Fusion (hyperplasia) of gill lamella	
Niacin	14–16 weeks	Muscle spasms, gastric edema	
Biotin	8–12 weeks	Skin lesions	
Vitamin B_{12}	12–16 weeks	Fragmented erythrocytes, increased level of immature erythrocytes	
Folacin	10–12 weeks	Anemia, absence of immature erythrocytes	
Choline	3–4 weeks	No specific clinical signs	
Myoinsitol	8–12 weeks	Distended abdomen, hemorrhages at base of fins	
Vitamin C	12–20 weeks	Spinal deformities (lordosis, scoliosis)	

[a]Only specific deficiency signs are listed. For most nutrient deficiencies, anorexia resulting in reduced growth is the first clinical sign that appears.

[b]Time to onset in fast-growing juvenile fish fed completely deficient diets. In larger fish fed deficient diets, time to onset will be longer for vitamins, essential fatty acids, and some minerals due to larger tissue reserves. Time to onset can be a useful tool to exclude certain deficiencies, especially when feed is changed.

[c]Primary clinical signs of deficiency only. Alterations in metabolism in many tissues are likely to be present with associated pathologies. Deficiency signs from Roberts, R. J., ed. 2001. Fish Pathology, 3rd Edition. New York: Harcourt Publishers, Ltd; Halver, J. E., and R. W. Hardy, eds. 2002. Fish Nutrition, 3rd Edition. Amsterdam, the Netherlands: Academic Press.

19

Feed Composition Tables

The tables of feed ingredient composition (Tables 19-1 to 19-6) provide information that makes it possible to define fish and shrimp feeds nutritionally and to formulate them economically for both research and commercial use. Data in the tables were compiled by the committee from a variety of sources. The primary sources were the previous edition of this publication and other NRC publications on the nutrient requirements of animals, the Atlas of Nutritional Data on United States and Canadian Feeds (NRC, 1972), the Handbook on Ingredients for Aquaculture Feeds (Hertrampf and Piedad-Pascual, 2000), the USDA Food Composition Data Base (USDA, 2005), and in the case of new or minor ingredients, published papers describing these ingredients. For some ingredients, data were obtained from other published sources considered by the committee to be reliable, and, in a few cases, unpublished data were incorporated. The committee used the most up-to-date values when more than one source was available. The values in the tables are average values; feed ingredients often vary in composition depending on source, method of processing, and other factors. The International Feed Number was included when it was known (because the International Network of Feed Information Centre is no longer active, feed numbers are not available for many ingredients). Values in the tables are on an "as-fed" basis.

REFERENCES

AAFCO (Association of American Feed Control Officials). 2009. Official Publication. Oxford, IN: AAFCO.

Hertrampf, J. W., and F. Piedad-Pascual. 2003. Handbook on Ingredients for Aquaculture Feeds. Dordrecht, The Netherlands: Kluwer Academic Publishers.

NRC (National Research Council). 1972. Atlas of Nutritional Data on United States and Canadian Feeds. Washington, DC: National Academy of Sciences.

USDA (United States Department of Agriculture). 2005. USDA National Nutrient Database for Standard Reference, Release 18. Nutrient Data Laboratory Home Page, *http://www.nal.usda.gov/fnic/foodcomp*. Accessed October 15, 2010.

TABLE 19-1 Typical Dry Matter and Proximate Composition Values for Natural and Chemically Defined Ingredients Commonly Used in Aquatic Animal Feeds (as-fed basis)[a]

Ingredient	International Feed Number[b]	Typical Dry Matter (%)	Proximate Composition (%)			
			Crude Protein	Crude Fat	Crude Fiber	Ash
Alfalfa meal, dried, 17% CP	1-00-023	92	17.1	2.8	24.1	9.8
Algae meal, *Spirulina*	5 19 931	93	57.5	7.7	3.6	6.2
Barley, whole grain ground	4-00-526	92	11.5	1.9	5.0	2.5
Blood meal, flash-dried porcine		92	87.6	1.6	—	—
Blood meal, spray dried	5-00-381	93	88.6	1.4	1.0	5.8
Bone meal, fish processing byproduct		92	39.5	3.5	—	45.0
Brewers' dried grains	5-02-141	92	23.1	6.4	13.7	3.7
Camelina meal		90	33.9	12	12.4	5.8
Canola (rapeseed) low erucic acid		89	32.9	3.2	—	—
Canola (rapeseed) meal, solvent extracted	5-03-871	93	38.0	3.8	11.1	6.8
Canola (rapeseed) protein concentrate meal		90	69.2	0.1	2.8	—
Casein, dried	5-01-162	91	84.0	0.2	0.2	3.5
Coconut meal, mechanically extracted	5-01-572	93	22.0	6.0	12.0	7.0
Corn, dent yellow, grain	4-02-935	88	8.5	3.6	2.3	1.3
Corn distillers grains with solubles, dried	5-02-843	91	27.0	9.3	9.1	6.4
Corn distillers solubles, dried	5-02-844	90	27.6	8.5	4.6	7.5
Corn grain meal		88	10.2	4.8	2.8	1.6
Corn gluten meal, dried, 60% CP min	5-09-538	91	63.7	2.2	1.5	1.6
Corn gluten feed, dried	5-02-903	90	21.5	3	9.4	7.0
Corn starch, gelatinized	4-02-889	90	0.4	0.4	0.1	0.1
Cottonseed meal, solvent extracted, 41% CP	5-07-872	92	41.7	1.8	11.3	6.4
Crab meal, process residue, dried	5-01-663	92	32.0	2.5	10.6	41.0
Fish solubles, condensed, 30% CP	5-01-969	50	31.5	6.1	0.5	9.6
Fish solubles, dried, 60% CP	5-01-971	93	64.1	8.2	1.3	2.5
Fish meal, anchovy, mechanically extracted	5-01-985	92	65.4	7.6	1.0	14.3
Fish meal, herring, mechanically extracted	5-02-000	92	72.0	8.4	0.6	10.4
Fish meal, menhaden, mechanically extracted	5-02-009	92	64.5	9.6	0.7	19.0
Fish meal, salmon, mechanically extracted	5-02-012	90	70.3	10.1	—	10.2
Fish meal, tuna, mechanically extracted	5-02-023	93	59.0	9.24	0.8	21.5
Fish meal, white, mechanically extracted	5-02-025	92	62.0	7.8	0.8	21.3
Gelatin	5-14-503	88	85.9	0.10	0.00	1.30
Kelp meal, dried		91	8.9	1.60	3.90	17.30
Krill meal, with shell	5-16-423	92	58.8	9.2	6.4	13.6
Linseed meal (flax meal), solvent extracted	5-02-048	90	35.0	1.6	8.9	5.7
Lupin meal (sweet white)	5-27-717	92	30.4	6.7	1.1	3.7
Meat and bone meal	5-00-388	94	50.9	9.7	2.4	29.2
Meat and bone meal, porcine		97	59.0	11.0	—	20.0
Meat meal	5-00-385	93	55.6	5.6	2.3	27.0
Molasses, cane, dried	4-04-695	94	9.6	0.8	6.2	12.5
Oats	4-03-309	90	11.2	5.4	10.6	2.6
Oat groats	4-03-331	90	15.5	6.1	2.5	2.0
Peanut meal (groundnut meal), solvent extracted	5-03-650	92	49.0	1.3	9.9	5.9
Peas (shelled and extruded)	5-03-600	90	25.3	1.4	6.9	3.3
Pea protein concentrate		90	55.0	2.0	—	6.0
Potato protein concentrate		93	81.1	2.7	—	0.5
Poultry byproduct meal, feed grade	5-03-798	89	55.9	13.6	2.1	14.5
Poultry byproduct meal, feed grade low ash		89	62.1	19.8	—	7.5
Poultry byproduct meal, pet food grade		91	57.7	—	—	—
Poultry feather meal, hydrolyzed, dried	5-03-795	93	83.3	5.4	1.2	2.9
Rice bran, with germ, solvent extracted	4-03-930	91	15.7	1.7	13.7	16.7
Rice bran	4-03-928	91	13.0	7.4	17.9	24.7
Rice polishings	4-03-943	90	13.6	14.5	4.2	8.3
Rice, broken with polishings (brewers' rice)	4-03-932	91	7.4	0.6	0.4	0.7
Rice protein concentrate		92	68.9	10.2	2.7	3.9
Safflower meal, mechanically extracted	5-04-109	91	20.0	6.7	35.4	4.1
Safflower meal, solvent extracted	5-08-501	90	22.0	1.4	—	—
Sesame meal, mechanically extracted	5-04-220	94	42.0	7.0	6.5	12.0
Shrimp meal, dried	5-04-226	88	39.5	3.2	12.8	27.2

continued

TABLE 19-1 Continued

Ingredient	International Feed Number[b]	Typical Dry Matter (%)	Proximate Composition (%)			
			Crude Protein	Crude Fat	Crude Fiber	Ash
Sorghum, milo, grain	4-04-444	89	9.9	2.8	2.3	1.8
Soybean seeds, extruded, full fat	5-04-597	90	35.2	18.0	5.0	4.5
Soybean meal, solvent extracted, 44% CP	5-04-604	89	44.0	1.5	7.3	6.3
Soybean meal, toasted, solvent extracted without hulls, 48% CP	5-04-612	90	48.5	0.9	3.4	5.8
Soybean protein concentrate		92	63.6	0.5	4.5	—
Soybean protein isolate	5-08-038	95	80.7	3.4	5.6	3.6
Squid liver meal	5-09-286	89	45.2	15.3	—	6.8
Squid meal	5-04-671	95	76.5	3.8	—	6.1
Sunflower meal, solvent extracted	5-04-737	90	32.3	2.3	21.0	7.0
Sunflower meal, partially dehulled, solvent extracted	5-04-739	93	46.5	0.5	13.0	7.1
Wheat, hard red winter, grain	4-05-268	88	14.8	1.8	2.5	2.1
Wheat, soft white, grain	4-05-336	86	10.8	1.7	2.8	2.0
Wheat bran	4-05-190	89	14.8	4.0	9.9	5.3
Wheat flour	4-05-199	88	11.7	1.2	1.3	0.4
Wheat gluten meal	5-05-220	89	80.7	1.5	0.5	0.7
Wheat middlings	4-05-205	89	16.6	4.0	7.5	4.5
Whey, dried, low-lactose	4-01-182	93	16.7	0.7	0.2	9.8
Yeast, brewers', dried	7-05-527	93	42.6	1.0	3.2	6.6
Yeast, Torula, dried	7-05-534	93	49.0	1.5	2.2	7.7
Cellulose powder		96	0	0	92.6	0

[a]Dash indicates data were not available.
[b]AAFCO (2009) has number and definitions.

TABLE 19-2 Amino Acid Composition of Ingredients (as-fed basis)[a]

Ingredient	International Feed Number	Dry Matter (%)	Typical Crude Protein (%)	Arginine (%)	Histidine (%)	Isoleucine (%)	Leucine (%)	Lysine (%)	Methionine (%)	Cystine (%)	Phenylalanine (%)	Tyrosine (%)	Threonine (%)	Tryptophan (%)	Valine (%)
Alfalfa meal, dried, 17% CP	1-00-023	92	17.10	0.77	0.33	0.81	1.28	0.85	0.27	0.29	0.80	0.54	0.71	0.34	0.88
Algae meal, *Spirulina*	5-19-931	93	57.50	4.15	1.09	3.21	4.95	3.03	1.15	0.66	2.78	2.58	2.97	0.93	3.51
Barley, whole grain ground	4-00-526	92	11.50	0.50	0.23	0.42	0.80	0.53	0.18	0.25	0.62	—	0.36	0.17	0.62
Blood meal, flash-dried porcine		92	87.60	3.70	7.00	1.51	12.30	8.22	1.67	0.82	5.48	2.80	3.18	1.00	7.70
Blood meal, spray dried	5-00-381	93	88.60	2.35	5.00	0.80	10.30	7.10	1.00	1.40	5.10	2.30	3.80	1.00	5.20
Bone meal, salmon processing byproduct		92	39.50	2.56	0.66	1.13	2.04	2.06	0.91	0.27	1.12	0.81	1.36	0.29	1.50
Brewers' dried grains	5-02-141	92	23.10	1.27	0.52	1.54	2.49	0.09	0.46	0.35	1.44	1.20	0.93	0.37	1.61
Camelina meal		90	33.90	2.62	0.75	1.20	2.13	1.54	0.61	0.66	1.40	—	1.30	0.42	1.61
Canola (rapeseed) meal, solvent extracted	5-03-871	93	38.00	2.32	1.10	1.51	2.60	2.02	0.77	0.97	1.50	0.99	1.50	0.46	1.94
Canola (rapeseed) meal, double low		93	38.00	2.32	0.64	1.09	2.89	0.65	0.50	0.46	1.39	0.99	0.98	0.10	1.50
Canola (rapeseed) meal, low erucic acid		89	32.90	2.06	0.99	1.35	2.50	1.98	0.71	0.30	1.41	0.79	1.56	0.43	1.79
Canola (rapeseed) meal, low glucosinolate		89	35.20	0.49	0.24	0.49	0.70	0.48	0.15	—	0.39	—	—	—	—
Canola (rapeseed) protein concentrate meal		90	69.20	4.20	1.70	2.82	4.92	3.10	1.26	1.27	2.81	0.99	2.49	0.20	3.33
Casein, dried	5-01-162	91	84.00	3.26	2.82	4.66	8.79	7.35	2.70	0.41	4.79	4.77	3.98	1.14	6.10
Coconut meal, mechanically extracted	5-01-572	93	22.00	2.30	0.30	1.00	1.49	0.54	0.33	0.20	0.80	—	0.60	0.20	1.00
Corn distillers grains with solubles, dried	5-02-843	91	27.00	1.10	0.65	1.00	2.80	0.90	0.51	0.50	1.20	0.31	0.92	0.20	1.33
Corn distillers solubles, dried	5-02-844	90	27.60	1.00	0.60	1.20	2.10	0.20	0.60	0.60	1.50	0.99	0.98	0.20	1.50
Corn gluten feed, dried	5-02-903	90	21.50	1.04	0.67	0.66	1.96	0.63	0.35	0.46	0.76	0.58	0.74	0.07	1.01
Corn gluten meal, dried, 60% CP	5-09-318	91	63.70	1.90	1.20	2.30	9.40	1.07	1.90	1.10	3.80	0.87	2.00	0.30	2.70
Corn grain meal		88	10.20	0.40	0.25	0.29	1.00	0.26	0.18	0.19	0.42	—	0.30	0.07	0.42
Corn, dent yellow, grain	4-02-935	88	8.50	0.40	0.25	0.29	1.00	0.24	0.18	0.18	0.42	—	0.29	0.07	0.42
Cottonseed meal, solvent extracted, 41% CP	5-07-872	92	41.70	4.18	1.07	1.45	2.32	1.60	0.58	0.73	2.18	0.94	1.34	0.53	1.90
Crab meal, process residue, dried	5-01-663	92	32.00	3.97	0.83	1.15	1.80	1.89	0.50	0.45	2.10	0.80	1.02	0.42	1.68
Fish solubles, condensed, 30% CP	5-01-969	50	31.50	1.66	1.09	0.70	1.54	1.38	0.53	0.24	0.70	0.40	1.00	0.29	1.00
Fish solubles, dried, 60% CP	5-01-971	93	64.10	3.05	2.10	2.05	2.97	3.51	1.18	0.62	1.53	0.85	1.35	0.59	2.10

continued

TABLE 19-2 Continued

Ingredient	International Feed Number	Dry Matter (%)	Typical Crude Protein (%)	Arginine (%)	Histidine (%)	Isoleucine (%)	Leucine (%)	Lysine (%)	Methionine (%)	Cystine (%)	Phenylalanine (%)	Tyrosine (%)	Threonine (%)	Tryptophan (%)	Valine (%)
Fish meal, anchovy, mechanically extracted	5-01-985	92	65.40	3.68	1.56	3.06	5.00	5.11	1.95	0.61	2.66	2.15	2.82	0.76	3.51
Fish meal, herring, mechanically extracted	5-02-000	92	72.00	3.73	1.53	3.64	4.69	7.30	2.20	1.60	2.68	2.10	2.49	0.67	3.26
Fish meal, menhaden, mechanically extracted	5-02-009	92	64.50	3.66	1.78	2.57	4.54	4.81	1.77	0.57	2.51	2.04	2.64	0.66	3.03
Fish meal, salmon byproduct, mechanically extracted	5-02-012	90	70.30	6.84	2.46	4.10	7.20	7.38	3.04	—	4.32	3.69	4.19		5.07
Fish meal, tuna byproduct, mechanically extracted	5-02-023	93	59.00	3.43	1.75	2.45	3.79	4.22	1.47	0.47	2.15	1.69	2.31	0.57	2.77
Fish meal, white byproduct, mechanically extracted	5-02-025	92	62.00	4.02	1.34	2.72	4.36	4.53	1.68	0.75	2.28	1.83	2.57	0.67	3.02
Gelatin	5-14-503	88	85.90	6.62	0.76	1.38	2.91	3.55	0.73	0.13	1.79	0.52	1.76	0.05	2.09
Kelp meal, dried		91	8.90	0.10	—	—	0.09	0.04	0.10	—	—	—	0.03	—	—
Krill meal, with shell	5-16-423	92	58.80	3.47	1.23	2.82	4.41	4.17	1.76	—	2.53	—	2.53	0.53	2.82
Linseed meal (flax meal), solvent extracted	5-02-048	90	35.00	2.97	0.68	1.56	2.06	1.24	0.59	0.59	1.57	1.03	1.26	0.52	1.74
Lupin meal (sweet white)	5-27-717	92	30.40	3.38	0.77	1.38	2.43	1.54	0.27	0.51	1.22	1.35	1.20	0.26	1.29
Meat and bone meal	5-00-388	94	50.90	3.60	0.96	1.70	3.20	2.60	0.67	0.33	1.70	1.30	1.70	0.26	2.25
Meat and bone meal, porcine		96	53.00	3.95	1.06	1.77	3.66	3.07	0.83	0.41	1.89	1.48	1.95	0.41	2.60
Meat meal	5-00-385	93	55.60	3.60	1.14	1.60	3.84	3.07	0.80	0.60	2.17	1.40	1.97	0.35	2.66
Oat groats	4-03-331	90	15.50	0.90	0.25	0.50	1.00	0.18	0.20	0.26	0.65	—	0.50	0.18	0.65
Oats	4-03-309	90	11.20	0.70	0.18	0.43	0.81	0.39	0.17	0.19	0.52	0.46	0.36	0.15	0.56
Pea protein concentrate		90	55.00	4.83	1.30	2.15	3.75	3.75	0.41	0.62	2.56	1.58	1.85	0.49	2.35
Peanut meal (groundnut meal), solvent extracted	5-03-650	92	49.00	3.37	0.96	1.43	3.00	2.67	0.65	0.50	1.70	1.09	1.65	0.30	2.45
Peas (shelled and extruded)	5-03-600	90	25.30	1.87	0.54	0.86	1.51	1.50	0.21	0.31	0.98	0.71	0.78	0.19	0.98
Potato protein concentrate		93	81.10	4.02	1.77	4.62	7.94	6.09	1.77	1.27	5.19	4.56	4.62	—	5.33
Poultry byproduct meal, feed grade	5-03-798	89	55.91	4.32	1.05	2.30	4.27	3.32	1.29	0.92	1.66	1.21	2.14	—	3.65
Poultry byproduct meal, feed grade low ash		89	62.13	4.99	1.17	2.38	4.14	3.44	1.24	—	1.63	1.30	2.39	—	3.03
Poultry byproduct meal, pet food grade		91	57.70	5.02	0.99	2.19	4.06	3.22	1.32	—	1.49	1.06	2.32	—	2.99
Poultry feather meal, hydrolyzed, dried	5-03-795	93	83.30	5.80	0.70	4.15	6.94	1.81	0.50	3.84	4.12	2.00	3.85	0.55	4.55
Rice bran	4-03-928	91	13.00	1.00	0.34	0.44	0.92	0.57	0.26	0.27	0.56	0.40	0.48	0.14	0.68
Rice bran, with germ, solvent extracted	4-03-930	91	15.70	0.85	0.29	0.51	1.01	0.54	0.21	0.20	0.56	0.54	0.45	0.21	0.65

Rice polishings	4-03-943	90	13.60	0.82	0.28	0.43	0.82	0.58	0.23	0.22	0.49	0.44	0.44	0.13	0.75
Rice, broken with polishings (brewers' rice)	4-03-932	91	7.37	0.55	0.18	0.36	0.64	0.27	0.18	0.13	0.36	—	0.27	0.09	0.46
Rice protein concentrate		92	68.90	5.58	1.38	5.51	2.75	2.41	1.77	1.45	3.52	3.32	2.54	0.81	4.34
Safflower meal, mechanically extracted	5-04-109	91	20.00	1.20	0.48	0.28	1.10	0.70	0.40	0.50	1.00	—	0.47	0.30	1.00
Safflower meal, solvent extracted	5-08-501	90	22.00	1.90	0.50	0.27	1.20	0.70	0.33	0.35	1.00	—	0.50	0.26	1.00
Sesame meal, mechanically extracted	5-04-220	94	42.00	5.06	1.16	2.28	3.30	1.37	1.48	0.60	2.32	—	1.71	0.82	2.53
Shrimp meal, dried	5-04-226	88	39.50	1.54	0.51	0.91	1.62	1.66	0.55	0.59	0.99	—	3.24	0.24	1.03
Sorghum, milo, grain	4-04-444	89	9.90	0.40	0.27	0.40	1.30	0.27	0.10	0.20	0.45	0.30	0.27	0.09	0.53
Soybean meal, solvent extracted without hulls, 48% CP	5-04-612	90	48.50	3.60	1.30	2.60	3.80	2.24	0.70	0.71	2.70	1.25	2.00	0.70	2.70
Soybean meal, solvent extracted, 44% CP	5-04-604	89	44.00	3.23	1.17	1.99	3.42	2.83	0.61	0.70	2.18	1.69	1.73	0.61	2.40
Soybean protein concentrate		92	63.63	4.64	1.58	2.94	4.92	3.93	0.81	0.89	3.28	2.30	2.47	0.84	3.06
Soybean protein isolate	5-08-038	95	80.70	4.15	1.08	3.21	4.95	3.02	1.15	0.66	2.78	2.58	2.97	0.93	3.51
Soybean seeds, extruded, full fat	5-04-597	90	35.20	2.60	0.96	1.61	2.75	2.22	0.53	0.55	1.83	1.32	1.41	0.48	1.68
Squid liver meal	5-09-286	89	45.20	2.48	—	—	—	2.56	0.97	—	—	—	1.74	0.49	
Squid meal	5-04-671	95	76.50	5.78	2.05	3.98	6.92	5.99	2.69	—	3.26	—	3.42	0.86	3.95
Sunflower meal, solvent extracted	5-04-737	90	32.30	2.93	0.92	1.44	2.31	1.20	0.82	0.66	1.66	1.03	1.33	0.44	1.74
Sunflower meal, solvent extracted without hulls	5-04-739	93	46.50	3.50	1.00	2.10	2.60	1.70	1.50	0.70	1.23	0.76	1.48	0.38	2.30
Wheat bran	4-05-190	89	14.80	0.64	0.30	0.51	0.89	0.36	0.21	0.27	0.63	0.43	0.37	0.17	0.59
Wheat flour	4-05-199	88	11.70	0.86	0.39	0.51	0.92	0.58	0.19	0.26	0.55	0.38	0.46	0.25	0.69
Wheat gluten meal	5-05-220	89	80.70	3.80	2.00	3.70	6.30	4.90	1.60	—	4.50	—	1.60	1.05	4.00
Wheat middlings	4-05-205	89	16.60	0.97	0.44	0.70	1.10	0.70	0.12	0.19	0.50	0.29	0.51	0.20	0.75
Wheat, hard red winter, grain	4-05-268	88	14.80	0.60	0.17	0.69	1.00	0.40	0.25	0.30	0.78	—	0.69	0.18	0.69
Wheat, soft, grain	4-05-336	86	10.80	0.40	0.20	0.43	0.60	0.12	0.14	0.20	0.49	—	0.28	0.12	0.48
Whey, dried, low-lactose	4-01-182	93	16.70	0.60	0.27	0.96	1.54	1.40	0.41	0.43	0.55	0.46	0.95	0.27	0.87
Yeast, brewers', dried	7-05-527	93	42.60	2.20	1.09	2.15	3.13	3.22	0.74	0.50	1.83	1.55	2.20	0.56	2.39
Yeast, Torula, dried	7-05-534	93	49.00	2.60	1.40	1.98	3.50	3.80	0.67	0.49	3.00	1.50	2.60	0.52	2.90

[a]Dash indicates data were not available.

TABLE 19-3 Mineral Composition of Ingredients Commonly Used in Aquatic Animal Feeds (as-fed basis)[a]

Ingredient	International Feed Number	Dry Matter (%)	Calcium (%)	Phosphorus (%)	Sodium (%)	Chlorine (%)	Potassium (%)	Magnesium (%)	Sulfur (%)	Copper (mg/kg)	Iron (mg/kg)	Manganese (mg/kg)	Selenium (mg/kg)	Zinc (mg/kg)
Alfalfa meal, dried, 17% CP	1-00-023	92	1.40	0.23	0.10	0.47	2.38	0.29	0.23	10.00	404.0	31.00	0.33	19.0
Algae meal, *Spirulina*	5-19-931	93	0.12	0.12	1.05	—	1.36	0.20	—	61.00	285.0	19.00	0.07	20.0
Barley, whole grain, ground	4-00-526	92	0.08	0.42	0.03	0.14	0.56	0.12	0.15	8.00	80.0	16.00	0.20	30.0
Blood meal, flash-dried porcine		92	0.04	0.17	0.60	—	0.58	0.02	—	7.00	2,200.0	0.80	—	20.0
Blood meal, spray dried	5-00-381	93	0.41	0.30	0.38	0.25	0.15	0.15	0.34	8.20	2,769.0	6.40	—	309.0
Bone meal, fish-processing byproduct		92	15.33	7.69	1.21	—	0.47	0.23	—	5.00	50.0	34.00	—	142.0
Brewers' dried grains	5-02-141	92	0.29	0.51	0.20	0.13	0.09	0.15	0.29	21.70	233.0	37.20	0.70	27.0
Camelina meal		90	0.33	0.94	0.10	—	1.24	0.40	—	9.00	—	45.00	—	85.0
Canola (rapeseed) meal, solvent extracted	5-03-871	93	0.63	1.01	0.07	0.11	1.22	0.51	0.85	6.00	142.0	49.00	1.10	69.0
Casein, dried	5-01-162	91	0.61	0.82	0.01	0.04	0.01	0.01	0.60	3.80	14.0	4.30	0.13	27.0
Coconut meal, mechanically extracted		93	0.17	0.60	0.06	0.03	0.60	—	—	—	—	—	—	—
Corn, dent yellow, grain	4-02-935	88	0.03	0.28	0.01	0.05	0.33	0.11	0.11	3.50	33.0	5.70	0.07	19.0
Corn distillers grains with solubles, dried	5-02-843	91	0.14	0.66	0.52	0.16	0.40	0.16	0.35	52.80	236.0	22.80	0.35	80.0
Corn distillers solubles, dried	5-02-844	90	0.30	1.44	0.23	0.26	1.64	0.59	0.36	81.00	555.0	72.80	0.36	84.0
Corn gluten meal, dried, 60% CP	5-09-318	91	0.07	0.44	0.05	0.07	0.19	0.07	0.57	26.10	229.0	6.30	0.83	31.0
Corn gluten feed, dried	5-02-903	90	0.22	0.83	0.15	0.22	0.92	0.33	0.22	48.00	460.0	24.00	0.27	70.0
Cottonseed meal, solvent extracted, 41% CP	5-07-872	92	0.17	1.17	0.04	0.04	1.39	0.41	0.30	19.00	208.0	21.00	0.06	61.0
Crab meal, process residue, dried	5-01-663	92	14.56	1.59	0.88	1.51	0.45	0.94	0.25	32.73	4,356.0	133.00	3.80	102.0
Fish solubles, condensed, 30% CP	5-01-969	50	0.16	0.57	2.45	2.93	1.64	0.03	0.12	46.60	276.0	13.20	1.97	43.0
Fish solubles, dried, 60% CP	5-01-971	93	0.40	1.20	0.40	2.50	—	0.27	0.45	20.00	948.0	10.00	2.70	76.0
Fish meal, anchovy, mechanically extracted	5-01-985	92	3.73	2.43	1.10	1.00	0.90	0.24	0.54	9.03	220.0	9.50	1.36	103.0
Fish meal, herring, mechanically extracted	5-02-000	92	2.20	1.67	0.59	0.99	1.08	0.14	0.46	5.60	114.0	4.80	1.95	125.0
Fish meal, menhaden, mechanically extracted	5-02-009	92	5.19	2.88	0.41	0.55	0.70	0.15	0.56	10.30	544.0	37.00	2.15	144.0
Fish meal, salmon, mechanically extracted	5-02-012	90	3.00	2.08	0.68	—	0.78	0.20	—	10.70	130.0	3.50	—	183.0
Fish meal, tuna, mechanically extracted	5-02-023	93	7.86	4.21	0.74	1.01	0.72	0.23	0.68	10.31	355.0	8.40	4.30	211.0
Fish meal, white, mechanically extracted	5-02-025	92	6.65	3.59	0.78	1.28	0.85	0.18	0.48	6.00	299.0	12.00	1.62	90.0
Fish meal, white, processing byproduct, presscake meal		91	7.31	3.81	0.78	0.50	0.83	0.18	0.48	5.90	181.0	12.40	1.62	90.0
Gelatin	5-14-503	88	0.06	0.04	0.20	—	0.02	0.02	—	21.60	11.1	1.05	0.40	1.4
Kelp meal, dried		91	1.20	0.16	2.40	—	2.30	0.85	0.73	5.00	566.0	62.00	0.40	46.0
Krill meal, with shell	5-30-288	92	0.46	0.29	0.73	—	0.09	0.23	—	10.00	10.0	2.70	—	47.0
Linseed meal (flax meal), solvent extracted	5-30-288	90	0.49	0.06	—	—	—	0.05	—	—	—	—	—	—
Lupin meal (sweet white)	5-27-717	92	0.22	0.51	0.02	0.03	1.10	0.19	0.24	6.00	54.0	1,390.00	0.07	32.0
Meat and bone meal	5-00-388	94	9.99	4.98	0.63	0.69	0.65	0.41	0.38	11.00	606.0	17.00	0.31	96.0

Meat and bone meal, porcine		96	8.70	4.30	0.64	—	0.64	0.23	—	10.00	320.0	22.00		140.0
Meat meal	5-00-385	93	9.00	3.88	0.80	0.97	0.57	0.35	0.45	10.00	440.0	10.00	0.37	94.0
Peanut meal (groundnut meal), solvent extracted	5-03-650	92	0.77	0.08	0.16	2.26	2.98	0.31	0.35	59.00	196.0	43.70	—	16.0
Peas (shelled and extruded)	5-03-600	90	0.11	0.39	0.04	0.05	1.02	0.12	0.20	9.00	65.0	23.00	0.38	23.0
Pea protein concentrate		90	0.09	0.84	0.01	—	1.96	0.22	0.39	15.00	85.0	25.00	0.08	59.0
Potato protein concentrate		93	0.15	0.45	0.05	—	0.80	0.10	—	—	40.0	—	—	—
Poultry byproduct meal, feed grade	5-03-798	89	0.27	0.61	0.07	0.03	1.16	0.27	0.31	15.00	142.0	26.70	—	20.0
Poultry feather meal, hydrolyzed, dried	5-03-795	93	3.51	1.83	0.82	0.54	0.39	0.18	0.52	14.12	442.0	11.00	0.78	121.0
Rice bran	4-03-928	91	0.25	0.66	0.69	0.28	0.28	0.20	1.47	6.40	74.0	12.50	0.82	68.0
Rice polishings	4-03-943	91	0.11	1.37	0.10	0.07	1.48	0.95	0.18	13.00	187.0	232.20	—	30.0
Rice, broken with polishings (brewers' rice)	4-03-932	90	0.08	0.13	0.06	0.04	0.15	0.02	0.15	3.30	12.2	12.30	—	2.1
Safflower meal, mechanically extracted	5-04-109	91	0.03	0.27	0.07	0.08	0.13	0.11	0.04	—	—	18.00	0.27	17.0
Safflower meal, solvent extracted	5-08-501	90	0.34	0.75	0.05	0.08	0.76	0.35	0.13	10.00	495.0	18.00	—	41.00
Sesame meal	5-04-220	94	2.00	1.30	0.04	0.06	1.39	0.80	0.40	—	95.0	48.00	—	100.00
Shrimp meal, dried	5-04-226	88	6.30	1.60	1.70	1.20	0.70	0.70	—	—	64.0	33.20	—	32.00
Sorghum, milo, grain	4-04-444	89	0.34	0.76	0.05	—	0.75	0.34	0.13	9.90	497.0	18.20	—	41.00
Soybean seeds, steam cooked, full fat	5-04-597	90	0.25	0.59	0.03	0.03	1.70	0.28	0.30	16.00	80.0	30.00	0.11	39.00
Soybean meal, solvent extracted, 44% CP	5-04-604	89	0.03	0.28	0.04	0.08	0.31	0.13	0.08	10.00	48.0	15.80	0.20	17.00
Soybean meal, solvent extracted without hulls, 48% CP	5-04-612	90	0.34	0.69	0.02	0.05	2.14	0.30	0.44	20.00	176.0	36.00	0.27	55.00
Soybean protein concentrate		92	0.30	0.65	0.04	0.04	2.11	0.29	0.42	23.00	140.0	30.60	0.10	52.00
Soybean protein isolate	5-08-038	95	0.18	0.78	1.00	—	0.08	0.04	—	15.99	145.0	390.00	0.00	40.00
Squid liver meal	5-09-286	89	—	—	—	—	—	—	—	—	—	—	—	—
Squid meal	5-04-671	95	0.09	0.99	—	—	—	—	—	—	—	—	—	—
Sunflower meal, solvent extracted	5-04-737	90	0.36	0.86	0.02	0.10	1.07	0.68	0.30	26.00	254.0	41.00	0.50	66.00
Sunflower meal, solvent extracted without hulls	5-04-739	93	0.42	0.94	0.22	0.16	1.19	0.69	0.21	4.00	31.0	18.90	2.13	98.00
Wheat, hard red winter, grain	4-05-268	88	0.04	0.37	0.02	0.05	0.43	0.12	0.14	5.10	35.0	30.40	0.29	35.00
Wheat, soft white, grain	4-05-336	86	0.05	0.30	0.06	0.07	0.40	0.10	0.10	9.70	43.0	30.00	0.10	28.00
Wheat bran	4-05-190	89	0.13	1.16	0.05	0.05	1.22	0.57	0.21	11.00	145.0	115.00	0.64	95.00
Wheat middlings	4-05-205	89	0.13	0.89	0.12	0.04	0.98	0.34	0.17	15.90	60.0	114.10	0.74	97.00
Wheat gluten meal	5-05-220	89	0.14	0.26	0.03	—	0.10	0.03	—	1.82	52.0	—	0.40	8.50
Whey, dried, low lactose	4-01-182	93	0.75	0.72	0.94	1.40	1.96	0.13	0.72	13.00	130.0	3.00	0.12	10.00
Yeast, brewers', dried	7-05-527	93	0.14	1.36	0.07	0.07	1.69	0.24	0.43	38.40	109.0	6.70	0.91	39.00
Yeast, Torula, dried	7-05-534	93	0.58	1.52	0.07	0.12	1.94	0.20	0.55	17.00	222.0	13.00	0.02	99.00

[a]Dash indicates data were not available.

TABLE 19-4 Vitamin Composition of Ingredients Commonly Used in Aquatic Animal Feeds (as-fed basis)[a]

Ingredient	International Feed Number	Biotin (mg/kg)	Choline (mg/kg)	Folacin (mg/kg)	Niacin (mg/kg)	Pantothenic Acid (mg/kg)	Pyridoxine (Vitamin B_6) (mg/kg)	Riboflavin (mg/kg)	Thiamin (mg/kg)	Vitamin A (mg/kg)	Vitamin D (mg/kg)	Vitamin B_{12} (mg/kg)	Vitamin E (mg/kg)	Vitamin K (mg/kg)
Alfalfa meal, dried, 17% CP	1-00-023	0.33	1,369	4.34	37.0	29.7	7.10	12.9	3.40	—	—	—	111.0	9.00
Algae meal, *Spirulina*	5-19-931	—	—	—	—	—	—	—	—	—	—	—	—	—
Barley, whole grain ground	4-00-526	0.20	1,027	0.40	57.2	6.4	—	2.0	5.00	—	—	—	36.0	—
Blood meal, spray dried	5-00-381	0.28			22.0	3.2	4.45	2.9	0.30	—	—	13.0	—	—
Brewers' dried grains	5-02-141	0.63	1,652	7.10	44.0	8.2	0.68	1.5	0.60	—	—	4.0	26.7	—
Camelina meal		—	—	—	—	—	—	—	—	—	—	—	—	—
Canola (rapeseed) meal, solvent extracted	5-03-871	0.98	6,700	0.83	160.0	9.5	7.2	5.8	5.20	—	—	0.0	13.4	—
Casein, dried	5-01-162	0.04	208	0.47	1.0	2.7	0.42	1.5	0.40	—	—	—	—	—
Coconut meal, mechanically extracted	5-01-572	—	1,100	—	28.6	6.1	—	3.3	0.66	—	—	—		—
Corn, dent yellow, grain	4-02-935	0.07	504	0.30	23.0	5.1	4.69	1.1	3.70	—	—	—	20.9	0.22
Corn distillers grains with solubles, dried	5-02-843	0.77	2,551	0.90	72.0	13.9	5.00	8.3	2.80	—	—	—	39.1	—
Corn distillers solubles, dried	5-02-844	1.63	4,687	1.30	122.0	22.9	8.85	17.0	6.60	—	—	3.0	50.5	—
Corn gluten meal, dried, 60% CP	5-09-318	0.19	352	0.30	60.0	3.5	6.90	2.0	0.30	—	—	—	23.4	—
Corn gluten feed, dried	5-02-903	0.14	1,518	0.28	66.0	17.0	13.00	2.4	2.00	—	—	0.0	8.5	—
Cottonseed meal, solvent extracted, 41% CP	5-07-872	0.97	2,764	1.40	41.0	13.7	7.00	3.3	6.60	—	—	0.0	16.0	—
Crab meal, process residue, dried	5-01-663	0.07	2,011	0.11	45.0	6.5	6.63	6.1	0.40	—	—	438.3	—	—
Fish solubles, condensed, 30% CP	5-01-969	0.14	3,370	0.20	176.0	35.7	12.14	12.9	5.50	—	—	507.0	—	—
Fish meal, anchovy, mechanically extracted	5-01-985	0.23	4,408	0.20	100.0	15.0	4.64	7.1	0.10	—	—	352.0	5.0	2.20
Fish meal, herring, mechanically extracted	5-02-000	0.49	5,266	0.30	85.0	17.3	4.77	9.7	0.40	—	—	430.0	22.1	—
Fish meal, menhaden, mechanically extracted	5-02-009	0.18	3,112	0.12	55.0	8.6	4.66	4.8	0.60	—	—	123.0	12.0	—

Fish meal, tuna, mechanically extracted	5-02-023	0.20	2,994	—	144.0	7.7	—	6.8	1.50	—	—	300.1	5.6	—
Fish meal, white, mechanically extracted	5-02-025	0.13	4,050	0.37	59.0	9.9	5.90	9.1	1.70	—	—	90.0	5.0	—
Fish meal, white, processing byproduct, presscake meal		0.08	3,099	0.35	59.0	9.9	5.92	9.1	1.70	—	—	89.5	8.9	—
Kelp meal, dried		0.10	275	0.10	23.0	7.0	—	5.0	1.00	66.00	—	0.0	150.0	—
Krill meal, with shell	5-16-423	—	—	—	—	—	—	—	—	—	—	—	—	—
Linseed meal (flax meal), solvent extracted	5-02-048	—	1,760	1.30	32.8	16.5	—	4.1	6.60	—	—	—	5.8	
Lupin meal (sweet white)	5-27-717	0.05	—	—	—	—	—	—	—	—	—	—	7.5	—
Meat and bone meal	5-00-388	0.08	1,996	0.41	49.0	4.1	—	4.7	0.40	—	—	90.0	1.6	—
Meat meal	5-00-385	0.08	2,077	0.50	57.0	5.0	2.40	4.7	0.60	—	—	80.0	1.2	—
Molasses, cane	4-04-696	0.11	1,922	0.50	53.0	4.9	4.60	5.3	0.20	—	—	91.0	1.0	—
Molasses, cane, dried	4-04-695	0.14	2,136	0.50	51.0	4.4	8.74	4.5	0.20	—	—	217.0	1.1	—
Oat groats	4-03-331	0.20	900	—	14.3	10.0	—	1.3	—	—	—	—	20.0	—
Oats	4-03-309	0.28	992	0.40	14.0	7.8	2.50	1.5	6.30	—	—	—	15.0	—
Pea protein concentrate		—	—	—	—	—	—	—	—	—	—	—	—	—
Peanut meal (groundnut meal), solvent extracted	5-03-650	0.67	704	0.10	36.0	37.5	4.22	2.8	0.90	—	—	—	5.4	—
Peas (shelled and extruded)	5-03-600	0.15	547	0.20	31.0	18.7	1.00	1.8	4.60	—	—	0.0	0.2	—
Potato protein concentrate		—	—	—	—	—	—	—	—	—	—	—	—	—
Poultry byproduct meal, feed grade	5-03-798	0.33	1,896	0.70	178.0	46.6	6.38	9.1	5.70	—	—	0.0	2.9	—
Poultry feather meal, hydrolyzed, dried	5-03-795	0.09	6,029	0.51	47.0	11.0	4.41	10.5	0.20	—	—	301.2	2.2	—
Rice bran	4-03-928	0.04	895	0.22	21.0	8.9	2.98	2.0	0.10	—	—	83.3	—	—
Rice polishings	4-03-943	0.42	1,128	2.20	284.0	23.0	29.12	2.9	22.60	—	—	—	60.7	—
Safflower meal, mechanically extracted	5-04-220	0.08	878	0.20	23.0	3.3	—	0.4	1.40	—	—	—	14.5	—
Safflower meal, solvent extracted	5-08-501	1.03	820	0.50	11.0	33.9	12.00	2.3	4.60	—	—	0.0	16.0	—
Sorghum, milo, grain	4-04-444	1.43	816	0.40	11.0	37.3	—	2.0	—	—	—	—	0.9	—
Soybean seeds, steam cooked, full fat	5-04-597	0.24	2,307	3.60	22.0	15.0	10.80	2.6	11.00	—	—	0.0	18.1	—
Soybean meal, solvent extracted, 44% CP	5-04-604	0.23	638	0.20	37.0	11.0	4.70	1.1	4.10	—	—	—	12.1	—

continued

TABLE 19-4 Continued

Ingredient	International Feed Number	Biotin (mg/kg)	Choline (mg/kg)	Folacin (mg/kg)	Niacin (mg/kg)	Pantothenic Acid (mg/kg)	Pyridoxine (Vitamin B_6) (mg/kg)	Riboflavin (mg/kg)	Thiamin (mg/kg)	Vitamin A (mg/kg)	Vitamin D (mg/kg)	Vitamin B_{12} (mg/kg)	Vitamin E (mg/kg)	Vitamin K (mg/kg)
Soybean meal, solvent extracted without hulls, 48% CP	5-04-612	0.26	2,731	1.37	22.0	15.0	6.40	3.1	3.20	—	—	0.0	2.3	—
Soybean protein concentrate		0.32	2,609	0.60	28.0	16.3	6.00	2.9	6.00	—	—	—	2.4	—
Soybean protein isolate	5-08-038	0.32	2,753	0.70	22.0	14.8	4.90	2.9	3.10	—	—	—	3.3	—
Sunflower meal, solvent extracted	5-04-737	1.40	3,791	1.14	264.0	29.9	11.10	3.0	3.00	—	—	0.0	9.1	—
Sunflower meal, solvent extracted without hulls	5-04-739	—	3,632	—	242.0	40.6	13.70	3.5	3.10	—	—	—	11.1	—
Wheat, hard red winter, grain	4-05-268	0.11	1,004	0.40	53.0	10.1	3.00	1.3	4.50	—	—	—	11.1	—
Wheat, soft winter, grain	4-05-268	—	929	0.40	52.0	9.6	3.20	1.5	4.50	—	—	—	16.0	—
Wheat bran	4-05-190	0.38	1,232	1.80	197.0	28.0	8.50	3.6	8.40	—	—	—	14.3	—
Wheat middlings	4-05-205	0.24	1,247	1.20	95.0	17.8	8.00	2.0	14.20	—	—	—	23.9	—
Whey, dried	4-01-182	0.27	1,820	0.85	10.0	47.0	4.00	27.1	4.10	—	—	23.0	0.3	—
Yeast, brewers', dried	7-05-527	1.04	3,847	9.70	443.0	110.7	37.10	34.1	85.20	—	—	1.0	2.1	—
Yeast, Torula, dried	7-05-534	0.58	2,881	22.40	492.0	84.2	36.30	49.9	6.20	—	—	—	—	—

[a]Dash indicates data were not available.

TABLE 19-5 Fatty Acid (Percentage of Total Fatty Acids) and Cholesterol Composition of Common Animal Fats, Fish Oils, and Vegetable Oils (as-fed basis)[a,b]

Lipid Source	International Feed Number	Sat ≤ C12	14:0	16:0	18:0	Total Sat	16:1	18:1	20:1	22:1	Total Mono	18:2 n-6	18:3 n-6	20:4 n-6	18:3 n-3	18:4 n-3	20:5 n-3
Animal Fat																	
Tallow	4-08-127	0.9	3.7	24.9	18.9	52.1	4.2	36.0	0.3	—	40.5	3.1	—	—	0.6	—	—
Pork fat (Lard)	4-04-790	0.3	1.3	23.8	13.5	41.1	2.7	41.2	1.0	—	44.9	10.2	—	—	1.0	—	—
Poultry fat	4-09-319	0.1	0.9	21.6	6.0	31.2	5.7	37.3	0.1	—	43.1	19.5	—	—	1.0	1.1	—
Yellow grease		—	1.9	16.2	10.5	29.9	2.5	47.5	—	—	50.0	17.5	—	—	1.9	—	—
Fish Oils																	
Anchovy	7-01-994	—	7.4	17.4	4.0	34.6	10.5	11.6	1.6	1.2	24.9	1.2	0.1	0.1	0.8	3.0	17.0
Capelin		—	7.9	11.1	1.0	20.0	11.1	17.0	18.9	14.7	61.7	1.7	t	0.1	0.4	2.1	4.6
Channel catfish, cultured		—	1.4	17.4	6.1	24.9	2.9	49.1	1.4	—	53.4	10.5	—	0.3	1.0	0.2	0.4
Cod liver	7-16-709	—	3.2	13.5	2.7	19.4	9.8	23.7	7.4	5.1	46.0	1.4	—	1.6	0.6	0.9	11.2
Copepod		—	10.6	14.5	1.7	26.8	4.3	10.3	7.0	11.4	33.8	3.3	—	0.4	2.5	11.8	7.6
Herring, Atlantic	7-08-048	0.2	6.4	12.7	0.9	22.8	8.8	12.7	14.1	20.8	56.4	1.1	0.2	0.3	0.6	1.7	8.4
Herring, Pacific		—	5.7	16.6	1.8	24.1	7.6	22.7	10.7	12.0	53.0	0.6	0.1	0.4	0.4	1.6	8.1
Krill		—	9.8	21.3	1.1	32.9	4.6	16.8	1.1	0.5	23.4	2.1	0.2	0.3	2.4	6.5	17.4
Mackerel		0.1	7.8	15.9	1.7	27.5	8.2	12.9	12.0	13.9	48.9	1.3	0.1	0.4	1.0	2.5	7.6
Menhaden	7-08-049	—	7.3	19.0	4.2	33.3	9.0	13.2	2.0	0.6	24.8	1.3	t	0.2	0.3	2.8	11.0
Pollock		—	4.0	13.3	2.5	20.1	6.8	22.7	7.2	4.9	42.5	0.9	—	0.4	0.7	2.4	12.6
Salmon, wild		—	3.7	10.2	4.7	18.6	8.7	18.6	8.4	5.5	41.2	1.2	—	0.9	0.6	2.1	12.0
Salmon, farmed byproduct		—	5.0	14.0	3.8	23.6	6.9	21.5	7.2	6.6	42.7	6.0	0.1	0.7	1.6	1.7	7.7
Sardine (Pacific-Californian)		—	7.6	16.2	3.5	28.6	9.2	11.4	3.2	3.6	27.4	1.3	—	1.6	0.9	2.0	16.9
Squid		—	4.9	15.5	3.0	24.3	6.7	15.6	7.2	5.0	34.5	2.8	—	t	—	3.0	14.1
Tuna (orbital oil)		—	2.9	17.2	4.9	24.4	5.2	15.3	t	1.4	25.7	8.0	—	1.6	1.2	1.0	6.6
Whitefish		—	—	—	—	—	—	—	—	—	—	—	—	—	—	—	—
Vegetable Oils																	
Canola (rapeseed)	4-06-144	—	—	3.1	1.5	7.4	—	60.0	1.3	1.0	62.3	20.2	—	—	12.0	—	—
Coconut	4-09-320	58.7	16.8	8.2	2.8	91.9	—	5.8	—	—	5.8	1.8	—	—	—	—	—
Corn	4-07-882	—	—	10.9	1.8	13.3	—	24.2	—	—	24.2	58.0	—	—	0.7	—	—
Cottonseed	4-20-836	—	0.8	22.7	2.3	27.1	0.8	17.0	—	—	17.8	51.5	—	—	0.2	—	—
Echium		—	—	7.0	4.0	11.0	t	16.0	t	t	16.0	15.0	11.0	—	32.0	14.0	—
Linseed	4-14-502	—	—	5.3	4.1	9.4	—	20.2	—	—	20.2	12.7	—	—	53.3	—	—
Olive		—	0.0	11.0	2.2	14.1	0.8	72.5	t	—	73.3	7.9	—	—	0.6	—	—
Palm		—	1.0	43.5	4.3	51.6	0.3	36.6	0.1	—	37.0	9.1	—	—	0.2	—	—
Palm kernel		56.4	16.0	8.2	2.2	82.8	—	15.5	—	—	15.5	2.2	—	—	—	—	—
Peanut	4-03-658	—	0.1	9.5	2.2	17.8	0.1	44.8	1.3	—	46.2	32.0	—	—	—	—	—
Safflower	4-20-526	—	0.1	6.2	2.2	9.5	0.4	11.7	—	—	12.1	74.1	—	—	0.4	—	—
Soybean	4-07-983	—	0.1	10.3	3.8	15.1	0.2	22.8	0.2	—	23.2	51.0	—	—	6.8	—	—
Sunflower	4-20-833	—	—	5.9	4.5	10.6	—	19.5	—	—	19.5	65.7	—	—	—	—	—

continued

TABLE 19-5 Continued

Lipid Source	International Feed Number	22:5 n-3	22:6 n-3	n3:n6 Ratio	Total n-3 LC-PUFA	U:S	Iodine Value	Vitamin A (mg/kg)	Vitamin D (mg/kg)	Vitamin E Activity (mg α-TE)	Cholesterol (mg/kg)	Energy Content (kcal/kg) DE[c]
Animal Fat												
Tallow	4-08-127	—	—	0.2	—	0.9	44	—	—	27	1,000	8,000
Pork fat (Lard)	4-04-790	—	—	0.1	—	1.4	64	90	—	15	950	8,285
Poultry fat	4-09-319	—	—	0.1	—	2.2	78	—	—	27	770	8,520
Yellow grease		—	—	0.1	—	2.3	75	—	—	—	—	8,550
Fish Oils												
Anchovy	7-01-994	1.6	8.8	24.0	27.4	1.9	183	—	—	—	—	8,445
Capelin		0.3	3.0	6.8	7.9	4.0	145	—	—	—	—	—
Channel catfish, cultured		0.3	1.3	0.3	2.0	3.0	—	—	—	—	—	—
Cod liver	7-16-709	1.7	12.6	9.0	25.5	4.2	—	300	2500	220	5700	—
Copepod		0.6	9.0	8.5	17.2	2.7	—	—	—	—	—	—
Herring, Atlantic	7-08-048	0.8	4.9	12.7	14.1	3.4	110	—	—	92		8,680
Herring, Pacific		0.8	4.8	15.7	13.7	3.1	140		—	—	7660	—
Krill		0.5	11.4	14.4	30.1	2.0	—	—	—	—	—	—
Mackerel		0.6	7.7	10.8	15.9	2.6	147	—	—	—	—	—
Menhaden	7-08-049	1.9	9.1	16.7	22.0	2.0	175	—	—	75	5210	8,475
Pollock		1.3	9.7	12.2	24.2	4.0	—	—	—	—	—	—
Salmon, wild		2.9	13.8	15.0	28.7	4.4	—	—	—	—	4850	—
Salmon, farmed byproduct		3.2	9.8	3.7	22.1	3.2	145	—	—	—	—	—
Sardine (Pacific-Californian)		2.5	12.9	12.1	32.3	2.5	162	—	—	—	710	—
Squid		1.1	17.0	12.9	33.1	3.1	190	—	—	—	13,000	—
Tuna (orbital oil)		1.4	23.5	3.1	31.4	3.1	—	—	—	—	—	—
Whitefish		—	—	—	—	—	—	—	—	—	—	—
Vegetable Oils												
Canola (rapeseed)	4-06-144	—	—	0.6	—	12.5	118	—	—	215	t	8,760
Coconut	4-09-320	—	—	0.0	—	0.1	10	—	—	7	t	8,405
Corn	4-07-882	—	—	0.0	—	6.5	125	—	—	198	t	8,755
Cottonseed	4-20-836	—	—	0.0	—	2.7	105	—	—	428	t	8,605
Echium		—	—	1.8	—	8.1	198	—	—	—	t	—
Linseed	4-14-502	—	—	4.2	—	9.6	185	—	—	18	t	—
Olive		—	—	< 0.1	—	6.1	86	—	—	120	t	8,750
Palm		—	—	0.0	—	0.9	50	—	—	335	t	8,010
Palm Kernel		—	—	0.0	—	0.2	18	—	—	62	t	—
Peanut	4-03-658	—	—	0.0	—	4.6	92	—	—	152	t	8,735
Safflower	4-20-526	—	—	0.0	—	9.5	140	—	—	349	—	8,760
Soybean	4-07-983	—	—	0.1	—	5.6	130	—	—	171	t	8,750
Sunflower	4-20-833	—	—	0.0	—	8.5	133	22	—	492	t	8,760

[a]Dash indicates that no data were available; t = trace amount.

[b]The fatty acid data were obtained from the following sources: Pearl (1995) of the Fats and Protein Research Foundation; USDA Food Composition Standard Release 11 (1997); Rossell, B., ed. 2009. Fish Oils. Oils and Fats Handbook, Volume 4. Oxford, UK: Leatherhead Food International, Leatherhead and Wiley-Blackwell; Padley, F. B., F. D. Gunstone, and J. L. Harwood. 1986. Occurrence and characteristics of oils and fats. Pp. 49–170 in The Lipid Handbook, F. D. Gunstone, J. L. Harwood, and F. B. Padley, eds. London, UK: Chapman Hall; Padley, F. B., F. D. Gunstone, and J. L. Harwood. 1994. Occurrence and characteristics of oils and fats. Pp. 47–223 in The Lipid Handbook, 2nd Edition, F. D. Gunstone, J. L. Harwood, and F. B. Padley, eds. London, UK: Chapman Hall; Gunstone, F. D., and J. L. Harwood. 2007. Occurrence and characterisation of oils and fats. Pp. 37–141 in The Lipid Handbook, 3rd Edition, F. D. Gunstone, J. L. Harwood, and A. J. Dijkstra, eds. London, UK: Chapman Hall. Values for fatty acid content do not always total 100% but represent means as obtained from various fat analysis conducted by gas-liquid chromatography.

[c]Calculated by the following relationship (Powles et al., 1995): $DE (kcal/kg) = 36.898 - (0.005 \times FFA) - (7.330 \times e^{-0.906 \times U:S})/4.184$ where FFA is the free fatty acid content in g/kg and U:S is the ratio of unsaturated to saturated fatty acids. In calculating DE, the free fatty acid concentrations of all fats were assumed to be 50 g/kg (or 5%).

TABLE 19-6 Chemical Composition of Some Purified Feed Ingredients Commonly Used for Aquatic Animal Research (as-fed basis)[a]

Ingredient	International Feed Number	Dry Matter (%)	DE (kcal/kg)	Crude Protein (%)	Crude Fat (%)	Calcium (%)	Phosphorus (%)	Lysine[b] (%)
Casein	5-01-162	91	4,135	84.0	0.2	0.61	0.82	7.35
Corn starch	4-02-889	90	4,000	0.4	0.4	0.00	0.03	—
Gelatin	5-14-503	88	2,800	85.9	0.1	0.60	0.4	3.55
Glucose monohydrate	4-02-125	90	3,360	0.3	—	—	—	—
Lactose	4-07-881	96	3,525	0.3	—	—	—	—
Sucrose	4-04-701	99	3,795	0.0	0.0	0.04	0.01	—

[a]Dash indicates that no data were available.

[b]Amino acid composition of casein is shown in Table 19-2. Other amino acids in gelatin: arginine, 6.60%; histidine, 0.66%; isoleucine, 1.42%; leucine, 2.91%; methionine, 0.76%; cystine, 0.12%; phenylalanine, 1.74%; tyrosine, 0.43%; threonine, 1.82%; tryptophan, 0.05%; and valine, 2.26%.

20

Common and Scientific Names of Species Discussed in This Report

TABLE 20-1 Common and Scientific Names of Species Discussed in This Report

Common Name	Scientific Name	Alternative Names
FISH		
African catfish	*Clarius gariepinus*	
Almaco jack	*Seriola rivoliana*	
American eel	*Anguilla rostrata*	
Angelfish	*Pterophyllum scalare*	
Arctic charr	*Salvelinus alpinus*	
Atlantic cod	*Gadus morhua*	
Atlantic halibut	*Hippoglossus hippoglossus*	
Atlantic salmon	*Salmo salar*	
Atlantic silverside	*Menidia menidia*	
Ayu	*Plecoglossus altivelis*	
Barramundi	*Lates calcarifer*	Asian sea bass
Basa	*Pangasius bocourti*	
Bighead carp	*Hypophthalmichthys nobilis*	
Black bullhead catfish	*Ameiurus melas*	
Black sea bream	*Mylio macrocephalus*	*Acanthopagrus schlegel* or *Sparus macrocephalus*
Blue tilapia	*Oreochromis aureus*	*Oreochromis aurea*
Bluegill	*Lepomis machrochirus*	
Brook trout	*Salvelinus fontinalis*	
Brown trout	*Salmo trutta linnaeus*	*Salmo trutta trutta*
Catfish	*Schilbeades molllis*	
Catla	*Catla catla*	
Channel catfish	*Ictalurus punctatus*	
Cherry salmon	*Oncorhynchus masou*	Japanese salmon
Chinese longsnout catfish	*Leiocassis longirostris*	
Chinese sucker	*Myxocyprinus asiaticus*	
Chinook salmon	*Oncorhynchus tshawytscha*	
Chum salmon	*Oncorhynchus keta*	
Cobia	*Rachycentron canadum*	Ling or Lemon fish
Coho salmon	*Oncorhynchus kisutch*	
Common carp	*Cyprinus carpio*	Fancy red carp
Crucian carp	*Carassius carassius*	
Cuneate drum	*Nibea miichthioides*	
Cutthroat trout	*Oncorhynchus clarki bouvieri*	
Estuary grouper	*Epinephelus striatus*	
Eurasian perch	*Perca fluviatilis*	
European bass	*Morone labrax*	
European eel	*Anguilla anguilla*	
European sea bass	*Dicentrarchus labrax*	
European whitefish	*Coregonus lavaretus*	

continued

TABLE 20-1 Continued

Common Name	Scientific Name	Alternative Names
Freshwater catfish	*Mystus nemurus*	
Giant sea perch	*Archoplites interruptus*	
Gibel carp	*Carassius auratus gibelio*	
Gilthead (sea) bream	*Sparus auratus*	*Sparus aurata*
Golden shiner	*Notemigonus crysoleucas*	
Goldfish	*Carassius auratus*	*Carassius auratus auratus*
Grass carp	*Ctenopharyngodon idella*	
Grey mullet	*Mugil cephalus*	
Grouper	*Epinephelus* spp.	
Guppy	*Poecilia reticulata*	
Haddock	*Melanogrammus aeglefinus*	
Hake	*Merluccius bilinearis*	
Herring	*Clupea harengus*	*Clupea harengus harengus*
Hybrid catfish	*Clarias gariepinus* × *Clarias macrocephalus*	
Hybrid striped bass	*Morone chrysops* × *Morone saxatilis*	Striped bass
Hybrid tilapia	*Oreochromis niloticus* × *Oreochromis aureus*	
Indian catfish	*Heteropneustes fossilis*	
Japanese (blue) parrotfish	*Scarus coeruleus*	
Japanese croaker	*Argyrosomus japonicus*	Mulloway
Japanese eel	*Anguilla japonica*	
Japanese flounder	*Paralichthys olivaceus*	
Japanese medaka	*Oryzias latipes*	
Japanese sea bass	*Lateolabrax japonicus*	Japanese sea perch
Jundia	*Rhamdia quelen*	
Lake trout	*Salvelinus namaycush*	
Largemouth bass	*Micropterus salmoides*	
Lingcod	*Ophiodon elongatus*	
Mahi mahi	*Coryphaena hippurus*	
Malabar grouper	*Epinephelus malabaricus*	
Mandarin fish	*Siniperca chuatsi*	
Milkfish	*Chanos chanos*	
Mossambique tilapia	*Oreochromis mossambicus*	Mozambique tilapia
Mrigal carp	*Cirrhinus mrigala*	Indian major carp
Mulloway	*Argyrosomus japonicus*	
Nile tilapia	*Oreochromis niloticus*	*Oreochromis nilotica*
Olive flounder	*Paralichthys olivaceus*	
Orange-spotted grouper	*Epinephelus coioides*	
Oriental weatherfish	*Misgurnus anguillcaudatus*	
Pacific salmon	*Oncorhynchus* spp.	
Pacu	*Piaractus mesopotamicus*	
Pangasius spp.	*Pangasius bocourti*, *P. hypohthalmus*	
Parrot fish	*Oplegnathus fasciatus*	Knife jaw
Pike	*Esox lucius*	
Pikeperch	*Sander lucioperca*	Zander
Pink salmon	*Oncorhynchus gorbuscha*	
Plaice	*Pleuronectes platessa*	
Plaice (American)	*Hippoglossoides platessoides*	
Platyfish (Southern)	*Xiphophorus maculatus*	*Xiphophorus hellerii*
Pollock	*Pollachius virens*	
Puffer (marbeled)	*Sphoeroides dorsalis*	
Rainbow trout	*Oncorhynchus mykiss*	
Red drum	*Sciaenops ocellatus*	
Red hybrid tilapia	*Oreochromis mossambicus* × *Oreochromis niloticus*	
Red pacu	*Piaractus brachypomus*	
Red sea bream	*Pagellus bogaraveo*	*Chrysophrys major* or *Pagrus major*
Red snapper	*Lutjanus argentimaculatus*	
Roach	*Rutilus rutilus*	
Rockfish (Japanese)	*Sebastes schulegeli*	Korean rockfish
Rohu carp	*Labeo rohita*	Indian major carp
Sabaki tilapia	*Oreochromis spilurus*	

TABLE 20-1 Continued

Common Name	Scientific Name	Alternative Names
Sablefish	*Anoplopoma fimbria*	
Sheatfish	*Silurus glanis*	
Siberian sturgeon	*Acipenser baerii*	
Silver bream	*Rhabdosargus sarba*	
Silver carp	*Hypophthalmichthys molitrix*	
Silver perch	*Bidyanus bidyanus*	
Skate	*Raja* spp.	
Skipjack tuna	*Katsuwonus pelamis*	
Smallmouth bass	*Micropterus dolomieu*	
Snakehead	*Channa striata*	
Sockeye salmon	*Oncorhynchus nerka*	
Starry flounder	*Paralichthys stellatus*	
Striped bass	*Morone saxatilis*	
Striped jack	*Longirostris delicatissimus*	*Pseudocaranx dentex*
Sturgeon	*Acipenser* spp.	
Sunfish (redbreast)	*Lepomis auritus*	
Swordfish (green)	*Xiphophorus helleri*	
Tambaqui	*Colossoma macropomum*	Black pacu
Tiger puffer	*Takifugu rubripes*	
Tra catfish	*Pangasianodon hypothalamus*	
Turbot	*Scophthalmus maximus*	*Psetta maxima*
Walleye	*Sander vitreum*	
White bass	*Morone chrysops*	
White grouper	*Epinephelus aeneus*	
White sturgeon	*Acipenser transmontanus*	
Whitefish	*Coregonus* spp.	
Winter flounder	*Pseudopleuronectes americanus*	
Yellow croaker	*Pseudosciaena crocea*	
Yellow perch	*Perca flavescens*	
Yellowtail	*Seriola lalandi*	*Seriola quinqueradiata* or Kingfish
Yellowtail flounder	*Pleuronectes ferrugineus*	
Zebrafish	*Danio rerio*	
Zillii's tilapia	*Tilapia zillii*	Zill's tilapia
SHRIMP and OTHER CRUSTACEANS		
American lobster	*Homarus americanus*	
Atlantic ditch shrimp	*Palaemonetes varians*	
Atlantic rock crab	*Cancer irroratus*	
Atlantic white shrimp	*Penaeus setiferus*	
Australian red claw crayfish	*Cherax quadricarinatus*	
Banana shrimp	*Fenneropenaeus merguiensis*	
Blue shrimp	*Litopenaeus stylirostris*	*Penaeus stylirostris*
Brown shrimp	*Penaeus aztecus*	
Common prawn	*Palaemon serratus*	
Common shrimp	*Crangon crangon*	Brown shrimp or Caridean shrimp
Fleshy prawn	*Fenneropenaeus chinensis*	*Penaeus chinensis* or *Penaeus orientalis*
Giant river prawn	*Macrobrachium rosenbergii*	Caridean shrimp
Indian white prawn	*Fenneropenaeus indicus*	*Penaeus indicus*
Kuruma prawn	*Marsupenaeus japonicus*	*Penaeus japonicus* or Kuruma shrimp
Louisiana crayfish	*Procambarus clarkii*	
Margaret River marron crayfish	*Cherax tenuimanus*	
Marine shrimp	*Penaeus penicillatus*	
Mexican cichlid	*Cichlasoma urophthalmus*	
Northern brown shrimp	*Penaeus aztecus*	
Oriental river prawn	*Macrobrachium nipponense*	
Pacific white shrimp	*Litopenaeus vannamei*	White-legged shrimp
Pink shrimp	*Pandalus borealis*	
Signal crayfish	*Pascifastacus leniusculus*	
Tiger shrimp	*Penaeus monodon*	Black tiger shrimp
Yellow leg shrimp	*Penaeus californiensis*	

Appendix A

Committee Statement of Task

A committee will prepare a report that evaluates the scientific literature on the nutrient requirements of fish and shrimp in all stages of life. The report will focus on the species that are most important commercially (e.g., catfish, tilapia, bass, trout, salmon, sea bass and sea bream, and shrimp), but other emerging species (e.g., halibut, Atlantic cod, and winter flounder) may be included.

The committee will examine estimates of nutrient requirements and signs of nutrient deficiencies as reported in the literature and will evaluate information on management techniques and feeding practices that influence requirements. The committee will also examine research findings on the use of various protein sources and review the effects of nutrition in commercial fish production on nutrient and waste excretion and on environmental pollution. Strategies to increase nutrient retention and thus reduce fecal and metabolic excretions that contribute to environmental pollution will be reviewed.

Based on its review, the committee will prepare a report that includes:

- a comprehensive analysis of recent research on feeding and nutrition of fish and shrimp, nutrient requirements, and physiological and environmental factors affecting requirements
- an update of the recommendations contained in the 1993 NRC publication "Nutrient Requirements of Fish," which currently serves as the authoritative source of information (in the U.S. and internationally) for feeding fish, with additional information on shrimp
- a description of feeding and production methods to reduce waste and environmental impacts
- a review of the benefits and detriments of including marine products in fish feeds
- information on the composition of feeds, feed additives (including substances such as antimicrobials and nutraceuticals), and other compounds routinely fed to fish and shrimp
- data on changes in the nutrient content of fish, such as omega-3 fatty acids, with changes in fish diet formulation.

Appendix B

Abbreviations and Acronyms

AA	Amino acid or Ascorbic acid (vitamin C)
AAFCO	Association of American Feed Control Officials
AASLP	Ascorbic acid stimulated lipid peroxidation
ACTH	Adrenocorticotropic hormone
ADC	Apparent digestibility coefficient
ADF	Acid detergent fiber
ADMD	Apparent dry matter digestibility
ADS	Absence of deficiency signs
AFB_1	Aflatoxin B_1
AFQ_1	Aflatoxin Q_1
AFSSA	Agence Française de Sécurité Sanitaire des Aliments (French Agency for Food Safety)
ALD	Apparent lipid digestibility
aNDF	Amylase-treated neutral detergent fiber
ANF	Antinutritional factor
ANOVA	Analysis of variance
AOAC	Association of Official Analytical Chemists
ARA	Arachidonic acid
ARG, Arg, R	Arginine
ATP	Adenosine tri-phosphate
BCAA	Branched-chain amino acids
BCAT	Branched-chain aminotransferase
BCKA	Branched-chain α-keto acid
BHA	Butylated hydroxylanisol
BHMT	Betaine-homocysteine methyltransferase
BHT	Butylated hydroxytoluene
BPL	Bonito egg polar lipid
BSE	Bovine spongiform encephalopathy
BW	Body weight
C/H	Collagen or hydroxyproline concentration
C2D	Ascorbate-2-glucose
C2MP	Ascorbyl-2-monophosphate
C2MP-Ca	Ascorbyl-2-monophosphate-Ca
C2MP-Mg	Ascorbyl-2-monophosphate-Mg
C2MP-Na	Ascorbyl-2-monophosphate-Na
C2MP-Na/Mg	Ascorbyl-2-monophosphate-Na/Mg
C2PP	Ascorbyl-2-polyphosphate
C2S	Ascorbyl-2-sulfate
CAA	Crystalline amino acids
cal	Calorie
CAST	Council for Agricultural Science and Technology
CCO	Cytochrome C oxidase
cDNA	Complementary deoxyribonucleic acid
CDP	Cytidine diphosphate
CF	Crude fiber

CGA	Chlorogenic acid
CHH	Crustacean hyperglycemic hormone
CLA	Conjugated linoleic acid
CMC	Carboxymethylcellulose
CP	Crude protein
CPL	Corn polar lipid
$cPLA_2$	Cytosolic phospholipase A_2
CPT	Cytidine diphosphate-choline phosphotransferases
CYS, Cys, C	Cystine
d	Days
D-AAO	D-amino acid oxidase
DAG	Diacylglycerol
DAS	Diacetoxyscirpenol
DE	Digestible energy
DEI	Digestible energy intake
DHA	Docosahexaenoic acid
DM	Dry matter
DNA	Deoxyribonucleic acid
DNase	Deoxyribonuclease
DNI	Digestible nitrogen intake
DON	Deoxynivalenol
DP	Digestible protein or Degree of polymerization
EAA	Essential amino acid
EAP	Estimated available phosphorus
ED	Enzyme data
EFA	Essential fatty acids
EFSA	European Food Safety Authority
EL	Chicken egg lecithin
Elovl	Elongation of very long chain fatty acids
ENL	Endogenous nitrogen losses
EO	Echium oil
EPA	Eicosapentaenoic acid
EPC	Purified egg phosphatidylcholine
EPT	Ethanolamine phosphotransferases
EU	European Union
FABP	Fatty acid binding proteins
Fad	Fatty acyl desaturase
FAD	Flavin adenine dinucleotide
FAO	Food and Agriculture Organization of the United Nations
FAS	Fatty acid synthetase
FBPase	Fructose-1,6-bisphosphatase
FCR	Feed conversion ratio
FDA	U.S. Food and Drug Administration
FE	Fecal energy losses or Feed efficiency
FFA	Free fatty acid
FFSM	Full-fat soybean meal
FIFO	Fish-in:fish-out
FM	Fish meal
FMN	Flavin mononucleotide
FO	Fish oil
FOM	Fecal organic matter
FOS	Fructo-oligosaccharide
FW	Freshwater
FWB	Final body weight
G	Growth
G6Pase	Glucose-6-phosphatase
GE	Gross energy or Enthalpy of combustion
GI	Gastrointestinal
GK	Glucokinase
GLA	Gamma (γ)-linolenic acid
GLUT1	Glucose transporter 1
GLUT2	Glucose transporter 2
GLUT4	Glucose transporter 4
GM	Genetically modified

G_{max}	Maximum growth
GPase	Glycogen phosphorylase
GPCR	G-Protein-coupled receptors
GPS	Global positioning system
GPx	Glutathione peroxidase
GRE	Growth rate exponential
GSase	Glycogen synthase
GSH	Glutathione
HCN	Hydrogen cyanide
H_dE	Heat of digestion and absorption processes
HDL	High-density lipoproteins
HE	Heat losses or heat production
H_eE	Basal metabolism or minimal metabolism
HE_f	Fasting heat production or Fasting heat losses
HE_m	Maintenance energy
H_iE	Heat increment of feeding
HIS, His, H	Histidine
H_jE	Heat of voluntary activity
HK	Hexokinases
HMB or OH-Met	2-Hydroxy-4-(methylthio)butanoic acid or Hydroxmethionine
H_rE	Heat of transformation and retention of substrates
HSI	Hepatosomatic index
H_wE	Heat of formation and excretion of metabolic wastes
H_xE	Heat of molting
IE	Intake of energy
IGF	Insulin-like growth factor
IGR	Instantaneous growth rate
ILE, Ile, I	Isoleucine
INRA	Institut National de la Recherche Agronomique (French National Institute for Agricultural Research)
IP	Intraperitoneal
IPCS	International Panel on Chemical Safety
IU	International units
IV	Intravenous
J	Joule
kcal	Kilocalorie
K_f	Efficiency of metabolizable energy utilization for lipid deposition
K_g	Efficiency of metabolizable energy utilization for growth or gain
kg	Kilogram
K_p	Efficiency of metabolizable energy utilization for protein deposition
K_{pf}	Efficiency of metabolizable utilization for production (whole body growth)
LAP	Land animal products
LC	Long-chain
LCA	Life cycle assessment/analysis
LC-PUFA	Long-chain polyunsaturated fatty acids
LD	Lipid deposition
LDL	Low-density lipoprotein
LEU, Leu, L	Leucine
LLC	Liver lipid content
LO	Linseed oil
LT	Leukotrienes
LXR	Liver X receptors
LYS, Lys, K	Lysine
MAG	Monoacylglycerol
MBS	Maximum body storage
MBW	Metabolic body weight
ME	Metabolizable energy
MEI	Metabolizable energy intake
MEM	Microextrusion marumerization
MET, Met, M	Methionine
MFA	Monosaturated fatty acids
MHA	Methionine hydroxyanalog
MKS	Maximum kidney storage

MLS	Maximum liver storage
MMS	Maximum muscle storage
MNB	Menadione nicotinamide bisulfite
MPB	Menadione dimethlypyrimidinol bisulfite
mRNA	Messenger ribonucleic acid
MSB	Menadione sodium bisulfite
MSBC	Menadione sodium bisulfite complex
MSU	Mississippi State University
Mt	Million tonnes (1 tonne = 1 metric ton = 1,000 kg)
NAD	Nicotinamide adenine dinucleotide
NADP	Nicotinamide adenine dinucleotide phosphate
NAS	National Academy of Sciences
NCP	Noncellulosic polysaccharides
ND	Not determined
NDF	Neutral detergent fiber
NE	Net energy
NEAA	Nonessential amino acids
NERC	Natural Environment Research Council
NHV	Normal hematocrit values
NR	No requirement determined
NR	Not required under practical conditions
NRC	National Research Council
NSP	Nonstarch polysaccharide
NT	Not tested
NVH	Norges Veterinærhøgskole (Norwegian School of Veterinary Science)
O	Oral
OSS	Optimum swimming stamina
OTA	Ochratoxin A
PA	Phosphatidic acid
PAF	Platelet-activating factor
PAH	Polycyclic aromatic hydrocarbons
PARA	Particle-assisted rotational agglomeration
PBDE	Polybrominated diphenyl ethers
PC	Phosphatidylcholine
PCB	Polychlorinated biphenyls
PD	Protein deposition
PD_{max}	Maximal protein deposition rate
PE	Phosphatidylethanolamine
PEPCK	Phosphoenolpyruvate carboxykinase
PER	Protein efficiency ratio
PFK-1	Phosphofructokinase 1
PFOS	Perfluorooctane sulfonate
PG	Prostaglandins
PHA	Phytohemagglutinin
PHE, Phe, F	Phenylalanine
PI	Phosphatidylinositol
PK	Pyruvate kinase
PKC	Protein kinase C
PKS	Polyketide
PL	Phospoholipid
PO	Palm oil
POP	Persistent organic pollutants
PPAR	Peroxisome proliferator-activated receptors
PPC	Potato protein concentrate
ppm	Parts per million
ppt	Parts per thousand (‰)
PS	Phosphatidylserine
PUFA	Polyunsaturated fatty acids
QIM	Quality index method
R	Required in diet but quantity not determined or Stress resistance
RE	Recovered energy
RE_l	Recovered energy as lipid
RE_p	Recovered energy as protein

RN	Recovered nitrogen
RNA	Ribonucleic acid
RNase	Ribonuclease
RO	Rapeseed oil (also known as canola oil)
S	Survival
SAH	S-Adenosylhomocysteine
SAM	S-Adenosylmethionine
SBA	Soybean agglutinin or Soybean lectin
SCD	Stearoyl CoA desaturase
SCO	Single cell oils
SDA	Specific dynamic action or Stearidonic acid
SDF	Soluble dietary fiber
SE	Surface energy loss
Sec or U	Selenocysteine
Se-met or Sem	Selenomethionine
SFA	Saturated fatty acids
SFM	Sunflower meal
SGA	Steroid glycoalkaloids
SGR	Specific growth rate
SIU	Système international des unites (International System of Units)
SKM	Saturation Kinetics Model
SL	Soybean lecithin
SOD	Superoxide dismutase
SPC	Soybean protein concentrate
SPE	Soybean protein extract
SPI	Soybean protein isolate
SREBP	Sterol regulatory element binding proteins
SUR	Survival
SW	Saltwater
t	Time
T	Temperature
TAG	Triacylglycerols
TBARS	Thiobarbituric acid reactive substance
TCA	Tricarboxylic acid
TDF	Total dietary fiber
TEQ	Toxic equivalent
TGC	Thermal-unit growth coefficient
THC	Total hemocyte count
THR, Thr, T	Threonine
TMA	Trimethylamine
TMAO	Trimethylamine oxide
TPP	Thiamin pyrophosphate
tRNA	Transfer RNA
TRP, Trp, W	Tryptophan
TSAA	Total sulfur amino acid
TTA	Tetradecylthioacetic acid
TYR, Tyr, Y	Tyrosine
UE	Urinary energy loss
UG/OMNR	University of Guelph/Ontario Ministry of Natural Resources
UN	Urinary nitrogenous wastes
USDA	United States Department of Agriculture
USFWS	United States Fish and Wildlife Service
VAL, Val, V	Valine
VC	Vertebral collagen content
Vitamin D_2	Ergocalciferol
Vitamin D_3	Cholecalciferol
Vitamin K_3	Menadione
VLDL	Very low-density lipoproteins
V_{max}	Maximum velocity
VO	Vegetable oil
VOT	Vinyl-2-oxazolidinethione
W_0	Initial weight
WG	Weight gain

Wt	Final weight
ZE	Branchial energy loss
ZN	Branchial nitrogenous waste

Appendix C

Committee Member Biographies

Ronald W. Hardy *(chair)* is professor and director of the Aquaculture Research Institute at the University of Idaho and affiliate professor at the University of Washington's School of Fisheries. His expertise is in fish nutrition and feeds, with a research focus in the areas of alternative feed ingredients for fish meal and fish oil for use in salmon and trout feeds. Research also includes molecular genetics in relation to somatic growth, nutrient partitioning, effects of diet on immune function, and ornamental fish production. He is also involved in aspects of conservation fisheries, such as diets for wild salmonids. Dr. Hardy was a member of the NRC Committee on Animal Nutrition (2000–2002) and the NRC's Subcommittee on Warm Water Fish Nutrition (1981–1983). Dr. Hardy received his B.S. in zoology from the University of Washington, his M.S. in nutrition from Washington State University, and his Ph.D. in fisheries from the University of Washington.

Delbert M. Gatlin, III *(vice-chair)* is a professor and University Faculty Fellow with the Department of Wildlife and Fisheries Sciences and a member of the Intercollegiate Faculty of Nutrition at Texas A&M University. He is an expert in basic nutrition of fishes and shellfishes, with primary emphasis on species cultured for human consumption, especially hybrid striped bass and red drum. Dr. Gatlin's research activities include determining dietary requirements for and metabolism of various nutrients with consideration given to such aspects as nutritional energetics, nutrition—disease interactions, and nutritional biochemistry of fish as it relates to human health. Dr. Gatlin sits on the board of directors of the Striped Bass Growers Association and is a Certified Fisheries Scientist and member of the American Fisheries Society, World Aquaculture Society, and American Society for Nutritional Sciences. He was a member of the NRC Committee on Animal Nutrition from 1996 to 1999. He has been Nutrition Section editor for the journal *Aquaculture* since 2009. Dr. Gatlin received his B.S. in fisheries and aquaculture from Texas A&M University and his Ph.D. in nutritional biochemistry from Mississippi State University.

Dominique P. Bureau is a professor in the Department of Animal and Poultry Science at the University of Guelph in Ontario, Canada. He is an expert on the nutrition of fish, mainly salmonid species (*Oncorhynchus* species, *Salvenius* species, and *Salmo salar*) for aquaculture or stock rehabilitation purposes. His research focuses on nutrient utilization, basic nutritional physiology, growth regulation, digestion, and feeding strategies as well as on ingredient quality and fish feed formulation. He also develops predictive models for growth, feed requirement, and waste output of fish. He supervises the activity of the UG/OMNR Fish Nutrition Research Laboratory, a joint venture between the Ontario Ministry of Natural Resources and the University of Guelph. In 2007, Dr. Bureau became a member of the Canada-Chile Sustainable Aquaculture Research & Training Network. He is a member of the editorial boards of Aquaculture Research, International AquaFeed, and Aquaculture Nutrition, and an associate editor of the Journal of the World Aquaculture Society. In 2005, he served as a Novelty Guidance Steering Committee member to the Canadian Food Inspection Agency and Health Canada. Dr. Bureau received his B.A.Sc. in bio-agronomy and his M.Sc. in animal science from the Université Laval in Québec and his Ph.D. in nutritional sciences from the University of Guelph.

Louis R. D'Abramo is the William L. Giles Distinguished Professor, an aquaculture biologist in the Department of Wildlife and Fisheries, and dean of the graduate school and associate vice president for academic affairs at Mississippi State University (MSU), where he has worked since 1984. Dr. D'Abramo's career at MSU has focused on development of management strategies to optimize production of fish, crustacean, and mollusc species for commercial aquaculture; nutritional physiology of larval and juvenile crustaceans, fish, and molluscs; development of formulated diets; and hormonal control of sex differentiation in crustaceans. One of the goals of his work is to develop sustainable commercial production practices based on the wise use of natural resources and environmental stewardship. Dr. D'Abramo has received awards for his research and teaching, including the Ralph E. Powe Research Excellence Award, the highest honor in faculty research from Mississippi State University, in March 2007; the Meritorious Award from the National Shellfisheries Association, Inc. in 2007; and the Highest Honor Bestowed by the World Aquaculture Society, the Exemplary Service Award in 2003. He is on the editorial boards of Aquaculture International, Aquaculture, and The Journal of Shellfish Research. He has participated in professional organizations by his service as president of the World Aquaculture Society (1992–1993) and of the National Shellfisheries Association (2005–2007). Dr. D'Abramo received his B.A. in natural sciences and mathematics from Assumption College and his M.Phil. and Ph.D. in ecology and evolutionary biology from Yale University.

D. Allen Davis is an associate professor of aquatic animal nutrition in Auburn University's Department of Fisheries and Allied Aquacultures. His primary research activities emphasize the determination of nutritional requirements for fish and shrimp. This includes the determination of dietary requirements, development of practical diet formulations, and the influence of diet formulations on nutrient loading of culture systems. His work aims to enhance understanding of nutrient requirements and to develop data needed by the commercial industry. He also maintains research interests geared toward the development of natural spawning techniques and the development of intensive culture techniques for marine fish and shrimp. Dr. Davis is a member of the American Fisheries Society and World Aquaculture Society, and he is an associate editor to the Journal of the World Aquaculture Society. He sits on the International Reviewing Board of the Egyptian Journal of Nutrition and Feeds. He is the recipient of the Alabama Agricultural Experiment Station of Auburn University's Directors Research Award for Assistant Professors in 2003, and the Aquaculture Engineering Society's Superior Paper Award. Dr. Davis received his B.S. in biology and chemistry from Northern Arizona University and his M.S. and Ph.D. in wildlife and fisheries sciences from Texas A&M University.

John E. Halver (NAS) is Professor Emeritus in Nutrition at the University of Washington's School of Aquatic and Fishery Sciences. He was a professor at the University of Washington for 24 years until his retirement in 1992. Prior to that, he was senior scientist in nutrition for the U.S. Fish and Wildlife Service (USFWS) in Seattle (1975–1978) and director of Western Fish Nutrition Laboratories of the USFWS (1950–1975). In 1976, he founded the Halver Corporation, an ecosystems management consultancy that serves the national and international arena, of which he is still the president and CEO. Dr. Halver is a member of the American Society for Nutrition, the American Fisheries Society, the World Aquaculture Society, and the American Institute of Fishery Research Biologists. He has membership in the National Academy of Sciences and the Hungarian Academy of Sciences. He has served as the chair of the first NRC publication on Nutrient Requirements of Fish; as Section 61 Liaison for Animal, Nutritional, and Applied Microbial Sciences; and on the Panel on Research and the Government Performance and Results Act. In 2005, the International Biographical Centre of Cambridge, England, named him one of the "Leading Scientists of the World." Dr. Halver received his B.Sc. in chemistry and his M.Sc. in organic chemistry from Washington State University and his Ph.D. in biochemistry from the University of Washington.

Åshild Krogdahl is a professor in the Department of Basic Sciences and Aquatic Medicine at the Norwegian School of Veterinary Science (NVH) and leader of the Nutrition Group at the Aquaculture Protein Centre, a Norwegian Centre of Excellence. She joined NVH in 1993 after 10 years of teaching and research in animal nutrition at the Norwegian University of Agriculture and 6 years at the associated Institute for Aquaculture Research. She is now responsible for the course in animal nutrition at NVH and teaches nutrition of nonruminant animals. Her research focuses on interactions between food components and intestinal physiology in various nonruminant animals (fish and humans included), with the goal of revealing the effects of antinutrients on digestive processes, transport and immune functions of the gut, and nutrient partitioning and requirement in the body. Presently, the Atlantic salmon is her main experimental animal. She is a referee for the journals Aquaculture, Aquaculture Nutrition, and Comparative Biochemistry and Physiology. She has served as an advisor or consultant to numerous committees and professional societies, including deputy president in the Senate of the Norwegian Research Council (1986–1989), representative for Norway in the EU's Cost Action 98 (1995–1999), leader of the Advisory Board of Directorate of Fisheries Institute of Nutrition (1998–2002), and leader of the Norwegian Scientific Committee for Food Safety (2004–present). Dr. Krogdahl earned her B.Sc., M.Sc., and Ph.D. at the University of Oslo.

Françoise Médale is director of research at the Joint Research Unit of the Aquaculture Nutrition and Genomics of INRA, the French National Institute for Agricultural Research. A specialist in the nutrition of farmed fish, Dr. Médale uses rainbow trout as her model organism to analyze the cellular and molecular mechanisms and regulators of energy use of nutrients. Her objectives are to reduce nitrogen use, thereby reducing nitrogen discharges, and to modulate fat deposits by nutrition in order to control the quality of characteristics in the final product, such as nutritional value, sensory characteristics, and yield. She also studies the relation of genetic variability to utilization of nutrients in different raw food ingredients. Dr. Médale also teaches fish nutrition at the University of Rennes and Agronomy Schools. She is an elected member of the Scientific Council of INRA, a member of the National Council of Universities Section 68, and an expert of the group on nutritional quality and health of aquatic products for AFSSA, the French Food Safety Agency. Dr. Médale received her degree in physiology and biochemistry from Toulouse III, her maitrise in animal physiology from Toulouse III, and her Ph.D. in nutrition and food science from the Université Paul Sabatier Toulouse.

Shi-Yen Shiau is the national endowed chair professor in the Department of Food Science at the National Taiwan Ocean University and the chair professor in the Department of Food and Nutrition at Providence University in Taiwan. Dr. Shiau is internationally renowned for his research, which is currently focused on the biological basis of micronutrient requirements and nutrient interactions in tilapia, grouper, and grass shrimp and their effects on immunity. He has received numerous awards for his research and teaching. Dr. Shiau serves on the editorial boards of the international journals Aquaculture, Aquaculture Nutrition, and Aquaculture Research and as editor-in-chief of Asian Fisheries Science. He is the past editor-in-chief of the Journal of the Chinese Nutrition Society and Journal of the Fisheries Society of Taiwan. Professor Shiau has been a member of a number of professional societies including the American Society for Nutritional Science, the World Aquaculture Society, and the Asian Fisheries Society. He has served as board director of the World Aquaculture Society and as president of the Taiwan Fisheries Society. Currently, he is the councilor of the Asian Fisheries Society. Since 1994, he has served on the International Union of Nutritional Science's Committee III/5: Nutrition and the Production of Fish and Shellfish and since 1995 on the National Committee for the Pacific Science Association. Dr. Shiau is also a scientific committee member of the International Symposium on Nutrition and Feeding of Fish. Dr. Shiau received his B.S. in food science from the National Taiwan Ocean University and both his M.S. and Ph.D. in nutrition from Texas Tech University.

Douglas R. Tocher is professor of molecular nutrition at the Institute of Aquaculture at the University of Stirling in Scotland where he has worked since 1998. His primary research has focused on the biochemistry, molecular biology, and genetic basis of regulation of lipid and fatty acid metabolism and nutrition in fish. Current research interests include transcription factors and the application of cell culture and post-genomic technologies to the study of fish lipid metabolism and nutrition. Prior to his current position, he worked for 16 years for the Natural Environment Research Council (NERC) at the Institute of Marine Biochemistry, Aberdeen (1982–1986) and the Unit of Aquatic Biochemistry of the University of Stirling (1986–1998). His main research interests during that period were in fish lipid and fatty acid metabolism, including embryonic and early larval development, eicosanoid metabolism, long-chain polyunsaturated fatty acid synthesis, and neural development. He is a member of the European Federation of Lipids, the American Oil Chemists' Society, and the European Aquaculture Society, and he is on the editorial boards of the journals Aquaculture, and Fish Physiology and Biochemistry. Dr. Tocher received a B.Sc. and Ph.D., both in biochemistry, from the University of Edinburgh in Scotland.

Appendix D

Recent Publications of the Board on Agriculture and Natural Resources

POLICY AND RESOURCES

Achievements of the National Plant Genome Initiative and New Horizons in Plant Biology (2008)
Achieving Sustainable Global Capacity for Surveillance and Response to Emerging Diseases of Zoonotic Origin: Workshop Report (2008)
Agricultural Biotechnology and the Poor: Proceedings of an International Conference (2000)
Agriculture, Forestry, and Fishing Research at NIOSH (2008)
Agriculture's Role in K-12 Education (1998)
Air Emissions from Animal Feeding Operations: Current Knowledge, Future Needs (2003)
Animal Biotechnology: Science-Based Concerns (2002)
Animal Care and Management at the National Zoo: Final Report (2005)
Animal Care and Management at the National Zoo: Interim Report (2004)
Animal Health at the Crossroads: Preventing, Detecting, and Diagnosing Animal Diseases (2005)
Biological Confinement of Genetically Engineered Organisms (2004)
California Agricultural Research Priorities: Pierce's Disease (2004)
Changes in the Sheep Industry in the United States: Making the Transition from Tradition (2008)
Countering Agricultural Bioterrorism (2003)
Critical Needs for Research in Veterinary Science (2005)
Designing an Agricultural Genome Program (1998)
Diagnosis and Control of Johne's Disease (2003)
Direct and Indirect Human Contributions to Terrestrial Carbon Fluxes (2004)
Ecological Monitoring of Genetically Modified Crops (2001)
Emerging Animal Diseases: Global Markets, Global Safety: Workshop Summary (2002)
Emerging Technologies to Benefit Farmers in Sub-Saharan Africa and South Asia (2009)
Ensuring Safe Food: From Production to Consumption (1998)
Environmental Effects of Transgenic Plants: The Scope and Adequacy of Regulation (2002)
Exploring a Vision: Integrating Knowledge for Food and Health (2004)
Exploring Horizons for Domestic Animal Genomics (2002)
Frontiers in Agricultural Research: Food, Health, Environment, and Communities (2003)
Future Role of Pesticides for U.S. Agriculture (2000)
Genetically Engineered Organisms, Wildlife, and Habitat: A Workshop Summary (2008)
Genetically Modified Pest-Protected Plants: Science and Regulation (2000)
Global Challenges and Directions for Agricultural Biotechnology (2008)
Incorporating Science, Economics, and Sociology in Developing Sanitary and Phytosanitary Standards in International Trade (2000)
National Capacity in Forestry Research (2002)
National Research Initiative: A Vital Competitive Grants Program in Food, Fiber, and Natural-Resources Research (2000)
Predicting Invasions of Nonindigenous Plants and Plant Pests (2002)
Professional Societies and Ecologically Based Pest Management (2000)
Publicly Funded Agricultural Research and the Changing Structure of U.S. Agriculture (2002)
Safety of Genetically Engineered Foods: Approaches to Assessing Unintended Health Effects (2004)
Scientific Advances in Animal Nutrition: Promise for a New Century (2001)
Scientific Criteria to Ensure Safe Food (2003)

Status of Pollinators in North America (2007)
The National Plant Genome Initiative (2002)
The Scientific Basis for Estimating Emissions from Animal Feeding Operations: Interim Report (2002)
The Scientific Basis for Predicting the Invasive Potential of Nonindigenous Plants and Plant Pests in the United States (2002)
The Use of Drugs in Food Animals: Benefits and Risks (2000)
Transforming Agricultural Education for a Changing World (2009)

ANIMAL NUTRITION PROGRAM—NUTRIENT REQUIREMENTS OF DOMESTIC ANIMALS SERIES AND RELATED TITLES

Mineral Tolerance of Animals: Second Revised Edition (2005)
Nutrient Requirements of Beef Cattle, Seventh Revised Edition, Update (2000)
Nutrient Requirements of Dairy Cattle, Seventh Revised Edition (2001)
Nutrient Requirements of Dogs and Cats (2006)
Nutrient Requirements of Horses, Sixth Revised Edition (2007)
Nutrient Requirements of Nonhuman Primates, Second Revised Edition (2002)
Nutrient Requirements of Small Ruminants: Sheep, Goats, Cervids, and New World Camelids (2007)
Nutrient Requirements of Swine, Tenth Revised Edition (1998)
Safety of Dietary Supplements for Horses, Dogs, and Cats (2009)
Scientific Advances in Animal Nutrition: Promise for a New Century (2001)
The First Seventy Years 1928–1998: Committee on Animal Nutrition (1998)
The Scientific Basis for Estimating Emissions from Animal Feeding Operations: Interim Report (2002)

Further information and prices are available from the National Academies Press website at *http://www.nap.edu/*. To order any of the titles above, go to *http://www.nap.edu/order.html* or contact the Customer Service Department at (888) 624-8373 or (202) 334-3313. Inquiries and orders may also be sent to the National Academies Press, 500 Fifth Street, NW, Lockbox 285, Washington, DC 20055.

Index

B

D

G

H

I

J

K

L

M

N

O

P

Q

R

S

T

U

V

W

Y

Z